MODULARITY IN DEVELOPMENT AND EVOLUTION

Modularity in Development and Evolution

EDITED BY
GERHARD SCHLOSSER and GÜNTER P. WAGNER

THE UNIVERSITY OF CHICAGO PRESS • CHICAGO AND LONDON

Gerhard Schlosser is scientific assistant at the Brain Research Institute, University of Bremen. He is interested in the evolution of development with a main research focus on neural development in vertebrates.

Günter P. Wagner is the Alison Richard Professor of Ecology and Evolutionary Biology at Yale University and recipient of a MacArthur Award. His interests are the theory of complex adaptations and the development and evolution of morphological characters. He most recently edited the book *The Character Concept in Evolutionary Biology.*

The University of Chicago Press, Chicago 60637
The University of Chicago Press, Ltd., London

Printed in the United States of America
13 12 11 10 09 08 07 06 05 04 1 2 3 4 5

ISBN: 0-226-73853-1 (cloth)
ISBN: 0-226-73855-8 (paper)

Library of Congress Cataloging-in-Publication Data
Modularity in development and evolution / edited by Gerhard Schlosser and Günter P. Wagner.
p. cm.
Includes bibliographical references.
ISBN 0-226-73853-1 (cloth : alk. paper)—ISBN 0-226-73855-8 (pbk. : alk. paper)
1. Developmental biology. 2. Evolution (Biology) I. Schlosser, Gerhard, 1963– II. Wagner, Günter P.
QH491.M59 2004
571.8—dc21

2003056415

∞ The paper used in this publication meets the minimum requirements of the American National Standard for Information Sciences—Permanence of Paper for Printed Library Materials, ANSI Z39.48-1992.

CONTENTS

SYNTHESIS

PREFACE

This book grew out of the symposium "Modularity in Development and Evolution," which one of us (G. Schlosser) had organized in May 2000 at the Hanse Institute for Advanced Study in Delmenhorst, Germany. It was an attempt to find common ground among participants from many different backgrounds, including developmental and evolutionary biology, philosophy, and computer modeling. Not surprisingly, the few days of the symposium did not culminate in a finely polished definition of "module," nor were we able to create consensus on how modularity should be conceptualized. However, our discussions certainly raised the awareness of the pervasiveness of modularity at different levels of organization (one participant remarked that while preparing his talk for the symposium he realized that he had been working on modules all along) and sharpened the perception of commonalities and differences between developmental and evolutionary modules. Since the symposium, several years of gestation have allowed the book to ripen and to absorb new influences. Whereas some of the participants of the symposium could not contribute to the book, several new contributors who had not attended the symposium were recruited. We wish to thank all who stayed on board from the beginning as well as all those who joined us en route for their willingness to look at their research from a new perspective and for their efforts to work out the role of modularity. We also wish to acknowledge the Stiftung Volkswagenwerk for generously funding the symposium and the Hanse Institute for Advanced Study for providing financial support and hosting the symposium in pleasant surroundings. Christie Henry and Erik Carlson of the University of Chicago Press were very helpful in turning a collection of essays into a stylistically coherent and visually pleasing whole. Finally, we are indebted to several anonymous reviewers of this book and to the many colleagues—too numerous to name—who discussed modularity and related topics with us over the years.

Gerhard Schlosser, Bremen
Günter Wagner, New Haven
December 2003

1 Introduction: The Modularity Concept in Developmental and Evolutionary Biology

GERHARD SCHLOSSER AND
GÜNTER P. WAGNER

Georges Cuvier was one of the foremost biologists of his time. He unearthed many spectacular fossils firmly establishing that organisms were replaced by different species throughout the ages. But he never turned into an evolutionist, despite the fact that the revolutionary climate in France was quite beneficial for the growth of such formerly heretical ideas, and two evolutionists—Jean-Baptiste Lamarck and Geoffroy Saint-Hilaire—worked more or less next door (Mayr 1982; Appel 1987). While religious motives may certainly have contributed to Cuvier's refusal to accept evolution, he had more rational reasons as well. Cuvier was fascinated by the harmonious integration of parts in an organism. He perceived organisms as complex and indecomposable wholes that could function properly only because of the perfect fit of one component with the other (his "correlation of parts"). Cuvier even thought that a small fragment of a bone should allow us to reconstruct the form of an entire organism. Such complex wholes cannot undergo evolution, because changing a single part results in the complete breakdown of the functional organization of the whole.

Why start a book about modularity with the views of a pre-Darwinian biologist? Because Cuvier's view of organisms is the antithesis of the perspective that motivated this book. His ideas about organisms do not allow for modularity. His important insight was that the absence of modularity makes evolution impossible. However, ample evidence clearly shows that Cuvier's premise is false. Organisms are not the tightly integrated wholes that he envisioned. Rather, they turn out to be modular—that is, they are composed of quasi-independent parts that are tightly integrated within themselves (so on a smaller scale Cuvier is right, after all) but develop or operate to a certain degree independently of each other. This modular organization, which is apparent during embryonic development as well as in other phases of the life cycle, may have contributed to the evolvability of organisms, because it reduces the probability of trade-offs for evolutionary change. Moreover, it quite likely channels evolution because it facilitates certain kinds of evolutionary changes more than others. As a consequence, modules that develop or operate as quasi-autonomous units during the life cycle may also serve as building blocks of mosaic evolution.

The idea that a modular organization of organisms facilitates and channels their evolutionary transformation was already suggested (from quite different perspectives) by Riedl (1975), Lewontin (1978), and Bonner (1988) but did not attract much attention until recent advances in developmental genetics brought to light that many gene regulatory networks and signaling cascades were highly conserved during evolution but were frequently reshuffled and recombined to generate novel structures (reviewed in Gerhart and Kirschner 1997; Gilbert 2000; Carroll et al. 2001). These findings have kindled hopes that a better understanding of modularity will be a major step toward a new synthesis between the long divorced fields of developmental and evolutionary biology (see, e.g., Zuckerkandl 1994; Wagner 1995, 1996; Wagner and Altenberg 1996; García-Bellido 1996; Raff 1996; Gilbert et al. 1996; Gerhart and Kirschner 1997; Gilbert 2000; Kirschner and Gerhart 1998; Hartwell et al. 1999; von Dassow and Munro 1999; Niehrs and Pollet 1999; Brandon 1999; Bolker 2000; Dover 2000; Stern 2000; Schlosser and Thieffry 2000; Gilbert and Bolker 2001; Schank and Wimsatt 2001; Carroll et al. 2001; Winther 2001; Schlosser 2002, in press).

A New Focus on an Old Idea

Modularity may have been in vogue in developmental and evolutionary biology only for a short time, but it is certainly not a new idea in other fields. Ever since humans began to erect buildings or construct machines, they have employed modular modes of construction. It is therefore not surprising that the notion of modularity has always played a prominent role in applied sciences or arts such as engineering, architecture, and—more recently—computer science. In cognitive sciences, which initially borrowed the modularity concept from computer science (Fodor 1983), controversy has been raging for decades over whether information processing in the brain occurs in a more modular or in a more distributed fashion (for some recent reviews see, e.g., Goldberg 1995; Goodale 1996; Mussa-Ivaldi 1999; Erickson 2001; Cohen and Tong 2001). Recently, this discussion infected the social sciences, when evolutionary psychologists put forward a modular model—sometimes characterized as the Swiss-pocket-knife model—of cognitive and social capacities (Cosmides and Tooby 1992), which is contested by alternative models suggesting that specialized social skills develop by learning from more generalized cognitive or social capacities (see collections of different views in Hirschfeld and Gelman 1995; Davies and Holcomb 1999).

We will avoid any discussion of these controversies in this book for several reasons. First, there is already a rather extensive literature on these subjects. Second, controversies in the cognitive and social sciences are sometimes conducted in a rather ideological spirit and along well-defined battle lines, because our picture of ourselves is at stake. Third, the notion of module that has been defended by some of the supporters and attacked by some of the critics of modularity in these fields seems to be a pretty narrow one. Conceptualizing modules as strictly autonomous units is a simplification that is easy to attack. It appears to be more fruitful to acknowledge that modularity comes in degrees. Some subprocesses have a higher degree of internal integration and independence from their surroundings than others. The fruitful questions to ask are, therefore, not *whether* there are modules or not, but *where* they are, under which conditions they show autonomous behavior, and what consequences this has for the developmental and evolutionary dynamics of the containing system. Fourth and finally, it seems more promising to explore the importance of modularity for developmental and evolutionary biology in some depth rather than trying to cover everything. The styles of thinking in developmental and evolutionary biology are so different that it requires a great deal of bridge building to even start a meaningful dialogue between these disciplines. Beginning to build this bridge is the purpose of this book.

There are certain other biases in this book. The editors are both zoologists, and plants have not gotten the share they deserve, nor have microorganisms. So this is largely a book about the role of modules during the development and evolution of metazoans. Nonetheless, we think that many of the discussions are of relevance for these other fields as well. Bringing together various perspectives on one group of organisms (sufficiently complex and diversified in itself) seemed more fertile to us than being "politically correct" in our choice of taxa. Moreover, although the examples of modules discussed in this book range from units of gene regulation to entire organisms, we treat intermediate-level modules such as organs and other classical morphological characters relatively briefly, because they have been discussed extensively in another recent book (Wagner 2001).

The Marks of Modules: Integration and Autonomy

Like many other important notions in biology (e.g., homology and function), the notion of modularity is intuitively clear and simple but conceptually very difficult to characterize. So it is not surprising that a generally accepted definition of a module does not exist and different authors use the concept in quite different ways, as will be evident from the contributions to this book. This is to a certain extent unavoidable, but it also involves the danger that modularity will degenerate into a fashionable but empty phrase unless its precise meaning is specified at least on a case-to-case basis. At present, retaining a pragmatic pluralism of different modularity concepts is probably a fruitful strategy for broadening our perspective and illuminating the importance of modularity at many different levels of organization. Finally, however, we should venture to make the deep relationships between these different concepts explicit in order to go beyond anecdotal storytelling and provide a firmer theoretical basis for analyzing the roles of modules in the dynamics of complex systems. Only then can we systematically explore the conditions under which modules of development also tend to act as units of evolutionary transformation.

Most generally, a module is conceived of in one of two senses: either as a component of a system that operates largely independently of other components or as a component of a system that is repeatedly used. Here, we understand a module very broadly in the first sense, as it is the more general and inclusive notion. Modules in the second sense can only be the iterated units they are, because they retain their integrity in different contexts—that is, because they are also modules in the first sense.

Moreover, modules may be characterized from structure- or process-oriented perspectives. From a structure-oriented perspective, modules are characterized as units that have more or stronger connections or interactions among their components than with their surroundings (see, e.g., McShea and Venit 2001) or that are defined by intertwined circuits of interactions (see, e.g., Mendoza et al. 1999; Thieffry and Sánchez 2004). From a process-oriented perspective, modules are characterized as units of interacting components that operate in an integrated (interdependent) but relatively context-insensitive manner and therefore behave relatively invariantly in different contexts (see, e.g., von Dassow and Munro 1999; Gilbert and Bolker 2001; Schlosser 2002, 2004; von Dassow and Meir 2004). Importantly, the process-oriented view implies that modules can be identified only in relation to some specified reference process (Wagner 1995, 1996; von Dassow and Munro 1999; Schlosser 2002).

Two aspects of modules are emphasized by both perspectives: their *integration* concerning their internal relations (between their components) and their *autonomy* concerning their external relations (to elements of the context). The advantage of the process-oriented approach is that it can specify integration and autonomy in operationally definable terms and specifically singles out those units that make an integrated and autonomous contribution to the *dynamics* of a containing system. This is not necessarily true for modules identified by structure-based criteria. However, in most cases, the modules identified by both approaches will match, because high internal and low external connectivity of a unit will often be reflected in an integrated but context-insensitive dynamic role. This was first pointed out by Herbert Simon (1962) in his seminal paper "The Architecture of Complexity."

Simon (1962) also was among the first to emphasize that integration and autonomy of modules are not absolute but come in degrees. Thus, modular systems are "nearly decomposable systems," whose behavior in the long run can be understood only when interactions between their subsystems are taken into account. In particular, the behavior of modules may be insensitive to some perturbations of the context but not to others. The degree to which modules behave in a context-insensitive fashion thus depends on the frequency with which these different types of perturbations occur in a given environment (i.e., a certain reference process). But even modules that behave in a highly context-independent fashion once they have been initiated remain connected to other modules via their inputs and outputs. Therefore, modules can be embedded in higher-order modules in a hierarchical fashion (Simon 1962; Raff 1996; von Dassow and Munro 1999; Schlosser 2002).

For example, during the development of organisms, components that operate as integrated and context-insensitive units and thus qualify as developmental modules can be recognized at many different levels ranging from molecular interactions to entire organisms (for reviews see Gerhart and Kirschner 1997; Gilbert 2000; Carroll et al. 2001). *Gene regulation* is often modular because the recruitment of the transcriptional machinery to a gene promoter due to the binding of a complex of transcription factors to a particular enhancer (regulatory region) proceeds independently of transcription factor binding at another enhancer. As important participants in these modular processes, enhancers themselves are often christened "modules" (Kirchhamer et al. 1996; Arnone and Davidson 1997; Yuh et al. 1998; Davidson 2001) but this structural application of the term is clearly derivative: calling enhancers modules is justified by the modular nature of the regulatory processes in which they are involved. *Gene regulatory networks* and *signaling cascades* (e.g., the Notch, Wnt, Hedgehog, TGF-β, and receptor tyrosine kinase pathways) are often modules whose capacity to behave normally even in atypical environments is reflected in their multiple use in development. Importantly, the same regulatory network or signaling cascade may activate different downstream processes in different tissues depending on which other tissue-specifically expressed factors they are combined with. *Cell types,* like muscle cells or neurons, are modules that, once determined, can differentiate normally in ectopic environments. *Organs,* like the limb bud, segments, or other tissue compartments, are even higher-level modules that can to a certain degree develop and function even if transplanted to abnormal locations. Finally, entire *organisms* may act as modules in symbiotic associations of different individuals that may form an even higher-level individual (e.g., superorganisms such as societies of social insects or the eukaryotic cell, which arose by endosymbiotic association of different prokaryotes).

Modules in Development and Evolution

As pointed out above, the recent excitement about modularity stems largely from accumulating evidence that some of the modular units of development were highly preserved but promiscuously recombined during evolution. These findings have fueled hopes that understanding developmental and behavioral modularity will give us deep insights into *constraints* on evolutionary processes (see, e.g., Gould and Lewontin 1979; Alberch 1980, 1982; Maynard Smith et al. 1985; Wake and Larson 1987; Wake 1991; Schwenk 1994; Shubin and Wake 1996; Arthur 1997; Wagner and Schwenk 2000) and thereby allow us to explain those evolutionary trends that are inexplicable by selection for environmental adaptations. However, these hopes are usually built on the assumption that modules of development will act as coherent and quasi-independent units of evolutionary transformations—that is, that modules of development and modules of evolution will normally coincide. But this assumption is not necessarily true (as pointed out by von Dassow and Munro 1999; Sterelny 2000), and it may hold only under certain conditions (for detailed discussion see Schlosser 2004). Two points in particular deserve to be emphasized.

First, there is an important asymmetry between the developmental and evolutionary autonomy of a unit: a network of interacting elements will behave relatively invariantly during development and hence qualify as a developmental module when it is little affected by the context in which it is embedded, regardless of whether it itself has strong and important effects on its surroundings or not. However, in order to make an independent fitness contribution and thus to be evolutionarily autonomous, a developmental module must also be relatively isolated from its surroundings regarding its *effects*—that is, it must be endowed with a separable function. Only if developmental modules are also dedicated to different adaptive functions is modularity expected to enhance evolvability. Otherwise evolvability in fact may be decreased. Hence, the generally expected positive effect of modularity on evolvability does not follow from the existence of developmental modules alone, but requires that the developmental modularity matches the modularity of adaptive functions the organism has to perform. Modular function might be seen as locomotion and sensory functions, for instance. Of course, eyes and legs all contribute to locomotory performance measured in the wild, but their adaptive functions are separable to a certain degree, because the environmental variables they have to adapt to can be independent.

Second, in contrast to developmental modules, modules of evolutionary transformation should be units of the so-called genotype-phenotype map (Riedl 1975; Cheverud 1984, 1996; Wagner 1996; Wagner and Altenberg 1996; Mezey et al. 2000; Cheverud 2004; Wagner and Mezey 2004) defined by their coherent and autonomous response to heritable variations. It is therefore possible that different developmental modules, which are independently perturbable during development (e.g., by environmental fluctuations or somatic gene mutations) but have an overlapping genetic basis (i.e., involve the expression of some of the same genes), are linked together into a single module of evolution by frequent pleiotropic effects of heritable variations, for example, due to germ line mutations in coding regions of pleiotropically employed genes (note that nonpleiotropic effects are also possible, e.g., due to mutations in domain specific cis-regulatory regions). Left and right limbs, for instance, are independent developmental modules, but they evolve in a coordinated fashion.

As a consequence, the relation between developmental modules and evolutionary modules is not so straightforward as is often thought and is not necessarily one to one. Insights into developmental modularity need to be supplemented with information on its genetic representation in order to draw any meaningful inferences about the units of evolutionary change. This said, it remains an attractive hypothesis that developmental modularity facilitates and channels evolutionary change and thereby contributes to the evolvability of complex systems. This book is dedicated to exploring this hypothesis from a variety of perspectives.

Outline of the Book

Part 1 of the book addresses the molecular and developmental basis of modularity. The chapters of this part bring examples of developmental modules at many different levels, review the evidence for their integrated and autonomous behavior during development, and discuss implications for their evolutionary fate. Part 2 then turns to problems of recognition and modeling of modules. The chapters of this section address the question of how developmental modules can be identified, for instance, from data on gene expression patterns and use computer models to analyze parameters that are important for the integrated and autonomous behavior of modules. Part 3 addresses the evolutionary dynamics and origin of modules. This section deals with both ideas about the origin of modules in evolution and the evolutionary modifications of modules during the evolution of specific organ systems. Part 4, finally, is dedicated to special kinds of modules—namely, individuals that act as modules in higher level units. The contribution of these modules is discussed in two basic scenarios: the symbiotic recruitment of individuals from different lineages and the organization of multiple individuals of the same lineage into a higher-level colony or multicellular organism.

We conclude the book with a chapter (by Gerhard Schlosser) that tries to integrate various aspects of modularity discussed in this book and provides a general overview of the roles of modules in development and evolution. It explores the relationship between the two concepts in some detail and discusses how modularity may have contributed to evolvability and the capacity to develop (developability), emphasizing a twofold contribution of modularity in allowing mosaic changes and facilitating the combinatorial generation of complexity. Examples of developmental modules at different levels ranging from gene regulation to organs and other higher-level modules are presented. On the basis of comparative studies it is argued that many of these developmental modules were indeed also modules of evolutionary transformation during vertebrate phylogeny.

References

Alberch, P. 1980. Ontogenesis and morphological diversification. *Am. Zool.* 20: 653–667.

Alberch, P. 1982. The generative and regulatory roles of development in evolution. In *Environmental adaptation and evolution,* ed. D. Mossakowski and G. Roth, 19–36. Stuttgart: Fischer.

Appel, T. A. 1987. *The Cuvier-Geoffroy debate: French biology in the decades before Darwin.* New York: Oxford University Press.

Arnone, M. I., and E. H. Davidson. 1997. The hardwiring of development: organization and function of genomic regulatory systems. *Development* 124:1851–1864.

Arthur, W. 1997. *The origin of animal body plans.* Cambridge: Cambridge University Press.

Bolker, J. A. 2000. Modularity in development and why it matters to evo-devo. *Am. Zool.* 40:770–776.

Bonner, J. T. 1988. *The evolution of complexity.* Princeton, N.J.: Princeton University Press.

Brandon, R. N. 1999. The units of selection revisited: the modules of selection. *Biol. Philos.* 14:167–180.

Carroll, S. B., J. K. Grenier, and S. D. Wearherbee. 2001. *From DNA to diversity.* Malden: Blackwell Science.

Cheverud, J. M. 1984. Quantitative genetics and developmental constraints on evolution by selection. *J. Theor. Biol.* 110:155–171.

Cheverud, J. M. 1996. Developmental integration and the evolution of pleiotropy. *Am. Zool.* 36:44–50.

Cheverud, J. M. 2004. Modular pleiotropic effects of quantitative trait loci on morphological traits. In *Modularity in development and evolution,* ed. G. Schlosser and G. P. Wagner. Chicago: University of Chicago Press.

Cohen, J., and F. Tong. 2001. The face of controversy. *Science* 293:2405–2407.

Cosmides, L., and J. Tooby. 1992. Cognitive adaptations for social exchange. In *The adapted mind,* ed. J. H. Barkow, L. Cosmides, and J. Tooby, 163–227. Oxford: Oxford University Press.

Davidson, E. H. 2001. *Genomic regulatory systems.* San Diego: Academic Press.

Davies, D. S., and H. Holcomb, eds. 1999. *The evolution of minds: psychological and philosophical perspectives.* Dordrecht: Kluwer.

Dover, G. 2000. How genomic and developmental dynamics affect evolutionary processes. *BioEssays* 22:1153–1159.

Erickson, R. P. 2001. The evolution and implications of population and modular neural coding ideas. *Prog. Brain Res.* 130:9–29.

Fodor, J. A. 1983. *The modularity of mind: an essay on faculty psychology.* Cambridge, Mass.: MIT Press.

García-Bellido, A. 1996. Symmetries throughout organic evolution. *Proc. Natl. Acad. Sci. U.S.A.* 93:14229–14232.

Gerhart, J., and J. Kirschner. 1997. *Cells, embryos, and evolution.* Malden: Blackwell Science.

Gilbert, S. F. 2000. *Developmental biology.* Sunderland: Sinauer.

Gilbert, S. F., and J. A. Bolker. 2001. Homologies of process and modular elements of embryonic construction. In *The character concept in evolutionary biology,* ed. G. P. Wagner, 559–579. San Diego: Academic Press.

Gilbert, S. F., J. M. Opitz, and R. A. Raff. 1996. Resynthesizing evolutionary and developmental biology. *Dev. Biol.* 173:357–372.

Goldberg, E. 1995. Rise and fall of modular orthodoxy. *J. Clin. Exp. Neuropsychol.* 17:193–208.

Goodale, M. A. 1996. Visuomotor modules in the vertebrate brain. *Can. J. Physiol. Pharmacol.* 74:390–400.

Gould, S. J., and R. C. Lewontin. 1979. The spandrels of San Marco and the Panglossian paradigm: a critique of the adaptationist programme. *Proc. R. Soc. Lond. B Biol. Sci.* 205:581–598.

Hartwell, L. H., J. J. Hopfield, S. Leibler, and A. W. Murray. 1999. From molecular to modular cell biology. *Nature* 402 (suppl.): C47–C52.

Hirschfeld, L., and S. Gelman, eds. 1995. *Mapping the mind: domain specificity in cognition and culture.* Cambridge: Cambridge University Press.

Kirchhamer, C. V., C.-H. Yuh, and E. H. Davidson. 1996. Modular cis-regulatory organization of developmentally expressed genes: two genes transcribed territorially in the sea urchin embryo, and additional examples. *Proc. Natl. Acad. Sci. U.S.A.* 93: 9322–9328.

Kirschner, M., and J. Gerhart. 1998. Evolvability. *Proc. Natl. Acad. Sci. U.S.A.* 95: 8420–8427.

Lewontin, R. C. 1978. Adaptation. *Sci. Am.* 239 (3): 156–169.

Maynard Smith, J., R. Burian, S. Kauffman, P. Alberch, J. Campbell, B. Goodwin, R. Lande, D. Raup, and L. Wolpert. 1985. Developmental constraints and evolution. *Q. Rev. Biol.* 60:265–287.

Mayr, E. 1982. *The growth of biological thought: diversity, evolution, and inheritance.* Cambridge, Mass.: Belknap Press.

McShea, D. W., and E. P. Venit. 2001. What is a part? In *The character concept in evolutionary biology,* ed. G. P. Wagner, 259–284. San Diego: Academic Press.

Mendoza, L., D. Thieffry, and E. R. Alvarez-Buylla. 1999. Genetic control of flower morphogenesis in *Arabidopsis thaliana:* a logical analysis. *Bioinformatics* 15:593–606.

Mezey, J. G., J. M. Cheverud, and G. P. Wagner. 2000. Is the genotype-phenotype map modular? a statistical approach using mock quantitative trait loci data. *Genetics* 156:305–311.

Mussa-Ivaldi, F. A. 1999. Modular features of motor control and learning. *Curr. Opin. Neurobiol.* 9:713–717.

Niehrs, C., and N. Pollet. 1999. Synexpression groups in eukaryotes. *Nature* 402: 483–487.

Raff, R. A. 1996. *The shape of life.* Chicago: University of Chicago Press.

Riedl, R. 1975. *Die Ordnung des Lebendigen.* Hamburg: Parey.

Schank, J. C., and W. C. Wimsatt. 2001. Evolvability: adaptation and modularity. In *Thinking about evolution,* ed. R. S. Singh, C. B. Krimbas, D. Paul, and J. Beatty, 322–335. Cambridge: Cambridge University Press.

Schlosser, G. 2002. Modularity and the units of evolution. *Theory Biosci.* 121:1–80.

Schlosser, G. In press. Amphibian variations: the role of modules in mosaic evolution. In *Modularity: understanding the development and evolution of natural complex systems,* ed. W. Callebaut and D. Rasskin-Gutman. Cambridge, Mass.: MIT Press.

Schlosser, G. 2004. The role of modules in development and evolution. In *Modularity in development and evolution,* ed. G. Schlosser and G. P. Wagner. Chicago: University of Chicago Press.

Schlosser, G., and D. Thieffry. 2000. Modularity in development and evolution. *BioEssays* 22:1043–1045.

Schwenk, K. 1994. A utilitarian approach to evolutionary constraint. *Zoology* 98: 251–262.

Shubin, N., and D. Wake. 1996. Phylogeny, variation, and morphological integration. *Am. Zool.* 36:51–60.

Simon, H. A. 1962. The architecture of complexity. *Proc. Am. Philos. Soc.* 106:467–482.

Sterelny, K. 2000. Development, evolution, and adaptation. *Philos. Sci.* 67 (suppl.): S369–S387.

Stern, D. L. 2000. Evolutionary developmental biology and the problem of variation. *Evolution* 54:1079–1091.

Thieffry, D., and L. Sánchez. 2004. Qualitative analysis of gene networks: toward the delineation of cross-regulatory modules. In *Modularity in development and evolution,* ed. G. Schlosser and G. P. Wagner. Chicago: University of Chicago Press.

von Dassow, G., and E. Meir. 2004. Exploring modularity with dynamical models of gene networks. In *Modularity in development and evolution,* ed. G. Schlosser and G. P. Wagner. Chicago: University of Chicago Press.

von Dassow, G., and E. Munro. 1999. Modularity in animal development and evolution: elements of a conceptual framework for evodevo. *J. Exp. Zool. (Mol. Dev. Evol.)* 285:307–325.

Wagner, G. P. 1995. The biological role of homologues: a building block hypothesis. *Neues Jahrb. Geol. Paläontol. Abh.* 19:36–43.

Wagner, G. P. 1996. Homologues, natural kinds and the evolution of modularity. *Am. Zool.* 36:36–43.

Wagner,G. P., ed. 2001. *The character concept in evolutionary biology.* San Diego: Academic Press.

Wagner, G. P., and L. Altenberg. 1996. Complex adaptations and the evolution of evolvability. *Evolution* 50:967–976.

Wagner, G. P., and J. G. Mezey. 2004. The role of genetic architecture constraints in the origin of variational modularity. In *Modularity in development and evolution,* ed. G. Schlosser and G. P. Wagner. Chicago: University of Chicago Press.

Wagner, G. P., and K. Schwenk. 2000. Evolutionarily stable configurations: functional integration and the evolution of phenotype stability. *Evol. Biol (N.Y.).* 31:155–217.

Wake, D. B. 1991. Homoplasy: the result of natural selection, or evidence of design limitations? *Am. Nat.* 138:543–567.

Wake, D. B., and A. Larson. 1987. Multidimensional analysis of an evolving lineage. *Science* 238:42–48.

Winther, R. G. 2001. Varieties of modules: Kinds, levels, origins, and behaviors. *J. Exp. Zool. (Mol. Dev. Evol.)* 291:116–129.

Yuh, C.-H., H. Bolouri, and E. H. Davidson. 1998. Genomic cis-regulatory logic: experimental and computational analysis of a sea urchin gene. *Science* 279:1896–1902.

Zuckerkandl, E. 1994. Molecular pathways to parallel evolution. 1. Gene nexuses and their morphological correlates. *J. Mol. Evol.* 39:661–678.

Part 1

The Molecular and Developmental Basis of Modularity

Angels and devils, mermaids and winged horses, dragons and unicorns: the fantastic beasts created by the human imagination are always new mixtures of familiar parts. However (as pointed out by R. A. Raff in *The Shape of Life* [1996]), these monstrous creations do not only attest the limits of human imagination; they are often composed of developmentally or functionally well-integrated units (wings, tails, etc.) in real animals, which are perceived as relatively autonomous—that is, which can easily be imagined to operate similarly in a different organic environment (e.g., wings on angels, devils, or horses). Such integrated but relatively context-insensitive, robust behavior is the hallmark of modules. Therefore, modules can be operationally identified by the fact that they develop or behave relatively normally in ectopic environments or that they are multiply employed in a single organism. Developmental modules can be recognized at many different levels, ranging from molecular interactions (e.g., between transcription factors and enhancers during gene regulation), gene regulatory networks, and signaling cascades to segments and complex organs.

The chapter by Craig Nelson discusses the genetic architecture of modules at the level of cell types, compartments, segments, and organs and how it may have evolved. Each of these modules is controlled by a selector gene encoding a transcription factor. In contrast to the traditional view of hierarchical control, selector genes regulate downstream genes at many different levels, often cooperating with signaling cascades to establish patterned gene expression domains. Because of their combinatorial interaction with selector genes, the same signaling cascades can be used over and over again to pattern different tissues in metazoan embryos (activating different target genes in each tissue depending on which selector they are combined with).

The promiscuous combinability of selector genes and signaling pathways and the multiple recruitment of the same target gene to different higher-level modules is greatly facilitated by the modularity of gene regulation itself. This is addressed in the chapter by Uwe Strähle and Patrick Blader, which shows that dimers of basic helix-loop-helix (bHLH) transcription factors act as modules that bind to a well-defined binding site and interact in a similar fashion (e.g., as activators) with the basal transcriptional apparatus in many different target genes, even though the bHLH binding sites form part of very different enhancers in each of these target genes. Furthermore, they show that enhancers themselves also qualify as modules because they can confer domain-specific expression independent of other enhancers.

The next chapter, by Gabrielle Kardon, Tiffany A. Heanue, and Clifford J. Tabin, then moves one step up in the hierarchy of modules and shows that entire networks of regulatory genes that crossregulate each other's expression can act as modules. The regulatory network involving *Pax, Six, Eya,* and *Dach* genes acts as a module, because the context insensitivity of regulatory interactions between these four players makes it apparently easy to activate this network in ectopic domains (e.g., after *Pax6* misexpression). This developmental robustness may also explain the ease with which the network is embedded into new contexts during evolution. The redeployment of the network for several novel developmental processes (such as ear and muscle development) during vertebrate evolution apparently was facilitated by the possibility of recruiting different paralogues of Pax or Six family transcription factors into the network.

Signaling cascades are another type of module that is characterized by extreme versatility. Apparently only five major signaling pathways are employed during metazoan embryogenesis, and these are used over and over again. The Notch module that José F. de Celis reviews in his contribution is involved in lateral inhibition or boundary formation during the development of many different structures derived from all three germ layers. While the pattern of interaction between its components and its logical role—to create differences between adjacent cells—seems to be conserved wherever it is employed and hence context insensitive, it nonetheless drives different cell fate decisions in each domain, because both the activation of the Notch module and the kind of target genes activated by it depend on the domain-specific availability of other transcription factors.

Domain-specific cooperation with other transcription factors at the cis-regulatory regions of target genes also explains how other signaling pathways, such as the sonic hedgehog and Wnt signaling pathways discussed in Anne-Gaëlle Borycki's chapter, can modulate different developmental decisions in each of their multiple domains, although the interactions between the components of a given signaling pathway are highly conserved. In addition, different signaling pathways may interact. One pathway may, for instance, control the availability of components for another pathway, thereby regulating the competence to respond to its particular ligand. Or two pathways may cooperate and form a higher-order module with a characteristic behavior. Cooperation of Shh and Wnt signaling pathways, for example, is specifically observed in processes of boundary formation or epithelial mesenchymal transitions.

Whereas all chapters so far characterize the context insensitivity of modules by their developmental autonomy, that is, by their ability to operate normally in different molecular and cellular environments, the following chapter, by James M. Cheverud, looks at modules from the perspective of quantitative genetics, addressing the modularity of pleiotropic effects of genetic variation. The chapter summarizes studies of quantitative trait loci (QTL) of many different traits in mice indicating that genetic variation on a given locus (QTL) typically has pleiotropic effects on different traits but that these pleiotropic effects are often restricted to modules of development or function (e.g., certain muscle attachment sites in the mouse mandible) and allow quasi-independent heritable variation of them. Because there is genetic variation for the degree of pleiotropic effects, such modular genotype-phenotype maps may be an evolved property of organisms.

The final chapter of part 1, by Christoph Redies and Luis Puelles, focuses on higher-level modules such as segments or compartments composed of many cells. The authors illustrate clearly that the modular organization of a developing organism can itself be reorganized. During development of the vertebrate central nervous system, embryonic modules (segments like prosomeres and rhombomeres) are transformed into the functional units of the adult nervous system. Cell adhesion molecules of the cadherin family play an important role during this process.

2 Selector Genes and the Genetic Control of Developmental Modules

CRAIG NELSON

Modules in Development and Evolution

Modularity pervades every level of biological organization. From proteins to populations, larger biological units are built of smaller, quasi-autonomous parts. This type of organization is essential to much of biological function. In particular, modular design enables evolutionary change. To quote Raff, "it is the property of modularity that allows the evolutionary dissociation of the developmental process and thus makes the evolution of development possible" (Raff and Sly 2000).

But what makes modularity possible? Specifically, how does the genetic architecture that controls the development of complex organisms encode the modular anatomical units that are the substrate of morphological evolution? To begin to address this question, this chapter will focus on the genetic control of the development of the primary anatomical modules which make up metazoan bodies, namely, segments, organs, compartments, and cell types. The genetic circuitry that controls the development of each of these modules shares common architectural features that facilitate their evolution. Specifically, the direct but contingent control of large regulatory hierarchies by a special class of transcription factors, known as selector genes, enables the dissociation, duplication, divergence, and co-option of modules that underlie morphological evolution.

In 1975 Riedl formalized the idea that anatomical complexity arises from the repetitive use of "standard parts" arranged in unique hierarchical combinations (Riedl 1978). This hierarchy is manifested as bodies made of organs, organs made of tissues, tissues made of cells, and so on. The modularity of these parts is evident in their deployment and development within a given animal, and in their continuity between species. Within any animal body, cell types, tissues, and organs occur repeatedly in multiple contexts. The overall form and function of any given body part is determined by the mix of these component parts. Complexity in animal body plans is derived from the arrangement and rearrangement of these component parts into larger functional units (Riedl 1978). The developmental autonomy of these units is apparent in the fact that, once induced, many of them can develop normally without regard to their precise surroundings. The fundamental evolutionary continuity of these parts between species is implicit in the study of comparative anatomy. As pointed out by Riedl, "the qualitative aspect of order lies . . . in the qualities of that component part of an organism which in morphology is named the homologue" (Riedl 1978). The ability of these modules to vary independently of one another is the substance of morphological evolution and allows for allometric, heterochronic, and neomorphic modification of these parts between lineages (Gilbert et al. 1996; Raff 1996; Wagner and Altenberg 1996; Raff and Sly 2000).

The impact of modular biological organization on the evolutionary process has become a topic of great interest to those who study the evolution of development and has been a central concept in a number of key contributions to the recent synthesis of developmental and evolutionary biology (Gilbert et al. 1996; Raff 1996; Wagner and Altenberg 1996; Gerhart and Kirschner 1997; von Dassow and Munro 1999; Carroll et al. 2001). Much of this work has stressed the evolutionary implications of modular organization, including the contribution of modular design to the generation of complexity, developmental robustness, and evolutionary flexibility or evolvability. Classical developmental concepts such as the morphogenetic field have been recast as the modules of development and the units of evolutionary change (Gilbert et al. 1996). Many of the common properties of modules at different levels of organization have been discussed, and these common properties have led to increasingly precise definitions of biological modularity (see Schlosser 2004).

The Genetic Control of Modularity

With a few notable exceptions (Carroll et al. 2001; Davidson 2001), relatively less explicit attention has been paid to common properties of the genetic architecture that controls the development of these modular units. How are developmental modules such as morphogenetic fields genetically differentiated from the surrounding tissue? What allows for the duplication of developmental modules? How do homologous developmental modules diverge over evolutionary time? What type of regulatory architecture controls the development of modules and how does this facilitate or constrain evolutionary change? What properties of this genetic architecture facilitate the generation of vast anatomical complexity from a finite set of modules? It is the goal of this chapter to begin to address some of these issues by describing unifying aspects of the genetic architecture controlling the development of the four main modules of anatomical organization—the segment, the organ, the compartment, and the cell—and discussing how this architecture effects the evolution of anatomical complexity.

Just as anatomy is composed of hierarchically nested modules, so, too, does the development of bilaterian animals proceed by repeatedly dividing the embryo into a suite of nested domains (Davidson 2001). These domains are established by cell-to-cell signaling processes that constitute the pattern formation engine of development and are marked by domains of cells with unique transcription profiles (these domains are sometimes referred to as equivalence groups). Within each transcriptional domain the patterning process is repeated (often with the same signaling pathways that patterned the preceding domain) and new transcriptional subdomains are established, within which the patterning process can be repeated yet again, leading to further subdivision, and so on, until the precise location of every cell type in the final body plan has been established and is marked by a unique profile of transcription factor expression (Davidson 2001). The establishment of these domains of unique transcriptional profiles is the first step in setting apart developmental modules from their neighbors.

While this process of patterning and subdivision is very deep, with many rounds of regionalization and patterning occurring between the initial division of the fertilized egg and the final differentiation of the cells making up the adult body, a few discrete levels of this regionalization stand out due to their correspondence with the primary modules of anatomical organization. Each of these levels of organization is controlled by the actions of a special class of transcription factors known as selector genes. The term "selector gene" was originally coined by García-Bellido to describe genes that define nonintermixing compartments of cells within developing organs (García-Bellido 1975). This definition has since been expanded to include transcription factors that define segmental, organ, and tissue-specific identities (Carroll et al. 2001). Some general features of selector genes include the fact that they are expressed in the structures whose identity they define, that they are required for the specification of that identity, and that under the appropriate conditions they are sufficient to specify that identity and all of the subsequent developmental processes that specification entails.

Thus, there exists a class of genes known as selector genes whose activity specifies the identity and subsequent development of the primary modules of anatomical organization. We might, therefore, reasonably expect to gain insight to the modular development and evolution of animal bodies by investigating the way in which these genes control the genetic regulatory architecture underlying these modules. Specifically, in this chapter we investigate the architecture of the regulatory networks controlled by selector genes which lead to the modular development and evolution of animal body parts and address how selector genes facilitate the generation of great anatomical complexity with a limited set of patterning cues.

Example of a Selector Hierarchy Controlling Development

Perhaps the best way to illustrate how selector genes act hierarchically to generate complex anatomy is through an example (fig. 2.1). In our lab we study the development of the flight appendages of the fruit fly, *Drosophila melanogaster.* The flight appendages of the adult fruit fly include two organs which develop in the second and third thoracic segments (T2 and T3) of the fly larva. T2 normally grows a wing, while T3 normally grows a winglike balancing organ called a haltere.

T2 and T3 and their flight appendages are differentiated from one another by the action of the Hox selector gene *Ubx. Ubx* is expressed in T3 and therefore the haltere primordium but not in T2 and the wing primordium. In the absence of *Ubx* activity, T3 develops as T2, and the flight organ of T3, the haltere, develops as a wing (Lewis 1978). Thus, we can see that the first layer of selector gene activity which specifies the identity of the segment, and therefore the organ it contains, is the activity of the Hox selector gene *Ubx.*

The second layer of selector gene activity controlling flight appendage development is the specification of the discrete organ identity within these segments by the organ-level selector complex VG-SD (the protein products of the *vestigial* and *scalloped* genes respectively). Within several segments of the larval fly, imaginal discs are set aside as primordia of the adult body wall and external organs. The development of each of these imaginal discs is controlled by one organ-specific selector gene, or by a combination of them. The development the flight organs, the wing and the haltere, is controlled by the activity of the wing selector gene complex VG-SD (Kim et al. 1996; Halder et al. 1998b). In the absence of this complex the fly lacks flight appendages entirely, while if this selector complex is misexpressed in one of the other imaginal discs, that disc can be converted from its normal identity into a wing (Kim et al. 1996). Thus, the VG-SD selector complex is both necessary and sufficient to induce the development of the flight appendage from an imaginal disc. The way in which this selector complex controls the development of these organs is the subject of later sections of this chapter.

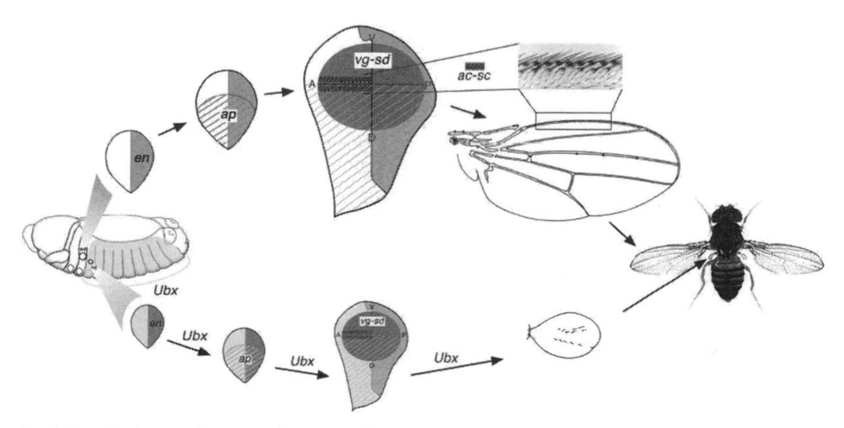

Fig. 2.1.—A schematic illustration of *Drosophila* flight appendage development from the second and third dorsal thoracic imaginal discs. Wing and haltere development proceeds from ectodermal imaginal discs derived from the second and third thoracic segments of the third-instar *Drosophila* larva, as illustrated on the far left of the figure. These discs are compartmentalized by the expression of *engrailed* in the posterior (*shaded*) and *apterous* (*diagonal lines*) in the dorsal halves of the disc. The portion of the disc that is to form the wing blade proper and the distal portion of the haltere is demarcated by VG-SD expression (*shaded*). Differentiated cell types are specified by the expression of cell type selector genes such as *achaete* and *scute* (*black stipples*), which direct the development of bristles in the mature structure. While development of the two structures (wing and haltere) are largely parallel, *Ubx* (*shaded*) represses the expression of target genes in the regulatory networks of each of the other selector genes, effectively transforming the ultimate morphology of the structure from a large flat wing blade into a small rounded haltere.

The third layer of selector gene activity defines the anteroposterior and dorsoventral compartments of the flight appendages through the activity of the compartmental selector genes *engrailed* and *apterous*. The definition of these compartments is essential to the patterning and normal development of the wing. In the absence of *engrailed* function the posterior half of the appendage develops with anterior identity, while in the absence of *apterous* function the appendage never grows out at all due to the loss of signaling centers whose specification is dependent on the normal division of the imaginal disc into dorsal and ventral territories (Blair 1995).

The final layer of selector gene activity deployed in wing development is the specification of cell types within the appendage by the activity of cell-type-specific selector genes such as the genes of the *achaete-scute* complex, which controls the development of neurons (Skeath and Carroll 1994). The expression of cell type selector genes is typically controlled by the signals patterning the appendage, ensuring the proper placement and arrangement of all of the component parts of the organ.

Thus, the unique identity of any cell within the flight appendage is marked by the combination of selector genes expressed within its nucleus. The expression of *Ubx* defines the segmental identity and therefore whether the cell will develop as a wing or haltere. The presence of the VG-SD selector complex defines the cell as a part of the flight organ and elicits within it the developmental program required to build that organ. The expression of *engrailed* and *apterous* defines the location of the cell within the organ and helps to pattern the structure. Finally, the expression of tissue-specific selector genes like *achaete-scute* defines the final differentiated phenotype of the individual cell in the mature appendage. The following section describes how selector genes control these developmental programs.

Selector Genes and the Control of Genetic Regulatory Hierarchies

With the exception of selector genes specifying terminally differentiated cell fates, selector genes specify structures whose development is controlled by deep genetic regulatory hierarchies characterized by many iterated rounds of patterning and transcriptional regionalization (Davidson 2001). This fact leads to the question of how selector genes influence these hierarchies. Because selector genes encode transcription factors, we can reasonably assume that they affect the transcription of downstream targets within these regulatory hierarchies. In principle, the influence of the selector gene could be limited to a few genes near the top of this hierarchy, with all else following from the activation of these few targets, or the selector could act throughout the hierarchy, infiltrating its influence in a very direct way on every step of the process. Many examples now suggest that selectors act by directly affecting the transcription of genes at many levels of the genetic regulatory hierarchies they control. A few examples of this are described in the following paragraphs.

Two selector genes act at distinct layers of the genetic hierarchy that controls flight appendage development in *Drosophila*. The first of these genes is *Ubx*, the Hox gene which differentiates the third thoracic segment from the second and therefore the haltere from the wing. The second is the VG-SD complex which controls the wing development program. Because the activity of *Ubx* appears to have been superimposed on the developmental program controlled by the VG-SD selector complex, we will begin by discussing what we have learned about the control of appendage development by VG-SD and then address the way in which this program is modified by *Ubx* activity.

By analyzing a number of wing-specific enhancers of genes known to be involved in wing development we hoped to begin to understand how the VG-SD complex controlled the expression of these genes and therefore, by extension, the wing development program as a whole (Guss et al. 2001). Genetic and molecular biological analysis revealed that each of these target enhancers is directly controlled by VG-SD through arrays of VG-SD binding sites (Halder et al. 1998b; Guss et al. 2001). None of the enhancers analyzed lacked direct input from the VG-SD selector complex. The fact that these targets genes act at multiple levels of the wing development program reveals that the VG-SD selector complex coordinates wing development by directly affecting the expression of many of the genes within the regulatory network it controls.

A similar scenario appears to be true for *Ubx*. As previously described, *Ubx* guides the T3 flight appendage to develop into a haltere rather than a wing. Genetic removal of *Ubx* activity from individual groups of cells in the developing haltere causes these cells to develop as wing cells (Weatherbee et al. 1998). The cellular autonomy of this effect suggests that *Ubx* directly regulates the wing program individually in each cell of the developing appendage. Expression analysis of a number of the individual genes that constitute the wing development program revealed that *Ubx* represses the expression of each of them in cells fated to develop as haltere (Weatherbee et al. 1998). These two findings suggest that *Ubx* modulates the wing development program controlled by VG-SD by directly repressing the expression of many individual genes in this program in cells of the developing haltere.

HNF3β and GATA-4 are expressed throughout the definitive endoderm of the vertebrate embryo and are essential for the proper development and differentiation of definitive endoderm derivatives, including the liver (Ang et al. 1993; Zaret 1996). In addition, HNF3β and GATA homologues in *Caenorhabditis elegans* and *Drosophila* appear to play a role in the specification of the definitive endoderm and can be sufficient to drive cells to endodermal fates (Weigel et al. 1989; Zhu et al. 1997; Zhu et al. 1998). Thus, HNF3β and GATA proteins appear to act as selector genes governing the development of definitive endoderm derivatives, including the liver. In vivo footprinting studies have shown that HNF3β and GATA factors bind to and directly regulate target genes expressed in the liver, including liver-specific serum albumin (Gualdi et al. 1996; Bossard and Zaret 1998). Because HNF3β and GATA proteins appear to act both very early in the specification of the endodermal derivatives and later, controlling the expression of liver-specific target genes, these studies suggest that HNF3β may directly regulate various levels of the endoderm development regulatory hierarchy in a way akin to the regulation of many target genes in the wing by the VG-SD selector complex.

Perhaps the best-known organ-level selector gene is the *Drosophila eyeless* gene and its vertebrate *Pax-6* homologues. One of the most remarkable stories of evolutionary conservation, the *Pax-6* family genes play an essential role in directing eye development in all organisms with eyes or photoreceptive cells (Gehring 1996). While the genetic regulatory networks controlled by *Pax-6* and its collaborators have not yet been exhaustively characterized, it has been suggested that *Pax-6* directly regulates crystallin expression in the lens (Sheng et al. 1997; Papatsenko et al. 2001) and rhodopsin expression in the photoreceptors (Sheng et al. 1997; Papatsenko et al. 2001). In addition to these very downstream markers of cellular differentiation, *Pax-6* must also directly regulate at least some target genes near the top of the regulatory program for eye development, as it is capable of directing the entire eye development program. Three such candidate targets of direct *Pax-6* regulation include *sine oculis, eyes absent,* and *dachshund,* each of which is required for *Pax-6* to direct the formation of an eye and each of which acts near the top of the eye development program (Shen and Mardon 1997; Halder et al. 1998a; see also Kardon et al. 2004). Thus it would appear that, like the other selector genes discussed here, *Pax-6* exerts its regulatory functions by binding directly to the enhancers of target genes throughout the regulatory program which it choreographs.

While the list of examples of selector genes which control their downstream regulatory networks by directly binding to the enhancers of target genes within these networks could be extended, these examples serve to prove the point that selector input to genetic regulatory hierarchies is not limited to the tops of these hierarchies. Rather, direct control by the selector gene appears to infiltrate all subordinate levels of the hierarchy. This type of control has some interesting potential explanations and consequences for the evolution of development.

Davidson has suggested that the primitive role for organ-level selector genes such as *eyeless/Pax-6, tinman,* and VG-SD was to direct the formation of terminally differentiated cell types with specialized functions, such as primitive photoreceptive cells or contractile cells (Davidson 2001). As these groups of cells evolved into increasingly complex organs, the selector gene which controlled their development served to mark the cells committed to that organ fate. Thus, as evolution selected for increasingly sophisticated organ structures composed of derivatives of the original cell group, the original cell type selector was recruited to control the expression of an increasingly complex genetic regulatory hierarchy. The assembly of a complex organ from a simple group of differentiated cells through the acquisition of new target genes should leave its evolutionary tracks in the direct control of each new target, in each new cell type, by the selector gene (fig. 2.2).

Because selector genes appear to directly regulate the majority of their targets, it is easy to envision the gradual acquisition and modification of these target networks over evolutionary time through the acquisition and loss of selector protein binding sites. Presumably the gradual acquisition of target genes would generally lead to gradual changes in the morphologies controlled by the selector in question. These changes could either be reinforced or eliminated by selective pressure, ultimately reshaping the animal without radical remodeling. This architecture would also make the network robust in the face of mutation. The distributed control of the network by the selector should keep much of the network functional even if portions of it are disrupted by the mutation of a component part, thus limiting the effects of mutation on the morphology of a structure.

Conversely, however, this type of architecture places great emphasis on the expression of the selector gene itself. Without the selector all is lost. The specific repression of selector gene expression therefore represents a potential strategy for quickly and dramatically reshaping the body plan of an animal with a minimum of genetic change. This is exactly the strategy that appears to have been employed in the repression of limb development in the posterior segments of primitive insects, leading to a limb-less abdomen. Primitive arthropods often have limbs on most of their body segments, including the abdominal segments. Insects, a derived arthropod group, have lost these abdominal limbs. The repression of abdominal limb growth appears to be due to the repression of the expression of the limb selector gene *Distalless* by the abdominal Hox genes (Panganiban et al. 1994; Warren et al. 1994; Warren and Carroll 1995; Palopoli and Patel 1998). Therefore, acquisition of binding sites for the abdominal Hox genes *AbdA* and *Ubx* in the *Distalless* (*Dll*) enhancer may have been the critical step for the evolution of the limb-less insect abdomen. While this suggests that a single locus was critically important to the evolution of a novel morphology, it does not suggest that the change in the regulation of this locus was necessarily sudden. In fact, recent data suggests that the acquisition of Hox regulation may be a stepwise and gradual process (Galant et al. 2002).

Repression of selector gene expression, however, while a temptingly rapid strategy for sculpting the body through the controlled loss of organs, does not appear to underlie all such morphological transformations. For instance, cave-dwelling populations of the fish *Astyanax mexicanus* have repeatedly lost their eyes. Perhaps the easiest and most complete way to have accomplished this loss would have been to abolish the expression of *Pax-6*, the selector gene for eye development. Molecular analysis of cave fish embryos however, shows that *Pax-6* expression has not been lost from the eye primordia of the embryo (Strickler et al. 2001), suggesting that the evolutionary alterations to eye development that result in the eyelessness of these fish have occurred in downstream portions of the eye development program.

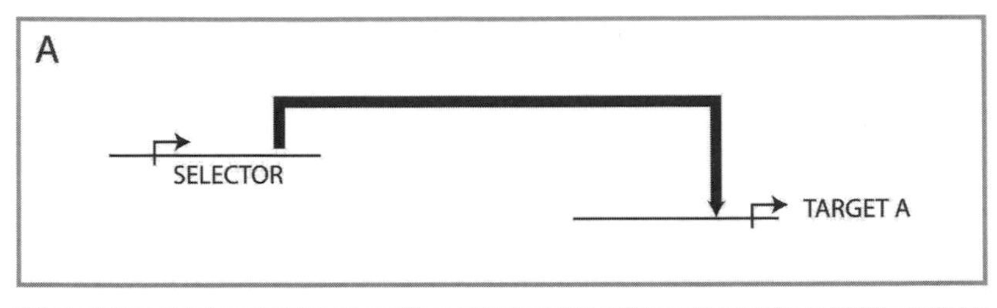
A
SELECTOR
TARGET A

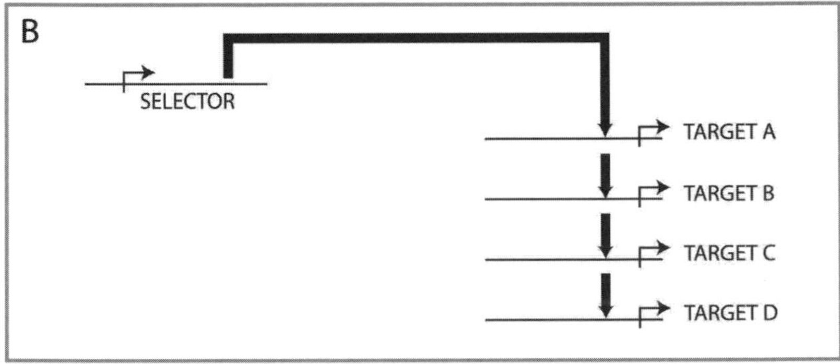
B
SELECTOR
TARGET A
TARGET B
TARGET C
TARGET D

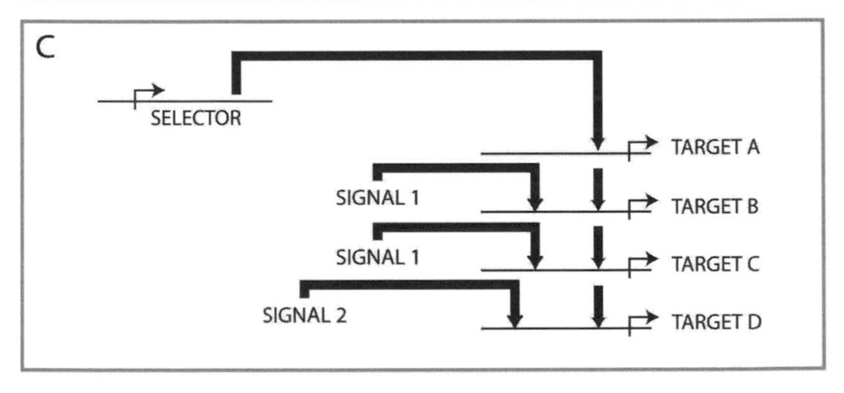
C
SELECTOR
TARGET A
SIGNAL 1
TARGET B
SIGNAL 1
TARGET C
SIGNAL 2
TARGET D

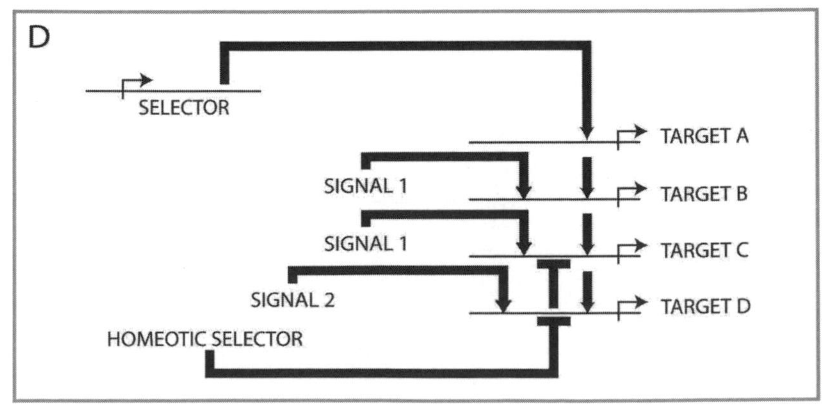
D
SELECTOR
TARGET A
SIGNAL 1
TARGET B
SIGNAL 1
TARGET C
SIGNAL 2
TARGET D
HOMEOTIC SELECTOR

Contingency and Selector Gene Action

The evolution of organ-level selector genes from primitive cell type selectors raises some interesting questions about the way in which the function of the selector itself would need to change over the course of this evolutionary process. Selectors of terminally differentiated phenotypes do not need to direct the differentiation of multiple cell types in precisely organized patterns. They simply need to turn on the differentiation program appropriate for the cell type in question. Organ-level and regional selectors, on the other hand, must orchestrate multiple downstream genetic hierarchies, including multiple cycles of regionalization and patterning and the induction of multiple terminally differentiated cell types. In order to do this, the selector must evolve contingency mechanisms such that it can coordinate multiple different genetic programs in the appropriate time and place. How is this contingency of selector target gene activation accomplished?

In general, transcriptional contingency is accomplished by combinatorial control of target cis-regulatory enhancers (Gerhart and Kirschner 1997). Multiple regulatory inputs must often be summed to elicit a transcriptional response. The activation of a target gene will often be contingent not only on the presence of one or more selector genes that define the identity of the developing organ but also on the intercellular signal transduction pathways patterning the organ. A few examples of this type of combinatorial regulation are discussed in the following paragraphs.

While we had shown that binding of the VG-SD selector complex was required for the activation of gene expression in the developing flight appendage in *Drosophila*, we suspected that it could not be sufficient for activation, since none of the target enhancers was activated throughout the region of VG-SD gene expression. Rather, each of the target enhancers we studied was expressed in a specific time and place within the wing disc, often in response to the intercellular signals known to be patterning the developing wing. This suggested that at least these two inputs (signal and selector) would be required to drive

Fig. 2.2.—Evolution of a selector regulatory network. This figure schematizes aspects of the "bottom-up" evolution of a selector regulatory network, as discussed in this chapter (see also Davidson 2001). Panel *A* illustrates the most primitive condition, in which the selector drives the expression of a single gene important for the differentiation of a specific cell type within the embryo. Panel *B* illustrates the modification of that cell type through the acquisition of other target genes by the selector through the evolution of new selector binding sites in the cis-regulatory regions of those target genes. Panel *C* illustrates the modification of certain cells within the structure through the dependence of specific target genes on selector input and cell-cell signaling pathways. Finally, panel *D* illustrates the differentiation of serial homologues by the influence of a homeotic selector on specific target genes within the regulatory network. This combinatorial regulation of target genes allows the generation of multiple cell types within a complex organ while the original selector gene retains control of organ development.

organ-specific patterned gene expression, and that perhaps these two inputs might be sufficient. We tested this possibility by building small synthetic enhancers composed of binding sites for the wing selector complex, VG-SD, and the transcriptional mediators of two intercellular signaling pathways, *Notch* and *Dpp*. We found that these synthetic enhancers were able to drive expression in the wing in response to the VG-SD complex and in patterns defined by the intercellular signaling pathways whose binding sites had been used in the constructs. In fact, the expression of these synthetic enhancers was very similar to that driven by the endogenous enhancers known to be responsive to these very same inputs. Arrays of selector complex or signal mediator binding sites alone, however, did not drive gene expression (Guss et al. 2001). This finding has two interesting consequences. First, the target enhancers of the wing selector complex integrate organ identity and signal responsiveness directly on the DNA. Second, the need for both signal and selector inputs to activate gene expression provides a mechanism whereby highly organ-specific responses can be generated in response to signals used throughout the developing body (Guss et al. 2001). This is one explanation for the way in which an entire complex body consisting of many segments, organs, and cell types can be patterned by a very limited set of intercellular signals used over and over but always eliciting the appropriate response in the receiving tissue.

The collaboration between signaling effectors and selector gene inputs on target enhancers is an example of the much broader phenomenon of combinatorial gene regulation. Combinatorial regulation of transcription is the rule rather than the exception and is known to refine the response of enhancers to the cellular environment in everything from phage to humans. What is uniquely interesting about the kind of combinatorial control exemplified by the synergy between signals and the VG-SD selector in the *Drosophila* wing is the developmental context and logic embedded in the enhancer. By requiring two discrete classes of input, of which one defines the organ type and the other patterns the organ, the enhancer elicits tissue-specific genetic responses to signals utilized throughout the body. The same signaling pathway which turns on eye-specific genes in the eye turns on wing-specific genes in the wing by synergizing with the selector genes which define those developmental fields. There are now several other examples of a similar developmental regulatory logic that combines tissue-specific transcription factors and signaling pathways in the activation of specific target genes in response to broadly utilized cell-cell signals. These include the activation of two target genes in the differentiating photoreceptor cell of the *Drosophila* eye and the activation of *even-skipped* (*eve*) in a subset of the *Drosophila* cardiac progenitor cells.

While *Pax-6* and its collaborators (*eyeless, twin of eyeless, eyes absent,* and *dachshund*) appear to act as selector genes for the eye developmental program as a whole (see Kardon et al. 2004), a different transcription factor, *lozenge,* is critical for the specification of cellular identity and terminal differentiation of a subset of cells in the *Drosophila* eye (Flores et al. 1998). The transcription of *lozenge* initiates just behind the morphogenetic furrow in a subset of the differentiating cells of the ommatidia. Two different laboratories have shown that the presence of *lozenge* in these cells defines the cell-type-specific transcription of a target gene in response to incoming intercellular signals. In the first study *prospero* expression is shown to be activated by the combination of cell-specific expression of *lozenge* and the receipt of an incoming EGF signal (Xu et al. 2000). In the second paper, activation of the *dPax2* gene is shown to require the presence of *lozenge* and incoming EGF and *Notch* signals (Flores et al. 2000). Thus, due to the additional requirement of *Notch* signaling, *dPax2* is expressed in only a subset of the cells in which *prospero* is expressed. This situation is clearly analogous to the collaboration of incoming signals with a cell-type-specific (or organ-type-specific) transcription factor that we find in the wing disc. The transcription factor(s) present in the cell collaborates with incoming signals to activate cell-specific targets of generic signaling pathways.

Another illustrative example of this phenomenon is found in the specification of heart progenitor cells during *Drosophila* development. A subset of the heart progenitor cells in the embryonic heart primordium of *Drosophila* is marked by the expressed of *eve*. By isolating and dissecting the heart progenitor cis-regulatory element of the *eve* gene, the Michelson lab was able to show that the expression of *eve* in these cells is dependent on five independent inputs of two discrete classes. The first class includes two transcription factors, *twist* and *Tinman,* which mark the cells and define them as dorsal mesoderm and prospective heart. The second set of inputs includes the wingless, *Decapentaplegic,* and MAP kinase signaling pathways which are patterning the tissue. Removal of any one of these inputs, either exogenously or from the enhancer, abrogates the transcription of *eve* in these cells (Halfon et al. 2000). This example of collaboration between cell-lineage-dependent transcription factors and generic signaling pathways is particularly interesting for two reasons. First, the transcription factors which mark the cell lineage are, like VG-SD, organ-level selector genes which orchestrate the development of an entire complex structure. Second, multiple selectors and multiple signals are required to impinge upon the enhancer in order to elicit target gene transcription.

Genetic Potentiation: Priming the Pump

While selector genes may not be able to activate transcription on their own, they may play important roles in the cell prior to the activation of specific target genes. For example, neither VG-SD nor HNF3β and GATA proteins are sufficient to activate transcription of many of their target genes, but both factors are expressed in many cells prior to the activation of specific target loci. What are these selector proteins doing, if anything, prior to the activation of these targets? HNF3β and GATA factors have been shown to bind to the enhancer of one of their target genes, liver-specific albumin, prior to the activation of albumin transcription and prior to the binding of any other transcription factors required for the activation of albumin transcription (Gualdi et al. 1996; Bossard and Zaret 1998). Furthermore, binding of purified HNF3 protein to DNA can alter nucleosome phasing and open chromatin structure near the HNF3 binding site (Cirillo et al. 1998; Shim et al. 1998). This suggests that one role for the early binding of selectors to the enhancers of their target genes prior to the activation of those genes may be decompaction of the chromatin structure at those enhancers thereby enabling the binding of subsequent transcription factors. This activity has been termed "genetic potentiation" (Zaret 1998) and represents an interesting paradigm for the action of selector genes as a group. Since the activation of target gene transcription by many selector genes is contingent upon other signals (as discussed above) it is possible that these factors have some preactivation role in readying the cell to respond to incoming signals. This may be reflected in the opening of target enhancers, as seen with HNF3β, and potentially readying the cell for the transcription of these targets in other ways such as positioning these loci within transcriptionally active regions of the nucleus (Cremer and Cremer 2001). Thus, while a transcriptional response to selector gene expression may not be apparent until other contingencies are met, the selector gene may be acting in other ways to specify the identity of the cell and ready it for action.

Conclusion

The modular architecture of metazoan body plans is generated by a similarly modular genetic regulatory hierarchy. In this hierarchy the identity of discrete anatomic and developmental modules is defined by a special class of transcription factors known as selector genes, which exert direct control over target genetic regulatory networks. Architectural complexity is achieved by the modification of downstream regulatory networks by upstream selector genes. Upstream selectors can exert their control either by affecting the expression of the downstream selector genes themselves or by insinuating their influence directly on multiple genes within the regulatory networks controlled by the downstream selector.

Selector genes are able to coordinate complex, deeply hierarchical processes in part because their control of target gene activation is contingent upon other developmentally relevant inputs, such as the cell-cell signaling pathways that pattern the embryo. The combinatorial control of gene expression by selector genes and signaling pathways results in context-specific genetic responses to widely utilized intercellular signals and suggests a solution to the question of how a very limited number of signal transduction pathways can be used to pattern all the diverse structures of the developing animal.

Finally, while selector genes depend on other inputs to activate gene expression, they appear to have activities that prime the genome of a cell to respond rapidly and specifically to these incoming signals. A better understanding of the way in which selector genes do this should lead to greater insight into the dynamics of the genomic regulation of development.

References

Ang, S. L., A. Wierda, D. Wong, K. A. Stevens, S. Cascio, J. Rossant, and K. S. Zaret. 1993. The formation and maintenance of the definitive endoderm lineage in the mouse: involvement of HNF3/forkhead proteins. *Development* 119:1301–1315.

Blair, S. S. 1995. Compartments and appendage development in *Drosophila. BioEssays* 17:299–309.

Bossard, P., and K. S. Zaret. 1998. GATA transcription factors as potentiators of gut endoderm differentiation. *Development* 125:4909–4917.

Carroll, S. B., J. K. Grenier, and S. D. Weatherbee. 2001. From DNA to diversity. Malden, Mass.: Blackwell Science.

Cirillo, L. A., C. E. McPherson, P. Bossard, K. Stevens, S. Cherian, E. Y. Shim, K. L. Clark, S. K. Burley, and K. S. Zaret. 1998. Binding of the winged-helix transcription factor HNF3 to a linker histone site on the nucleosome. *EMBO J.* 17:244–254.

Cremer, T., and C. Cremer. 2001. Chromosome territories, nuclear architecture and gene regulation in mammalian cells. *Nat. Rev. Genet.* 2:292–301.

Davidson, E. H. 2001. Genomic regulatory systems: development and evolution. San Diego: Academic Press.

Flores, G. V., A. Daga, H. R. Kalhor, and U. Banerjee. 1998. Lozenge is expressed in pluripotent precursor cells and patterns multiple cell types in the *Drosophila* eye through the control of cell-specific transcription factors. *Development* 125:3681–3687.

Flores, G. V., H. Duan, H. Yan, R. Nagaraj, W. Fu, Y. Zou, M. Noll, and U. Banerjee. 2000. Combinatorial signaling in the specification of unique cell fates. *Cell* 103: 75–85.

Galant, R., C. M. Walsh, and S. B. Carroll. 2002. Hox repression of a target gene: extradenticle-independent, additive action through multiple monomer binding sites. *Development* 129:3115–3126.

García-Bellido, A. 1975. Genetic control of wing disc development in *Drosophila. Ciba Found. Symp.* 29:161–182.

Gehring, W. J. 1996. The master control gene for morphogenesis and evolution of the eye. *Genes Cells* 1:11–15.

Gerhart, J., and M. Kirschner. 1997. Cells, embryos, and evolution. Malden, Mass.: Blackwell Science.

Gilbert, S. F., J. M. Opitz, and R. A. Raff. 1996. Resynthesizing evolutionary and developmental biology. *Dev. Biol.* 173:357–372.

Gualdi, R., P. Bossard, M. Zheng, Y. Hamada, J. R. Coleman, and K. S. Zaret. 1996. Hepatic specification of the gut endoderm in vitro: cell signaling and transcriptional control. *Genes Dev.* 10:1670–1682.

Guss, K. A., C. E. Nelson, A. Hudson, M. E. Kraus, and S. B. Carroll. 2001. Control of a genetic regulatory network by a selector gene. *Science* 292:1164–1167.

Halder, G., P. Callaerts, S. Flister, U. Walldorf, U. Kloter, and W. J. Gehring. 1998a. Eyeless initiates the expression of both sine oculis and eyes absent during *Drosophila* compound eye development. *Development* 125:2181–2191.

Halder, G., P. Polaczyk, M. E. Kraus, A. Hudson, J. Kim, A. Laughon, and S. Carroll. 1998b. The Vestigial and Scalloped proteins act together to directly regulate wing-specific gene expression in *Drosophila. Genes Dev.* 12:3900–3909.

Halfon, M. S., A. Carmena, S. Gisselbrecht, C. M. Sackerson, F. Jimenez, M. K. Baylies, and A. M. Michelson. 2000. Ras pathway specificity is determined by the integration of multiple signal-activated and tissue-restricted transcription factors. *Cell* 103: 63–74.

Kardon, G., T. A. Heanue, and C. J. Tabin. 2004. The *Pax/Six/Eya/Dach* network in development and evolution. In *Modularity in development and evolution,* ed. G. Schlosser and G. P. Wagner. Chicago: University of Chicago Press.

Kim, J., A. Sebring, J. J. Esch, M. E. Kraus, K. Vorwerk, J. Magee, and S. B. Carroll. 1996. Integration of positional signals and regulation of wing formation and identity by *Drosophila* vestigial gene. *Nature* 382:133–138.

Lewis, E. B. 1978. A gene complex controlling segmentation in *Drosophila. Nature* 276:565–570.

Palopoli, M. F., and N. H. Patel. 1998. Evolution of the interaction between Hox genes and a downstream target. *Curr. Biol.* 8:587–590.

Panganiban, G., L. Nagy, and S. B. Carroll. 1994. The role of the Distal-less gene in the development and evolution of insect limbs. *Curr. Biol.* 4:671–675.

Papatsenko, D., A. Nazina, and C. Desplan. 2001. A conserved regulatory element present in all *Drosophila* rhodopsin genes mediates Pax6 functions and participates in the fine-tuning of cell-specific expression. *Mech. Dev.* 101:143–153.

Raff, R. A. 1996. The shape of life: genes, development, and the evolution of animal form. Chicago: University of Chicago Press.

Raff, R. A., and B. J. Sly. 2000. Modularity and dissociation in the evolution of gene expression territories in development. *Evol. Dev.* 2:102–113.

Riedl, R. 1978. Order in living organisms: a systems analysis of evolution. Chichester: John Wiley and Sons.

Schlosser, G. 2004. The role of modules in development and evolution. In *Modularity in development and evolution,* ed. G. Schlosser and G. P. Wagner. Chicago: University of Chicago Press.

Shen, W., and G. Mardon. 1997. Ectopic eye development in *Drosophila* induced by directed dachshund expression. *Development* 124:45–52.

Sheng, G., E. Thouvenot, D. Schmucker, D. S. Wilson, and C. Desplan. 1997. Direct regulation of rhodopsin 1 by Pax-6/eyeless in *Drosophila:* evidence for a conserved function in photoreceptors. *Genes Dev.* 11:1122–1131.

Shim, E. Y., C. Woodcock, and K. S. Zaret. 1998. Nucleosome positioning by the winged helix transcription factor HNF3. *Genes Dev.* 12:5–10.

Skeath, J. B., and S. B. Carroll. 1994. The achaete-scute complex: generation of cellular pattern and fate within the *Drosophila* nervous system. *FASEB J.* 8:714–721.

Strickler, A. G., Y. Yamamoto, and W. R. Jeffery. 2001. Early and late changes in Pax6 expression accompany eye degeneration during cavefish development. *Dev. Genes Evol.* 211:138–144.

von Dassow, G., and E. Munro. 1999. Modularity in animal development and evolution: elements of a conceptual framework for *evodevo*. *J. Exp. Zool.* 285:307–325.

Wagner, G. P., and L. Altenberg. 1996. Perspective: complex adaptations and the evolution of evolvability. *Evolution* 50:967–976.

Warren, R., and S. Carroll. 1995. Homeotic genes and diversification of the insect body plan. *Curr. Opin. Genet. Dev.* 5:459–465.

Warren, R. W., L. Nagy, J. Selegue, J. Gates, and S. Carroll. 1994. Evolution of homeotic gene regulation and function in flies and butterflies. *Nature* 372:458–461.

Weatherbee, S. D., G. Halder, J. Kim, A. Hudson, S. and Carroll. 1998. Ultrabithorax regulates genes at several levels of the wing-patterning hierarchy to shape the development of the *Drosophila* haltere. *Genes Dev.* 12:1474–1482.

Weigel, D., G. Jurgens, F. Kuttner, E. Seifert, and H. Jackle. 1989. The homeotic gene fork head encodes a nuclear protein and is expressed in the terminal regions of the *Drosophila* embryo. *Cell* 57:645–658.

Xu, C., R. C. Kauffmann, J. Zhang, S. Kladny, and R. W. Carthew. 2000. Overlapping activators and repressors delimit transcriptional response to receptor tyrosine kinase signals in the *Drosophila* eye. *Cell* 103:87–97.

Zaret, K. 1998. Early liver differentiation: genetic potentiation and multilevel growth control. *Curr. Opin. Genet. Dev.* 8:526–531.

Zaret, K. S. 1996. Molecular genetics of early liver development. *Annu. Rev. Physiol.* 58:231–251.

Zhu, J., T. Fukushige, J. D. McGhee, and J. H. Rothman. 1998. Reprogramming of early embryonic blastomeres into endodermal progenitors by a *Caenorhabditis elegans* GATA factor. *Genes Dev.* 12:3809–3814.

Zhu, J., R. J. Hill, P. J. Heid, M. Fukuyama, A. Sugimoto, J. R. Priess, and J. H. Rothman. 1997. end-1 encodes an apparent GATA factor that specifies the endoderm precursor in *Caenorhabditis elegans* embryos. *Genes Dev.* 11:2883–2896.

3 The Basic Helix-Loop-Helix Proteins in Vertebrate and Invertebrate Neurogenesis

UWE STRÄHLE AND PATRICK BLADER

The fundamental molecular mechanisms underlying neurogenesis in the fruit fly *Drosophila melanogaster* and in vertebrates have been conserved during evolution despite obvious differences in their body plans (Salzberg and Bellen 1996; Arendt and Nübler-Jung 1997; Hassan and Bellen 2000). Basic helix-loop-helix (bHLH) transcription regulatory proteins play pivotal roles in these developmental processes (Lewis 1996; Chan and Jan 1999; Guillemot 1999). Here, we will introduce the structure of bHLH proteins in general, provide an overview of the conserved use of bHLH proteins in neurogenesis, and assess to what extent bHLH proteins can be regarded as modules. The conserved use of bHLH proteins in vertebrate and invertebrate neurogenesis will then be used to illustrate the concept of modularity in development and evolution. In the last part of this review we will turn to a different type of regulatory module. We will describe the regulatory architecture of the proneural bHLH gene *neurogenin1* in the zebrafish. In contrast to amniotes, lower vertebrates are characterized by two phases of neurogenesis, the first (primary) neurogenesis taking place at the neural plate stage. Regulatory regions controlling expression of *neurogenin1* during primary neurogenesis are also active in the mouse neural tube, thus providing a different illustration of modularity in development and evolution.

As detailed by G. Schlosser (2004), the theory of modularity defines a module as an entity that serves as a building block of a function that operates in an integrated and relatively autonomous manner. Function and context-insensitivity are essential parameters of a module. The functional description is in our minds a crucial prerequisite for a precise delineation of modules and is very much determined by perspective. A chemist working at atomic resolution would certainly define module differently from a developmental biologist. For the purpose of this review we define a module as an assembly of biological structures that fulfills a function in an integrated and context-insensitive manner. Function as defined here is not merely the interaction of molecules but an interaction that yields a biological output which is characteristic of the module. Furthermore, application of the module is flexible. To be recognized as a module, it has to be used either in different processes in the same organisms or in different organisms, exploiting its invariant functional properties in the same or different processes. A module is therefore also characterized by its reiterated use. We regard a module as an integrated assembly of biological structures that is structurally and functionally conserved and is context independent in operational terms. If modules are even partially context dependent, their operational definition is wrong or imprecise, and they therefore fail to qualify as modules entirely.

The bHLH Motif Is an Ancient Protein Domain

Transcription regulators utilizing a bHLH domain have been identified in organisms as diverse as yeast and man (Atchley and Fitch 1997) and control a vast range of physiological and developmental processes including phosphate metabolism, myogenesis, and neurogenesis (Brand et al. 1993; Lee 1997; Robinson et al. 2000). The bHLH domain is thus an ancient and, in evolutionary terms, highly successful protein structure.

The 12–15-amino-acid basic region of the bHLH domain mediates binding to specific DNA sequences (Davis et al. 1990). The DNA recognition sequences of bHLH proteins are derivatives of the core hexamer CANNTG, the so called E-box (Ghysen and Dambly-Chaudiere 1989; Blackwell et al. 1990; Blackwell and Weintraub 1990; Gradwohl et al. 1996; Atchley and Fitch 1997). The larger HLH subdomain which is located immediately C-terminally to the basic region is composed of two amphipathic α-helices separated by a loop of variable length (Ferre-D'Amare et al. 1993, 1994; Ma et al. 1994; Atchley and Fitch 1997). This domain permits protein-protein interaction. Some HLH proteins can form homodimers (Amati and Land 1994). However, heterodimerization with different HLH proteins appears to be the physiologically preferred association (Cabrera and Alonso 1991; Sun and Baltimore 1991). Heterodimeric partners can exert either a positive or a negative effect on the transcriptional activity of the complex. For example, bHLH partners such as *Drosophila* Daughterless and the vertebrate E12 and E47 proteins act as positive cofactors (Murre et al. 1989; Giebel et al. 1997), whereas vertebrate Id or *Drosophila* Extramacrochaetae (Emc) proteins, which lack the basic domain (fig. 3.1),

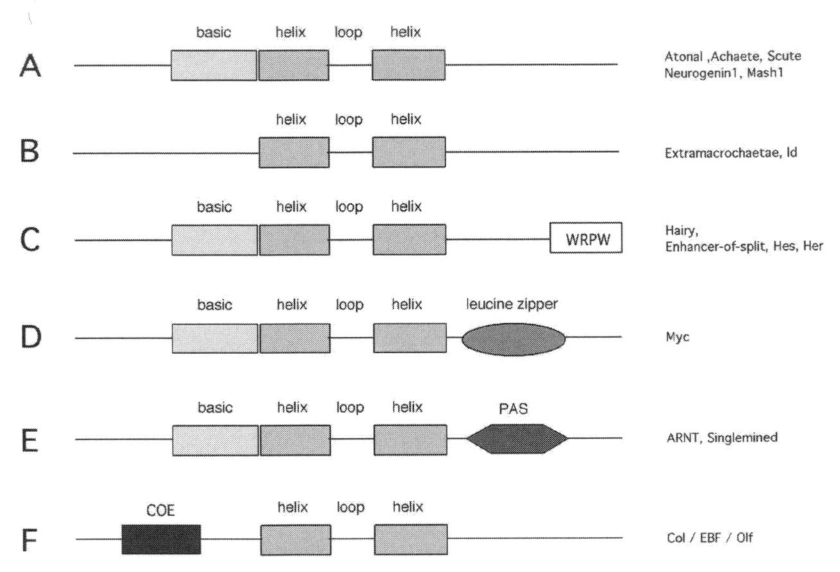

Fig. 3.1.—Domain structure of bHLH proteins. *A,* Domain structure of prototype bHLH proteins typically found in proneural genes. The basic domain serves as a DNA binding surface, while the HLH domain is involved in protein-protein interaction with other HLH domains. *B,* Negatively acting *extramacrochaetae* and *Id* genes lack the basic domain. These proteins inhibit activity of other bHLH protein by forming heterodimers unable to bind to DNA. *C,* bHLH genes of the *enhancer-of-split* and *hairy* class act as repressors. They interact with the corepressor Groucho via the WRPW motif. *D,* bHLH genes can contain a leucine zipper as a second protein-protein interaction domain. The protooncogene *myc* is an example of this class of bHLH gene. *E,* The bHLH domain can be also combined with a PAS protein-protein interaction module. Examples of this class of bHLH proteins are the transcription factors ARNT and Singleminded. (F) The Col/EBF/Olf proteins have an HLH domain but lack the basic domain. These proteins utilize the COE domain for DNA binding.

act negatively by preventing binding of the complex to DNA (Sun et al. 1991; Van Doren et al. 1991, 1992). Thus, the activity of bHLH proteins is context-sensitive, depending on which dimerization partners are available.

In certain other cases, further versatility of function is achieved by combination of the bHLH domain with other, distinct protein interaction motifs. In addition to the bHLH domain, the bHLH repressor Hairy contains a WRPW motif (fig. 3.1) which allows it to interact with the repressor Groucho (Paroush et al. 1994; Van Doren et al. 1994; Fisher et al. 1996; Fisher and Caudy 1998). Other bHLH proteins were found to contain a leucine zipper or a PAS domain (Atchley and Fitch 1997; Bouchard et al. 1998; Crews 1998), increasing the spectrum of interacting protein partners further. Members of the Col/EBF/Olf transcription factors have an HLH motif but bind DNA using the COE domain (Dubois et al. 1998) (fig. 3.1). Reasons for the evolutionary success of the HLH domain may thus have been on the one hand its flexibility to interact with different HLH partners and on the other hand its ability to be combined intramolecularly with distinct protein interaction domains to allow the integration of distinct signals on the same transcription regulator.

Drawing from our definition of a module in the introduction, neither the bHLH domain per se nor the entire protein qualifies as module. They function only in combination with heteromeric partners, and are therefore not context insensitive. However, the heterodimeric assembly of a specific bHLH protein with the coactivator Daughterless, for example, forms a module. It has the well-definable function of binding to the E-box and activating the basic transcription machinery. We will hereafter refer to this dimeric assembly as the bHLH module.

A Cascade of bHLH Proteins Controls Neurogenesis in *Drosophila*

As mentioned already, some bHLH proteins play pivotal roles in neurogenesis. We will therefore investigate next how bHLH modules are utilized in these processes in the fruit fly. The development of the external sensory organs of *Drosophila* has served as a paradigm of neurogenesis in general and exemplifies the reiterated use of the bHLH proteins in a cascade of regulatory genes (fig. 3.2). The nervous system of *Drosophila* is derived from the ectoderm and involves a hierarchy of developmental steps. Prepattern genes define first the position of cell groups in the ectoderm which have the equivalent potential to develop into neurons. These groups of cells are characterized by expression of the proneural genes *achaete* and *scute* and are referred to as proneural clusters (Ghysen and Dambly-Chaudiere 1989; Ghysen et al. 1993; Jimenez and Modolell 1993; Campos-Ortega 1995, 1997; Vervoort et al. 1997; Dambly-Chaudiere and Vervoort 1998). In the next step, a single cell is selected from the proneural cluster to become the sensory organ precursor. The remaining cells within the cluster take up the fate of epidermoblasts. Once specified, the sensory organ precursor cell undergoes stereotypic asymmetric cell divisions which generate the distinct cell types characteristic of the mature external sense organ (Ghysen and Dambly-Chaudiere 1989; Campuzano and Modolell 1992; Jan and Jan 1993; Gho et al. 1999).

HLH proteins play crucial roles at each step of external sense organ formation (fig. 3.2; Fisher and Caudy 1998). The negatively acting *hairy* and *emc* genes act as prepattern genes, which are required among other genes to define the position of proneural clusters. They are repressors of external sense organ formation and act by inhibiting the activity of proneural genes (Botas et al. 1982; Cubas and Modolell 1992; Martinez et al. 1993). Proneural genes such as *achaete* and *scute* (Cubas et al. 1991) belong also to the bHLH superfamily and are defined by two criteria. First, they are expressed prior to neuronal differentia-

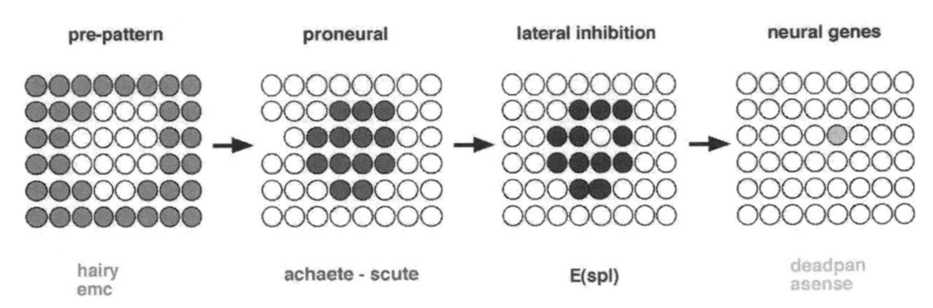

Fig. 3.2.—A cascade of HLH genes controls external sensory organ development in *Drosophila.* The repressors *hairy* and *extramacrochaetae* act as prepattern genes by inhibiting proneural gene expression and activity, respectively. Expression of the proneural genes *achaete* and *scute* marks the proneural cluster. Via Delta/Notch-mediated lateral inhibition, *enhancer-of-split* genes (*E(spl)*) are activated in cells destined to become epidermoblasts. They act as repressors of proneural and neural genes as well as Delta. The selected neuroblast expresses the neural bHLH genes *deadpan* and *asense* which are involved in differentiation control. (This figure was adapted from Fisher and Caudy 1998.)

tion and mark a neural precursor state. Second, they are necessary and sufficient for the specification and development of neural lineages in the ectoderm (Ghysen et al. 1993; Hassan and Bellen 2000).

The sensory organ precursor cell is selected from the proneural cluster through a process called lateral inhibition which involves the Notch/Delta signaling system (Heitzler and Simpson 1991, 1993; Artavanis-Tsakonas et al. 1995, 1999). *achaete* and *scute* genes control expression of the ligand Delta initially in every cell of the proneural cluster (Kunisch et al. 1994). Activation of the Notch receptor by Delta causes activation of the genes of the *Enhancer-of-split* (*E(spl)*) complex (fig. 3.3) (Bailey and Posakony 1995; Lecourtois and Schweisguth 1995, 1997, 1998). The *E(spl)* genes, which contain a bHLH in association with a WRPW motif, are negative regulators of *Delta* and proneural genes (Klambt et al. 1989; Oellers et al. 1994; Paroush et al. 1994). As a consequence cells with strong activation of Notch turn off *Delta* and *achaete-scute* expression, drop out of the neuronal program, and instead develop as epidermal cells (Kunisch et al. 1994; Heitzler et al. 1996).

Further differentiation of the neuron from the sensory organ precursor cell requires the activity of panneural genes—yet another subgroup of bHLH proteins. The neural bHLH transcription factors *deadpan* and *asense* are widely expressed in the *Drosophila* nervous system (fig. 3.2). They are expressed later than proneural genes, and their mutant phenotypes are rather weak, suggesting that they control certain aspects of neuronal differentiation rather than the initial establishment of the neuronal program (Bier et al. 1992; Brand et al. 1993; Jarman et al. 1993a; Roark et al. 1995).

Thus, neurogenesis is controlled by a cascade of distinct bHLH modules which have different functions in neuronal specification. Heterodimers of Achaete or Scute with the repressor Emc define negatively acting bHLH modules, while distinct and positively acting bHLH modules are formed in heterodimeric association with Daughterless. This exemplifies also that components of modules such as Daughterless or Emc can associate with other protein partners to form distinct modules. This raises the question of whether these differently composed associations have different functions. Distinct bHLH genes act as proneural genes in other sense organs or in the central nervous system (CNS) (Jimenez and Campos-Ortega 1990; Martin-Bermudo et al. 1991, 1993; Goulding et al. 2000; Huang et al. 2000). For example, the bHLH factor Atonal is crucial for the development of most chordotonal organs, a subset of olfactory sense organs, some multidendritic neurons, and the eye (Jarman et al. 1993b; Baker et al. 1996; Gupta and Rodrigues 1997). Misexpression of *atonal* induces preferentially ectopic chordotonal organs (Jarman et al. 1993b; Jarman and Ahmed 1998). In contrast, forced expression of *scute* results predominantly in the formation of additional external sensory organs, suggesting that proneural genes convey specificity to the neuronal programs that they

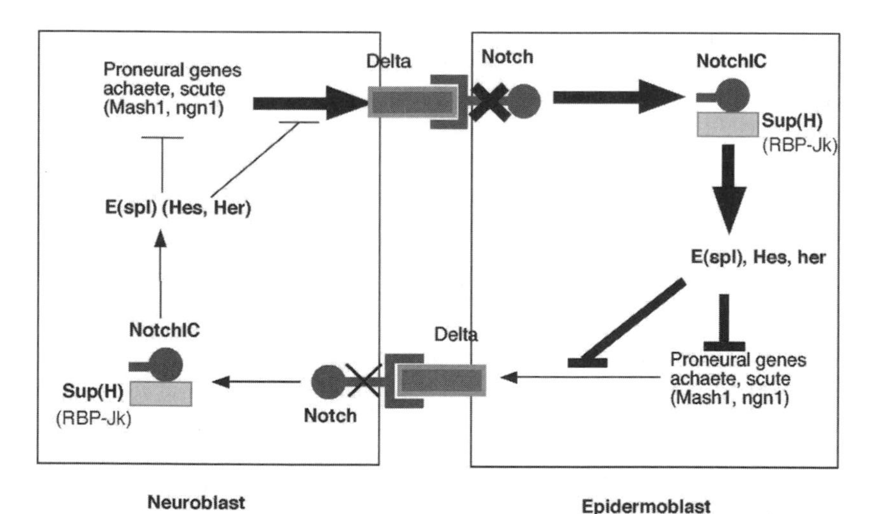

Fig. 3.3.—Delta/Notch-mediated selection of a neuroblast from a proneural cluster. *Delta* is under positive control of *achaete* and *scute* genes, which leads to expression of *Delta* in all cells of the proneural cluster. Delta activates the Notch receptor, which becomes proteolytically processed into NotchIC. Notch IC forms a complex with Suppressor-of-hairless and activates the genes of the *enhancer-of-split* complex *E(spl)*. The *E(spl)* genes act as repressors which inhibit expression of proneural genes and of *Delta*. As a consequence, these cells produce less Delta. They are less able to activate Notch in adjacent cells, which can therefore express higher levels of *Delta*. This mechanism leads ultimately to a complete repression of *Delta* and proneural genes in most of the cells of the proneural cluster. A central issue in this model of lateral inhibition is how the initial differences in levels of *Delta* expression arise. This could be due to inherent variation of *Delta* activation in individual cells of a proneural cluster. There is also evidence that extrinsic factors can cause unequal levels of *Delta* expression within a cluster. Notch/Delta signaling employs highly related genes in vertebrates and invertebrates. Vertebrate gene names are indicated in brackets in cases where they differ from the *Drosophila* homologues.

control. Domain swap experiments between *atonal* and *scute* identified the basic domain as a crucial determinant for specificity (Chien et al. 1996). Thus, proneural bHLH modules can differ in function.

Variations on a Common Theme: bHLH Genes in Vertebrate Neurogenesis

The vertebrate genomes encode bHLH genes which are highly related to *Drosophila* proneural genes (fig. 3.4). Many of these homologues are expressed in the peripheral and central nervous system of vertebrates, suggesting also functions of bHLH modules in the neuronal development of vertebrates.

Neurogenesis in amphibian and teleost embryos begins toward the end of gastrulation. The first neurons have already become postmitotic at this stage and become organized in functional neuronal circuits soon after the neural tube has formed. These large, early born neurons are referred to as primary neurons to distinguish them from the later born, smaller secondary neurons (Kimmel and Westerfield 1989). It is commonly held that primary neurogenesis is a process specific to lower vertebrates. Amniotes like the mouse and the chicken appear not to form primary neurons but rather immediately employ secondary neurogenesis to build their nervous system.

In *Xenopus* and zebrafish embryos, the bHLH factor Neurogenin1 (Ngn1) (Ma et al. 1996; Blader et al. 1997; Korzh et al. 1998), most closely related to *Drosophila biparous* or *tap* (Bush et al. 1996; Gautier et al. 1997), is expressed in the neural plate in broad domains prior to neural differentiation (fig. 3.5). Misexpression of *ngn1* by injection of synthetic RNA into *Xenopus* and zebrafish embryos induces the formation of ectopic neurons in the epidermis (Ma et al. 1996; Blader et al. 1997; Perron et al. 1999; Takke et al. 1999). Similar results are obtained when the three related genes of the mouse, *ngn1, ngn2,* and *ngn3,* are ectopically expressed in the zebrafish (fig. 3.6), suggesting conserved action of the distinct vertebrate bHLH factors. Thus, like the *Drosophila* proneural genes, vertebrate bHLH modules are able to respecify the fate of epidermal cells to become neurons.

The domains of *ngn1* expression in the neural plate correspond to sites of primary neurogenesis (fig. 3.5). The number of primary neurons that develop from these areas is, however, much smaller than the number of cells that express *ngn1*. This suggests that not all *ngn1*-expressing cells develop into neurons and that *ngn1* expression, like that of the *Drosophila* proneural homologues, demarcates precursor cell groups in the neural plate from which individual cells are selected to enter a neuronal differentiation pathway (Ma et al. 1996; Blader et al. 1997; Korzh et al. 1998). Proneural genes of *Drosophila* are

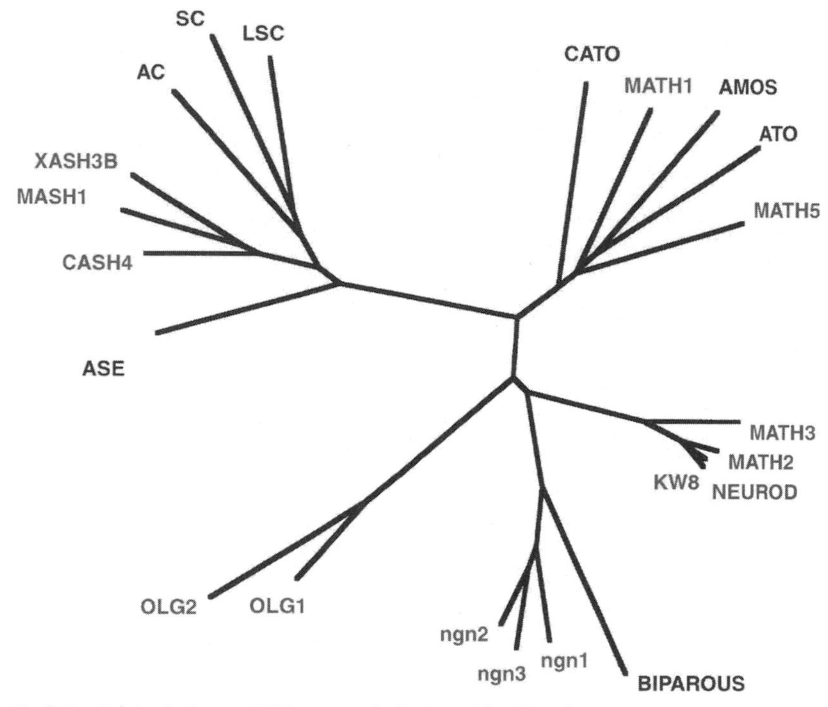

Fig. 3.4.—Relationship between bHLH genes involved in *Drosophila* and vertebrate neurogenesis. AC, Achaete; SC, Scute; LSC, Lethal-of-scute; OLG, Oligo; ATO, Atonal; ASE, Asense; ngn, Neurogenin; MASH, mouse achaete scute homologue; CASH, chicken achaete scute homologue; XASH, *Xenopus* achaete scute homologue; MATH, mouse atonal homologue. (Kindly provided by E. Cau and G. Gradwohl.)

under the control of Notch/Delta-mediated lateral inhibition. There is evidence that similar mechanisms act in the neural plate of *Xenopus* and zebrafish to select neuronal precursors from a larger set of initially equivalent cells (Chitnis et al. 1995; Henrique et al. 1995; Chitnis and Kintner 1996; Dornseifer et al. 1997; Appel and Eisen 1998; Haddon et al. 1998). *ngn1*, like its *Drosophila* proneural counterpart, appears to control *delta* gene expression, as *delta* becomes ectopically activated when *ngn1* is misexpressed (Ma et al. 1996; Blader et al. 1997). Moreover, expression territories of *delta* genes and *ngn1* appear to coincide in the neural plate of *Xenopus* and zebrafish embryos (Chitnis et al. 1995; Dornseifer et al. 1997; Appel and Eisen 1998; Haddon et al. 1998), and targeted mutation of the related factors *Ngn1* or *Ngn2* in the mouse leads to loss of expression of a *delta-1-like* gene in the cranial sensory placodes (Fode et al. 1998; Ma et al. 1998)—underscoring a functional dependence of the two gene classes, as in *Drosophila*. Furthermore, forced expression of *delta* genes in zebrafish and *Xenopus* inhibits *ngn1* expression (Ma et al. 1996; Blader et al. 1997), and mutations in *delta* genes cause an increased number of neurons in ze-

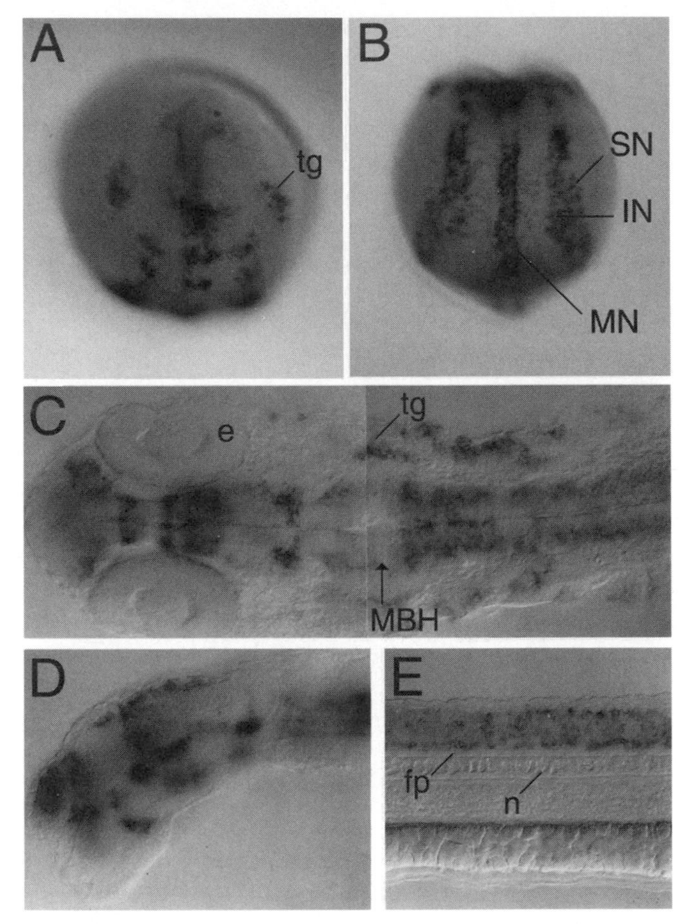

Fig. 3.5.—The neural determination gene *ngn1* is expressed in many neurogenic regions of the developing zebrafish embryo. *A*, and *B*, 3-somite-stage embryo with view of anterior and posterior neural plate, respectively. *C–E*, Brain in dorsal (*C*) and lateral view (*D*) and lateral view of spinal cord (*E*) of 26 hour embryos. Embryos were hybridized to zebrafish antisense *ngn1* mRNA probe. e, eye; fp, floor plate; IN, interneuron; MBH, midbrain-hindbrain boundary; MN, motor neurons; n, notochord; SN, Rohon Beard sensory neurons; tg, trigeminal ganglion.

brafish embryos (Appel et al. 1999; Riley et al. 1999; Holley et al. 2000), as do *delta* mutations in *Drosophila*. Activated Notch in vertebrates appears to transduce via downstream proteins similar to those in the fruit fly, suggesting that the entire pathway has been conserved (de la Pompa et al. 1997; Wettstein et al. 1997; Kao et al. 1998; Shimizu et al. 2000). Moreover, *E(spl)*-like genes named *Hes* or *her* in vertebrates for their sequence similarity to both *hairy* and *E(spl)* genes of *Drosophila* function downstream of Notch signaling (Kageyama and Ohtsuka 1999; Ohtsuka et al. 1999; Takke and Campos-Ortega 1999; Takke et al. 1999). In summary, this shows that vertebrate Ngn1 modules like

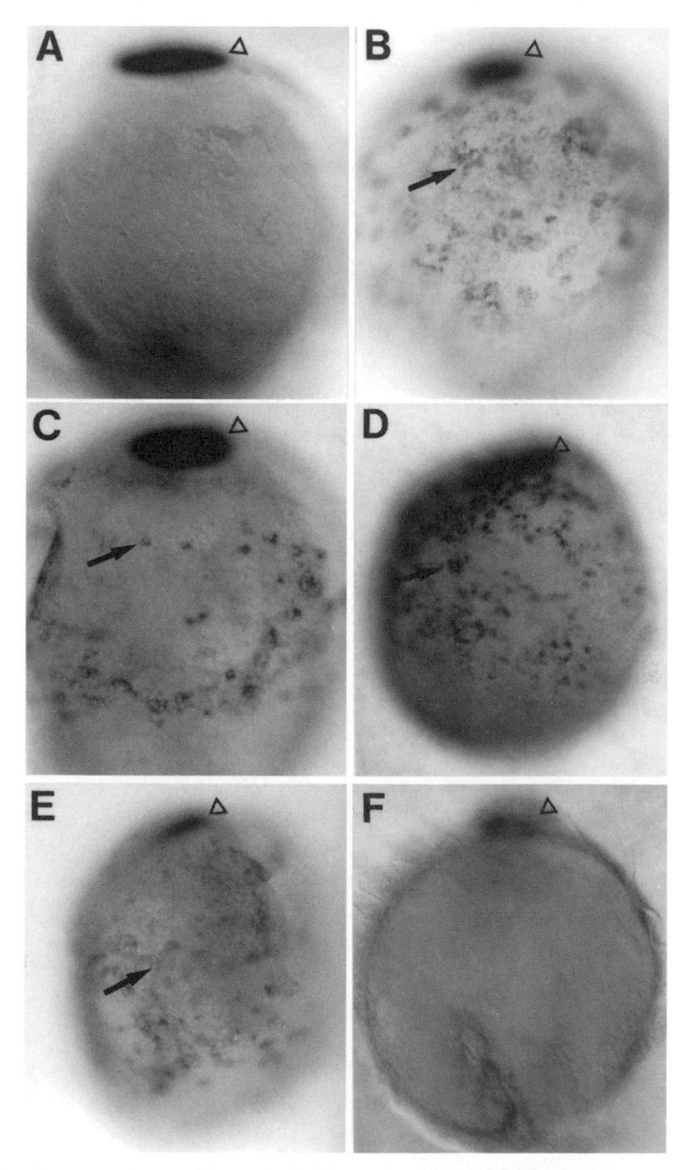

Fig. 3.6.—Mouse *Ngn1, Ngn2, and Ngn3* and zebrafish *ngn1* but not mouse *Mash1* induce ectopic sensory neurons in the nonneural ectoderm of zebrafish embryos. Uninjected control embyos (*A*) and embryos injected with mouse *Ngn2* (*B*), mouse *Ngn1* (*C*), mouse *Ngn3* (*D*), zebrafish *Ngn1* (*E*), and *Mash1* (*F*) mRNA were stained with a probe directed against the neural marker isl-1 at the 3-somite stage. Strong ectopic activation of *isl-1* (*arrows* in *B–D*) is detected in the yolk sac ectoderm of *ngn*-injected embryos but not in uninjected controls (*A*) or *Mash1*-injected embryos (*F*). The polster that expresses *isl-1* also at this stage is indicated by an open triangle. Embryos are shown in ventral view, with anterior up. (P. Blader, G. Gradwohl, F. Guillemot, and U. Strähle, unpublished data.)

the homologous modules in *Drosophila* are controlled by Delta/Notch signaling. Thus, the Ngn1 module is not only structurally similar to the proneural modules of the fruit fly but is also regulated in a comparable way.

The Delta/Notch system is involved in many other differentiation processes which do not involve regulation of proneural genes. Hence, proneural bHLH proteins are not integral part of a Notch/Delta module. They have to be regarded as separate modules according to our definition, as their linkage to the Delta/Notch signaling system is context dependent. The high conservation of individual signaling components within the Notch/Delta pathway suggests that the entire signal transduction system forms a conserved module which can be combined with distinct downstream targets such as the Ngn1 module that controls the cell-specific responses.

Vertebrates Contain Multiple bHLH Factors with Distinct Functions

bHLH genes involved in vertebrate neurogenesis have been grouped into determination and differentiation genes to indicate their temporal role in specification of neurons (fig. 3.7). While determination genes, which are regarded as equivalent to the proneural genes of *Drosophila,* act early on neural precursors, the differentiation genes control the differentiation of neurons in a way reminiscent of neural genes of *Drosophila.*

As already discussed, *Drosophila* proneural genes show specificity in the neuronal programs that they control. Similarly, it was proposed that vertebrate bHLH factors display specificity in their activity. When expressed in the zebrafish yolk sac ectoderm, *ngn1* induced mostly neurons with characteristics of primary sensory neurons (Blader et al. 1997). Studies in *Xenopus* have arrived at similar conclusions (Perron et al. 1999). Also, *ngn1* expressed in the chicken premigratory neural crest by retroviral infection biased neurogenesis toward the formation of sensory neurons (Perez et al. 1999). *Mash1* does not elicit ectopic activation of sensory neurons in the zebrafish yolk sac epithelium (fig. 3.6; P. Blader and U. Strähle, unpublished data). In contrast to *ngns, Mash1* controls differentiation of the autonomous nervous system, another neural-crest-derived class of peripheral neurons. These neurons fail to form in *Mash1* knock-out mice (Guillemot et al. 1993). Thus, the vertebrate bHLH genes *Mash1* and *Ngn1* control distinct neuronal differentiation programs in the neural crest. Similarly, distinct populations of neurons are controlled by the two factors in the CNS (Fode et al. 2000).

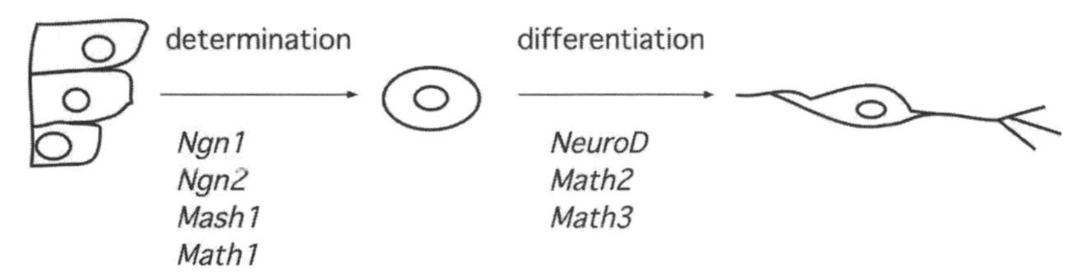

Fig. 3.7.—A hierarchy of determination and differentiation genes is involved in vertebrate neurogenesis.

In *Drosophila,* a hierarchy of bHLH modules is involved in neurogenesis in external sense organs (fig. 3.2). There is also evidence in vertebrates that Ngn modules control the expression of other bHLH genes such as *NeuroD* (fig. 3.6). Although misexpression of *NeuroD* induces ectopic neurons in nonneural ectoderm in a fashion similar to that of *ngns*, its true role is as a differentiation rather than a determination gene (Lee et al. 1995; Ma et al. 1996). The role of vertebrate differentiation genes such as *NeuroD,* but also the *atonal* homologues *Math2* and *Math3* in the mouse, has been proposed to be similar to that of the bHLH factors *asense* and *deadpan* of *Drosophila* in that they control aspects of the neuronal differentiation program (Ahmad et al. 1998; Schwab et al. 1998; Miyata et al. 1999). Thus, as in *Drosophila,* vertebrate bHLH modules are organized in cascades to control consecutive steps in the specification of neurons.

To complicate things further, although *Ngns* play a central role as determination factors of neural progenitors, *Ngns* can act also as differentiation genes. In the olfactory epithelium, *Ngn1* expression is controlled by *Mash1* (Cau et al. 1997). It thus functions downstream of the proneural gene *Mash1* in olfactory neurons and hence appears to act as a differentiation factor in this context (Cau et al. 1997). The closest relative of *ngns* in *Drosophila* is *biparous* (or *tap*). In chemosensory and CNS lineages, *biparous* is expressed quite late, during differentiation, suggesting that it is a differentiation gene (Bush et al. 1996; Gautier et al. 1997). It is, however, also expressed in the antennal discs at the time when the precursors of olfactory organs form and may thus have proneural activity in this context (Ledent et al. 1998). Similarly, *atonal* acts as a determination (proneural) gene during chordotonal organ development (Jarman et al. 1993b) and at the same time as a differentiation factor in a group of neurons that innervate the optic lobe in the brain (Hassan et al. 2000). Furthermore, the *Drosophila* bHLH factor *asense* and the vertebrate *NeuroD* are utilized as differentiation factors in the respective organisms, yet, when misexpressed at an earlier stage, these genes have proneural gene activity (Brand et al. 1993; Lee et al. 1995). Thus, the redeployment of proneural bHLH modules in differentiation processes is a strategy common to both invertebrates and vertebrates. The timing of bHLH module expression (i.e., the wiring of the control regions to prepattern genes and upstream bHLH factors) may define its function in the regulatory cascade controlling neural differentiation. Hence, whether a particular bHLH module has proneural (neuronal determination) or differentiation activity is context dependent.

These examples indicate how critical the correct evaluation of the operational description of modules needs to be. The bHLH modules described here have been highly conserved during evolution and are employed in neurogenesis in both invertebrates and vertebrates. However, their application may be as a determination or differentiation factor, and it may not be restricted to neurogenesis, as some are also deployed in other processes such as pancreas differentiation (Gradwohl et al. 2000). One can thus define the function of these modules only in the following general way: they form heterodimers and interact specifically with E-box-like DNA recognition sequences and activate the transcriptional machinery in a cell-specific manner. Their function cannot be exclusively linked to neurogenesis, as it is context dependent.

Spatiotemporal Control of the Expression of the bHLH Gene *ngn1* in the Neural Plate of Zebrafish Embryos

ngn1 is not uniformly expressed in the neuroectoderm of zebrafish (Blader et al. 1997; Korzh et al. 1998) (fig. 3.5, *A, B*). In the anterior neural plate of zebrafish embryos, which gives rise to the brain, *ngn1* is expressed in distinct cell groups, which prefigure the primordia of early differentiating brain neurons (Fig 3.5, *A, C, D*). In addition, the olfactory placodes and the trigeminal ganglia which arise at the boundary between neural plate and nonneural ectoderm also express *ngn1* mRNA (Blader et al. 1997; Korzh et al. 1998). In the prospective spinal cord (fig. 3.5, *B*), longitudinal domains of expression are discernible, forecasting the location of the three main primary neuronal types: the primary motor, inter-, and sensory neurons (or Rohon Beard neurons). The two medial stripes flanking the floor plate will give rise to motor neurons. They are separated by a gap of nonexpressing cells from a more laterally located broad stripe. This lateral stripe is in fact composed of the two closely spaced territories corresponding to interneurons and sensory neurons, as can be seen in more posterior aspects of the neural plate, where the differentiation of primary interneurons is delayed relative to that of the lateral Rohon Beard sensory neurons. *ngn1* mRNA marks the territories of all major primary neurons in the neural plate of the zebrafish embryo (Blader et al. 1997; Korzh et al. 1998).

The pattern of expression in the neural plate suggests that *ngn1* is under complex spatial control. Motor neuron development requires Sonic hedgehog (Shh) in amniote embryos (Roelink et al. 1994; Marti et al. 1995; Roelink et al. 1995; Chiang et al. 1996). This appears to be also the case for primary motor neurons in zebrafish embryos even though only indirect evidence supports this notion so far: Overexpression of Shh or Hedgehog (Hh) signal transducers leads to ectopic differentiation of motor neurons (Hammerschmidt et al. 1996; Blader et al. 1997). This increase of motor neurons in injected embryos is anticipated by an expansion of *ngn1* expression, suggesting that the medial domain of *ngn1* expression depends on Shh signals (Blader et al. 1997). Development of primary neurons is, however, not impaired in the *shh* mutant *sonic-you* (Schauerte et al. 1998). At least two other closely related *hh* genes are, however, expressed in the notochord and the ventral neural tube (Krauss et al. 1993; Ekker et al. 1995; Currie and Ingham 1996). These Hhs have similar inducing activities and are thus likely to compensate for the lack of *shh* activity. In agreement, mutants that lack both notochord and floor plate as possible source of Hh signals fail to form motor neurons (Beattie et al. 1997), as do embryos incapable of receiving the Hh signal (Lewis and Eisen 2001).

Embryos with impaired BMP2b/BMP7 signaling show defects in the development of neural crest and sensory neurons (Kishimoto et al. 1997; Nguyen et al. 1998; Barth et al. 1999; Dick et al. 2000; Nguyen et al. 2000; Schmid et al. 2000). BMP2b/7 signals are necessary for ventralization of both mesoderm and ectoderm during gastrulation. In the ectoderm, the mutant phenotype is characterized by an expansion of the neural plate into ventral regions on the expense of epidermis. The territory of both the interneurons and the motor neurons is expanded in BMP mutants, as determined by staining with neuron-specific markers (Barth et al. 1999; Nguyen et al. 2000). While Rohon Beard sensory neurons are absent in BMP2b/BMP7-deficient embryos, it is not clear whether their absence is paralleled by lack of *ngn1* expression due to the expansion and distortion of the neural plate in the mutant. In contrast, mutations in the molecularly uncharacterized *narrowminded* locus affect development of sensory neurons and neural crest but do not affect the size of the neural plate (Artinger et al. 1999). Moreover, the lateral domain of *ngn1* expression is reduced in the lateral aspects of the mutant neural plate, indicating a possible prepattern function of *narrowminded* in the control of *ngn1* in the lateral proneural domain (P. Blader and U. Strähle, unpublished data).

This dependence of *ngn1* expression on *narrowminded* gene activity indicates a positive mode of *ngn1* regulation. However, it is also possible that negative regulatory mechanisms operate in parallel to restrict *ngn1* expression to subdomains of the neural plate by inhibition in intervening regions. In amphibians, the zinc-finger transcription factor *Zic2,* which is expressed in regions between proneural domains, inhibits *ngn1* expression when expressed uniformly (Brewster et al. 1998). Several bHLH inhibitors, such as the vertebrate *emc* and *hairy/enhancer-of-split* homologues, are expressed in the neural plate (Muller et al. 1996; Sawai and Campos-Ortega 1997; Takke et al. 1999). No clear involvement of these genes in establishing the pattern of *ngn1* in the neural plate has yet been demonstrated. It is rather unlikely that the Notch/Delta signaling system plays a crucial role in specifying the position of proneural domains in the neural plate, since misexpression of dominant negative forms of Delta do not lead to ectopic activation of *ngn1* expression (Blader et al. 1997).

In summary, this shows that the complex pattern of the expression of *ngn1* is the result of multiple, converging signal transduction pathways. It involves the Hh and BMP pathway to control medial-lateral differentiation and other, yet unidentified pathways which interact along the A-P axis to specify the complex pattern in the anterior neural plate. Thus expression of the Ngn1 module in the neural plate is under complex control, which is probably a mechanism to restrict primary neurogenesis spatially.

Regulatory Regions Driving *ngn1* Expression Are Conserved between Fish and Mouse

Here we turn to a different issue, asking whether cis-regulatory elements could qualify as modules in the same way as bHLH modules. Regulatory regions evolve much faster than protein-protein interaction domains (Tautz and Eisen 2000). This, together with our limited understanding of the regulatory process, prevents a systematic comparison between *Drosophila* and vertebrate regulatory elements. We will restrict our discussion to the vertebrate lineage.

To investigate the mechanisms controlling the expression of *ngn1* in the zebrafish, we embarked on an analysis of the cis-regulatory elements. The *ngn1* gene contains one intron in its 5′ nontranslated region (P. Blader and U. Strähle, unpublished data). This structure is shared with the highly related *ngn1* genes of mouse and human. Sequence comparisons between the human and zebrafish locus revealed significant sequence conservation outside the coding region. Not only the promoter, that is, the region immediately upstream of the transcription start site, but also two other regions display sequence conservation between zebrafish and human (fig. 3.8). Reporter constructs in which the coding region of the *green fluorescent protein* (*gfp*) was placed under control of a 3.1 kb *ngn1* upstream sequence drive *gfp* expression in primary Rohon Beard sensory neurons in 1-day-old trans-

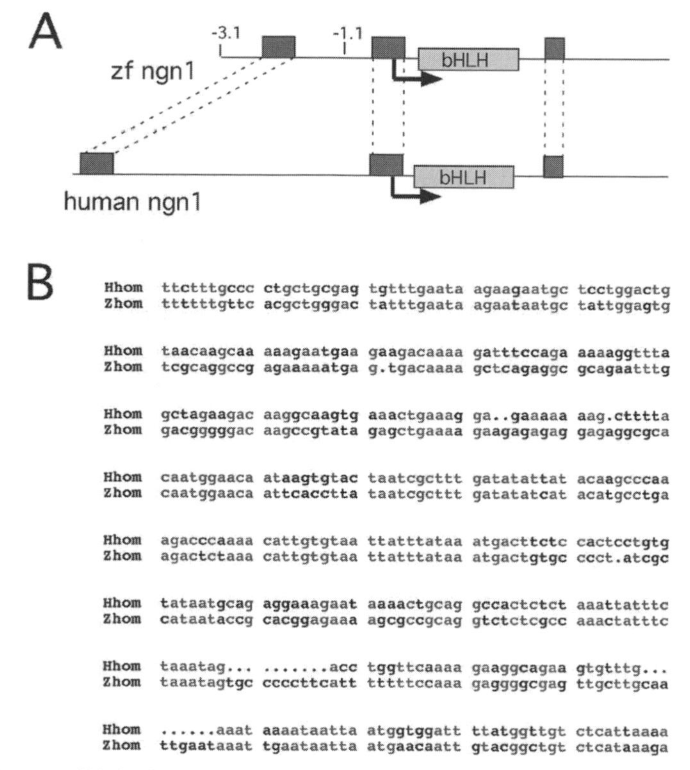

Fig. 3.8.—Several blocks of conserved sequence are located in the 5′ and 3′ region of the zebrafish and human *ngn1* gene. *A*, Schematic overview. Conserved sequence blocks in noncoding regions are indicated in black. The coding region is shown in grey. *B*, Sequence comparison of the upstream homology between human (Hhom) and zebrafish (Zhom) *ngn1*.

genic embryos. Constructs in which the upstream sequence was deleted to 1.1 kb failed to direct *gfp* expression in sensory neurons (P. Blader and U. Strähle, unpublished data). This suggests that the *ngn1* promoter is not sufficient and that sequences upstream of 1.1 kb are required for control of reporter activity in sensory neurons.

Rohon Beard sensory neurons are transient in the zebrafish and degenerate toward the end of the first week of development (Williams et al. 2000). Mammalian embryos do not form this transient sensory neuron population. Instead, they develop only the neural-crest-derived dorsal root ganglia. The strong sequence conservation of the human *ngn1* gene in a region which is required for expression in primary sensory neurons in the zebrafish neural tube prompted us to test the activity of the zebrafish upstream region in mouse embryos. The 3.1 kb zebrafish upstream regulatory sequence drives expression of the reporter in dorsal aspects of the mouse neural tube, corresponding to the D2A population of interneurons (fig. 3.9; P. Blader, R. Scardigli, F. Guillemot, and

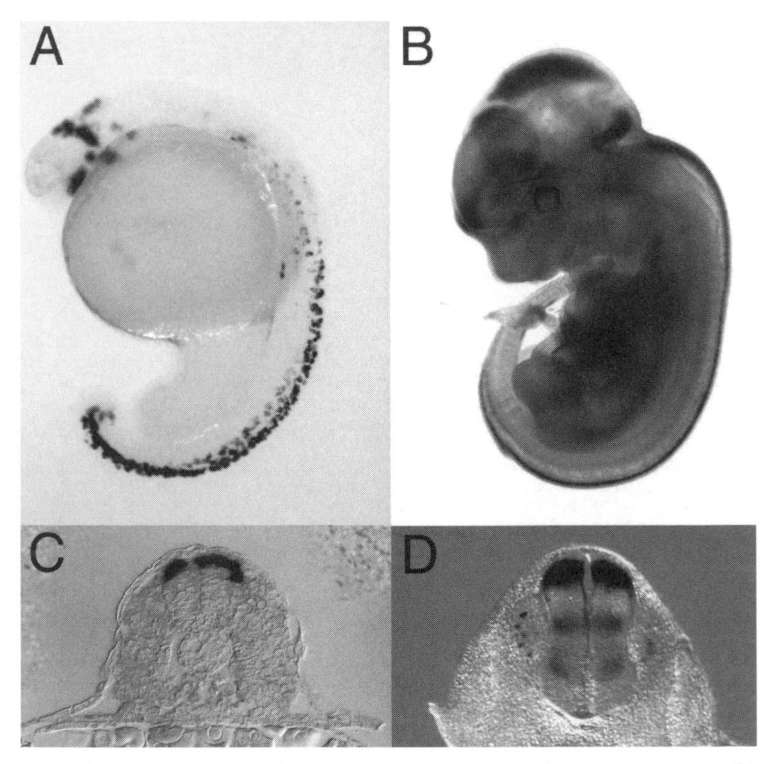

Fig. 3.9.—Zebrafish *3.1ngn1* transgene drives reporter gene expression in dorsal primary sensory neurons of the zebrafish (*A, C*) and in dorsal aspects of the spinal cord of mouse embryos (*B, D*), suggesting conserved regulatory mechanisms in the two species. *A* and *B*, Whole mounts. *C*, and *D*, Transverse sections through spinal cord at trunk levels.

U. Strähle, unpublished data). These results strongly suggest that regulatory mechanisms controlling *ngn1* expression have been conserved during evolution. Furthermore, these data indicate that common regulatory mechanisms underlie primary and secondary neurogenesis. It is clear that these 3.1 kb upstream sequences regulate reporter expression in only a subset of the cells that normally express *ngn1*. Thus, other regulatory sites are necessary for all aspects of *ngn1* expression. The conserved promoter and sequence downstream of the coding region (fig. 3.8, *A*) may be other regions contributing to the many facets of *ngn1* expression during CNS development.

Cis-regulatory elements such as the upstream enhancer identified between −3.1 kb and −1.1 kb of the *ngn1* gene act relatively independently of context, as they can be transferred to other promoters. Furthermore, the enhancer is structurally conserved between zebrafish and mouse and functions in the mouse neural tube. Even though we have not directly tested context independence, that is, activity of the enhancer in front of a heterologous promoter, it is a good candidate for a cis-regulatory module.

Conclusion

Mechanisms underlying neurogenesis have been highly conserved during evolution. bHLH modules play important roles in these process as both neuronal determinants and genes that control specific neuronal differentiation processes. bHLH modules are linked to other regulatory modules in a conserved fashion, leading to maintenance of complex regulatory pathways in organisms as diverse as the fruit fly and vertebrates. It appears that mechanisms—once they have been invented in evolution, such as those which segregate neurons from common precursor cell pools—have been maintained in modifications during evolution.

The control of expression of bHLH genes is crucial for their function, as the very same bHLH genes act at differing levels within these regulatory cascades. Analysis of the regulatory elements that drive expression of *ngn*s in vertebrates suggests that not only have the bHLH structures been conserved, but also the regulatory regions have been maintained, and this conserved use of regulatory sequences highlights enhancers as a distinct class of modules.

Acknowledgments

We thank Elise Cau, Carlos Parras, Thomas Dickmeis, and François Guillemot for their critical reading of this manuscript and D. Biellmann for artwork. We are grateful to the Institute National de la Santé et de la Recherche Medicale, the Centre National de la Recherche Scientifique, the Hôpital Universitaire de Strasbourg, AFM, ACI, and ARC for financial support.

References

Ahmad, I., H. R. Acharya, J. A. Rogers, A. Shibata, T. E. Smithgall, and C. M. Dooley. 1998. The role of NeuroD as a differentiation factor in the mammalian retina. *J. Mol. Neurosci.* 11:165–178.

Amati, B., and H. Land. 1994. Myc-Max-Mad: a transcription factor network controlling cell cycle progression, differentiation and death. *Curr. Opin. Genet. Dev.* 4: 102–108.

Appel, B., and J. S. Eisen. 1998. Regulation of neuronal specification in the zebrafish spinal cord by Delta function. *Development* 125:371–380.

Appel, B., A. Fritz, M. Westerfield, D. J. Grunwald, J. S. Eisen, and B. B. Riley. 1999. Delta-mediated specification of midline cell fates in zebrafish embryos. *Curr. Biol.* 9: 247–256.

Arendt, D., and K. Nübler-Jung. 1997. Dorsal or ventral: similarities in fate maps and gastrulation patterns in annelids, arthropods and chordates. *Mech. Dev.* 61: 7–21.

Artavanis-Tsakonas, S., K. Matsuno, and M. E. Fortini. 1995. Notch signaling. *Science* 268:225–232.

Artavanis-Tsakonas, S., M. D. Rand, and R. J. Lake. 1999. Notch signaling: cell fate control and signal integration in development. *Science* 284:770–776.

Artinger, K. B., A. B. Chitnis, M. Mercola, and W. Driever. 1999. Zebrafish narrow-minded suggests a genetic link between formation of neural crest and primary sensory neurons. *Development* 126:3969–3979.

Atchley, W. R., and W. M. Fitch. 1997. A natural classification of the basic helix-loop-helix class of transcription factors. *Proc. Natl. Acad. Sci. U.S.A.* 94:5172–5176.

Bailey, A. M., and J. W. Posakony. 1995. Suppressor of hairless directly activates transcription of enhancer of split complex genes in response to Notch receptor activity. *Genes Dev.* 9:2609–2622.

Baker, N. E., S. Yu, and D. Han. 1996. Evolution of proneural atonal expression during distinct regulatory phases in the developing *Drosophila* eye. *Curr. Biol.* 6: 1290–1301.

Barth, K. A., Y. Kishimoto, K. B. Rohr, C. Seydler, S. Schulte-Merker, and S. W. Wilson. 1999. Bmp activity establishes a gradient of positional information throughout the entire neural plate. *Development* 126:4977–4987.

Beattie, C. E., K. Hatta, M. E. Halpern, H. Liu, J. S. Eisen, and C. B. Kimmel. 1997. Temporal separation in the specification of primary and secondary motoneurons in zebrafish. *Dev. Biol.* 187:171–182.

Bier, E., H. Vaessin, S. Younger-Shepherd, L. Y. Jan, and Y. N. Jan. 1992. deadpan, an essential pan-neural gene in *Drosophila,* encodes a helix-loop-helix protein similar to the hairy gene product. *Genes Dev.* 6:2137–2151.

Blackwell, T. K., L. Kretzner, E. M. Blackwood, R. N. Eisenman, and H. Weintraub. 1990. Sequence-specific DNA binding by the c-Myc protein. *Science* 250:1149–1151.

Blackwell, T. K., and H. Weintraub. 1990. Differences and similarities in DNA-binding preferences of MyoD and E2A protein complexes revealed by binding site selection. *Science* 250:1104–1110.

Blader, P., N. Fischer, G. Gradwohl, F. Guillemont, and U. Strahle. 1997. The activity of neurogenin1 is controlled by local cues in the zebrafish embryo. *Development* 124:4557–4569.

Botas, J., J. Moscoso del Prado, and A. Garcia-Bellido. 1982. Gene-dose titration analysis in the search of trans-regulatory genes in *Drosophila. EMBO J.* 1:307–310.

Bouchard, C., P. Staller, and M. Eilers. 1998. Control of cell proliferation by Myc. *Trends Cell Biol.* 8:202–206.

Brand, M., A. P. Jarman, L. Y. Jan, and Y. N. Jan. 1993. asense is a *Drosophila* neural precursor gene and is capable of initiating sense organ formation. *Development* 119: 1–17.

Brewster, R., J. Lee, and A. Ruiz i Altaba. 1998. Gli/Zic factors pattern the neural plate by defining domains of cell differentiation. *Nature* 393:579–583.

Bush, A., Y. Hiromi, and M. Cole. 1996. Biparous: a novel bHLH gene expressed in neuronal and glial precursors in *Drosophila. Dev. Biol.* 180:759–772. Erratum appears in *Dev. Biol.* 205 (2) (1999): 332.

Cabrera, C. V., and M. C. Alonso. 1991. Transcriptional activation by heterodimers of the achaete-scute and daughterless gene products of *Drosophila. EMBO J.* 10: 2965–2973.

Campos-Ortega, J. A. 1995. Genetic mechanisms of early neurogenesis in *Drosophila melanogaster. Mol. Neurobiol.* 10:75–89.

Campos-Ortega, J. A. 1997. Neurogenesis in *Drosophila:* an historical perspective and some prospects. *Perspect. Dev. Neurobiol.* 4:267–271.

Campuzano, S., and J. Modolell. 1992. Patterning of the *Drosophila* nervous system: the achaete-scute gene complex. *Trends Genet.* 8:202–208.

Cau, E., G. Gradwohl, C. Fode, and F. Guillemot. 1997. Mash1 activates a cascade of bHLH regulators in olfactory neuron progenitors. *Development* 124:1611–1621.

Chan, Y. M., and Y. N. Jan. 1999. Conservation of neurogenic genes and mechanisms. *Curr. Opin. Neurobiol.* 9:582–588.

Chiang, C., Y. Litingtung, E. Lee, K. E. Young, J. L. Corden, H. Westphal, and P. A. Beachy. 1996. Cyclopia and defective axial patterning in mice lacking Sonic hedgehog gene function. *Nature* 383:407–413.

Chien, C. T., C. D. Hsiao, L. Y. Jan, and Y. N. Jan. 1996. Neuronal type information encoded in the basic-helix-loop-helix domain of proneural genes. *Proc. Natl. Acad. Sci. U.S.A.* 93:13239–13244.

Chitnis, A., D. Henrique, J. Lewis, D. Ish-Horowicz, and C. Kintner. 1995. Primary neurogenesis in *Xenopus* embryos regulated by a homologue of the *Drosophila* neurogenic gene Delta. *Nature* 375:761–766.

Chitnis, A., and C. Kintner. 1996. Sensitivity of proneural genes to lateral inhibition affects the pattern of primary neurons in *Xenopus* embryos. *Development* 122: 2295–2301.

Crews, S. T. 1998. Control of cell lineage-specific development and transcription by bHLH-PAS proteins. *Genes Dev.* 12:607–620.

Cubas, P., J. F. de Celis, S. Campuzano, and J. Modolell. 1991. Proneural clusters of achaete-scute expression and the generation of sensory organs in the *Drosophila* imaginal wing disc. *Genes Dev.* 5:996–1008.

Cubas, P., and J. Modolell. 1992. The extramacrochaetae gene provides information for sensory organ patterning. *EMBO J.* 11:3385–3393.

Currie, P. D., and P. W. Ingham. 1996. Induction of a specific muscle cell type by a hedgehog-like protein in zebrafish. *Nature* 382:452–455.

Dambly-Chaudiere, C., and M. Vervoort. 1998. The bHLH genes in neural development. *Int. J. Dev. Biol.* 42:269–273.

Davis, R. L., P. F. Cheng, A. B. Lassar, and H. Weintraub. 1990. The MyoD DNA binding domain contains a recognition code for muscle-specific gene activation. *Cell* 60: 733–746.

de la Pompa, J. L., A. Wakeham, K. M. Correia, E. Samper, S. Brown, R. J. Aguilera, T. Nakano, T. Honjo, T. W. Mak, J. Rossant, and R. A. Conlon. 1997. Conservation of the Notch signalling pathway in mammalian neurogenesis. *Development* 124:1139–1148.

Dick, A., M. Hild, H. Bauer, Y. Imai, H. Maifeld, A. F. Schier, W. S. Talbot, T. Bouwmeester, and M. Hammerschmidt. 2000. Essential role of Bmp7 (snailhouse) and its prodomain in dorsoventral patterning of the zebrafish embryo. *Development* 127: 343–354.

Dornseifer, P., C. Takke, and J. A. Campos-Ortega. 1997. Overexpression of a zebrafish homologue of the *Drosophila* neurogenic gene Delta perturbs differentiation of primary neurons and somite development. *Mech. Dev.* 63:159–171.

Dubois, L., L. Bally-Cuif, M. Crozatier, J. Moreau, L. Paquereau, and A. Vincent. 1998. XCoe2, a transcription factor of the Col/Olf-1/EBF family involved in the specification of primary neurons in *Xenopus. Curr. Biol.* 8:199–209.

Ekker, S. C., A. R. Ungar, P. Greenstein, D. P. von Kessler, J. A. Porter, R. T. Moon, and P. A. Beachy. 1995. Patterning activities of vertebrate hedgehog proteins in the developing eye and brain. *Curr. Biol.* 5:944–955.

Ferre-D'Amare, A. R., P. Pognonec, R. G. Roeder, and S. K. Burley. 1994. Structure and function of the b/HLH/Z domain of USF. *EMBO J.* 13:180–189.

Ferre-D'Amare, A. R., G. C. Prendergast, E. B. Ziff, and S. K. Burley. 1993. Recognition by Max of its cognate DNA through a dimeric b/HLH/Z domain. *Nature* 363: 38–45.

Fisher, A., and M. Caudy. 1998. The function of hairy-related bHLH repressor proteins in cell fate decisions. *BioEssays* 20:298–306.

Fisher, A. L., S. Ohsako, and M. Caudy. 1996. The WRPW motif of the hairy-related basic helix-loop-helix repressor proteins acts as a 4-amino-acid transcription repression and protein-protein interaction domain. *Mol. Cell. Biol.* 16:2670–2677.

Fode, C., G. Gradwohl, X. Morin, A. Dierich, M. LeMeur, C. Goridis, and F. Guillemot. 1998. The bHLH protein NEUROGENIN 2 is a determination factor for epibranchial placode-derived sensory neurons. *Neuron* 20:483–494.

Fode, C., Q. Ma, S. Casarosa, S. L. Ang, D. J. Anderson, and F. Guillemot. 2000. A role for neural determination genes in specifying the dorsoventral identity of telencephalic neurons. *Genes Dev.* 14:67–80.

Gautier, P., V. Ledent, M. Massaer, C. Dambly-Chaudiere, and A. Ghysen. 1997. tap, a *Drosophila* bHLH gene expressed in chemosensory organs. *Gene* 191:15–21.

Gho, M., Y. Bellaiche, and F. Schweisguth. 1999. Revisiting the *Drosophila* microchaete lineage: a novel intrinsically asymmetric cell division generates a glial cell. *Development* 126:3573–3584.

Ghysen, A., and C. Dambly-Chaudiere. 1989. Genesis of the *Drosophila* peripheral nervous system. *Trends Genet* 5:251–255.

Ghysen, A., C. Dambly-Chaudiere, L. Y. Jan, and Y. N. Jan. 1993. Cell interactions and gene interactions in peripheral neurogenesis. *Genes Dev.* 7:723–733.

Giebel, B., I. Stuttem, U. Hinz, and J. A. Campos-Ortega. 1997. Lethal of scute requires overexpression of daughterless to elicit ectopic neuronal development during embryogenesis in *Drosophila. Mech. Dev.* 63:75–87.

Goulding, S. E., P. zur Lage, and A. P. Jarman. 2000. amos, a proneural gene for *Drosophila* olfactory sense organs that is regulated by lozenge. *Neuron* 25:69–78.

Gradwohl, G., A. Dierich, M. LeMeur, and F. Guillemot. 2000. neurogenin3 is required for the development of the four endocrine cell lineages of the pancreas. *Proc. Natl. Acad. Sci. U.S.A* 97:1607–1611.

Gradwohl, G., C. Fode, and F. Guillemot. 1996. Restricted expression of a novel murine atonal-related bHLH protein in undifferentiated neural precursors. *Dev. Biol.* 180: 227–241.

Guillemot, F. 1999. Vertebrate bHLH genes and the determination of neuronal fates. *Exp. Cell Res.* 253:357–364.

Guillemot, F., L. C. Lo, J. E. Johnson, A. Auerbach, D. J. Anderson, and A. L. Joyner. 1993. Mammalian achaete-scute homolog 1 is required for the early development of olfactory and autonomic neurons. *Cell* 75:463–476.

Gupta, B. P., and V. Rodrigues. 1997. Atonal is a proneural gene for a subset of olfactory sense organs in *Drosophila. Genes Cells* 2:225–233.

Haddon, C., L. Smithers, S. Schneider-Maunoury, T. Coche, D. Henrique, and J. Lewis. 1998. Multiple delta genes and lateral inhibition in zebrafish primary neurogenesis. *Development* 125:359–370.

Hammerschmidt, M., M. J. Bitgood, and A. P. McMahon. 1996. Protein kinase A is a common negative regulator of Hedgehog signaling in the vertebrate embryo. *Genes Dev.* 10:647–658.

Hassan, B. A., and H. J. Bellen. 2000. Doing the MATH: is the mouse a good model for fly development? *Genes Dev.* 14:1852–1865.

Hassan, B. A., N. A. Bermingham, Y. He, Y. Sun, Y. N. Jan, H. Y. Zoghbi, and H. J. Bellen. 2000. atonal regulates neurite arborization but does not act as a proneural gene in the *Drosophila* brain. *Neuron* 25:549–561.

Heitzler, P., M. Bourouis, L. Ruel, C. Carteret, and P. Simpson. 1996. Genes of the Enhancer of split and achaete-scute complexes are required for a regulatory loop between Notch and Delta during lateral signalling in *Drosophila. Development* 122:161–171.

Heitzler, P., and P. Simpson. 1991. The choice of cell fate in the epidermis of *Drosophila. Cell* 64:1083–1092.

Heitzler, P., and P. Simpson. 1993. Altered epidermal growth factor-like sequences provide evidence for a role of Notch as a receptor in cell fate decisions. *Development* 117:1113–1123.

Henrique, D., J. Adam, A. Myat, A. Chitnis, J. Lewis, and D. Ish-Horowicz. 1995. Expression of a Delta homologue in prospective neurons in the chick. *Nature* 375: 787–790.

Holley, S. A., R. Geisler, and C. Nusslein-Volhard. 2000. Control of her1 expression during zebrafish somitogenesis by a delta-dependent oscillator and an independent wave-front activity. *Genes Dev.* 14:1678–1690.

Huang, M. L., C. H. Hsu, and C. T. Chien. 2000. The proneural gene amos promotes multiple dendritic neuron formation in the *Drosophila* peripheral nervous system. *Neuron* 25:57–67.

Jan, Y. N., and L. Y. Jan. 1993. HLH proteins, fly neurogenesis, and vertebrate myogenesis. *Cell* 75:827–830.

Jarman, A. P., and I. Ahmed. 1998. The specificity of proneural genes in determining *Drosophila* sense organ identity. *Mech. Dev.* 76:117–125.

Jarman, A. P., M. Brand, L. Y. Jan, and Y. N. Jan. 1993a. The regulation and function of the helix-loop-helix gene, asense, in *Drosophila* neural precursors. *Development* 119:19–29.

Jarman, A. P., Y. Grau, L. Y. Jan, and Y. N. Jan. 1993b. atonal is a proneural gene that directs chordotonal organ formation in the *Drosophila* peripheral nervous system. *Cell* 73:1307–1321.

Jimenez, F., and J. A. Campos-Ortega. 1990. Defective neuroblast commitment in mutants of the achaete-scute complex and adjacent genes of *D. melanogaster. Neuron* 5:81–89.

Jimenez, F., and J. Modolell. 1993. Neural fate specification in *Drosophila. Curr. Opin. Genet. Dev.* 3:626–632.

Kageyama, R., and T. Ohtsuka. 1999. The Notch-Hes pathway in mammalian neural development. *Cell Res.* 9:179–188.

Kao, H. Y., P. Ordentlich, N. Koyano-Nakagawa, Z. Tang, M. Downes, C. R. Kintner, R. M. Evans, and T. Kadesch. 1998. A histone deacetylase corepressor complex regulates the Notch signal transduction pathway. *Genes Dev.* 12:2269–2277.

Kimmel, C. B., and M. Westerfield. 1989. Primary neurons of the zebrafish. In *Signal and sense,* ed. G. M. Edelman and M. W. Cowan, 561–588. New York: Wiley-Liss.

Kishimoto, Y., K. H. Lee, L. Zon, M. Hammerschmidt, and S. Schulte-Merker. 1997. The molecular nature of zebrafish swirl: BMP2 function is essential during early dorsoventral patterning. *Development* 124:4457–4466.

Klambt, C., E. Knust, K. Tietze, and J. A. Campos-Ortega. 1989. Closely related transcripts encoded by the neurogenic gene complex enhancer of split of *Drosophila melanogaster. EMBO J.* 8:203–210.

Korzh, V., I. Sleptsova, J. Liao, J. He, and Z. Gong. 1998. Expression of zebrafish bHLH genes ngn1 and nrd defines distinct stages of neural differentiation. *Dev. Dyn.* 213: 92–104.

Krauss, S., J.-P. Concordet, and P. W. Ingham. 1993. A functionally conserved homolog of the *Drosophila* segment polarity gene *hedgehog* is expressed in tissues with polarizing activity in zebrafish embryos. *Cell* 75:1431–1444.

Kunisch, M., M. Haenlin, and J. A. Campos-Ortega. 1994. Lateral inhibition mediated by the *Drosophila* neurogenic gene delta is enhanced by proneural proteins. *Proc. Natl. Acad. Sci. U.S.A.* 91:10139–10143.

Lecourtois, M., and F. Schweisguth. 1995. The neurogenic suppressor of hairless DNA-binding protein mediates the transcriptional activation of the enhancer of split complex genes triggered by Notch signaling. *Genes Dev.* 9:2598–2608.

Lecourtois, M., and F. Schweisguth. 1997. Role of suppressor of hairless in the delta-activated Notch signaling pathway. *Perspect. Dev. Neurobiol.* 4:305–311.

Lecourtois, M., and F. Schweisguth. 1998. Indirect evidence for Delta-dependent intracellular processing of notch in *Drosophila* embryos. *Curr. Biol.* 8:771–774.

Ledent, V., F. Gaillard, P. Gautier, A. Ghysen, and C. Dambly-Chaudiere. 1998. Expression and function of tap in the gustatory and olfactory organs of *Drosophila. Int. J. Dev. Biol.* 42:163–170.

Lee, J. E.. 1997. Basic helix-loop-helix genes in neural development. *Curr. Opin. Neurobiol.* 7:13–20.

Lee, J. E., S. M. Hollenberg, L. Snider, D. L. Turner, N. Lipnick, and H. Weintraub. 1995. Conversion of *Xenopus* ectoderm into neurons by NeuroD, a basic helix-loop-helix protein. *Science* 268:836–844.

Lewis, J. 1996. Neurogenic genes and vertebrate neurogenesis. *Curr. Opin. Neurobiol.* 6:3–10.

Lewis, K. E., and J. S. Eisen. 2001. Hedgehog signaling is required for primary motoneuron induction in zebrafish. *Development* 128:3485–3495.

Ma, P. C., M. A. Rould, H. Weintraub, and C. O. Pabo. 1994. Crystal structure of MyoD bHLH domain-DNA complex: perspectives on DNA recognition and implications for transcriptional activation. *Cell* 77:451–459.

Ma, Q., Z. Chen, I. del Barco Barrantes, J. L. de la Pompa, and D. J. Anderson. 1998. neurogenin1 is essential for the determination of neuronal precursors for proximal cranial sensory ganglia. *Neuron* 20:469–482.

Ma, Q., C. Kintner, and D. J. Anderson. 1996. Identification of neurogenin, a vertebrate neuronal determination gene. *Cell* 87:43–52.

Marti, E., D. A. Bumcrot, R. Takada, and A. P. McMahon. 1995. Requirement of 19K form of Sonic hedgehog for induction of distinct ventral cell types in CNS explants. *Nature* 375:322–325.

Martin-Bermudo, M. D., F. Gonzalez, M. Dominguez, I. Rodriguez, M. Ruiz-Gomez, S. Romani, J. Modolell, and F. Jimenez. 1993. Molecular characterization of the lethal of scute genetic function. *Development* 118:1003–1012.

Martin-Bermudo, M. D., C. Martinez, A. Rodriguez, and F. Jimenez. 1991. Distribution and function of the lethal of scute gene product during early neurogenesis in *Drosophila. Development* 113:445–454.

Martinez, C., J. Modolell, and J. Garrell. 1993. Regulation of the proneural gene achaete by helix-loop-helix proteins. *Mol. Cell. Biol.* 13:3514–3521.

Miyata, T., T. Maeda, and J. E. Lee. 1999. NeuroD is required for differentiation of the granule cells in the cerebellum and hippocampus. *Genes Dev.* 13:1647–1652.

Muller, M., E. von Weizsacker, and J. A. Campos-Ortega. 1996. Expression domains of a zebrafish homologue of the *Drosophila* pair-rule gene hairy correspond to primordia of alternating somites. *Development* 122:2071–2078.

Murre, C., P. S. McCaw, H. Vaessin, M. Caudy, L. Y. Jan, Y. N. Jan, C. V. Cabrera, J. N. Buskin, S. D. Hauschka, A. B. Lassar, et al. 1989. Interactions between heterologous helix-loop-helix proteins generate complexes that bind specifically to a common DNA sequence. *Cell* 58:537–544.

Nguyen, V. H., B. Schmid, J. Trout, S. A. Connors, M. Ekker, and M. C. Mullins. 1998. Ventral and lateral regions of the zebrafish gastrula, including the neural crest progenitors, are established by a bmp2b/swirl pathway of genes. *Dev. Biol.* 199: 93–110.

Nguyen, V. H., J. Trout, S. A. Connors, P. Andermann, E. Weinberg, and M. C. Mullins. 2000. Dorsal and intermediate neuronal cell types of the spinal cord are established by a BMP signaling pathway. *Development* 127:1209–1220.

Oellers, N., M. Dehio, and E. Knust. 1994. bHLH proteins encoded by the Enhancer of split complex of *Drosophila* negatively interfere with transcriptional activation mediated by proneural genes. *Mol. Gen. Genet.* 244:465–473.

Ohtsuka, T., M. Ishibashi, G. Gradwohl, S. Nakanishi, F. Guillemot, and R. Kageyama. 1999. Hes1 and Hes5 as notch effectors in mammalian neuronal differentiation. *EMBO J.* 18:2196–2207.

Paroush, Z., R. J. Finley, Jr., T. Kidd, S. M. Wainwright, P. W. Ingham, R. Brent, and D. Ish-Horowicz. 1994. Groucho is required for *Drosophila* neurogenesis, segmentation, and sex determination and interacts directly with hairy-related bHLH proteins. *Cell* 79:805–815.

Perez, S. E., S. Rebelo, and D. J. Anderson. 1999. Early specification of sensory neuron fate revealed by expression and function of neurogenins in the chick embryo. *Development* 126:1715–1728.

Perron, M., K. Opdecamp, K. Butler, W. A. Harris, and E. J. Bellefroid. 1999. X-ngnr-1 and Xath3 promote ectopic expression of sensory neuron markers in the neurula ectoderm and have distinct inducing properties in the retina. *Proc. Natl. Acad. Sci. U.S.A.* 96:14996–15001.

Riley, B. B., M. Chiang, L. Farmer, and R. Heck. 1999. The deltaA gene of zebrafish mediates lateral inhibition of hair cells in the inner ear and is regulated by pax2.1. *Development* 126:5669–5678.

Roark, M., M. A. Sturtevant, J. Emery, H. Vaessin, E. Grell, and E. Bier. 1995. scratch, a pan-neural gene encoding a zinc finger protein related to snail, promotes neuronal development. *Genes Dev.* 9:2384–2398.

Robinson, K. A., J. I. Koepke, M. Kharodawala, and J. M. Lopes. 2000. A network of yeast basic helix-loop-helix interactions. *Nucleic Acids Res.* 28:4460–4466.

Roelink, H., A. Auggsburger, J. Heemskerk, V. Kortzh, S. Norlin, A. Ruiz i Altaba, Y. Tanabe, M. Placzek, T. Edlund, T. Jessell, and J. Dodd. 1994. Floor plate and motor neuron induction by vhh-1, a vertebrate homolog of hedgehog expressed by the notochord. *Cell* 76:761–775.

Roelink, H., J. A. Porter, C. Chiang, Y. Tanabe, D. T. Chang, P. A. Beachy, and T. M. Jessell. 1995. Floor plate and motor neuron induction by different concentrations of the amino-terminal cleavage product of sonic hedgehog autoproteolysis. *Cell* 81: 445–455.

Salzberg, A., and H. J. Bellen. 1996. Invertebrate versus vertebrate neurogenesis: variations on the same theme? *Dev. Genet.* 18:1–10.

Sawai, S., and J. A. Campos-Ortega. 1997. A zebrafish Id homologue and its pattern of expression during embryogenesis. *Mech. Dev.* 65:175–185.

Schauerte, H. E., F. J. M. van Eden, C. Fricke, J. Odenthal, U. Strähle, and P. Haffter. 1998. Sonic hedgehog is not required for floor plate induction in the zebrafish embryo. *Development* 125:2983–2993.

Schlosser, G. 2004. The role of modules in development and evolution. In *Modularity in development and evolution,* ed. G. Schlosser and G. P. Wagner. Chicago: University of Chicago Press.

Schmid, B., M. Furthauer, S. A. Connors, J. Trout, B. Thisse, C. Thisse, and M. C. Mullins. 2000. Equivalent genetic roles for bmp7/snailhouse and bmp2b/swirl in dorsoventral pattern formation. *Development* 127:957–967.

Schwab, M. H., S. Druffel-Augustin, P. Gass, M. Jung, M. Klugmann, A. Bartholomae, M. J. Rossner, and K. A. Nave. 1998. Neuronal basic helix-loop-helix proteins (NEX, neuroD, NDRF): spatiotemporal expression and targeted disruption of the NEX gene in transgenic mice. *J. Neurosci.* 18:1408–1418.

Shimizu, K., S. Chiba, N. Hosoya, K. Kumano, T. Saito, M. Kurokawa, Y. Kanda, Y. Hamada, and H. Hirai. 2000. Binding of Delta1, Jagged1, and Jagged2 to Notch2 rapidly induces cleavage, nuclear translocation, and hyperphosphorylation of Notch2. *Mol. Cell. Biol.* 20:6913–6922.

Sun, X. H., and D. Baltimore. 1991. An inhibitory domain of E12 transcription factor prevents DNA binding in E12 homodimers but not in E12 heterodimers. *Cell* 64: 459–470. Erratum in *Cell* 66 (3) (1991): 423.

Sun, X. H., N. G. Copeland, N. A. Jenkins, and D. Baltimore. 1991. Id proteins Id1 and Id2 selectively inhibit DNA binding by one class of helix-loop-helix proteins. *Mol. Cell. Biol.* 11:5603–5611.

Takke, C., and J. A. Campos-Ortega. 1999. her1, a zebrafish pair-rule like gene, acts downstream of notch signalling to control somite development. *Development* 126: 3005–3014.

Takke, C., P. Dornseifer, E. von Weizsacker, and J. A. Campos-Ortega. 1999. her4, a zebrafish homologue of the *Drosophila* neurogenic gene E(spl), is a target of NOTCH signalling. *Development* 126:1811–1821.

Tautz, D., and J. S. Eisen. 2000. Evolution of transcriptional regulation. *Curr. Opin. Genet. Dev.* 10:575–579.

Van Doren, M., A. M. Bailey, J. Esnayra, K. Ede, and J. W. Posakony. 1994. Negative regulation of proneural gene activity: hairy is a direct transcriptional repressor of achaete. *Genes Dev.* 8:2729–2742.

Van Doren, M., H. M. Ellis, and J. W. Posakony. 1991. The *Drosophila* extramacrochaetae protein antagonizes sequence-specific DNA binding by daughterless/achaete-scute protein complexes. *Development* 113:245–255.

Van Doren, M., P. A. Powell, D. Pasternak, A. Singson, and J. W. Posakony. 1992. Spatial regulation of proneural gene activity: auto- and cross-activation of achaete is antagonized by extramacrochaetae. *Genes Dev.* 6:2592–2605.

Vervoort, M., C. Dambly-Chaudiere, and A. Ghysen. 1997. Cell fate determination in *Drosophila. Curr. Opin. Neurobiol.* 7:21–28.

Wettstein, D. A., D. L. Turner, and C. Kintner. 1997. The *Xenopus* homolog of *Drosophila* Suppressor of Hairless mediates Notch signaling during primary neurogenesis. *Development* 124:693–702.

Williams, J. A., A. Barrios, C. Gatchalian, L. Rubin, S. W. Wilson, and N. Holder. 2000. Programmed cell death in zebrafish rohon beard neurons is influenced by TrkC1/NT-3 signaling. *Dev. Biol.* 226:220–230.

4 The *Pax/Six/Eya/Dach* Network in Development and Evolution

GABRIELLE KARDON, TIFFANY A. HEANUE,
AND CLIFFORD J. TABIN

Introduction

An emerging theme in developmental and evolutionary biology is the conservation of networks of regulatory genes working together during the development of a wide range of metazoan taxa. In many cases, these networks of genes, often referred to as regulatory cassettes, are used in conserved processes for the development of homologous structures. However, these cassettes are often deployed in different temporal and spatial developmental contexts and expanded to include different gene family members and downstream targets. In fact, such modification of regulatory cassettes may be an important mechanism for generating evolutionary novelty.

In this chapter, we examine one evolutionarily conserved cassette of transcriptional regulators. The network of *eyeless* (*ey*), *sine oculis* (*so*), *eyes absent* (*eya*), and *dachshund* (*dac*) was first identified in *Drosophila* as a critical regulator of eye development (reviewed in Wawersik and Maas 2000). Their respective homologues, the *Pax, Six, Eya,* and *Dach* genes, have also been found in vertebrates (reviewed in Wawersik and Maas 2000) and are coexpressed in a variety of developmental contexts, including the developing eye. The expression patterns of these genes often overlap in a manner suggesting that these genes may indeed be functioning as a network and implying that this network has acquired new functions in vertebrate development. Recently, Heanue and colleagues (Heanue et al. 1999) have demonstrated that the *Pax/Six/Eya/Dach* network plays a critical role in myogenesis. Data from several labs also indicate that this network is important for eye and ear development (Torres et al. 1996; Xu et al. 1999; reviewed in Wawersik and Maas 2000). Comparison of the networks used during vertebrate myogenesis, eye and ear development reveals that different members of the *Pax, Six, Eya,* and *Dach* gene families have been employed and that they are differently regulated. The evolutionary expansion and inclusion of other members of these four gene families appears to have been critical for the deployment of the cassette in novel developmental processes. Not only are the new family members expressed in different temporal and spatial contexts, but also they interact with different partner proteins and activate different downstream targets.

The *Pax, Six, Eya,* and *Dach* Gene Families

The *Pax/Six/Eya/Dach* network is composed of interactions between four different families of genes. Two of the families, the *Pax* and *Six* genes, encode DNA-binding transcription factors, while the other two families, the *Eya* and *Dach* genes, encode transcriptional coactivators. With the evolution of vertebrates, all four gene families have expanded, and new family members have been co-opted into the *Pax/Six/Eya/Dach* network.

The *Pax* genes constitute a large and relatively diverse family of transcription factors (reviewed by Miller et al. 2000). *Pax* genes are defined by the presence of a paired domain, a 128-amino-acid DNA-binding domain. In addition, some *Pax* genes contain a complete or partial homeodomain and/or a distinctive octapeptide motif. Based on a comparison of domain structure and sequence, *Pax* genes are divided into four classes, *PaxA–PaxD*. In *Drosophila,* there are single members of the *A* and *B* classes, but two members of the *C* and three members of the *D* family (fig. 4.1). In vertebrates, no members of the *A* class exist. However, there has been an expansion of the other classes: there are three members of the *B* class, two members of the *C* class, and four members of the *D* class (Miller et al. 2000; fig. 4.1).

The *Six* genes comprise a family of transcription factors that contain a homeodomain and a Six domain (reviewed by Kawakami et al. 2000). The homeodomain is unique because it lacks two highly conserved amino acid residues typical of most homeodomains. Both the Six domain and the homeodomain are necessary for specific DNA binding. In addition to its DNA-binding role, the Six domain is essential for binding to Eya proteins (Pignoni et al. 1997). Members of the *Six* family have been divided into three subfamilies based on the lengths of the region C-terminal to the homeodomain (Kawakami et al. 2000). In *Drosophila,* each of the subfamilies contains one family member. In vertebrates, the *Drosophila sine oculis, optix,* and *six4* subfamilies have been expanded to include two orthologues each (Kawakami et al. 2000; fig. 4.1).

The *Eya* genes constitute a family of transcriptional coactivators, each of which contains an Eya domain. The Eya domain is a highly

Drosophila	Vertebrate
PaxA: pox neuro	
PaxB: sparkling	*Pax 2,5,8*
PaxC: twin of eyeless *eyeless*	*Pax 4,6*
PaxD: pox meso	*Pax 1,9*
gooseberry *gooseberry neuro* *paired*	*Pax 3,7*

Drosophila	Vertebrate
sine oculis	*Six 1,2*
optix	*Six 3, Optx 2*
six 4	*Six 4,5*

Drosophila	Vertebrate
eyes absent	*Eya 1,2,3,4*

Drosophila	Vertebrate
dachshund	*Dach 1,2*

Fig. 4.1.—Comparison of the members of the *Pax, Eya, Six,* and *Dach* genes found in *Drosophila* and vertebrates. The *Pax* and *Six* families contain several subfamilies. The *Pax* family is divided into *PaxA, PaxB, PaxC,* and *PaxD* subfamilies. The *Six* family is divided into the *Six1, Six2, Six3,Optx2,* and *Six4,5* subfamilies. See references in text.

conserved region at the C-terminus of the Eya proteins (Xu et al. 1997b). This domain has been shown to be the site of protein-protein interactions between the *Drosophila* Eya and So proteins and between Eya and Dac proteins (Pignoni et al. 1997; Chen et al. 1997). At the N-terminus of Eya proteins is a nonconserved proline-, serine-, threonine-rich region capable of functioning as a transcriptional activator (Xu et al. 1997a). Eya proteins do not contain known DNA-binding motifs, suggesting that Eya must act in concert with DNA-binding proteins to regulate transcription (Wawersik and Maas 2000). In *Drosophila,* there is a single *eyes absent* gene, and in vertebrates, the family has expanded to include four members (Bonini et al. 1993; Borsani et al. 1999; Xu et al. 1997a; fig. 4.1).

The other family of transcriptional coactivators in the network is the *Dach* genes. Two regions of high similarity, an N-terminal region termed DD1/Dachbox-N and a C-terminal region termed DD2/Dachbox-C, have been identified in all known members of the Dach family (Hammond et al. 1998; Davis et al. 1999). The N-terminal domain has been shown to be necessary for transcriptional activation in yeast (Chen et al. 1997), while the C-terminus has been demonstrated to be critical for binding to Eya proteins in both *Drosophila* and chick (Chen et al. 1997; Heanue et al. 1999). Because there is no known DNA-binding domain in Dach proteins, the transcriptional activation function of Dach must be mediated by interactions with DNA-binding proteins. In *Drosophila,* there is a single *dachshund* gene, and in vertebrates, the family has expanded to two members (Mardon et al. 1994; Hammond et al. 1998; Davis et al. 1999; Heanue et al. 1999; Kozmik et al. 1999; Caubit et al. 1999; fig. 4.1).

Drosophila Eye Development

The *Drosophila* eye consists of a hexagonal array of approximately 750 ommatidia, each containing photoreceptor and accessory cells (reviewed in Wolff and Readt 1993). The eye develops from a small number of cells set aside in the eye imaginal disc (Younoussi-Hartenstein et al. 1993). During the third instar of larval development, ommatidia are generated as a wave of differentiation, known as the morphogenetic furrow, moves from posterior to anterior across the eye disc (Tomlinson and Ready 1987). Anterior to the furrow, cells are undifferentiated, whereas posterior to it, cells are recruited into ommatidia and differentiate into photoreceptors. The initiation and propagation of the morphogenetic furrow is necessary for the proper formation of ommatidia, and the *eyeless/sine oculis/eyes absent/dachshund* network is essential for this process.

The importance of the *ey, so, eya,* and *dac* genes for *Drosophila* eye development was revealed through both loss- and gain-of-function studies. In *ey, so, eya,* and *dac* mutants, the eye anlagen initially form normally. However, during the third instar the eyes fail to develop in all four mutant backgrounds because of lack of morphogenetic furrow initiation and massive cell death in the eye disc (Quiring et al. 1994; Halder et al. 1998; Cheyette et al. 1994; Bonini et al. 1993; Mardon et al. 1994). Thus, each of these four genes is necessary for proper eye development. Conversely, gain-of-function studies have shown that these genes are also sufficient to initiate eye development. In particular, targeted misexpression of *ey, eya,* or *dac* in the antennal imaginal disc induces the formation of ectopic eyes (Halder et al. 1995; Bonini et al. 1997; Shen and Mardon 1997). Interestingly, these genes are not equally potent inducers of ectopic eyes: *ey* is capable of inducing large ectopic eyes with high frequency, while *eya* and *dac* induce smaller eyes and at a much lower frequency.

Consistent with their role in eye development, all four genes are expressed in the developing eye. *ey* is expressed first in the earliest eye anlagen and subsequently becomes restricted to the cells anterior to the morphogenetic furrow (Quiring et al. 1994). *so* and *eya* have similar patterns of expression (Cheyette et al. 1994; Bonini et al. 1993). Prior to morphogenetic furrow formation, both are expressed along the posterior and lateral edges of the eye disc with decreasing levels towards the central region. After morphogenetic furrow propagation, the two genes are expressed anterior to, within, and posterior to the furrow. Prior to morphogenetic furrow formation, *dac* is expressed at the posterior margin of the eye, and subsequently, during furrow propagation, it becomes restricted to cells anterior to the furrow (Mardon et al. 1994).

The initial expression and regulation of *ey, so, eya,* and *dac* is primarily linear (fig. 4.2, *A*). *ey* is the earliest expressed component of the

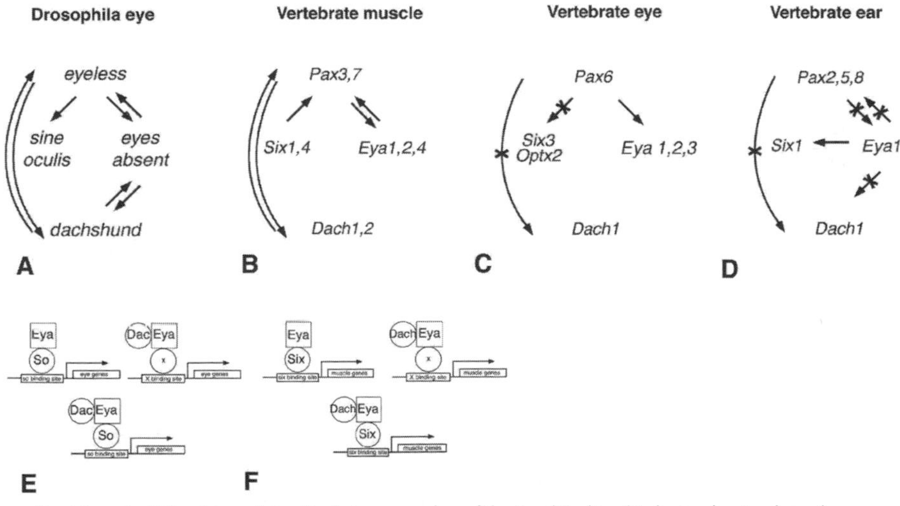

Fig. 4.2.—*A–D,* Regulatory relationships between members of the *Pax/Six/Eya/Dach* network acting during the development of the *Drosophila* eye, vertebrate muscle, eye, and ear. *A,* The known sufficient interactions between *ey, so, eya,* and *dac* in the *Drosophila* eye as established by experiments ectopically misexpressing these genes. Not shown are the necessary interactions between these genes as established by analysis of mutants. Analysis of mutants has determined that *so, eya,* and *dac* are not necessary for *ey* expression and that *dac* is not necessary for *eya* expression. In addition, mutant analysis has determined that *ey* is necessary for *so* and *eya* expression, *eya* is necessary for *so* and *dac* expression, and *so* is necessary for *eya* expression. *B,* The known sufficient interactions between members of the network in vertebrate muscle as determined by ectopic misexpression studies in the chick using *Pax3, Six1, Eya2,* and *Dach2.* Not indicated is the fact that *Pax3* is not necessary for *Dach2* expression, as revealed by normal *Dach2* expression in *Splotch* mice. *C,* Known necessary interactions in the vertebrate eye between members of the network as determined by analysis *Six3, Eya1,* and *Dach1* expression in mice with mutations in *Pax6. D,* Known necessary interactions in the vertebrate ear between members of the network as determined by analysis of *Pax2* and *8, Six1,* and *Dach1* expression in *Eya1* mutant mice and analysis of *Eya1* and *Dach1* expression in *Pax2* mutant mice. *E* and *F,* Proposed interactions between Six, Eya, and Dach proteins functioning in the development of the *Drosophila* eye (*E*) and vertebrate muscle (*F*). See references in text.

ey/so/eya/dac network (Quiring et al. 1994) and is the most potent inducer of ectopic eyes (Halder et al. 1995). In addition, *ey* is expressed in the absence of *so, eya,* or *dac* (Halder et al. 1998; Bonini et al. 1997; Shen and Mardon 1997). *ey,* in turn, is required for the expression of *eya* and *so* and is sufficient to induce *so* and *eya* (Chen et al. 1997; Halder et al. 1998). Recently, Niimi and colleagues have shown that *ey* regulation of *so* is direct, as the Ey protein binds and activates *so* through an eye-specific enhancer (Niimi et al. 1999). *eya* and *so* regulate each other's expression; in the absence of *eya,* no *so* is expressed, and in the absence of *so, eya* is downregulated (Halder et al. 1998). Finally, *dac* is downstream of *ey, so,* and *eya.* For instance, ectopic *ey* induces *dac,* and *ey* is expressed in the absence of *dac* (Shen and Mardon 1997). Also, in the absence of *eya,* no *dac* is expressed, but in the absence of *dac, eya* is expressed normally (Chen et al. 1997).

Although *ey, so, eya,* and *dac* appear initially to be regulated in a linear manner, these genes ultimately function in a nonlinear network, and all four are required for eye development (fig. 4.2, *A*). For instance, gain-of-function studies demonstrate that both *eya* and *dac* are able to induce expression of genes initially upstream in the network. In particular, ectopic *eya* induces *ey* expression, and ectopic *dac* induces both *eya* and *ey* expression (Bonini et al. 1997; Chen et al. 1997; Shen and Mardon 1997). Also, experiments combining loss- and gain-of-function approaches show that the ability of *so, eya,* or *dac* to induce ectopic eyes requires the function of the initially upstream *ey.* In *ey* mutants, ectopic *so, eya,* and *dac* are unable to induce ectopic eyes either singly or in combination (Bonini et al. 1998; Chen et al. 1997; Pignoni et al. 1997). Conversely, *so, eya,* and *dac* are as critical for eye development as *ey.* In *eya, so,* or *dac* mutants ectopic *ey* is unable to induce ectopic eyes (Chen et al. 1997; Halder et al. 1998).

The network is further complicated by the synergistic relationship between *so* and *eya* and between *eya* and *dac.* As mentioned previously, ectopic expression of *eya* or *dac* is able to induce ectopic eyes, albeit small ones and at a low frequency. However, ectopic expression of *so* and *eya* or *eya* and *dac* is able to induce larger ectopic eyes and at a higher frequency (Pignoni et al. 1997; Chen et al. 1997). This synergistic relationship between *so* and *eya* and between *eya* and *dac* is underlain by interactions between the proteins they encode. In particular, GST pull-down and yeast two-hybrid assays establish that So and Eya proteins and Eya and Dac proteins physically interact. Several models may explain the relationship between So, Eya, and Dac proteins and their downstream target genes (fig. 4.2, *E*). The transcriptional coactivator Eya may bind to the transcription factor So, which in turn binds to *so* binding sites upstream of target genes. The transcriptional coactivators Dac and Eya may bind to a third unidentified transcription factor which binds upstream of target genes. Alternatively, a third, as yet undemonstrated, possibility is that Dac, Eya, and So together form a transcriptional complex which activates downstream targets.

Recently, a second *Pax* family member, *twin of eyeless* (*toy*), has been found to play an important role in *Drosophila* eye development (Czerny et al. 1999). Like *ey, toy* is a member of the *PaxC* family and is more likely orthologous to *Pax6* than *ey. toy* is expressed earlier in development than *ey* in the developing head ectoderm. Later, *toy* is found in the cells anterior to the morphogenetic furrow within the eye disc. *toy* appears to act upstream of the entire network, and of *ey* in particular. Ectopic Toy induces both ectopic eyes and *ey* expression. *toy* may in fact directly regulate *ey* expression; several *toy*-binding sites essential for eye-specific expression have been identified in the *ey* enhancer (Hauck et al. 1999). This expansion of the *ey/so/eya/dac* network to include a second member of the *Pax* family appears to be unique to *Drosophila.*

Interestingly, a second *Six* gene, *optix,* has been identified in *Drosophila* and appears to be involved in eye morphogenesis by an *ey*-independent mechanism (Seimiya and Gehring 2000). *optix* is a member of the *Six3* subfamily and is expressed in a pattern different from *so* in the developing eye. Early in eye development, *optix* is expressed throughout the eye disc and later becomes restricted to the region anterior to the morphogenetic furrow (more like the expression pattern of *ey* or *toy*). Unlike *so,* ectopic *optix* can induce ectopic eyes. In addition, unlike *so, optix* does not function synergistically with *eya* in eye development; ectopic *optix* with ectopic *eya* does not induce ectopic eyes at a higher frequency. Moreover, unlike *eya* or *dac,* ectopic *optix* can induce ectopic eyes in an *ey* mutant background. In total, these results suggest that in the context of ectopic eye formation, *optix* acts in a partially different pathway from the one regulated by *ey.* However, since no loss-of-function mutants for *optix* are currently available, the functional role of *optix* in the eye disc remains uncertain.

Vertebrate Myogenesis

In vertebrates, somites are segmentally organized mesodermal structures that are the embryonic precursors of all skeletal muscle (reviewed in Christ and Ordahl 1995). Somites initially form as epithelial balls and, in response to patterning signals from surrounding tissues, acquire distinct fates. The dorsal region of the somite forms the dermamyotome, which gives rise to the dermatome and myotome. The myotome, in turn, gives rise to the epaxial (deep back) muscles that differentiate in situ and to the hypaxial muscles that form from cells that migrate away from the somites and differentiate into body wall and limb muscles. In both epaxial and hypaxial muscle cells the expression of the muscle-specific helix-loop-helix transcription factors, *Myf5* and *MyoD,* marks the initiation of the myogenic differentiation program.

The role of the *Pax/Six/Eya/Dach* network in myogenesis was suggested by the expression of several gene family members in the somite. In particular, *Pax3* and *Pax7; Six1* and *Six4; Eya1, Eya2,* and *Eya4;* and *Dach1* and *Dach2* are all expressed in the dorsal somite prior to expression of the myogenic genes (Williams and Ordahl 1994; Jostes et al. 1990; Oliver et al. 1995b; Esteve and Bovolenta 1999; Borsani et al. 1999; Mishima and Tomarev 1998; Xu et al. 1997b; Heanue et al. 1999; Heanue et al. 2002). In addition, some of these genes are also expressed in the undifferentiated myogenic precursors migrating into the limbs. These genes therefore appear to be good candidates for genes acting upstream of the myogenic regulatory factors.

Recent studies in the chick have revealed some of the regulatory relationships between *Pax3, Six1, Eya2,* and *Dach2* within the somite and their role during myogenesis. Previous somite culture experiments and analysis of mouse splotch (*Pax3*) mutants had firmly established the importance of *Pax3* for induction of myogenesis (Maroto et al. 1997; Tajbakhsh and Buckingham 1999). However, the regulatory relationships of *Six, Eya,* and *Dach* to *Pax* and the roles of these three gene families in myogenesis were unknown. Somite culture and in vivo misexpression experiments in chick now have established that *Pax3* positively regulates *Dach2* and *Eya2* expression (Heanue et al. 1999; Kardon et al. 2002; fig. 4.2, *B*). Interestingly, analysis of *Dach2* expression in splotch mutants reveals that at least in the mouse, *Pax3* is not necessary for *Dach2* expression (Davis et al. 2001). Conversely, *Six1, Eya2,* and *Dach2* are able to weakly induce *Pax3*. However, misexpression of *Dach2* with *Eya2* or *Eya2* with *Six1* in somite culture is able to strongly upregulate *Pax3*. Thus, *Dach2* and *Eya2* or *Eya2* with *Six1* synergistically regulate *Pax3* expression. In addition, these culture experiments also have established that these pairs of genes synergistically regulate myogenesis. In particular, while misexpression of *Dach2, Eya2,* or *Six1* alone was unable to induce *MyoD* and *Myosin heavy chain,* misexpression of *Dach2* with *Eya2* or *Eya2* with *Six1* was able to induce these myogenic genes.

The synergistic regulation of myogenesis by *Dach2* with *Eya2* and by *Eya2* with *Six1* is underlain by specific protein-protein interactions (fig. 4.2, *F*). GST pull-down and yeast two-hybrid assays demonstrate that Dach2 and Eya2 proteins physically interact and also Eya2 and Six1 proteins interact (Heanue et al. 1999). The protein-protein interactions appear to be specific for particular family members. In particular, while Eya2 and Six1 proteins strongly interact, Eya2 does not appear to interact with Six3 (Heanue et al. 1999). The importance of

Eya/Six function in regulating the myogenic regulatory gene *myogenin* has been demonstrated in several studies. Gel mobility shift assays have demonstrated that Six1 and Six4 proteins bind to the MEF3 binding site of *myogenin* (Spitz et al. 1998). More recently, Ohto and colleagues (Ohto et al. 1999) have shown that Six and Eya proteins synergistically activate *myogenin.* Furthermore, the magnitude of cooperative activation of *myogenin* transcription depends on particular combinations of Six and Eya proteins. Although not yet tested, it is possible that Dach2 participates directly in this Six-Eya transcriptional complex.

Many aspects of the *Pax/Six/Eya/Dach* network functioning in *Drosophila* eye development are strikingly conserved in the *Pax/Six/Eya/Dach* network functioning in vertebrate myogenesis. In both networks, *Pax* and *Dach* act in a positive feedback loop, and *Pax* positively regulates *Eya.* Similar to the synergistic regulation of eye development by *dac* and *eya* and by *eya* and *so,* myogenesis is regulated synergistically by *Dach* and *Eya* and by *Eya* and *Six.* Moreover, in both *Drosophila* and chick, this synergism is underlain by interactions between Dach and Eya proteins and between Eya and Six proteins. There is also evidence that the interaction domains of these proteins have been conserved between *Drosophila* and chick. When chick *Dach2* is ectopically expressed in *Drosophila dac* null mutants, the mutant eye phenotype is rescued. This suggests that chick *Dach2* can functionally interact with *Drosophila eya* and that the interaction domain of the Dac and Dach2 proteins is similar (Heanue et al. 1999). In addition, yeast two-hybrid assays have directly shown that chick Six1 can physically interact with *Drosophila* Eya demonstrating that the interaction domains of Six1 and So are conserved (Heanue et al. 1999).

One notable dissimilarity between the fly and the chick *Pax/Six/Eya/Dach* networks is the employment of *Pax3,* and potentially *Six4,* in vertebrate myogenesis. *Pax3* is a member of the *PaxD* subfamily and therefore is not orthologous to *ey,* a *Pax*C subfamily gene (Miller et al. 2000), and likewise *Six4* is not orthologous to *so* (Kawakami et al. 2000). Thus, it appears that during vertebrate evolution, different members of the *Pax* and *Six* families have been substituted in the network.

Vertebrate Eye Development

The morphological development of the vertebrate eye differs dramatically from the development of the fly eye (reviewed in Grainger 1992). Vertebrate eye development begins with the outpouching of the diencephalic portion of the neural tube. This outpouching, the optic vesicle, subsequently contacts the head ectoderm and interacts with the overlying ectoderm as it thickens into the lens placode. The lens placode then invaginates, detaches from the adjacent ectoderm, forms a lens vesicle, and eventually lengthens to form the lens of the eye. Concurrently, the optic vesicle folds inward on itself and surrounds the developing lens. The cells of this optic cup proliferate and differentiate into the neural and pigmented layers of the adult retina.

Pax6, Six3, Optx2, Eya1 and *Eya2,* and *Dach1* are all expressed in the developing eye. *Pax6* is initially expressed in the optic vesicle and in the head surface ectoderm in both the lens and otic regions prior to placode formation. As development proceeds, *Pax6* becomes localized to the lens placode and the neural retina (Grindley et al. 1995; Li et al. 1994; Walther and Gruss 1991). *Six3* is also expressed in the developing optic vesicle and later in both the neural retina and the lens (Ohto et al. 1998; Oliver et al. 1995a). *Optx2* has a slightly different expression pattern: it is found in the developing optic vesicle and later in the neural retina but appears to be absent in the developing lens (Jean et al. 1999; Lopez-Rios et al. 1999; Ohto et al. 1998; Toy and Sundin 1999; Toy et al. 1998). Two members of the *Eya* family, *Eya1* and *Eya2,* are expressed in complementary patterns in the developing eye (Xu et al. 1997b). Early in development, *Eya1* is expressed in the lens vesicle and the peripheral region of the optic vesicle, and it later is localized to the anterior epithelium of the lens and the retinal pigmented epithelium. In contrast, *Eya2* is never found in the lens or pigmented epithelium, but instead is expressed in the neural retina. Finally, *Dach1* is expressed in the developing optic vesicle and later in the neural retina (Hammond et al. 1998; Kozmik et al. 1999; Heanue et al. 2002).

The critical role of the *Pax, Six,* and *Eya* genes in vertebrate eye development has been revealed primarily by analysis of mouse and human mutations (reviewed in Wawersik and Maas 2000). Mutations in *Pax6, Six3,* and *Optx2* all lead to severe eye defects. Mutations in mouse *Pax6* cause the *Small eye* (*Sey*) phenotype (Hill et al. 1991; Hogan et al. 1986). *Sey/+* heterozygotes have lens and cornea abnormalities, while *Sey/Sey* homozygotes lack eyes altogether. Similarly, in humans haploinsufficiency for *Pax6* leads to aniridia, and homozygous *Pax6* mutations lead to anophthalmia (Glaser et al. 1994a, 1994b; Ton et al. 1991). In addition, mutations in human *Six3* cause microphthalmia, while mutations in *Optx2* result in anophthalmia (Gallardo et al. 1999; Wallis et al. 1999). Analysis of mice with null mutations in *Eya1* has not revealed any major defects in eye development (Xu et al. 1999). However, a subset of human *Eya1* mutations does result in cataracts and anterior segment abnormalities (Azuma et al. 2000).

Gain-of-function studies also confirm the importance of the *Pax* and *Six* genes for vertebrate eye development. Ectopic expression, via RNA injection, of *Pax6* in *Xenopus* or *Six3* in the teleost medaka results in ectopic retina and lenslike structures (Altmann et al. 1997; Chow et al.

1999; Loosli et al. 1999; Oliver et al. 1996). In addition, misexpression of *Optx2* in chick retinal pigmented epithelium induced cells to express neural-retina-specific markers (Toy et al. 1998). Overexpression of *Optx2* in *Xenopus* induced proliferation of retinal cells (Zuber et al. 1999).

At present, little is known about the regulatory relationships between the *Pax, Six, Eya,* and *Dach* genes functioning during eye development. However, some data on the relationship of *Pax6* to *Six, Eya,* and *Dach* genes have been gathered from analysis of early *Sey* embryos. In the *Sey/Sey* mice, *Eya1* is downregulated in the developing lens in the absence of functional *Pax6* (Xu et al. 1997a). The expression of *Six3* and *Dach1* in the optic vesicle and neural retina is unaffected in the *Sey/Sey* mice (Oliver et al. 1995a; Heanue et al. 2002). Therefore, in the developing eye *Eya1, Six3,* and *Dach1* are differentially regulated by *Pax6*. Understanding of the regulatory relationships among the *Eya, Six,* and *Dach* genes awaits the further analysis of mouse single and double knockouts.

The *Pax/Six/Eya/Dach* network appears to be critical for eye development in both vertebrates and *Drosophila,* despite the radically different structure and morphogenesis of their eyes. In *Drosophila* these genes are important for the initiation and propagation of the morphogenetic furrow, while in vertebrates these genes are required for the development of the lens and the retina. Members of the *Pax, Six, Eya,* and *Dach* gene families are all expressed in both vertebrate and *Drosophila* eyes, but there are some interesting differences in the particular genes expressed and their regulation. *ey* in *Drosophila* and its orthologue, *Pax6,* in vertebrates are key regulators of eye development. However, while *so* is critical for *Drosophila* eye development, there is no evidence that its orthologues, *Six1* and *Six2,* are used in vertebrate eyes. Instead, members of the *optix/Six3* subfamily, *Six3* and *Optx2,* are important for vertebrate eye development. Comparison of the regulation of these genes in fly and vertebrate eyes reveals that some, but not all, aspects of the regulatory network are conserved. In both the vertebrate and *Drosophila* eye, *Eya* gene expression is dependent on *Pax* genes. Interestingly, both *Six3* expression is independent of *Pax6* in the vertebrate eye and *optix* expression is independent of *ey* in the *Drosophila* eye. The regulation of *Dach* genes appears to differ between the two systems; in vertebrates expression of *Dach1* is independent of *Pax6,* while in *Drosophila dac* expression is dependent on *ey.*

Vertebrate Ear Development

The vertebrate inner ear derives from a thickened area of ectoderm, the otic placode, localized close to the hindbrain (reviewed in Torres and

Giraldez 1998). The otic placode invaginates to give rise successively to the otic cup and then to the otic vesicle. From the early otic cup, cells delaminate to give rise to the cochlear and vestibular neurons. The other components of the inner ear derive from the otic vesicle. The vesicle undergoes intense proliferative growth and differentiates into the endolymphatic duct, semicircular canals, vestibule, and cochlea. The expression of *Pax, Six, Eya,* and *Dach* genes in the developing otic cup and vesicle, and the ear defects resulting from mutations in some of these genes, indicate that the *Pax/Six/Eya/Dach* network is also critical for vertebrate ear development.

Pax2, Pax5, and *Pax8, Six1, Eya1,* and *Dach1* are expressed in various temporal and spatial patterns in the developing inner ear. *Pax8* is the earliest *Pax* gene to be expressed in the developing otic region. *Pax8* is expressed in the prospective otic placode and in the developing otic vesicle but is downregulated as the vesicle differentiates (Heller and Brändli 1999; Plachov et al. 1990). *Pax2* is associated with the auditory region of the inner ear. It begins to be expressed in the otic cup and is restricted to the ventral half of the otic vesicle that will give rise to the cochlea and adjacent sacculus (Nornes et al. 1990). In addition in *Xenopus, Pax5* is transiently expressed in the invaginating otic vesicle (Heller and Brändli 1999). *Six1* is found in the otic placode, vesicle, and facioacoustic ganglion (Oliver et al. 1995b). Finally, both *Eya1* and *Dach1* are expressed in the otic vesicle. *Eya1* is initially expressed in the ventromedial wall of the otic vesicle, which is the site of the future sensory epithelia of the cochlea (Xu et al. 1997b; Kalatzis et al. 1998). *Dach1* is also expressed in the ventromedial wall of the otic vesicle and in the vestigial ganglia (Heanue et al. 2002).

Loss-of-function studies have demonstrated that at least *Pax2* and *Eya1* are required for normal ear development. Analysis of *Pax2* mutant mice shows that *Pax2* is necessary for differentiation of the auditory regions of the inner ear. In these mutants, the otic vesicle invaginates and the cochlear neurons segregate normally from the vesicle. However, in the subsequent morphogenesis of the otic vesicle, neither the cochlea nor the cochlear ganglion differentiates. *Eya1* mutant mice have an even more dramatic ear phenotype (Xu et al. 1999). These mice have defects in their inner, middle, and outer ears. With regard to the inner ear, the otic vesicle forms but fails to develop further, and no inner structures form. The critical role of *Eya1* in ear development is also found in humans. Haploinsufficiency in human *Eya1* results in branchio-oto-renal syndrome (Abdelhak et al. 1997a, 1997b; Kumar et al. 1998), which is characterized by hearing loss. Mice with null mutations in another gene, *Pax8,* expressed in the developing ear have also been generated (Mansouri et al. 1998). However, these mutant mice do not have ear phenotypes. It is possible, although not yet tested, that upregulation of *Pax2* and/or *Pax5* compensates for the loss of functional *Pax8*.

Analysis of *Eya1* and *Pax2* mutant mice reveals some of the regulatory relationships between *Pax2* and *Pax8, Six1, Eya1*, and *Dach1*. In mice lacking functional *Eya1, Six1* expression is lost (Xu et al. 1999). This demonstrates that *Six1* expression is regulated by *Eya1*. In contrast, *Pax2, Pax8*, and *Dach1* expression is unaffected in the *Eya1* mutant, indicating that these genes are regulated independently of *Eya1* (Heanue et al. 2002; Xu et al. 1999). In *Pax2* mutant mice expression of *Eya1* and *Dach1* is unaffected, suggesting that their expression is regulated independently of *Pax2*. However, their expression may be regulated by *Pax5* and/or *Pax8* (Heanue et al. 2002).

In summary, the coexpression of *Pax, Six, Eya*, and *Dach* genes in the developing ear together with the ear phenotypes in mice mutant for *Pax2* and *Eya1* strongly suggests that the *Pax/Six/Eya/Dach* network is functionally important in vertebrate ear development. In the employment of this network for ear development the *sine oculis* orthologue, *Six1*, has been used, but the *eyeless PaxC* orthologues have not. Instead, *PaxB* subfamily members *Pax2, Pax5*, and *Pax8* have been utilized. Although there has been little analysis of the regulatory relationships between the genes, some aspects of the regulation found in the *Drosophila* eye have been conserved in the vertebrate ear, while others have not. For instance, as in *Drosophila, Six1* expression is dependent on *Eya1*. However, unlike the fly eye, *Dach1* expression appears to be independent of *Eya1*.

Comparison of the *Pax/Six/Eya/Dach* Networks Employed in *Drosophila* and Vertebrate Development

The *Pax/Six/Eya/Dach* network has been employed in *Drosophila* and vertebrates in a variety of different developmental contexts (summarized in fig. 4.2). In *Drosophila* the network in the eye primarily consists of *eyeless, sine oculis, eyes absent*, and *dachshund*. Although not explicitly tested, it is possible that the network is employed in *Drosophila* in other developmental contexts, perhaps using other members of the *Pax* and *Six* families. For example, in the *Drosophila* larval eye (Bolwig's organ) both *sine oculis* and *eyes absent* are required for its proper development, but *eyeless* and *twin of eyeless* are not (Suzuki and Saigo 2000). Potentially, another *Pax* family member is important in this developmental context. In the future it will be interesting to see whether in other parts of the *Drosophila* embryo the network is found to function with members of the *Pax A, B*, and *D* subfamilies or with *optix* or *six4*.

With the evolution of vertebrates, each of the gene families has undergone duplications, and the network has been employed in several different developmental contexts. As might be expected, *ey* and *so* orthologues have been used in the vertebrate *Pax/Six/Eya/Dach* networks. *ey* and *toy* and their orthologue, *Pax6*, have been used in *Drosophila* and vertebrate eye development, respectively. Similarly, *so* and its orthologue, *Six1*, have been employed in *Drosophila* eye development and in vertebrate muscle and ear development. However, the vertebrate *Pax* and *Six* families are complex and include multiple subfamilies. Nonorthologous members of the *Pax* and *Six* families have been employed in the *Pax/Six/Eya/Dach* network. Members of the *PaxB* (*Pax2/Pax5/Pax8*) and *D* (*Pax3* and *Pax7*) subfamilies have been used in vertebrate ear and muscle development, respectively. In vertebrate eye development, *Six3* and *Optx2* and *not* the *so* orthologues *Six1* or *Six2* are used. In addition, all known members of the *Eya* and *Dach* families appear to have been employed in the network in different developmental contexts. Comparison of the particular members of the *Pax, Six, Eya,* and *Dach* families used suggests that there is no necessary relationship between which particular family members must be used together in concert. For example, *Six1* can work in a network with either *Eya1* or *Eya2*, and *Eya1* can work in a network with either *Six3* (*Optx2*) or *Six1*.

Many aspects of the regulatory relationships between *Pax, Six, Eya,* and *Dach* genes have been conserved between *Drosophila* and vertebrates. Initially *Pax* genes are most upstream, followed by *Six, Eya,* and *Dach* genes. Subsequently, positive regulatory loops are established between the components to form a complex network. How have these tight regulatory relationships been maintained? In *Drosophila,* part of this regulation is direct; the Ey protein binds to an eye-specific enhancer of *so* and activates *so* (Niimi et al. 1999). Although it has not yet been demonstrated, *Pax* genes may bind to *Six* upstream regions. Another possibility is that the *Six* transcription factors may directly bind to and transactivate *Pax, Eya,* and *Dach* genes. The tight regulatory relationships between *Six* and *Eya* and between *Eya* and *Dach* may be indirect yet made necessary by the physical interactions between the proteins they encode. A third possibility is that the entire network of genes is maintained by common regulatory regions upstream of the *Pax, Six, Eya,* and *Dach* genes. The upstream regions may be important for restricting the temporal and spatial distribution of these genes.

Although many of the regulatory relationships have been conserved in the *Pax/Six/Eya/Dach* network, there are significant instances of nonconservation. Within vertebrates, some genes have been decoupled from the tight network of internal regulation. For instance, within the vertebrate eye, *Dach1* expression is independent of *Pax6* (Heanue et al.

2002). In fact, there are multiple documented cases where members of the *Pax, Six, Eya,* and *Dach* families are operating entirely independently of the network. In the *Drosophila* wing and vertebrate limb *dac* and *Dach1* (Mardon et al. 1994; Hammond et al. 1998; Kozmik et al. 1999; Davis et al. 1999; Heanue et al. 1999), respectively, are clearly functioning independently of the network, as no *Pax, Six,* or *Eya* genes are coexpressed in these regions (LeClair et al. 1999; Xu et al. 1997a, 1997b; Oliver et al. 1995a, 1995b). In vertebrates *Pax1* is strongly expressed in the sclerotomal region of the somites, but no *Six, Eya,* or *Dach* genes are expressed in this region (Hammond et al. 1998; Kozmik et al. 1999; Davis et al. 1999; Heanue et al. 1999). Another interesting evolutionary novelty has been the expansion of the network in *Drosophila* to include both *ey* and *toy* (Czerny et al. 1999; Hauck et al. 1999). So far, no such similar expansion has been discovered in vertebrates.

Evolution of the *Pax/Six/Eya/Dach* Regulatory Network

The evolution of the *Pax, Six, Eya,* and *Dach* genes is characterized by the expansion of each of these gene families (fig. 4.3). In the lower Metazoa, the only gene family that has been currently identified is the *Pax* family. In cnidarians, four *Pax* genes, *A–D,* are present (Miller et al. 2000). The identification of cnidarian *Six, Eya,* and *Dach* genes awaits further research. On the basis of the distribution of genes in *Drosophila* and vertebrates, it appears that members of all four families are present before the protostome-deuterostome split. At this node,

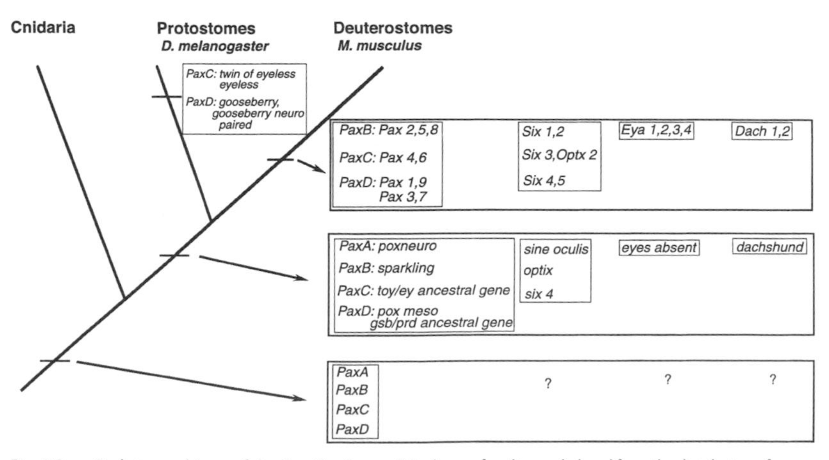

Fig. 4.3.—Evolutionary history of the *Pax, Six, Eya,* and *Dach* gene families as deduced from the distribution of genes in cnidarians, *Drosophila,* and mouse (Miller et al. 2000; Kawakami et al. 2000).

the *Pax* family had undergone duplication and contained at least five members. Also, three *Six* family genes, one *Eya,* and one *Dach* gene were present. With the evolution of protostomes, there were further duplications in the *Pax* family, so that *Drosophila* has eight *Pax* genes. Within the evolution of deuterostomes, there has been an expansion of all four gene families. Vertebrates have at least nine *Pax* genes, six *Six* genes, four *Eya* genes, and two *Dach* genes.

The phylogenetic distribution and the developmental expression of the *Pax, Six, Eya,* and *Dach* genes provide some insights into the origin of the *Pax/Six/Eya/Dach* network. The presence of all four gene families in the protostome/deuterostome common ancestor suggests that the network may have been present and functional very early in animal phylogeny. Moreover, if cnidarians are found to have *Six, Eya,* and *Dach* genes, the network may have originated even earlier. The ancestral developmental function of *Pax* genes, perhaps in the context of the *Pax/Six/Eya/Dach* network, may have been in the development of the nervous system. Although *Pax* genes are expressed in a variety of tissues in higher animals, most are expressed in the nervous system during development (Miller et al. 2000). Intriguingly, the one *Pax* gene, *Pax-Cam,* examined in the anthozoan cnidarian *Acoropora* is found in the presumptive developing neurons (Miller et al. 2000). It will be interesting to see whether *Six, Eya,* and *Dach* genes are coexpressed in the developing neurons and whether all four genes function in a regulatory network important for nervous system development.

With the evolution of protostomes and deuterostomes, the *Pax/Six/Eya/Dach* network has acquired a diversity of functions in the developing embryo. This diversification of function has been accompanied by the expansion and use of different members of the *Pax, Six, Eya,* and *Dach* families. In fact, it could be argued that it is the deployment of other gene family members that has allowed the *Pax/Six/Eya/Dach* network to be successfully used and functional in so many developmental contexts. The use of different family members may allow for diversification of function via three different mechanisms. First, the use of genes with different temporal and spatial patterns of expression may permit the network to operate in novel developmental contexts. Second, the differences in the DNA-binding specificities of different Pax and Six proteins may allow activation of different downstream targets. Finally, variation in protein-protein interactions between Six and Eya proteins and between Eya and Dach proteins may allow activation of different target genes. Such protein-protein specificity and its importance for target gene activation have been clearly demonstrated with the Six/Eya activation of *myogenin* (Ohto et al. 1999; Spitz et al. 1998). Overall, the important and diverse functions of the *Pax/Six/Eya/Dach* network in the developing embryo are striking and may have served a critical role in the evolution of both protostomes and deuterostomes. The expansion and use of different members of the *Pax, Six, Eya,* and *Dach* families in the network has allowed for developmental modification of the *Pax/Six/Eya/Dach* network. Such modification of developmental cassettes may be an important mechanism for generating evolutionary novelty in the animal lineage.

Evolution of Regulatory Cassettes

In this chapter we have examined one example of a regulatory cassette, a network of genes working together in many different developmental contexts. In general, the continued maintenance of regulatory cassettes in different developmental contexts in a wide variety of taxa suggests that there is some developmental and evolutionary utility to these cassettes. Here we have examined in detail the *Pax/Six/Eya/Dach* network. Another example of such a cassette is the *tinman/dmef2/pannier* and *Nkx/Mef2c/Gata4* networks important for heart development in both *Drosophila* and vertebrates, respectively (reviewed in Harvey and Rosenthal 1999). In the case of the *Pax/Six/Eya/Dach* network, the tight regulatory relationships between *Six* and *Eya* and between *Eya* and *Dach* probably originated and were maintained by selection for the necessary physical interactions between the proteins these genes encode. The origin and maintenance of the relationship between *Pax* and the other three genes is less clear. In *Drosophila, ey* directly regulates *so*. It is possible that *Pax* genes, in general, directly regulate *Six* (excluding members of the *optix* subfamily) and perhaps also *Eya* and *Dach* genes. Potentially, transcriptional regulation by *Pax* genes of *Six, Eya,* and *Dach* genes has allowed *Six, Eya,* and *Dach* genes to be coselected as a gene network. Over the course of evolution, regulatory cassettes have proved to be extremely versatile. In the case of the *Pax/Six/Eya/Dach* network, each of the gene families has expanded, and different members of these families have been used in the network. The use of different family members may have allowed the *Pax/Six/Eya/Dach* network to be used in different spatial and temporal developmental contexts and to activate different downstream targets. In addition, in some developmental contexts the internal regulatory relationships within the *Pax/Six/Eya/Dach* network have been modified and may have made the network more versatile. Both the developmental utility and the evolutionary flexibility of regulatory cassettes may make these cassettes central to animal development and evolution. Future examination of a broad array of animal taxa will reveal how universally these regulatory cassettes have been used.

References

Abdelhak S., V. Kalatzis, R. Heilig, S. Compain, D. Samson, C. Vincent, F. Levi-Acobas, C. Cruaud, M. Le Merrer, M. Mathieu, R. Konig, J. Vigneron, J. Weissenbach, C. Petit, and D. Weil. 1997a. Clustering of mutations responsible for branchio-oto-renal (BOR) syndrome in the eyes absent homologous region (eyaHR) of eya1. *Hum. Mol. Genet.* 6:2247–2255.

Abdelhak, S., V. Kalatzis, R. Helig, S. Compain, D. Samson, C. Vincent, D. Weil, C. Cruad, I. Sahly, M. Leibovici, M. Bitner-Glindzicz, M. Francis, D. Lacombe, J. Vigeron, R. Charachon, K. Boven, P. Bedbeder, N. Van Regemorter, J. Weissenbach, and C. Petit. 1997a. A human homologue of the *Drosophila* eyes absent gene underlies branchio-oto-renal (BOR) syndrome and identifies a novel gene family. *Nat. Genet.* 15:157–164.

Altmann, C. R., R. L. Chow, R. A. Lang, and A. Hemmati-Brivanlou. 1997. Lens induction by pax6 in *Xenopus laevis. Dev. Biol.* 185:119–123.

Azuma, N., A. Hirakiyama, T. Inoue, A. Asaka, and M. Yamada. 2000. Mutations of a human homologue of the *Drosophila* eyes absent gene (eya1) detected in patients with congenital cataracts and ocular anterior segment anomalies. *Hum. Mol. Genet.* 9:363–366.

Bonini, N. M., Q. T. Bui, G. L. Gray-Board, and J. M. Warrick. 1997. The *Drosophila* eyes absent gene directs ectopic eye formation in a pathway conserved between flies and vertebrates. *Development* 124 (23): 4819–4826.

Bonini, N. M., W. M. Leiserson, and S. Benzer. 1993. The eyes absent gene: genetic control of cell survival and differentiation in the developing *Drosophila* eye. *Cell* 72 (3): 379–395.

Bonini, N. M., W. M. Leiserson, and S. Benzer. 1998. Multiple roles of the eyes absent gene in *Drosophila. Dev. Biol.* 196 (1): 42–57.

Borsani, G., A. DeGrandi, A. Ballabio, A. Bulfoni, L. Bernard, S. Banfi, C. Gattuso, N. Mariani, M. Dixon, D. Donnai, K. Metcalfe, R. Winter, M. Robertson, R. Axton, A. Brown, V. van Heyningen, and I. Hanson. 1999. EYA4, a novel vertebrate gene related to *Drosophila* eyes absent. *Hum. Mol. Genet.* 8 (1): 11–23.

Caubit, X., R. Thankgarah, T. Theil, J. Wirth, H. G. Nothwang, U. Ruther, and S. Krauss. 1999. Mouse Dac, a novel nuclear factor with homology to *Drosophila* dachshund shows a dynamic expression in the neural crest, the eye, the neocortex, and the limb bud. *Dev. Dyn.* 214 (1): 66–80.

Chen, R., M. Amoui, Z. Zhang, and G. Mardon. 1997. Dachshund and eyes absent proteins form a complex and function synergistically to induce ectopic eye development in *Drosophila. Cell* 91 (7): 893–903.

Cheyette, B. N., P. J. Green, K. Martin, H. Garren, V. Hartenstein, and S. L. Zipursky. 1994. The *Drosophila* sine oculis locus encodes a homeodomain-containing protein required for the development of the entire visual system. *Neuron* 12 (5): 977–996.

Chow, R. L., C. R. Altmann, R. A. Lang, and A. Hemmati-Brivanlou. 1999. Pax6 induces ectopic eyes in a vertebrate. *Development* 126:4213–4222.

Christ, B., and C. P. Ordahl. 1995. Early stages of chick somite development. *Anat. Embryol.* 191 (5): 381–396.

Czerny, T., G. Halder, U. Kloter, A. Souabni, W. J. Gehring, and M. Busslinger. 1999. twin of eyeless, a second Pax-6 gene of *Drosophila,* acts upstream of eyeless in the control of eye development. *Mol. Cell* 3 (3): 297–307.

Davis, R. J., W. Shen, T. A. Heanue, and G. Mardon. 1999. Mouse Dach, a homologue of *Drosophila* dachshund, is expressed in the developing retina, brain and limbs. *Dev. Genes Evol.* 209 (9): 526–536.

Davis, R. J., W. Shen, Y. I. Sandler, T. A. Heanue, and G. Mardon. 2001. Characterization of mouse *Dach2,* a homologue of *Drosophila dachshund. Mech. Dev.* 102: 169–179.

Esteve, P., and P. Bovolenta. 1999. cSix4, a member of the six family of transcription factors, is expressed during placode and somite development. *Mech. Dev.* 85 (1–2): 161–165.

Gallardo, M. E., J. Lopez-Rios, I. Fernaud-Espinosa, B. Granadino, R. Sanz, C. Ramos, C. Ayuso, M. J. Seller, H. G. Brunner, P. Bovolenta, and S. Rodriguez de Cordoba. 1999. Genomic cloning and characterization of the human homeobox gene six6 reveals a cluster of six genes in chromosone 14 and associates six6 hemizygosity with bilateral anophthalmia and pituitary anomalies. *Genomics* 61:82–91.

Glaser, T., L. Jepeal, J. G. Edwards, S. R. Young, J. Favor, and R. L. Maas. 1994a. Pax6 gene dosage effect in a family with congenital cataracts, aniridia, anophthalmia and central nervous system defects. *Nat. Genet.* 7:463–471.

Glaser, T., D. S. Walton, and R. L. Maas. 1994b. Genomic structure, evolutionary conservation and aniridia mutations in the human pax6 gene. *Nat. Genet.* 2:232–238.

Grainger, R. M. 1992. Embryonic lens induction: shedding light on vertebrate tissue determination. *Trends Genet.* 8:349–355.

Grindley, J. C., D. R. Davidson, and R. E. Hill. 1995. The role of pax 6 in eye and nasal development. *Development* 121:1422–1442.

Halder, G., P. Callaerts, S. Flister, U. Walldorf, U. Kloter, and W. J. Gehring. 1998. Eyeless initiates the expression of both sine oculis and eyes absent during *Drosophila* compound eye development. *Development* 125 (12): 2181–2191.

Halder, G., P. Callaerts, and W. J. Gehring. 1995. Induction of ectopic eyes by targeted expression of the eyeless gene in *Drosophila. Science* 267 (5205): 1788–1792.

Hammond, K. L., I. M. Hanson, A. G. Brown, L. A. Lettice, and R. E. Hill. 1998. Mammalian and *Drosophila* dachshund genes are related to the Ski proto-oncogene and are expressed in eye and limb. *Mech. Dev.* 74 (1–2): 121–131.

Harvey, R. P., and N. Rosenthal. 1999. *Heart development.* San Diego: Academic Press.

Hauck, B., W. J. Gehring, and U. Walldorf. 1999. Functional analysis of an eye specific enhancer of the eyeless gene in *Drosophila. Proc. Natl. Acad. Sci. U.S.A.* 96:564–569.

Heanue, T. A., R. J. Davis, D. H. Rowitch, A. Kispert, A. P. McMahon, G. Mardon, and C. J. Tabin. 2002. *Dach1,* a vertebrate homologue of *Drosophila dachshund,* is expressed in the developing eye and ear of both chick and mouse and is regulated independently of *Pax* and *Eya* genes. *Mech. Dev.* 111 (1–2): 75–87.

Heanue, T. A., R. Reshef, R. J. Davis, G. Mardon, G. Oliver, S. Tomarev, A. B. Lassar, and C. J. Tabin. 1999. Synergistic regulation of vertebrate muscle development by Dach2, Eya2, and Six1, homologs of genes required for *Drosophila* eye formation. *Genes Dev.* 13 (24): 3231–3243.

Heller, N., and A. W. Brändli. 1999. *Xenopus* pax 2/5/8 orthologues: novel insights into pax gene evolution and identification of pax 8 as the earliest marker for otic and pronephric cell lineages. *Dev. Genet.* 24:208–219.

Hill, R. E., J. Favor, B. L. Hogan, C. C. T. Ton, G. F. Saunders, I. M. Hansom, J. Prosser, T. Jordan, et al. 1991. Mouse Small eye results from mutations in a paired-like homeobox containing gene. *Nature* 354:522–525.

Hogan, B. L., G. Horsburgh, J. Cohen, C. M. Hetherington, G. Fisher, and M. F. Lyon. 1986. Small eyes (Sey): a homozygous lethal mutation on chromosome 2 which affects the differentiation of both lens and nasal placodes in the mouse. *J. Embryol. Exp. Morphol.* 97:95–110.

Jean, D., G. Bernier, and P. Gruss. 1999. Six6 (Optx2) is a novel murine Six3-related homeobox gene that demarcates the presumptive pituitary/hypothalmic axis and the ventral optic stalk. *Mech. Dev.* 84:31–40.

Jostes, B., C. Walther, and P. Gruss. 1990. The murine paired box gene, Pax7, is expressed specifically during the development of the nervous and muscular system. *Mech. Dev.* 33 (1): 27–37.

Kalatzis, V., I. Sahly, A. El-Amraoui, and C. Petit. 1998. Eya1 expression in the developing ear and kidney: towards the understanding of the pathogenesis of branchio-oto-renal (BOR) Syndrome. *Dev. Dyn.* 213 (4): 486–499.

Kardon, G., T. A. Heanue, and C. J. Tabin. 2002. Pax3 and Dach2 positive regulation in the developing somite. *Dev. Dyn.* 224 (3): 350–355.

Kawakami, K., S. Sato, H. Ozaki, and K. Ikeda. 2000. Six family genes: structure and function as transcription factors and their roles in development. *BioEssays* 22 (7): 616–626.

Kozmik, Z., P. Pfeffer, J. Kralova, J. Paces, V. Paces, A. Kalousova, and A. Cvekl. 1999. Molecular cloning and expression of the human and mouse homologues of the *Drosophila* dachshund gene. *Dev. Genes Evol.* 209 (9): 537–545.

Kumar, S., H. A. M. Marres, C. W. R. J. Cremers, and W. J. Kimberling. 1998. Identification of three novel mutations in human EYA1 protein associated with branchio-oto-renal syndrome. *Hum. Mutat.* 11:443–449.

LeClair, E. E., L. Bonfiglio, and R. S. Tuan. 1999. Expression of the paired box genes Pax1 and Pax9 in limb skeleton development. *Dev. Dyn.* 214 (4): 101–116.

Li, H. S., J. M. Yang, R. D. Jacobson, D. Pasko, and O. Sundin. 1994. Pax 6 is first expressed in a region of ectoderm anterior to the early neural plate: implications for stepwise determination of the lens. *Dev. Biol.* 162:181–194.

Loosli, F., S. Winkler, and J. Wittbrodt. 1999. Ectopic retina in response to six3 overexpression. *Genes Dev.* 13:649–654.

Lopez-Rios, J., M. E. Gallardo, S. Rodriguez de Cordoba, and R. Bovolenta. 1999. Six9 (Optx2), a new member of the six gene family of transcription factors, is expressed at early stages of vertebrate ocular and pituitary development. *Mech. Dev.* 83:155–159.

Mansouri, A., K. Chowdhury, and P. Gruss. 1998. Follicular cells of the thyroid gland require Pax8 gene function. *Nat. Genet.* 19 (1): 87–90.

Mardon, G., N. M. Solomon, and G. M. Rubin. 1994. dachshund encodes a nuclear protein required for normal eye and leg development in *Drosophila. Development* 120 (12): 3473–3486.

Maroto, M., R. Reshef, A. E. Munsterberg, S. Koester, M. Goulding, and A. B. Lassar. 1997. Ectopic Pax-3 activates MyoD and Myf-5 expression in embryonic mesoderm and neural tissue. *Cell* 89 (1): 139–148.

Miller, D. J., D. C. Hayward, J. S. Reece-Hoyes, I. Scholten, J. Catmull, W. J. Gehring, P. Callaerts, J. E. Larsen, and E. E. Ball. 2000. Pax gene diversity in the basal cnidarian *Acropora millepora* (Cnidaria, Anthozoa): implications for the evolution of the Pax gene family. *Proc. Natl. Acad. Sci. U.S.A.* 97 (9): 4475–4480.

Mishima, N., and S. Tomarev. 1998. Chicken Eyes absent 2 gene: isolation and expression pattern during development. *Int. J. Dev. Biol.* 42 (8): 1109–1115.

Niimi, T., M. Seimiya, U. Kloter, S. Flister, and W. J. Gehring. 1999. Direct regulatory interaction of the eyeless protein with an eye-specific enhancer in the sine oculis gene during eye induction in *Drosophila. Development* 126:2253–2260.

Nornes, H. O., G. R. Dressler, E. W. Knapik, U. Deutsch, and P. Gruss. 1990. Spatially and temporally restricted expression of Pax2 during murine neurogenesis. *Development* 109 (4): 797–809.

Ohto, H., S. Kamada, K. Tago, S. I. Tominaga, H. Ozaki, S. Sato, and K. Kawakami. 1999. Cooperation of Six and Eya in activation of their target genes through nuclear translocation of Eya. *Mol. Cell. Biol.* 19 (10): 6815–6824.

Ohto, H., T. Takizawa, T. Saito, M. Kobayashi, K. Ikeda, and K. Kawakami. 1998. Tissue and developmental distribution of Six family gene products. *Int. J. Dev. Biol.* 42: 667–677.

Oliver, G., F. Loosli, R. Koster, J. Wittbrodt, and P. Gruss. 1996. Ectopic lens induction in fish in response to the murine homeobox gene six3. *Mech. Dev.* 60:233–239.

Oliver, G., A. Mailhos, R. Wehr, N. G. Copeland, N. A. Jenkins, and P. Gruss. 1995a. Six3, a murine homologue of the sine oculis gene, demarcates the most anterior border of the developing neural plate and is expressed during eye development. *Development* 121 (12): 4045–4055.

Oliver, G., R. Wehr, N. A. Jenkins, N. G. Copeland, B. N. Cheyette, V. Hartenstein, S. L. Zipursky, and P. Gruss. 1995b. Homeobox genes and connective tissue patterning. *Development* 121 (3): 693–705.

Pignoni, F., B. Hu, K. H. Zavitz, J. Xiao, P. A. Garrity, and S. L. Zipursky. 1997. The eye-specification proteins So and Eya form a complex and regulate multiple steps in *Drosophila* eye development. *Cell* 91 (7): 881–891. (Published erratum appears in *Cell* 92 (4, 20 February 1998), following 585.)

Plachov, D., K. Chowdhury, C. Walther, D. Simon, J. L. Guenet, and P. Gruss. 1990. Pax8, a murine paired box gene expressed in the developing excretory system and thyroid gland. *Development* 110 (2): 643–651.

Quiring, R., U. Walldorf, U. Kloter, and W. J. Gehring. 1994. Homology of the eyeless gene of *Drosophila* to the Small eye gene in mice and Aniridia in humans. *Science* 265 (5173): 785–789.

Seimiya, M., and W. J. Gehring. 2000. The *Drosophila* homeobox gene optix is capable of inducing ectopic eyes by an eyeless-independent mechanism. *Development* 127: 1879–1886.

Shen, W., and G. Mardon. 1997. Ectopic eye development in *Drosophila* induced by directed dachshund expression. *Development* 124 (1): 45–52.

Spitz, F., J. Demignon, A. Porteu, A. Kahn, J. P. Concordet, D. Daegelen, and P. Maire. 1998. Expression of myogenin during embryogenesis is controlled by Six/sine oculis homeoproteins through a conserved MEF3 binding site. *Proc. Natl. Acad. Sci. U.S.A.* 95 (24): 14220–14225.

Suzuki, T., and K. Saigo. 2000. Transcriptional regulation of *atonal* required for *Drosophila* larval eye development by concerted action of Eyes absent, Sine oculis and Hedgehog signaling independent of Fused kinase and Cubitus interrupus. *Development* 127:1521–1540.

Tajbakhsh, S., and M. Buckingham. 1999. The birth of muscle progenitor cells in the mouse: spatiotemporal consideration. *Curr. Top. Dev. Biol.* 48:225–268.

Tomlinson, A., and D. F. Ready. 1987. Neuronal differentiation in the *Drosophila* ommatidium. *Dev. Biol.* 120:366–376.

Ton, C. C. T., H. Hirovenen, H. Miwa, M. W. Weil, A. P. Monaghan, T. Jordan, V. van Heynigen, N. D. Hastie, H. Meijers-Heijboer, M. Dreschler, et al. 1991. Positional cloning and characterization of a paired box and homeobox containing gene from the aniridia region. *Cell* 67:1059.

Torres, M., and F. Giraldez. 1998. The development of the vertebrate inner ear. *Mech. Dev.* 71:5–21.

Torres, M., E. Gomez-Pardo, and P. Gruss. 1996. Pax2 contributes to inner ear patterning and optic nerve trajectory. *Development* 122 (11): 3381–3391.

Toy, J., and O. H. Sundin. 1999. Expression of the optx2 homeobox gene during mouse development. *Mech. Dev.* 83:183–186.

Toy, J., J. M. Yang, G. S. Leppert, and O. H. Sundin. 1998. The optx2 homeobox gene is expressed in early precursors of the eye and activates retina-specific genes. *Proc. Natl. Acad. Sci. U.S.A.* 95 (18): 10643–10648.

Wallis, D. E., E. Roessler, U. Hehr, L. Nanni, T. Wiltshire, A. Richieri-Costa, G. Gillessen-Kaesbach, E. H. Zackai, J. Rommens, and M. Muenke. 1999. Mutations in the homeodomain of six3 gene cause holoprosencephaly. *Nat. Genet.* 22:196–198.

Walther, C., and P. Gruss. 1991. Pax-6, a murine paired box gene, is expressed in the developing CNS. *Development* 113 (4): 1435–1449.

Wawersik, S., and R. L. Maas. 2000. Vertebrate eye development as modeled in *Drosophila. Hum. Mol. Genet.* 9 (6): 917–925.

Williams, B. A., and C. P. Ordahl. 1994. Paxs expression in segmental mesoderm marks early stages in myogenic cell specification. *Development* 120 (4): 785–796.

Wolff, T., and D. F. Readt. 1993. Pattern formation in the *Drosophila* retina. In *The development of Drosophila melanogaster,* edited by A. Martinez Arias and M. Bate. Cold Spring Harbor, N.Y.: Cold Spring Harbor Laboratory Press.

Xu, P. X., J. Adams, H. Peters, M. C. Brown, S. Heaney, and R. Maas. 1999. Eya1-deficient mice lack ears and kidneys and show abnormal apoptosis of organ primordia. *Nat. Genet.* 23 (1): 113–117.

Xu, P. X., J. Cheng, J. A. Epstein, and R. L. Maas. 1997a. Mouse Eya genes are expressed during limb tendon development and encode a transcriptional activation function. *Proc. Natl. Acad. Sci. U.S.A.* 94 (22): 11974–11979.

Xu, P. X., I. Woo, H. Her, D. R. Beier, and R. L. Maas. 1997b. Mouse Eya homologues of the *Drosophila* eyes absent gene require Pax6 for expression in lens and nasal placode. *Development* 124 (1): 219–231.

Younoussi-Hartenstein, A., U. Tepass, and V. Hartenstein. 1993. Embryonic origin of the imaginal discs of the head of *Drosophila melanogaster. Roux's Arch. Dev. Biol.* 203:60–73.

Zuber, M. E., M. Peroon, A. Philpott, A. Bang, and W. A. Harris. 1999. Giant eyes in *Xenopus laevis* by overexpression of Xoptx2. *Cell* 98:341–352.

5 The Notch Signaling Module

JOSÉ F. DE CELIS

In the context of developmental biology, cell-to-cell communication includes the mechanisms by which cells interchange molecular signals that affect and/or direct the acquisition of particular cell fates. Cell signaling influences many aspects of cell behavior, such as cell division, growth, and polarity. In many instances the immediate consequences of signaling are changes in gene expression triggered by the transcriptional regulators that constitute the final elements of each signaling pathway. Signaling pathways consist of a number of proteins that are functionally related in a hierarchical manner. Each pathway generally includes extracellular ligands, membrane receptors, and a chain of intracellular transducers that modify the activity of a transcription factor. The interactions between members of a given pathway are determined by molecular recognition and therefore are context independent. This makes each signaling pathway a "module" of interacting proteins that contributes to the regulation of gene expression.

The Notch signaling pathway influences many cell fate choices during the development of multicellular organisms (Artavanis-Tsakonas et al. 1999). The elements that constitute the pathway are conserved in vertebrates and invertebrates. Furthermore, Notch affects similar developmental operations in all organisms where its functional requirements have been analyzed, including lateral inhibition during cell fate choice and local induction in the establishment of developmental boundaries (Artavanis-Tsakonas et al. 1999). For these reasons, the Notch pathway can be considered a signaling module that regulates gene expression. According to this view, the elements of the pathway constitute a conserved set of interacting proteins that modify the activity of a transcriptional regulator. There are two fundamental aspects that provide functionality and versatility to the Notch module: the regulation in space and time of Notch activation and the integration between Notch and other developmental signals affecting gene expression. Recent data in both vertebrates and invertebrates suggest that these two biological aspects of Notch functionality rely on the combinatorial regulation of the expression of Notch ligands and target genes.

Elements of the Notch Signaling Pathway

The conserved components of the Notch signaling module include the ligands, receptors, and the transducer molecule Suppressor of Hairless (Su(H)) (fig. 5.1) (Muskavitch 1994; Fleming 1998; Artavanis-Tsakonas et al. 1999). These "core components" of the "Notch module" have been shown to be required for Notch activity in organisms suitable for genetic analysis, such as the mouse, chicken, *Drosophila,* and *Caenorhabditis elegans* (Lewis 1998; Greewald 1998). In addition, several proteins are required for the proteolytic processing of the receptor that both precedes and follows ligand activation, and for posttranscriptional modifications of the receptor affecting its interactions with the ligands (Chan and Jan 1998; Wu and Rao 1999; Blair 2000; fig. 5.1). Other genes affecting Notch signaling have also been identified in genetic screens carried out in flies (Brand and Campos-Ortega 1990; Klein and Campos-Ortega 1993; Go and Artavanis-Tsakonas 1998; Royet et al. 1998). Some of these genes, such as *deltex* and *Suppressor of deltex,* interact with Notch and are phylogenetically conserved (Busseau et al. 1994; Diederich et al. 1994; Matsuno et al. 1998; Fostier et al. 1998). It is unclear, however, to what extent they contribute exclusively to regulating Notch activity in different organisms, and they are not considered further in this review. Finally, although in most instances Notch signaling requires Su(H) function, several examples of Notch activities appear to be independent of Su(H) (Shawber et al. 1996; Matsuno et al. 1997; Rusconi and Corbin 1998, 1999; Dumont et al. 2000; Martinez Arias 1998; Zecchini et al. 1999). These observations suggest that Notch can be engaged in interactions with other downstream transducers, although its nature is still unknown.

The ligands of the Notch receptor (*D*elta, *S*errate, and *L*AG-2 [DSL] proteins) are transmembrane proteins containing a variable number of repeats (1–16) resembling those of the epidermal growth factor (EGF) and a conserved DSL domain (fig. 5.2). Members of the Ser family also contain a cysteine-rich domain in their extracellular domain. All DSL proteins have a unique transmembrane domain and a short cytoplas-

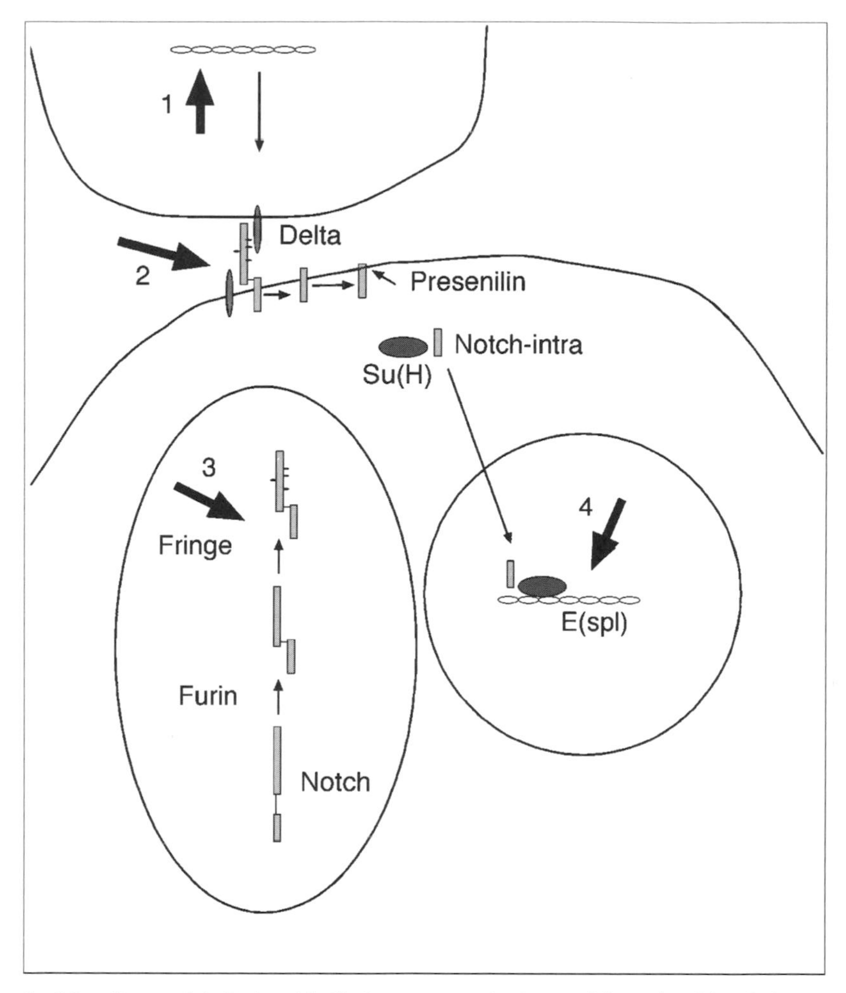

Fig. 5.1.—Elements of the Notch module. The figure represents the elements of the Notch module and other proteins involved in Notch functionality in their corresponding subcellular compartments. The events leading to Notch signaling are indicated by numbered arrows: 1, regulated expression of *D*elta, *S*errate, and *L*AG-2 (DSL) proteins (Delta); 2, activation of the Notch receptor in the cellular membrane; 3, posttranscriptional modifications of Notch in the Golgi apparatus; and 4, transcriptional regulation of target genes (*E(spl)*) by Su(H)/N-intra protein complexes in the nucleus.

mic tail whose sequences are highly divergent between different members of the family (fig. 5.2) (Lendahl 1998; Lissemore and Starmer 1999). Most data are compatible with Notch's being activated by DSL proteins expressed in adjacent cells by a mechanism that requires trans-endocytosis of the Notch extracellular domain into DSL-expressing cells (Heitzler and Simpson 1991; Muskavitch 1994; Parks et al. 2000).

The Notch proteins have an overall similar structure consisting of several functional domains formed by repetitions of conserved motifs

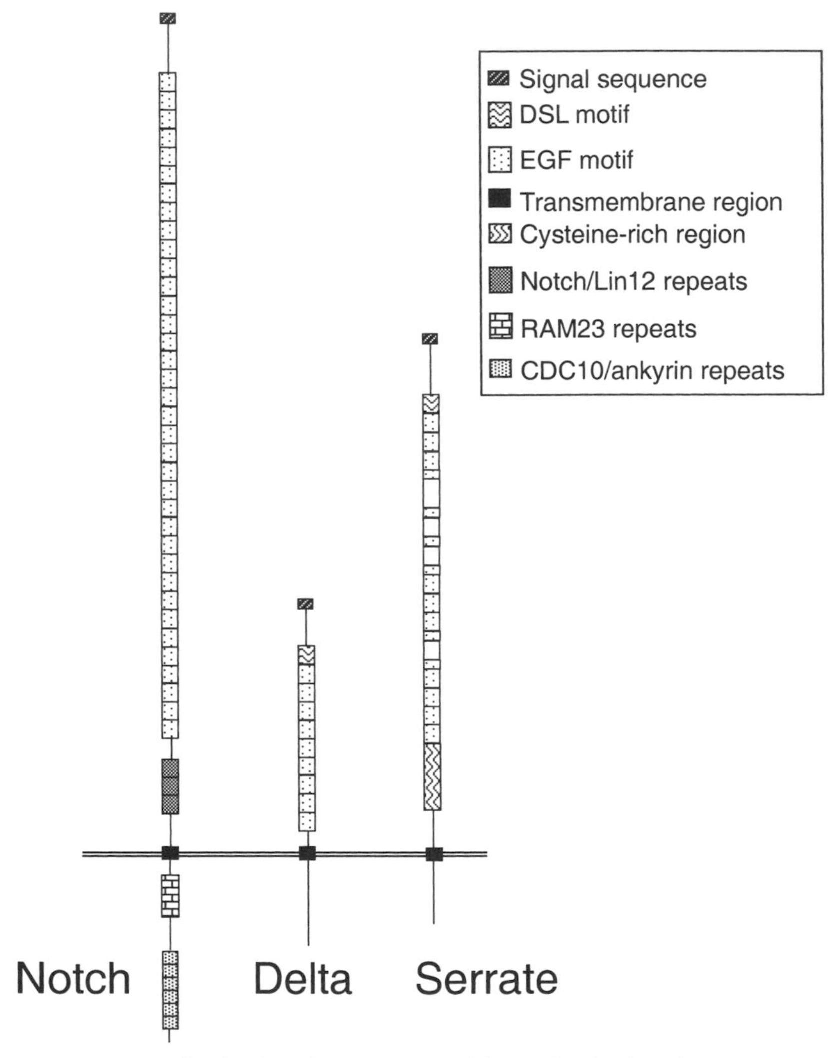

Fig. 5.2.—Structure of Notch, Delta, and Ser proteins. Structural domains of Notch, Delta, and Ser proteins are represented by filled squares.

(Mumm and Kopan 2000; fig. 5.2). In the extracellular domain there is a variable number (ranging from 10 in *C. elegans* LP-1 to 36 in *Drosophila* Notch) of a 40-amino-acid repeat homologous to the EGF (Johansen et al. 1989; Kidd et al. 1989; Roehl and Kimble 1993). Closer to the transmembrane domain, there are three copies of a characteristic cysteine-rich repeat called Notch/Lin-2 (LN repeats). The transmembrane domain links the extracellular region with a long cytoplasmic tail whose major characteristics are the presence of a 50-amino-acid-long domain that mediates interactions with Su(H), termed the RAM23 domain (Tamura et al. 1995), and seven CDC10/ankyrin repeats flanked by nuclear localization signals (Stifani et al. 1992; Fleming 1998; Mumm and Kopan 2000; fig. 5.2). The ankyrin motif was previously found in Cdc10 and SWI6 proteins of yeast and is thought to mediate protein-protein interactions (Breeden and Nasmyth 1987; Bennet 1992).

Notch proteins are highly conserved throughout evolution, and, interestingly, individual EGF repeats show higher homology to the corresponding repeats in other orthologues than to repeats of the same protein (Lardelli et al. 1995; Fleming 1998). This suggests that individual EGF repeats or groups of repeats within the protein have specific and distinct functions (Lawrence et al. 2000). However, only in some cases has it been possible to assign separate functions to particular repeats. For example, the EGF repeats 11 and 12 are necessary and sufficient for interaction with ligands, and a single amino acid substitution in a cysteine of repeat 11 abolishes Notch activity in *Drosophila* (Fehon et al. 1990; Rebay et al. 1991; de Celis et al. 1993). Mutations in the EGF repeats 24–29 of *Drosophila* Notch affect the activity of the protein and the efficiency of interactions between Notch and Delta (Dl) (Kelley et al. 1987; Lieber et al. 1992). These repeats appear to mediate the interaction of Notch with the glycosyl transferase protein Fringe (de Celis and Bray 2000; Ju et al. 2000). Interestingly, these repeats are also necessary for the binding of Wingless to Notch observed in vitro (Wesley 1999).

Mechanism and Regulation of Notch Activation

Notch signaling involves ligand-dependent proteolytic processing of the receptor, translocation to the nucleus of the processed form (N-intra), and transcriptional activation of downstream genes by Su(H)/N-intra complexes (Mumm and Kopan 2000; fig. 5.1). During its transit to the cell membrane, the 300 kDa Notch protein undergoes a proteolytic event in the Golgi apparatus by a Furin-convertase to generate two polypeptides (p120 and p180) that remain together bound by calcium (Logeat et al. 1998; Blaumueller et al. 1997; Chan and Jan 1998; Rand et al. 2000). The molecular details of Notch-ligand interactions between neighboring cells resulting in Notch activation are not known. However, the activation of Notch requires clathrin-mediated endocytosis, where the Notch extracellular domain is internalized into Dl-expressing cells (Seugnet et al. 1997; Parks et al. 2000). This observation suggests that once Notch looses its extracellular domain, the remaining protein is now available for further proteolytic processing to generate an active N-intra fragment (fig. 5.1). The cascade of proteolytic events leading to cytoplasmic N-intra is now beginning to be unraveled. First, the Notch membrane-bound fragment is cleaved to generate a C-terminal fragment susceptible to proteolytic processing in its transmembrane domain by Presenilins to generate the membrane-free intracellular fragment N-intra (Kidd et al. 1998; Parks et al. 2000; Mumm et al. 2000; Ye and Fortini 1998; Ye et al. 1999; Steiner et al. 1999; Ray et al. 1999; De Strooper et al. 1999; Struhl and Greenwald 1999; Lecourtois and Schweisguth 1998). The N-intra fragment interacts with Su(H), forming the active complex that mediates most aspects of Notch signaling.

Transcriptional Responses to Notch: The Su(H) Molecule

Su(H) in flies and its orthologues Lag-1 in *C. elegans* and RBP-Jk/KBF2/CBF1 in mammals are known as CSL proteins. These proteins have a sequence-specific DNA binding domain and are located both in the nucleus and in the cytoplasm. In cultured cells CSL proteins are associated, in the absence of Notch activation, with a transcriptional repressor complex including Skip and a histone-deacetylase complex (Hsieh and Hayward 1995; Gho et al.1996; Kao et al. 1998; Zhou et al. 2000). This complex has an intrinsic activity as a transcriptional repressor (Dou et al. 1994; Kao et al. 1998). Upon binding of N-intra to CSL the complex changes from a transcriptional repressor to a transcriptional activator (Morel and Schweisguth 2000; Klein et al. 2000; Furriols and Bray 2000). This transition appears to be mediated by the elimination of the histone-deacetylase complex and the incorporation of histone acetylases and the transcriptional activator Lag3/mastermind (fig. 5.3) (Kurooka and Honjo 2000; Petcherski and Kimble 2000a, 2000b; Mumm and Kopan 2000).

Transcriptional Responses to Notch: The *E(spl)* Gene Complex

The best-characterized N-intra/CSL targets are a group of related genes known as *E(spl)* (ESR in *Xenopus* and HES in mammals; Takke et al. 1999; Wurmbach et al. 1999; Kageyama and Ohtsuka 1999; Ohtsuka et al. 1999). The regulatory regions of these genes contain CSL binding sites whose presence is needed to confer responsiveness to Notch signaling (Bailey and Posakony 1995; Lecourtois and Schweisguth 1995). The *E(spl)* genes encode basic helix-loop-helix proteins that repress the expression of a number of genes, among others the proneural genes during lateral inhibition (Oelles et al. 1994). The regulation of *Drosophila E(spl)* genes is complex, with individual genes of the complex being expressed in specific temporal and spatial patterns (de Celis et al. 1996a; Nellesen et al. 1999; Cooper et al. 2000; fig. 5.4). The regulatory regions of individual *E(spl)* genes contain a variable number of

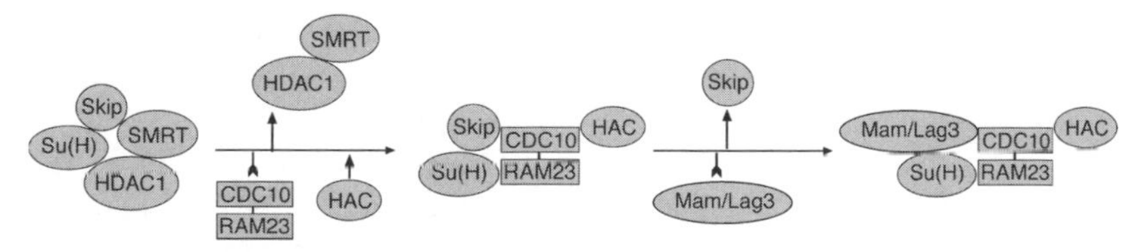

Fig. 5.3.—Elements implicated in transcriptional activation mediated by Notch signaling. Named circles represent the proteins identified in vitro that participate in the regulation of Notch target genes. HDAC1, histone-deacetylase complex 1; HAC, histone-acetylase complex; SMRT, transcriptional corepressor; Mam/lag3, mastermind and lag3 transcriptional activators; Su(H), Suppressor of Hairless; CDC10-RAM23, part of the intracellular domain of Notch; Skip, Ski-interacting protein. (Data are taken from Mumm and Kopan 2000.)

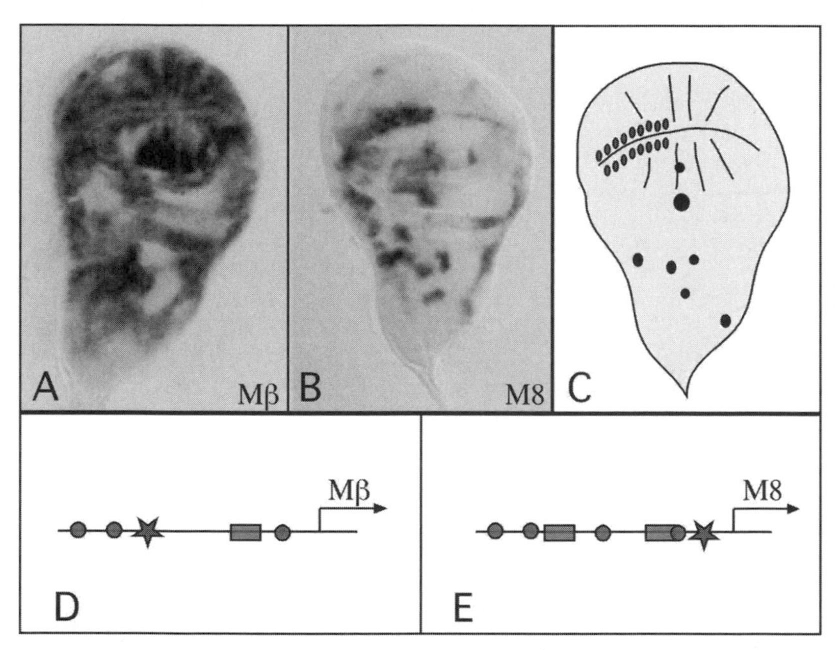

Fig. 5.4.—Transcriptional regulation of two *E(spl)* genes. *A*, Expression of *Drosophila E(spl)mβ* (*A*) in the wing imaginal disc. *B*, Expression of *Drosophila E(spl)m8* in the wing imaginal disc. *C*, Representation of the wing imaginal disc showing the position of the wing veins (*vertical dark lines*), wing margin (*horizontal dark line*), and sensory organs (*black circles*). The expression of *E(spl)mβ* is related to the developing veins and wing margin, whereas expression of *E(spl)m8* is related to sensory organs and to the wing margin. *D* and *E*, Representation of the 5′ regulatory regions of *E(spl)mβ* (*D*) and *E(spl)m8* (*E*), indicating the position of Su(H) binding sites (*squares*), basic helix-loop-helix (bHLH) activators (*star*), and bHLH repressors (*circles*). (Data for *D* and *E* are taken from Nellesen et al. 1999.)

Su(H) binding sites that are necessary for the response to Notch (Bailey and Posakony 1995). In addition, it is likely that *E(spl)* contains gene-specific sequences mediating the interaction with other tissue-specific transcriptional regulators that confer a particular expression domain on each individual gene (Nellesen et al. 1999; Cooper et al.

2000; fig. 5.4). Thus, it appears that Notch signaling is one component of a combination of signals that regulate the expression of individual *E(spl)* genes. Other Notch target genes share this mode of regulation. For example, in the case of the *Drosophila* gene *vestigial,* Notch participates in the activation of one of its enhancers, the dorsoventral boundary enhancer that also contains Su(H) binding sites and additional elements that restrict its activation to the dorsoventral boundary of the wing disc (Kim et al. 1996, 1997).

A common picture emerges when we consider the effect of Notch signaling in the regulation of target genes, where the Notch/CSL transcriptional activator in combination with other transcriptional regulators affects the expression of target genes. According to this view Notch signaling converts a transcriptional repressor complex into a transcriptional activation complex that, together with other transcription factors, regulates the expression of target genes. This implies that genes will be receptive to Notch activation only when they contain appropriate binding sites for Su(H), and that individual targets will respond to Notch only when other, context-specific transcriptional regulators are present in the same cell. The combination of common Su(H)/CSL binding sites and binding sites for other transcriptional regulators could be important in determining specific cellular responses to Notch activation. Thus, the variety of responses to Notch activity observed in different cell types will be determined by the structure of the regulatory region of downstream genes and the presence of context-specific transcriptional regulators.

Spatiotemporal Regulation of Notch Activity

Inappropriate activation of Notch has dramatic effects on cell proliferation and patterning, and this implies that the time, pattern, and intensity of Notch signaling are tightly regulated. Several mechanisms have been identified which influence Notch activity. They include the localized expression of the ligands, posttranscriptional modifications to the receptor that modify the interaction with the ligands, interactions between the ligands and receptor that restrict the domain of activation, and the expression of other transcription factors that attenuate or antagonize Notch activity (Panin and Irvine 1998; fig. 5.5).

The expression of the *Dl* and *Ser* genes is highly regulated and constitutes the first level of control of the time course and spatial pattern of Notch activation (Haenlin et al. 1990; Henrique et al. 1995; Kooh et al. 1993; Thomas et al. 1991; Fleming et al. 1990). This transcriptional control relies on the existence of long and complex regulatory regions organized in a modular manner that direct the expression of

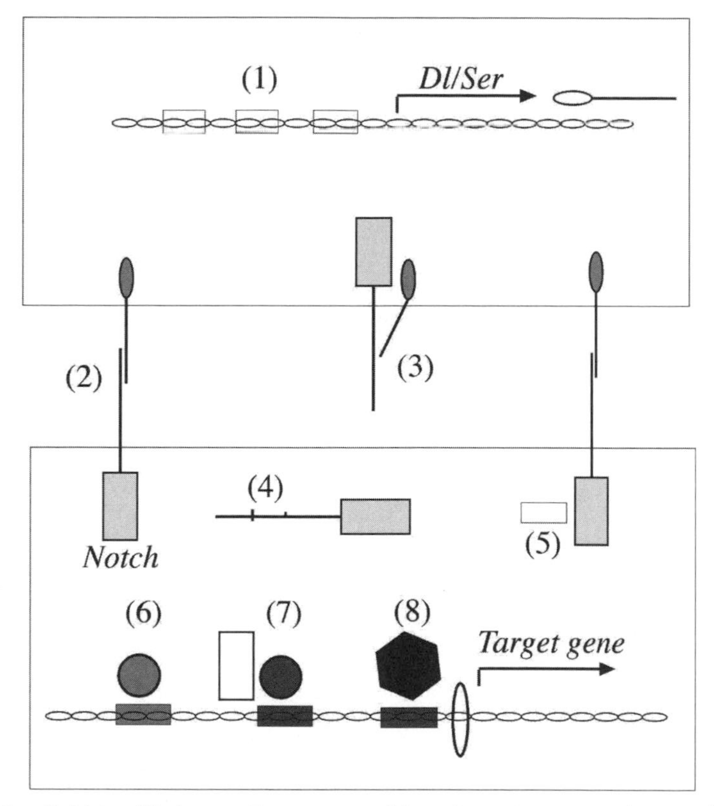

Fig. 5.5.—Modulation of Notch activity. The main aspects of the regulation in space, time, and intensity of Notch signaling include (1) transcriptional control of Notch ligands by an array of enhancers (*empty boxes*), (2) Dl and Ser interactions with Notch between neighboring cells, (3) unproductive interactions between the ligands and Notch taking place in the same cell, (4) posttranscriptional modifications of Notch influencing its interactions with Dl and Ser, (5) binding of proteins (*empty rectangle*) that reduce Notch activation to the Notch intracellular domain, (6) cell or tissue transcriptional coactivators (*light gray circle*) required for the expression of Notch target genes, (7) the presence of Su(H)/N-intra binding sites (*medium gray box*) or alternatively of E(spl) binding sites (not shown), (8) transcriptional repressors that negatively influence the response to Su(H)/N-intra (*dark hexagon*). Steps 1–5 refer to Notch activation, and steps 6–8 to regulation of Notch target genes.

these genes in particular subsets of their expression domains (fig. 5.5) (Haenlin et al. 1994; Bachmann and Knust 1998). The characterization of these regulatory sequences has so far been studied only in *Drosophila,* where individual enhancers directing expression of *Dl* and *Ser* in the developing nervous system (*Dl*) or in specific regions of the imaginal discs and embryo (*Ser*) have been identified (Haenlin et al. 1994; Bachmann and Knust 1998). This organization implies that multiple tissue-specific regulatory inputs can be integrated at the level of *Dl* and *Ser* promoters (fig. 5.5). In addition, the expression of the ligands is regulated by Notch itself in some biological processes, generating a feedback loop that stabilizes the domain of ligand expression and receptor activation (de Celis and Bray 1997; Micchelli et al. 1997; Papayannopoulos et al. 1998).

The expression of the ligands not only determines the places where Notch is activated, but also helps to restrict this activation to cells expressing low levels of ligand. This is based on interactions taking place in the cell expressing both the ligand and the receptor (cis-interactions) that reduce the activity of the receptor and the efficiency of signaling to neighboring cells (fig. 5.5; de Celis and Bray 1997; Jacobsen et al. 1998). The antagonism between the ligands and the receptor depends on the integrity of the Notch repeats 24–29 and the activity of the glycosyl transferase Fringe (Klein and Martinez-Arias 1998; Panin et al. 1997; Wu and Rao 1999; de Celis and Bray 2000). The combination of cis and trans interactions between the ligands and receptor are important for determining the polarity of signaling, from cells expressing higher levels of ligands to cells expressing higher levels of the receptor. The maintenance of the polarity of signaling is also reinforced by a positive feedback loop that increases the expression of Notch in cells where it has been activated (de Celis and Bray 1997; de Celis et al. 1997). This mechanism operates mainly during the formation of developmental boundaries by Notch.

Other elements affecting the regulation of Notch activity, at least in *Drosophila,* include a family of genes that antagonize or reduce the efficiency of Notch signaling. Thus, nested within the *E(spl)* complex, and also in other genomic regions of *Drosophila,* there are several genes regulated by N-intra/Su(H) whose main activity seems to be to antagonize E(spl) function. These genes constitute the *Bearded* (*Brd*) family, which in *Drosophila* is formed by seven members that share high sequence similarity (Lai et al. 2000a, 2000b; Leviten et al. 1997). The activation of *Brd* by Notch constitutes a common theme in signaling, where the activity of a signaling pathway triggers negative feedback mechanisms that attenuate the consequences of signaling (Perrimon and McMahon 1999). Interestingly, a novel gene downstream of Notch that antagonizes Notch function has been recently identified in *Xenopus* (Lamar et al. 2001), suggesting that negative feedbacks affecting Notch activity are conserved during evolution. In addition, other proteins whose expression is independent of Notch also contribute to attenuating Notch activity in *Drosophila.* The best examples are Disheveled (Dsh) and Nubbin (Nub). Disheveled is a multidomain cytoplasmic protein involved in Wingless signaling (Klingensmith et al. 1994; Cadigan and Nusse 1997). In addition, Dsh can bind to the cytoplasmic region of Notch and appears to reduce the efficiency of Notch activation (fig. 5.5) (Axelrod et al.

1996). The interaction between Notch and Dsh could be particularly relevant in places where Notch and Wingless signaling act in the same or adjacent groups of cells, as happens during the formation of the *Drosophila* wing dorsoventral boundary (Blair 1995). The nuclear protein Nub belongs to the POU family of transcriptional regulators and binds to specific sequences that are present in some Notch downstream genes such as *vestigial* (Neumann and Cohen 1998). The expression of *nub* in the *Drosophila* wing is independent of Notch and seems to participate in restricting high levels of Notch activity to the dorsoventral boundary (Neumann and Cohen 1998). It is unclear whether the activities of Dsh and Nub in antagonizing Notch signaling are conserved functions of these proteins in other organisms. In any case, the effects of these proteins observed in *Drosophila* illustrate the existence of multiple levels of regulation of the efficiency of Notch signaling. The relationships between proteins acting on Notch and on its target genes could be another mechanism that affects the consequences of signaling, contributing to cell-specific responses to a generic signal.

Conserved Aspects of Notch Signaling: Lateral Inhibition and Boundary Formation

Two processes where Notch plays a significant and conserved role are lateral inhibition during neurogenesis and the establishment of developmental boundaries during organogenesis (fig. 5.6). Neurogenesis in a variety of organisms implies selection of neural precursors within a field of cells competent to follow a neural fate. The competence to adopt the neural fate is conferred by proneural proteins, which are basic helix-loop-helix proteins that regulate the expression of subordinate genes through E-box sequences (Campuzano and Modolell 1992; Modolell and Campuzano 1998). During lateral inhibition Notch prevents cells from entering into the neural fate, and this function is mediated by the repression of proneural gene expression by E(spl)/HES proteins (fig. 5.6) (Nakao and Campos-Ortega 1996). The activation of Notch during neurogenesis depends on *Dl,* whose regulatory region, interestingly, contains E-box sequences (Kunisch et al. 1994). This allows *Dl* activation by proneural proteins and results in high levels of *Dl* expression in cells entering the neural fate (Kunisch et al. 1994). These cells, in consequence, activate Notch signaling in the neighboring cells and prevent them from following the same developmental pathway (fig. 5.6). Interestingly, *Drosophila E(spl)* genes also contain E-box sequences that are necessary for their expression in proneural fields (Nellesen et al. 1999). In this way, proneural proteins not only

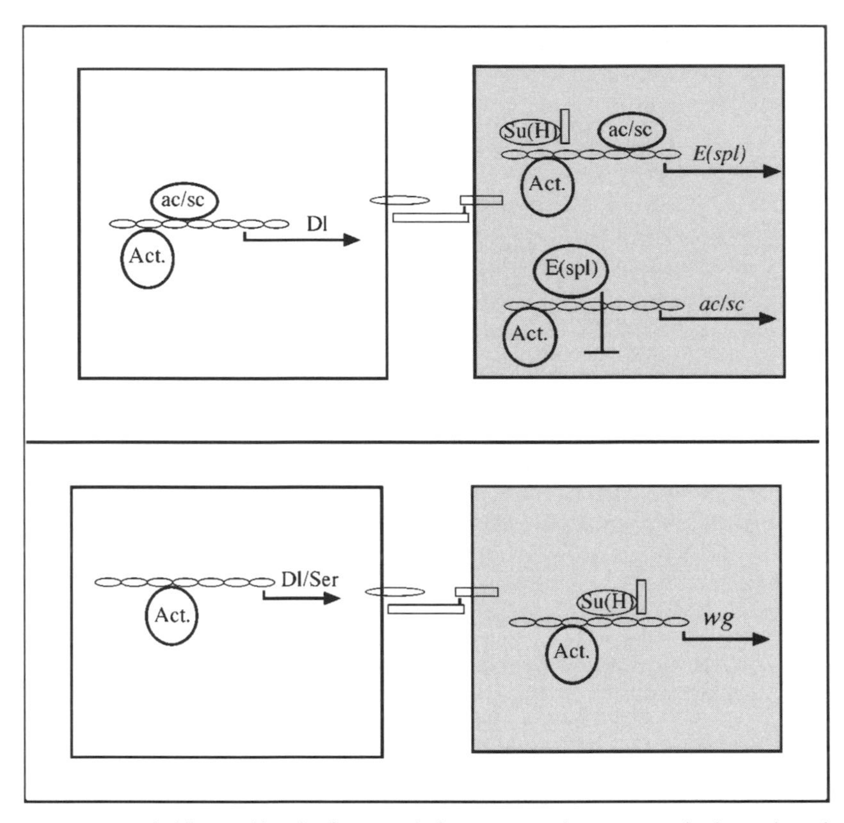

Fig. 5.6.—Lateral inhibition and boundary formation. The figure represents the main events related to Notch signaling taking place during lateral inhibition (*upper panel*) and boundary formation (*lower panel*). The expression of the proneural genes (*ac/sc*) is repressed by E(spl) proteins during lateral inhibition. Proneural genes, in turn, contribute to regulating the expression of *E(spl)* genes and *Dl*. In contrast, during boundary formation, E(spl) proteins are not required. In these processes both Dl and Ser ligands are required, and their interactions with Notch are modified by Fringe. Su(H)/N-intra complexes regulate the expression of several target genes (*wg*), which in turn may contribute to the maintenance of *Dl* and *Ser* expression. In these regulatory interactions other transcriptional regulators participate in the regulation of *Dl, Ser,* proneural, and *E(spl)* gene expression (indicated by "Act." in circles).

participate in the regulation of *Dl,* but also contribute to the localized expression of some *E(spl)* genes. This circuitry of interactions coordinates the expression of proneural genes with Notch signaling through Notch/E(spl) activities and permits the reinforcement and extinction of proneural gene expression (fig. 5.6).

A development boundary is a cellular domain that separates a field of cells into two different populations that usually have different genetic specifications. Notch has been implicated in the creation of developmental boundaries in different biological processes in several organisms, including the formation of the dorsoventral boundary in developing imaginal discs and vertebrate limbs and the separation of adjacent somites during somitogenesis (Irvine and Vogt 1997). In contrast to neurogenesis, where the role of E(spl)/HES proteins is determinant, during boundary formation these genes do not contribute to Notch activity (de Celis et al. 1996a, 1996b). Thus, the effects of Notch are exclusively mediated by N-intra/Su(H) transcriptional activation of target genes that are insensitive to E(spl) repression (fig. 5.6). The place and timing of developmental boundary formation are also initiated by the expression of the Notch ligands Dl and Ser. In addition, the effectiveness of ligand-receptor interactions plays a central role in determining which ligand activates Notch and where this activation takes place (Irvine and Vogt 1997; Blair 2000).

The sensitivity of Notch to its ligands is regulated by Fringe (Fng), a glycosyl transferase that is localized in the Golgi apparatus, where it binds to Notch and catalyzes the incorporation of GlcNAc to O-fucose on specific EGF repeats (Munro and Freeman 2000; Ju et al. 2000). The activity of Fng is critical for the establishment of Notch activation domains. Thus, Fng potentiates activation of Notch by Dl and prevents activation of Notch by Ser. In the *Drosophila* wing and eye discs, and in vertebrate limbs, the expression of *fng* is restricted to dorsal cells, where it prevents Ser from signaling in the dorsal compartment (Irvine and Vogt 1997; Blair 2000). The absence of *Fng* in ventral cells diminishes the signaling activity of Dl in this compartment. In this manner, dorsal cells signal to ventral cells through Ser, while ventral cells signal to dorsal cells through Dl. Most processes requiring *fng* activity also involve the formation of developmental boundaries, such as the segmentation of the legs in flies and somitogenesis in vertebrates. In this last process one vertebrate orthologue of *fng, lunatic fringe,* is expressed in stripes in the presomitic mesoderm. There are no described effects for *fng* independent of Notch, suggesting that *fng* is a genuine member of the Notch module. For example, in mice lacking *lunatic fringe* the failures in somite formation are very similar to those observed in the absence of Notch or Notch ligands (Wu and Rao 1999).

Concluding Remarks

The Notch module of interacting proteins is used throughout animal development in processes where the regulation of target gene expression is complex and combinatorial. Although the Notch signaling pathway is a context-independent module, in the sense that the molecular interactions between its members are conserved and invariant, the outcome of Notch signaling is highly context dependent. This seems to be determined by the nature of the regulatory regions of both Notch ligands and target genes. It is in these regulatory regions that the places of Notch activation and the responses to Notch are encoded. These cis-regulatory regions determine the requirement for other cell-type transcription factors and the capability of responding to either E(spl), N-intra/Su(H) or both. The context-dependent activation and outcome of Notch signaling is probably a general aspect of other signaling pathways where the specificity is not in the pathway itself, but rather in the combination of active transcriptional regulators present in a cell at a given time.

Acknowledgments

I thank R. Barrio, C. Extavour, A. García-Bellido, A. Martínez-Arias, and M. Ruiz-Gomez for critical reading of the manuscript.

References

Artavanis-Tsakonas, S., M. D. Rand, and R. J. Lake. 1999. Notch signaling: Cell fate control and signal integration in development. *Science* 284:770–776.

Axelrod, J. D., K. Matsuno, S. Artavanis-Tsakonas, and N. Perrimon. 1996. Interaction between wingless and Notch signaling pathways mediated by dishevelled. *Science* 271:1826–1832.

Bachmann, A., and E. Knust. 1998. Dissection of *cis*-regulatory elements of the *Drosophila* gene *Serrate. Dev. Genes Evol.* 208:346–351.

Bailey, A. M., and J. W. Posakony. 1995. Suppressor of Hairless directly activates transription of *Enhancer of split* Complex genes in response to Notch receptor activity. *Genes and Development* 9:2609–2622.

Bennet, V. 1992. Ankyrins: adaptors between diverse plasma membrane proteins and the cytoplasm. *J. Biol. Chem.* 267:8703–8706.

Blair, S. S. 1995. Compartments and appendage development in *Drosophila. BioEssays* 4:299–309.

Blair, S. S. 2000. Notch signalling: Fringe really is a glycosystransferase. *Curr. Biol.* 10: 608–612.

Blaumueller, C. M., P. Qui, P. Zagouras, and S. Artavanis-Tsakonas. 1997. Intracellular cleavage of Notch leads to a heterodimeric receptor on the plasma membrane. *Cell* 90:281–291.

Brand, M., and J. A. Campos-Ortega. 1990. Second site modifiers of the *split* mutation of *Notch* define genes involved in neurogenesis in *Drosophila melanogaster. Roux's Arch. Dev. Biol.* 198:275–285.

Breeden, L., and K. Nasmyth. 1987. Similarity between cell-cycle genes of budding yeast and fission yeast and the *Notch* locus of *Drosophila. Nature* 329:651–654.

Busseau, I., R. J. Diederich, T. Xu. and S. Artavanis-Tsakonas. 1994. A member of the *Notch* group of interacting loci, *deltex* encodes a cytoplasmic basic protein. *Genetics* 136:585–596.

Cadigan, K. M., and R. Nusse. 1997. Wnt signalling: a common theme in animal development. *Genes Dev.* 11:3286–3305.

Campuzano, S., and J. Modolell. 1992. Patterning of the *Drosophila* nervous system—the *achaete-scute* gene complex. *Trends Genet.* 8:202–208.

Chan, Y. M., and Y. N. Jan. 1998. Roles for proteolysis and trafficking in Notch maturation and signal transduction. *Cell* 4:423–426.

Cooper, M. T., D. M. Tyler, M. Furriols, A. Chalkiadaki, C. Delidakis, and, S. Bray. 2000. Spatially restricted factors cooperate with notch in the regulation of Enhancer of split genes. *Dev. Biol.* 221:390–403.

de Celis, J. F., R. Barrio, A. del Arco, and A. Garcia-Bellido. 1993. Genetic and molecular analysis of a *Notch* mutation in its Delta- and Serrate- binding domain. *Proc. Natl. Acad. Sci. U.S.A.* 90:4037–4041.

de Celis, J. F., and S. Bray. 1997. Feedback mechanisms affecting Notch activation at the dorso-ventral boundary in the *Drosophila* wing. *Development* 124:3241–3251.

de Celis, J. F., and S. Bray. 2000. The *Abruptex* domain of Notch regulates negative interactions between Notch, its ligands and Fringe. *Development* 127:1291–1302.

de Celis, J. F., S. Bray, and A. Garcia-Bellido. 1997. Notch signalling regulates *veinlet* expression and establishes boundaries between veins and interveins in the *Drosophila* wing. *Development* 124:1919–1928.

de Celis, J. F., J. de Celis, P. Ligoxiars, A. Preiss, C. Delidakis, and S. Bray. 1996a. Functional relationships between *Notch, Su(H)* and the bHLH genes of the *E(spl)* complex: the *E(spl)* genes mediate only a subset of *Notch* activities during imaginal development. *Development* 122:2719–2728.

de Celis, J. F., A. Garcia-Bellido, and S. Bray. 1996b. Activation and function of *Notch* at the dorsoventral boundary in the *Drosophila* wing imaginal disc. *Development* 122:359–369.

De Strooper, B., W. Annaert, P. Cupers, P. Saftig, K. Craessaerts, J. S. Mumm, E. H. Schroeter, V. Schrijvers, M. S. Wolfe, W. J. Ray, A. Goate, and R. Kopan. 1999. A presenilin-1-dependent gamma-secretase-like protease mediates release of Notch intracellular domain. *Nature* 389:518–522.

Diederich, R. J., K. Matsuno, H. Hing, and S. Artavanis-Tsakonas. 1994. Cytosolic interaction between deltex and Notch ankyrin repeats implicates deltex in the Notch signal. *Development* 120:473–481.

Dou, S., X. Zeng, P. Cortes, B. H. Erdjument, P. Tempst, T. Honjo, and L. D. Vales. 1994. The recombination signal sequence-binding protein RBP-2N functions as a transcriptional repressor. *Mol. Cell. Biol.* 14:3310–3319.

Dumont, E., K. P. Fuchs, G. Bommer, B. Christoph, E. Kremmer, and B. Kempkes. 2000. Neoplastic transformation by Notch is independent of transcriptional activation by RBP-J signalling. *Oncogene* 19:556–561.

Fehon, R. G., P. J. Kooh, I. Rabay, C. L. Regan, T. Xu, M. A. T. Muskavitch, and S. Artavanis-Tsakonas. 1990. Molecular interactions between the protein products of the neurogenic loci Notch and Delta, two EGF-homologous genes in *Drosophila*. *Cell* 61:523–534.

Fleming, R. J. 1998. Structural conservation of Notch receptors and ligands. *Semin. Cell Dev. Biol.* 9:599–607.

Fleming, R. J., T. N. Scottgale, R. J. Diederich, and S. Artavanis-Tsakonas. 1990. The gene *Serrate* encodes a putative EGF-like transmembrane protein essential for proper ectodermal development in *Drosophila melanogater. Genes Dev.* 4:2188–2201.

Fostier, M., D. A. Evans, S. Artavanis-Tsakonas, and M. Baron. 1998. Genetic characterization of the *Drosophila melanogaster Suppressor of deltex* gene: A regulator of Notch signaling. *Genetics* 150:1477–1485.

Furriols, M., and S. Bray. 2000. Dissecting the mechanisms of Suppressor of Hairless function. *Dev. Biol.* 227:520–532.

Gho, M., M. Lecourtais, G. Geraud, J. Posakony, and F. Schweisguth. 1996. Subcellular localization of Suppressor of Hairless in *Drosophila* sense organ cells during Notch signalling. *Development* 122:1673–1682.

Go, M. J., and S. Artavanis-Tsakonas. 1998. A genetic screen for novel components of the Notch signaling pathway during *Drosophila* bristle development. *Genetics* 150: 211–220.

Greewald, I. 1998. LIN12/Notch signaling: lessons from worms and flies. *Genes Dev.* 12:1751–1762.

Haenlin, M., B. Kramatschek, and J. A. Campos-Ortega. 1990. The pattern of transcription of the neurogenic gene *Delta* of *Drosophila melanogaster. Development* 110: 905–914.

Haenlin, M., M. Kunisch, B. Kramatschek, and J. A. Campos-Ortega. 1994. Genomic regions regulating early embryonic expression of the *Drosophila* neurogenic gene *Delta. Mech. Dev.* 47:99–110.

Heitzler, P., and P. Simpson. 1991. The choice of cell fate in the epidermis of *Drosophila. Cell* 64:1032–1042.

Henrique, D., J. Adam, A. Myat, A. Chitnis, J. Lewis, and D. Ish-Horovitch. 1995. Expression of a *Delta* homologue in prospective neurons in the chick. *Nature* 375: 787–790.

Hsieh, J. J., and S. D. Hayward. 1995. Masking of the CBF1/RBPJ kappa transcriptional repression domain by Epstein-Barr virus EBNA2. *Science* 268:560–563.

Irvine, K. D., and T. F. Vogt. 1997. Dorso-ventral signalling in limb development. *Curr. Opin. Cell Biol.* 9:867–876.

Jacobsen, T. L., K. Brennan, A. Martinez-Arias, and M. A. T. Muskavitch. 1998. *Cis*-interactions between Delta and Notch modulate neurogenic signalling in *Drosophila. Development* 125:4531–4550.

Johansen, K. M., R. G. Fehon, and S. Artavanis-Tsakonas. 1989. The Notch gene product is a glicoprotein expressed on the cell surface of both epidermal and neural precursor cells during *Drosophila* development. *J. Cell. Biol.* 109:2427–2440.

Ju, B. G., S. Jeong, E. Bae, S. Hyun, S. Carroll, J. Yim, and J. Kim. 2000. Fringe forms a complex with Notch. *Nature* 405:191–195.

Kageyama, R., and T. Ohtsuka. 1999. The Notch-Hes pathway in mammalian neural development. *Cell Res.* 9:179–188.

Kao, H. Y., P. Ordentlich, N. Koyano-Nakagawa, Z. Tang, M. Downes, C. R. Kintner, R. M. Evans, and T. Kadesch. 1998. A histone deacetylase corepressor complex regulates the Notch signal transduction pathway. *Genes Dev.* 15:2269–2277.

Kelley, M. R., S. Kidd, W. A. Deutsh, and M. W. Young. 1987. Mutations altering the structure of epidermal growth factor-like coding sequences at the *Drosophila Notch* locus. *Cell* 51:539–548.

Kidd, S., M. K. Baylies, G. P. Gasic, and M. W. Young. 1989. Structure and distribution of the Notch protein in developing *Drosophila. Genes Dev.* 3:1113–1129.

Kidd, S., T. Lieber, and M. W. Young. 1998. Ligand-induced cleavage and regulation of nuclear entry of Notch in *Drosophila* melanogaster embryos. *Genes Dev.* 12: 3728–3740.

Kim, J., K. Johnson, H. J. Chen, S. Carroll, and A. Laughon. 1997. *Drosophila* Mad binds to DNA and directly mediates activation of *vestigial* by Decapentaplegic. *Nature* 388:304–308.

Kim, J., A. Sebring, J. J. Esch, M. E. Kraus, K. Vorwerk, J. Magee, and S. B. Carroll. 1996. Integration of positional signals and regulation of wing formation by *Drosophila vestigial* gene. *Nature* 382:133–138.

Klein, T., and J. A. Campos-Ortega. 1993. Second-site modifiers of the Delta wing phenotype in *Drosophila melanogaster. Roux's Arch. Dev. Biol.* 202:49–60.

Klein, T., and A. Martinez-Arias. 1998. Interactions among Delta, Serrate and Fringe modulate Notch activity during *Drosophila* wing development. *Development* 125: 2951–2962.

Klein, T., L. Seugnet, M. Haenlin, and A. Martinez Arias. 2000. Two different activities of Suppressor of Hairless during wing development in *Drosophila. Development* 127:3553–3566.

Klingensmith, J., R. Nusse, and N. Perrimon. 1994. The *Drosophila* segment polarity gene *dishevelled* encodes a novel protein required for response to the *wingless* signal. *Genes Dev.* 8:118–130.

Kooh, P. J., R. G. Fehon, and M. A. T. Muskavitch. 1993. Implications of dynamic patterns of Delta and Notch expression for cellular interactions during *Drosophila* development. *Development* 117:493–507.

Kunisch, M., M. Haenlin, and J. A. Campos-Ortega. 1994. Lateral inhibition mediated by the *Drosophila* neurogenic gene-*Delta* is enhanced by proneural proteins. *Proc. Natl. Acad. Sci. U.S.A.* 91:10139–10143.

Kurooka, H., and T. Honjo. 2000. Functional interactions between the mouse Notch1 intracellular region and histone acetyltransferases PCAF and GCN5. *J. Biol. Chem.* 275:17211–17220.

Lai, E. C., R. Bodner, J. Kavaler, G. Freschi, and J. W. Posakony. 2000a. Antagonism of Notch signaling activity by members of a novel protein family encoded by the *Bearded* and *Enhancer of split* gene complexes. *Development* 127:291–306.

Lai, E. C., R. Bodner, and J. W. Posakony. 2000b. The *Enhancer of split* complex of *Drosophila* includes four Notch-regulated members of the *Bearded* gene family. *Development* 127:3441–3455.

Lamar, E., G. Deblandre, D. Wettstein, V. Gawantka, N. Pollet, C. Niehrs, and C. Kintner. 2001. Nrarp is a novel intracellular component of the Notch signaling pathway. *Genes Dev.* 15:1885–1899.

Lardelli, M., R. Willians, and U. Lendahl. 1995. *Notch*-related genes in animal development. *Int. J. Dev. Biol.* 39:769–780.

Lawrence, N., T. Klein, K. Brennan, and A. Martinez Arias. 2000. Structural requirements for Notch signalling with Delta and Serrate during the development and patterning of the wing disc of *Drosophila. Development* 14:3185–3195.

Lecourtois, M., and F. Schweisguth. 1995. The neurogenic Suppressor of Hairless DNA-binding protein mediates the transcriptional activation of the *Enhancer of split* Complex genes triggered by Notch signalling. *Genes Dev.* 9:2598–2608.

Lecourtois, M., and F. Schweisguth. 1998. Indirect evidence for Delta-dependent intracellular processing of Notch in *Drosophila* embryos. *Curr. Biol.* 8:771–774.

Lendahl, U. 1998. A growing family of Notch ligands. *BioEssays* 2:103–107.

Leviten, M. W., E. C. Lai, and J. W. Posakony. 1997. The *Drosophila* gene *Bearded* encodes a novel small protein and shares 3′ UTR sequence motifs with multiple Enhancer of split Complex genes. *Development* 20:4039–4051.

Lewis, J. 1998. Notch signalling and the control of cell fate choices in vertebrates. *Semin. Cell Dev. Biol.* 9:583–589.

Lieber, T., C. S. Wesley, E. Alcamo, B. Hassel, J. F. Krane, J. A. Campos-Ortega, and M. W. Young. 1992. Single amino acid sustitutions in EGF-like elements of Notch and Delta modify *Drosophila* development and affect cell adhesion *in vitro. Neuron* 9:847–859.

Lissemore, J. L., and W. T. Starmer. 1999. Phylogenetic analysis of vertebrate and invertebrate Delta/Serrate/LAG-2 (DSL) proteins. *Mol. Phylogenet. Evol.* 11: 308–319.

Logeat, F., C. Bessia, C. Brou, O. LeBail, S. Jarriault, N. G. Seidah, and A. Israel. 1998. The Notch1 receptor is cleaved constitutively by a furin-like convertase. *Proc. Natl. Acad. Sci. U.S.A.* 14:8108–8112.

Martinez-Arias, A. 1998. Interactions between Wingless and Notch during the assignment of cell fates in *Drosophila. Int. J. Dev. Biol.* 42:325–333.

Matsuno, K., D. Eastman, T. Mitsiades, A. M. Quinn, M. L. Carcanciu, P. Ordentlich, T. Kadesch, and S. Artavanis-Tsakonas. 1998. Human *deltex* is a conserved regulator of Notch signalling. *Nat. Genet.* 19:74–78.

Matsuno, K., M. J. Go, X. Sun, D. S. Eastman, and S. Artavanis-Tsakonas. 1997. Suppressor of Hairless-independent events in Notch signaling imply novel pathway elements. *Development* 124:4265–4273.

Micchelli, C. A., E. J. Rulifson, and S. Blair. 1997. The function and regulation of *cut* expression on the wing margin of *Drosophila:* Notch, Wingless and a dominant negative role for Delta and Serrate. *Development* 124:1485–1495.

Modolell, J., and S. Campuzano. 1998. The *achaete-scute* complex as an integrating device. *Int. J. Dev. Biol.* 42:275–282.

Morel, V., and F. Schweisguth. 2000. Repression by Suppressor of Hairless and activation by Notch are required to define a single row of single-minded expressing cells in the *Drosophila* embryo. *Genes Dev.* 14:377–388.

Mumm, J. S., and R. Kopan. 2000. Notch signaling: from the outside in. *Dev. Biol.* 228:151–165.

Mumm, J. S., E. H. Schroeter, M. T. Saxena, A. Griesemer, X. L. Tian, D. J. Pan, W. J. Ray, and R. Kopan. 2000. A ligand induced extracellular cleavage regulates gamma-secretase-like proteolytic activation of Notch1. *Mol. Cell* 5:197–206.

Munro, S., and M. Freeman. 2000. The Notch signalling regulator Fringe acts in the Golgi apparatus and requires the glicosyltransferase signature motif DxD. *Curr. Biol.* 10:813–820.

Muskavitch, M. A. T. 1994. Delta-Notch signaling and *Drosophila* cell fate choice. *Dev. Biol.* 166:415–430.

Nakao, K., and J. A. Campos-Ortega. 1996. Persistent expression of genes of the *Enhancer of split* complex suppresses neural development in *Drosophila. Neuron* 16:275–286.

Nellesen, D. T., E. C. Lai, and J. W. Posakony. 1999. Discrete enhancer elements mediate selective responsiveness of *Enhancer of split* complex genes to common transcriptional activators. *Dev. Biol.* 213:33–53.

Neumann, C. J., and S. M. Cohen. 1998. Boundary formation in *Drosophila* wing: Notch activity attenuated by the POU protein Nubbin. *Science* 281:409–413.

Oelles, N., M. Dehio, and E. Knust. 1994. bHLH proteins encoded by the *Enhancer of split* complex of *Drosophila* negatively interferes with transcrptional activation mediated by proneural genes. *Mol. Gen. Genet.* 224:2743–2755.

Ohtsuka, T., M. Ishibashi, G. Gradwohl, S. Nakanishi, F. Guillemot, and, R. Kageyama. 1999. Hes1 and Hes5 as Notch effectors in mammalian neuronal differentiation. *EMBO J.* 18:2196–2207.

Panin, V. M., and K. D. Irvine. 1998. Modulators of Notch signaling. *Semin. Cell Dev. Biol.* 9:609–617.

Panin, V. M., V. Papayannopoulos, R. Wilson, and K. D. Irvine. 1997. Fringe modulates Notch-ligand interactions. *Nature* 387:908–912.

Papayannopoulos, V., A. Tomlinson, V. M. Panin, C. Rauskolb, and K. D. Irvine. 1998. Dorsal-ventral signaling in the *Drosophila* eye. *Science* 281:2031–2034.

Parks, A. L., K. M. Klueg, J. R. Stout, and M. A. T. Muskavitch. 2000. Ligand endocytosis drives receptor dissociation and activation in the Notch pathway. *Development* 127:1373–1385.

Perrimon, N., and A. P. McMahon. 1999. Negative feedback mechanisms and their roles during pattern formation. *Cell* 97:13–16.

Petcherski, A. G., and J. Kimble. 2000a. LAG-3 a putative transcrptional activator in the *C. elegans* Notch pathway. *Nature* 405:364–368.

Petcherski, A. G., and J. Kimble. 2000b. mastermind is a putative activator for Notch. *Curr. Biol.* 10:471–473.

Rand, M. D., L. M. Grimm, A. Artavanis-Tsakonas, V. Patriub, S. C. Blacklow, J. Sklar, and J. C. Astar. 2000. Calcium depletion dissociates and activates heterodimeric Notch receptors. *Mol. Cell. Biol.* 20:1825–1835.

Ray, W. J., M. Yao, J. Mumm, E. H. Schroeter, P. Saftig, M. Wolfe, D. J. Selkoe, R. Kopan, and A. M. Goate. 1999. Cell surface presenilin-1 participates in the gamma-secretase-like proteolysis of Notch. *J. Biol. Chem.* 274:36801–36807.

Rebay, I., R. J. Fleming, R. G. Fehon, P. Chervas, and S. Artavanis-Tsakonas. 1991. Specific EGF repeats of Notch mediate interactions with Delta and Serrate: implications for Notch as a multifunctional receptor. *Cell* 67:687–699.

Roehl, H., and J. Kimble. 1993. Control of cell fate in *C. elegans* by a GLP-1 peptide consisting primarily of ankyrin repeats. *Nature* 364:632–635.

Royet, J., T. Bouwmeester, and S. M. Cohen. 1998. Notchless encodes a novel WD40-repeat-containing protein that modulates Notch signaling activity. *EMBO J.* 24: 7351–7360.

Rusconi, J. C., and V. Corbin. 1998. Evidence for a novel Notch pathway required for muscle precursor selection in *Drosophila. Mech. Dev.* 79:39–50.

Rusconi, J. C., and V. Corbin. 1999. A widespread and early requirement for a novel Notch function during *Drosophila* embryogenesis. *Dev. Biol.* 215:388–398.

Seugnet, L., P. Simpson, and M. Haenlin. 1997. Requirements for dynamin during Notch signalling in *Drosophila* neurogenesis. *Dev. Biol.* 192:585–598.

Shawber, C., D. Nofziger, J. J. Hsieh, C. Lindsell, O. Bogler, D. Hayward, and G. Weinmaster. 1996. Notch signaling inhibits muscle cell differentiation through a CBF1-independent pathway. *Development* 122:3765–73.

Steiner, H., K. Duff, A. Capell, H. Romig, M. G. Grim, S. Lincoln, J. Hardy, X. Yu, M. Picciano, K. Fechteler, M. Citron, R. Kopan, B. Pesold, S. Keck, M. Baader, T. Tomita, T. Iwatsubo, R. Baumeister, and C. Haass. 1999. A loss of function mutation of presenilin-2 interferes with amyloid beta-peptide production and notch signaling. *J. Biol. Chem.* 274:28669–28673.

Stifani, S., C. M. Blaumueller, N. J. Redhead, R. H. Hill, and S. Artavanis-Tsakonas. 1992. Human homologs of a *Drosophila Enhancer of split* gene product define a novel family of nuclear proteins. *Nat. Genet.* 2:119–127.

Struhl, G., and I. Greenwald. 1999. Presenilin is required for activity and nuclear access of Notch in *Drosophila. Nature* 398:522–525.

Takke, C., P. Dornseifer, E. von Weizsacker, and J. A. Campos-Ortega. 1999. *her4,* a zebrafish homologue of the *Drosophila* neurogenic gene *E(spl),* is a target of NOTCH signalling. *Development* 126:1811–1821.

Tamura, K., Y. Taniguchi, S. Minoguchi, T. Sakai, TTun, T. Furukawa, and T. Honjo. 1995. Physical interactions between a novel domain of the receptor Notch and the transcription factor RBP-Jk/Su(H). *Curr. Biol.* 5:1416–1423.

Thomas, U., S. A. Speicher, and E. Knust. 1991. The *Drosophila* gene *Serrate* encodes an EGF-like transmembrane protein with complex expression patterns in embryos and wing disc. *Development* 111:749–761.

Wesley, C. S. 1999. Notch and wingless regulate expression of cuticle patterning genes. *Mol. Cell. Biol.* 8:5743–5758.

Wu, J. Y., and Y. Rao. 1999. Fringe: defining borders by regulating the Notch pathway. *Curr. Opin. Neurobiol* 9:537–543.

Wurmbach, E., I. Wech, and A. Preiss. 1999. The *Enhancer of split* complex of *Drosophila melanogaster* harbors three classes of Notch responsive genes. *Mech. Dev.* 80:171–180.

Ye, Y., and M. E. Fortini. 1998. Characterization of *Drosophila* Presenilin and its colocalization with Notch during development. *Mech. Dev.* 79:199–211.

Ye, Y., N. Lukinova, and M. E. Fortini. 1999. Neurogenic phenotypes and altered Notch processing in *Drosophila Presenilin* mutants. *Nature* 6727:525–529.

Zecchini, V., K. Brennan, and A. Martinez-Arias. 1999. An activity of Notch regulates JNK signalling and affects dorsal closure in *Drosophila*. *Curr. Biol.* 9:460–469.

Zhou, S. F., M. Fujimuro, J. Hsieh, L. Chen, A. Miyamoto, G. Weinmaster, and S. D. Hayward. 2000. SKIP, a CBF1-associated protein, interacts with the ankyrin repeat domain of NotchIC to facilitate NotchIC function. *Mol. Cell. Biol.* 20:2400–2410.

6 Sonic Hedgehog and Wnt Signaling Pathways during Development and Evolution

ANNE-GAËLLE BORYCKI

Introduction

The diversity of morphologies observed throughout the animal kingdom may suggest the existence of as many molecular mechanisms as distinct shapes and developmental processes identified during evolution. However, the knowledge gained by developmental biologists over the past 20 years contradicts this intuitive concept and instead emphasizes that few growth factors and pathways are utilized by adult cells and, of those, fewer still participate in embryonic patterning. Indeed, work so far has found that establishing the vertebrate body axis and coordinating cell fate specification with cell growth and cell death during organogenesis requires mainly the activity of five signaling pathways: Notch/Delta, receptor tyrosine kinase-mediated (such as FGF [Fibroblast Growth Factor]), Wnt, TGFβ (Transforming Growth Factor), and Shh (Sonic hedgehog). This observation has led to the new concept that signaling pathways and gene networks define "genetic modules" that act autonomously and are redeployed during evolution to introduce variations on a common theme, allowing the generation of novel complexity, specificity, and morphologies necessary to the genesis of organ diversity. This suggests an extraordinary conservation of proteins and signaling pathways that control the developmental processes underlying the formation of otherwise morphologically diverse creatures. This concept has multiple implications for our understanding of the developmental processes controlled by these signaling pathways. In particular, a corollary to this concept is the necessary variation in the regulation and the biochemical properties of the components of these signaling pathways, in order to integrate the necessary changes. Another issue is the use of these modules in networks during vertebrate development and evolution, implying cross-talks between signaling pathways with various intersection points. This certainly implies that signaling pathways also act in a nonautonomous manner and rely on other pathways to be active.

In this chapter, I describe the components of the wingless and hedgehog signaling pathways as elucidated in *Drosophila* and review the degree of conservation of these signaling pathways in vertebrates, which suggests that these pathways constitute autonomous modules. Recent data from several labs suggest that genetic modules have also conserved function during development and evolution, therefore defining modules of higher degree, which I call "functional modules." To illustrate this notion of "functional module," I will discuss the conserved function of Shh signaling during myogenesis in chordates. Subtle alterations in the biochemical properties of the components of the signaling pathway or in their regulation provide a mechanism for introducing diversity into the temporal and spatial way these genetic modules act in distinct organisms. Examples of these alterations related to the Shh signaling pathway will be provided, and in particular the nature of the interaction between Shh and Wnt signaling pathway in the context of muscle development will be discussed. Finally, I will compare the degree of conservation of Shh-Wnt interaction and cooperation in different tissues and animal models and refer to these observations as nonautonomous characteristics of signaling modules.

The Wingless Signaling Pathway: A Context-Independent Genetic Module?

The wingless (Wg) signaling pathway was first elucidated in *Drosophila* (Nüsslein-Volhard and Wieschaus 1980). Mutations in *wg* have both short- and long-range effects on anteroposterior patterning of the epidermis (Zecca et al. 1996; Neumann and Cohen 1997). Components of the Wg signaling pathway were subsequently characterized because mutants for these proteins either mimicked *wg* mutations or had opposite phenotypes. Wingless is a secreted glycoprotein that activates the Wg signaling pathway following binding to its putative receptor Frizzled 2 (dFz2) (fig. 6.1, *A*) (Bhanot et al. 1996; Zhang and Carthew 1998). *fz* genes encode seven-transmembrane receptor proteins with an extracellular cystein-rich domain (Vinson et al. 1989). In *Drosophila,* only dFz2 mediates Wg signaling through the canonical pathway described here (Cadigan and Nusse 1997), whereas signaling through Fz diverges downstream of Dishevelled (Dsh) and functions in tissue polarity (Shulman et al. 1998; Mlodzik 1999). The mechanism by which dFz2 transduces Wg signals from the membrane to the cytoplasm is not yet known. However, the fact that seven-transmembrane

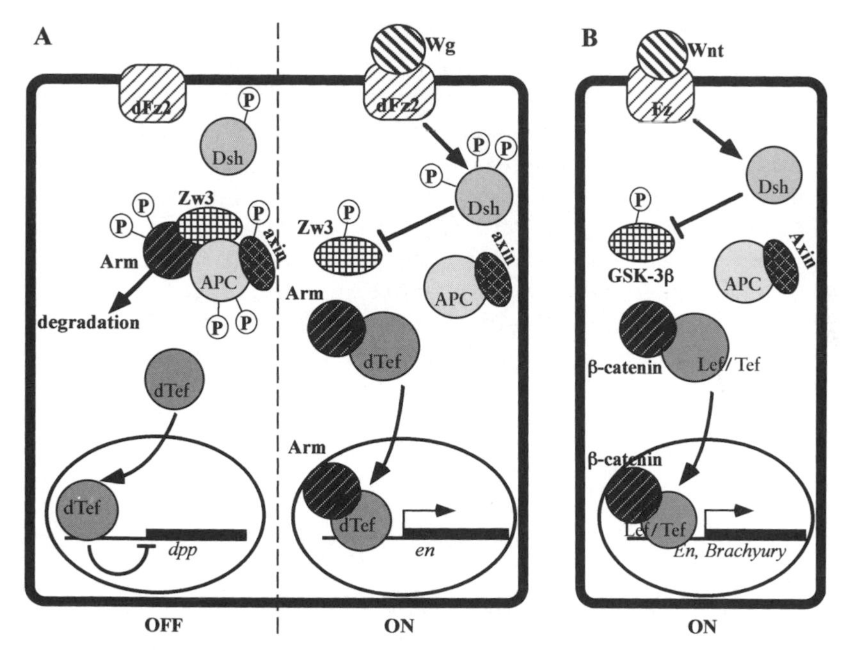

Fig. 6.1.—Conservation of the wingless signaling pathway from invertebrates to vertebrates. Two states characterize the wingless (Wg) signaling pathway. In the absence of Wg (*OFF*), Wg target genes are repressed, whereas in the presence of Wg (*ON*), Wg target genes are activated. Activation of the Wg signaling pathway involves the disruption of a multicomponent cytoplasmic complex, resulting in the nuclear translocation of dTcf. Components of this pathway are conserved from *Drosophila* (*A*) to vertebrates (*B*). Fz, Frizzled; Wg, wingless; Dsh, Dishevelled; Zw3, zeste-white 3; Arm, armadillo; APC, adenomatous polyposis coli; Tcf, T-cell factor; *dpp*, decapentaplegic; *En*, Engrailed; GSK-3*β*, Glycogen synthase kinase 3 beta; P inside circle, phosphorylated site.

domain receptors usually activate G proteins and that G-protein inhibitors block signaling through rat Fz2 may point to the possible involvement of G proteins in dFz2 signaling (Slusarski et al. 1997). Immediately downstream of dFz2 is dishevelled (Dsh), a cytoplasmic phosphoprotein with highly conserved PDZ (*P*SD-95, *D*isc-large, *Z*O-1) and DEP (*D*ishevelled, *E*gl-10, *P*leckstrin) domains, both found in proteins interacting with G proteins (Klingensmith et al. 1994). Wg signaling leads to the hyperphosphorylation of Dsh, which is thought to be mediated by casein kinase II (Willert et al. 1997). Activation of Dsh, directly or indirectly, inhibits Zeste-white 3 (Zw3) (fig. 6.1), which otherwise is constitutively active in the absence of Wg (Siegfried et al. 1992). Again, modulation of Zw3 activity occurs via its phosphorylation on a serine residue (Ruel et al. 1999). Inhibition of Zw3 ultimately results in the stabilization of Armadillo (Arm) (Riggleman et al. 1990), a protein that contains twelve internal repeats in the centre of the protein, several serine residues in its N terminal domain, and a transcriptional activator domain in the C terminal domain (Peifer and Wieschaus 1990). Stabilized Arm proteins accumulate in the cytoplasm, translocate to the nucleus, where they form complexes with the HMG-box transcription factor, dTcf, also called pangolin (Brunner et al. 1997), and activate transcription of target genes. Members of the Lef/Tcf family of proteins (Lymphoid enhancer-binding factor/T-cell factor) are poor transcriptional activators by themselves (Eastman and Grosschedl 1999; Roose and Clevers 1999), but association with Arm greatly increases their transcriptional activity (van de Wetering et al. 1997).

The canonical Wg signaling pathway is characterized by two states, the active state (or ON, described above) and the inactive state (OFF), found in the absence of Wg (fig. 6.1, *A*). In the absence of Wg, hypophosphorylated Dsh does not inhibit Zw3. A multicomponent complex forms that includes Zw3 associated with Arm, the product of the adenomatous polyposis coli gene (APC) and Axin (Hayashi et al. 1997; Hamada et al. 1999). Axin plays a central role within this complex by facilitating Zw3-induced phosphorylation of Arm and APC (Ikeda et al. 1998; Yanagawa et al. 2000), a process that subsequently targets Arm to the ubiquitine-proteasome-mediated degradation pathway (Aberle et al. 1997).

dTcf, in the absence of Arm, enters the nucleus, and, in association with corepressor proteins, represses target gene expression. Several proteins exhibit dTcf corepressor activity. Among them, Groucho binds a central domain of dTcf (Cavallo et al. 1998), the C-terminal-binding protein (CtBP) binds the C-terminal domain of dTcf (Nibu et al. 1998; Roose and Clevers 1999), and the CREB-binding protein (CBP) binds the HMG domain of dTcf (Waltzer and Bienz 1998).

In the presence of Wg, Axin is rapidly dephosphorylated, possibly through the action of protein phosphatase 2A (PP2A) (Seeling et al. 1999). This results in the destabilization of Axin and the decrease of its affinity for Arm (Willert et al. 1999; Yamamoto et al. 1999). Thus, unphosphorylated Arm escapes recognition by Slimb, a component of E3 ubiquitine ligase (Jiang and Struhl 1998).

Components of the Wg signaling pathway are remarkably conserved in *Caenorhabditis elegans* and vertebrates (summarized in table 6.1). Several homologues of Wg have been isolated in vertebrates, and this growing family consists of no fewer than sixteen members, whereas in *C. elegans* five homologues have been isolated, called Ce-Wnt1, Ce-Wnt2, lin-44, mom-2, and egl-20 (Cadigan and Nusse 1997). Similarly, several homologues of the Fz family have been isolated in *C. elegans* (lin-14, mom-5, mig-1) (Sawa et al. 1996; Thorpe et al. 1997), and in vertebrates (11 members) (Wang et al. 1996). Three Dsh homologues have been isolated in *C. elegans* and in vertebrates (Sussman et al. 1994; Klingensmith et al. 1996; Koelle and Horvitz 1996; Tsang et al. 1996). The vertebrate homologue of Zw3 is Glycogen synthase kinase 3 (GSK-3) (Dominguez et al. 1995; He et al. 1995), which, like

Table 6.1: Wnt/Wg signaling pathway in invertebrates and vertebrates

DROSOPHILA	C. ELEGANS	VERTEBRATES
Wg	Mom-2, egl-20, lin-44, Ce-Wnt1, Ce-Wnt2	Wnt
dFfz2	Mom-5, lin-14, mig-1	Fz
Dsh		Dsh
Zw3		GSK-3
APC	APR-1	APC
Axin		Axin
Arm	Wrm-1, Bar-1	β-catenin
dTcf	Pop-1	Lef1/Tcf

Zw3, forms a complex with vertebrate APC and Axin proteins (fig. 6.1, *B*). This complex also includes the homologue of Arm, called β-catenin in vertebrates, and either Bar-1 or Wrm-1, the only β-catenin homologues involved in Wnt signaling out of the three homologues isolated in *C. elegans* (McCrea et al. 1991; Butz et al. 1992; Korswagen et al. 2000). As in *Drosophila,* Wnt signaling in vertebrates results in the stabilization of β-catenin, its translocation to the nucleus, and its binding to the HMG-box transcription factors of the Lef1/Tcf family (fig. 6.1, *B*). In vertebrates, four homologues of dTcf have been characterized, called Lef1, Tcf1, Tcf3, and Tcf4 (Eastman and Grosschedl 1999), whereas in *C. elegans* the homologue of Dtcf is Pop-1 (Lin et al. 1995).

This conservation of the core elements of the Wg signaling pathway in several species of the taxa including *Drosophila, C. elegans, Xenopus,* chick, mouse, and human may therefore define a conserved genetic module. The Wg/Wnt signaling module is abundantly utilized during embryonic development, arguing for the view that it constitutes a relatively autonomous module reused at diverse sites and stages of embryonic development. For instance, in the *Drosophila* embryo, Wg signaling has short- and long-range functions during epidermal cell specification. Its short-range action is required for the production of naked cuticle, whereas its long-range action specifies the pattern of denticle diversity (Zecca et al. 1996; Neumann and Cohen 1997). In the *C. elegans* embryo, Mom-2 signaling is essential for mesodermal cell fate specification associated with the successive cell divisions of the four-cell-stage embryo along the anteroposterior axis (Thorpe et al. 1997). In the *Xenopus* embryo, Wnt activity was first demonstrated because injection in ventral blastomeres induced dorsal mesoderm and axis duplication (Sokol et al. 1991). In the mouse neural tube, Wnt signaling is essential for the maintenance of Engrailed expression at the midbrain-hindbrain boundary (Danielian and McMahon 1996). Targeted mutations in the mouse embryo of *Wnt* genes have identified specific roles for *Wnt1* in midbrain/cerebellum formation (McMahon and Bradley 1990; McMahon et al. 1992), *Wnt3a* in paraxial mesoderm formation (Takada et al. 1994), *Wnt4* in kidney formation (Stark et al. 1994), and *Wnt7a* in limb formation (Parr and McMahon 1995). Together, these observations support the idea that Wnt signaling may function in a context-independent manner for the specification, survival, and growth of various cell types.

The Hedgehog Signaling Pathway: A Context-independent Genetic Module?

The hedgehog (hh) signaling pathway was also first described in *Drosophila,* where mutations in *hh* disrupt the polarity of larval segments, resulting in larvae covered with spiky denticles like a hedgehog (Nüsslein-Volhard and Wieschaus 1980). Hedgehog is synthesized as a precursor protein, which contains a signal peptide, a highly conserved N-terminal domain, and a more divergent C-terminal domain (Hall et al. 1995, 1997). Posttranslational modifications of hh involve an autoproteolytic cleavage, which generates a 19 kDa membrane-bound N-terminal peptide, which is slowly released in the extracellular compartment, and a diffusible C-terminal peptide (Lee et al. 1994; Perler 1998). The C-terminal peptide catalyzes the cleavage of hh and the transfer of cholesterol on the N-terminal peptide (Porter et al. 1996a, 1996b), which alone carries all biological activities (Porter et al. 1995).

In the presence of N-terminal hh peptide, the signaling pathway is activated (ON) (fig. 6.2, *A*). hh binds to Patched (Ptc), a multipass transmembrane protein (Ingham et al. 1991; Marigo et al. 1996a), and relieves the inhibitory action of Ptc on Smoothened (Smo), a seven-pass transmembrane protein that belongs to the family of Frizzled receptors (Alcedo et al. 1996; van den Heuvel and Ingham 1996). Recent data suggest that Ptc normally destabilizes Smo, but in the presence of hh, Ptc is internalized, and Smo, following phosphorylation, stabilizes and accumulates at the cell surface (Denef et al. 2000; Ingham et al. 2000).

In the cytoplasm, hh signaling induces the dissociation of a multicomponent complex from microtubules (Ruiz i Altaba 1997). This cytoplasmic complex of approximately 400–800 kDa contains Costal 2 (Cos2), a novel protein that has homology with the kinesin protein superfamily in its N-terminal domain (Robbins et al. 1997; Sisson et al. 1997), Fused (Fu), a serine/threonine kinase (Therond et al. 1996), and Cubitus interruptus (Ci), a zinc finger transcription factor (Aza-Blanc and Kornberg 1999). Biochemical analyses have shown that Cos2 mediates the association of the complex with microtubules and its retention in the cytoplasm (Robbins et al. 1997; Sisson et al. 1997), whereas Fu is required for the transmission of hh signaling (Therond et al. 1996). Dissociation of the Cos2-Fu-Ci complex from micro-

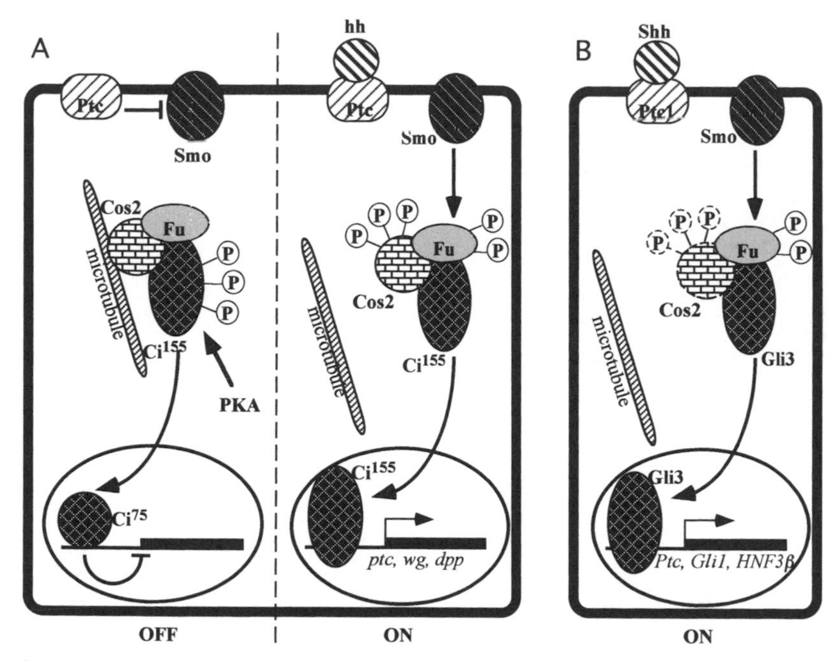

Fig. 6.2.—Conservation of the hedgehog signaling pathway from invertebrates to vertebrates. Two states characterize the hedgehog (hh) pathway, which depend on the presence (*ON*) or the absence (*OFF*) of hh and its binding to the receptor Patched (Ptc). Most components of the hh pathway are conserved from *Drosophila* (*A*) to vertebrates (*B*), except Cos2 (in dashes in vertebrates), which has not yet been isolated in vertebrates. Also conserved is the hh-dependent regulation of Ci proteolytic cleavage in *Drosophila* and Gli3 cleavage in vertebrates, which generates a truncated protein that acts as a transcriptional repressor. Ptc, Patched; Smo, Smoothened; Fu, Fused; Cos2, Costal 2; Ci, Cubitus interruptus; Shh, Sonic hedgehog; *wg*, wingless; *dpp*, decapentaplegic; PKA, protein kinase A; HNF-3β, hepatocyte nuclear factor 3 beta; P inside circle, phosphorylated site.

tubules is followed by the hyperphosphorylation of both Cos2 and Fu and the stabilization of full-length Ci (Ci^{155}), which then translocates to the nucleus and induces transcription of target genes (G. Wang et al. 2000).

In the absence of N-terminal hh peptide, Ptc represses Smo, and the signaling pathway is inactivated (OFF) (fig. 6.2, *A*). The complex Cos2-Fu-Ci is then tightly associated with microtubules, and Ci is proteolytically cleaved into a 75 kDa N-terminal fragment (Ci^{75}) that contains a repressor domain and the zinc finger DNA binding domain but is deprived of the C-terminal activation domain (Aza-Blanc et al. 1997). Ci^{75} translocates to the nucleus, where it represses the transcription of target genes. Cleavage of Ci^{155} into the 75 kDa transcriptional repressor is thought to occur via the proteasome pathway (Maniatis 1999), following phosphorylation of Ci^{155} by Protein Kinase A (PKA) (Price and Kalderon 1999; Wang et al. 1999).

Target genes of hh signaling include *wg*, *dpp*, and *ptc* itself (Hidalgo

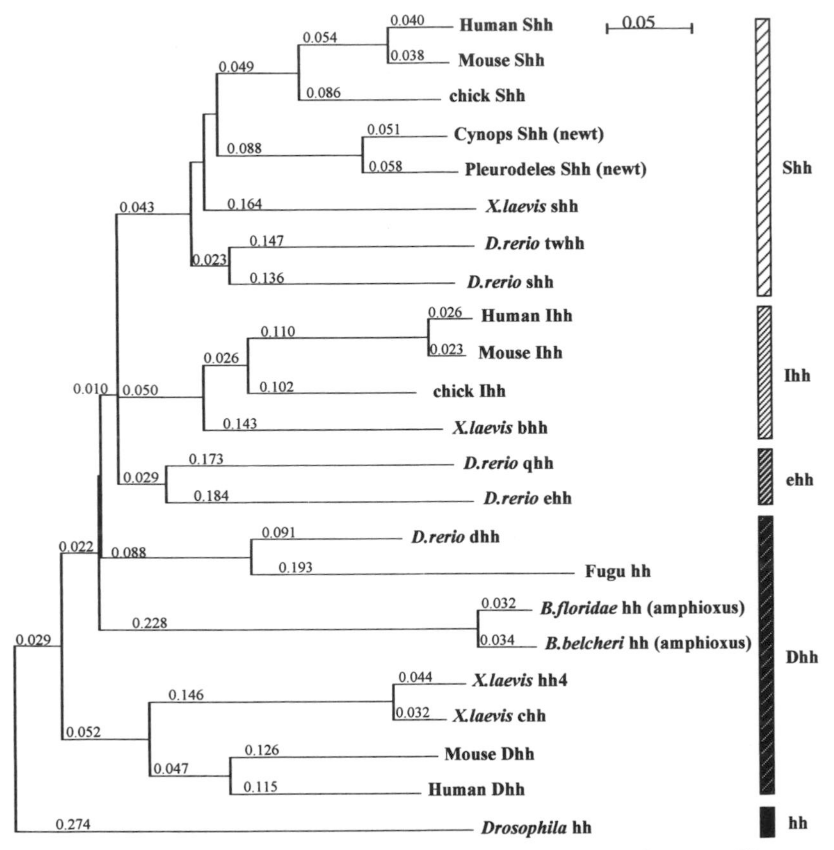

Fig. 6.3.—Neighbor joining tree of the hedgehog family of signaling molecules. Amino acid sequences of hh proteins were aligned using the CLUSTAL multiple alignment program. A neighbor joining tree analysis (http://pbil.univ-lyon1.fr/software/njplot.html) was performed to emphasize the evolutionary relationship between proteins. Branch lengths are proportional to distances. Groups of related proteins are indicated by shaded bars on the right side of the figure. Numbers above branches indicate branch length. Ihh: Indian hedgehog; Shh: Sonic hedgehog; Dhh: Desert hedgehog; bhh: banded hedgehog; chh: cephalic hedgehog; ehh: echina hedgehog; twhh: tiggy-winkle hedgehog; qhh: qiqihar hedgehog.

and Ingham 1990; Basler and Struhl 1994; Tabata and Kornberg 1994). This creates a negative feedback loop where hh signaling locally upregulates *ptc* expression, resulting in the sequestration of free hh molecules, a mechanism that limits the diffusion of hh signaling (Chen and Struhl 1996).

Vertebrate homologues of hh have been isolated in many species (Hammerschmidt et al. 1997). Figure 6.3 shows a unique *hh* gene in *Drosophila* and amphioxus, and several genes in vertebrates, indicating that new members of the family arose by gene duplication immediately after the bifurcation of vertebrates from cephalochordates. *Dhh* (Desert hedgehog) is likely to have arisen from an earlier duplication event, as it appears the most closely related to *Drosophila hh*. A second duplication event, which occurred more recently, generated Shh (Sonic hedgehog) and *Ihh* (Indian hedgehog). Interestingly, the zebrafish *Danio rerio* has two *Shh*-related genes, *shh* and *twhh* (tiggy-winkle hedgehog), and two other *hh*-related genes, *ehh* (echina hedgehog) and *qhh* (qiqihar hedgehog) (P. W. Ingham, unpublished observation), which appear to have arisen before the *Ihh*-related genes, and finally one *Dhh* gene more closely related to amphioxus and *Fugu hh* than to other *Dhh*-related genes. This suggests that an additional duplication event occurred in teleost *hh* genes, a view in agreement with previous observations on *Hox* genes (Aparicio 2000). Therefore, we may expect that additional *hh* genes, related to *Ihh* or *Dhh,* will be identified by the zebrafish genome project, unless the genome duplication was followed by specific *hh* gene loss.

Although most studies have focused on Sonic hedgehog (Shh), it is assumed that other members of the family transduce their signal to the nucleus through the same pathway (fig. 6.2, *B*). Shh and hh share similarities in their structure, and both are activated by autoproteolytic cleavage of a precursor protein. Two vertebrate homologues of Ptc have been described, Ptc1 and Ptc2 (Goodrich et al. 1996; Marigo et al. 1996c; Motoyama et al. 1998). Of the two receptors, Ptc1 appears to have conserved the characteristics of Ptc: (1) it binds Shh; (2) it is not expressed in Shh-producing cells; (3) it is up-regulated in cells in response to Shh signaling (Ingham 1998). Smo and Fu appear to be the unique vertebrate homologues of *Drosophila* Smo and Fu, respectively (Stone et al. 1996; Murone et al. 2000). *Drosophila* Ci has three vertebrate homologues called Gli1, Gli2, and Gli3 (Hui et al. 1994; Marigo et al. 1996b; Marine et al. 1997). Of those, only Gli3 has been shown to be posttranslationally regulated in a Shh-dependent manner via proteolytic cleavage and therefore, like Ci, to act as a repressor or activator (Ruiz i Altaba 1999; Sasaki et al. 1999). No vertebrate homologue of Cos2 has been isolated so far, but it is likely that Gli3 is also involved in a multicomponent complex with Fu and a Cos2 homologue that may tether the complex to microtubules.

Thus, *Drosophila* and vertebrate hedgehog signaling pathways appear remarkably conserved. In addition, both pathways are extensively used during development to trigger a wide range of effects in diverse tissues (table 6.2). These effects range from anteroposterior or dorsoventral patterning to cell fate determination or tissue outgrowth, suggesting that hedgehog signaling acts as a context-independent module. The hh/Shh module is then used in an opportunistic manner by various cells and tissues to induce the expression of transcription factors or signaling molecules essential for cell fate determination, cell growth, or patterning during development.

Table 6.2: Role of hedgehog proteins during *Drosophila* and vertebrate development

Tissue/ Process	Activity	Reference
Drosophila:		
Epidermis	Maintenance of parasegmental borders: Hh, produced by the anterior cell of each segment, maintains *wg* expression in the posterior parasegment border	Lee et al. 1992
Wing disc	Antero-posterior patterning: Hh, produced by posterior cells, induces *dpp* expression at the antero-posterior margin	Tabata and Kornberg 1994
Leg disc	Antero-posterior patterning: Hh, produced by posterior cells, induces *dpp* expression at the dorsal antero-posterior margin and *wg* at the ventral antero-posterior margin	Diaz-Benjumea et al. 1994
Eye disc	Anterior progression of the morphogenetic furrow: Hh, produced by posterior differentiated cells, induces *dpp* expression in the morphogenetic furrow	Heberlein et al. 1993
Vertebrate:		
Left-right	Left-right asymmetry: Shh asymmetrically expressed on the left of the node induces left-specific expression of cNr1, a TGF-β related protein	Levin et al. 1995; Ramsdell and Yost 1998
Neural tube	Dorso-ventral patterning: Shh produced by the notochord induces ventral cell fate (floor plate, motor neuron, and interneuron formation at posterior levels, and dopaminergic and cholinergic neurons at midbrain and hindbrain levels)	Roelink et al. 1994; Jessell 2000
Somite	Dorso-ventral patterning: Shh, produced by the notochord, induces ventral cell fate (*Pax1* in sclerotome) and dorsal cell fate (*Myf5* in epaxial skeletal muscles)	Fan et al. 1995; Munsterberg et al. 1995; Borycki and Emerson 2000
Limb	Antero-posterior and proximo-distal patterning: Shh, produced by the ZPA, induces	Riddle et al. 1993; Ng et al. 1999

Does the Hedgehog Signaling Pathway Define a Conserved Functional Module?

I discussed previously how the signaling cascade activated following the binding of hh/Shh to its receptor has such a degree of conservation that it could be considered an autonomous genetic module. Could this concept be further developed by demonstrating the conserved use of this genetic module in specific aspects of embryonic development? An illustration of this statement is the conserved role of Shh signaling during vertebrate embryonic skeletal muscle development.

In vertebrates, all trunk skeletal muscles originate from somites,

Table 6.2 *continued*

Tissue/ Process	Activity	Reference
	antero-posterior digit identity; it also maintains *Fgf4* expression in the AER, which regulates proximo-distal limb outgrowth	
Eye	Eye-field subdivision: Shh, produced in the forebrain, induces *Pax2* and represses *Pax6* to generate two separate eye fields	Chiang et al. 1996
Gut	Hindgut formation: Shh, expressed in the endoderm, induces visceral mesoderm specification through the induction of *Bmp4* and *HoxD10* expression	Roberts et al. 1995
Pancreas	Pancreas formation and growth, insulin-producing islet number: down-regulation of endoderm Shh expression allows pancreas formation; Shh and Ihh also regulate pancreas growth and endocrine cell number	Hebrok et al. 2000
Lung	Lung growth and morphogenesis: Shh, produced by the lung bud epithelium, induces growth of epithelial and mesenchymal cells of the lung bud	Bellusci et al. 1997
Tooth	Tooth bud growth: Shh, produced by the dental ectoderm, stimulates growth of epithelial and mesenchymal cells of the tooth bud	Mo et al. 1997; Dassule et al. 2000
Hair follicle	Placode growth: Shh, produced by epidermal cells, causes the proliferation of epidermal and mesodermal cells, which are responsible for placode growth	Bitgood and McMahon 1995; St.-Jacques et al. 1998
Long bones	Bone growth: Ihh, produced by prehypertrophic chondrocytes, induces the production of PTHrP in the perichondrium, which blocks differentiation of chondrocytes	Vortkamp et al. 1996
Testis	Spermatogenesis: Dhh, produced by Sertoli s cells, is required for the differentiation of Leydig cells and peritubular cells that support spermatogenesis	Bitgood et al. 1996

which are mesodermal structures that form in an anterior-to-posterior, segmentally arranged manner on either side of the neural tube/notochord complex (Christ and Ordahl 1995). Cells from newly formed somites are pluripotent and become restricted to a specific cell fate under the influence of signaling molecules produced by surrounding tissues (Bumcrot and McMahon 1995; Lassar and Munsterberg 1996; Borycki and Emerson 2000). More specifically, the first skeletal muscle cells to become evident in the dorsomedial domain of the somite of 8.0–8.5-day-old mouse embryos give rise to epaxial muscles, which will form deep back muscles (fig. 6.4). Another well-characterized

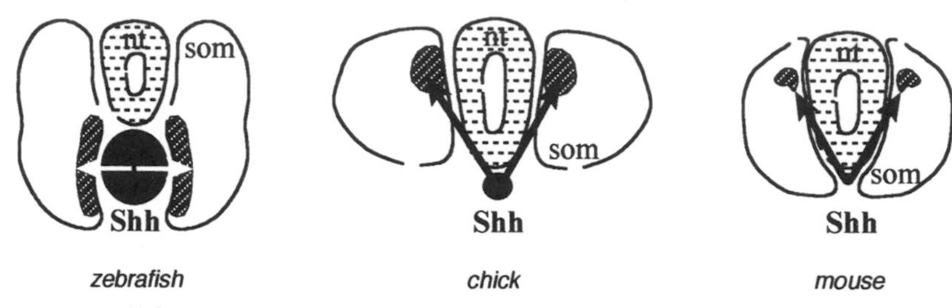

Fig. 6.4.—Shh function during myogenesis is conserved among vertebrates. Shh, produced by the notochord (*dark grey*), specifies mesodermal cells to the skeletal muscle lineage through the induction of *Myf5* and *MyoD* expression in adaxial cells in zebrafish (*grey*) and in epaxial muscle progenitor cells in chick and mouse (*grey*). nt, Neural tube; som: somite.

muscle cell population, which develops two days after the epaxial lineage has formed, gives rise to hypaxial muscles, which will form body wall and limb muscles (Tajbakhsh and Buckingham 2000). Both epaxial and hypaxial muscle progenitor cells enter the myogenic program following the induction of *Myf5* and *MyoD*, two basic helix-loop-helix (bHLH) transcription factors whose function is essential for skeletal muscle formation (Rudnicki et al. 1993). In the mouse embryo, *Myf5* is the first of these transcription factors to be detected in somites (Ott et al. 1991; Tajbakhsh et al. 1996). Targeted inactivation of *Shh* results in the specific loss of *Myf5* induction in epaxial muscle progenitor cells (Borycki et al. 1999), demonstrating that Shh, produced by the adjacent notochord, is required for the determination of somitic cells to the epaxial muscle lineage (fig. 6.4). In contrast, Shh signaling does not appear to be essential for the induction of *Myf5* in hypaxial muscle progenitor cells and for the determination of somitic cells to the hypaxial muscle lineage (Borycki et al. 1999; Kruger et al. 2001).

Noticeably, this essential role of Shh in the determination of the epaxial muscle lineage appears to be conserved among vertebrates (fig. 6.4), as ablation of the Shh-expressing notochord in avian embryos results in the specific loss of epaxial, but not hypaxial, muscles (Rong et al. 1992), and its replacement by Shh-coated beads restores epaxial myogenesis (Borycki et al. 1998).

Shh also plays an essential role in the development of a subpopulation of skeletal muscle cells in zebrafish embryos. The first population of *Myf5*- and *MyoD*-expressing muscle progenitor cells is observed in the paraxial mesoderm of the zebrafish embryo prior to the formation of somites (Weinberg et al. 1996; Chen et al. 2001). The cells of this group are located around the notochord and are called adaxial cells (fig. 6.4). Adaxial cells differentiate into mononucleated slow muscle fibers that migrate to the surface of the somite and into a population of Engrailed-expressing cells that extend from the notochord to the surface of the embryo, called muscle pioneer cells (Devoto et al. 1996). Most of the remaining somitic cells form multinucleated fast muscle fibers. Genetic analyses of zebrafish mutants have shown that mutations in genes of the Shh signaling pathway result in a characteristic phenotype called the U-shaped type, which invariably corresponds to the lack or the reduction of muscle pioneer cells and slow muscle fibers (Blagden et al. 1997; Lewis et al. 1999; Barresi et al. 2000). This phenotype is also associated with the loss of *Myf5* and *MyoD* expression in adaxial cells, but not in somitic fast muscle progenitor cells (Blagden et al. 1997; Lewis et al. 1999; Barresi et al. 2000), indicating that Shh signaling is required for slow but not fast muscle development in zebrafish embryos.

Similarly, *Myf5* and *MyoD* are first expressed in the prospective dorsal mesoderm at the blastula stage of *Xenopus* embryos (Ho and Hanawalt 1991; Hopwood et al. 1991, 1992), and this expression correlates with the mesodermal expression pattern of a member of the FGF family of growth factors, eFGF (Isaacs et al. 1992). In agreement with this observation, eFGF is required in vivo for early mesodermal expression of *Myf5* and *MyoD* (Fisher et al. 2002), and in vitro plays an essential role in the myogenic community effect (Standley et al. 2001). However, Shh signaling appears essential for later expression of *MyoD* and *Myf5* in the paraxial mesoderm of neurula-stage *Xenopus* embryos. Indeed, injection of *Ptc* mRNA on one side of the embryo or treatments with cyclopamine or forskolin, two pharmaceutical compounds that specifically inhibit Shh activity, results in the specific down-regulation of *Myf5* and *MyoD* in the paraxial mesoderm at the neurula stage, without any effect on the early mesodermal expression at the blastula stage (M.-E. Pownall, personal communication).

Together, these data argue for the conservation of Shh function in the determination and the differentiation of one muscle lineage that also correspond to the earliest population of muscle cells observed during embryonic development in all vertebrates examined to date (fig. 6.4). We may therefore define "functional module" as a module of higher degree, which has conserved function in the formation of specific tissues during evolution. For instance, the genetic module "Shh signaling pathway" would define a conserved "functional module" during vertebrate myogenesis. The generalization of this concept of signaling pathways constituting functional modules awaits further analyses of the role of Shh signaling in the formation of other tissues across the phyla. Of immediate interest would be the analysis of the function of Shh signaling in muscle formation in other species, and in particular in the closer urochordates and cephalochordates. Urochordates, such as ascidians, have a large notochord flanked by muscles in the tail region, which consist of a primary and a secondary muscle lineage that differentially express splicing products of a unique myogenic bHLH gene (Meedel et al. 1997). Interestingly, whereas secondary muscle cells require cell-cell interaction for their specification, primary muscle cells are specified autonomously under the control of a maternal cytoplasmic determinant called macho-1, which encodes a zinc finger transcription factor of the Zic/Gli family (Nishida and Sawada 2001). The cephalochordates, such as amphioxus, have two genes of the myogenic bHLH family, and their muscles, lateral to the notochord, are organized into somites (Araki et al. 1996). Despite the existence of a single amphioxus hedgehog gene, which is specifically expressed in the notochord and the ventral neural tube (Shimeld 1999), little is known concerning the mechanisms of specification of muscle cells in amphioxus embryos.

A direct consequence of the conservation of Shh function in the induction of *Myf5* and *MyoD* is the necessary selective pressure during evolution to preserve Gli-binding sites in their regulatory sequences. Direct evidence that Shh signaling induces the transcription of myogenic bHLH genes was facilitated by the characterization of the mouse *Myf5* locus, which has a modular organization of distinct enhancers controlling individual sites of embryonic expression (Hadchouel et al. 2000; Summerbell et al. 2000). Recently, it was shown that a unique Gli-binding site in the *Myf5* epaxial enhancer is necessary for epaxial *Myf5* expression (Gustafsson et al. 2001), demonstrating unequivocally that *Myf5* is a primary target of Shh signaling in epaxial muscle progenitor cells of the mouse embryo. Although still awaiting a direct demonstration, the prediction, based on the conserved function of Shh in *Myf5* and *MyoD* induction in vertebrates, would be that Gli-binding sites are present in the regulatory sequences of *Myf5* and *MyoD* of other vertebrates and that their activity is essential to transduce Shh signals. This observation has interesting implications for the molecular evolution of members of the myogenic bHLH transcription factor family. In vertebrates, there are four members of this family, MyoD, Myf5, Myogenin, and MRF4, which, because of the divergence existing between the two amphioxus and the four vertebrate myogenic bHLH genes, are thought to have derived from two successive duplications after the bifurcation from the cephalochordates (Atchley et al. 1994; Araki et al. 1996). The first duplication event would have generated a common ancestor to *MyoD* and *Myf5* and a common ancestor to *Myogenin* and *MRF4* (Atchley et al. 1994). Gene duplication events involve the duplication of both coding and regulatory regions. Regulatory sequences of duplicate genes may subsequently diverge via a DDC (duplication-degeneration-complementation) mechanism (Force et al. 1999), in which deleterious mutations introduced in enhancers are preserved because of the complementation provided by the duplicate gene. Consistent with this model, the structural organization of mouse *Myf5* regulatory sequences is strikingly different from that of *MyoD* regulatory sequences. Whereas mouse *Myf5* regulatory sequences display a complex organization of about ten enhancers sequentially arranged along 190 kb, the mouse *MyoD* gene contains two unrelated, short enhancers, one proximal and one distal (Goldhamer et al. 1992; Asakura et al. 1995; Hadchouel et al. 2000; Summerbell et al. 2000). However, the avian *MyoD* enhancer shares little homology with the mouse *MyoD* enhancer (Goldhamer et al. 1992; Pinney et al. 1995), and a single 300 bp region upstream of zebrafish *Myf5,* instead of 190 kb in the mouse, is sufficient to drive expression of a reporter construct in zebrafish somites and presomitic mesoderm (Hadchouel et al. 2000; Summerbell et al. 2000; Chen et al. 2001). This suggests that further alterations in the regulatory sequences of *Myf5* and *MyoD* occurred, probably selected during evolution to integrate differences in muscle development among vertebrates. In this context, Shh response elements are to be found displaced and relocalized within these new regulatory regions.

Together, these observations indicate that Shh signaling is a context-independent module with conserved function during vertebrate muscle development. To achieve this degree of functional conservation, despite alterations in regulatory sequences of the myogenic regulatory bHLH genes following their duplication, a selection pressure must be invoked in favour of the maintenance of Shh response elements within otherwise divergent regulatory sequences.

Interaction between Hedgehog and Wingless Signaling Pathways: A Context-Dependent Mechanism

Despite the conservation of the hh signaling pathway module during evolution, there is evidence that diversity was introduced in the hh signaling cascade during evolution through multiple mechanisms. These mechanisms include

1. The cooperation between the hh mediators Ci or Gli with other tissue-specific transcription factors on promoter regions. The multiplicity of *hh* related genes with tissue-specific expression pattern raises the question as to how distinct functions are achieved if all hh proteins transduce their signal via an identical signaling pathway with no more than three Gli transcription factors to transmit these effects. This may require the cooperation of additional transcription factors acting together with Gli on enhancers of target genes to create tissue-specific gene expression.

2. The control of the diffusion of hh via its interaction with extracellular matrix components, such as heparan-sulfate-containing proteoglycans (HSPG). HSPGs were recently shown to be involved in the diffusion of hh via the identification of a *Drosophila* mutant, *tout-velu* (*Ttv*), as the fly homologue of a human gene *EXT* (Exostose), which encodes an enzyme essential for the synthesis of heparan sulfate glycos-

aminoglycans (Bellaiche et al. 1998). *Ttv* is required in the fly for the diffusion of hh away from its source (The et al. 1999). *EXT,* its human homologue, is a mutation involved in hereditary exostoses, a skeletal disorder (Lind et al. 1998), and may be similarly involved in the diffusion of Ihh, as Ihh regulates the growth and differentiation of chondrocytes (Vortkamp et al. 1996). This mechanism is likely to introduce diversity within the same tissue, as, depending on the presence of these extracellular matrix components, cells will receive variable concentrations of hh. This may participate in setting up gradients of hh concentration, which are involved in the determination of distinct cell types along the dorsoventral axis of the neural tube, for instance (Jessell 2000).

3. The specific expression of inhibitory proteins that prevent the interaction ligand-receptor, such as the recently characterized hh-inhibitory protein (HIP), which was isolated based on its high affinity with Shh (Chuang and McMahon 1999).

4. The generation of multiple orthologues of transcription factors that act downstream of the signaling cascade, with distinct biochemical properties. For instance, Gli1, Gli2, and Gli3, the three vertebrate homologues of Ci, differ structurally in the absence of PKA-phosphorylation sites in Gli1, but not in Gli2 or Gli3 (Ruiz i Altaba 1999; B. Wang et al. 2000), rendering Gli1 insensitive to PKA-dependent proteolytic cleavage. Consequently, only Gli2 and Gli3 have retained the ability to act as both transcriptional repressor and activator, whereas Gli1 acts as a transcriptional activator. This difference is likely to have consequences in the overall transcriptional activity of *Gli1*-expressing versus *Gli2/3*-expressing cells at a given time.

5. The temporally and spatially regulated competence of cells to respond to the signaling molecule. This mechanism is widely used to modulate the activity of a signaling pathway during evolution and consists in controlling temporally and spatially the availability of one or several components of the signaling pathway. One such regulation was recently described during the control of somite myogenesis by Shh (Borycki et al. 2000).

Myogenesis in birds and mammals proceeds in tight coordination with the formation of somites, which occurs following segmentation and epithelialization of the presomitic mesoderm. This is intriguing in view of the constitutive production of Shh by the notochord along the anteroposterior axis of the embryo (Marti et al. 1995) and indicates that presomitic mesodermal cells are not competent to respond to Shh signals. Consistent with this observation, *Gli1, Gli2,* and *Gli3* are not expressed in presomitic mesoderm of avian embryos, but are induced at the time somites form (Borycki et al. 2000), suggesting that the competence to respond to Shh is acquired by presomitic

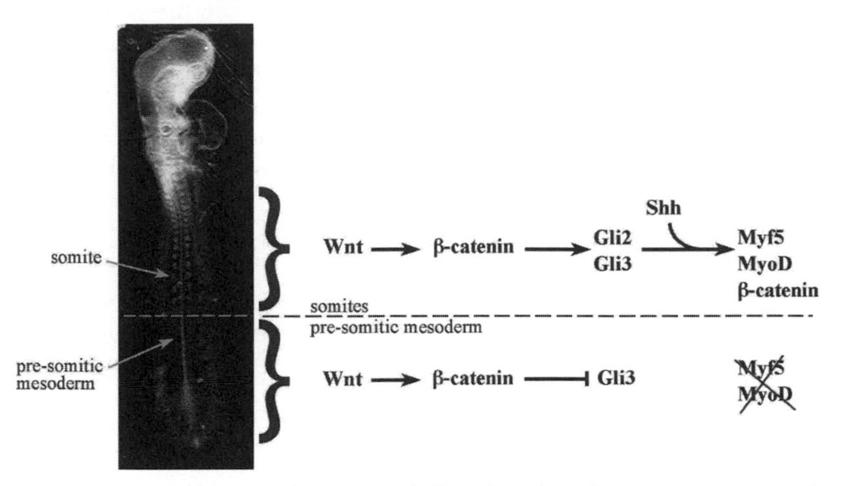

Fig. 6.5.—An example of intersection between Wnt and Shh signaling pathways during somitogenesis. Wnt signaling controls the initiation of Shh response during somite formation by controlling the transcription of *Gli2* and *Gli3,* mediators of Shh signals. Activation of *Gli2* and *Gli3* at the time somites form provides somitic cells with a set of transcription factors to transduce Shh signals, leading to the activation of *Myf5* and *MyoD.* In contrast, Wnt signaling prevents premature activation of the Shh signaling pathway in the paraxial mesoderm by repressing *Gli3* transcription in the presomitic mesoderm.

mesodermal cells with the induction of *Gli* gene expression. In contrast to birds and mammals, myogenesis in frog and fish is not coordinated with somite formation, but is initiated prior to somite formation (Hopwood et al. 1989, 1991; Weinberg et al. 1996), indicating that the competence to respond to Shh signals is acquired much earlier in lower vertebrates, at the presomitic mesoderm level. This observation is consistent with the early expression of *Gli1, Gli2,* and *Gli3* in the presomitic mesoderm of zebrafish and *Xenopus* embryos (Marine et al. 1997).

The mechanism controlling the coordinated induction of *Gli* gene expression with somite formation has been elucidated in avian embryos and involves Wnt-dependent repression of *Gli3* transcription in presomitic mesodermal cells (fig. 6.5) (Borycki et al. 2000). In contrast, acquisition of the competence to respond to Shh at the time somites form involves Wnt-mediated induction of *Gli2* and *Gli3* transcription (fig. 6.5) (Borycki et al. 2000). Both repression of *Gli3* expression in the presomitic mesoderm and induction of *Gli2* and *Gli3* in somites are controlled by the canonical Wnt signaling pathway (see above) (Borycki et al. 2000). This suggests that distinct transcriptional complexes involving Lef/Tcf transcription factors and corepressors mediate the differential control of *Gli* gene expression by Wnt signaling between presomitic mesodermal and somitic cells (fig. 6.5). One prediction from this model is that *Gli* gene expression in *Xenopus* and zebrafish, which is observed in the presomitic mesoderm earlier than in amniotes, would be controlled by distinct mechanisms. If this is true, we would expect that regulatory sequences of lower and higher vertebrate *Gli* genes have evolved differently.

Interestingly, at later stages of avian muscle development, a reverse relationship between Shh and Wnt signaling is observed, where Shh signaling controls the expression of *β*-catenin, a component of the Wnt signaling pathway, in the somitic dermomyotome (fig. 6.5) (Schmidt et al. 2000).

Thus, mechanisms modulating the activity of the conserved Shh signaling pathway have been established during evolution. Although acting at different levels of the signaling cascade, they specifically affect the spatial and temporal regulation of the competence of cells to respond to Shh signals. In some instances, the acquisition of the competence to respond to Shh is controlled by another signaling pathway. This is the case during myogenesis, where a direct interaction between Wnt and Shh signaling pathways has been demonstrated, establishing that these signaling pathways act also in a context-dependent manner, by establishing a network with feedback loops and cross-regulations that affect the activity of the signaling pathway in a tissue-specific manner.

Hedgehog-Wingless Interactions: A Common Theme Redeployed in Diverse Tissues and Organisms

Although evidence for direct transcriptional control of components of one signaling pathway by another signaling pathway is still lacking in other systems, the cooperation between Shh and Wnt signaling has been reported in several cases.

In the *Drosophila* embryo, pair-rule gene expression in stripes prefigures the anteroposterior organization of epidermal segments. Segment polarity genes, such as *hh* and *wg,* then act to refine the anteroposterior patterning of epidermal segments. Specifically, *hh* and *wg* are involved in the maintenance of the parasegmental border. A parasegment is defined as the posterior domain of one segment and the anterior domain of the next segment (Martinez-Arias and Lawrence 1985). Hh is expressed in cells posterior to the border, whereas Wg is expressed in cells anterior to the border (fig. 6.6, *A*) (van den Heuvel et al. 1989; Mohler and Vani 1992). Wg induces the transcription of *en* in posterior cells, which results in the secretion of Hh (Martinez-Arias et al. 1988; Heemskerk et al. 1991; Lee et al. 1992; Tabata et al. 1992). Hh acts on anterior cells to maintain *wg* expression (Lee et al. 1992), therefore defining a feedback loop where Hh and Wg mutually regulate each other.

A similar relationship between Hh and Wg exists in the leg imaginal disc (fig. 6.6, *B*). Hh is expressed in cells of the posterior compart-

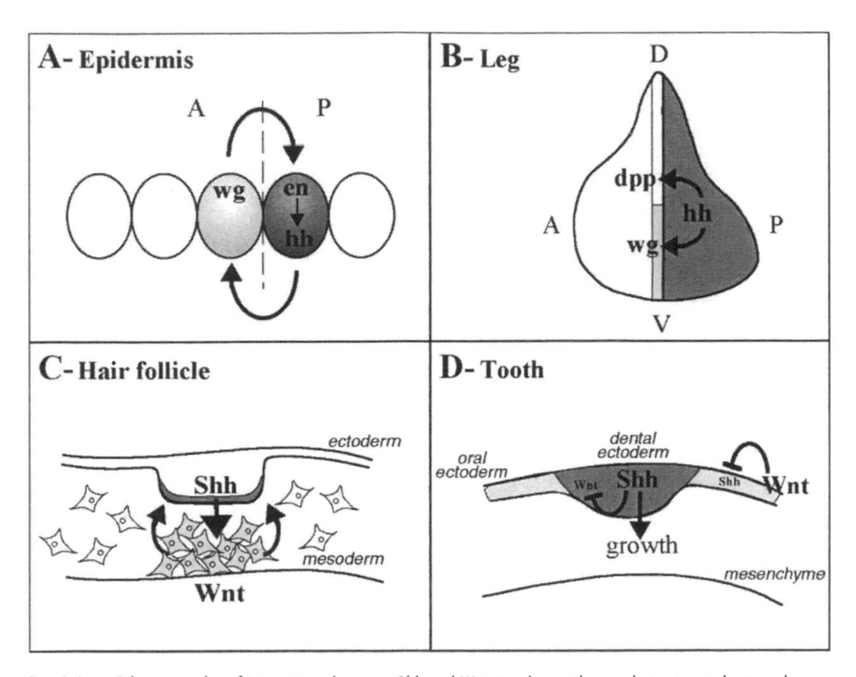

Fig. 6.6.—Other examples of interactions between Shh and Wnt signaling pathways during invertebrate and vertebrate development. *A,* Wg and Hh function during anteroposterior patterning of the *Drosophila* epidermis. *B,* Wg and Hh function during anteroposterior patterning of the *Drosophila* leg imaginal disc. *C,* Wnt and Shh function during hair follicle formation. *D,* Wnt and Shh function during tooth development. For a detailed explanation, see the text. Hh, Shh expressing compartments are indicated in dark grey, and Wg/Wnt expressing compartments are indicated in light grey. En, Engrailed; wg, wingless; dpp, decapentaplegic; hh, hedgehog.

ment of the leg imaginal disc and induces the production of Wg in ventral cells of the anteroposterior border of the disc (Diaz-Benjumea et al. 1994).

In vertebrates, Wnt and Shh signaling are associated with multiple processes that involve epithelial-mesenchymal cell interaction such as hair follicle and tooth bud development. Hair follicle development is a multistep process that involves first the formation of placodes regularly spaced on the skin and second the differentiation of placodes into hair follicles (fig. 6.6, C). Placodes are formed following the induction by mesodermal cells on the overlying ectoderm to produce a signal that stimulates the condensation of mesodermal cells to form a placode (Hardy 1992). Further stimulation of epidermal cells on themselves as well as on mesodermal cells results in the growth of the placode. Later, epithelial cells move away from the mesoderm and begin to differentiate into hair follicle. Shh is expressed by epithelial cells following placode formation as they invaginate into the dermis, whereas Wnt is produced by mesodermal cells (Bitgood and McMahon 1995; Saitoh et al. 1998; St.-Jacques et al. 1998; Millar et al. 1999). Evidence for a possible interaction of both signaling pathways comes from the discovery that overexpression of β-catenin or overexpression of Shh or mutation in Ptc results in an abnormal increase in hair follicle number and the formation of skin tumors (Dahmane et al. 1997; Oro et al. 1997; Gat et al. 1998). However, the basal cell carcinomas caused by Shh overexpression differ from the follicular tumors caused by β-catenin overexpression (Oro and Scott 1998), indicating that each signaling pathway may have additional independent functions in hair follicle formation.

Like hair follicle development, tooth development initiates with a thickening of the oral ectoderm to form the dental ectoderm and continues with a subsequent invagination into the underlying mesenchyme. This process is correlated with the condensation of the mesenchyme and the formation of a tooth bud (Thesleff et al. 1995) (fig. 6.6, *D*). Shh is specifically expressed by epithelial cells from the dental ectoderm, whereas *Wnt* genes are expressed by epithelial cells of the surrounding oral ectoderm (Dassule and McMahon 1998; Sarkar and Sharpe 1999; Dassule et al. 2000). In this context, Wnt signaling controls the boundary between oral and dental ectoderm by down-regulating Shh expression in the oral ectoderm (Sarkar et al. 2000).

Although all these examples await the formal demonstration that either Gli or Lef/Tcf proteins control the expression of Wnt or Shh themselves or one component of their signaling pathways, they are consistent with the idea that Wnt-Shh interactions are largely used throughout development and during evolution. Examination of the processes involving Wnt and Shh reported in the literature points to a common theme, which is the establishment or maintenance of boundaries and/or epithelial-mesenchymal interactions leading to a change in cell fate. This indicates that Wnt and Shh signaling may be part of a network that is used in some precise aspects of embryonic development.

Conclusion

With the study of multiple developmental organisms, the remarkable conservation of entire signaling pathways across the phyla became evident. This led to the consideration that signaling pathways are autonomous genetic modules, which have been maintained during evolution and redeployed at diverse sites during embryogenesis. Although such modules may appear at first to introduce constraints and therefore to prevent evolutionary diversification, detailed analysis of signaling pathways involved in fundamental developmental processes has emphasized that instead, modules constitute an ideal framework to introduce controlled variability in order to assign new functions to a signaling pathway. A direct implication of this is the necessary selective pressure, following the alteration of one element of the module during evolution, to initiate complementing modifications on downstream components of the signaling pathway in order to preserve the cascade of interactions that characterizes the module. The difficulty of achieving such coordination in the generation of variability at the protein level is overcome in many instances by the use of gene and/or genome duplication as a means of generating diversity. This mechanism was largely employed in the case of Wg and Hh signaling pathways, as attested by the large number of paralogues of each of these signaling molecules identified in vertebrates. Following the duplication event, variability was introduced in duplicated components of the Hh and Wg signaling pathway either via the partitioning of ancestral functions between members of the newly formed family (i.e., the dual activator/repressor function of *Drosophila* Ci shared between vertebrate Gli1 and Gli3), or via the alteration of regulatory sequences in order to generate new embryonic expression sites. The former mechanism can be elegantly demonstrated in knock-in experiments where a family member gene is inserted into the locus of its paralogue to assess its ability to rescue the mutant phenotype. Although such experiments have not been performed for members of the Gli family or the Hh family yet, they proved to be invaluable in identifying divergent functions among members of the MyoD family of myogenic regulatory factors by showing that defects were only partially rescued (Wang and Jaenisch 1997). The latter mechanism is illustrated by the adaptability of genetic modules in order to reiterate a cascade of gene activation in a new location, implying that changes in cis or trans regulation must occur to integrate transcriptional regulators provided by the new environment. I presented evidence in this review that additional regulatory mechanisms control temporally and spatially the competence of cells and tissues to respond to signaling molecules and are differently used among vertebrates to control the timing of Shh response in the paraxial mesoderm. Perhaps this is the most widely used mechanism to control and introduce versatility in genetic modules during evolution. This regulation delineates how, despite their apparent autonomy, genetic modules are subject to and part of a larger signaling network, which remodels them into context-dependent modules.

Acknowledgments

Work in my laboratory is supported by the Wellcome Trust, the BBSRC, and the Royal Society. I would like to thank Hilary Ashe and Jorge Caamaño for their critical reading of this manuscript. I am grateful to Philip Ingham, Mary-Elizabeth Pownall, and Charles Emerson for sharing sequences and results prior to publication.

References

Aberle, H., A. Bauer, J. Stappert, A. Kispert, and R. Kemler. 1997. beta-catenin is a target for the ubiquitin-proteasome pathway. *EMBO J.* 16:3797–3804.

Alcedo, J., M. Ayzenzon, T. Von Ohlen, M. Noll, and J. E. Hooper. 1996. The *Drosophila* smoothened gene encodes a seven-pass membrane protein, a putative receptor for the hedgehog signal. *Cell* 86:221–232.

Aparicio, S. 2000. Vertebrate evolution: recent perspective from fish. *Trends Genet.* 16: 54–56.

Araki, I., K. Terazawa, and N. Satoh. 1996. Duplication of an amphioxus myogenic bHLH gene is independent of vertebrate myogenic bHLH gene duplication. *Gene* 171:231–236.

Asakura, A., G. E. Lyons, and S. J. Tapscott. 1995. The regulation of MyoD gene expression: conserved elements mediate expression in embronic axial muscles. *Dev. Biol.* 171:386–398.

Atchley, W. R., W. M. Fitch, and M. Bronner-Fraser. 1994. Molecular evolution of the MyoD family of transcription factors. *Proc. Natl. Acad. Sci. U.S.A.* 91:11522–11526.

Aza-Blanc, P., and T. B. Kornberg. 1999. Ci: a complex transducer of the hedgehog signal. *Trends Genet.* 15:458–462.

Aza-Blanc, P., F. A. Ramirez-Weber, M. P. Laget, C. Schwartz, and T B. Kornberg. 1997. Proteolysis that is inhibited by hedgehog targets Cubitus interruptus protein to the nucleus and converts it to a repressor. *Cell* 89:1043–1053.

Barresi, M. J., H. L. Stickney, and S. H. Devoto. 2000. The zebrafish slow-muscle-omitted gene product is required for Hedgehog signal transduction and the development of slow muscle identity. *Development* 127:2189–2199.

Basler, K., and G. Struhl. 1994. Compartment boundaries and the control of *Drosophila* limb pattern by hedgehog protein. *Nature* 368:208–214.

Bellaiche, Y., I. The, and N. Perrimon. 1998. Tout-velu is a *Drosophila* homologue of the putative tumour suppressor EXT-1 and is needed for Hh diffusion. *Nature* 394: 85–88.

Bellusci, S., Y. Furuta, M. G. Rush, R. Henderson, G. Winnier, and B. L. Hogan. 1997. Involvement of Sonic hedgehog (Shh) in mouse embryonic lung growth and morphogenesis. *Development* 124:53–63.

Bhanot, P., M. Brink, C. H. Samos, J. C. Hsieh, Y. Wang, J. P. Macke, D. Andrew, J. Nathans, and R. Nusse. 1996. A new member of the frizzled family from *Drosophila* functions as a Wingless receptor. *Nature* 382:225–230.

Bitgood, M. J., and A. P. McMahon. 1995. Hedgehog and Bmp genes are coexpressed at many diverse sites of cell-cell interaction in the mouse embryo. *Dev. Biol.* 172: 126–138.

Bitgood, M. J., L. Shen, and A. P. McMahon. 1996. Sertoli cell signaling by Desert hedgehog regulates the male germline. *Curr. Biol.* 6:298–304.

Blagden, C. S., P. D. Currie, P. W. Ingham, and S. M. Hughes. 1997. Notochord induction of zebrafish slow muscle mediated by Sonic hedgehog. *Genes Dev.* 11: 2163–2175.

Borycki, A.-G., A. M. C. Brown, and C. P. Emerson, Jr. 2000. Shh and Wnt signaling pathways converge to control Gli gene activation in avian somites. *Development* 127:2075–2087.

Borycki, A.-B., B. Brunk, S. Tajbakhsh, M. Buckingham, C. Chiang, and C. P. Emerson, Jr. 1999. Sonic hedgehog controls epaxial muscle determination through Myf5 activation. *Development* 126:4053–4063.

Borycki, A.-G., and C. P. Emerson, Jr. 2000. Multiple tissue interactions and signal transduction pathways control somite myogenesis. In *Somitogenesis,* ed. C. Ordahl, 48:165–224. San Diego, Academic Press.

Borycki, A.-G., L. Mendham, and C. P. Emerson, Jr. 1998. Control of somite patterning by Sonic hedgehog and its downstream signal response genes. *Development* 125: 777–790.

Brunner, E., O. Peter, L. Schweizer, and K. Basler. 1997. Pangolin encodes a Lef-1 homologue that acts downstream of Armadillo to transduce the Wingless signal in *Drosophila. Nature* 385:829–833.

Bumcrot, D. A., and A. P. McMahon. 1995. Somite differentiation: Sonic signals somites. *Curr. Biol.* 5:612–614.

Butz, S., J. Stappert, H. Weissig, and R. Kemler. 1992. Plakoglobin and beta-catenin: distinct but closely related. *Science* 257:1142–1144.

Cadigan, K. M., and R. Nusse. 1997. Wnt signaling: a common theme in animal development. *Genes Dev.* 11:3286–3305.

Cavallo, R. A., R. T. Cox, M. M. Moline, J. Roose, G. A. Polevoy, H. Clevers, M. Peifer, and A. Bejsovec. 1998. *Drosophila* Tcf and Groucho interact to repress Wingless signalling activity. *Nature* 395:604–608.

Chen, Y., and G. Struhl. 1996. Dual roles for patched in sequestring and transducing hedgehog. *Cell* 87:553–563.

Chen, Y. H., W. C. Lee, C. F. Liu, and H. J. Tsai. 2001. Molecular structure, dynamic expression, and promoter analysis of zebrafish (*Danio rerio*) myf-5 gene. *Genesis* 29: 22–35.

Chiang, C., Y. Litingtung, E. Lee, K. E. Young, J. L. Corden, H. Westphal, and P. A. Beachy. 1996. Cyclopia and defective axial patterning in mice lacking sonic hedgehog gene function. *Nature* 383:407–413.

Christ, B., and C. P. Ordahl. 1995. Early stages of chick somite development. *Anat. Embryol.* 191:381–396.

Chuang, P. T., and A. P. McMahon. 1999. Vertebrate Hedgehog signalling modulated by induction of a Hedgehog-binding protein. *Nature* 397:617–621.

Dahmane, N., J. Lee, P. Robins, P. Heller, and A. Ruiz i Altaba. 1997. Activation of the transcription factor Gli1 and the Sonic hedgehog signalling pathway in skin tumours. *Nature* 389:876–881.

Danielian, P. S., and A. P. McMahon. 1996. Engrailed-1 as a target of the Wnt-1 signalling pathway in vertebrate midbrain development. *Nature* 383:332–334.

Dassule, H. R., P. Lewis, M. Bei, R. Maas, and A. P. McMahon. 2000. Sonic hedgehog regulates growth and morphogenesis of the tooth. *Development* 127:4775–4785.

Dassule, H. R., and A. P. McMahon. 1998. Analysis of epithelial-mesenchymal interactions in the initial morphogenesis of the mammalian tooth. *Dev. Biol.* 202:215–227.

Denef, N., D. Neubüser, L. Perez, and S. M. Cohen. 2000. Hedgehog induces opposite changes in turnover and subcellular localization of patched and smoothened. *Cell* 102:521–531.

Devoto, S. H., E. Melancon, J. S. Eisen, and M. Westerfield. 1996. Identification of separate slow and fast muscle precursor cells in vivo, prior to somite formation. *Development* 122:3371–3380.

Diaz-Benjumea, F. J., B. Cohen, and S. M. Cohen. 1994. Cell interaction between compartments establishes the proximal-distal axis of *Drosophila* legs. *Nature* 372: 175–179.

Dominguez, I., K. Itoh, and S. Y. Sokol. 1995. Role of glycogen synthase kinase 3 beta as a negative regulator of dorsoventral axis formation in *Xenopus* embryos. *Proc. Natl. Acad. Sci. U.S.A.* 92:8498–8502.

Eastman, Q., and R. Grosschedl. 1999. Regulation of LEF-1/TCF transcription factors by Wnt and other signals. *Curr. Opin. Cell Biol.* 11:233–240.

Fan, C. M., J. A. Porter, C. Chiang, D. T. Chang, P. A. Beachy, and M. Tessier-Lavigne. 1995. Long-range sclerotome induction by sonic hedgehog: direct role of the amino-terminal cleavage product and modulation by the cyclic AMP signaling pathway. *Cell* 81:457–465.

Fisher, M. E., H. V. Isaacs, and M. E. Pownall. 2002. eFGF is required for activation of XmyoD expression in the myogenic cell lineage of *Xenopus laevis*. *Development* 129:1307–1315.

Force, A., M. Lynch, F. B. Pickett, A. Amores, Y. L. Yan, and J. Postlethwait. 1999. Preservation of duplicate genes by complementary, degenerative mutations. *Genetics* 151:1531–1545.

Gat, U., R. DasGupta, L. Degenstein, and E. Fuchs. 1998. De novo hair follicle morphogenesis and hair tumors in mice expressing a truncated beta-catenin in skin. *Cell* 95: 605–614.

Goldhamer, D. J., A. Faerman, M. Shani, and C. P. Emerson, Jr. 1992. Regulatory elements that control the lineage-specific expression of myoD. *Science* 256:538–542.

Goodrich, L. V., R. L. Johnson, L. Milenkovic, J. McMahon, and M. Scott. 1996. Conservation of the hedgehog/patched signaling pathway from flies to mice: induction of a mouse patched gene by hedgehog. *Genes Dev.* 10:301–312.

Gustafsson, M. K., H. Pan, D. F. Pinney, Y. Liu, A. Lewandowski, D. J. Epstein, and C. P. Emerson, Jr. 2001. Myf5 is a direct target of long-range Shh signaling and Gli regulation for muscle specification. *Genes Dev.* 16:114–126.

Hadchouel, J., S. Tajbakhsh, M. Primig, T. H. Chang, P. Daubas, D. Rocancourt, and M. Buckingham. 2000. Modular long-range regulation of Myf5 reveals unexpected heterogeneity between skeletal muscles in the mouse embryo. *Development* 127:4455–4467.

Hall, T. M., J. A. Porter, P. A. Beachy, and D. J. Leahy. 1995. A potential catalytic site revealed by the 1.7-A crystal structure of the amino-terminal signalling domain of Sonic hedgehog. *Nature* 378:212–216.

Hall, T. M., J. A. Porter, K. E. Young, E. V. Koonin, P. A. Beachy, and D. J. Leahy. 1997. Crystal structure of a Hedgehog autoprocessing domain: homology between Hedgehog and self-splicing proteins. *Cell* 91:85–97.

Hamada, F., Y. Tomoyasu, Y. Takatsu, M. Nakamura, S. Nagai, A. Suzuki, F. Fujita, M. Shibuya, H. Toyoshima, N. Ueno, and T. Akiyama. 1999. Negative regulation of Wingless signaling by D-axin, a *Drosophila* homolog of axin. *Science* 283:1739–1742.

Hammerschmidt, M., A. Brook, and A. P. McMahon. 1997. The world according to hedgehog. *Trends Genet.* 13:14–21.

Hardy, M. H. 1992. The secret life of the hair follicle. *Trends Genet.* 8:55–61.

Hayashi, S., B. Rubinfeld, B. Souza, P. Polakis, E. Wieschaus, and A. J. Levine. 1997. A *Drosophila* homolog of the tumor suppressor gene adenomatous polyposis coli down-regulates beta-catenin but its zygotic expression is not essential for the regulation of Armadillo. *Proc. Natl. Acad. Sci. U.S.A.* 94:242–247.

He, X., J. P. Saint-Jeannet, J. R. Woodgett, H. E. Varmus, and I. B. Dawid. 1995. Glycogen synthase kinase-3 and dorsoventral patterning in *Xenopus* embryos. *Nature* 374: 617–622.

Heberlein, U., T. Wolff, and G. M. Rubin. 1993. The TGF beta homolog dpp and the segment polarity gene hedgehog are required for propagation of a morphogenetic wave in the *Drosophila* retina. *Cell* 75:913–926.

Hebrok, M., S. K. Kim, B. St. Jacques, A. P. McMahon, and D. A. Melton. 2000. Regulation of pancreas development by hedgehog signaling. *Development* 127: 4905–4913.

Heemskerk, J., S. DiNardo, R. Kostriken, and P. H. O'Farrell. 1991. Multiple modes of engrailed regulation in the progression towards cell fate determination. *Nature* 352: 404–410.

Hidalgo, A., and P. Ingham. 1990. Cell patterning in the *Drosophila* segment: spatial regulation of the segment polarity gene patched. *Development* 110:291–301.

Ho, L., and P. C. Hanawalt. 1991. Gene-specific DNA repair in terminally differentiating rat myoblasts. *Mutat. Res.* 255:123–141.

Hopwood, N. D., A. Pluck, and J. B. Gurdon. 1989. MyoD expression in the forming somites is an early response to mesoderm induction in *Xenopus* embryos. *EMBO J.* 8:3409–3417.

Hopwood, N. D., A. Pluck, and J. B. Gurdon. 1991. *Xenopus* Myf-5 marks early muscle cells and can activate muscle genes ectopically in early embryos. *Development* 111: 551–560.

Hopwood, N. D., A. Pluck, J. B. Gurdon, and S. M. Dilworth. 1992. Expression of XMyoD protein in early *Xenopus laevis* embryos. *Development* 114:31–38.

Hui, C. C., D. Slusarski, K. A. Platt, R. Holmgren, and A. L. Joyner. 1994. Expression of three mouse homologs of the *Drosophila* segment polarity gene cubitus interruptus, Gli, Gli-2, and Gli-3, in ectoderm- and mesoderm-derived tissues suggests multiple roles during postimplantation development. *Dev. Biol.* 162:402–413.

Ikeda, S., S. Kishida, H. Yamamoto, H. Murai, S. Koyama, and A. Kikuchi. 1998. Axin, a negative regulator of the Wnt signaling pathway, forms a complex with GSK-3beta and beta-catenin and promotes GSK-3beta-dependent phosphorylation of beta-catenin. *EMBO J.* 17:1371–1384.

Ingham, P. W. 1998. The patched gene in development and cancer. *Curr. Opin. Genet. Dev.* 8:88–94.

Ingham, P. W., S. Nystedt, Y. Nakano, W. Brown, D. Stark, M. van den Heuvel, and A. M. Taylor. 2000. Patched represses the Hedgehog signalling pathway by promoting modification of the Smoothened protein. *Curr. Biol.* 10:1315–1318.

Ingham, P. W., A. M. Taylor, and Y. Nakano. 1991. Role of the *Drosophila* patched gene in positional signalling. *Nature* 353:184–187.

Isaacs, H. V., D. Tannahill, and J. M. Slack. 1992. Expression of a novel FGF in the *Xenopus* embryo: a new candidate inducing factor for mesoderm formation and anteroposterior specification. *Development* 114:711–720.

Jessell, T. M. 2000. Neuronal specification in the spinal cord: inductive signals and transcriptional codes. *Nat. Rev. Genet.* 1:20–29.

Jiang, J., and G. Struhl. 1998. Regulation of the Hedgehog and Wingless signalling pathways by the F-box/WD40-repeat protein Slimb. *Nature* 391:493–496.

Klingensmith, J., R. Nusse, and N. Perrimon. 1994. The *Drosophila* segment polarity gene dishevelled encodes a novel protein required for response to the wingless signal. *Genes Dev.* 8:118–130.

Klingensmith, J., Y. Yang, J. D. Axelrod, D. R. Beier, N. Perrimon, and D. Sussman. 1996. Conservation of dishevelled structure and function between flies and mice: isolation and characterization of Dvl2. *Mech. Dev.* 58:15–26.

Koelle, M. R., and H. R. Horvitz. 1996. EGL-10 regulates G protein signaling in the *C. elegans* nervous system and shares a conserved domain with many mammalian proteins. *Cell* 84:115–125.

Korswagen, H. C., M. A. Herman, and H. C. Clevers. 2000. Distinct beta-catenins mediate adhesion and signalling functions in *C. elegans*. *Nature* 406:527–532.

Kruger, M., D. Mennerich, S. Fees, R. Schafer, S. Mundlos, and T. Braun. 2001. Sonic hedgehog is a survival factor for hypaxial muscles during mouse development. *Development* 128:743–752.

Lassar, A. B., and A. E. Munsterberg. 1996. The role of positive and negative signals in somite patterning. *Curr. Opin. Neurobiol.* 6:57–63.

Lee, J. J., S. C. Ekker, D. P. von Kessler, J. A. Porter, B. I. Sun, and P. A. Beachy. 1994. Autoproteolysis in hedgehog protein biogenesis. *Science* 266:1528–1537.

Lee, J. J., D. P. von Kessler, S. Parks, and P. A. Beachy. 1992. Secretion and localized transcription suggest a role in positional signaling for products of the segmentation gene hedgehog. *Cell* 71:33–50.

Levin, M., R. L. Johnson, C. D. Stern, M. Kuehn, and C. Tabin. 1995. A molecular pathway determining left-right asymmetry in chick embryogenesis. *Cell* 82:803–814.

Lewis, K. E., P. D. Currie, S. Roy, H. Schauerte, P. Haffter, and P. W. Ingham. 1999. Control of muscle cell-type specification in the zebrafish embryo by Hedgehog signalling. *Dev. Biol.* 216:469–480.

Lin, R., S. Thompson, and J. R. Priess. 1995. pop-1 encodes an HMG box protein required for the specification of a mesoderm precursor in early *C. elegans* embryos. *Cell* 83:599–609.

Lind, T., F. Tufaro, C. McCormick, U. Lindahl, and K. Lidholt. 1998. The putative tumor suppressors EXT1 and EXT2 are glycosyltransferases required for the biosynthesis of heparan sulfate. *J. Biol. Chem.* 273:26265–26268.

Maniatis, T. 1999. A ubiquitin ligase complex essential for the NF-kappaB, Wnt/Wingless, and Hedgehog signaling pathways. *Genes Dev.* 13:505–510.

Marigo, V., R. A. Davey, Y. Zuo, J. M. Cunningham, and C. J. Tabin. 1996a. Biochemical evidence that patched is the Hedgehog receptor. *Nature* 384:176–179.

Marigo, V., R. L. Johnson, A. Vortkamp, and C. J. Tabin. 1996b. Sonic hedgehog differentially regulates expression of GLI and GLI3 during limb development. *Dev. Biol.* 180:273–283.

Marigo, V., M. P. Scott, R. L. Johnson, L. V. Goodrich, and C. J. Tabin. 1996c. Conservation in hedgehog signaling: induction of a chicken patched homolog by Sonic hedgehog in the developing limb. *Development* 122:1225–1233.

Marine, J.-C., E. J. Bellefoid, H. Pendeville, J. A. Martial, and T. Pieler. 1997. A role for *Xenopus* Gli-type zinc-finger proteins in the early embbryonic paterning of the mesoderm and neuroectoderm. *Mech. Dev.* 63:211–225.

Marti, E., R. Takada, D. A. Bumcrot, H. Sasaki, and A. P. McMahon. 1995. Distribution of Sonic hedgehog peptides in the developing chick and mouse embryo. *Development* 121:2537–2547.

Martinez-Arias, A., N. E. Baker, and P. W. Ingham. 1988. Role of segment polarity genes in the definition and maintenance of cell states in the *Drosophila* embryo. *Development* 103:157–170.

Martinez-Arias, A., and P. A. Lawrence. 1985. Parasegments and compartments in the *Drosophila* embryo. *Nature* 313:639–642.

McCrea, P. D., C. W. Turck, and B. Gumbiner. 1991. A homolog of the armadillo protein in *Drosophila* (plakoglobin) associated with E-cadherin. *Science* 254:1359–1361.

McMahon, A. P., and A. Bradley. 1990. The Wnt-1 (int-1) proto-oncogene is required for development of a large region of the mouse brain. *Cell* 62:1073–1085.

McMahon, A. P., A. L. Joyner, A. Bradley, and J. A. McMahon. 1992. The midbrain-hindbrain phenotype of Wnt-1-/Wnt-1- mice results from stepwise deletion of engrailed-expressing cells by 9.5 days postcoitum. *Cell* 69:581–595.

Meedel, T. H., S. C. Farmer, and J. J. Lee. 1997. The single MyoD family gene of *Ciona intestinalis* encodes two differentially expressed proteins: implications for the evolution of chordate muscle gene regulation. *Development* 124:1711–1721.

Millar, S. E., K. Willert, P. C. Salinas, H. Roelink, R. Nusse, D. J. Sussman, and G. S. Barsh. 1999. WNT signaling in the control of hair growth and structure. *Dev. Biol.* 207:133–149.

Mlodzik, M. 1999. Planar polarity in the *Drosophila* eye: a multifaceted view of signaling specificity and cross-talk. *EMBO J.* 18:6873–6879.

Mo, R., A. M. Freer, D. L. Zinyk, M. A. Crackower, J. Michaud, H. H. Heng, K. W. Chik, X. M. Shi, L. C. Tsui, S. H. Cheng, A. L. Joyner, and C. Hui. 1997. Specific and redundant functions of Gli2 and Gli3 zinc finger genes in skeletal patterning and development. *Development* 124:113–123.

Mohler, J., and K. Vani. 1992. Molecular organization and embryonic expression of the hedgehog gene involved in cell-cell communication in segmental patterning of *Drosophila. Development* 115:957–971.

Motoyama, J., T. Takabatake, K. Takeshima, and C. Hui. 1998. Ptch2, a second mouse Patched gene is co-expressed with Sonic hedgehog. *Nat. Genet.* 18:104–106.

Munsterberg, A. E., J. Kitajewski, D. A. Bumcrot, A. P. McMahon, and A. B. Lassar. 1995. Combinatorial signaling by sonic hedgehog and wnt family members induces myogenic bHLH gene expression in the somite. *Genes Dev.* 9:2911–2922.

Murone, M., S. M. Luoh, D. Stone, W. Li, A. Gurney, M. Armanini, C. Grey, A. Rosenthal, and F. J. de Sauvage. 2000. Gli regulation by the opposing activities of fused and suppressor of fused. *Nat. Cell Biol.* 2:310–312.

Neumann, C. J., and S. M. Cohen. 1997. Long-range action of Wingless organizes the dorsal-ventral axis of the *Drosophila* wing. *Development* 124:871–880.

Ng, J. K., K. Tamura, D. Buscher, and J. C. Izpisua-Belmonte. 1999. Molecular and cellular basis of pattern formation during vertebrate limb development. *Curr. Top. Dev. Biol.* 41:37–66.

Nibu, Y., H. Zhang, and M. Levine. 1998. Interaction of short-range repressors with *Drosophila* CtBP in the embryo. *Science* 280:101–104.

Nishida, H., and K. Sawada. 2001. macho-1 encodes a localized mRNA in ascidian eggs that specifies muscle fate during embryogenesis. *Nature* 409:724–729.

Nüsslein-Volhard, C., and E. Wieschaus. 1980. Mutations affecting segment number and polarity in *Drosophila. Nature* 287:795–801.

Oro, A. E., K. M. Higgins, Z. Hu, J. M. Bonifas, E. H. Epstein, Jr., and M. P., Scott. 1997. Basal cell carcinomas in mice overexpressing sonic hedgehog. *Science* 276: 817–821.

Oro, A. E., and M. P. Scott. 1998. Splitting hairs: dissecting roles of signaling systems in epidermal development. *Cell* 95:575–578.

Ott, M. O., E. Bober, G. Lyons, H. Arnold, and M. Buckingham. 1991. Early expression of the myogenic regulatory gene, myf-5, in precursor cells of skeletal muscle in the mouse embryo. *Development* 111:1097–1107.

Parr, B. A., and A. P. McMahon. 1995. Dorsalizing signal Wnt-7a required for normal polarity of D-V and A-P axes of mouse limb. *Nature* 374:350–353.

Peifer, M., and E. Wieschaus. 1990. The segment polarity gene armadillo encodes a functionally modular protein that is the *Drosophila* homolog of human plakoglobin. *Cell* 63:1167–1176.

Perler, F. B. 1998. Protein splicing of inteins and hedgehog autoproteolysis: structure, function, and evolution. *Cell* 92:1–4.

Pinney, D. F., F. C. de la Brousse, A. Faerman, M. Shani, K. Maruyama, and C. P. Emerson, Jr. 1995. Quail myoD is regulated by a complex array of cis-acting control sequences. *Dev. Biol.* 170:21–38.

Porter, J. A., S. C. Ekker, W. J. Park, D. P. Von Kessler, K. E. Young, C. H. Chen, Y. Ma, A. S. Woods, R. J. Cotter, E. V. Koonin, and P. A. Beachy. 1996a. Hedgehog patterning activity: role of a lipophilic modification mediated by the carboxy-terminal autoprocessing domain. *Cell* 86:21–34.

Porter, J. A., D. P. von Kessler, S. C. Ekker, K. E. Young, J. J. Lee, K. Moses, and P. A. Beachy. 1995. The product of hedgehog autoproteolytic cleavage active in local and long-range signalling. *Nature* 374:363–366.

Porter, J. A., K. E. Young, and P. A. Beachy. 1996b. Cholesterol modification of hedgehog signaling proteins in animal development. *Science* 274:255–259.

Price, M. A., and D. Kalderon. 1999. Proteolysis of cubitus interruptus in *Drosophila* requires phosphorylation by protein kinase A. *Development* 126:4331–4339.

Ramsdell, A. F., and H. J. Yost. 1998. Molecular mechanisms of vertebrate left-right development. *Trends Genet.* 14:459–465.

Riddle, R. D., R. L. Johnson, E. Laufer, and C. Tabin. 1993. Sonic hedgehog mediates the polarizing activity of the ZPA. *Cell* 75:1401–1416.

Riggleman, B., P. Schedl, and E. Wieschaus. 1990. Spatial expression of the *Drosophila* segment polarity gene armadillo is posttranscriptionally regulated by wingless. *Cell* 63:549–560.

Robbins, D. J., K. E. Nybakken, R. Kobayashi, J. C. Sisson, J. M. Bishop, and P. P. Therond. 1997. Hedgehog elicits signal transduction by means of a large complex containing the kinesin-related protein costal2. *Cell* 90:225–234.

Roberts, D. J., R. L. Johnson, A. C. Burke, C. E. Nelson, B. A. Morgan, and C. Tabin. 1995. Sonic hedgehog is an endodermal signal inducing Bmp-4 and Hox genes during induction and regionalization of the chick hindgut. *Development* 121:3163–3174.

Roelink, H., A. Augsburger, J. Heemskerk, V. Korzh, S. Norlin, A. Ruiz i Altaba, Y. Tanabe, M. Placzek, T. Edlund, T. M. Jessell, and J. Dodd. 1994. Floor plate and motor neuron induction by vhh-1, a vertebrate homolog of hedgehog expressed by the notochord. *Cell* 76:761–775.

Rong, P. M., M. A. Teillet, C. Ziller, and N. M. Le Douarin. 1992. The neural tube/notochord complex is necessary for vertebral but not limb and body wall striated muscle differentiation. *Development* 115:657–672.

Roose, J., and H. Clevers. 1999. TCF transcription factors: molecular switches in carcinogenesis. *Biochim. Biophys. Acta* 1424:M23–M37.

Rudnicki, M. A., P. N. Schnegelsberg, R. H. Stead, T. Braun, H. H. Arnold, and R. Jaenisch. 1993. MyoD or Myf-5 is required for the formation of skeletal muscle. *Cell* 75:1351–1359.

Ruel, L., V. Stambolic, A. Ali, A. S. Manoukian, J. R. and Woodgett. 1999. Regulation of the protein kinase activity of Shaggy (Zeste-white3) by components of the wingless pathway in *Drosophila* cells and embryos. *J. Biol. Chem.* 274:21790–21796.

Ruiz i Altaba, A. 1999. Gli proteins encode context-dependent positive and negative functions: implications for development and disease. *Development* 126:3205–3216.

Ruiz i Altaba, A. 1997. Catching a Gli-mpse of hedgehog. *Cell* 90:193–196.

Saitoh, A., L. A. Hansen, J. C. Vogel, and M. C. Udey. 1998. Characterization of Wnt gene expression in murine skin: possible involvement of epidermis-derived Wnt-4 in cutaneous epithelial-mesenchymal interactions. *Exp. Cell Res.* 243:150–160.

Sarkar, L., M. Cobourne, S. Naylor, M. Smalley, T. Dale, and P. T. Sharpe. 2000. Wnt/Shh interactions regulate ectodermal boundary formation during mammalian tooth development. *Proc. Natl. Acad. Sci. U.S.A.* 97:4520–4524.

Sarkar, L., and P. T. Sharpe. 1999. Expression of Wnt signalling pathway genes during tooth development. *Mech. Dev.* 85:197–200.

Sasaki, H., Y. Nishizaki, C.-C. Hui, N. Nakafuku, and H. Kondoh. 1999. Regulation of Gli2 and Gli3 activities by an amino-terminal repression domain: implication of Gli2 and Gli3 as primary mediators of Shh signaling. *Development* 126:3915–3924.

Sawa, H., L. Lobel, and H. R. Horvitz. 1996. The *Caenorhabditis elegans* gene lin-17, which is required for certain asymmetric cell divisions, encodes a putative seven-transmembrane protein similar to the *Drosophila* frizzled protein. *Genes Dev.* 10: 2189–2197.

Schmidt, M., M. Tanaka, and A. Munsterberg. 2000. Expression of beta-catenin in the developing chick myotome is regulated by myogenic signals. *Development* 127: 4105–4113.

Seeling, J. M., J. R. Miller, R. Gil, R. T. Moon, R. White, and D. M. Virshup. 1999. Regulation of beta-catenin signaling by the B56 subunit of protein phosphatase 2A. *Science* 283:2089–2091.

Shimeld, S. M. 1999. The evolution of the hedgehog gene family in chordates: insights from amphioxus hedgehog. *Dev. Genes Evol.* 209:40–47.

Shulman, J. M., N. Perrimon, and J. D. Axelrod. 1998. Frizzled signaling and the developmental control of cell polarity. *Trends Genet.* 14:452–458.

Siegfried, E., T. B. Chou, and N. Perrimon. 1992. wingless signaling acts through zeste-white 3, the *Drosophila* homolog of glycogen synthase kinase-3, to regulate engrailed and establish cell fate. *Cell* 71:1167–1179.

Sisson, J. C., K. S. Ho, k. Suyama, and M. P. Scott. 1997. Costal2, a novel kinesin-related protein in the Hedgehog signaling pathway. *Cell* 90:235–245.

Slusarski, D. C., V. G. Corces, and R. T. Moon. 1997. Interaction of Wnt and a Frizzled homologue triggers G-protein-linked phosphatidylinositol signalling. *Nature* 390: 410–413.

Sokol, S., J. L. Christian, R. T. Moon, and D. A. Melton. 1991. Injected Wnt RNA induces a complete body axis in *Xenopus* embryos. *Cell* 67:741–752.

Standley, H. J., A. M. Zorn, and J. B. Gurdon. 2001. eFGF and its mode of action in the community effect during *Xenopus* myogenesis. *Development* 128:1347–1357.

Stark, K., S. Vainio, G. Vassileva, and A. P. McMahon. 1994. Epithelial transformation of metanephric mesenchyme in the developing kidney regulated by Wnt-4. *Nature* 372:679–683.

St.-Jacques, B., H. R. Dassule, I. Karavanova, V. A. Botchkarev, J. L. Li, P. S. Danielian, J. A. McMahon, P. M. Lewis, R. Pans, and A. P. McMahon. 1998. Sonic hedgehog signaling is essential for hair development. *Curr. Biol.* 8:1058–1068.

Stone, D. M., M. Hynes, M. Armanini, T. A. Swanson, Q. Gu, R. L. Johnson, M. P. Scott, D. Pennica, A. Goddard, H. Phillips, M. Noll, J. E. Hooper, F. de Sauvage, and A. Rosenthal. 1996. The tumour-suppressor gene patched encodes a candidate receptor for Sonic hedgehog. *Nature* 384:129–134.

Summerbell, D., P. R. Ashby, O. Coutelle, D. Cox, S. Yee, and P. W. Rigby. 2000. The expression of Myf5 in the developing mouse embryo is controlled by discrete and dispersed enhancers specific for particular populations of skeletal muscle precursors. *Development* 127:3745–3757.

Sussman, D. J., J. Klingensmith, P. Salinas, P. S. Adams, and R. Nusse, and N. Perriman. 1994. Isolation and characterization of a mouse homolog of the *Drosophila* segment polarity gene dishevelled. *Dev. Biol.* 166:73–86.

Tabata, T., S. Eaton, and T. B. Kornberg. 1992. The *Drosophila* hedgehog gene is expressed specifically in posterior compartment cells and is a target of engrailed regulation. *Genes Dev.* 6:2635–2645.

Tabata, T., and T. B. Kornberg. 1994. Hedgehog is a signaling protein with a key role in patterning *Drosophila* imaginal discs. *Cell* 76:89–102.

Tajbakhsh, S., E. Bober, C. Babinet, S. Pournin, H. Arnold, and M. Buckingham. 1996. Gene targeting the myf-5 locus with nlacZ reveals expression of this myogenic factor in mature skeletal muscle fibres as well as early embryonic muscle. *Dev. Dyn.* 206:291–300.

Tajbakhsh, S., and M. Buckingham. 2000. The birth of muscle progenitor cells in the mouse: spatiotemporal considerations. *Curr. Top. Dev. Biol.* 48:225–268.

Takada, S., K. L. Stark, M. J. Shea, G. Vassileva, J. A. McMahon, and A. P. McMahon. 1994. Wnt-3a regulates somite and tailbud formation in the mouse embryo. *Genes Dev.* 8:174–189.

The, I., Y. Bellaiche, and N. Perrimon. 1999. Hedgehog movement is regulated through tout velu-dependent synthesis of a heparan sulfate proteoglycan. *Mol. Cell* 4: 633–639.

Therond, P. P., J. D. Knight, T. B. Kornberg, and J. M. Bishop. 1996. Phosphorylation of the fused protein kinase in response to signaling from hedgehog. *Proc. Natl. Acad. Sci. U.S.A.* 93:4224–4228.

Thesleff, I., A. Vaahtokari, P. Kettunen, and T. Aberg. 1995. Epithelial-mesenchymal signaling during tooth development. *Connect. Tissue Res.* 32:9–15.

Thorpe, C. J., A. Schlesinger, J. C. Carter, and B. Bowerman. 1997. Wnt signaling polarizes an early *C. elegans* blastomere to distinguish endoderm from mesoderm. *Cell* 90:695–705.

Tsang, M., N. Lijam, Y. Yang, D. R. Beier, A. Wynshaw-Boris, and D. J. Sussman. 1996. Isolation and characterization of mouse dishevelled-3. *Dev. Dyn.* 207:253–262.

van den Heuvel, M., and P. W. Ingham. 1996. smoothened encodes a receptor-like serpentine protein required for hedgehog signalling. *Nature* 382:547–551.

van den Heuvel, M., R. Nusse, P. Johnston, and P. A. Lawrence. 1989. Distribution of the wingless gene product in *Drosophila* embryos: a protein involved in cell-cell communication. *Cell* 59:739–749.

van de Wetering, M., R. Cavallo, D. Dooijes, A. van Beest, J. van Es, J. Loureiro, A. Ypma, D. Hursch, T. Jones, A. Bejsovec, M. Peifer, M. Mortin, and H. Clevers. 1997. Armadillo coactivates transcription driven by the product of the *Drosophila* segment polarity gene dTCF. *Cell* 88:789–799.

Vinson, C. R., S. Conover, and P. N. Adler. 1989. A *Drosophila* tissue polarity locus encodes a protein containing seven potential transmembrane domains. *Nature* 338: 263–264.

Vortkamp, A., K. Lee, B. Lanske, G. V. Segre, H. M. Kronenberg, and C. J. Tabin. 1996. Regulation of rate of cartilage differentiation by Indian hedgehog and PTH-related protein. *Science* 273:613–622.

Waltzer, L., and M. Bienz. 1998. *Drosophila* CBP represses the transcription factor TCF to antagonize Wingless signalling. *Nature* 395:521–525.

Wang, B., J. F. Fallon, and P. A. Beachy. 2000. Hedgehog-regulated processing of Gli3 produces an anterior/posterior repressor gradient in the developing vertebrate limb. *Cell* 100:423–434.

Wang, G., K. Amanai, B. Wang, and J. Jiang. 2000. Interactions with Costal2 and suppressor of fused regulate nuclear translocation and activity of cubitus interruptus. *Genes Dev.* 14:2893–2905.

Wang, G., B. Wang, and J. Jiang. 1999. Protein kinase A antagonizes Hedgehog signaling by regulating both the activator and repressor forms of Cubitus interruptus. *Genes Dev.* 13:2828–2837.

Wang, Y., and R. Jaenisch. 1997. Myogenin can substitute for Myf5 in promoting myogenesis but less efficiently. *Development* 124:2507–2513.

Wang, Y., J. P. Macke, B. S. Abella, K. Andreasson, P. Worley, D. J. Gilbert, N. G. Copeland, N. A. Jenkins, and J. Nathans. 1996. A large family of putative transmembrane receptors homologous to the product of the *Drosophila* tissue polarity gene frizzled. *J. Biol. Chem.* 271:4468–4476.

Weinberg, E. S., M. L. Allende, C. S. Kelly, A. Abdelhamid, T. Murakami, P. Andermann, O. G. Doerre, D. J. Grunwald, and B. Riggleman. 1996. Developmental regulation of zebrafish MyoD in wild-type, no tail and spadetail embryos. *Development* 122:271–280.

Willert, K., M. Brink, A. Wodarz, H. Varmus, and R. Nusse. 1997. Casein kinase 2 associates with and phosphorylates dishevelled. *EMBO J.* 16:3089–3096.

Willert, K., S. Shibamoto, and R. Nusse. 1999. Wnt-induced dephosphorylation of axin releases beta-catenin from the axin complex. *Genes Dev.* 13:1768–1773.

Yamamoto, H., S. Kishida, M. Kishida, S. Ikeda, S. Takada, and A. Kikuchi. 1999. Phosphorylation of axin, a Wnt signal negative regulator, by glycogen synthase kinase-3beta regulates its stability. *J. Biol. Chem.* 274:10681–10684.

Yanagawa, S., J. S. Lee, Y. Matsuda, and A. Ishimoto. 2000. Biochemical characterization of the *Drosophila* axin protein. *FEBS Lett.* 474:189–194.
Zecca, M., K. Basler, and G. Struhl. 1996. Direct and long-range action of a wingless morphogen gradient. *Cell* 87:833–844.
Zhang, J., and R. W. Carthew. 1998. Interactions between Wingless and DFz2 during *Drosophila* wing development. *Development* 125:3075–3085.

7 Modular Pleiotropic Effects of Quantitative Trait Loci on Morphological Traits

JAMES M. CHEVERUD

An important aspect of the genetic architecture of complex traits is the structure of the pleiotropic effects of loci contributing to trait variation. Functionally and developmentally related morphological traits are often inherited together and therefore tend to evolve as integrated units. This morphological integration has been evident in the inheritance and evolution of mammalian cranial characters (Cheverud 1982, 1995, 1996; Ackermann and Cheverud 2000; Marroig and Cheverud 2001). The coinheritance of traits is measured by their genetic correlation (Falconer and Mackay 1997). Genetic correlations between traits can arise through linkage disequilibrium among separate genes affecting different characters or through pleiotropy, when a single gene affects multiple characters. The underlying genotypic basis of genetic correlations can be examined through quantitative trait locus (QTL) studies, where random markers are used to identify portions of the genome affecting morphological traits. A less-than-perfect correlation among traits can occur either by modular pleiotropic effects, where different sets of traits are affected by different QTLs, or by antagonistic pleiotropy, where the same QTL has opposite effects on different characters. The distinction between modular and antagonistic pleiotropy can have important evolutionary consequences with regard to direct and correlated responses to selection (Gromko 1995).

Riedl (1978) hypothesized that in response to natural selection "the system of reciprocal gene effects will imitate the essential aspects of the phene system" (196–197), so that the epigenetic system imitates functional phenotypic dependences. "The selection of interdependence therefore consistently increases the genetic interconnection of those features which have a basic, immediate, and long-lasting functional connection" (Riedl 1978, 196). Thus, in order for evolution to proceed, Riedl (1978) hypothesizes that natural selection will result in modules of functionally interacting phenotypes coming under control of separate gene systems and that the genome should imitate functional phenotypic dependences. He predicts modularity rather than antagonistic pleiotropy. Wagner and Altenberg (1996; Wagner 1996) come to a similar conclusion. They suggest that specific genes affect specific character complexes while different character complexes are affected by different sets of genes. Thus, the genotype-phenotype relationship will be modular. In contrast, one would expect antagonistic pleiotropy in situations involving allocation trade-offs due to competition among tissues for the common resources required for growth. Riska's (1986) variable part and variable proportion models of development predict antagonistic pleiotropy. It is unknown whether modularity or antagonistic pleiotropy is responsible for the lack of perfect correlation among parts.

Quantitative Trait Locus Studies

Over the past ten years advances in molecular and statistical genetics (Lander and Botstein 1989; Haley and Knott 1992) have made it possible to measure the effects of small genomic regions on a wide range of phenotypes. In most applications of this approach two distinct inbred lines are crossed, producing a genetically homogenous, heterozygous F_1 hybrid population. This hybrid population is then intercrossed to produce an F_2 hybrid population segregating for all genetic differences between the parental inbred lines or backcrossed to one of the parental lines. Measuring marker phenotypes in F_2 individuals specifies the segregation of a set of polymorphic genetic markers that can then be correlated with the segregation of phenotypes of interest. Genomic regions of genotype-phenotype correlation are referred to as quantitative trait loci (QTLs).

The most commonly used statistical approach to mapping QTL effects across the genome is interval mapping (Lander and Botstein 1989; Haley and Knott 1992). In interval mapping, genetic markers are scored throughout the genome spaced every 10–20 centiMorgans (cM; Darvasi and Soller 1994) along each chromosome. The segregation of these markers in the F_2 or backcross generation is used to construct a genetic map placing the markers on chromosomes relative to one another.

Genotypic scores are then calculated every 2 cM in the intervals between the markers, using the probability that an individual is homozygous for either parent allele or heterozygous at the specified location. At marker loci where the genotype is directly measured, additive genotypic scores are assigned as -1, 0, and $+1$ for the first parental homozygote, the heterozygote, and the alternate parental homozygote, respectively. Dominance genotypic scores are 0 for both parental homozygotes and 1 for the heterozygote. Genotypic scores in the intermarker interval are imputed using the genotypes measured at the flanking markers and the rates of recombination between the location of interest and the flanking markers (Haley and Knott 1992).

Genotypic scores along the chromosomes are then serially correlated with the phenotypes of interest. The map position with the smallest probability of obtaining the observed result by chance when there is no genetic effect is taken as the most likely position of a QTL affecting the phenotype. The QTL support region is identified as that region with less than a 10-fold increase in this probability. In an F_2 intercross experiment these regions are usually about 20 cM in length. It should be noted that there are probably many genes in such a confidence region. Thus, it is possible that what appears to be a single QTL in an F_2 generation may actually reflect the joint effects of several genes. For example, given 30,000 genes in the mouse genome, there would be about 400 genes in a typical QTL support interval. Narrowing the genomic region in question requires accumulating additional recombination, most often by generating a randomly mated advanced intercross (AI) line in which QTL mapping resolutions of 1 cM are possible (Darvasi and Soller 1995). At this finer resolution each region would contain between 10 and 20 genes.

It is possible to measure the pleiotropic effects of QTLs by using multiple phenotypes in a single analysis. When different traits map to the same location, the QTL is said to have pleiotropic effects on those traits. The theoretical considerations of evolutionary modularity discussed above lead to the prediction that pleiotropy will be most common among modules of functionally and developmentally related traits. Thus, QTL studies allow a critical test of modularity in the genotype-phenotype system. However, it must be remembered that the joint effects measured in a QTL study also may be due to separate closely linked genes in linkage disequilibrium. Discriminating between pleiotropy and linkage disequilibrium as causes of joint mapping of traits requires finer mapping resolution, such as is obtained in AI lines, to break down linkage disequilibrium in the mapping population.

The Structure of Pleiotropic QTL Effects

We have undertaken a series of QTL studies mapping the genomic location of effects on a wide range of morphological and growth characters in mice. These include analyses of cranial morphology (Leamy et al. 1999), mandibular morphology (Cheverud et al. 1997; Cheverud 2000a; Mezey et al. 2000), growth in body weight (Cheverud et al. 1996; Vaughn et al. 1999), and body composition (Cheverud et al. 2001). In all of these studies, multiple traits have been analyzed with a view to measuring pleiotropic effects and determining whether pleiotropic modules reflect basic functional and developmental relationships.

Our studies utilize the F_2 intercross of two inbred mouse strains, Large (LG/J) and Small (SM/J). LG/J was generated by inbreeding a line that had been selected for large body size (Chai 1956a, 1956b, 1957, 1961, 1968; Goodale 1938, 1941; Wilson et al. 1971). SM/J was generated by inbreeding a line selected for small body size in a separate experiment (Chai 1956a, 1956b, 1957, 1961, 1968; MacArthur 1944). Chai's (1956a, 1956b) studies indicated that these two strains differed for a relatively large number of genes of small effect on adult body size. Such effects are thought to be important in evolution but have not usually been measured in gene mapping studies.

In two separate experiments (Cheverud et al. 1996; Kramer et al. 1998; Vaughn et al. 1999), we crossed LG/J females to SM/J males, producing approximately 50 F_1 hybrid animals in each experiment. These hybrid animals were intercrossed to produce 535 F_2 animals in the first experiment and 510 animals in the second experiment. Sample sizes of at least 500 are recommended for QTL studies seeking to measure small effects without substantial statistical bias (Beavis 1994; Lynch and Walsh 1998). Animals were weaned from their mothers at 3 weeks of age and housed with three or four other animals in single-sex cages. They were fed a standard rodent chow (PicoLab Rodent Chow 20 [#5053]) ad libitum. Individual mice were weighed weekly from 1 to 10 weeks using a digital scale. At 10 weeks animals were either bred or sacrificed. Animals that were bred were sacrificed after they had successfully produced offspring. Organs, including the liver, heart, spleen, and kidneys, were removed and weighed at necropsy, with the tissue being saved for DNA extraction. In the second experiment, the reproductive fat pad was also weighed and the tail length measured at necropsy. After necropsy, mouse carcasses were macerated by dermestid beetles to prepare the skeletons for measurement.

DNA was extracted from the spleen or liver and used to score microsatellite polymorphisms. Over 10,000 microsatellite polymorphisms have been mapped in mice (Mouse Genome Database 2001; Dietrich et al. 1992, 1996). Approximately half of these are expected to be polymorphic between the LG/J and SM/J strains (Routman and Cheverud 1995). Polymorphic microsatellite loci have different lengths of dinucleotide repeats in the strains compared. PCR was used to amplify microsatellite sequences, and length polymorphisms were scored on high-density agarose gels (Dietrich et al 1992; Routman and Cheverud 1994, 1995). Each animal was then scored as SM/J homozygous, het-

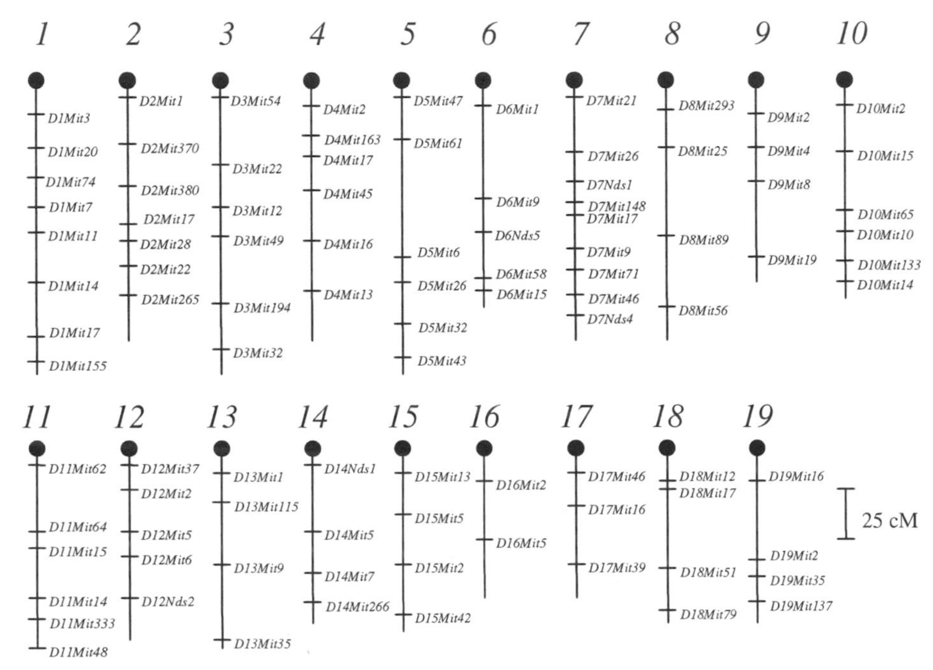

Fig. 7.1.—Genetic map positions of microsatellite markers used in QTL mapping studies of the LG/J-by-SM/J intercross (Cheverud et al. 1996; Vaughn et al. 1999).

erozygous, or LG/J homozygous at each marker locus. Seventy-six marker loci were scored in the first experiment (Cheverud et al. 1996), with an additional 21 being scored in the second experiment (Vaughn et al. 1999). Marker positions for these loci are illustrated in figure 7.1.

Somatic Growth

Mammalian growth occurs in two major stages, an early growth stage, where growth occurs largely by cell division, and a later growth period, during which organs increase in weight primarily through cell enlargement (Atchley and Zhu 1997; Atchley et al. 2000). Different organs also grow differentially during these two major stages. For example, brain growth occurs early, while somatic growth is greater during the later period. These growth periods are also controlled through different physiological systems, with the growth hormone axis being responsible for growth during the later period. Biometrical studies had indicated the possibility of negative pleiotropy between early and later growth (Riska et al. 1984).

In a replicated set of QTL studies of postnatal growth in the cross between LG/J and SM/J mice, Cheverud et al. (1996) and Vaughn et al. (1999) found that early body weight growth (prenatal to week 3)

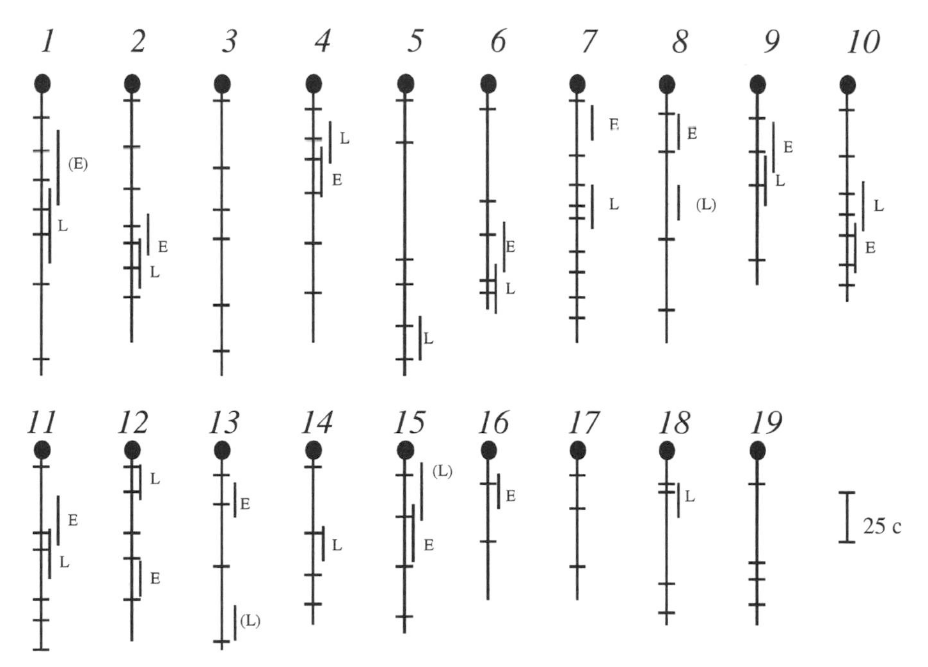

Fig. 7.2.—Genetic map of early (E) and late (L) growth QTLs (Cheverud et al. 1996; Vaughn et al. 1999) shown with 95% confidence regions. Parenthetical entries represent locations with exclusive effects on early or late weights but not on growth itself.

and later body weight growth (week 3 to week 10) were generally under separate genetic control (see fig. 7.2). These two growth periods showed a modular genetic relationship. Genes affecting growth affected either early or later growth.

Mandibular Morphology

The mandible is formed from several different mesenchymal condensations derived from neural crest cells invading the first branchial arch (Atchley and Hall 1991; Atchley 1993), although it is a single fused structure in adult mammals. The condensations produce the various processes of the mandible that coalesce to form the definitive structure, including the condyloid, coronoid, angular, and alveolar processes and the mandibular corpus (see fig. 7.3, *A*). After the condensations begin to ossify, their development depends on interactions with other tissues. Specifically, the ascending ramus of the mandible serves as a region for attachment of the muscles of mastication (see fig. 7.3, *B*), including the temporalis muscle inserted into the coronoid process, the lateral pterygoid muscle inserted into the condyloid process, and the masseter and medial pterygoid muscles inserted into the angular process and the posterior portion of the mandibular corpus. The definitive size and shape

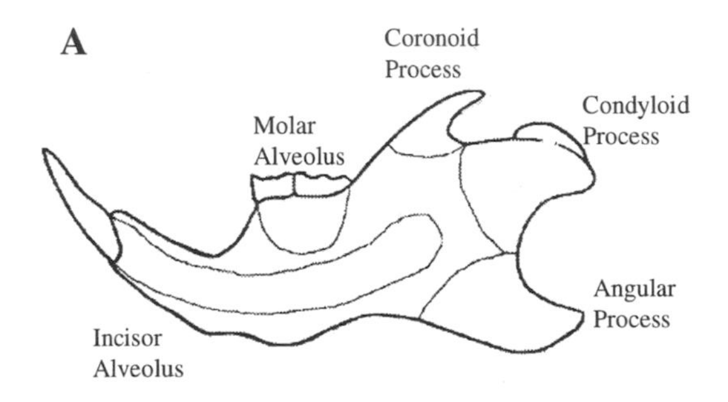

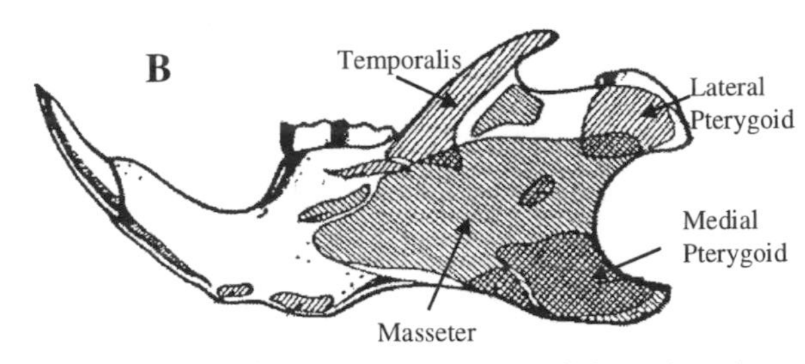

Fig. 7.3.—*A,* Mandibular condensations that coalesce to form the adult mandible. *B,* Mandibular muscle attachment areas. (Modified from Atchley and Hall 1991.)

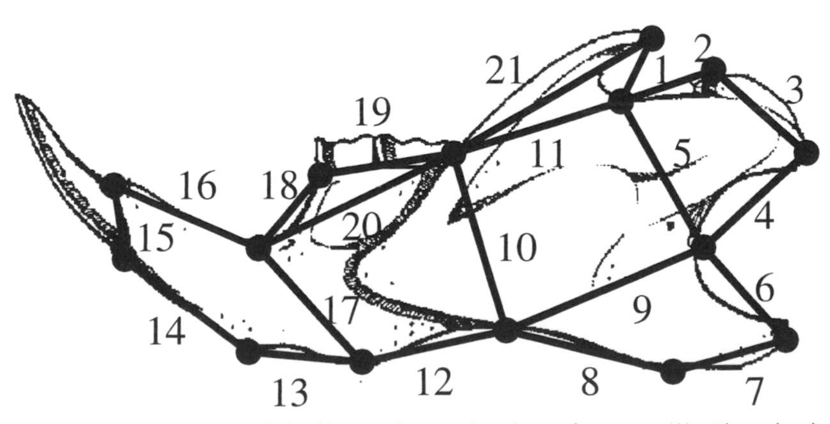

Fig. 7.4.—Linear measurements calculated from two-dimensional coordinates of murine mandibles (Cheverud et al. 1997; Cheverud 2000a). The dots indicate landmark locations.

of these processes depends on muscle function (Moore 1981; Herring and Lakers 1981; Atchley et al. 1984). Likewise, the anterior portion of the mandibular corpus serves as the alveolus supporting the incisor tooth with the molar alveolus lying superior and posterior to the incisor alveolus. These areas form under the influence of interactions with the adjacent, enamel-forming epithelium (Atchley and Hall 1991).

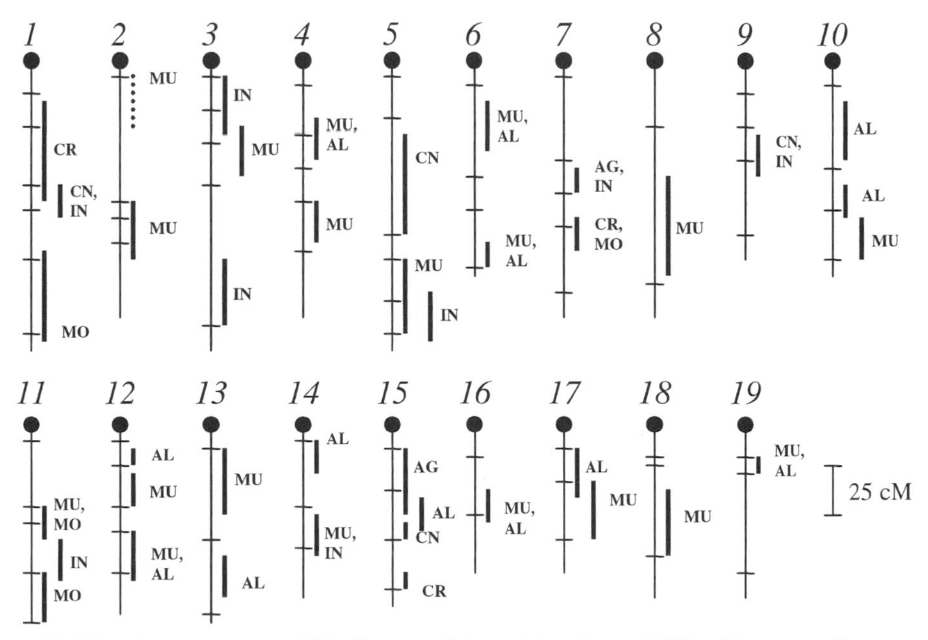

Fig. 7.5.—Genetic map positions of QTLs affecting mandibular morphology shown with 95% confidence regions. CR indicates that the QTL affects the coronoid process; CN, the condyloid process; AG, the angular process; MU, all the muscle attachment regions of the ascending ramus; IN, the incisor alveolus; MO, the molar alveolus; and AL, the total alveolar region. The dashed region on chromosome 2 indicates a region that was not sufficiently mapped in the first F_2 intercross to allow precise QTL placement. (After Cheverud 2000a.)

Maintenance of alveolar bone depends on continuing interaction with the embedded tooth. If the tooth is removed, the alveolus will resorb. A series of 21 individual mandibular measurements delineating the functional and developmental regions of the mandible were collected on each individual (see fig. 7.4).

In our analyses of QTLs for murine mandibular morphology, we located 41 QTLs affecting various parts of the mandible (Cheverud et al. 1997; Cheverud 2000a; fig. 7.5). We found that most of these QTLs affected functionally and developmentally distinct portions of the mandible, 27% affecting individual subregions such as the condyloid, coronoid, or angular processes or the incisor or molar alveolus, 44% having effects restricted to either the alveolar region or ascending ramus, and 29% affecting the whole mandible. None of the whole-mandible QTLs displayed antagonistic pleiotropy with opposite effects on different mandibular regions. Therefore, most of these QTLs affected the size of local regions, with pleiotropy restricted to modules. Examples of each kind of QTL are provided in figure 7.6. On occasion, a QTL affecting primarily one module will have a measured effect on a trait that is part of another module, such as in the condyloid QTL in figure 7.6, C,

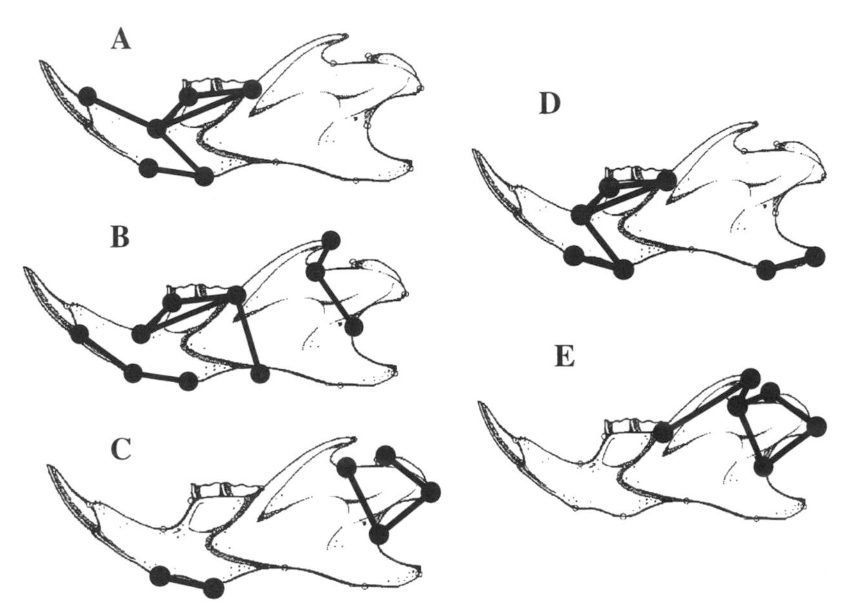

Fig. 7.6.—Examples of mandibular QTLs with different levels of modular effects. The lines drawn on the mandible indicate individual measures with significant QTL effects. *A,* QTL affecting alveolar processes on proximal chromosome 6; *B,* QTL affecting the entire mandible on distal chromosome 6; *C,* QTL affecting the condyloid process on chromosome 9; *D,* QTL affecting the alveolar process on proximal chromosome 10; *E,* QTL affecting muscular processes of the ascending ramus on distal chromosome 10. (After Cheverud et al. 1997.)

which includes inferior corpus length along with the condyloid measurements, and in the alveolar QTL in figure 7.6, *D,* which includes inferior angular length along with a series of alveolar measures. However, the concentration of significant results in single modules in each of these cases is greater than expected by chance (Cheverud et al. 1997; Cheverud 2000a). In an analysis of overall levels of modularity in mandibular QTLs, Mezey et al. (2000) found that the mandibular regions were significantly modular.

Cranial Morphology

The size and shape of the elements of the skull are strongly affected by their associated soft tissues. Growth of the related soft tissues results in growth and remodeling of cranial bones. A particularly sharp contrast exists between those parts of the cranium that surround and protect the brain and those that serve the face. In eutherian mammals the brain grows first, reaching its full size at a relatively early age, so that the neurocranium also completes growth early (Moore 1981). In rodents most of this growth is prenatal. The face enlarges during a later growth period, beginning at about 3 postnatal weeks. While the integrity of the cranium as a whole is required for efficient function and to withstand the forces of mastication, it seems likely that there is a certain degree of separability in the development and function of these two cranial regions.

Leamy et al. (1999) mapped QTLs for a series of murine cranial measurements (see fig. 7.7) and discovered 26 QTLs affecting cranial morphology (see fig. 7.8). Most of these QTLs affected specific cranial regions: 8 QTLs affected the neurocranium; 8 affected the face; and 10 affected the whole cranium. As in the analysis of mandibular morphology, all module-specific QTLs had the same effects on all module components. Only two of the whole-cranium QTLs provided evidence for antagonistic pleiotropy, with opposite effects on the neurocranium and face.

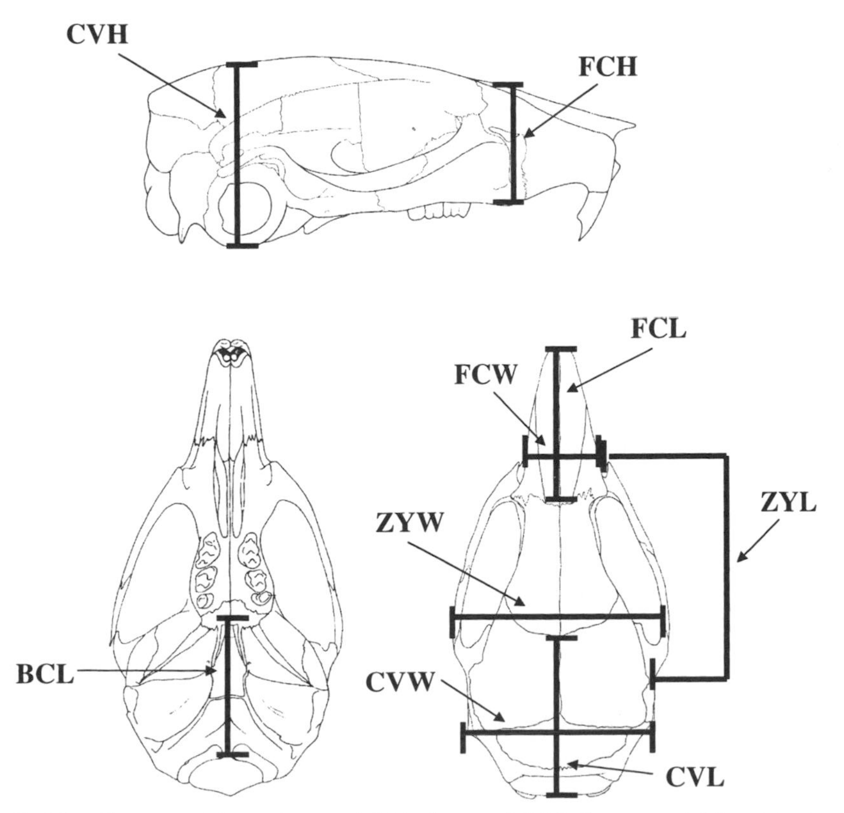

Fig. 7.7.—Measurements taken on the mouse skull (after Leamy et al. 1999). Measurements include cranial vault height (CVH), facial height (FCH), basicranial length (BCL), facial width (FCW), facial length (FCL), cranial vault width (CVW), cranial vault length (CVL), zygomatic arch width (ZYW), and zygomatic arch length (ZYL).

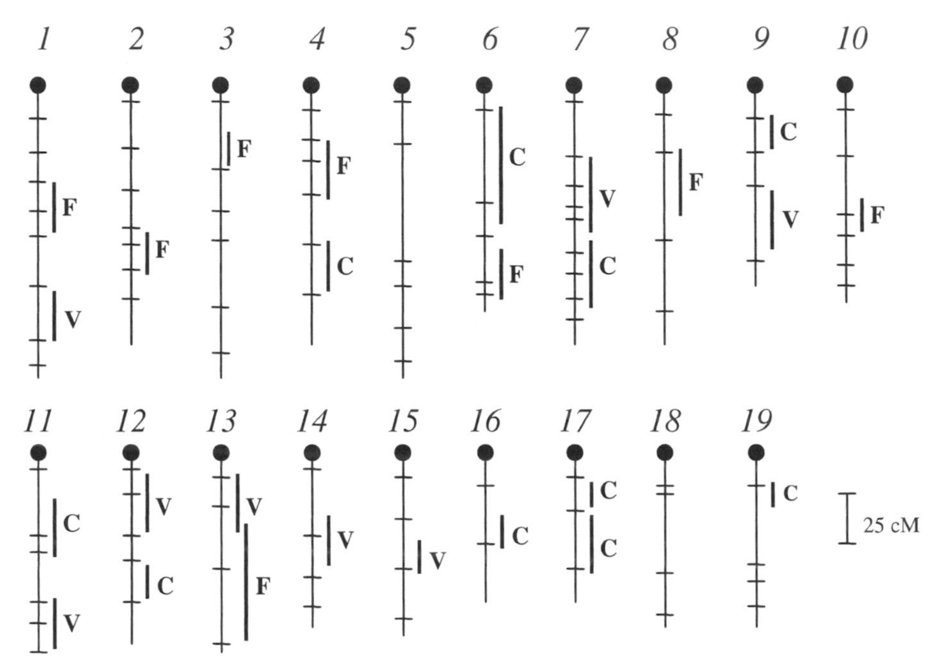

Fig. 7.8.—Genetic map positions of QTLs affecting cranial morphology shown with 95% confidence regions (after Leamy et al. 1999). A 'C' indicates a QTL affecting the whole cranium, 'V' a QTL affecting the cranial vault, and 'F' a QTL affecting the face.

Body Composition

Cheverud et al. (2001) reported gene mapping for body composition traits in mice, including adiposity (the weight of the reproductive fat-pad relative to the total body weight) and tail length (a measure of skeletal size). While both aspects of body composition are affected by the growth hormone axis, the timing and physiology of growth are different. The tail has essentially completed growth by 10 weeks of age, by which time the vertebral epiphyses have fused. However, adiposity continues to increase up to at least 20 weeks of age, with much of its growth occurring after tail length has reached its adult size. Furthermore, the genetic correlation between tail length and adiposity is relatively low in the LG/J-by-SM/J intercross population (Kramer et al. 1998) and drops to zero when their common correlation with overall body weight is controlled.

Cheverud et al. (2001) mapped eight QTLs affecting adiposity and nine QTLs affecting tail length (see fig. 7.9). Of these QTLs, the two aspects of body composition share only two, those on chromosomes 8 and 12. Fat deposition and skeletal size seem to have a modular relationship to one another, with each being affected by different sets of genes.

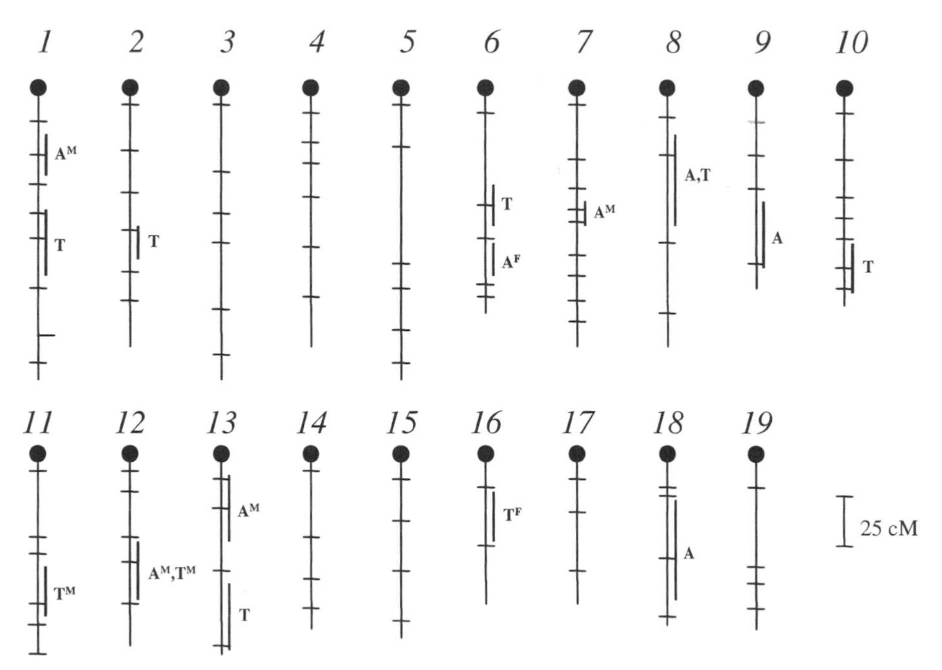

Fig. 7.9.—Map locations and 95% confidence regions for adiposity (A; reproductive fat pad weight/body weight) and tail length (T). A superscript indicates a sex-specific effect, with M for males and F for females.

The results obtained for all of these body size and morphological traits are consistent with one another. Physiologically related traits tend to be affected by a common set of genes, separate from sets of genes affecting complementary traits. Such results support the view that pleiotropy has a modular structure (Riedl 1978; Wagner 1996; Wagner and Altenberg 1996). Of course, this structure is not absolute. Within each set of traits, genes were also discovered that had effects spanning the whole system, such as genes affecting the whole mandible or the whole skull. However, these system-wide QTLs remain in the minority.

A more inclusive view of the structure of pleiotropy considers it a nested hierarchy. QTLs at the finest structural level of pleiotropy only affect a single low-level module, such as the coronoid process of the mandibular ascending ramus. At the next higher level, several submodules, such as the coronoid, condyloid, and angular processes of the mandible, are united in a single ascending ramus module united by the common effect of the masticatory muscles on bony process development. We discovered many QTLs with modular effects at this intermediate level, affecting the muscular processes without affecting the alveolar regions of the mandible. Finally, we also discovered QTLs affecting the whole mandible, which in turn can be seen as a module in higher-order cranial morphology. Thus, the modular structure of pleiotropy will be manifested in a hierarchical nesting of modules, with each module at any particular level associated with specific QTLs.

Genetic Variation in Pleiotropy and Differential Epistasis

The review of QTL effects provided above supports Riedl's (1978) concept of the imitatory genetic system in which the organization of patterns of gene effects imitates the organization of functional and developmental phenotypic patterns. If Riedl's (1978) ideas are correct, the organization of gene effects must have evolved and must continue to evolve to match phenotypic patterns. In order for this evolution to occur, there has to be heritable genetic variation in the morphological patterns of gene effects. In other words, there must be genetic variation in the pleiotropic effects of a gene locus.

One way in which genetic variation in pleiotropy can occur is by differential epistasis, in which epistatic interactions between loci differ for different phenotypes. These interactions may quash genetic variance for one trait affected by a gene while leaving the variance of a second trait unaffected. For example, genotypic values for traits X and Y at loci A and B are presented in table 7.1. Genotypic values are the average phenotype of individuals carrying the specified two-locus genotype. There is no epistasis for trait X at these two loci, since a value of +1 is added for every A or B allele, starting with −2 for the aabb homozygote. This is a perfectly additive set of genotypic values. On the other hand, strong epistasis exists for trait Y because the contrast in genotypic values at one locus varies depending on the genotype present at the second locus. For example, in the population of individuals with the bb genotype, each A allele adds a value of one to trait Y so that the two homozygotes differ by 2 units. The two alleles are codominant because the genotypic value of the heterozygote is midway between the homozygotes. However, among those with the Bb genotype, the A allele is dominant to a because the Aa and AA genotypes are equal to one another. Furthermore, unlike the homozygotes in the bb row, the homozygotes only differ by 1 instead of 2 units. Finally, among those with the BB genotype, there is overdominance, with the alternate homozygotes having the same value. Thus, the genotypic values for the A locus vary depending on the genotype present at the B locus.

As with any interaction, the reverse must also be true: genotypic values at the B locus must vary depending on the A locus genotype present, as seen here for trait Y. Note that locus B has pleiotropic effects on traits X and Y when the A locus genotype is aa or Aa but affects trait X only in the AA subpopulation and thus within this subpopulation displays no pleiotropy. When there are different epistatic patterns for different traits, there is genetic variation for pleiotropy at the B locus,

Table 7.1: Differential epistasis: genotypic values for traits X and Y at loci A and B

	Genotype		
Genotype	aa	Aa	AA
Trait X:			
bb	−2	−1	0
Bb	−1	0	1
BB	0	1	2
Trait Y:			
bb	−2	−1	0
Bb	−1	0	0
BB	0	1	0

Note: There is no epistasis for trait X, but there is epistasis for trait Y because differences in genotypic values at any one locus change depending on the genotype at the second locus.

providing fodder for the evolution of pleiotropy. Pleiotropy at the B locus becomes weaker as the A allele increases in frequency. When the population is fixed for the AA genotype, the B locus has no pleiotropic effects because it affects trait X but not trait Y.

Our QTL studies have often found evidence for epistasis, including epistasis for adult body weight and for elements of body composition (Routman and Cheverud 1997; Cheverud 2000b; Cheverud et al. 2001). The magnitude of epistatic gene effects, about 0.2 standard deviation units, is similar to the level of main effects found in the QTL experiments. There are many epistatic interactions, each of small effect, for each trait examined. Whether these epistatic interactions are differential or not cannot be fully determined from the data presented in these papers, although the two loci affecting both elements of body composition, on chromosomes 8 and 12, have significant epistatic interactions for tail length but not for adiposity (Cheverud et al. 2001). Thus, there should be differential epistasis for these two traits at these two loci.

Another indication of the possible presence of differential epistasis can be seen in situations in which the relationship between two phenotypic traits varies with the genotype at a particular locus. For example, Sing et al. (1988) report that the usual positive correlation between blood levels of triglyceride and cholesterol depends on the genotype at the *ApoE* locus. Individuals with genotypes carrying the *e*4 allele show no correlation, in contrast to the other genotypes present in the population. This suggests the possibility that the *ApoE* locus is involved in differential epistatic interactions with another locus, or loci, for these two traits. The *ApoE* locus would play the same role as the A locus in the example in table 7.1, where pleiotropic effects of other loci on triglyceride and cholesterol levels vary depending on the presence or absence of particular genotypes at the *ApoE* locus.

Genotypic variation in pleiotropic QTL effects on mandibular morphology provides further evidence of differential epistasis. Pleiotropic patterns for mandibular traits were described above (Cheverud et al. 1997; Cheverud 2000a). To detect differential epistasis, we examined QTL effects on the phenotypic correlation and allometry between individual mandibular measurements and total mandibular length (J. M. Cheverud, T. H. Ehrich, T. T. Vaughn, S. Koreishi, R. B. Linsey, and L. S. Pletscher, unpublished manuscript). Allometry describes the relationship between the size of one part of an organ and the size of the entire organ or body as a whole (Gould 1966). If different genotypes at a QTL specify different regressions of part on whole, then different forms of epistasis are likely to exist between that QTL and other, unspecified loci for the part and whole phenotypes. This is exemplified in table 7.1, where the regression of Y on X is 1.0 for the aa and Aa genotypes and 0.0 for the AA genotype.

We performed a whole-genome scan for regions with an effect on the level of covariance or correlation between individual mandibular traits and total mandibular length. This was accomplished by fitting a least squares regression model regressing an individual trait on the interaction between genotype scores and total mandibular length, holding the main effects of genotype scores and mandibular length constant, every 2 cM along the genome.

Most chromosomes have regions that cause significant variation in the relationship between individual mandibular traits and total mandible length, including chromosomes 1–4, 6, 7, 10, 11, 14, 15, and 17–19 (Cheverud et al., unpublished manuscript). All individual mandibular measurements participated in at least two QTLs, with a wide range of effects indicating ubiquity of differential epistasis for mandibular traits in relation to total mandible length.

A few examples of such interactions are presented in figure 7.10. Part *A* presents the relationship between posterior mandibular corpus height and mandible length for each genotype at an interaction QTL on chromosome 17 located at marker *D17Mit16*. The graph shows that the relationship between posterior corpus height and total length is strongest for LL homozygotes ($r = 0.54$) and weakest for the SS homozygotes ($r = 0.15$). Indeed, the regression slope for the SS genotype is not significantly greater than zero, indicating independence of these two traits in individuals carrying this genotype. Thus, the SS genotype at *D17Mit16* is like the AA genotype in table 1 in that the regression of part (Y) on whole (X) is 0.0 while other genotypes at this locus have a positive slope. Variation in the strength of the relationship between the two traits is most likely due to differential epistatic interactions involving a QTL at *D17Mit16* and other gene loci, although the possibility of genotype-by-environment interactions involving this QTL cannot be ruled out at present. Another interesting aspect of this graph is that the QTL has no overall effect on either trait, as illustrated by a lack of spread among the regression lines at the joint mean of the population (the mean is 11.5 mm for mandible length and 3.05 mm for posterior corpus height). A QTL screen for either trait alone or the two in combination would detect no QTL effect at *D17Mit16*, because the effect of this QTL on posterior corpus height is conditional on the length of the mandible. For small mandibles, to the left of the graph, the SS genotype produces a larger-than-average posterior corpus height, while in large mandibles the LL genotype has this effect. The effect of this QTL on posterior corpus height depends on the overall size of the mandible, and specific alleles have opposite effects at contrasting mandibular sizes. Presumably, if the S allele were fixed in this population so that everyone had the SS genotype, there would be no correlation between posterior corpus height and mandible length. Alternatively, if the L allele were fixed, the population would show a strong correlation between these two traits.

A second example (fig. 7.10, *B*) is presented for a QTL 2 cM downstream from marker *D10Mit2* on chromosome 10. This QTL affects the relationship between the anterior inferior mandibular base length and total mandible length. This example is similar to the one discussed above, but the roles of the L and S alleles are reversed. Those with an SS genotype show a relatively strong correlation ($r = 0.38$), while there is no significant correlation among those with the LL genotype. However, unlike the genotypes in the previous example, the genotypes are quite distinct at the joint trait means (11.5 mm, 2.1 mm), indicating that this QTL has detectable effects on both traits. Even so, the regression lines cross within the interval of trait sizes represented in the population, so that the effect of the QTL on anterior inferior base length changes sign depending on whether short (LL is associated with a longer anterior inferior basal length) or long mandibles (LL is associated with a shorter anterior inferior basal length) are considered.

Finally, a third example is given in figure 7.10, *C*, for a QTL 12 cM downstream from marker *D10Mit133* on chromosome 10 affecting the relationship between posterior angular height and total mandibular length. At this QTL, the LL and SS homozygotes display very similar, weak relationships between the two traits ($r_{SS} = 0.18$; $r_{LL} = 0.20$). In contrast, the relationship is relatively strong for the LS heterozygotes. As in the first example, this QTL has no average effect on either trait, as indicated by a lack of divergence between the regression lines at the joint mean (11.5 mm, 2.1 mm) of the traits, and the QTL produces underdominance for posterior angular height in small mandibles and overdominance in large mandibles.

As can be seen from the examples given above, it is possible that

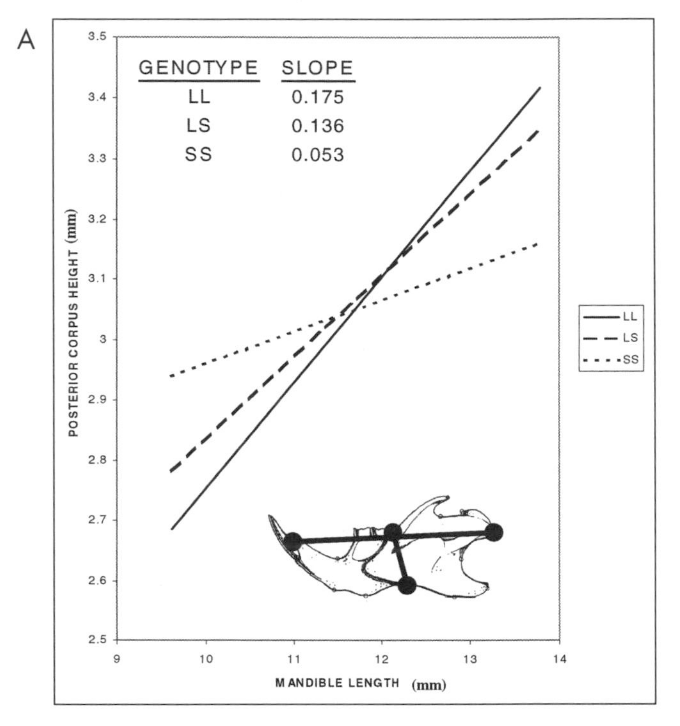

Fig. 7.10.—Differential epistasis examples from the mouse mandible. *A,* Genotype-specific regressions at a QTL on chromosome 17 at marker *D17Mit16,* affecting the relationship between posterior corpus height and total mandible length. *B,* Genotype-specific regressions at a QTL on chromosome 10 at 2 cM distal to marker *D10Mit2,* affecting the relationship between anterior inferior basal length and total mandible length. *C,* Genotype-specific regressions at a QTL on chromosome 10 at 12 cM distal to marker *D10Mit133* affecting the relationship between posterior angular height and total mandible length.

differential epistasis will be commonly found when the multivariate effects of genes on morphology are investigated more fully. In each of the systems discussed, there is evidence for genetic variance in the level of pleiotropic effects between traits. This variance is required for the evolution of pleiotropy and hence for the evolution of modularity. However, it is as yet unclear what the structure of this variation in pleiotropy will be like. It is usually argued that pleiotropy has become more restricted over evolutionary time as organs have become individuated (Riska 1986; Wagner 1996; Wagner and Altenberg 1996). Does this suggest that we should expect stronger genetic variance in pleiotropy at higher hierarchical levels of the modular system so that modularity can more easily evolve at these higher levels? Neither the theoretical underpinnings of such expectations nor an empirical understanding of patterns is yet available. However, further investigations of differential epistasis will be important for our understanding of the evolution of modularity.

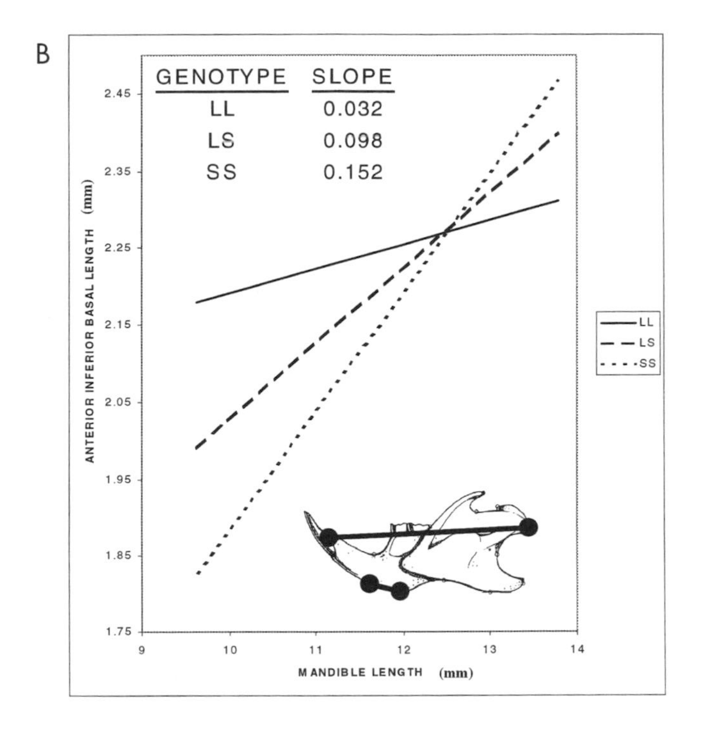
B
GENOTYPE
SLOPE
LL
0.032
LS
0.098
SS
0.152
2.45
2.35
2.25
2.15
2.05
1.95
1.85
1.75
ANTERIOR INFERIOR BASAL LENGTH (mm)
9
10
11
12
13
14
MANDIBLE LENGTH (mm)
LL
LS
SS

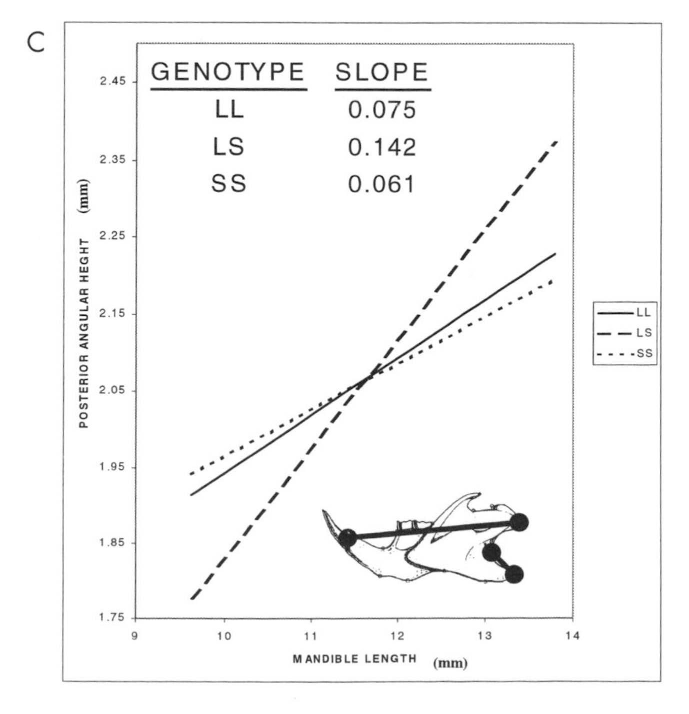
C
GENOTYPE
SLOPE
LL
0.075
LS
0.142
SS
0.061
2.45
2.35
2.25
2.15
2.05
1.95
1.85
1.75
POSTERIOR ANGULAR HEGHT (mm)
9
10
11
12
13
14
MANDIBLE LENGTH (mm)
LL
LS
SS

Summary

We have examined patterns of pleiotropic effects of quantitative trait loci (QTL) in a wide range of morphological systems, including the mouse mandible, skull, body size growth, and body composition. In each instance we found that pleiotropic effects tend to be restricted to sets of developmentally and/or physiologically related traits, although a few loci may affect each system as a whole. We have measured the effects of individual genes on mandibular morphology, skull morphology, body size growth, and body composition in F_2 hybrid animals from the intercross of the LG/J and SM/J inbred mouse strains. In general, we found that most QTLs affect discrete modules, such as the muscular processes or alveolus of the mandible, the neurocranium or face of the skull, early or later postnatal growth, and adiposity or skeletal size. A minority of the genes affect the whole organ, such as the mandible as a whole, and only a very few genes were found with antagonistic pleiotropic effects on different cranial modules. Thus, patterns of pleiotropic effects are modular in that functionally and developmentally related traits are generally affected by unique quantitative trait loci. The structure of pleiotropic relationships was further investigated by examining between-genotype variation in allometric relationships between individual mandibular measurements and total mandibular length. A large number of instances of varying pleiotropic relations were mapped across the mouse genome. Variation at these loci provides the potential for evolution of pleiotropy at other sets of gene loci.

Acknowledgments

I wish to thank Andrea Peripato, Reinaldo Alves de Brito, Jane Kenney, Susan Cropp, Thomas Erich, and Jason Wolf for comments on the manuscript. This work was supported by National Science Foundation grant DEB-9726433 and National Institutes of Health grant DK52514.

References

Ackermann, R., and J. Cheverud. 2000 Phenotypic covariance structure in tamarins (genus *Saguinus*): a comparison of variation patterns using matrix correlations and common principal component analysis. *Am. J. Phys. Anthropol.* 111:489–501.

Atchley, W. R. 1993. Genetic and developmental aspects of variability in the mammalian mandible. In *The skull: development,* vol. 1, ed. J. Hanken and B. K. Hall, 207–247 Chicago: University of Chicago Press.

Atchley, W. R., and B. K. Hall. 1991. A model for the development and evolution of complex morphological structures. *Biol. Rev.* 66:101–157.

Atchley, W. R., S. W. Herring, B. Riska, and A. Plummer. 1984. Effects of the muscular dysgenesis gene on developmental stability in the mouse mandible. *J. Craniofac. Genet. Dev. Biol.* 4:179–189.

Atchley, W. R., R. Wei, and P. Crenshaw. 2000. Cellular consequences in the brain and liver of age-specific selection for rate of development in mice. *Genetics* 155: 1347–1357.

Atchley W. R., and J. Zhu. 1997. Developmental quantitative genetics, conditional epigenetic variability and growth in mice. *Genetics* 147:765–776.

Beavis, W. D. 1994. The power and deceit of QTL experiments: lessons from comparative QTL studies. In *49th Annual Corn and Sorghum Research Conference,* 252–268. Washington, D.C.: American Seed Trade Association.

Chai, C. 1956a. Analysis of quantitative inheritance of body size in mice. 1. Hybridization and maternal influence. *Genetics* 41:157–164.

Chai, C. 1956b. Analysis of quantitative inheritance of body size in mice. 2. Gene action and segregation. *Genetics* 41:167–178.

Chai, C. 1957. Analysis of quantitative inheritance of body size in mice. 3. Dominance. *Genetics* 42:601–607.

Chai, C. 1961. Analysis of quantitative inheritance of body size in mice. 4. An attempt to isolate polygenes. *Genet. Res.* 2:25–32.

Chai, C. 1968. Analysis of quantitative inheritance of body size in mice. 5. Effects of small numbers of polygenes on similar genetic backgrounds. *Genet. Res.*11:239–246.

Cheverud, J. 1982. Phenotypic, genetic, and environmental morphological integration in the cranium. *Evolution* 36:499–516.

Cheverud, J. 1995. Morphological integration in the saddle-back tamarin (*Saguinus fuscicollis*) cranium. *Am. Nat.,* 145:63–89.

Cheverud, J. 1996. Quantitative genetic analysis of cranial morphology in the cotton-top (*Saguinus oedipus*) and saddle-back (*S. fuscicollis*) tamarins. *J. Evol. Biol.* 9:5–42.

Cheverud, J. 2000a. The genetic architecture of pleiotropic relations and differential epistasis. In *The character concept in evolutionary biology,* ed. G. Wagner, 411–434. New York: Academic Press.

Cheverud, J. 2000b. Detecting epistasis among quantitative trait loci. In *Epistasis and the evolutionary process,* ed. J. Wolf, E. Brodie III, and M. Wade, 58–81. New York: Oxford University Press.

Cheverud, J., E. Routman, F. M. Duarte, B. van Swinderen, K. Cothran, and C. Perel. 1996. Quantitative trait loci for murine growth. *Genetics* 142:1305–1319.

Cheverud, J. M., E. J. Routman, and D. K. Irschick. 1997. Pleiotropic effects of individual gene loci on mandibular morphology. *Evolution* 51:2004–2014.

Cheverud, J., T. Vaughn, L. S. Pletscher, A. Peripato, E. Adams, C. Erickson, and K. King-Ellison. 2001. Genetic architecture of adiposity in the cross of Large (LG/J) and Small (SM/J) inbred mice. *Mamm. Genome* 12:3–12.

Darvasi, A., and M. Soller. 1994. Optimum spacing of genetic markers for determining linkage between marker loci and quantitative trait loci. *Theor. Appl. Genet.* 89: 351–357.

Darvasi, A., and M. Soller. 1995. Advanced intercross lines, an experimental population for fine genetic mapping. *Genetics* 141:1199–1207.

Dietrich, W., H. Katz, S. Lincoln, H.-S. Shin, J. Friedman, N. Dracopoli, and E. S. Lander. 1992. A genetic map of the mouse suitable for typing intraspecific crosses. *Genetics* 131:423–447.

Dietrich, W., J. Miller, R. Steen, M. Merchant, D. Damron-Boles, Z. Husain, R. Dredge, M. Daly, K. Ingalls, T. O'Connor, C. Evans, M. DeAngelis, D. Levison, L. Kruglyak, N. Goodman, N. Copeland, N. Jenkins, T. Hawkins, L. Stein, D. Page, and E. Lander. 1996. A comprehensive genetic map of the mouse genome. *Nature* 380:149–152.

Falconer, D. S., and T. F. C. Mackay. 1997. *Introduction to quantitative genetics.* New York: Longman Press.

Goodale, H. 1938. A study of the inheritance of body weight in the albino mouse by selection. *J. Hered.* 29:101–112.

Goodale, H. 1941. Progress report on possibilities in progeny test breeding. *Science* 94:442–443.

Gould, S. J. 1966. Allometry and size in ontogeny and phylogeny. *Biol. Rev.* 41: 587–640.

Gromko, M. H. 1995. Unpredictability of correlated response to selection: pleiotropy and sampling interact. *Evolution* 49:685–693.

Haley, C. S., and S. A. Knott. 1992. A simple regression method for mapping quantitative trait loci in line crosses using flanking markers. *Heredity* 69:315–324.

Herring, S. W., and T. C. Lakars. 1981. Craniofacial development in the absence of muscle contraction. *J. Craniofac. Genet. Dev. Biol.* 1:341–357.

Kramer, M. G., T. T. Vaughn, L. S. Pletscher, K. King-Ellison, E. Adams, C. Erickson, and J. M. Cheverud. 1998. Genetic variation in body weight growth and composition in the intercross of Large (LG/J) and Small (SM/J) inbred strains of mice. *Genet. Mol. Biol.* 21:211–218.

Lander, E. S., and D. Botstein. 1989. Mapping Mendelian factors underlying quantitative traits using RFLP linkage maps. *Genetics* 121:185–199.

Leamy, L., E. Routman, and J. Cheverud. 1999. Quantitative trait loci for early and late developing skull characters in mice: a test of the genetic independence model of morphological integration. *Am. Nat.* 153:201–214.

Lynch, M., and B. Walsh. 1998. *Genetics and analysis of quantitative traits.* Sunderland, Mass.: Sinauer Associates.

MacArthur, J. 1944. Genetics of body size and related characters. 1. Selection of small and large races of the laboratory mouse. *Am. Nat.* 78:142–157.

Marroig, G., and J. Cheverud. 2001. A comparison of phenotypic variation and covariation patterns and the role of phylogeny, ecology and ontogeny during cranial evolution of New World monkeys. *Evolution* 55:2576–2600.

Mezey, J., J. Cheverud, and G. Wagner. 2000. Is the genotype-phenotype map modular? a statistical approach using mouse QTL data. *Genetics* 156:305–311.

Moore, W. 1981. *The mammalian skull.* Cambridge: Cambridge University Press.

Mouse Genome Database (MGD). 2001. Mouse genome informatics Web site Bar Harbor, Me.: Jackson Laboratory. World Wide Web (URL: http://www.informatics.jax.org/). (September 2001).

Riedl, R. 1978. *Order in living organisms.* New York: Wiley Press.

Riska, B. 1986. Some models for development, growth, and morphometric correlation. *Evolution* 40:1303–1311.

Riska B., W. R. Atchley, and J. J. Rutledge. 1984. A genetic analysis of targeted growth in mice. *Genetics* 107:79–101.

Routman, E., and J. Cheverud. 1994. A rapid method of scoring simple sequence repeat polymorphisms with agarose gel electrophoresis. *Mamm. Genome* 5:187–188.

Routman, E., and J. Cheverud. 1995. Polymorphism for PCR-analyzed microsatellites: Data for two additional inbred mouse strains and the utility of agarose gel electrophoresis. *Mamm. Genome* 6:401–404.

Routman, E. J., and J. M. Cheverud. 1997. Gene effects on a quantitative trait: two-locus epistatic effects measured at microsatellite markers and at estimated QTL. *Evolution* 51:1654–1662.

Sing, C., E. Boerwinkle, P. Moll, and A. Templeton. 1988. Characterization of genes affecting quantitative traits in humans. In *Proceedings of the 2nd International Conference on Quantitative Genetics,* ed. B. Weir, E. Eisen, M. Goodman, and G. Namkoong, 250–269. Sunderland, Mass.: Sinauer Associates.

Vaughn, T. T., L. S. Pletscher, A. Peripato, K. King-Ellison, E. Adams, C. Erikson, and J. M. Cheverud. 1999. Mapping quantitative trait loci for murine growth: a closer look at genetic architecture. *Genet. Res.* 74:313–322.

Wagner, G. 1996. Homology, natural kinds, and the evolution of modularity. *Am. Zool.* 36:36–43.
Wagner, G., and L. Altenberg. 1996. Perspective: complex adaptations and the evolution of evolvability. *Evolution* 50:967–976.
Wilson, S. P., H. D. Goodale, W. H. Kyle, and E. F. Godfrey. 1971. Long term selection for body weight in mice. *J. Hered.* 62:228–234.

8 Central Nervous System Development: From Embryonic Modules to Functional Modules

CHRISTOPH REDIES AND LUIS PUELLES

The central nervous system (CNS) of vertebrates displays an immensely complex three-dimensional architecture that evolves during development from a relatively simple two-dimensional sheet of tissue, the neuroepithelium. The neuroepithelium is the wall of the neural tube, a hollow structure that is located at the center of the body and extends along the entire length of the body axis. While the anteriormost part of the neural tube gives rise to the brain, the rest develops into the spinal cord. We will describe two types of modules in the vertebrate CNS. The first type of module is most obvious early in embryogenesis. It is represented by distinct adjacent domains of neuroepithelial tissue that are systematically arranged along both the longitudinal and transverse axes of the neural tube. The second type of module is functional and encountered in the mature brain. It is represented by neural circuits that are composed of nerve cell aggregates and their connections. The neural circuits provide the basis for information processing, which is the major biological function of the CNS. The question that we will address in this chapter is how the early embryonic modules of the CNS give rise to the later functional modules. In particular, we will review evidence that a family of cell-cell adhesion molecules, the cadherins, is involved in this transformation and provides an adhesive code for the diverse structures of both the embryonic and the mature CNS.

Embryonic CNS Modules

The early embryonic CNS has a prominent modular structure. It consists of blocks of neuroepithelial tissue that form a mosaic-like pattern along the longitudinal (anterior-posterior) axis and transverse (dorsal-ventral) axis of the CNS (reviewed in Puelles and Rubenstein 1993;

Lumsden and Krumlauf 1996; Rubenstein et al. 1998; Redies and Puelles 2001). Each embryonic domain represents a relatively homogeneously specified histogenetic field, in which neural cells proliferate, migrate, and differentiate into characteristic neurons and glia (Rendahl 1924; Puelles et al. 1987). Although there is some exchange of signaling molecules and also of cells between these modules, histogenesis in each embryonic field is largely independent of neighboring units once the embryonic modules have been established. In each module, specific gene constellations operate, and the cells derived from each module share functional characteristics in the mature brain. The embryonic structure of the vertebrate brain is thus rather similar to the structure of other parts of the body that are divided into independent histogenetic fields, which later differentiate into specialized parts of the body or into individual organs.

The major longitudinal domains of the neural tube are the floor plate, the basal plate, the alar plate, and the roof plate (see figs. 8.1, *A*, 8.6, *C*). In the transverse dimension, complete segments are called neuromeres. There are seven more or less morphologically distinct neuromeres in the hindbrain (rhombencephalon), called rhombomeres 1–7 (for review, see Lumsden and Krumlauf 1996), while other transverse units called isthmus and pseudorhombomeres 8–11 form the less overtly segmented, rostral and caudal parts of the hindbrain (Cambronero and Puelles 2000). In the forebrain (prosencephalon), 6–8 transverse segments have been postulated, and they are called prosomeres (Bulfone et al. 1993; for review, see Puelles and Rubenstein 1993; Puelles 1995; Hauptmann and Gerster 2000). Figure 8.1 shows an overview of the neuromeres of the vertebrate brain. According to the prosomeric model proposed by Puelles, Rubenstein, and coworkers, the conventional diencephalon consists of parts of all six prosomeres, whereas the telencephalon is an enlargement mostly of the dorsal part of prosomere 5. Many of these major embryonic domains can be further divided into smaller, secondary domains. For example, the pretectum (alar prosomere 1) contains at least three transverse secondary subdivisions, and the dorsal thalamus (alar prosomere 2) at least five subdivisions (fig. 8.2; Redies et al. 2000). The telencephalon consists of the subpallium and the pallium, both of which can be further subdivided (fig. 8.3; Puelles et al. 2000).

Many of the genes that are involved in patterning the embryonic neural tube have been identified. They belong to several families of transcription factors or gene regulatory proteins. Many of these genes are also involved in pattern formation in other organs and organisms, in vertebrates and invertebrates. In the hindbrain, the expression and role of the Hox genes has been particularly well studied (for review, see Wilkinson 1993). In the mid- and forebrain, a number of other gene

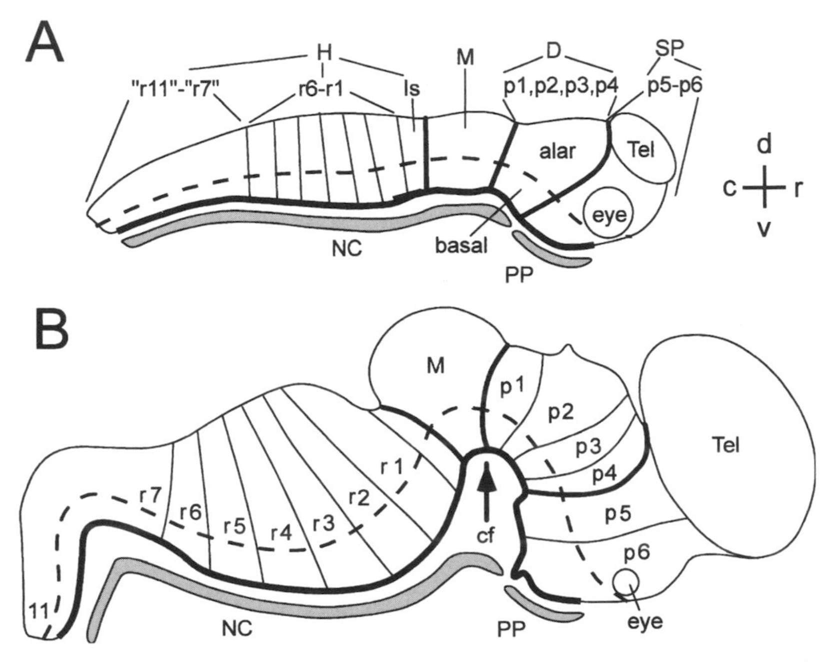

Fig. 8.1.—Embryonic modules of the mouse brain. *A*, Schematic representation of the early embryonic brain. The hindbrain (H) consists of seven rhombomeres (r1–r7) and four pseudorhombomeres (r8–r11). Rostral to the hindbrain, the isthmic region (Is), the midbrain (M), and prosomeres p1–p4 of the diencephalon (D) are found, in a caudal-to-rostral sequence. The anlage of the telencephalon (Tel) and the eye are outpouchings of the secondary prosencephalon (SP). *B*, The neuromeres at a more advanced stage of development. The midbrain and telencephalon have enlarged considerably. Also note the differential growth in the alar regions of prosomeres p1–p6. The dashed line indicates the alar–basal plate boundary and the thick black line indicates the floor of the brain, which is situated just dorsal to the notochord (NC) and the prechordal plate (PP). Both lines run in parallel to the longitudinal (rostral-caudal) axis of the brain. Note that this axis bends at the cephalic flexure (cf) during development. The lines orthogonal to the alar–basal plate boundary represent transverse (ventral-dorsal) boundaries between the neuromeres. c, caudal; d, dorsal; r, rostral; v, ventral.

families are expressed, such as members of the Otx, Pax, Dlx, Gbx, Emx, Wnt, Sox, Six, and Nkx families of genes. Most of these genes have in common that they are expressed in restricted regions of the brain. Some genes are expressed in relatively large regions that correspond to entire embryonic divisions. For example, the expression of five transcription factors (Pax-6, Tbr1, Emx1, Dlx2, Nkx2.1) was used to delineate the telencephalic divisions of the chicken and mouse brain (fig. 8.3; Puelles et al. 2000). The expression domains of such patterning factors are often nested, but, like the Hox genes in the hindbrain, they show sharp expression borders that often coincide with postulated transverse and longitudinal boundaries. At some of the transverse boundaries, for example, in the isthmic region and the zona limitans intrathalamica of the thalamus, signaling centers are found that are known or proposed to induce additional regionalization in the surrounding segments (Puelles and Rubenstein 1993; Marín and Puelles 1994; Shimamura and Rubenstein 1997).

In the brain, the number and the topological relation of the neuromeres are well conserved between species, from the lamprey brain to the human brain (Puelles et al. 1996; Puelles and Verney 1998; Pombal and Puelles 1999; Wullimann and Puelles 1999; Davila et al. 2000; Hauptmann and Gerster 2000; Milan and Puelles 2000; Redies et al. 2000). Despite the large morphological differences between the mature brains of the different vertebrate species, the neuromeric organization represents a basic scheme that is common to all vertebrates investigated so far. This allows studying how brain diversity is generated during evolution and development and what brain regions in the different species derive from which embryonic division (Puelles 1995; Redies and Puelles 2001). Thus, the mapping of gene expression in the embryonic brain provides a powerful additional tool for assessing—with due precautions—which brain regions in the different species are homologous to each other (Holland and Garcia-Fernandez 1996; Smith-

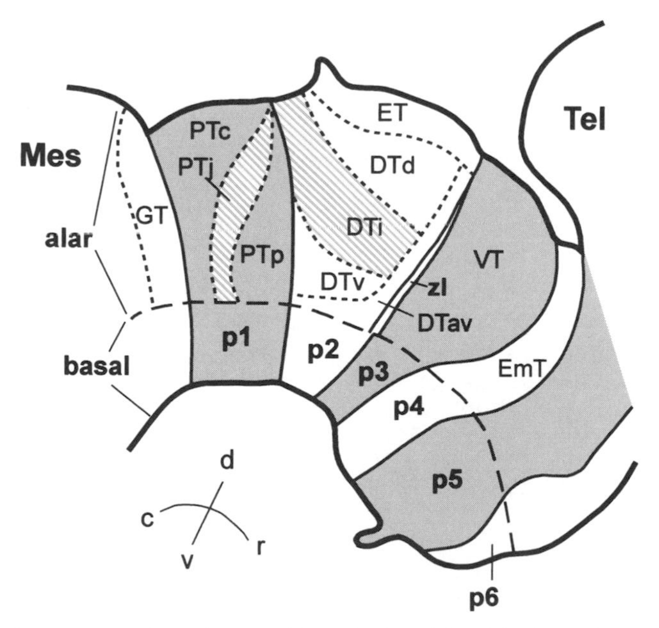

Fig. 8.2.—Primary and secondary embryonic modules in the diencephalon of the chicken. The pretectum (alar prosomere 1) can be divided into three secondary subdivisions (PTc, PTj, and PTp) and the dorsal thalamus (alar prosomere 2) into five secondary subdivisions (ET, DTd, DTi, DTv, and DTav). Like the prosomeres, their secondary subdivisions persist as coherent blocks of gray matter in the mature brain. Different shadings mark the different (sub)divisions. The dashed line indicates the alar–basal plate boundary. c, caudal; d, dorsal; EmT, eminentia thalami; GT, griseum tectale; Mes, mesencephalon; p1–p6, prosomeres 1–6; r, rostral; Tel, telencephalon; v, ventral; zl, intrathalamic zona limitans. (Reproduced with permission from Redies et al. [2000], Copyright © 2000, Wiley-Liss, Inc., a subsidiary of John Wiley & Sons, Inc.)

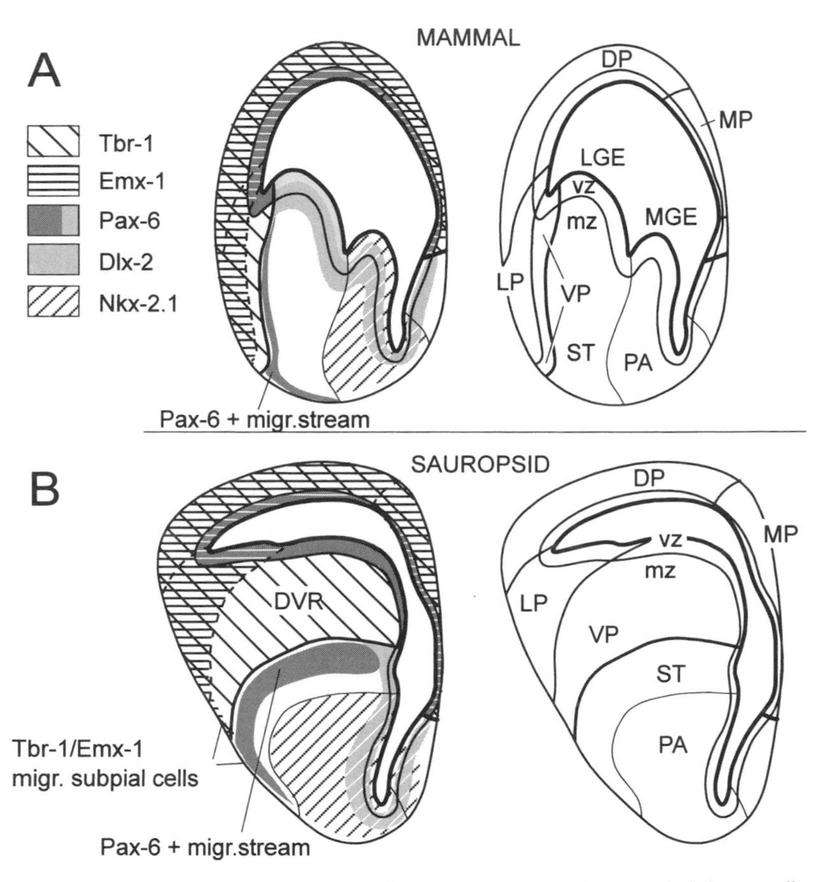

Fig. 8.3.—Embryonic modules in the telencephalon of a mammal (mouse, *A*) and a sauropsid (chicken, *B*). Different shadings, as indicated on the left side of panel *A*, indicate the expression domains of the five transcription factors Tbr-1, Emx-1, Pax-6, Dlx-2, and Nkx-2.1. The light shading for Pax-6 indicates overlapping expression with Dlx-2. The expression domains define divisions in the pallium (MP, medial pallium; DP, dorsal pallium; LP, lateral pallium; and VP, ventral pallium) and in the subpallium (ST, striatum; and PA, pallidum). Note that the arrangement of the different embryonic divisions is similar in both species but their relative size differs. DVR, dorsal ventricular ridge; LGE, lateral ganglionic eminence; MGE, medial ganglionic eminence; migr., migratory; mz, mantle layer; vz, ventricular layer. (Reproduced with permission from figure 1 of the study by Puelles et al. [1999], Copyright © 1999, Swets & Zeitlinger Publishers.)

Fernandez et al. 1998; Puelles et al. 2000). Moreover, many of the genes studied play a role in the patterning and/or stabilization of the embryonic divisions.

Functional Modules

In the mature CNS, a different type of modularity is found that is unique to the CNS. This type of module is functional and forms the basis for information processing. Functional modules can process information relatively independently of each other and are typically composed of separate nerve cell aggregates (brain nuclei or regions) that are connected to each other in a highly ordered fashion by specific fiber tracts. Exchange of information is high within the modules and low between the modules. Like the embryonic modules, each functional module has a physical location in the CNS and is embedded between other functional modules. Moreover, each module can be characterized by the type of information that it processes. For example, sensory information is conveyed to the CNS from the sensory organs, such as the eye, the inner ear, the olfactory epithelium, or the sensory receptors of the viscera, muscles, ligaments, and skin. One of the advantages of having spatially segregated functional modules in the CNS is that the different types of information (visual, auditory, olfactory, etc.) can be processed in parallel and simultaneously in specialized neural circuits (sensory systems). A typical example of a functional system, the ascending auditory system, is shown in figure 8.4. Most functional systems can be divided into a number of neural subcircuits (i.e., X and Y pathways of the visual system, or the macular and semicircular organ pathways of the vestibular system, or the mechanosensitive, thermosensitive, and nociceptive parts of the somatosensory system). There are also multisensory regions in the brain, which integrate information from different sensory systems. Moreover, motor systems generate hierarchically appropriate body responses to environmental or internal stimuli at different neural levels. Another advantage of having specialized functional modules is that the neural architecture of each module can be adapted independently during evolution to carry out the type of information processing that is required under environmental pressure in each case. In addition, the neural architecture in each individual module can be fine-tuned independently by plasticity and experience during the lifetime of an organism.

Typically, each functional system is composed of brain gray matter areas that are derived from several of the embryonic modules. For example, the visual, auditory, and motor systems each comprise specialized areas or subsets of brain nuclei in the cerebral cortex, basal ganglia, thalamus, brainstem, nerves, and so on (see schematic diagram in fig. 8.5). It is generally believed that the gray matter areas that are derived from each embryonic module carry out a similar type of computation. For example, the type of computation that is carried out in the different areas of the cerebral cortex is similar for the visual, auditory, and motor cortical areas. The same applies to the functionally specialized regions of the basal ganglia and thalamus. It has been shown that some cortical or dorsal thalamic regions can even replace each other's function if the regions or their input are exchanged early enough in development (Molnar and Blakemore 1991; Schlaggar and O'Leary

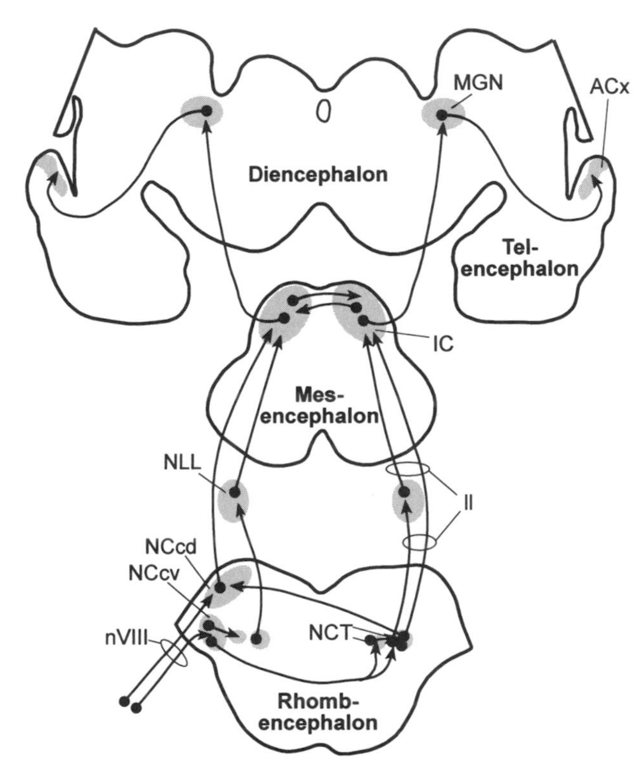

Fig. 8.4.—Functional modularity in the vertebrate brain. A simplified, schematic drawing of the ascending auditory system in man is shown. Auditory information reaches the rhombencephalon via the auditory nerve (nVIII), which terminates in the cochlear nuclei (NCcd and NCcv). From here, auditory information is conveyed to both sides of the brain along specific fiber connections to other hindbrain nuclei (NCT, NLL) and along the projection of the lateral lemniscus (ll) to the midbrain auditory center (inferior colliculus [IC]). Auditory information is transmitted further to a relay nucleus of the diencephalon (medial geniculate nucleus [MGN]) and, finally, to the auditory cortex (ACx) of the telencephalon. Note that the components of this system are widely distributed throughout the brain and are derived from several of the embryonic divisions.

1991). The fact that each functional system is composed of derivatives of several embryonic modules may thus relate to the need to carry out diverse types of computation within each system.

From Embryonic to Functional Modularity

At first glance, embryonic modularity and functional modularity seem rather different principles of CNS organization. On the one hand, embryonic modules represent local units of tissue lying side by side. On the other hand, functional modules are neural circuits that globally interconnect specialized regions located in many different parts of the CNS. How do the two types of modularity relate to each other? The approaches to answering this question have included fate-mapping studies, transplantation experiments, mapping of developmentally regulated genes, and other techniques. From these studies, a number of principles have emerged that are summarized in the following sections.

Embryonic Divisions Are Transformed into Coherent Domains of Mature Gray Matter

Studies on chick/quail chimeras have shown that the neuromeres of the hindbrain differentiate into transverse blocks of gray matter (Hallonet and Le Douarin 1993; Marín and Puelles 1995; Wingate and Lumsden 1996; Cambronero and Puelles 2000). Many of the hindbrain nuclei have a multisegmental origin and fuse across the neuromere boundaries. A developmental study of cadherin expression, patterns of histogenesis, radial glial topology, and known neuronal migration patterns in the chicken forebrain (see below) suggests that the prosomeres and many of their subdivisions also persist in the mature brain (Redies et al. 2000, 2001; Yoon et al. 2000). In the avian and mammalian forebrain, most brain nuclei originate in a single embryonic division, and they remain in this division, with few exceptions (Puelles and Medina 1994; Puelles and Verney 1998; Redies et al. 2000, 2001; Verney et al. 2001).

The hindbrain boundaries and at least some of the prosomeric boundaries restrict migration of early neurons (Puelles et al. 1992; Figdor and Stern 1993; Martínez et al. 1995). Extensive subpial migration tangential to the surface of the brain occurs in some of the prosomeric subdivisions, but this migration also seems to be restricted within the divisional borders in the cases studied so far (e.g., the anteroventral

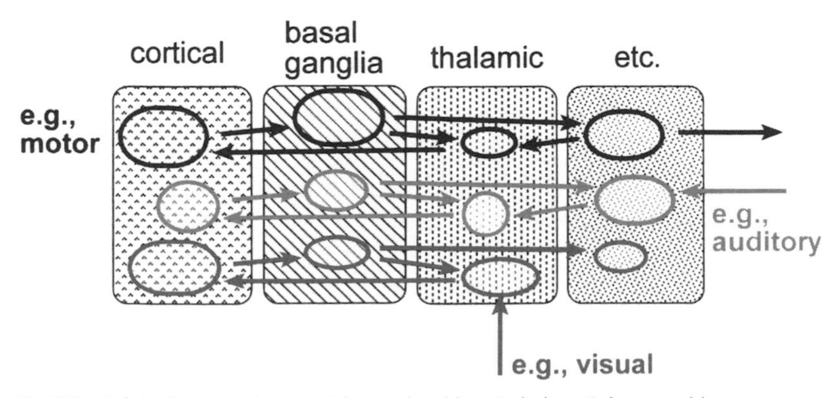

Fig. 8.5.—Relation between embryonic and functional modularity in the brain. Embryonic modules represent spatially separate, largely independent histogenetic fields. Each field gives rise to a coherent domain of gray matter that is later characterized by a particular way of information processing (e.g., information processing in a cortical, basal ganglia, thalamic way). Each domain contains several brain nuclei or regions that are connected to nuclei or regions in other domains by fiber tracts, thus forming neural circuits that are parts of different functional systems (e.g., visual, auditory, motor).

domain of the dorsal thalamic division; Puelles et al. 1992; Uchikawa et al. 1999). The exchange of neural cells between the embryonic modules is thus limited (Tan and Le Douarin 1991; Birgbauer and Fraser 1994; Marín and Puelles 1995; Anderson et al. 1997; Lavdas et al. 1999; Cambronero and Puelles 2000), and most neurons remain in the embryonic fields where they are born (Fraser et al. 1990; Smith-Fernandez et al. 1998; Inoue et al. 2000). Some specific populations of neurons, however, have been shown to cross divisional boundaries. For example, most cortical interneurons have their origin in the ganglionic eminences of the telencephalic subpallium and migrate across the (probably longitudinal) pial-subpial boundary to reach the cortex (Anderson et al. 1997; Lavdas et al. 1999; Cobos et al. 2001). Likewise, oligodendrocytes, a glial cell type that forms the myelin sheets in the CNS, spread from a few restricted regions over the entire brain during development (Timsit et al. 1995; Perez Villegas et al. 1999). Migration of specific populations of early neurons across longitudinal (dorsoventral) boundaries has also been observed in the hindbrain (Tan and Le Douarin 1991). Nevertheless, the vast majority of neural cells remain in the division where they are born.

In summary, it is evident that embryonic modularity is not a transient phenomenon that is observed only early in development, to be replaced later by a different type of brain organization. Rather, the neuromeres and their subdivisions are transformed into functional structures. Embryonic modularity thus represents one of the bases for mature brain architecture. This conclusion, although disregarded for a large part of the twentieth century, had already been reached by some of the early developmental neurobiologists (Tello 1923; Rendahl 1924; Bergquist and Källén 1954; Vaage 1969).

Embryonic Divisions and Derived Gray Matter Domains Form Complete Radial Units

The third dimension of the CNS is the radial one. It runs orthogonal to the transverse and longitudinal axis in most CNS regions. In the wall of the neural tube, the radial dimension extends from the inner (ventricular) surface to the outer (pial) surface of the neural tube. Ventricular neuroepithelial cells and, later in development, radial glia cells extend their processes along the radial dimension, irrespective of morphogenetic deformations, so that each point at the brain pial surface is connected to another point (or small area) at the ventricular surface by the apical and basal processes, respectively, of such radial cells. Early in development, when the wall of the neural tube is still thin, the radial processes are short and straight and are only minimally divergent as they approach the surface. At later stages, radial glia processes are much longer and can be variously deflected, compressed, or forced apart by differential growth phenomena and differentiating intrinsic or extrinsic elements in the brain wall (nuclei, fiber tracts, blood vessels, as well as sulci, adhesions, outpouchings, prominences, or recesses of the ventricular and pial surfaces). At the divisional boundaries, changes in radial glial density are often observed. Some of the boundaries in the chicken and mouse brain are delineated by dense glial structures (Mai et al. 1998; Redies et al. 2000). Radial glial processes seem to extend all the way from the ventricular to the pial surface until rather late stages of development (Striedter and Beydler 1997; Redies et al. 2000). Their final, more or less deformed course reveals the primordial topologically radial dimension.

The radial glial processes guide the migration of early postmitotic neurons from the inner (ventricular) layer, where cells proliferate, to the outer (mantle) layer, where neurons differentiate. In the mantle layer, more and more neurons accumulate as development proceeds, and the mantle layer increases in thickness to a considerable degree. The neurons aggregate in the various nuclei and pronuclei and layers of the mantle layer. The proliferative activity of the embryonic divisions varies (Puelles et al. 1987; Bayer and Altmann 1995). Moreover, in many brain divisions, a varying number of cells may come to lie at different levels of radial stratification. As a consequence, the gray matter domains that are derived from the embryonic divisions can show considerable differences in their final size and form, as shown in detail for forebrain morphogenesis in the chicken (Puelles et al. 1987; De Castro et al. 1998; Redies et al. 2000; Yoon et al. 2000).

The radial glial cells thus represent an array of positional cues, which define unique positions relative to the two-dimensional sheet of neuroepithelial cells that make up the neural tube wall early in development. This positional scheme is fully translated into the radial dimension, that is, the thickness of the neural tube wall, by the positional cues that are carried by the processes of the radial glia. As a consequence, the molecularly (genetically) specified embryonic divisions of the early neuroepithelium are extended up to the pial surface in terms of differential adhesiveness and histogenetic properties, possibly influencing such diverse phenomena as proliferation, cell migration, and axonal navigation, leading to gradual implementation of differential neuronal fates within the fundamental context of the primary modules.

Other indications of the persistence of the radial embryonic units in the mature brain come from fate-mapping studies in the hindbrain of quail/chick chimeras. These studies (see above) demonstrate the persistence of the rhombomeres and pseudorhombomeres as complete radial units. Similar conclusions were reached for the prosomeres in the embryonic chicken forebrain (Redies et al. 2000, 2001; Yoon et al.

2000). In other vertebrate species such as the lamprey, frog, and zebrafish, the persistence of the embryonic divisions in the adult brain is even more obvious, because most postmitotic neuroblasts do not migrate far from the ventricular ependyma, differentiating instead into mature neurons largely in a periventricular stratum of the mantle layer (Puelles et al. 1996; Pombal and Puelles 1999; Wullimann and Puelles 1999).

The concept of persisting radial units in the vertebrate brain does not necessarily imply a persistence of radial glia in the mature or adult brain. In fact, some radial glial populations were shown to transform into astrocytes late in development (Voigt 1989), and some mature ependymocytes extend their apical process to contact intraneural blood vessels, instead of the pial surface. This concept also does not imply a complete separation of all neuronal cell groups between neighboring units (see above). It rather refers to the fact that the majority of the neurons born in a given division remain in it and form a coherent gray matter domain that fills the space in between the corresponding parts of the ventricular and pial surfaces of the brain.

Derived Gray Matter Domains Retain Their Embryonic Topological Relationships

Detailed studies of the chicken hindbrain and forebrain have shown that a given gray matter domain that has emerged dorsal, ventral, rostral, or caudal to another embryonic division is always found at that same topological position in the mature brain, despite the sometimes marked differences in growth and the morphogenetic deformation that take place in adjacent embryonic divisions (Puelles et al. 1987; Marín and Puelles 1995; Cambronero and Puelles 2000; Redies et al. 2000, 2001). As a frame of reference for this analysis, the longitudinal axis of the brain has to be identified in each brain region. Originally, the neural tube is oriented along the longitudinal body axis (fig. 8.1, *A*), but, as development proceeds, it forms three flexures, of which the cephalic flexure is the most prominent one (cf. in fig. 8.1, *B*). As a consequence, the longitudinal axis changes direction in the mature brain as one moves from the spinal cord to the hypothalamus, where the longitudinal axis ends rostrally (Puelles et al. 1987; Puelles and Rubenstein 1993; Shimamura et al. 1995). The rostralmost part of the longitudinal brain axis is flipped almost 180 degrees compared to the axial orientation of the spinal cord.

The same topological consistency between the embryonic divisions and the gray matter domains that are derived from them is found for some of the secondary forebrain subdivisions, such as those of the pretectum (alar prosomere 1) and the dorsal thalamus (alar prosomere 2;

fig. 8.2, *B;* Redies et al. 2000). The telencephalic pallium in tetrapods can be divided into four gray matter domains (medial, dorsal, lateral, and ventral pallium; fig. 8.3; Smith-Fernandez et al. 1998; Puelles et al. 2000). These domains also keep the same relative topological positions as the embryonic divisions they derive from. Since all these gray matter domains extend from the ventricular to the pial surface (see above), the topological arrangement of the embryonic divisions remains the same at both the ventricular and the pial surfaces, irrespective of any differences in size and form of the surface areas (one-to-many, or divergent, versus many-to-one, or compressed, topological transformations; Kuhlenbeck 1967).

Cadherins Provide a Combinatorial Code of Potentially Adhesive Cues for CNS Modules

It is clear that the initial patterning of the neural tube by gene regulatory proteins must be translated, in one way or another, into the expression of genes which regulate the various aspects of brain morphogenesis and circuit formation, such as cell migration, cell sorting and aggregation, axon outgrowth and fasciculation, target recognition, and synapse formation and stabilization. Several of the gene families that regulate these processes have been identified. They include diffusible molecules that set up molecular gradients for cell and axon migration, molecules that mediate cell-cell and cell-substrate adhesion, molecules that mediate short-range or long-range attraction and repulsion between the neural cells and their processes, and other types of molecules (for reviews, see Tessier-Lavigne and Goodman 1996; Redies 1997; Stoeckli and Landmesser 1998; O'Leary and Wilkinson 1999; Kaprielian et al. 2000). Here, we will focus on a family of cell-cell adhesion molecules, the cadherins. Studies of this class of molecules have provided insight into some of the developmental mechanisms that eventually lead to the emergence of mature functional architecture from the segmental structure of the embryonic CNS. Other types of molecules involved in this transformation will also be mentioned.

Cadherins are transmembrane glycoproteins that are known to confer adhesive specificity to cell surface membranes (for review, see Takeichi 1988; Gumbiner 1996). Several dozen cadherins are expressed in the vertebrate CNS (for review, see Redies 2000). They are all characterized by repetitive so-called cadherin repeats in their extracellular domains but show considerable differences in their cytoplasmic tails. On the basis of these differences, they have been classified into several subfamilies (for review, see Nollet et al. 2000). Adhesive specificity has been shown to be mediated by domains at the N-terminal end of the extracellular domain.

In general, two populations of cells that express the same subtype of cadherin tend to aggregate selectively with each other, both in vivo and in vitro. In contrast, two populations of cells that express different subtypes of cadherins tend to segregate from each other. This finding has been obtained for many combinations of cadherins and by numerous studies (for review, see Takeichi 1988; Redies 2000). It is referred to as "preferentially homotypic binding" and implies that each cadherin subtype confers some adhesive specificity on the cellular structures that express it. For some combinations of cadherins, heterotypic binding is also observed, but it is usually weaker than homotypic binding, with few exceptions (Shimoyama et al. 2000).

The expression of almost all cadherins that have been studied in the CNS to date is spatially restricted to a high degree (for review, see Redies 2000). The expression pattern of each cadherin is unique, although there is partial overlap between the expression domains of many cadherins. Coexpression of cadherin subtypes has been observed both at the regional and cellular levels. The cadherin-mediated adhesive code is thus a combinatorial one, and the combined expression patterns are highly complex. Interestingly, it is possible to relate them to several of the developmental processes that are observed during the emergence of functional modularity from embryonic modularity, as discussed in the following sections.

The concept of preferentially homotypic binding between cadherin-expressing cell somata can be generalized to neuronal processes. It has been observed in numerous in vivo studies that neurites associate with each other according to which cadherin subtype they express (for review, see Redies 2000). This association of nerve fibers is called "fasciculation," and it may play a role in the pathfinding of neurites that grow along preestablished routes, following pioneering fibers. A role for N-cadherin in this fasciculation has been experimentally established in vivo (Honig and Rutishauser 1996; Iwai et al. 1997). Moreover, the neurites that are growing into a target area and the neurons in that area often coexpress the same cadherin subtype. It has therefore been suggested that some aspects of the interaction between the ingrowing neurites and the neurons in the target center are at least partly mediated by cadherins (Redies et al. 1993; for review, see Redies 2000), jointly with other molecules (semaphorins, ephrins, netrins, etc.). Last but not least, it has been proposed that the specific association of pre- and postsynaptic neuronal membranes at the synapse is mediated by cadherins (Yamagata et al. 1995; Fannon and Colman 1996; Uchida et al. 1996). In summary, cadherins seem to participate not only in morphogenetic processes but also in so-called morphostatic processes (Wagner and Misof 1993; Wagner 1994). The latter type of processes tends to fix histogenetic patterns, which are dynamic in space and time, into permanent anatomical structure, at different levels of histogenesis.

The Divisional Pattern of the Embryonic CNS Is Translated into a Pattern of Potentially Adhesive Cues

Despite their histological homogeneity, the neuroepithelium and the radial glia show a considerable molecular diversity at the regional level, as a result of the patterning processes that take place early in development. This regional molecular diversity is also reflected in the expression of morphogenetic and/or morphostatic molecules, such as cadherins (Gänzler and Redies 1995; for review, see Redies and Takeichi 1996; Redies 2000). At the divisional borders, the subtype of cadherin that the radial glia and the other neuroepithelial cells express often changes so that many of the boundaries coincide with a change of adhesiveness. Since the differential adhesive cues are carried by the radial glial processes, the adhesive boundaries extend all the way from the ventricular to the pial surface (see, e.g., the cadherin-7-positive longitudinal domain of the chicken spinal cord [*arrows* in fig. 8.6]).

It has been proposed that changes in adhesiveness at the divisional boundaries play a role in restricting cell mixing and migration across such boundaries (Gänzler and Redies 1995). Cell culture experiments show that neural cells that are derived from adjacent embryonic subdivisions generally do not mix freely with each other in a calcium-dependent manner (Götz et al. 1996; Stoykova et al. 1997). Moreover, Espescth et al. (1998) demonstrated that F-cadherin plays a role in localizing the neural cells that express this molecule to specific regions in

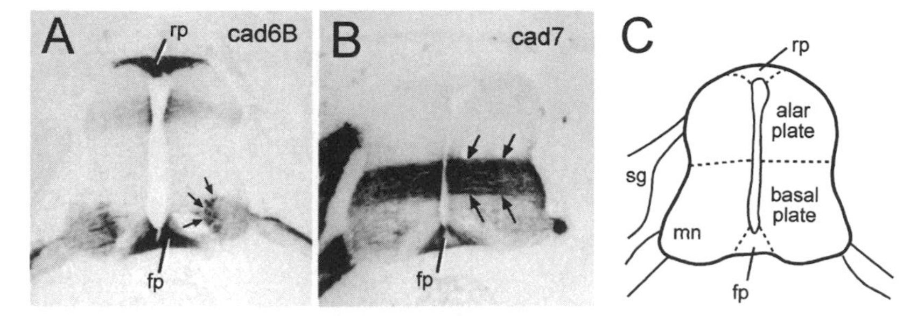

Fig. 8.6.—Longitudinal (ventral-to-dorsal) domains of the embryonic chicken spinal cord, as revealed by the expression of cadherin-6B (cad6B, *A*) and cadherin-7 (cad7, *B*). *C* shows a schematic diagram of the major ventral-to-dorsal subdivisions. Cadherin-6B is expressed by the roof plate (rp), a subdivision of the alar plate, a subgroup of motor neurons (mn), and the floor plate (fp). Cadherin-7 expression is especially prominent in a dorsal subdivision of the basal plate and in the floor plate. sg, spinal ganglion. (Reprinted from Redies [2000], with permission from Elsevier Science, Copyright © 2000.)

the developing neural tube of *Xenopus*. In the mouse telencephalon, cadherin-6 and R-cadherin were shown to be involved in maintaining the pallial-subpallial boundary (Inoue et al. 2001). Another group of molecules, the ephrins and their receptors, have also been implicated in separation of embryonic divisions in the vertebrate brain. This group of molecules was shown to mediate mutual adhesion and repulsion between neural cells and their processes in several systems (for review, see Mellitzer et al. 2000). In the hindbrain of the chicken, Eph receptors and ephrins were implicated in the segmental restriction of cell intermingling (Xu et al. 1999). Like the cadherins (see below), ephrins and their receptors also regulate neurite outgrowth and target recognition in the developing brain (for review, see O'Leary and Wilkinson 1999).

In general, the regional expression of cadherins in the neuroepithelium sets in at a stage of development when the regional patterning by gene regulatory proteins is already well advanced. This is in accordance with the general assumption that the expression of morphogenetic genes is regulated downstream of gene regulatory proteins. The details of how this translation is regulated at the genetic level are unclear at present. To our knowledge, there has been no report so far of a complete and precise overlap between the expression of a gene regulatory protein and a morphogenetic and/or morphostatic molecule. However, there are several studies that clearly show a relation between the two types of genes. For example, Stoykova et al. (1997) showed that, in the forebrain of the mouse, there is a regional coexpression of R-cadherin and Pax-6, and the two molecules share a common expression boundary that coincides with a divisional boundary, the pallio-subpallial limit. Wild-type neuroepithelial cells that are derived from either side of the boundary segregate to some extent in vitro. In Pax-6-deficient mice, the boundary of R-cadherin expression disappears and greater cell mixing is observed, possibly due to some repatterning of the ventral pallium (Stoykova et al. 2000).

Early embryonic pattern formation does not stop when the neuromeres and the major longitudinal subdivisions have been formed. In many of the primary divisions, secondary regionalization is observed (see also above). For example, in the cerebellum, an extensive secondary patterning gives rise to multiple and discrete parasagittal compartments of the cerebellar cortex and the deep cerebellar nuclei. Interestingly, this secondary patterning process involves some molecular players that already play a role in the primary patterning of the neural tube, such as sonic hedgehog and members of the BMP, Wnt, Pax, cadherin, Eph, and ephrin families of molecules (Millen et al. 1995; Arndt et al. 1998; Lin and Cepko 1998; Karam et al. 2000).

Adhesive Patterning Takes Place at Multiple Levels of Gray Matter Organization

Following early embryonic pattern formation, one of the next steps in CNS development is the generation of early neurons that translocate from the ventricular to the mantle layer in several waves, basically by radial migration (see above). Depending on the region involved, outside-in, inside-out, or mixed patterns of neuronal stratification can be observed in the mantle layer with regard to the cell birth dates. From the beginning of mantle layer formation, specific populations of differentiating neurons express cadherins (Gänzler and Redies 1995; Korematsu and Redies 1997; Wöhrn et al. 1998, 1999; Yoon et al. 2000). As the mantle layer grows in thickness, these populations of neurons seem to sort out progressively, according to which cadherin they express, forming early brain nuclei (fig. 8.7) and cortical layers. There seems to be no clear-cut relation between the expression of cadherin subtypes by the radial glia and their expression by the early neurons that are generated in a given embryonic division. In the chicken forebrain, many regions show coexpression of a cadherin subtype by radial glia and by at least some postmitotic neurons, but there are also examples of regions that generate neurons that express a given cadherin without expression of this molecule by the radial glia in the same region. Also, not all neurons born in a given embryonic division express the same cadherin. Cadherin expression may change over time in a given area during neurogenesis. Nevertheless, it is clear that the regionalization of adhesive specificity (and of other molecular determinants of morphogenesis) takes place not only in the neuroepithelium, but also in the mantle layer that derives from it. How far the two processes depend on each other, or are based on other mechanisms, is unclear at present. The adhesive patterning along the radial dimension that takes place during the generation of successive mantle layer strata further increases the heterogeneity of cadherin expression by the brain gray matter in each region (fig. 8.7).

In the hindbrain, neurogenesis follows a repetitive pattern in the rhombomeres and pseudorhombomeres, with similar types of neurons generated in each module (Clarke and Lumsden 1993; Marín and Puelles 1995; Cambronero and Puelles 2000). Perhaps as a consequence of this repetitiveness, similar populations of neurons come to lie at the boundaries of adjacent rhombomeres, and these populations often fuse across the boundaries at later stages of development to form the conventional hindbrain sensory and motor brain nuclei of multisegmental origin. In accordance with this observation, many of the multisegmental hindbrain nuclei show a relatively uniform expression of cadherins

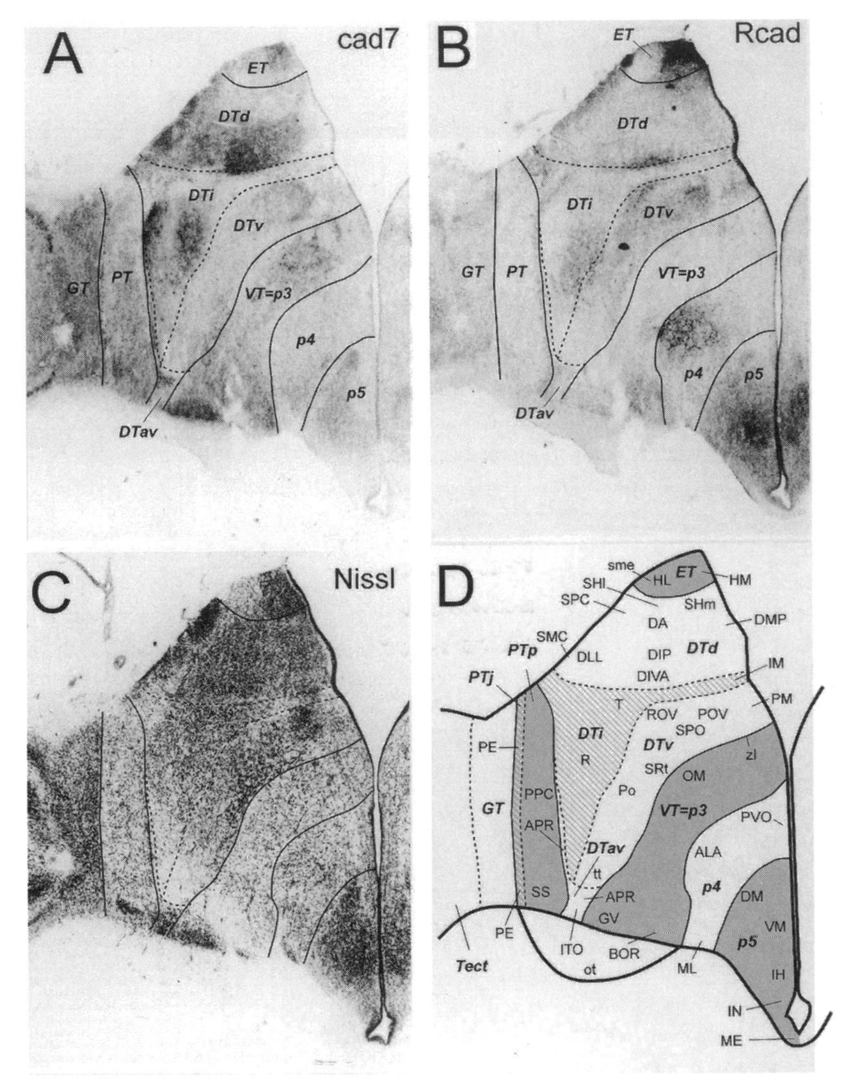

Fig. 8.7.—Expression of cadherin-7 (cad7, *A*) and R-cadherin (Rcad, *B*) in the chicken diencephalon at 11 days of incubation. A Nissl stain of an adjacent section is shown in *C*, and a schematic diagram of the embryonic divisions for this level of transverse sectioning is shown in *D*. The embryonic divisions (*different shadings in D*) have given rise to gray matter domains that contain specific sets of diencephalic brain nuclei, which differentially express cadherin-7 and R-cadherin. ALA, anterior nucleus of the ansa lenticularis; APR, perirotundic area; BOR, nucleus of the basal optic root; DA, dorsal anterior nucleus; DIP, dorsointermediate posterior nucleus; DIVA, dorsal intermedial ventral anterior nucleus; DLL, dorsolateral lateral nucleus; DM, dorsomedial nucleus; DMP, dorsomedial posterior nucleus; DTav, anteroventral subdivision of dorsal thalamus; DTd, dorsal tier of dorsal thalamus; DTi, intermediate tier of dorsal thalamus; DTv, ventral tier of dorsal thalamus; ET, epithalamus; GT, griseum tectale; GV, ventral geniculate nucleus; HL, lateral habenula; HM, medial habenula; IH, inferior hypothalamic nucleus; IM, intermediomedial nucleus; IN, infundibular nucleus of hypothalamus; ITO, prospective interstitial nucleus of the optic tract; ME, median eminence; ML, lateral mamillary nucleus; OM, nucleus of the occipito-mesencephalic tract; ot, optic tract; p3–p5, prosomeres 3–5; PE, external pretectal nucleus; PM, medial pontine nucleus; Po, posterior nucleus; POV, periovoidal nucleus; PPC, principal precommissural

(C. Redies, unpublished observations). However, detailed analysis of the connectivity of some multisegmental hindbrain complexes, for example, of the vestibular sensory column, has revealed distinct specializations in the fiber connections within this column (e.g., contralateral versus ipsilateral targets, ascending versus descending fiber projections, central versus peripheral topography of tracts; Glover 2000). These specializations clearly relate to the rhombomeric origins of the different parts (Diaz et al. 1998). The afferent (vestibular) information is thus distributed to a longitudinal "column" of gray matter, but this column can be divided into transverse units that differ in their segmental origin, histological architecture, and functional specialization.

The forebrain boundaries, which coincide with adhesive changes in the neuroepithelial cells and their processes during early embryogenesis (see above), are gradually replaced by the borders of the brain nuclei that derive from each division (see fig. 8.7). At these late stages of development, most cadherins are not expressed along the entire border from the ventricular to the pial surface, but a given border is often delineated by several different types of adhesive changes along its radial course, in a complex pattern that needs careful tracing in three dimensions (Redies et al. 2000, 2001). Perhaps as a consequence of this heterogeneity, a large majority of the conventionally named brain nuclei in the forebrain of the chicken seem to originate in one division and remain in it; that is, unlike many hindbrain nuclei, they are of unisegmental origin.

Nevertheless, note that the optic tract, a longitudinal system of afferents, crosses the series of prosomeres on its way to the midbrain and connects with a series of segmentally distributed visual centers with different functions in visual processing (Martínez et al. 1991; Puelles et al. 1991; Puelles 1995) in a way that is similar to the columnar and transverse organization mentioned above for the hindbrain. The difference between hindbrain and forebrain nuclei therefore lies not so much in the observed functional organization as in the gross histological appearance of the nuclei. This has led anatomists, on the one hand, to split nuclei in the forebrain and to give them different names, and, on the other hand, to lump functionally distinct units under rougher, unifying nuclear concepts in the hindbrain.

Most cadherins studied so far are expressed in multiple brain re-

---

nucleus; PT, pretectum; PVO, periventricular organ; R, nucleus rotundus; ROV, retroovoidal nucleus; SHl, lateral subhabenular region; SHm, medial subhabenular region; SMC, superficial microcellular nucleus; sme, stria medullaris; SPC, superfical parvocellular nucleus; SPO, nucleus semilunaris periovoidalis; SRt, subrotundic nucleus; SS, superficial synencephalic nucleus; Tect, optic tectum; T, triangular nucleus; tt, tectothalamic tract; VM, ventromedial hypothalamic nucleus; VT, ventral thalamus; zl, intrathalamic zona limitans. (Modified and reproduced with permission from Redies et al. [2000], Copyright © 2000, Wiley-Liss, Inc., a subsidiary of John Wiley & Sons, Inc.)

gions, and each histogenetic primary or secondary division usually gives rise to gray matter structures that express several cadherins. This is in contrast to some of the gene regulatory factors that are expressed during early pattern formation, because these usually define large regions in the CNS, such as entire pallial, alar, or basal domains (fig. 8.5). In the next section, we address the question of what role the scattered expression of multiple cadherins throughout the CNS may play in the establishment of functional connectivity between parts of the larger regions.

Adhesive Specificity as a Basis for Neural Circuit Formation and Synaptogenesis

As the neurons settle in the mantle layer (and sometimes before), they start to extend neurites to other neurons that are located in the same or in other regions of the CNS (interneurons and projection neurons). The first (pioneer) axons are usually extending in early development when the embryonic brain is still small and the distances to be covered are relatively short. At this early stage of development, diffusible substances, such as members of the netrin and slit families of genes, set up regional molecular gradients in the CNS that guide axonal growth by attractive and repulsive mechanisms. Gradients of attractive or repulsive cues (e.g., of ephrins and their receptors) are established in the target areas, and the ingrowing axons often carry corresponding gradients of matching receptors that help to correctly position the terminals within the target areas (for review, see Drescher et al. 1997; O'Leary and Wilkinson 1999). The expression patterns of many of these molecules are highly heterogeneous in the CNS, but in general, their relation to the early embryonic divisional pattern has been studied in less detail than that of the cadherins (but see, e.g., Puelles et al. 1987 for acetylcholinesterase, which is thought to function in early development as a cell adhesion protein; see Chédotal et al. 1995 for SC1/BEN/DM-GRASP; see Allendoerfer et al. 1999 for FORSE-1; and see Xu et al. 1999 for ephrins and Eph receptors).

Molecules such as cadherins seem to play only a minor role (if any) in the initial establishment of axonal connectivity by pioneer axons within the CNS, because the early areas expressing a given cadherin subtype are often separated from each other by areas that do not express the same cadherin subtype. Members of the immunoglobulin superfamily of adhesion molecules, which show more extensive heterophilic binding, are more likely to be involved in the outgrowth of pioneer axons (for review, see Tessier-Lavigne and Goodman 1996). Some of the early fiber tracts were initially described as preferentially growing along divisional boundaries (e.g., see Wilson et al. 1993; Puelles

1995), which often contain glial cell populations that express specific adhesion molecules. However, recent analysis of this point suggests that forebrain tracts like the posterior commissure, retroflex tract, mammillothalamic tract and fornix tract lie topographically "adjacent to" segmental boundaries and not "in" the boundaries (Hjörth and Key 2001). Many fiber systems clearly cross regional boundaries to project from one division to another. It is possible that adhesive cues that are expressed at a boundary guide or restrict growth of fibers at one side of the boundary or across it, in concert with other attractive and repulsive mechanisms (see, e.g., axon guidance at the midline; Stoeckli and Landmesser 1995; for review, see Kaprielian et al. 2000).

After initial fiber connections have formed, many more axons usually follow the path that is set up by the pioneer axons in the vertebrate brain. For this tracking of axons along preexisting axon pathways, adhesive cues could possibly be used. Many cadherins were shown to be differentially expressed by the very early axonal pathways, and cadherin expression usually persists during active neurite outgrowth and fasciculation (Redies et al. 1992; Shimamura et al. 1992; for review, see Redies 2000). What remains to be shown experimentally is that this fiber outgrowth is selectively induced by a homotypic adhesive mechanism, that is, that fibers that express a cadherin subtype preferentially grow along neurites that express the same cadherin subtype.

Both growing neurites and their target often express the same cadherin subtype, in a matching fashion (Redies et al. 1993; for review, see Redies 1997). For example, figure 8.8 shows that specific brain nuclei in the visual system (fig. 8.8, *B*) and in the motor system (fig. 8.8, *A*) express R-cadherin and are connected by fiber tracts that also express this molecule (Arndt and Redies 1996). Subcircuits of the other functional systems also express R-cadherin. Such a relation of cadherin expression with functional connectivity patterns has been observed for most other classic cadherins, and also for some protocadherins (for review, see Redies 2000). Each cadherin is expressed in subcircuits of several functional systems, although one or the other system may dominate in the expression pattern. In turn, the functional systems can be subdivided into a number of neural subcircuits, each of which is marked by the expression of a specific cadherin. Cadherins thus provide an adhesive code not only for the divisional structure of the embryonic brain and its derived gray matter but also for the diverse functional connections of the mature brain.

The matching expression of cadherins by neurites and their targets raises the question of whether the adhesive specificity mediated by cadherins is involved in some aspects of target recognition, such as growth arrest, defasciculation, or branching of axons, as well as synaptogenesis, or synaptic plasticity. A role for N-cadherin and a few other

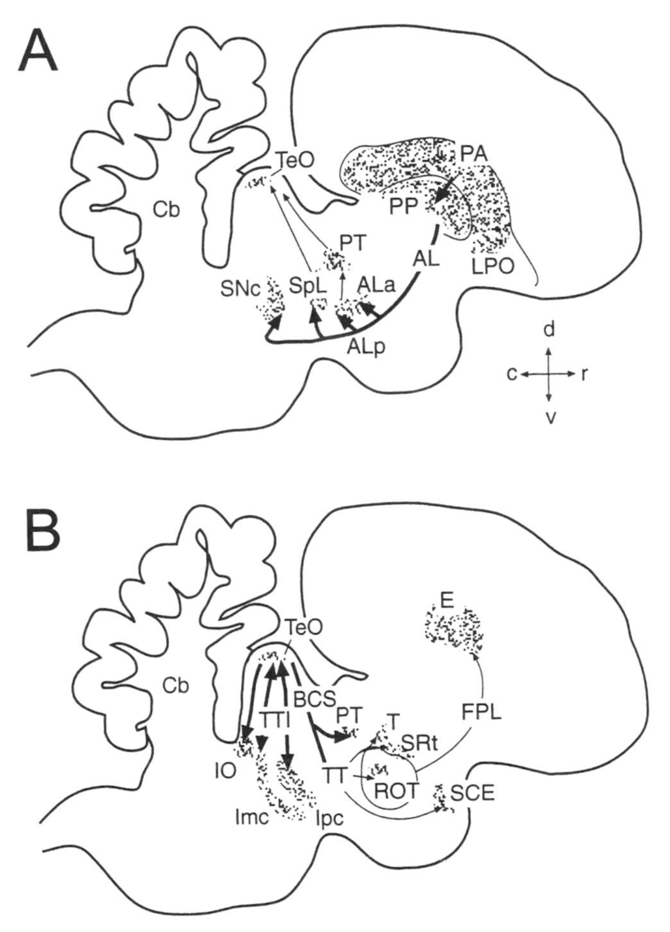

Fig. 8.8.—Schematic overview of R-cadherin expression by neural circuits in the motor system (*A*) and the visual system (*B*) of the chicken embryo at 11 days of incubation. The abbreviations denote brain structures that express R-cadherin. The arrows represent fiber tracts that connect the R-cadherin-positive gray matter structures. The fiber tracts represented by the thick arrows were found to express R-cadherin by immunostaining. Subcircuits of other functional systems, such as the auditory and somatosensory systems, also express R-cadherin. AL, ansa lenticularis; ALa, anterior nucleus of the ansa lenticularis; ALp, posterior nucleus of the ansa lenticularis; BCS, brachium of the superior colliculus; c, caudal; Cb, cerebellum; d, dorsal; E, ectostriatum; FPL, lateral forebrain bundle; Imc, isthmic nucleus, principal magnocellular part; IO, isthmooptic nucleus; Ipc, isthmic nucleus, principal parvocellular part; LPO, parolfactory lobe; PA, paleostriatum augmentatum; PP, paleostriatum primitivum; PT, principal pretectal nucleus; r, rostral; ROT, nucleus rotundus; SCE, external cellular stratum; SNc, substantia nigra, pars compacta; SpL, lateral spiriform nucleus; SRt, subrotundic nucleus; TeO, optic tectum; T, triangular nucleus; TT, tectothalamic tract; TTI, tectoisthmic tract; v, ventral. (Reproduced with permission from Arndt and Redies [1996], Copyright © 1996, Wiley-Liss, Inc., a subsidiary of John Wiley & Sons, Inc.)

cadherins has been demonstrated for some of these developmental mechanisms, such as for layer-specific termination of axons (Inoue and Sanes 1997) and synaptic plasticity (Yamagata et al. 1999; Manabe et al. 2000). It has also become clear in recent years that cadherins are expressed during the formation and stabilization of synapses (Yamagata et al. 1995; Fannon and Colman 1996; Uchida et al. 1996; Tanaka et al. 2000), sometimes in specific synapses (Huntley and Benson 1999), suggesting a role for cadherins in synaptic specificity. However, the precise role of cadherin-mediated adhesiveness in target recognition and its relation to other developmental mechanisms remains to be studied in more detail.

General Conclusion and Outlook

We reviewed evidence for two types of modularity in the vertebrate CNS. The morphological embryonic divisions of the early neural tube represent the first type of module. The embryonic divisions are well conserved in evolution. They form largely independent histogenetic fields that are specified by position-dependent expression of members of various gene regulatory proteins. These proteins regulate the expression of genes involved in patterning and morphogenesis. Each division takes up a topologically unique location in relation to the other divisions along the longitudinal and transverse axes of the neural tube. This topological relation does not change during development, irrespective of distortions that take place when the brain grows in size. As development proceeds, secondary and tertiary patterning occurs within each division and along the radial dimension. The mechanisms regulating these additional rounds of patterning are not well known at present. It has been speculated that secondary and tertiary patterning may depend on the extent to which each brain division grows during development (Redies and Puelles 2001). As a result of the patterning processes, diverse gray matter structures emerge within each division. Gray matter areas from different divisions become connected by fiber tracts, in a precise and orderly fashion, to form neural circuits. The neural circuits represent the other type of modularity in the CNS, that is, functional modularity. The neural circuits are computational units that simultaneously and independently carry out specific information-processing tasks. Evolutionary divergence of brain regions may be based on altered gene expression, which, in turn, may lead to differences between species in the growth of an individual neuromere, in the migratory behavior of the derived cell populations, or in cell fate determination (Redies and Puelles 2001).

The embryonic divisional pattern is not lost during development but is translated into an array of potentially adhesive cues, which is manifested by the expression of cadherins and other cell adhesion proteins. The cadherin-based adhesive cues possibly regulate the emergence of functional gray matter structures, such as brain nuclei, or other functional subdivisions within gray matter. Moreover, it is often found that the same cadherin subtype is present in different regions of the CNS, which become functionally connected to form parts of specific neural circuits. One of the general mechanisms that may underlie these diverse developmental processes is the preferentially homotypic adhesiveness mediated by cadherins. In general, the cytoplasmic membranes of two cells that express the same cadherin subtypes at the same time selectively associate with each other (for review, see Takeichi 1988; Redies 2000). Inversely, the cytoplasmic membranes of two cells that express different cadherin combinations segregate from each other. In CNS development, the concept of homotypic adhesive specificity can be applied to the association of neuroepithelial cells, neurons, neurites, and synaptic membranes (Takeichi et al. 1990; Redies et al. 1993; Redies 1995; Redies 2000). It has recently been proposed (Redies and Puelles 2001) that this concept may perhaps be as basic as the concept of the "Hebbian synapse" (Hebb 1949), which also regulates the establishment of functional connectivity in the CNS. The Hebbian concept implies that the functional connection between two neurons that repeatedly fire at the same time becomes stronger. Both concepts describe the functional binding of neurons. Adhesive specificity may provide one of the bases for structural, morphogenic, and/or morphostatic binding, while the Hebbian rule regulates functional binding at the level of plastic neural activity. Recent results suggest that these two concepts meet each other at the synapse (Yamagata et al. 1999; Manabe et al. 2000; Tanaka et al. 2000).

Acknowledgments

The research described in this chapter was supported by DFG Re 616/4-3 (to C. Redies) and DGES grant PB98-0397 (to L. Puelles).

References

Allendoerfer, K. L., A. Durairaj, G. A. Matthews, and P. H. Patterson. 1999. Morphological domains of Lewis-X/FORSE-1 immunolabeling in the embryonic neural tube are due to developmental regulation of cell surface carbohydrate expression. *Dev. Biol.* 211:208–219.

Anderson, S. A., D. D. Eisenstat, L. Shi, and J. L. R. Rubenstein. 1997. Interneuron migration from basal forebrain to neocortex: dependence on *Dlx* genes. *Science* 278: 474–476.

Arndt, K., S. Nakagawa, M. Takeichi, and C. Redies. 1998. Cadherin-defined segments and parasagittal cell ribbons in the developing chicken cerebellum. *Mol. Cell. Neurosci.* 10:211–228.

Arndt, K., and C. Redies. 1996. Restricted expression of R-cadherin by brain nuclei and neural circuits of the developing chicken brain. *J. Comp. Neurol.* 373:373–399.

Bayer, S. A., and J. Altmann. 1995. Principles of neurogenesis, neuronal migration, and neural circuit formation. In *The rat nervous system*, ed. G. Paxinos, 1079–1098. San Diego: Academic Press.

Bergquist, H., and B. Källén. 1954. Notes on the early histogenesis and morphogenesis of the central nervous system in vertebrates. *J. Comp. Neurol.* 100:627–659.

Birgbauer, E., and S. E. Fraser. 1994. Violation of cell lineage restriction compartments in the chick hindbrain. *Development* 120:1347–1356.

Bulfone, A., L. Puelles, M. H. Porteus, M. A. Frohmann, G. R. Martin, and J. L. R. Rubenstein. 1993. Spatially restricted expression of *Dlx-1, Dlx-2 (Tes-1), Gbx-2,* and *Wnt-3* in the embryonic day 12.5 mouse forebrain defines potential transverse and longitudinal segmental boundaries. *J. Neurosci.* 13:3155–3172.

Cambronero, F., and L. Puelles. 2000. Rostrocaudal nuclear relationships in the avian medulla oblongata: A fate map with quail chick chimeras. *J. Comp. Neurol.* 427: 522–545.

Chédotal, A., O. Pourquié, and C. Sotelo. 1995. Initial tract formation in the brain of the chick embryo: selective expression of the BEN/SC1/DM-GRASP cell adhesion molecule. *Eur. J. Neurosci.* 7:198–212.

Clarke, J. D., and A. Lumsden. 1993. Segmental repetition of neuronal phenotype sets in the chick embryo hindbrain. *Development* 118:151–162.

Cobos, I., L. Puelles, and S. Martínez. 2001. The avian telencephalic subpallium originates tangentially migrating inhibitory neurons that invade the dorsal ventricular ridge and the cortical areas. *Dev. Biol.* 239:30–45.

Davila, J. C., S. Guirado, and L. Puelles. 2000. Expression of calcium-binding proteins in the diencephalon of the lizard *Psammodromus algirus. J. Comp. Neurol.* 427: 67–92.

De Castro, F., I. Cobos, L. Puelles, and S. Martinez. 1998. Calretinin in pretecto- and olivocerebellar projections in the chick: immunohistochemical and experimental study. *J. Comp. Neurol.* 397:149–162.

Diaz, C., L. Puelles, F. Marín, and J. C. Glover. 1998. The relationship between rhombomeres and vestibular neuron populations as assessed in quail-chicken chimeras. *Dev. Biol.* 202:14–28.

Drescher, U., F. Bonhoeffer, and B. K. Müller. 1997. The Eph family in retinal axon guidance. *Curr. Opin. Neurobiol.* 7:75–80.

Espeseth, A., G. Marnellos, and C. Kintner. 1998. The role of *F-cadherin* in localizing cells during neural tube formation in *Xenopus* embryos. *Development* 125:301–312.

Fannon, A. M., and D. R. Colman. 1996. A model for central synaptic junctional complex formation based on the differential adhesive specificities of the cadherins. *Neuron* 17:423–434.

Figdor, M. C., and C. D. Stern. 1993. Segmental organization of embryonic diencephalon. *Nature* 363:630–634.

Fraser, S., R. Keynes, and A. Lumsden. 1990. Segmentation in the chick embryo hindbrain is defined by cell lineage restrictions. *Nature* 344:431–435.

Gänzler, S. I. I., and C. Redies. 1995. R-cadherin expression during nucleus formation in chicken forebrain neuromeres. *J. Neurosci.* 15:4157–4172.

Glover, J. C. 2000. Neuroepithelial "compartments" and the specification of vestibular projections. *Prog. Brain Res.* 124:3–21.

Götz, M., A. Wizenmann, S. Reinhardt, A. Lumsden, and J. Price. 1996. Selective adhesion of cells from different telencephalic regions. *Neuron* 16:551–564.

Gumbiner, B. M. 1996. Cell adhesion: the molecular basis of tissue architecture and morphogenesis. *Cell* 84:345–357.

Hallonet, M. E., and N. M. Le Douarin. 1993. Tracing neuroepithelial cells of the mesencephalic and metencephalic alar plates during cerebellar ontogeny in quail-chick chimaeras. *Eur. J. Neurosci.* 5:1145–1155.

Hauptmann, G., and T. Gerster. 2000. Regulatory gene expression patterns reveal transverse and longitudinal subdivisions of the embryonic zebrafish forebrain. *Mech. Dev.* 91:105–118.

Hebb, D. O. 1949. *The organization of behaviour.* New York: John Wiley.

Hjörth, E. T., and B. Key. 2001. Are pioneer axons guided by regulatory gene expression domains in the zebrafish forebrain? High-resolution analysis of the patterning of the zebrafish brain during axon tract formation. *Dev. Biol.* 229:271–286.

Holland, P. W. H., and J. Garcia-Fernandez. 1996. Hox genes and chordate evolution. *Dev. Biol.* 173:382–395.

Honig, M. G., and U. S. Rutishauser. 1996. Changes in the segmental pattern of sensory neuron projections in the chick hindlimb under conditions of altered cell adhesion molecule function. *Dev. Biol.* 175:325–337.

Huntley, G. W., and D. L. Benson. 1999. Neural (N)-cadherin at developing thalamocortical synapses provides an adhesion mechanism for the formation of somatotopically organized connections. *J. Comp. Neurol.* 407:453–471.

Inoue, A., and J. R. Sanes. 1997. Lamina-specific connectivity in the brain: regulation by N-cadherin, neurotrophins, and glycoconjugates. *Science* 276:1428–1431.

Inoue, T., S. Nakamura, and N. Osumi. 2000. Fate mapping of the mouse prosencephalic neural plate. *Dev. Biol.* 219:373–383.

Inoue, T., T. Tanaka, M. Takeichi, O. Chisaka, S. Nakamura, and N. Osumi. 2001. Role of cadherins in maintaining the compartment boundary between the cortex and striatum during development. *Development* 128:561–569.

Iwai, Y., T. Usui, S. Hirano, R. Steward, M. Takeichi, and T. Uemura. 1997. Axon patterning requires *D*N-cadherin, a novel neuronal adhesion receptor, in the *Drosophila* embryonic CNS. *Neuron* 19:77–89.

Kaprielian, Z., R. Imondi, and E. Runko. 2000. Axon guidance at the midline of the developing CNS. *Anat. Rec.* 261:176–197.

Karam, S. D., R. C. Burrows, C. Logan, S. Koblar, E. B. Pasquale, and M. Bothwell. 2000. Eph receptors and ephrins in the developing chick cerebellum: relationship to sagittal patterning and granule cell migration. *J. Neurosci.* 20:6488–6500.

Korematsu, K., and C. Redies. 1997. Restricted expression of cadherin-8 in segmental and functional subdivisions of the embryonic mouse brain. *Dev. Dyn.* 208:178–189.

Kuhlenbeck, H. 1967. *The central nervous system of vertebrates.* Vol. 1, *Propaedeutics to comparative neurology.* Basel: Karger.

Lavdas, A. A., M. Grigoriou, V. Pachnis, and J. G. Parnavelas. 1999. The medial ganglionic eminence gives rise to a population of early neurons in the developing cerebral cortex. *J. Neurosci.* 19:7881–7888.

Lin, J. C., and C. L. Cepko. 1998. Granule cell raphes and parasagittal domains of Purkinje cells: complementary patterns in the developing chick cerebellum. *J. Neurosci.* 18:9342–9353.

Lumsden, A., and R. Krumlauf. 1996. Patterning the vertebrate neuraxis. *Science* 274:1109–1115.

Mai, J. K., C. Andressen, and K. W. Ashwell. 1998. Demarcation of prosencephalic regions by CD15-positive radial glia. *Eur. J. Neurosci.* 10:746–751.

Manabe, T., H. Togashi, N. Uchida, S. C. Suzuki, Y. Hayakawa, M. Yamamoto, H. Yoda, T. Miyakawa, M. Takeichi, and O. Chisaka. 2000. Loss of cadherin-11 adhesion receptor enhances plastic changes in hippocampal synapses and modifies behavioral responses. *Mol. Cell. Neurosci.* 15:534–546.

Marín, F., and L. Puelles. 1994. Patterning of the embryonic avian midbrain after experimental inversions: a polarizing activity from the isthmus. *Dev. Biol.* 163:19–37.

Marín, F., and L. Puelles. 1995. Morphological fate of rhombomeres in quail/chick chimeras: a segmental analysis of hindbrain nuclei. *Eur. J. Neurosci.* 7:1714–1738.

Martínez, S., R. M. Alvarado-Mallart, M. Martínez-de-la-Torre, and L. Puelles. 1991. Retinal and tectal connections of embryonic nucleus superficialis magnocellularis and its mature derivatives in the chick. *Anat. Embryol.* 183:235–243.

Martínez, S., F. Marín, M. A. Nieto, and L. Puelles. 1995. Induction of ectopic *engrailed* expression and fate change in avian rhombomeres: intersegmental boundaries as barriers. *Mech. Dev.* 51:289–303.

Mellitzer, G., Q. Xu, and D. G. Wilkinson. 2000. Control of cell behaviour by signalling through Eph receptors and ephrins. *Curr. Opin. Neurobiol.* 10:400–408.

Milan, F. J., and L. Puelles. 2000. Patterns of calretinin, calbindin, and tyrosine-hydroxylase expression are consistent with the prosomeric map of the frog diencephalon. *J. Comp. Neurol.* 419:96–121.

Millen, K. J., C. C. Hui, and A. L. Joyner. 1995. A role for *En-2* and other murine homologues of *Drosophila* segment polarity genes in regulating positional information in the developing cerebellum. *Development* 121:3935–3945.

Molnar, Z., and C. Blakemore. 1991. Lack of regional specificity for connections formed between thalamus and cortex in coculture. *Nature* 351:475–477.

Nollet, F., P. Kools, and F. van Roy. 2000. Phylogenetic analysis of the cadherin superfamily allows identification of six major subfamilies besides several solitary members. *J. Mol. Biol.* 299:551–572.

O'Leary, D. D., and D. G. Wilkinson. 1999. Eph receptors and ephrins in neural development. *Curr. Opin. Neurobiol.* 9:65–73.

Perez Villegas, E. M., C. Olivier, N. Spassky, C. Poncet, P. Cochard, B. Zalc, J. L. Thomas, and S. Martínez. 1999. Early specification of oligodendrocytes in the chick embryonic brain. *Dev. Biol.* 216:98–113.

Pombal, M. A., and L. Puelles. 1999. Prosomeric map of the lamprey forebrain based on calretinin immunocytochemistry, Nissl stain and ancillary markers. *J. Comp. Neurol.* 414:391–422.

Puelles, L. 1995. A segmental morphological paradigm for understanding vertebrate forebrains. *Brain Behav. Evol.* 46:319–337.

Puelles, L., J. A. Amat, and M. Martínez-de-la-Torre. 1987. Segment-related, mosaic neurogenetic pattern in the forebrain and mesencephalon of early chick embryos. 1. Topography of AChE-positive neuroblasts up to stage HH18. *J. Comp. Neurol.* 266: 247–268.

Puelles, L., M. Guillén, and M. Martínez-de-la-Torre. 1991. Observations on the fate of nucleus superficialis magnocellularis of Rendahl in the avian diencephalon, bearing on the organization and nomenclature of neighboring retinorecipient nuclei. *Anat. Embryol.* 183:221–233.

Puelles, L., E. Kuwana, E. Puelles, A. Bulfone, K. Shimamura, J. Keleher, S. Smiga, and J. Rubenstein. 2000. Pallial and subpallial derivatives in the embryonic chick and mouse telencephalon, traced by the expression of the genes Dlx-2, Emx-1, Nkx-2.1, Pax-6 and Tbr-1. *J. Comp. Neurol.* 424:409–438.

Puelles, L., E. Kuwana, E. Puelles, and J. L. Rubenstein. 1999. Comparison of the mammalian and avian telencephalon from the perspective of gene expression data. *Eur. J. Morphol.* 37:139–150.

Puelles, L., and L. Medina. 1994. Development of neurons expressing tyrosine hydroxylase and dopamine in the chicken brain: a comparative segmental analysis. In *Phylogeny and development of catecholamine systems in the CNS of vertebrates,* ed. W. J. A. J. Smeets and A. Reiner, 381–401. Cambridge: Cambridge University Press.

Puelles, L., F. J. Milán, and M. Martínez-de-la-Torre. 1996. A segmental map of architectonic subdivisions in the diencephalon of the frog *Rana perezi:* acetylcholinesterase-histochemical observations. *Brain Behav. Evol.* 47:279–310.

Puelles, L., and J. L. R. Rubenstein. 1993. Expression patterns of homeobox and other putative regulatory genes in the embryonic mouse forebrain suggest a neuromeric organization. *Trends Neurosci.* 16:472–479.

Puelles, L., M. P. Sanchez, R. Spreafico, and A. Fairen. 1992. Prenatal development of calbindin immunoreactivity in the dorsal thalamus of the rat. *Neuroscience* 46: 135–147.

Puelles, L., and C. Verney. 1998. Early neuromeric distribution of tyrosine-hydroxylase-immunoreactive neurons in human embryos. *J. Comp. Neurol.* 394:283–308.

Redies, C. 1995. Cadherin expression in the developing vertebrate brain: from neuromeres to brain nuclei and neural circuits. *Exp. Cell Res.* 220:243–256.

Redies, C. 1997. Cadherins and the formation of neural circuitry in the vertebrate CNS. *Cell Tissue Res.* 290:405–413.

Redies, C. 2000. Cadherins in the central nervous system. *Prog. Neurobiol.* 61: 611–648.

Redies, C., M. Ast, S. Nakagawa, M. Takeichi, M. Martínez-de-la-Torre, and L. Puelles. 2000. Morphological fate of diencephalic neuromeres and their subdivisions revealed by mapping cadherin expression. *J. Comp. Neurol.* 421:481–514.

Redies, C., K. Engelhart, and M. Takeichi. 1993. Differential expression of N- and R-cadherin in functional neuronal systems and other structures of the developing chicken brain. *J. Comp. Neurol.* 333:398–416.

Redies, C., H. Inuzuka, and M. Takeichi. 1992. Restricted expression of N- and R-cadherin on neurites of the developing chicken CNS. *J. Neurosci.* 12:3525–3534.

Redies, C., L. Medina, and L. Puelles. 2001. Cadherin expression by embryonic divisions and derived gray matter structures in the telencephalon of the chicken. *J. Comp. Neurol.* 438:253–285.

Redies, C., and L. Puelles. 2001. Modularity in vertebrate brain development and evolution. *BioEssays* 23:1100–1111.

Redies, C., and M. Takeichi. 1996. Cadherins in the developing central nervous system: an adhesive code for segmental and functional subdivisions. *Dev. Biol.* 180: 413–423.

Rendahl, H. 1924. Embryologische und morphologische Studien über das Zwischenhirn beim Huhn. *Acta Zool. (Stockh.)* 5:241–344.

Rubenstein, J. L. R., K. Shimamura, S. Martínez, and L. Puelles. 1998. Regionalization of the prosencephalic neural plate. *Annu. Rev. Neurosci.* 21:445–477.

Schlaggar, B. L., and D. D. M. O'Leary. 1991. Potential of visual cortex to develop an array of functional units unique to somatosensory cortex. *Science* 252:1556–1560.

Shimamura, K., D. J. Hartigan, S. Martínez, L. Puelles, and J. L. R. Rubenstein. 1995. Longitudinal organization of the anterior neural plate and neural tube. *Development* 121:3923–3933.

Shimamura, K., and J. L. R. Rubenstein. 1997. Inductive interactions direct early regionalization of the mouse brain. *Development* 124:2709–2718.

Shimamura, K., T. Takahashi, and M. Takeichi. 1992. E-cadherin expression in a particular subset of sensory neurons. *Dev. Biol.* 152:242–254.

Shimoyama, Y., G. Tsujimoto, M. Kitajima, and M. Natori. 2000. Identification of three human type-II classic cadherins and frequent heterophilic interactions between different subclasses of type-II classic cadherins. *Biochem. J.* 349:159–167.

Smith-Fernandez, A., C. Pieau, J. Reperant, E. Boncinelli, and M. Wassef. 1998. Expression of the Emx-1 and Dlx-1 homeobox genes define three molecularly distinct domains in the telencephalon of mouse, chick, turtle and frog embryos: implications for the evolution of telencephalic subdivisions in amniotes. *Development* 125:2099–2111.

Stoeckli, E. T., and L. T. Landmesser. 1995. Axonin-1, Nr-CAM, and Ng-CAM play different roles in the in vivo guidance of chick commissural neurons. *Neuron* 14: 1165–1179.

Stoeckli, E. T., and L. T. Landmesser. 1998. Axon guidance at choice points. *Curr. Opin. Neurobiol.* 8:73–79.

Stoykova, A., M. Götz, P. Gruss, and J. Price. 1997. *Pax6*-dependent regulation of adhesive patterning, R-cadherin expression and boundary formation in developing forebrain. *Development* 124:3765–3777.

Stoykova, A., D. Treichel, M. Hallonet, and P. Gruss. 2000. Pax6 modulates the dorsoventral patterning of the mammalian telencephalon. *J. Neurosci.* 20:8042–8050.

Striedter, G. F., and S. Beydler. 1997. Distribution of radial glia in the developing telencephalon of chicks. *J. Comp. Neurol.* 387:399–420.

Takeichi, M. 1988. The cadherins: cell-cell adhesion molecules controlling animal morphogenesis. *Development* 102:639–655.

Takeichi, M., H. Inuzuka, K. Shimamura, T. Fujimori, and A. Nagafuchi. 1990. Cadherin subclasses: differential expression and their roles in neural morphogenesis. *Cold Spring Harbor Symp. Quant. Biol.* 55:319–325.

Tan, K., and N. Le Douarin. 1991. Development of the nuclei and cell migration in the medulla oblongata: application of the quail-chick system. *Anat. Embryol.* 183: 321–343.

Tanaka, H., W. Shan, G. R. Phillips, K. Arndt, O. Bozdagi, L. Shapiro, G. W. Huntley, D. L. Benson, and D. R. Colman. 2000. Molecular modification of N-cadherin in response to synaptic activity. *Neuron* 25:93–107.

Tello, J. F. 1923. Les différenciations neuronales dans l'embryon du poulet, pendant les premiers jours de l'incubation. *Travaux Lab. Rech. Biol. (Madr.)* 21:1–93.

Tessier-Lavigne, M., and C. S. Goodman. 1996. The molecular biology of axon guidance. *Science* 274:1123–1133.

Timsit, S., S. Martinez, B. Allinquant, F. Peyron, L. Puelles, and B. Zalc. 1995. Oligodendrocytes originate in a restricted zone of the embryonic ventral neural tube defined by DM-20 mRNA expression. *J. Neurosci.* 15:1012–1024.

Uchida, N., Y. Honjo, K. R. Johnson, M. J. Wheelock, and M. Takeichi. 1996. The catenin/cadherin adhesion system is localized in synaptic junctions bordering transmitter release zones. *J. Cell Biol.* 135:767–779.

Uchikawa, M., Y. Kamachi, and H. Kondoh. 1999. Two distinct subgroups of Group B Sox genes for transcriptional activators and repressors: their expression during embryonic organogenesis of the chicken. *Mech. Dev.* 84:103–120.

Vaage, S. 1969. The segmentation of the primitive neural tube in chick embryos (*Gallus domesticus*). *Ergeb. Anat. Entwicklungsgesch.* 41:1–88.

Verney, C., N. Zecevic, and L. Puelles. 2001. Structure of longitudinal brain zones that provide the origin for the substantia nigra and ventral tegmental area in human embryos, as revealed by cytoarchitecture and tyrosine hydroxylase, calretinin, calbindin, and GABA immunoreactions. *J. Comp. Neurol.* 429:22–44.

Voigt, T. 1989. Development of glial cells in the cerebral wall of ferrets: direct tracing of their transformation from radial glia into astrocytes. *J. Comp. Neurol.* 289:74–88.

Wagner, G. P. 1994. Homology and the mechanisms of development. In *Homology: the hierarchic basis of comparative biology,* ed. B. K. Hall, 274–301. San Diego: Academic Press.

Wagner, G. P., and B. Y. Misof. 1993. How can a character be developmentally constrained despite variation in developmental pathways? *J. Evol. Biol.* 6:449–455.

Wilkinson, D. G. 1993. Molecular mechanisms of segmental patterning in the vertebrate hindbrain and neural crest. *BioEssays* 15:499–505.

Wilson, S. W., M. Placzek, and A. J. Furley. 1993. Border disputes: do boundaries play a role in growth-cone guidance? *Trends Neurosci.* 16:316–323.

Wingate, R. J. T., and A. Lumsden. 1996. Persistence of rhombomeric organization in the postnatal hindbrain. *Development* 122:2143–2152.

Wöhrn, J.-C. P., S. Nakagawa, M. Ast, M. Takeichi, and C. Redies. 1999. Combinatorial expression of cadherins and the sorting of neurites in the tectofugal pathways of the chicken embryo. *Neuroscience* 90:985–1000.

Wöhrn, J.-C. P., L. Puelles, S. Nakagawa, M. Takeichi, and C. Redies. 1998. Cadherin expression in the retina and retinofugal pathways of the chicken embryo. *J. Comp. Neurol.* 396:20–38.

Wullimann, M. F., and L. Puelles. 1999. Postembryonic neural proliferation in the zebrafish forebrain and its relationship to prosomeric domains. *Anat. Embryol.* 199: 329–348.

Xu, Q., G. Mellitzer, V. Robinson, and D. G. Wilkinson. 1999. In vivo cell sorting in complementary segmental domains mediated by Eph receptors and ephrins. *Nature* 399:267–271.

Yamagata, K., K. I. Andreasson, H. Sugiura, E. Maru, M. Dominique, Y. Irie, N. Miki, Y. Hayashi, M. Yoshioka, K. Kaneko, H. Kato, and P. F. Worley. 1999. Arcadlin is a neural activity-regulated cadherin involved in long term potentiation. *J. Biol. Chem.* 274:19473–19479.

Yamagata, M., J. P. Herman, and J. R. Sanes. 1995. Lamina-specific expression of adhesion molecules in developing chick optic tectum. *J. Neurosci.* 15:4556–4571.

Yoon, M.-S., L. Puelles, and C. Redies. 2000. Formation of cadherin-expressing brain nuclei in diencephalic alar plate subdivisions. *J. Comp. Neurol.* 421:461–480.

Part 2

Recognition and Modeling of Modules

One of the major challenges in the analysis of complex systems like organisms is the recognition of modules. With the advent of functional genomics the awareness of this problem has become increasingly acute. While it is already possible to monitor the expression of thousands of genes using DNA microarrays, reliable methods for predicting relatively autonomous networks of tightly interacting genes from these data still have to be elaborated. Moreover, identifying modules in such large data sets is not just an empirical problem. In order to draw the boundarics between one module and another we need precise criteria for delineating modules. The general characterization of modules as units of elements that exhibit relatively invariant, robust behavior due to a high degree of cooperative interactions between their components and a low degree of perturbability by other elements does not suffice for this purpose. How can we measure degrees of robustness? And when is the contribution of a unit of interacting elements to a particular biological process sufficiently robust for the unit to qualify as a module? Theoretical models are helpful in exploring such questions. Due to their easy modifiability in silico these models also allow us to identify parameters that are critical for ensuring high degrees of robustness in a module.

In the chapter by Christof Niehrs, "synexpression groups" are introduced as an operationally characterized and heuristically useful concept to identify modules. Synexpression groups are defined by the close temporal or spatial coexpression of a cluster of genes that are also functionally related. In contrast to other concepts of modules, synexpression groups by definition include only components of modules that are coexpressed in multiple domains. Therefore, synexpression groups exclude multifunctional components of a developmental module that play additional roles independent of the module as well as domain-specific components.

The next chapter, by Roland Somogyi, Stefanie Fuhrman, Gary Anderson, Chris Madill, Larry D. Greller, and Bernard Chang, addresses the important question of how the modular structure of gene interaction networks can be reverse engineered from the flood of gene expression data created by novel high-throughput methods of functional genomics. Using an extensive data set of gene expression at various developmental stages of rat spinal cord development, the authors introduce several methods for identifying clusters of genes that are spatially or temporally expressed in a coordinated manner and reconstructing their interactions.

How modules can be delineated in complex networks of known regulatory interactions is further detailed in the chapter by Denis Thieffry and Lucas Sánchez. The focus of their logical approach, which considers only general and qualitative aspects of regulatory interactions (i.e., whether they are activating or inhibitory), is on closed regulatory circuits (feedback loops) because these have predictable effects on dynamics. Negative feedback circuits allow the buffering of perturbations, whereas positive feedback circuits may act as developmental switches between alternative developmental pathways. A cross-regulatory module is defined as the set of genes participating in a series of intertwined circuits. Such a module is predicted to play a particular and relatively invariant dynamical role, which should facilitate its repeated use during development. Examples from flower morphogenesis in *Arabidopsis* and pattern formation in *Drosophila* illustrate this approach.

The intrinsic dynamics of modules are also central to the argument of George von Dassow and Eli Meir. They even define a module by its intrinsic dynamics (rather than by the connectivity of the network underlying its invariant behavior) and discuss how such modules can be combined into higher-order modules in a hierarchical fashion. Using a dynamical systems approach that allows nonlinear interactions between various genes and proteins to be included and is thus biologically quite realistic, they model interactions of the segment polarity network in *Drosophila*. Their model shows surprisingly robust behavior and results in the formation of segment boundaries under a wide range of initial conditions and parameter variations with important implications for the evolutionary modifiability of the segment polarity network.

The final chapter in part 2, by Andrew Wuensche, discusses the relation between network architecture and the modularity of network dynamics on a more general and abstract level concentrating on random Boolean networks. Due to the limited number of states that are possible in such networks, their dynamics inevitably lead to a sequence of repeated states, or an "attractor." Only a limited number of attractors exist for each network, and the same attractor may be reached from many different initial conditions (jointly constituting its "basin of attraction"). Random Boolean networks have, for instance, been used to model genetic regulatory networks, where each node is interpreted as a gene and attractors correspond to different cell types characterized by stable differences in gene activity patterns. Exploration of such models can uncover important connections between network architecture and dynamical modularity. For instance, attractors turn out to be more stable (context insensitive) in networks containing subnetworks than in homogeneously connected networks.

9 Synexpression Groups: Genetic Modules and Embryonic Development

CHRISTOF NIEHRS

Summary

Synexpression groups are clusters of coexpressed genes that share the biological process in which they are functioning. They have recently been identified by global gene expression analysis in various eukaryotes. Synexpression groups function in diverse processes, from cell signaling and cell cycling to protein and lipid biosynthesis. They can be used to assemble molecular pathways and to predict gene function. Compartmentalization of the genetic potential in synexpression groups may be a key determinant facilitating evolutionary change leading to animal diversity by enhancing developmental modularity.

Introduction

Biology can be broken down into modules, that is, distinct ensembles consisting of interacting elements which are employed in a combinatorial fashion interacting with other units or processes. Modules occur at all levels of biological organization and can encompass physical clusters (e.g., biomolecular complexes of many kinds) as well as processes (e.g., genetic pathways; table 9.1). Modules are redeployed in many different contexts. In development, for example, a handful of modularly organized growth factor signaling pathways (e.g., Wnt, Shh, and TGF-β) are reutilized over and over again during ontogeny, from axis formation until organogenesis and in all animals analyzed.

In this chapter I will focus my discussion on one kind of molecular module, a genetic network that we call a "synexpression group" and that may play an important role during embryonic development (Niehrs and Pollet 1999).

Table 9.1: Modules in biology

Level of Organization	Module	Examples
Molecular	Building block	Amino acids, nucleotides
	Protein domain	Kinase domains, homeodomains
	Protein	No examples, since obvious
	Genetic network	Map kinase pathway, synexpression groups
Supramolecular	Multiprotein complex	Ribosome, proteasome, microtubules
Subcellular	Organelle	Mitochondria, nucleus
Supracellular	Cell	Epithelial-, mesenchymal cells
	Tissue	Somites, neuromuscular units, osteomuscular units, hair follicles, feather follicles
Anatomical	Organ	Insect segments, vertebrae, appendages
Organismic	Individual	Worker ants

Examples of Synexpression Groups

Differential gene transcription is thought to be a central mechanism underlying cell differentiation and embryonic development. Hence, where and when a gene is expressed can give important clues to its function and is a tag characterizing the expressing cell type. Furthermore, knowledge about concomitant transcriptional changes of many genes provides insight into genetic networks. Recent technical advances in gene expression monitoring have permitted global surveys of gene activity at high resolution during development and differentiation. Such surveys have been most massively done in yeast using DNA microarrays, and with more limited numbers of genes in rat and *Xenopus* embryos as well as human cells. A recurring theme in the studies that surveyed gene expression at high resolution (i.e., in many different cells or experimental conditions) was the observation of expression clusters (Cho et al. 1998; Chu et al. 1998; DeRisi et al. 1997; Hughes et al. 2000; Spellman et al. 1998; Vishwanath et al. 1999; Wen et al. 1998). Surprisingly, genes within a cluster are typically also functionally interacting. This implies that for uncharacterized members of such a cluster the mere fact of tight coexpression allows prediction of possible function by a guilt-of association paradigm.

We discovered synexpression groups in a gene expression survey in *Xenopus* embryo, using whole-mount in situ hybridization to identify differentially expressed genes during early embryogenesis. Of the 270 genes identified, 15% could be grouped into four gene sets, each of

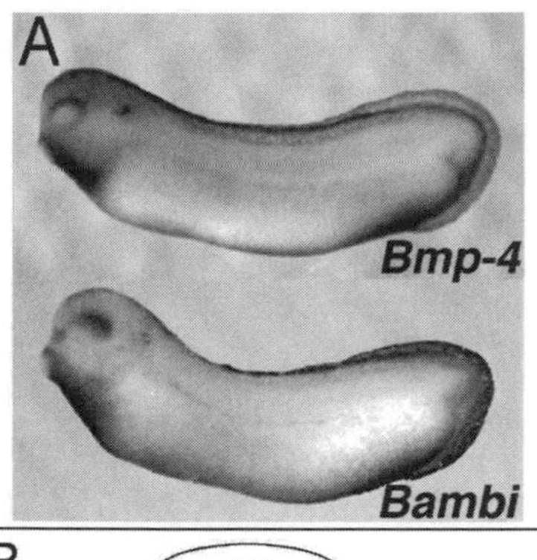

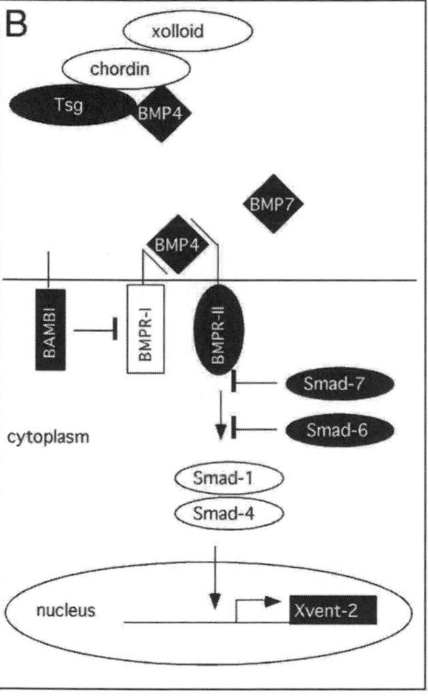

Fig. 9.1.—The BMP4 synexpression group. *A*, Whole-mount in situ hybridization of *Bmp4* and *BAMBI* in *Xenopus* tailbud embryos, illustrating the expression of members of the BMP4 synexpression group (Gawantka et al. 1998). Expression patterns are highly similar, including dorsal eye, heart, proctodaeum, and lateral plate mesoderm. *B*, Schematic drawing of the BMP signaling pathway. Bmp4 and Bmp7 are ligands for type I and II BMP transmembrane receptors, encoding Ser/thr kinases. Upon ligand binding, type II receptors phosphorylate Smad1 transcriptional activators, which complex with Smad4 and enter the nucleus, activating the transcription of immediate early target genes such as *Xvent2*, which in turn mediates the transcriptional repression of downstream targets. Inhibitory Smads such as Smad6 and Smad7 as well as BAMBI interfere with signal transduction. Chordin binds and inhibits BMPs and is itself cleaved by the Xolloid protease. Twisted gastrulation (Tsg) dislodges latent BMPs bound to Chordin BMP-binding fragments generated by Xolloid cleavage, providing a permissive signal that allows high BMP signaling in the embryo. Components of the BMP signal transduction pathway coexpressed in the pattern shown in *A* are highlighted in black. (Adapted from Niehrs and Pollet 1999.)

which has a very distinctive, complex expression pattern (Gawantka et al. 1998). An example is the BMP4 group (fig. 9.1, *A*), members of which are expressed like this growth factor dorsally in the eye, the heart, the tailbud, and lateral plate mesoderm of tailbud stage *Xenopus* embryos. This group consists of seven members, which all encode components of the BMP signaling pathway, as studied in early dorsoventral patterning of mesoderm (fig. 9.1, *B*), including ligands, receptor, and downstream components of the pathway. Such sets of coordinately expressed genes which act in the same process are referred to as "synexpression groups" (Gawantka et al. 1998; Niehrs 1997). The expression pattern of these genes is similar to the growth factor itself, and BMP4 indeed coordinately induces them (Bhushan et al. 1998; Frisch and Wright 1998; Gawantka et al. 1998; Hata et al. 1998). This group illustrates three other points. (1) The proteins encoded by genes of synexpression groups are not homologous (e.g., receptor, ligand, transcription factor) and hence coexpression is not the consequence of gene duplication; (2) while the coregulation is relatively tight, there are minor differences in the pattern—for example, unlike the other members of the group, BMP4 itself is expressed in the otic vesicle; (3) not all components of a pathway are coordinately expressed—for example, Smad1 and Smad4, which also participate in BMP4 signaling, are ubiquitously expressed during early embryogenesis (Lagna et al. 1996; Meersman et al. 1997).

Another example is the endoplasmic reticulum (ER) synexpression group, members of which are highly expressed in tissues active in secretion. Genes of this group act in the early steps of secretion, either in translocation (e.g., translocon subunits) or in protein folding in the ER (protein disulphide isomerase) (Schatz and Dobberstein 1996). The common regulatory mechanism of this group is unknown, but it suggests a transcriptional feedback between the secretory load of a cell and the expression of key components involved in protein translocation across the ER. The Delta1 synexpression group includes mostly basic helix-loop-helix (bHLH) genes that are expressed in a pattern characteristic for this ligand of the Notch receptor, for example, the CNS and the forming somites. The possibility that members of this group function in the Notch pathway (Artavanis-Tsakonas et al. 1995) has been confirmed by functional analysis of three novel members of this group, *ESR4*, *ESR5*, and *Nrarp* (Jen et al. 1999; Lamar et al. 2001). The shared expression is probably due to Delta1-responsive elements in the genes' promoters. The largest synexpression group identified in *Xenopus* is the chromatin synexpression group. Characteristic for these genes is their repression in tissues cells of which become postmitotic. Most of these genes are known to encode chromatin proteins (e.g., histones, high mobility group [HMG] proteins), or genes indirectly interacting with chromatin, such as ornithine decarboxylase, a key enzyme in spermidine synthesis. The common regulatory mechanism of this group is also unknown, but it is probably cell cycle related.

Gene groups akin to synexpression groups were also identified by producing a high-resolution temporal map of mRNA expression of 112 genes during spinal cord development (Wen et al. 1998). Using RT-PCR analysis, five basic waves of expression were discovered. A number of genes acting in the same process mapped to particular expression profiles. For example, genes encoding glutamate decarboxylases, GABA transporter, and six GABA receptor subunits function in

GABAergic signaling and are coexpressed in one characteristic wave, suggesting a synexpression group of components involved in GABA signaling.

Strong evidence for synexpression groups has been obtained in yeast, where gene-chip-based expression analyses have now been carried out with all 6,000 genes. One such study revealed that genes grouped on the basis of their common expression profile also frequently share a common role (DeRisi et al. 1997). For example, 17 genes were induced by overexpression of *YAP1*, encoding a transcription factor conferring resistance to various noxious agents. Of these, nine genes encode different kinds of oxidoreductases thought to act in detoxication. Using the same technique, the transcriptional program of sporulation in yeast was analyzed, and seven distinct temporal profiles of induction were observed (Chu et al. 1998). Genes expressed in a given profile typically are involved in the same process. For example, 11 of 17 genes with known function coexpressed in the "early I induction" temporal profile play a role in meiosis, mostly chromosome pairing, suggesting a synexpression group. As a common regulator of these genes, the *URS1* transcription factor complex has been implicated (Chu et al. 1998).

Using DNA chip technology, the transcriptional program underlying the serum response of fibroblasts has been investigated. Again, coordinated regulation of groups of genes whose products act at different steps in a common process was a recurring theme. For example, five genes involved in the biosynthesis of cholesterol are coordinately expressed, sharing a common expression profile following serum addition to starved fibroblasts (Vishwanath et al. 1999), which defines a synexpression group.

Taken together, these results reveal that coordinate expression of genes acting in the same cell biological process is a widespread phenomenon in eukaryotes. Synexpression groups are clusters of genes that are coexpressed *and* share the biological process in which they are involved, that is, where there is tight correlation between function and expression. It is the phenomenon of tight correlation between expression and function which is novel and which distinguishes it from regulons (see below). Table 9.2 summarizes major synexpression groups in vertebrates. Such synexpression groups may relate both to temporal and spatial coregulation. Identifying synexpression groups requires expression profiling at high resolution, correlating either multiple temporal or spatial data points. For example, it is not useful to consider genes exclusively expressed in muscle as a synexpression group, as these may be involved in diverse yet muscle-specific biological processes, for example, respiration, contraction, and cell adhesion. Further, it may not be safe to assume that whenever a gene is expressed in a cell it is there

Table 9.2: Examples of synexpression groups in vertebrates

Synexpression Group (Animal Model)	No. of Genes	Significance	References
BMP4 (*Xenopus*)	7	BMP signaling	Bhushan et al. 1998; Frisch and Wright 1998; Gawantka et al. 1998; Hata et al. 1998; Oelgeschlager et al. 2000; Onichtchouk et al. 1999
DeltaI (*Xenopus*)	5	Delta-Notch signaling	Gawantka et al. 1998
GABA (rat)	9	GABA neurotransmission	Wen et al. 1998
FGF8 (zebrafish)	6	FGF signaling	Niehrs and Meinhardt 2002
sonic hedgehog (mouse, *Xenopus*)	4	Hedgehog signaling	Chuang and McMahon 1999; Motoyama et al. 1998; Platt et al. 1997; Takabatake et al. 2000
Chromatin (*Xenopus*)	25	Cell-cycle-regulated DNA scaffold components	Gawantka et al. 1998
ER-import (*Xenopus*)	7	Components of ER/ secretion machinery	Gawantka et al. 1998
Ribosome biogenesis (*Xenopus*)	6	Nuclear import	Wischnewski et al. 2000
Cholesterol biosynthesis (human fibroblasts)	5	Membrane biosynthesis	Vishwanath et al. 1999

for a functional reason (Miklos and Rubin 1996). This could be particularly true for synexpression group genes, which may be required at some but not all times or in some but not all tissues showing characteristic expression of the group. Also, little is known about protein expression, and mRNAs may remain untranslated. Mutations leading to the loss of gratuitous expression would probably affect essential expression as well and can therefore expected to be selected against.

Synexpression Groups versus Operons and Regulons

The lac operon was the first example of tight integration of gene expression and function in bacteria and consists of *lacZ*, a hydrolase cleaving lactose, *lacY*, a galactoside permease, and *lacA*, a galactoside transacetylase. These genes are organized in a cistron, a linear transcription unit where the genes are coregulated by a repressor (Jacob 1997). Operons abound in prokaryotes, and remarkably, there appear to be operons for cell division and protein export (Dandekar et al. 1998; Lawrence 1997), similar to the chromatin and ER import groups of *Xenopus*.

In eukaryotes operons are generally missing. Genes are not transcribed as a polycistronic mRNA and are typically arranged on chromosomes without correlation of function or expression, although there are exceptions, for example, *rRNA, histone,* and *HOX, immunoglobulin,* and *globin* genes, as well as certain genes in *C. elegans* (Blumenthal 1998). Eukaryotic genes instead are individually regulated by complex promoters. Hence, coordinate expression is not likely to be achieved by cis-regulation of physically associated transcription units, as many genes of a synexpression group reside on different chromosomes. More likely, expression is coordinated by trans-acting factors regulating common enhancer elements (although posttranscriptional mechanisms are also conceivable). This type of coordinated gene expression was first discovered in bacteria and later in yeast and is called a regulon. A regulon is a group of operons regulated by a common regulator. In a regulon, a battery of target genes contains common promoter elements, which are recognized by one or a few transcription factors that integrate transcriptional responses. Often, different operons are integrated by one transcription factor. The transcriptional responses can hence be global, for example, in the case of sigma factor S, which integrates more than 30 *E. coli* genes that have a diverse functions, from virulence and osmoprotection to thermotolerance (Loewen and Hengge-Aronis 1994); or the response can be more restricted, as in case of the galactose regulon involved in galactose transport and metabolism (Weickert and Adhya 1993). Microarray analysis followed by clustering motif analysis has identified many regulons in yeast (Tavazoie et al. 1999). The concepts of regulons and synexpression groups are clearly related. Yet, unlike most regulons, synexpression groups consist of member genes whose role is typically restricted to one process, for example, ER import, while most regulons integrate genes with distinct roles. For example, the leucine LRP regulon controls operons involved in isoleucine-valine biosynthesis, oligopeptide transport, and serine and threonine catabolism (Newman et al. 1992). Unlike the genes of a regulon, by definition a gene that is member of one synexpression group cannot be part of another.

A practical application of synexpression groups is that they allow genes to be assembled in groups that define molecular pathways and strong predictions to be made about the role played by constituent members with unknown functions. Such prognoses were already successful for six genes from our initial in situ screen (Jen et al. 1999; Lahaye et al. 2002; Lamar et al. 2001; Onichtchouk et al. 1996, 1999; Weissman et al. 2001) and for yeast *Spo70*, an effector of meiosis (Chu et al. 1998). Clustering of genes permitted very specific testable hypotheses to be formulated, and these were rapidly confirmed experimentally. Using filter-arrayed cDNA libraries, robotic processing of DNA/RNA probes and automated whole-mount in situ hybridization or gene expression screening with gene chip technology can be largely automated (Houston and Banks 1997; Plickert et al. 1997). It can be expected that such temporal and spatial high-resolution maps of thousands of genes will reveal many more synexpression groups. Synexpression groups promise thus to help elucidate molecular pathways during development and differentiation and to become an important prognostic tool in the postgenome era in systematic prediction of unknown gene function.

Evolutionary Implications: Functional Compartmentalization of the Genome

One explanation for the pervasive occurrence of modules in general is that once the low-probability genotype promoting a set of favorable interacting elements arises during evolution, it becomes locked ("Never change a winning team"). The combination of a hierarchical assembly of biological modules and a network-like connection of them is thought to enhance evolvability, that is, to promote the capacity for rapid evolutionary change and diversification (Cheverud 1996; Gerhart and Kirschner 1997; Raff 1996; Riedl 1978; Wagner 1995).

The coevolution of function and expression observed in synexpression groups implies that there is strong selective pressure promoting this type of genetic organization. There are two possible ways of explaining the adaptiveness of synexpression groups, one in terms of their benefit to individual organisms and the other in terms of their capacity to evolve (Niehrs and Pollet 1999). The latter type of forward-looking adaptation is problematic, though, because of the difficulty of explaining how this may be coming about by selection in individuals, as has been discussed elsewhere (Barton and Partridge 2000; Gerhart and Kirschner 1997; Wagner and Altenberg 1996). Selection for modules may have evolved primarily due to their conferring increased fitness on individual organisms, with enhanced evolvability as a by-product. Alternatively, there may be direct group selection for modular organization and synexpression groups. These different scenarios are inherently difficult to prove.

Synexpression Groups and Benefits to the Individual Organism

Operons, regulons, and synexpression groups may be adaptive for the individual organism because they allow a set of genes to be coordinately expressed only when required, for example, the Lac operon in the presence of galactose or the ER import group in tissues with a secretory load. Apart from energetic economy, interacting gene products frequently need to assemble stoichiometrically or may require cotranslation to form a complex (Dandekar et al. 1998), which is promoted by coexpression. Also, different cell states may require expression of mutually exclusive groups of interacting genes; for example, expression of the chromatin group is probably incompatible with a nondividing cell state. This may enhance fitness and allow selection at the level of the individual. Furthermore, the autoregulation of expression via positive and negative feedback loops of a signal transduction pathway such as the BMP4 synexpression group may enhance the robustness of the signaling output and protect against physiological fluctuations (Becskei and Serrano 2000).

Synexpression Groups and Genetic Networks

Synexpression groups may be also adaptive because they enhance the modularity of the genetic potential and promote rapid evolution. The completion of various genome projects highlighted once more that the primary source of animal diversity is not differences in gene products. *Drosophila* has only about twice as many genes as yeast despite its much greater complexity. Instead, the regulatory linkages that make the function of gene products or whole genetic modules contingent on extracellular conditions are thought to be a major factor driving evolutionary change (Britten and Davidson 1969; Duboule and Wilkins 1998; Gerhart and Kirschner 1997; Huang 1998).

Compartmentalizing an organism's genetic potential in synexpression groups may produce evolutionary change requiring few mutational steps (Niehrs and Pollet 1999). Synexpression groups are modules performing an elementary cell biological function, such as forming a signaling circuit or a secretion module. They do not perform higher-order functions, such as building vertebrae, because such higher-level morphological modules are made up of lower-level modules, such as synexpression groups. Through promoter shuffling in a master regulator controlling a battery of target genes, an entire synexpression battery may be redeployed in a new developmental context (such as a new time or site of expression; for example, an ER import group in a novel secretory cell type) or connected to other regulatory pathways. Higher-order integration of multiple synexpression groups into regulons can be also envisaged, permitting regulatory networks of enormous complexity to be built. The classic Britten-Davidson model of coordinated gene regulation has anticipated these and other adaptive features of integrated gene expression (Britten and Davidson 1969).

Synexpression Groups and Developmental Integration

Developmental integration is thought to underlie harmonious changes during morphological diversification in evolution (Cheverud 1996; Gerhart and Kirschner 1997; Raff 1996; Riedl 1978; Wagner 1995). For example, the ancestors of fore- and hindlimbs were genetically linked together early in their history by the common expression of Hox genes. This may explain why digits arose at the same time in hand and foot. These appendages have evolved together presumably because their development has been brought under the control of a common enhancer for the Hox complex (Shubin et al. 1997). Correlated characters may coevolve when properties such as size, shape, and enzyme composition are genetically or epistatically coupled and if this coupling results in covariation. Character coupling during individual development will thus affect population-level evolutionary processes (Cheverud 1996).

Likewise, tissues sharing the same synexpression group may coevolve through genetic integration. Function and regulation of synexpression group genes will be subject to genetic variation, and this variation will be manifested in various tissues controlled by a synexpression group through pleiotropy. For example, BMP4 heterozygous null mice display haploinsufficient phenotypes such as cystic kidney, craniofacial malformations, microphthalmia, and preaxial polydactyly of the hindlimbs (Dunn et al. 1997). Hence, these tissues experience genetic covariation. In turn, such correlated characters will have the tendency to coevolve. However, while it is reasonable that the genetic linkage of appendages may lead to adaptive codiversification of these functionally related structures, it is less obvious what properties are shared by kidney, head skeleton, and limbs that make it advantageous to have them cocontrolled by BMP4. Possibly, it is the need for an autoregulatory signaling module during the embryonic patterning process. However, a combination of tissues expressing a synexpression group may also simply reflect evolutionary history, the chance acquisition of a functional module proving adaptive. It has been tacitly assumed that genetic covariation in synexpression groups will lead to phenotypic covariation. However, while this is plausible, it may well be that much of genetic variation will remain phenotypically silent, because the genetic changes are phenotypically neutral or because they are canalized (Schlichting and Pigliucci 1998).

Developmental integration has a price, which is the developmental constraint it generates, that is, restrictions on potential evolutionary trajectories due to a requirement of internal functional coherence of the module. Evolutionarily highly conserved features such as phylotypic stages (e.g., pharyngula in vertebrates) are thought to result from developmental constraints (Hall 1992; Raff 1996; Riedl 1978; Schlichting and Pigliucci 1998). Hence, besides developmental integration, developmental parcellation—that is, uncoupling of characters—has to be considered to be important during evolution (Schlichting and Pigliucci 1998; Wagner 1995). An example is the uncoupling of fore- and hindlimb development during evolution, as is most obviously the case in bird wings.

Evolutionary Dynamics of Synexpression Groups

Only a subset of functionally interacting genes are part of synexpression groups. For example, only some but not all genes involved in methionine biosynthesis are part of the synexpression group (Spellman et al. 1998), and *Smad1* and *Smad4*, which also participate in BMP4 signaling, are ubiquitously expressed during early embryogenesis (Lagna et al. 1996; Meersman et al. 1997) and hence are not part of the BMP4 group. One possible explanation for such nonselected genes is that they are evolutionarily younger and may not have had enough time to be recruited. Another reason may be that nonrecruited components may serve in multiple processes (e.g., Smad4 in BMP, TGF-β, and activin pathways) and therefore are not tightly coupled in expression to any one synexpression group. This leads to the prediction that genes with a role in multiple pathways may rarely be members of a synexpression group.

Once a gene is recruited into a synexpression group, this should be very stable due to the connectivity in multiple tissues, and a prediction is that synexpression groups should be conserved even between different animal phyla. Indeed, (1) there are chromatin synexpression groups both in yeast and in *Xenopus*, and (2) in the homologous *Drosophila* dpp and *Xenopus* BMP4 pathways, inhibitory Smads (*dad*; *Smad6/7*) are coexpressed with the ligands (Niehrs and Pollet 1999). In contrast, the region-specific deployment of individual synexpression groups, for example, by changing the promoter of a master regulator (Britten and Davidson 1969), may be subject to much greater variability, as this may require few genetic changes.

Conclusions

Global gene expression analysis has revealed an unexpected principle of organization in eukaryotes—namely, groups of genes showing highly coordinated expression and interacting in a common functional pathway. Such synexpression groups reflect the functional compartmentalization of the eukaryote genome and reveal a striking formal parallel to the prokaryote operon. They cover a remarkably diverse functional range, from cell signaling and the cell cycle to protein biosynthesis. A major adaptive advantage may be to integrate such pathways via master regulators and to contribute to the modularity of the genetic potential, promoting evolvability. We can make the following testable predictions regarding synexpression groups:

1. Components of supramolecular complexes will probably be organized in synexpression groups, since they often require synthesis in stoichiometric amounts.

2. Once a gene is recruited into a synexpression group expressed in multiple tissues, this should be evolutionarily stable due to the connectivity, and hence, such synexpression groups should be conserved even between different animal phyla. In contrast, expression of the master regulator may be less conserved.

3. Genes with a role in multiple pathways will rarely be members of a synexpression group.

4. The promoters of synexpression group members share common regulatory elements.

5. There are adaptive reasons to be found for functionally integrating seemingly unrelated tissues through shared synexpression groups.

6. In evolutionary computer simulations virtual organisms using a synexpression mode will fare better than competitors without modular gene expression; that is, synexpression modes enhance evolvability.

Acknowledgments

I thank Emil Karaulanov for critical reading of the manuscript.

References

Artavanis-Tsakonas, S., K. Matsuno, and M. E. Fortini. 1995. Notch signaling. *Science* 268:225–232.

Barton, N., and L. Partridge. 2000. Limits to natural selection. *BioEssays* 22:1075–1084.

Becskei, A., and L. Serrano. 2000. Engineering stability in gene networks by autoregulation. *Nature* 405:590–593.

Bhushan, A., Y. Chen, and W. Vale. 1998. Smad7 inhibits mesoderm formation and promotes neural cell fate in *Xenopus* embryos. *Dev. Biol.* 200:260–268.

Blumenthal, T. 1998. Gene clusters and polycistronic transcription in eukaryotes *BioEssays* 20:480–487.

Britten, R. J., and E. H. Davidson. 1969. Gene regulation for higher cells: a theory. *Science* 165:349–357.

Cheverud, J. M. 1996. Developmental integration and the evolution of pleiotropy. *Am. Zool.* 36:44–50.

Cho, R. J., M. J. Campbell, E. A. Winzeler, L. Steinmetz, A. Conway, L. Wodicka, T. G. Wolfsberg, A. E. Gabrielian, D. Landsman, D. J. Lockhart, and R. W. Davis. 1998. A genome-wide transcriptional analysis of the mitotic cell cycle. *Mol. Cell* 2:65–73.

Chu, S., J. DeRisi, M. Eisen, J. Mulholland, D. Botstein, P. O. Brown, and I. Herskowitz.

1998. The transcriptional program of sporulation in budding yeast. *Science* 282: 699–705.
Chuang, P. T., and A. P. McMahon. 1999. Vertebrate Hedgehog signalling modulated by induction of a Hedgehog-binding protein. *Nature* 397:617–621.
Dandekar, T., B. Snel, M. Huynen, and P. Bork. 1998. Conservation of gene order: a fingerprint of proteins that physically interact. *Trends Biochem. Sci.* 273:324–328.
DeRisi, J. L., R. I. Vishwanath, and P. O. Brown. 1997. Exploring the metabolic and genetic control of gene expression on a genomic scale. *Science* 278:680–686.
Duboule, D., and A. S. Wilkins. 1998. The evolution of "bricolage." *Trends Genet.* 14: 54–59.
Dunn, N. R., G. E. Winnier, L. K. Hargett, J. J. Schrick, A. B. Fogo, and B. L. Hogan. 1997. Haploinsufficient phenotypes in Bmp4 heterozygous null mice and modification by mutations in Gli3 and Alx4. *Dev. Biol.* 188:235–247.
Frisch, A., and C. V. E. Wright. 1998. XBMPRII, a novel *Xenopus* type II receptor mediating BMP signalling in embryonic tissues. *Development* 125:431–442.
Gawantka, V., N. Pollet, H. Delius, R. Pfister, M. Vingron, R. Nitsch, C. Blumenstock, and C. Niehrs. 1998. Gene expression screening in *Xenopus* identifies molecular pathways, predicts gene function and provides a global view of embryonic patterning. *Mech. Dev.* 77:95–141.
Gerhart, J., and M. Kirschner. 1997. *Cells, embryos and evolution*. Malden: Blackwell.
Hall, B. K. 1992. *Evolutionary developmental biology*. London: Chapman and Hall.
Hata, A., G. Lagna, J. Massague, and A. Hemmati-Brivanlou. 1998. Smad6 inhibits BMP/Smad1 signaling by specifically competing with the Smad4 tumor suppressor. *Genes Dev.* 12:186–197.
Houston, J. G., and M. Banks. 1997. The chemical-biological interface: developments in automated and miniaturised screening technology. *Curr. Opin. Biotechnol.* 8: 734–740.
Huang, F. 1998. Syntagms in development and evolution. *Int. J. Dev. Biol.* 42:487–494.
Hughes, T. R., M. J. Marton, A. R. Jones, C. J. Roberts, R. Stoughton, C. D. Armour, H. A. Bennett, E. Coffey, H. Dai, Y. D. He, M. J. Kidd, A. M. King, M. R. Meyer, D. Slade, P. Y. Lum, S. B. Stepaniants, D. D. Shoemaker, D. Gachotte, K. Chakraburtty, J. Simon, M. Bard, and S. H. Friend. 2000. Functional discovery via a compendium of expression profiles. *Cell* 102:109–126.
Jacob, F. 1997. The operon—25 years later. *C. R. Acad. Sci. Paris* 320:199–206.
Jen, W.-C., V. Gawantka, N. Pollet, C. Niehrs, and C. Kintner. 1999. Periodic repression of Notch pathway genes governs the segmentation of *Xenopus* embryos. *Genes Dev.* 13:1486–1499.
Lagna, G., A. Hata, A. Hemmati-Brivanlou, and J. Massague. 1996. Partnership between DPC4 and SMAD proteins in TGF-beta signalling pathways. *Nature* 383:832–836.
Lahaye, K., S. Kricha, and E. Bellefroid. 2001. XNAP, a conserved ankyrin repeat-containing protein with a role in the Notch pathway during *Xenopus* primary neurogenesis. *Mech. Dev.* 110:113–124.
Lamar, E., G. Deblandre, D. Wettstein, V. Gawantka, N. Pollet, C. Niehrs, and C. Kintner. 2001. Nrarp is a novel intracellular component of the Notch signaling pathway. *Genes Dev.* 15:1885–1899.
Lawrence, J. G. 1997. Selfish operons and speciation by gene transfer. *Trends Microbiol.* 5:355–359.
Loewen, P. C., and R. Hengge-Aronis. 1994. The role of the sigma factor sigma S (KatF) in bacterial global regulation. *Annu. Rev. Microbiol.* 48:53–80.
Meersman, G., K. Verschueren, L. Nelles, C. Blumenstock, H. Kraft, G. Wuytens, J. Remacle, C. A. Kozak, P. Tylzanowsky, C. Niehrs, and D. Huylebroeck. 1997. The C-terminal domain of Mad-like signal transducers is sufficient for biological activity in vivo and transcriptional activation. *Mech. Dev.* 61:127–140.

Miklos, G. L., and G. M. Rubin. 1996. The role of the genome project in determining gene function: insights from model organisms. *Cell* 86:521–529.

Motoyama, J., H. Heng, M. A. Crackower, T. Takabatake, K. Takeshima, L. C. Tsui, and C. Hui. 1998. Overlapping and non-overlapping Ptch2 expression with Shh during mouse embryogenesis. *Mech. Dev.* 78:81–84.

Newman, E. B., R. D'Ari, and R. T. Lin. 1992. The leucine-Lrp regulon in *E. coli:* a global response in search of a raison d'etre. *Cell* 68:617–619.

Niehrs, C. 1997. Gene-expression screens in vertebrate embryos: more than meets the eye. *Genes Funct.* 1:229–231.

Niehrs, C., and H. Meinhardt. 2002. Modular feedback. *Nature* 417:35–36.

Niehrs, C., and N. Pollet. 1999. Synexpression groups in eukaryotes *Nature* 402: 483–487.

Oelgeschlager, M., J. Larrain, D. Geissert, and E. M. De Robertis. 2000. The evolutionarily conserved BMP-binding protein Twisted gastrulation promotes BMP signalling. *Nature* 405:757–763.

Onichtchouk, D., Y.-G. Chen, R. Dosch, V. Gawantka, H. Delius, J. Massagué, and C. Niehrs. 1999. Silencing of TGF-beta signalling by the pseudoreceptor BAMBI. *Nature* 401:480–485.

Onichtchouk, D., V. Gawantka, R. Dosch, H. Delius, K. Hirschfeld, C. Blumenstock, and C. Niehrs. 1996. The Xvent-2 homeobox gene is part of the BMP-4 signaling pathway controling dorsoventral patterning of *Xenopus* mesoderm. *Development* 122:3045–3053.

Platt, K. A., J. Michaud, and A. L. Joyner. 1997. Expression of the mouse Gli and Ptc genes is adjacent to embryonic sources of hedgehog signals suggesting a conservation of pathways between flies and mice. *Mech. Dev.* 62:121–135.

Plickert, G., M. Gajewski, G. Gehrke, H. Gausepohl, J. Schlossherr, and H. Ibrahim. 1997. Automated in situ detection (AISD) of biomolecules. *Dev. Genes Evol.* 207: 362–367.

Raff, R. A. 1996. *The shape of life.* Chicago: University of Chicago Press.

Riedl, R. 1978. *Order in living organisms: a systems analysis of evolution.* Chichester: John Wiley and Sons.

Schatz, G., and B. Dobberstein. 1996. Common principles of protein translocation across membranes. *Science* 271:1519–1526.

Schlichting, C. D., and M. Pigliucci. 1998. *Phenotypic evolution: a reaction norm perspective.* Sunderland: Sinauer Associates.

Shubin, N., C. Tabin, and S. Carroll. 1997. Fossils, genes and the evolution of animal limbs. *Nature* 388:639–648.

Spellman, P. T., G. Sherlock, M. Q. Zhang, V. R. Iyer, K. Anders, M. B. Eisen, P. O. Brown, D. Botstein, and B. Futcher. 1998. Comprehensive identification of cell cycle–regulated genes of the yeast *Saccharomyces cerevisiae* by microarray hybridization. *Mol. Biol. Cell* 9:3273–3297.

Takabatake, T., T. C. Takahashi, Y. Takabatake, K. Yamada, M. Ogawa, and K. Takeshima. 2000. Distinct expression of two types of *Xenopus* Patched genes during early embryogenesis and hindlimb development. *Mech. Dev.* 98:99–104.

Tavazoie, S., J. D. Hughes, M. J. Campbell, R. J. Cho, and G. M. Church. 1999. Systematic determination of genetic network architecture. *Nat. Genet.* 22:281–285.

Vishwanath, R. I., M. B. Eisen, D. T. Ross, G. Schuler, T. Moore, J. C. F. Lee, J. M. Trent, L. M. Staudt, J. Hudson, M. S. Boguski, D. Lashkari, D. Shalon, D. Botstein, and P. O. Brown. 1999. The transcriptional program in the response of human fibroblasts to serum. *Science* 283:83–87.

Wagner, G. P. 1995. Adaptation and the modular design of organisms. In *Advances in artificial life,* ed. F. Moran, A. Moreno, J. J. Merelo, and P. Chacon, 317–328. Berlin: Springer.

Wagner, G. P., and L. Altenberg. 1996. Perspective: Complex adaptations and the evolution of evolvability. *Evolution* 50:967–976.
Weickert, M. J., and S. Adhya. 1993. The galactose regulon of *Escherichia coli*. *Mol. Microbiol.* 10:245–251.
Weissman, J. T., H. Plutner, and W. E. Balch. 2001. The mammalian guanine nucleotide exchange factor mSec12 is essential for activation of the Sar1 GTPase directing endoplasmic reticulum export. *Traffic* 2:465–475.
Wen, X., S. Fuhrman, G. S. Michaels, D. B. Carr, S. Smith, J. L. Barker, and R. Somogyi. 1998. Large-scale temporal gene expression mapping of central nervous system development. *Proc. Natl. Acad. Sci. U.S.A.* 95:334–339.
Wischnewski, J., M. Solter, Y. Chen, T. Hollemann, and T. Pieler. 2000. Structure and expression of *Xenopus* karyopherin-beta3: definition of a novel synexpression group related to ribosome biogenesis. *Mech. Dev.* 95:245–248.

10 Systematic Exploration and Mining of Gene Expression Data Provides Evidence for Higher-Order, Modular Regulation

ROLAND SOMOGYI, STEFANIE FUHRMAN, GARY ANDERSON, CHRIS MADILL, LARRY D. GRELLER, AND BERNARD CHANG

Introduction

How do gene products interact to produce tissues, organs, and appendages? How does an organism remain stable, on the one hand, and differentiate into a fully developed organism, on the other hand? How can we identify regulatory modules and major sets of regulatory connections that control these modules in development, health, and disease? In this chapter we show how exploratory and advanced methods of analysis can be used systematically to examine gene expression data from experiments on central nervous system (CNS) development and provide evidence of modularity in development.

Modularity can be observed at an anatomical level, where the structures of cells define the boundaries of tissues and organs, and also at a deeper, molecular level, within cells, where molecular regulatory networks control cellular function. Many examples in development, for example, major insect anatomical structures and their developmental regulatory and segmentation patterns, demonstrate the action of a set of master genes that act as switches to activate the genetic programs that generate these modules (Gehring 1998). These types of modules are conserved in more complex organisms, such as vertebrates and mammals, but the overall regulation is much more complex and intertwined there.

Given the advances in molecular biological tools and high-throughput measurement technologies, we now have the capacity not only to acquire full genomic sequences, but also to profile globally the activity patterns of genes captured in RNA and protein expression. However, while data acquisition, storage, and management are being carried out more or less routinely, we are still faced with major challenges in the analysis and interpretation of molecular activity data.

Using high-throughput biology in combination with advanced data analysis and modeling approaches, we can systematically identify complex regulatory networks, and the major interactions that constitute and control particular modules.

Exploratory Analysis of Expression Data Provides Evidence for Regulatory Modules in CNS Development

Large-scale expression profiling offers the scale of data from which we can infer structural features and principles of the overall architecture of genetic regulatory networks. Here we will examine time series of gene expression in rat central nervous development, specifically cervical spinal cord and hippocampus (Wen et al. 1998; Chang et al. 2001). Expression was measured using reverse-transcription polymerase chain reaction (RT-PCR), which delivers the highest sensitivity (down to 10 copies of RNA per sample), dynamic range (6–8 orders of magnitude), and fidelity (Somogyi et al. 1995) of all gene expression measurement methods. Measurements were carried out on triplicate animals, and the values were averaged before analysis.

One may begin with exploratory analysis and visualization of these data to discover qualitative relationships among the genes studied here. As a first step, cluster analysis methods help us to see coexpression groups and major temporal regulatory profiles.

Hierarchical Clustering

For example, hierarchical, agglomerative clustering orders genes within a tree structure that reflects increasing similarity of the expression patterns (fig. 10.1, *top panel*). This method is based on the principle that each gene or cluster at the end of a branch is united with its closest gene or cluster into a new node on the tree, which defines a new cluster that in turn is combined with its closest cluster into a new node. All clusters or nodes are ultimately united into a single branch. While this resulting dendrogram is a useful visualization tool, it has several limitations; for example, it presumes that each gene can be part of only one cluster (not an appropriate assumption for expression data), it does not answer how many clusters there are in the data, and it forces each gene into a cluster (some genes may be "outliers" and not belong to a "natural" cluster at all).

From another example of hierarchical clustering of this same data set (Wen et al. 1998), we have visually isolated four clusters and an outlier group (fig. 10.2). The expression profiles are colored according to functional gene family, and under each column of color matrix plots we show the average expression profile or centroid of the cluster. We

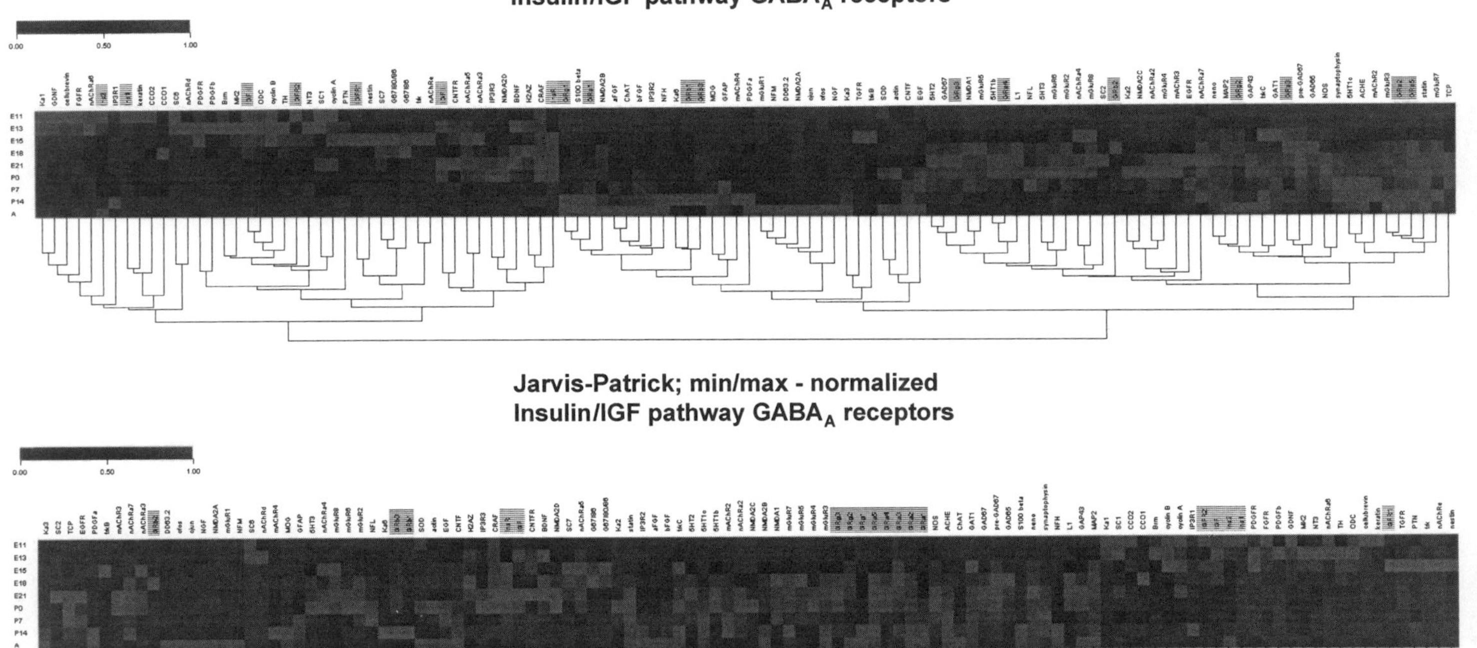

Fig. 10.1.—Temporal gene expression color matrix maps and cluster dendrograms for rat central nervous development containing min-max-normalized expression patterns using Euclidean hierarchical clustering and average linkage (*top*) and Jarvis-Patrick clustering (*bottom*) (see http://www.biosystemix.com/ModularityChapter/ModularityFiguresWWW or http://www.press.uchicago.edu/books/schlosser for color graphics). Low to high expression is represented as blue to red. The genes involved in the Insulin/IGF pathway are highlighted in light blue and the GABA receptors are highlighted in pink. The time points are embryonic days 11, 13, 15, 18, and 21 (E11, E13, E15, E18, and E21), postnatal days 0, 7, and 14 (P0, P7, and P14), and adult (A). Data analysis and graphics were generated using the GeneLinker™ Gold 2.0 software from Predictive Patterns Software (www.predictivepatterns.com), distributed by Biosystemix (www.biosystemix.com).

have identified here four major waveforms of progressive developmental expression, representing more general cases than the more detailed clusters of the 11 Jarvis-Patrick clusters (see "Jarvis-Patrick Clustering" below) in the lower panel of figure 10.1. Interestingly, there are clear constraints in the representation of functional gene families within these waveforms. For example, wave 3 (fig. 10.2) is almost entirely composed of neurotransmitter receptor genes (*red*) and absolutely no growth factor signaling genes. The latter (*green*) are restricted to early developmental expression during neurogenesis (wave 1) and late developmental expression (wave 4) during gliogenesis. It is now obvious that the expression modules also correspond to known functional gene groups, therefore qualifying cluster analysis as a useful method for expression module identification.

Partitional Clustering

Differing from hierarchical clustering, partitional clustering divides data into disjoint, flat clusters without imposing a hierarchical structure. The Jarvis-Patrick method is deterministic, works well when data partitions into nonspherical or nonconvex clusters, and is noniterative, while the *k*-means method is nondeterministic but allows the user to specify a priori the number of clusters desired.

Jarvis-Patrick Clustering

The Jarvis-Patrick method of clustering (Jarvis and Patrick 1973) is a prediction method based on similarity between neighbors, as measured by a number of different distance metrics (Euclidean, Manhattan, Pearson, Chebychev, Spearman, and so on). It is useful when tight clusters might be discovered in larger loose clusters, or when clustering speed is an issue.

When using the Jarvis-Patrick method, the user specifies the number of closest neighbors to examine and the number of these neighbors required to be in common. For each of the data points, the specified number of nearest neighbors is identified. Two data points are placed into the same cluster if they are in each other's nearest neighbor list and they share a specified number of common neighbors.

Using Jarvis-Patrick clustering on the same data (fig. 10.1, *lower panel*), we are presented with a number of clusters (the number of clusters is not determined ahead of time) and a number of outlier genes that do not cluster. Within each of the 11 clusters, we observe a very high degree of similarity of the expression profiles and characteristic waveforms of developmental gene activity indicative of specific expression modules. For example, the cluster on the extreme right of figure 10.1 captures genes that show early developmental activity that peaks at

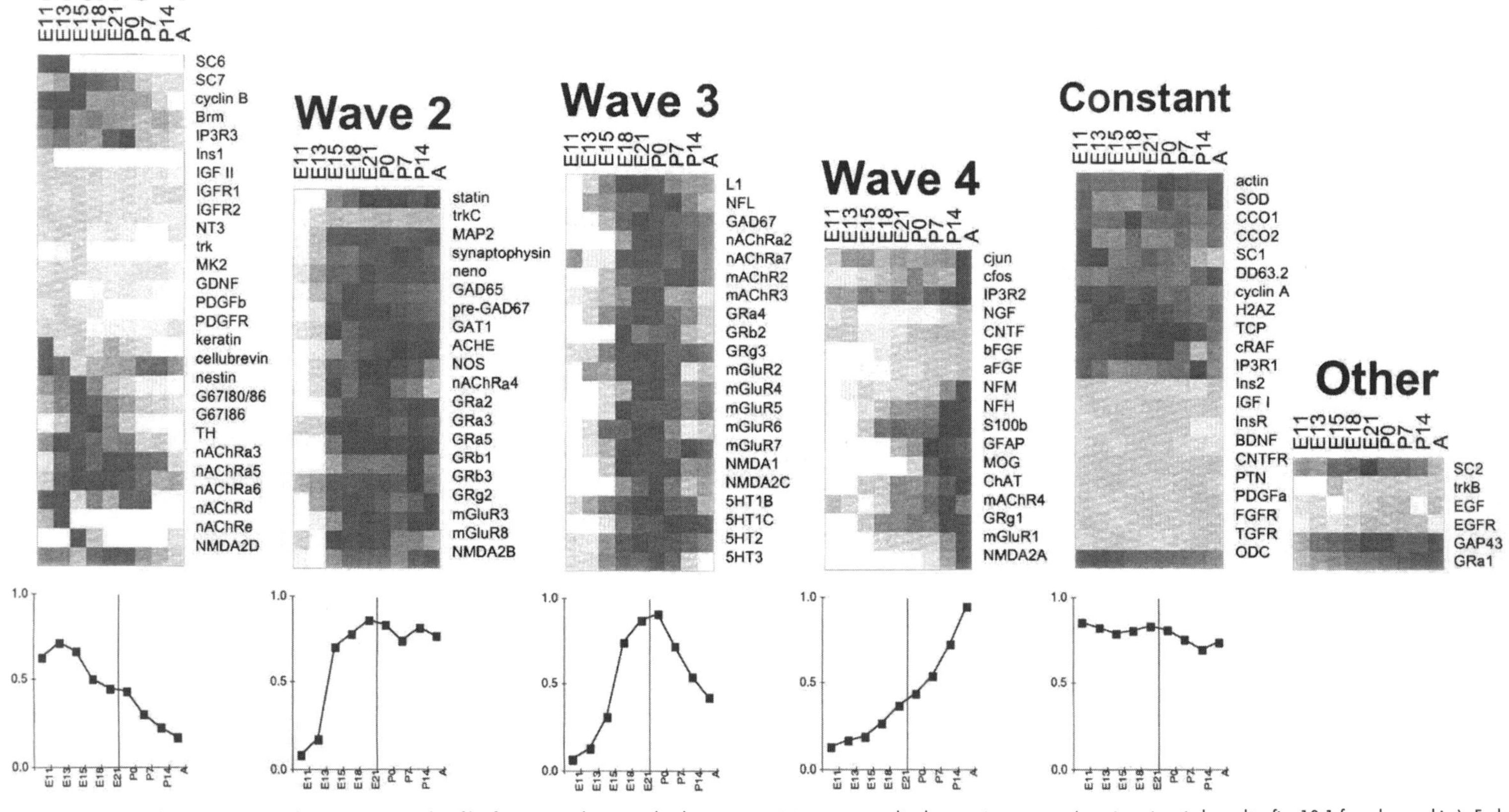

Fig. 10.2.—Temporal gene expression color matrix maps and profiles for rat central nervous development containing max-normalized expression patterns (see sites given in legend to fig. 10.1 for color graphics). Each color matrix map contains genes clustered by using the FITCH clustering method (Felsenstein 1993). Each of the profiles at the bottom of the graph is the average of the expressions for all the genes in that particular cluster. The time points and abbreviations of gene names are the same as in figure 10.1. Genes are colored according to functional families: green, growth factors and growth factor receptors; red, neurotransmitter receptors; purple, neuroglial markers; blue, other gene families.

embryonic day 15 (E15), the cluster to its left shows genes that peak at E11 and E13, and the other clusters capture specific transient and late activation profiles. In one example, genes from the same gene family, the GABA receptors (highlighted in pink), tend to group together in the third cluster from the right. This underlines the modular structure identified here—genes from the same gene family are expected to share some genetic control elements.

If modular structures persist at a deeper level of regulatory architectures, one would expect similar findings from the study of different related biological processes, also using different analysis methods. We asked ourselves whether some of the same expression modules found in spinal cord development could be identified in the development of the hippocampus. Hippocampal development is delayed with respect to cervical spinal cord development, and we therefore began our expression measurements at embryonic day 15 (E15) versus E11 in spinal cord development. We compared partitional to hierarchical clustering methods, and whether these identify some of the same patterns or modules that we have found in spinal cord development (Wen et al. 1998; Chang et al. 2001).

K-Means Clustering

When carrying out the partitional method of *k*-means clustering, the user must first specify the number of clusters. Then the algorithm iteratively optimizes the distribution of the gene expression vectors, so that each expression pattern is closer to the centroid (the average expression pattern of all cluster members) of its own cluster than to the centroid of any of the other cluster. Note that the procedure is generally set such that the initial centroids are selected randomly, and therefore multiple runs of *k*-means on the same data set may result in different clusterings; these may provide examples of alternate analyses that provide interesting, complementary views of the data.

We used *k*-means to allocate the hippocampal development expression profiles into seven clusters. The output of this clustering is shown in figure 10.3; for each cluster, we show the average pattern or centroid on the left, and all of its component expression profiles superimposed on the right. Note that the hippocampal cluster centroids represent waveforms similar to those shown in figure 10.2 for spinal cord development; for example, K1 is similar to wave 1, K2 to wave 3, K3 to wave 2, K6 to wave 1, and K5 to a delayed wave 1 pattern. K4 and K7 represent exceptional profiles outside the typical waveform patterns, similar to the "Other" group in figure 10.2. When comparing the results of the partitional clustering to hierarchical clustering of the same hippocampal data set (fig. 10.4), we observe much overlap between both clustering methods. Cluster boundaries for seven hierarchical

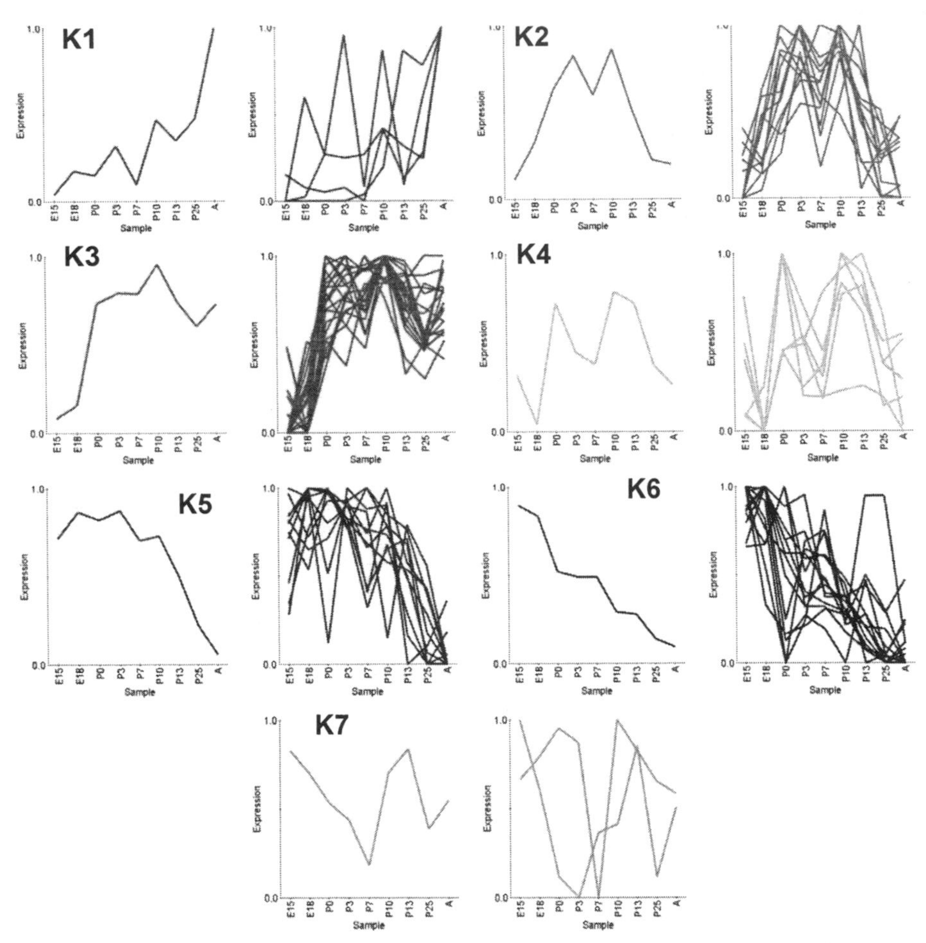

Fig. 10.3.—Line graphs of normalized temporal expression patterns for genes involved in hippocampal development, analyzed using Euclidean distance as the similarity measure, and then clustered using the *k*-means method (see sites given in legend to fig. 10.1 for color graphics). Clusters, shown as multiple-line plots, are color-coded as in figure 10.4. Single-line plots shown in the same colors are the corresponding centroids of these clusters. Data analysis and graphics were generated using the GeneLinker™ Gold 2.0 software from Predictive Patterns Software (www.predictivepatterns.com), distributed by Biosystemix (www.biosystemix.com).

clusters were determined visually, marked by brackets, and labeled H1 through H7 on the graph. There are three singletons, S1, S2, and S3, identified as well. Gene label background colors reflect the color scheme of the *k*-means clusters as shown in figure 10.3. Note that, for example, hierarchical cluster H1 contains exactly the same members as *k*-means cluster K1, H2 largely corresponds to K2, H3 largely corresponds to K3, and so forth. Comparison of partitional and hierarchical clustering of hippocampal development with spinal cord development clusters essentially shows the same representative expression

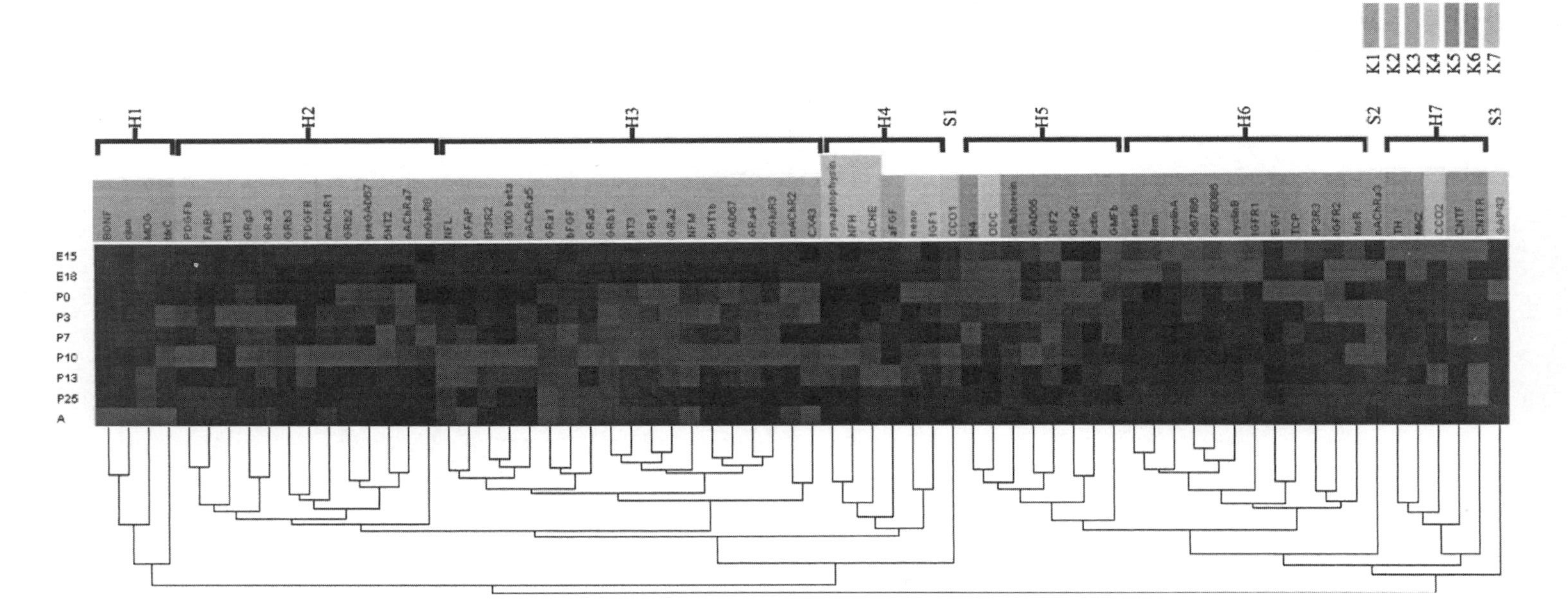

Fig. 10.4.—Temporal gene expression color matrix maps for hippocampal development containing min-max-normalized expression patterns for seventy genes (see sites given in legend to fig. 10.1 for color graphics). Low to high expression is represented as blue to red. A dendrogram from Euclidean hierarchical clustering with average linkage is shown at the bottom of the map. The Euclidean hierarchical clusters are shown bracketed at the top of the map. Highlight colors correspond to *k*-means clusters 1–7 (key at upper right of each map). The time points are embryonic days 15 and 18 (E15 and E18), postnatal days 0, 3, 7, 10, 13, and 25 (P0, P3, P7, P10, P13, and P25), and adult (A). Data analysis and graphics were generated using the GeneLinker™ Gold 2.0 software from Predictive Patterns Software (www.predictivepatterns.com), distributed by Biosystemix (www.biosystemix.com).

waves, providing evidence for generalizable modules of expression control in CNS development.

Self-Organizing Map

The self-organizing map (SOM) is essentially a clustering method with roots in artificial neural networks (Kohonen 1997). This unsupervised machine learning method maps the original low- or high-dimensional data that populate the "source" space into a more readily interpretable low-dimensional "target" space. An important property of SOMs is that proximity relationships (similarities and distances) between the original data elements in the source space are well preserved in the easily visualized low-dimensional target space. SOMs have been applied to many domains, including exploration of gene expression data sets (Golub et al. 1999; Tamayo et al. 1999; Toronen et al. 1999; Hill et al. 2000).

Broadly speaking, SOM algorithms work by following an initialization phase with an iterative training phase and a final reassignment or calibration phase. Reference vectors are assigned to the target space nodes during initialization. In the training phase, for each data vector, the node closest the vector (using any reasonable distance metric) is identified and moved toward it by an amount proportional to their separation. A window centered on the closest node can be employed in determining which neighboring nodes in the target space are also moved toward the data vector, and by how much. The target space node-movement process is performed over all the data vectors until some reasonable machine learning convergence criterion is met. Then, in the final phase, each data vector is reassigned to the node in the target space to which it is closest. Thus, proximity relationships between the data vectors populating the original source space are preserved in the low-dimensional target space, for the most part.

Thus, a SOM works similarly to *k*-means clustering but is richer. With *k*-means, one chooses the number of clusters to fit the data. For a SOM, one chooses the shape and size of a network of nodes into which to map the data. Like *k*-means, a SOM initially populates its nodes with clusters by randomly sampling the data and then refines the nodes in a systematic fashion. Unlike *k*-means clustering, however, it is possible for a node of a SOM to end up without any associated cluster items when the map is complete. A further difference with *k*-means clustering is that a SOM automatically provides some information on the similarity between nodes, that is, how strongly certain nodes resemble each other.

We have performed a SOM analysis on the hippocampal development data set, and the results are presented in figures 10.5, 10.6, and

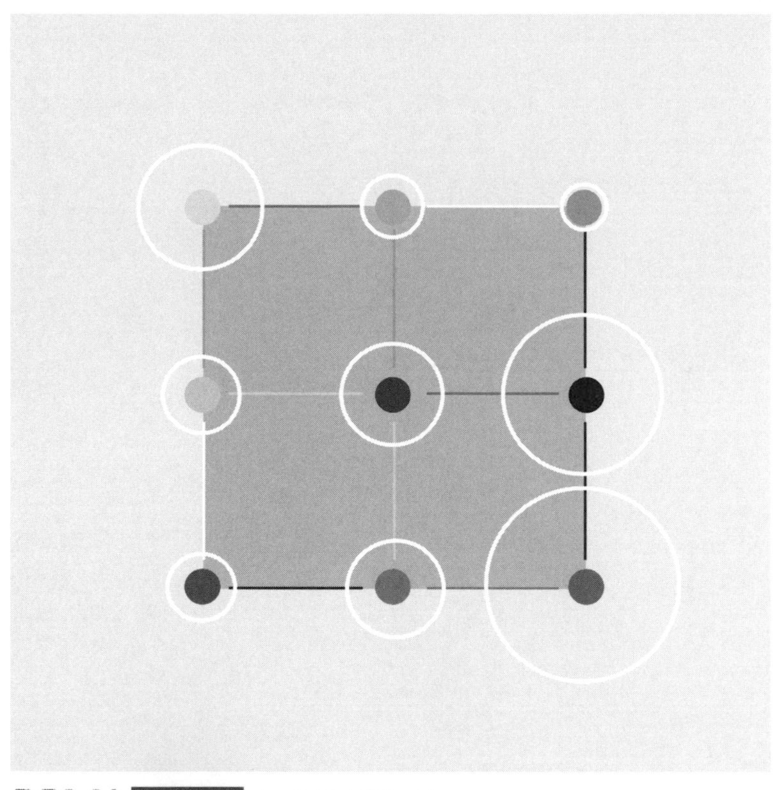

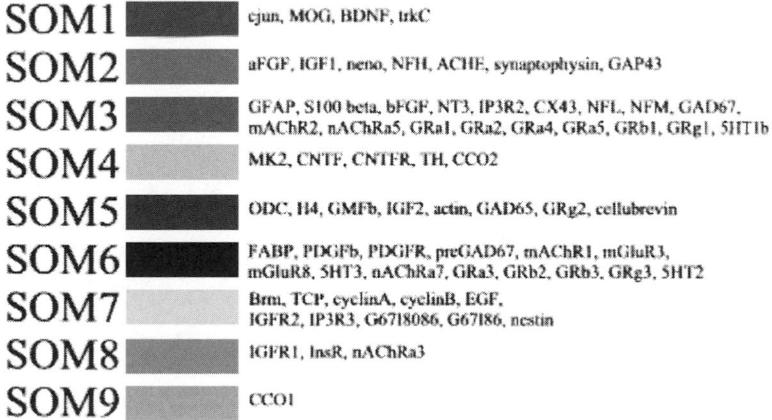

Fig. 10.5.—Self-organizing map with nine nodes for the seventy genes involved in hippocampal development (see sites given in legend to fig. 10.1 for color graphics). A proximity-gradient map is shown in blue with a darker color indicating a higher similarity among the nodes. Each node in the map is depicted as a small solid circle and represents a cluster. The white circles around the nodes are the cardinality rings. A larger ring indicates more items in the cluster. The vertical and horizontal lines that connect adjacent nodes are collectively referred to as the proximity grid. A darker red indicates a higher similarity between adjacent nodes. Nine nodes were used in this particular map. Data analysis and graphics were generated using the GeneLinker™ Gold 2.0 software from Predictive Patterns Software (www.predictivepatterns.com), distributed by Biosystemix (www.biosystemix.com).

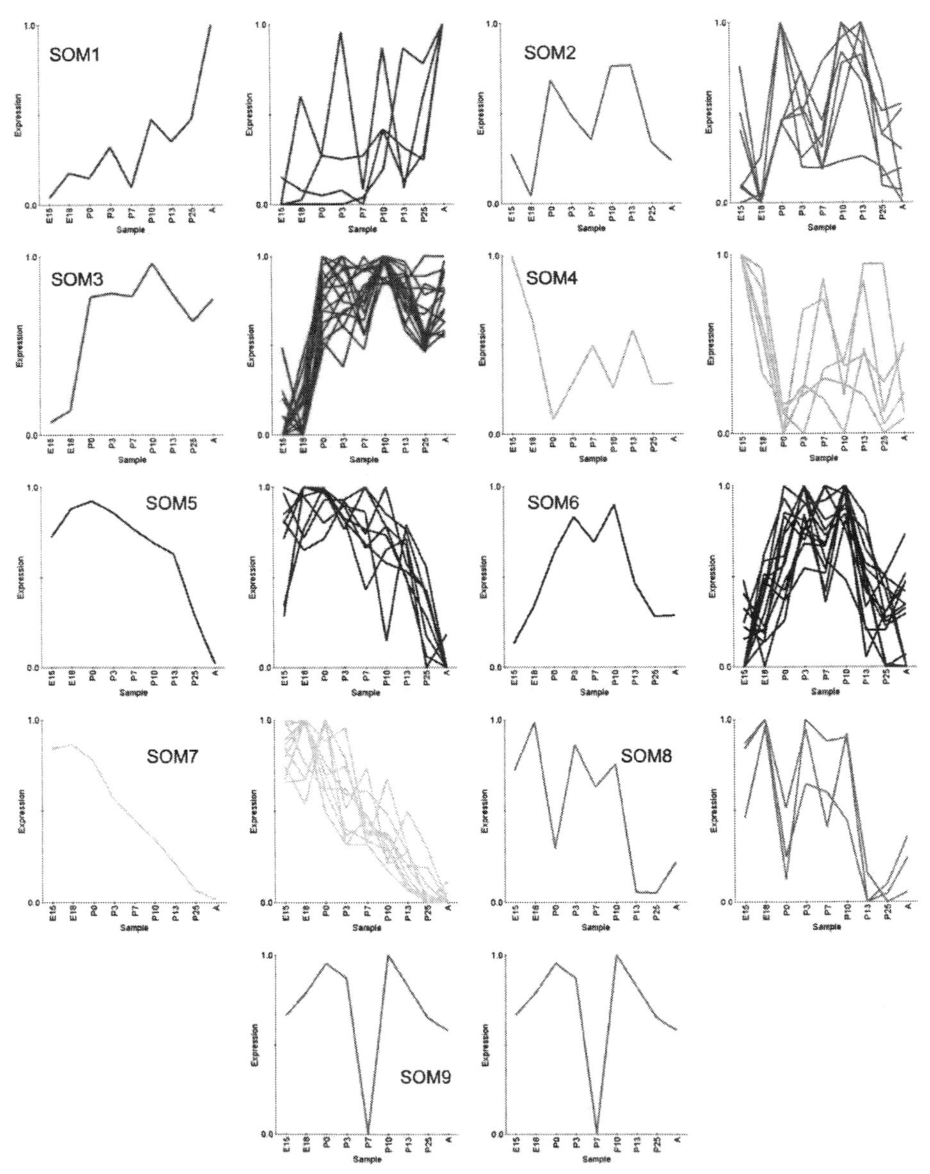

Fig. 10.6.—Line graphs of temporal expression patterns for genes involved in hippocampal development clustered using a self-organizing map (see sites given in legend to fig. 10.1 for color graphics). The clusters (SOM1–SOM9) are color-coded as in figures 10.5 and 10.7. Single-line plots are the centroids that correspond to multiple-line plots (clusters) of the same color. Data analysis and graphics were generated using the GeneLinker™ Gold 2.0 software from Predictive Patterns Software (www.predictivepatterns.com), distributed by Biosystemix (www.biosystemix.com).

10.7. Figure 10.5 shows the proximity-gradient map, which is a high-level view of the average proximity (or similarity) among the nodes of the SOM, with a darker blue indicating higher similarity. Each node in the map is depicted as a small solid circle and represents a cluster. The white circles around the nodes are the cardinality rings—the larger the ring, the more items in the cluster. The vertical and horizontal lines that connect adjacent nodes are collectively referred to as the proximity grid. It shows more accurately the similarity between adjacent nodes. Color renderings of the connections between nodes reflect the underlying proximity gradient; for example, a darker red indicates higher similarity.

Figure 10.6 shows the clustering results from a SOM. For each cluster, we show the average pattern or centroid on the left, and all of its component expression profiles superimposed on the right. Figure 10.7 shows that the clustering obtained from a SOM is very similar to that from hierarchical clustering. Again, the cluster boundaries for the seven hierarchical clusters were determined visually and labeled H1 through H7 on the graph. Gene label background colors reflect the color scheme of the SOM clusters as shown in figure 10.6. SOM cluster numbers SOM1, SOM4, SOM3, SOM2, SOM5, SOM7 + SOM8, and SOM4 correspond exactly to hierarchical cluster numbers H1, H2, H3, H4 + S3, H5, H6 + S2, and H7, respectively, with the exception of mGluR3. The singleton S1 from hierarchical clustering is matched by the SOM cluster number SOM9, which contains the same single gene (CCO1).

Principal Component Analysis

Principal component analysis (PCA) is used to reduce a large number of interrelated variables to a typically much smaller set of uncorrelated variables. The principal components are new variables comprising different linear combinations of the variables from the original data. In machine learning, it is useful as a direct unsupervised computational approach to delineating the most influential independent features in a data set. It has been used to analyze gene expression data (Wen et al. 1998; Alter et al. 2000).

We performed a PCA on the different time points for the hippocampal development data set. Figure 10.8 represents the projection of the genes on the plane formed by the first two principal components. The coloring of the genes indicates the result from the hierarchical clustering. It is evident that genes that cluster together in the hierarchical clustering also fall into subregions on this projection. It is worthwhile to point out that genes in clusters 3 and 6 from the hierarchical clustering exhibit anticorrelation, as they are at opposite ends of the

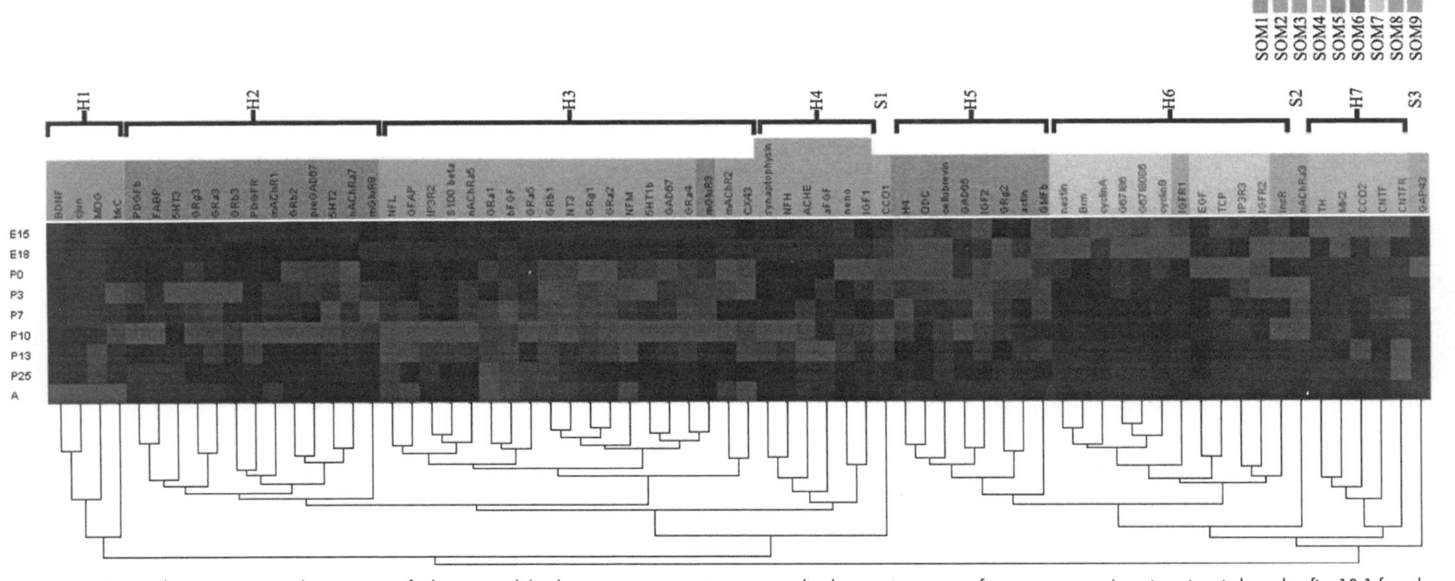

Fig. 10.7.—Temporal gene expression color matrix map for hippocampal development containing min-max-normalized expression patterns for seventy genes (see sites given in legend to fig. 10.1 for color graphics). Low to high expression is represented as blue to red. A dendrogram from a Euclidean hierarchical clustering with average linkage is shown at the left of the map. The Euclidean hierarchical clusters are shown bracketed at the right of the map. Highlight colors correspond to self-organizing map clusters 1–9 (key at upper right of each map). The time points and abbreviations of gene names are the same as in figure 10.4. Data analysis and graphics were generated using the GeneLinker™ Gold 2.0 software from Predictive Patterns Software (www.predictivepatterns.com), distributed by Biosystemix (www.biosystemix.com).

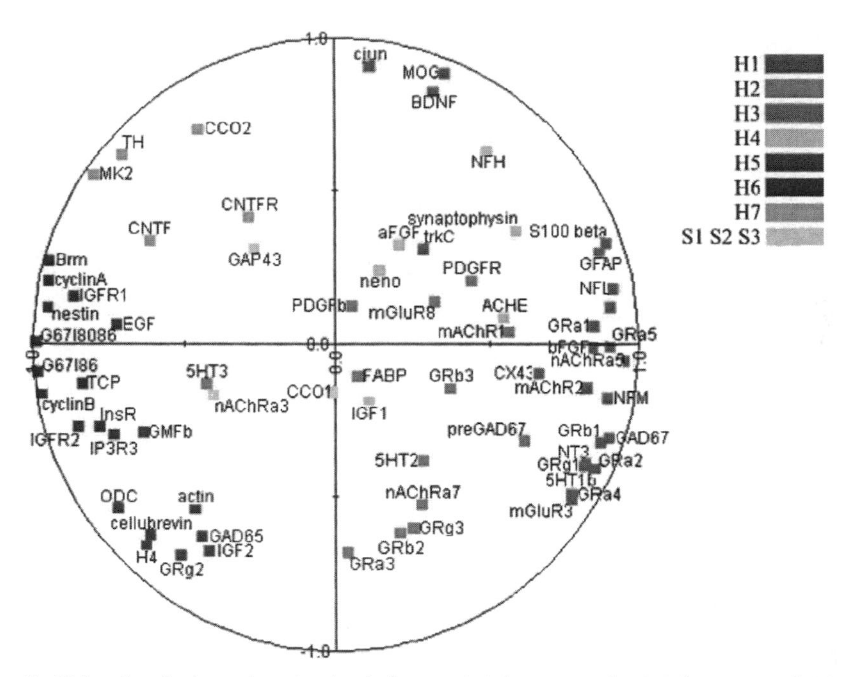

Fig. 10.8.—Normalized score plot projected on the first two principal components of a principal component analysis using the nine time points for hippocampal development (see sites given in legend to fig. 10.1 for color graphics). The abbreviations of the gene names follow those in figure 10.1. The coloring of the genes indicates the clusters they belong to from a Euclidean hierarchical clustering (H1–H7, S1–S3) and correspond to the bracketing shown at the top of figures 10.4 and 10.7. Data analysis and graphics were generated using the GeneLinker™ Gold 2.0 software from Predictive Patterns Software (www.predictivepatterns.com), distributed by Biosystemix (www.biosystemix.com).

axis for the first principal component. This is also evident from the color matrix plot of the hierarchical clustering in figure 10.4.

Advanced Analysis of Expression Data for the Concrete Determination of Network Wiring

While exploratory analysis provides qualitative evidence for regulatory modules of genes which define particular functional stages of development, we ultimately seek in-depth answers on the specific context-dependent roles of each gene and how these genes interact in networks within and among these modules. To answer these questions we need to apply advanced data-mining and modeling methods to infer the regulatory connections and to interpret them within a plausible, abstract model architecture. We recently reviewed the contributions advanced data-mining approaches can make in a variety of biomedical research areas (Somogyi and Greller 2001). We concluded that the main challenge lies in discovering the combinatorial and nonlinear regulatory interactions in these networks, for which the key enabling requirements are sophisticated and efficient computational pattern recognition methods.

We will discuss here an example of how regulatory gene-gene interactions can be inferred or "reverse engineered" directly from gene activity data, and what we may learn from such a gene network diagram in terms of the complexity and substructure of the network. The idea of gene network reverse engineering was first established on model Boolean networks (Somogyi et al. 1997; Liang et al. 1998) and has been explored quantitatively (D'haeseleer et al. 1999) and qualitatively on experimental data (for general review, see D'haeseleer et al. 2000).

The time series data shown above in the context of exploratory analysis (nine time points that capture signature waves of developmental phases) may prompt an investigator to look for gene activity patterns that are suggestive of causal interactions. For example, what if we repeatedly find that an elevation in "gene A" is followed by an elevation of "gene B" one hour later in a proportional fashion? Could A be an activator of B? In a more complex scenario, what if we discovered that elevation of both "gene A" and "gene B" together (not individually) was followed by a decrease in the "gene C" expression level? Could we infer that A and B together repress gene C? Is there a systematic way to establish plausible hypotheses of gene interactions based on such inferences?

We have used a linear mathematical model to carry out this type of analysis quantitatively on the time series data of spinal cord development, and hippocampal development and injury (D'haeseleer et al. 1999), covering a total of 70 genes. Essentially, the linear model posits that the expression level of a gene at a particular time point equals the sum of the expression levels of all genes in the network at an earlier time (an arbitrarily short time before the time point in question), each gene multiplied by a positive or negative constant. To more plausibly reflect the trends in the data, all time series were subject to a cubic spline interpolation procedure, which fills out the gaps between the measured time points while smoothing out abrupt changes. The challenge of the analysis is to fill in the gene-gene interaction matrix of linear factors with the optimal set of values that can most accurately reproduce the measured training data. For this purpose, we applied genetic algorithms, neural nets, and least-squares minimizers to optimally fit these parameters. Remarkably, the model was able to fit the training data exceedingly well (fig. 10.9), passing the first hurdle of the reverse engineering challenge, that is, determining that a mathematical network architecture is sufficient to reflect the complexity of the data.

However, more research is required to determine the uniqueness of

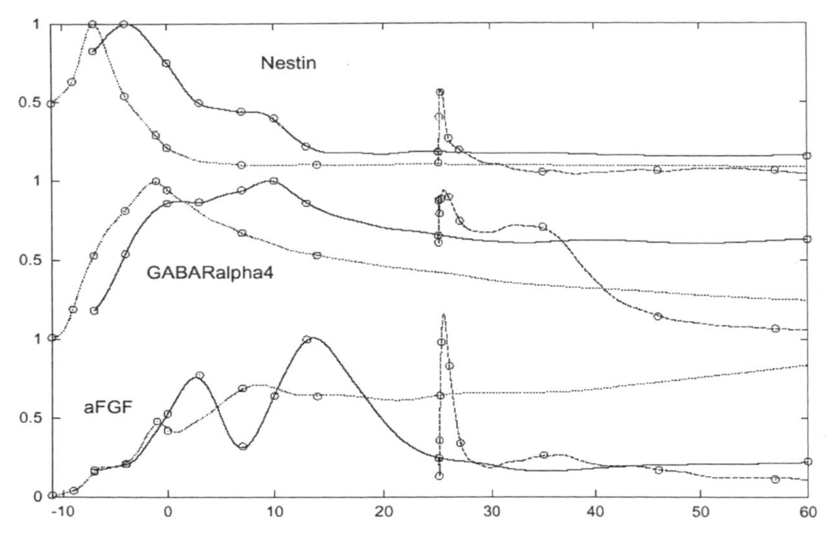

Fig. 10.9.—Gene network reverse engineering: fitting the experimental data with a linear model. The points represent development and injury, and the lines represent a simulation using a linear model. The model faithfully reproduces the time series of the training data sets. The dotted lines represent the spinal cord, starting 11 days before birth. The solid lines represent the development of the hippocampus, starting 7 days before birth. The dashed lines represent the hippocampus kainite injury, starting at postnatal day 25. GABaRa4, γ-aminobutyric acid receptor; aFGF, acidic fibroblast growth factor.

these network solutions. Indirect optimizers such as genetic algorithms (GAs) and artificial neural networks (ANNs) can be more prone to variations in outcome parameter values (e.g., connection weights in ANNs) with respect to procedural starting conditions than are the more direct optimizers such as maximum likelihood or least squares. But, such vagaries in optimal parameter estimation by GAs or ANNs can be overcome in practice by employing established training methods including regularization, cross-validation, and committees of learners. When these kinds of training regimens are employed, consistent themes in the relative strengths of network interconnection weights frequently emerge, even though there are variations in outcome parameters and weights across the individual instances of trained GAs or ANNs (Bishop 1995; Duda et al. 2001).

What can we say about the global features of the network architecture from our reverse engineering study of CNS development? First of all, we discovered that the gene-gene interaction matrix of this network is sparse; that is, there is only a small number of major interaction weights that are not zero or near zero. This affirms our general intuition about biological networks, that is, that all genes are not equally connected: Each gene receives only a small number of decisive regula-

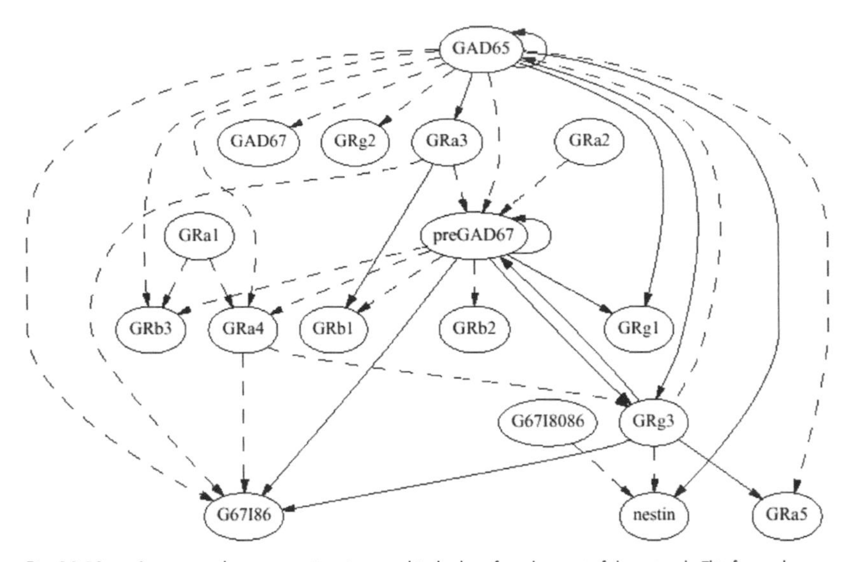

Fig. 10.10.—Gene network reverse engineering: graphic display of a select part of the network. This figure shows a hypothetical gene interaction diagram for the γ-aminobutyric acid (GABA) signaling family inferred from developmental gene expression data (spinal cord and hippocampus data). Although individual proposed interactions have not yet been experimentally verified, the predicted high connectivity within this gene family appears biologically plausible. The positive feedback interaction of the glutamic acid decarboxylase (GAD) species has been proposed independently in another study (Somogyi et al. 1995). Solid lines correspond to positive interactions; broken lines suggest inhibitory relationships.

tory inputs, and there is a relatively small subset of genes that performs most of the regulation.

We also found indications for modular structures in the network graph, that is, groups of genes from particular functional families that are tightly connected among themselves (the complete 70 × 70 network graph is not shown here due to size limitations). For example, in figure 10.10, we show such a group of 17 genes, 16 of which are involved in the signal transduction of the neurotransmitter GABA (GABA receptors [GRs] and glutamic acid decarboxylase [GAD], which synthesizes GABA). This is again clear evidence for the presence of dynamic regulatory modules in these networks that overlap with existing gene function families.

Concluding Remarks

Many questions remain to be explored in the study of genetic network modularity. For example, we have not yet fully defined a module and have yet to discover how many there are. The size of modules will depend on the scope of the biological events. It would be reasonable to define a module as a functional unit of genes necessary for tissue development. The genes in a module would all receive inputs from the same master gene or genes, although not necessarily direct inputs; that is, some would receive master gene inputs via other genes within the same module. In the context of organ development, each of the individual tissues that make up the organ could itself be considered a module. We may therefore need to consider the idea of nested or hierarchical modules.

The dynamics of interaction among modules has not yet been determined; this is an important consideration for the study of communication between tissues and will be important in understanding organ development. It is reasonable to assume that each module contains outputs to communicate with other modules. This leads to the question of the accessibility of one module by another; that is, are some modules open to communication through only one particular gene, or do they contain a number of different genes that can act as inputs? In the former case, a genetic program involving such a module could be activated only under very specific circumstances. In some cases, certain combinations of inputs may be required to activate a module.

The possibility that a gene could be a member of more than one module needs to be addressed, since some genes change functions over the course of development. In addition, some genes may play multiple roles at the same developmental stage, for example, one role in normal development and another in response to injury during development. The possibility of multiple roles also exists for modules within tissues. Detailed gene activity data, combined with gene network reverse engineering tools and experimental validation, will help us build the models we need to answer these questions.

So far, large-scale gene expression studies have generated more questions than they have answered about genetic networks. Further time series data sets from studies of organ development, tissue regeneration, and degenerative diseases will be needed for a greater understanding of this subject. The application of combinations of computational approaches for analyzing the data should provide further information concerning the organization of gene modules and their dynamics. These studies will be critical for the development of therapeutic approaches to tissue regeneration, including the reversal of damage from spinal cord injuries and of brain damage caused by strokes and degenerative diseases. Although reverse engineering will require a large amount of experimental and computational work, it should eventually become the basis for making reasonably accurate predictions about tissue responses to therapeutic drugs for use in tissue regeneration studies.

Acknowledgments

The authors wish to state their appreciation for the support of the Molecular Mining software development team, specifically the product managers, Steve Misener and Ross Dickson, in building and fine-tuning the analysis tools used here, and to Xiling Wen and Millicent Dugich-Djordjevich for carrying out the experimental measurements.

References

Alter, O., P. O. Brown, and D. Botstein. 2000. Singular value decomposition for genome-wide expression data processing and modeling. *Proc. Natl. Acad. Sci. U.S.A.* 97: 10101–10106.

Bishop, C. M. 1995. *Neural networks for pattern recognition.* Oxford: Oxford University Press, Clarendon Press.

Chang, B., R. Somogyi, and S. Fuhrman. 2001. Evidence for shared genetic programs from cluster analysis of hippocampal gene expression dynamics in development and response to injury. *Restor. Neurol. Neurosci.* 18:115–125.

D'haeseleer, P., S. Liang, and R. Somogyi. 2000. Genetic network inference: from co-expression clustering to reverse engineering. *Bioinformatics.* 16:707–726.

D'haeseleer, P., X. Wen, S. Fuhrman, and R. Somogyi. 1998. Mining the gene expression matrix: inferring gene relationships from large scale gene expression data. In *Proceedings of the International Workshop on Information Processing in Cells and Tissues* (IPCAT97), ed. M. Holcombe and R. Paton, 203–212. New York: Plenum Press.

D'haeseleer, P., X. Wen., S. Fuhrman, and R. Somogyi. 1999. Linear modeling of mRNA expression levels during CNS development and injury. In *Proceedings of Pacific Symposium on Biocomputing '99* (PSB99), ed. R. B. Altman, A. K. Dunker, L. Hunter, and T. E. Klein, 41–52. Singapore: World Scientific Publishing Co..

Duda, R. O., P. E. Hart, and D. G. Stork. 2001. *Pattern classification.* New York: John Wiley and Sons.

Felsenstein, J. 1993. PHYLIP (Phylogeny Inference Package) v3.5c. Seattle: University of Washington, Department of Genetics.

Gehring, W. J. 1998. Master control genes in development and evolution. New Haven, Conn.: Yale University Press.

Golub, T. R., D. K. Slonim, P. Tamayo, C. Huard, M. Gaasenbeek, J. P. Mesirov, H. Coller, M. L. Loh, J. R. Downing, M. A. Caligiuri, C. D. Bloomfield, and E. S. Lander. 1999. Molecular classification of cancer: class discovery and class prediction by gene expression monitoring. *Science* 286:531–537.

Hill, A., C. P. Hunter, B. T. Tsung, G. Tucker-Kellogg, and E. L. Brown. 2000. Genomic analysis of gene expression in *C. elegans. Science* 290:809–812.

Jarvis, R. A., and Edward A. Patrick. 1973. Clustering using a similarity measure based on shared nearest neighbors. *IEEE Trans. Comput.* C-22:1025–1034.

Kohonen, T. 1997. *Self-organizing maps.* 2d ed. Berlin: Springer.

Liang, S., S. Fuhrman, and R. Somogyi. 1998. REVEAL, a general reverse engineering algorithm for inference of genetic network architectures. In *Proceedings of the Pacific Symposium on Biocomputing '98* (PSB98), ed. R. B. Altman, A. K. Dunker, L. Hunter, and T. E. Klein, 18–29. Singapore: World Scientific Publishing Co.

Somogyi, R., S. Fuhrman, M. Askenazi, and A. Wuensche. 1997. The gene expression matrix: Towards the extraction of genetic network architectures. In *Proceedings of the Second World Congress of Nonlinear Analysts* (WCNA96), Athens, 1996. Published in *Nonlinear Anal.* 30 (3): 1815–1824.

Somogyi, R., and L. D. Greller. 2001. The dynamics of molecular networks: applications to therapeutic discovery. *Drug Discov. Today* 6:1–11.

Somogyi, R., X. Wen, W. Ma, and J. L. Barker. 1995. Developmental kinetics of GAD family mRNAs parallel neurogenesis in the rat spinal cord. *J. Neurosci.* 15: 2575–2591.

Tamayo, P., D. Slonim, J. Mesirov, Q. Zhu, S. Kitareewan, E. Dmitrovsky, E. S. Lander, and T. R. Golub. 1999. Interpreting patterns of gene expression with self-organizing maps: methods and application to hematopoietic differentiation. *Proc. Natl. Acad. Sci. U.S.A.* 96:2907–2912.

Toronen, P., M. Kolehmainen, G. Wong, and E. Castren. 1999. Analysis of gene expression data using self-organizing maps. *FEBS Lett.* 451:142–146.

Wen, X., S. Fuhrman, G. S. Michaels, D. B. Carr, S. Smith, J. L. Barker, and R. Somogyi. 1998. Large-scale temporal gene expression mapping of central nervous system development. *Proc. Natl. Acad. Sci. U.S.A.* 95:334–339.

11 Qualitative Analysis of Gene Networks: Toward the Delineation of Cross-Regulatory Modules

DENIS THIEFFRY AND LUCAS SÁNCHEZ

Introduction: Toward a Computational Representation of the Regulome

The interplay between genetic and molecular methodologies is very fruitful in the comprehension of biological processes. The genetic approach allows identification of the genes through the induction of mutations affecting the process. In addition, the genetic analysis of the epistatic relationships between these mutations permits delineation of the genetic network, which is at the basis of the biological process in question. The molecular characterization of mutations and recent techniques based on the expression of reporter genes lead to various types of data, such as the organization of genes in terms of sequences coding for the gene products (protein or RNA), the cis-acting regulatory sequences that interact with the factors controlling the expression of genes, and the identification of domains endowed with specific functions exerted by the gene products.

With the development of automated sequencing and PCR techniques, most of these types of studies have been thoroughly enhanced, leading to a change of focus from the level of individual molecular mechanisms to whole cells or organisms. As a first outcome, dozens of genomes, ranging from bacteria to human, have been or are being fully sequenced. Moreover, systematic sequencing of variable regions in hundreds of individuals of a single species are being performed to evaluate genetic variability. Mapping and expression studies have also been profoundly modified, since, for example, PCR-amplified fragments are increasingly used in place of genetic or enzymatic markers to obtain high-resolution molecular maps. Similarly, 2D protein gels (coupled with mass spectroscopy) and DNA chips (or "microarrays," encompassing thousands of PCR-amplified cDNA probes distributed on a single glass slide) are now used to characterize the expression of hundreds or thousands of genes simultaneously, in various cells types, mutant backgrounds, and culture conditions (for an overview of DNA chips, see Granjeaud et al. 1999). Finally, automation of gene cloning, of in situ expression analyses, and of loss-of-function studies is also underway. This genomic emphasis is well attested by the increasing use of terms such as "genome," "transcriptome," "proteome," and "metabolome" to speak of DNA sequences, mRNA, proteins, and metabolism at the level of the whole organism. Similarly, the term "regulome" can be used to cover the regulatory mechanisms controlling gene expression and enzyme activities at the genomic level.

To cope with this explosion of data, many groups are working on the development of appropriate forms of storage and querying, leading to numerous and diverse types of databases (for an overview, see the first issue of *Nucleic Acid Research* of this year, covering over hundred different databases, or see Misener and Krawetz 2000). A long-term prospect of these developments is to support the delineation of dynamical models accounting for the spatiotemporal behavior of cells or organisms. However, even in the most documented cases, we are still far from reaching a solid basis for such model building, particularly in the case of multicellular eukaryotic organisms. On the one hand, relatively little is known about the organization and the regulation of plant and animal genomes. On the other hand, the information already gathered seems difficult to subsume under the traditional regulatory schemes or simple extensions thereof. Indeed, regulatory regions are much more spread out in plants and metazoans than in bacteria or simple eukaryotes such as yeast. In addition, many factors are often involved in the regulation of a single plant or animal gene. Furthermore, many of these factors appear to be poorly specific and are found in a wide variety of cellular types. Finally, posttranscriptional regulation mechanisms seem both more frequent and more sophisticated in high eukaryotes (e.g., alternative splicing, mRNA maturation, posttranslational modifications of proteins such as acetylation, phosphorylation, or glycosylation, and proteolysis).

In any case, gene regulation is increasingly conceived in terms of the combinatorics of regulatory factors whose varying associations across space and time would determine the outcome in terms of gene expression. Intercellular interactions, however, have also been implicated in gene regulation. Beyond the characterization of the concentrations of all regulatory factors, some knowledge of the precise localization in the cell of these regulatory factors (i.e., on the chromosomes in the case of DNA binding factors), of the states of the neighboring cells, and of the temporal sequence of regulatory events would be required to infer the transcriptional status of a given cell.

If a full model of gene regulation at the level of a eukaryotic cell or organism is probably still out of reach, recent studies on specific networks and subnetworks already provide interesting clues about how one could proceed to disentangle large genomic networks into "cross-regulatory modules" (see below for the definition of "cross-regulatory modules"). Indeed, several chapters of this book offer formal methods of dealing with the dynamical analysis of the expression of groups of genes cooperating to drive some particular cellular or developmental process. In the present chapter, we take advantage of a series of specific studies dealing with cell differentiation and pattern formation in model organisms such as *Arabidopsis thaliana* and *Drosophila melanogaster*. Using a logical multilevel and asynchronous formalism (Thomas 1991), these studies emphasize the relationships between the dynamics of gene expression and the structure of the corresponding regulatory network, in particular, the occurrence of "feedback circuits" (i.e., closed chains of regulatory interactions, also often called "feedback loops") (Thomas et al. 1995).

More specifically, positive circuits (i.e., circuits involving an even number of negative interactions) have been associated with the generation of alternative regimes of gene expression (Monod and Jacob 1961; Lewis et al. 1977; Meinhardt 1978; Thomas 1978). Since then, much progress has been made in the molecular analysis of gene regulation and development. In the process, dozens of examples of direct or indirect positive autoregulation of key regulatory genes involved in cell differentiation have been found (e.g., *Myo*D, *Wg*). In parallel, the requirement of positive feedback circuits for multistationarity, as well as that of negative circuits for sustained oscillations, has been formally proved by several authors (Plahte et al. 1995; Gouzé 1998; Snoussi 1998). Consequently, the biological roles of positive and negative feedback circuits have been further clarified. Whereas negative circuits allow the buffering of gene dosage effects, as well as tight control of the expression of key regulatory genes, positive regulatory circuits may constitute developmental switches, allow the occurrence of alternative developmental pathways, and/or encode positional information. We still need to understand, however, how such elementary circuits behave in the presence of additional regulatory inputs or when several such circuits are intertwined. In the next section, we will review some of the results already obtained in the context of a logical analysis of several networks involved in cell differentiation and pattern formation in *Arabidopsis thaliana* and *Drosophila melanogaster*. In each case, we will see that it is possible to consistently isolate, model, analyze, and simulate subsets of cross-regulating genes while keeping track of the way these networks are embedded in larger genetic systems. Altogether, these studies lead to the delineation of the notion of "cross-regulatory module," defined in terms of intertwined regulatory circuits.

Genetic Control of Flower Morphogenesis in *Arabidopsis thaliana*

Our first example deals with the problem of flower morphogenesis in *Arabidopsis thaliana*. Developed in collaboration with Luis Mendoza and Elena Alvarez-Buylla (Instituto de Ecología, UNAM, Mexico), our analysis relies on the various genetic and molecular data accumulated during the last twenty years (for reviews, see Parcy et al. 1998; Ma 1998). A dozen genes directly implicated in the differentiation of the four types of flower organs (sepals, petals, stamens, and carpels) have been partly characterized. These were initially identified for the drastic effects of some of their mutations on flower morphology, leading to the loss of structures or transformation of a type of flower organ into another ("homeotic mutations"). Most of these genes have been found to code for transcriptional factors. To what extent these genes constitute the whole set of relevant regulatory genes remains unclear. In addition, although some data have already been gathered on the timing and the level of expression of these genes (using in situ hybridization or labeling experiments), little is known yet on the molecular details of the regulatory mechanisms involved.

This system thus constituted a good opportunity to model and analyze a complex but still incompletely characterized developmental network. The challenge was to find ways to extract the maximum from the data already available in order to gain new biological insights into the flowering process. An extensive analysis of the literature led us to propose a ten-element regulatory network, encompassing both known and hypothetical interactions (fig. 11.1, *A*). Although it is fairly well established that most of these genes interact with several other genes of the network, available data do not allow the direct derivation of quantitative or even qualitative parameters for these interactions. Consequently, we focused on a qualitative model of the network, using a logical formalism developed by René Thomas and collaborators in Brussels, Belgium (see, e.g., Thomas 1991). Briefly, for each gene of a network, this formalism associates

- A logical variable, representing the concentration or activity of the final product
- A logical function, representing the rate of gene expression, for example, transcriptional rates
- A set of logical parameters, qualifying the strength of individual interactions or groups of interactions controlling a given gene

The logical parameters allow the specification of various types of functions of classical Boolean algebra (e.g., AND, OR, NOR). All logical variables, functions, and parameters can take a limited number of

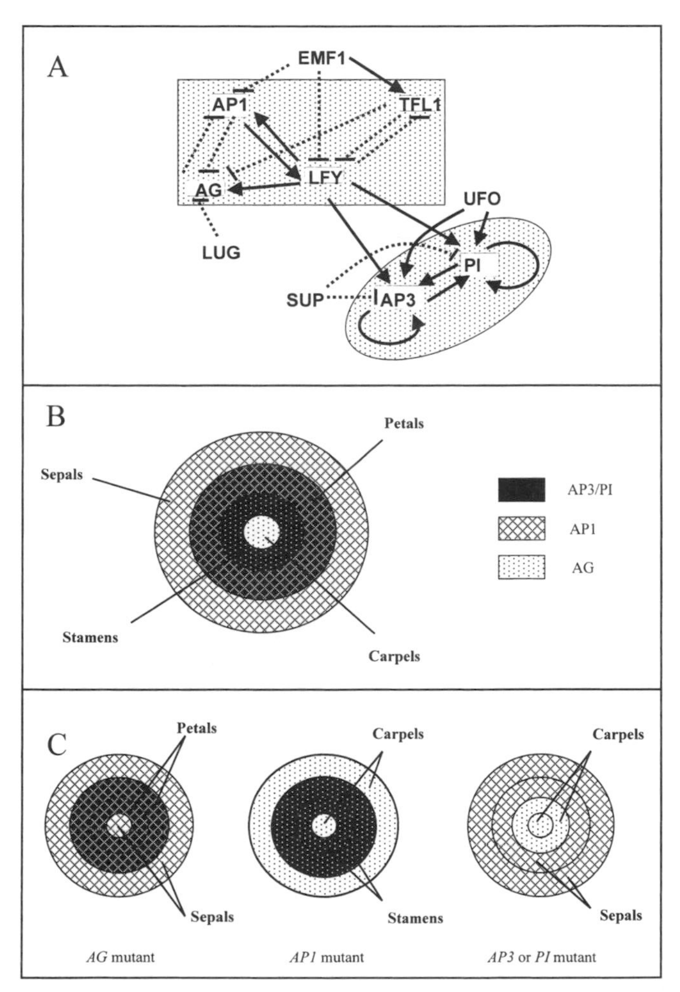

Fig. 11.1.—Genetic network involved in the control of flower morphogenesis in *Arabidopsis thaliana. A,* Diagram describing the cross-regulatory interactions between the most important genes controlling flower morphogenesis in *Arabidopsis.* Activatory and repressory relationships are indicated by arrows and blunt lines, respectively. Focusing on the genes involved in feedback circuits, two subnetworks, or "cross-regulatory modules" can be readily distinguished (*highlighted boxes*), the first formed by *AP3* and *PI* cross-regulations, the second involving the genes *TFL1, LFY, AP1,* and *AG. B,* Schematic representation of the concentric organization of the flower bud, with indication of the domains of expression of four organ-specific genes. *C,* Simulation of the patterns of gene expression in three homeotic mutants, with a description of resulting phenotype. Gene symbols: AG, *AGAMOUS;* AP1, *APETALA 1;* AP3, *APETALA 3;* EMF1, *EMBRYONIC FLOWER 1;* LFY, *LEAFY;* LUG, *LEUNIG;* PI, *PISTILLATA;* SUP, *SUPERMAN;* TFL1, *TERMINAL FLOWER 1;* UFO, *UNUSUAL FLORAL ORGANS.*

integral values according to a gene-specific scale. In the simple "Boolean" case, these can take two values, 0 or 1. Note that, in contrast to most modeling studies referring to the Boolean formalism or extensions thereof, Thomas's approach is asynchronous and consequently associates a specific time delay to each variable commutation (i.e., a qualitative change in some regulator concentration). This asynchronous treatment eliminates the most problematic artifacts of the synchronous logical approaches (e.g., spurious logical cycles) (see Thomas and D'Ari 1990 for further details).

In the context of our *Arabidopsis* model, inspection of the network led us to identify two subnetworks, or "modules," each formed by several intertwined regulatory circuits. This observation suggested a strategy that consisted of analyzing the two modules separately in a first step and then combining the results obtained for each module in order to characterize the dynamical behavior of the whole system (Mendoza et al. 1999).

In the case of the first module, formed by the genes *PISTILLATA* and *APETALA 3,* available molecular information supports the selection of specific values for all logical parameters (recall that in the Boolean case, these logical parameters can take only the values 0 or 1). This module was shown to determine the choice between two regimes of gene expression, ensuring the cooperative expression of *PISTILLATA* and *APETALA 3* needed for the formation of petals and stamens.

The analysis of the second module, involving genes *APETALA 1, AGAMOUS, LEAFY,* and *TERMINAL FLOWER 1,* proved to be somewhat more complicated but still tractable. In this case, qualitative information about functional patterns of gene expression was used to select values for most logical parameters. Our analysis shows that this network is mainly involved in the choice between two stables states of gene expression, the first allowing sepal and petal formation, the second leading to stamen and carpel development. Moreover, we showed that in addition to these four genes, another gene, *EMBRYONIC FLOWER 1,* has to be taken into account to go from the vegetative state to the flowering pathways. As this gene is found to regulate but not to be regulated by other genes in the network, its expression could be represented by a simple input variable.

Combining the parameter constraints obtained for the two modules, we could assign values to most of the logical parameters. Moreover, the alternative regimes of gene expression of the whole system were obtained as a straightforward combination of those of the two subsystems. This leads to expression regimes corresponding to the four types of organs (sepals, petals, stamens, carpels; represented in fig. 11.1, *B*), the vegetative state, and a puzzling state combining the expression of flower inhibition genes (*TERMINAL FLOWER 1* and *EMBRYONIC FLOWER 1*) with that of *PISTILLATA* and *APETALA 3*. This last pattern was automatically obtained as long as the constraints needed to obtain the other—physiologically understood—states were fulfilled. Although it is probable that this expression state is never reached during normal development, one could try to induce it experimentally, for example, through a transient induction of *PISTILLATA* and *APETALA 3* in vegetative cells. This constitutes a prediction of the model, still awaiting experimental confirmation.

Another important prediction emerges as a result of a careful analysis of the parameter values selected. Indeed, to obtain the right regimes of gene expression, we have to assign zero values to all parameters accounting for *LEAFY* expression. This means that this gene would never be significantly expressed, a situation itself incompatible with the fact that experimentalists have shown that its expression is needed for the transition from the vegetative state to flowering (see, e.g., Busch et al. 1999; Wagner et al. 1999). As we obtained the same result even in the context of the analysis of multilevel models (unpublished results), we were led to predict the existence of at least one additional regulator of *LEAFY,* which would play a key role in the transition from the vegetative state to the flowering pathways. This prediction finds some support in a recent experimental paper reporting the identification of a new gene, *LUMINIDEPENDENS,* coding for a nuclear protein and probably implicated in the regulation of *LEAFY* (Aukerman et al. 1999).

The two predictions just mentioned derive directly from unexpected but nevertheless robust results of our model analysis. This clearly suggests that, even when dealing with complex systems with limited constraining data, a qualitative but still rigorous formal treatment may lead the modeler to reconsider some of his initial assumptions. Indeed, as just showed, the integration of different types of constraints, dealing for example with some parameter values or with some of the relevant regime states of gene expression, often leads to unforeseen consequences. It is precisely in such situations that formal modeling can be the most useful, leading to clear predictions amenable to experimental test and thus potentially contributing to the ongoing characterization of the system under study.

Though still crude, our model already allows the simulation of different types of mutations. Most straightforward is the simulation of loss-of-function mutations, which consists in setting the corresponding variables and parameters to zero (see the examples given in fig. 11.1, C). More sophisticated cases, such as cis-regulatory mutations, gains of function, or even mutations of specific protein domain, can also be simulated through appropriate value changes for some of the logical parameters.

An important and original outcome of our logical analysis consists in the delineation of the roles of two regulatory circuits, one involving the genes *APETALA 1* and *AGAMOUS* and the other involving the genes *APETALA 3* and *PISTILLATA*. Formed by the cross-inhibitory interactions between *APETALA 1* and *AGAMOUS,* the first circuit constitutes a switch forcing a choice between the expression of one or the other of these genes. Made of the mutual activation of genes *APETALA 3* and *PISTILLATA,* the second circuit also constitutes a switch, this time coordinating the expression of both genes. Furthermore, our analysis leads to the conclusion that the other circuits (involving up to four genes) found in the network are not fully functional from a dynamical point of view. By "functional," we mean here that the corresponding circuit generates multistationary properties (i.e., alternative regimes of gene expression) in the case of a positive circuit, or homeostatic properties (buffering or oscillation of gene expression) in the case of a negative circuit. Saying that a circuit is *not* functional does not imply, however, that the constitutive interactions are dispensable for the proper functioning of the network, but rather that the corresponding sequences of interactions do not generate multistationary or homeostatic properties.

Our model proves to be completely consistent with the famous combinatory "ABC" model (Coen and Meyerowitz 1991). In this respect, one of the achievements of our analysis consists in providing a more explicit molecular basis for the heuristic power of the ABC model. In particular, the comparison of various mutant phenotypes and gene expression patterns (under wild-type and mutant backgrounds) led to the delineation of new interactions. Our analysis further identifies the most important regulatory interactions (those forming the functional feedback circuits), accounting for the crucial properties of the ABC model. Finally, our analysis reveals some gaps in our present molecular understanding of the system (e.g., a missing regulator of *LEAFY*).

This is certainly not the end of the story, as recent experimental data suggest the existence of several additional regulators—in particular, a series of factors involved in the induction of flowering, most likely via *LEAFY* activation (Blazquez and Weigel 2000). Moreover, we still lack a proper understanding of the details of the kinetics of gene expression, even in the case of the genes already taken into account in our model. Finally, up to now, the spatial dimension was treated only in an implicit way, thus forbidding us to account for the specific numbers and shapes of each type of floral organ, which are characteristic of flowering plant species.

Pattern Formation during *Drosophila melanogaster* Development

One of the most extensively studied developmental processes is certainly the early embryogenesis of *Drosophila melanogaster.* A saturated mutagenesis followed by the careful screening of many embryonic mutant phenotypes led to the identification of most (if not all) regulatory genes controlling the patterning processes along the dorsoventral and the anteroposterior axes (see Nüsslein-Volhardt and Wieschaus 1980; Nüsslein-Volhardt and Roth 1989).

The present picture can be summarized as follows (for a review, see Lawrence 1992). Initial development takes place in a syncytial environment. The zygotic nuclei divide synchronously and migrate to the periphery, where they form the layer of blastoderm cells, which gives rise to the somatic component of the animal. When the eggs are laid, these contain maternal products that are asymmetrically distributed in them. After fertilization, these maternal products lead to the formation of concentration gradients along the dorsoventral and anteroposterior axes. These gradients in turn act upon the zygotic genome, activating a series of interacting genes and leading to a progressive definition of specific spatial patterns of gene expression along the dorsoventral and the anteroposterior axes. These expression patterns ultimately determine the fates of the cells of the embryo, leading to differentiated tissues and segments, first in the larva and later in the adult fly.

On the basis of an extensive review of the literature, we are working on the development of multilevel logical models encompassing the most relevant genes and their cross-regulations. The sets of genes involved in the establishment of the dorsoventral and anteroposterior patterns are analyzed separately, as their expressions are largely independent during the segmentation process. Their expressions are coupled, however, in the head and tail, which are not covered by the present study.

Dorsoventral Patterning

The dorsoventral pattern results from the different responses of cells to the Dorsal protein morphogen, which is distributed according to a nuclear gradient along the dorsoventral axis, with the highest concentration on the ventral side (reviewed in Chasan and Anderson 1993). As described in Sánchez et al. (1997), about a dozen genes have been clearly implicated in this process. The analysis of the feedback structure of this network leads to the identification of two regulatory modules (fig. 11.2, *A*). The first module is simply constituted by the (indirect) positive autoregulatory gene *decapentaplegic,* and the second by the genes *twist* and *snail.* The other genes apparently behave as input variables

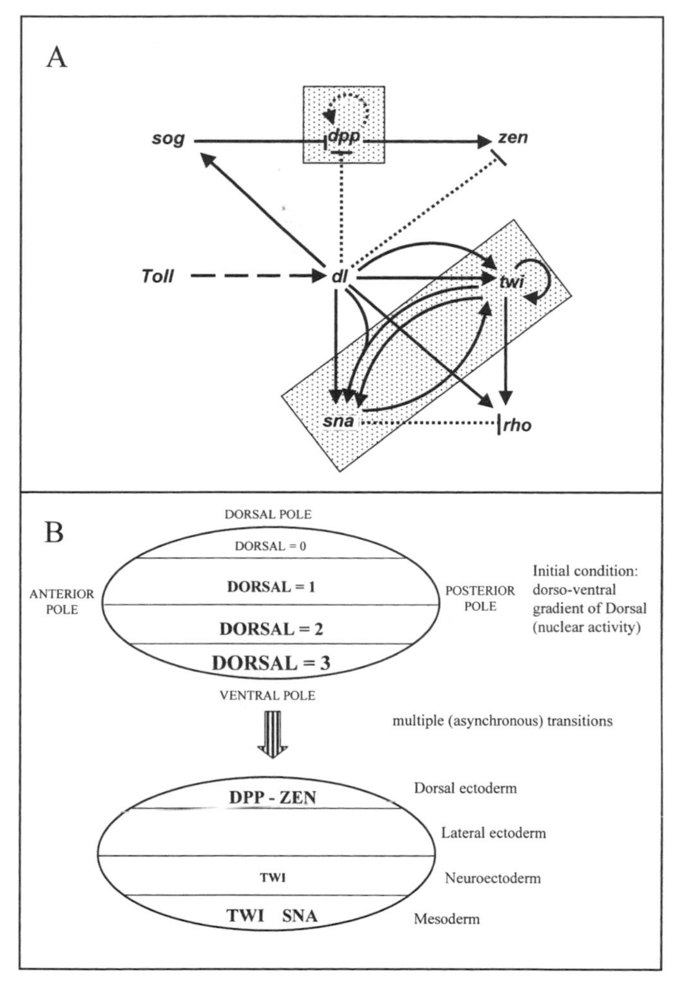

Fig. 11.2.—Dorsoventral patterning in the trunk of *Drosophila melanogaster* embryo (several downstream genes are omitted for the sake of clarity). *A,* Functional relationships among the most important genes involved in the determination of this pattern. Activatory and repressory relationships are indicated by arrows and blunt lines, respectively. *B,* Qualitative simulation of the main gene activity patterns across the dorsoventral axis. The initial gradient of Dorsal activity is represented by a multilevel logical variable. In the resulting gene expression pattern, the size of the symbol TWI for Twist is higher in the mesoderm than in the neuroectoderm to reflect the higher amount of TWI protein in the former tissue. The establishment of the dorsoventral pattern is as follows. The Dorsal protein is the same for all cells along the dorsoventral axis. The protein is localized in the cytoplasm and forms a complex with the protein Cactus. A signal, released ventrally, interacts with the receptor Toll, present in all cells along the dorsoventral axis. The activation of this receptor produces a second signal, which is internalized and provokes the dissociation of the Dorsal-Cactus complex. This results in the migration of the free Dorsal protein into the nucleus. Since the signal that activates Toll is released ventrally, the amount of Dorsal protein in the nuclei has a peak of higher concentration in the ventral-most side of the embryo, decreasing gradually toward the dorsal side where no Dorsal protein is present. Gene symbols: *dl* = *dorsal, dpp* = *decapentaplegic, rho* = *rhomboid, sna* = *snail, twi* = *twist, sog* = *short gastrulation, zen* = *zerknüllt.*

or output functions. Here also, our strategy consisted in analyzing these modules separately to later reconstitute the global dynamics of gene expression. But in the present case, detailed genetic and molecular data are available. These suggest the existence of multiple (qualitatively different) functional levels for several regulatory factors (above all for the morphogen Dorsal) and thus call for the use of multilevel variables and functions to represent these factors.

In addition, several context-sensitive or synergic regulatory effects are found. For example, higher concentrations of Dorsal protein by itself is able to activate gene *snail,* but lower concentrations of Dorsal protein requires the concerted action of Twist protein to activate gene *snail.*

For proper parameter values, our multi-level logical model accounts for the essential qualitative effects of the Dorsal gradient during the dorsoventral determination process (fig. 11.2, *B*). In particular, our analysis of the *twist-snail* module shows how different transient levels of Dorsal protein can lead to the durable selection of one specific regime of gene expression among three alternative ones.

At the basis of this multistable behavior, we find two positive circuits formed by *twist* auto-regulation, and by *snail* and *twist* positive cross-regulations. These circuits behave like switches acted upon by different concentrations of Dorsal, ultimately determining the different cell types that make up the embryonic dorsoventral pattern. The resulting gene expression regimes are found to consistently represent the differentiation of three distinct regions along the dorsoventral axis (ectoderm, neuroectoderm and mesoderm).

Other genes are involved to refine the resulting pattern, in particular *decapentaplegic* whose (indirect) auto-regulation is found to allow a durable choice between the most dorsal fate (leading to dorsal ectoderm and amniosera) and the lateral ectoderm fate.

The known facts are well accounted by our model. More important, several predictions emerge from this analysis. First, we predict that the protein Twist must have two distinct functional thresholds, one for auto-activation, the other for the activation of the gene *snail.* In addition, the auto-activation threshold must be smaller than that of *snail* activation. Finally, the role of Snail protein is crucial for the maintenance of *twist* expression, allowing a stable choice between the mesoderm or neuroectoderm developmental pathways. None of these could have been derived from existing experimental data without a formal analysis. They constitute predictions of our model analysis, all susceptible to be tested experimentally.

In addition, we predict that if the gene *snail* shows auto-regulation (as might be concluded from the detection of binding sites for the Snail protein in the promoter region of its gene, see Mauhin et al. 1993), either positive or negative, this auto-regulation should not be crucial for the determination of the embryonic dorsoventral pattern.

Finally, our model also allows the simulation of various types of mutants, such as loss or gain of function, or cis-regulatory mutations, leading to results which are so far consistent with available data (for a recent review, see, e.g., Podos and Ferguson 1999).

Anteroposterior Patterning

An analysis of the gene network controlling the formation of the anteroposterior pattern at the origin of the body segmentation is under way. Here also, the process is initiated by a series of maternal genes transcribed during oogenesis. Their products are deposited asymmetrically in the oocyte. This initial pattern directly stimulates the expression of several zygotic segmentation genes, leading to the progressive definition of the typical stripe pattern of gene expression associated with the different segments of the larvae (see Nüsslein-Volhardt and Roth 1989; Lawrence 1992; Rivera-Pomar and Jäckle 1996).

As the genetic control of head and tail formation are not yet very well characterized, our analysis primarily focuses on the formation of the trunk segments. In this context, our approach leads to a rational decomposition of the segmentation network into well-defined sets of feedback circuits sharing one or several elements. The modules obtained correspond exactly to the different classes of segmentation genes ("gap," "pair-rule," "segment polarity") defined by the molecular geneticists on the basis of mutant phenotypes (fig. 11.3, *A*).

Focusing on the gap module, we proposed a multilevel model (fig. 11.3, *B*), allowing the simulation of the kinetics of gene expression observed in the wild-type fly (fig. 11.3, *C*), as well as the prediction of the phenotype of various mutants (see table 11.1 for a selection of single and multiple loss-of-function mutations) (Sánchez and Thieffry 2001). Here also, our analysis emphasizes the roles of the various feedback circuits present in the regulatory matrix and specifies the range of parameter values allowing their proper functioning. In particular, we find that the positive circuit made of the cross-inhibition between the genes *Krüppel* and *giant* predominantly contributes to the refinement of segmental boundaries.

The fascinating establishment of alternate stripes of gene expression along the anteroposterior axis has already stimulated a wealth of theoretical studies (see e.g., Meinhardt 1986; Hunding et al. 1990; Lacalli 1990; Burstein 1995; Bodnar 1997; Reinitz et al. 1998; von Dassow et al. 2000). Strikingly, when comparing the regulatory structures corresponding to the different models for the sole gap regulatory module, many differences appear. Focusing on qualitative aspects, indeed, as

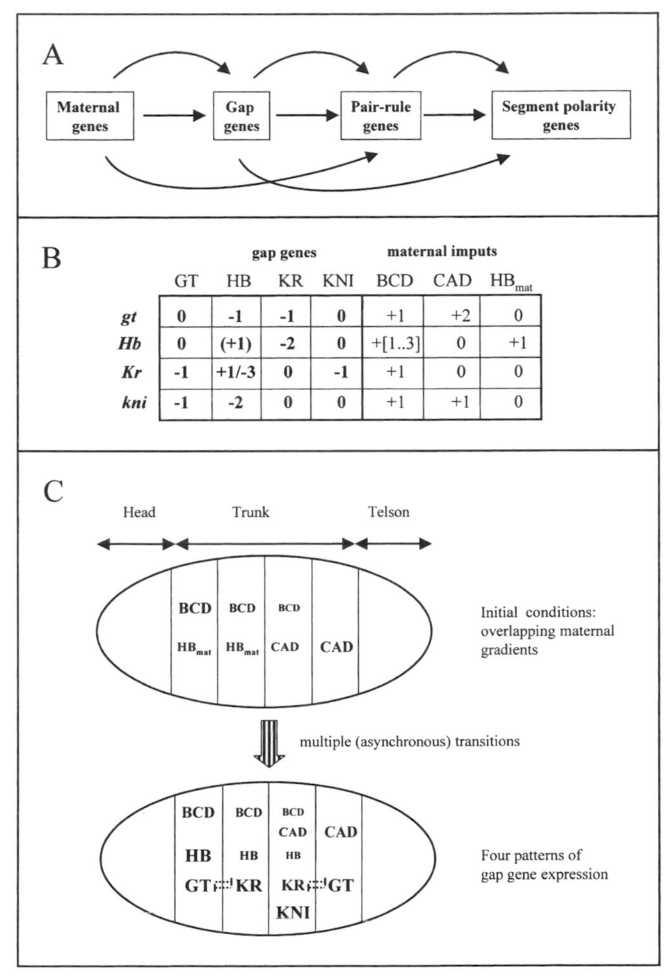

	GT	HB	KR	KNI	BCD	CAD	HB_{mat}
gt	0	-1	-1	0	+1	+2	0
Hb	0	(+1)	-2	0	+[1..3]	0	+1
Kr	-1	+1/-3	0	-1	+1	0	0
kni	-1	-2	0	0	+1	+1	0

Fig. 11.3.—Anteroposterior patterning in the trunk region of *Drosophila melanogaster* embryo. *A,* Segmentation originates from the combined action of maternal organizers and the zygotic genome. The zygotic segmentation genes can be divided into three categories depending on the number of segments affected by their mutations: the gap genes affect several contiguous segments; the pair-rule genes affect complete alternate segments; the segment polarity genes affect each segment. The segmentation genes constitute a hierarchical system in which interactions take place in temporal order: the maternal organizers act upon the zygote genome, resulting in activation of the gap genes; the combined action of maternal and gap gene products determines the activation of the pair-rule genes; finally, the pair-rule genes determine the activity of the segment polarity genes. *B,* Qualitative matrix modeling the interactions between the maternal products and the gap genes (including cross-regulations between these). The signs + and − represent positive and negative interactions, whereas integers specify the functional threshold concentrations when a gene regulates several target genes. Note that we take into account a dual interaction from *Hb* on *Kr,* as well as a multilevel interaction from *Bcd* on *Hb*. Brackets indicate synergic effects. *C,* Qualitative simulation of the gap gene expression. *Upper section:* initial maternal product pattern. *Bottom section:* resulting expression pattern for the gap genes. The size of the gene symbols accounts for the different amounts of maternal products and for the different activity levels of the gap genes. BCD, Bicoid; CAD, Caudal; HB_{mat}, maternal Hunchback; GT, Giant; HB_{zyg}, zygotic Hunchback; KNI, Knirps; KR, Krüppel.

Table 11.1: Qualitative simulation of loss-of-function mutations at maternal and gap segmentation genes

Genetic Background (Loss of Function)	Final State (GT, HB, KR, KNI) A	B	C	D	Observations and Predictions
Wild type	1300	0220	0111	1000	
Bicoid	0001	0001	0001	1000	Loss of GT in A; loss of HB in ABC and KR in BC; KNI expands anteriorly into AB
Hunchback$_{mat}$	1300	0220	0111	1000	
caudal	1300	0220	0120	0000	Increase of KR in C; loss of KNI in C; loss of GT in D
giant	0300	0220	0111	0001	KNI expands posteriorly into D;
Krüppel	1300	1200	1100	1000	GT expands into B and C; loss of KNI in C
knirps	1300	0220	0120	1000	Increase of KR in C;
Hunchback$_{mat\&zyg}$	1000	1000	1000	1000	GT expands into B and C; loss of KR in B and C; loss of KNI in C
giant-Krüppel	0300	0200	0101	0001	KNI expands posteriorly into D
Krüppel-knirps	1300	1200	1100	1000	GT expands into B and C
giant-knirps	0300	0220	0120	0000	Increase of KR in C

Note: Four different domains are distinguished in the trunk of the embryo: A, B, C, and D, from the anterior to the posterior pole. In each of these regions, the levels of activity of the four gap genes (*giant, Hunchback, Krüppel,* and *knirps*) are given in the form of a logical vector. Note that we consider only two different functional activity levels for Giant (GT) and Knirps (KNI) (0 and 1), but three different levels for Krüppel (KR) (0, 1, and 2) and four for Hunchback (HB) (0, 1, 2, and 3). Thus, "1300" stands for high activity levels for Giant and Hunchback, but no activity of Krüppel and Knirps; whereas "0220" means no Giant, a medium level of Hunchback, a high level of Krüppel, and no Knirps activity. For each mutant and for each region, indirect effects are emphasized (underscored values) and further described in the last column. Whereas the simple mutants have all already been well characterized experimentally (and this information was indeed used in the course of the modeling process), double mutants still await full experimental characterization. The patterns given in the bottom four lines of the table thus constitute predictions of our model analysis.

few as four interactions are assigned the same signs (i.e., repression versus activation) in all models, on a total of sixteen potential interactions between the four gap genes. These discrepancies reflect the difficulty of interpreting the numerous genetic and molecular data obtained with different in vitro or in vivo methodologies.

It is this confusing situation which in fact motivated Reinitz to choose a reverse-engineering approach, favoring wild-type in vivo expression patterns against promoter dissection experiments. This strategy, however, also has a price. Indeed, a series of assumptions have to be made to render the problem computationally tractable. In particular, the generic equations retained by Reinitz exclude beforehand potential context-sensitive interactions, since they formalize all regulatory contributions affecting the expression of a given gene as a sum of nonlinear but monotonous terms. Similarly, these equations do not allow multiple interactions between two factors or, a fortiori, multiple interactions with different signs.

To a large extent, our theoretical approach and that of Reinitz complement each other. Relying on very different assumptions and using different formalisms, our model analyses and their confrontation should lead to new insights about the dynamical properties of the segmentation network, beyond the known individual properties of each regulatory factor. Our logical analysis should help to delineate the roles of each of the feedback circuits found. But quantitative spatiotemporal simulations are also needed for a fine account of the patterns of expression observed experimentally. Comparisons between the two approaches should lead to a reconsideration of conflicting assumptions in light of the new experimental data, including those produced by the group of Reinitz.

Conclusions

The Role of Feedback Circuits

The examples discussed above deal with regulatory networks from very different organisms and rely on different kinds of experimental data. In the case of flower morphogenesis in *Arabidopsis*, only limited information on mutant phenotypes and on some patterns of gene expressions was available, but a consistent qualitative model could still be elaborated, leading to several new predictions. In the process, we identified two modules, one involving two and the other four genes. Though the largest of these modules contains several intertwined circuits involving up to four genes, one specific two-element positive circuit could be singled out and associated with crucial differentiative properties.

Similarly, our analysis of the network controlling the dorsoventral patterning in *Drosophila* leads to the delineation of two small modules, the first involving one (*decapentaplegic*) and the second two genes (*twist* and *snail*). With respect to the anteroposterior patterning in *Drosophila*, among the seven feedback circuits within the gap module, the feedback circuit between *giant* and *Krüppel* proved to be the most instrumental for the final expression pattern of the gap genes.

To be sure, in the lack of precise and complete data, our qualitative model analyses leave out potentially important genes, interactions, or processes and furthermore involve several simplifying assumptions. In particular, the alternative "stable states" in the phase space of the different cross-regulatory modules considered represent in fact regimes of gene expression which are found only transiently during development. Each of these models should thus be considered a provisional approximation but still leads to insights into what can already be rigorously inferred from our present partial and qualitative knowledge of the corresponding genetic system. Taken together, these studies particularly emphasize the dynamical and biological roles of specific regulatory circuits embedded in more complex networks. Note that though some of these networks contain several large circuits (i.e., involving three or more elements), only short, positive circuits (involving one or two genes) were found to be functional. Seeing a circuit in an interaction graph, of course, does not give enough evidence to infer the generation of the typical corresponding dynamical properties (i.e., multistationarity in the case of a positive circuit, homeostasis or oscillations of gene expression in the case of a negative circuit). In addition, specific parameter constraints have to be fulfilled, and, most often, only a subset of circuits can be simultaneously functional for a fully specified logical model (including the determination of the parameter values). The larger a circuit is, the more parameter constraints have to be fulfilled to ensure dynamical functionality, as more nodes and connections are involved (for a proper mathematical estimation of the parameter constraints for circuit functionality in the case of fully connected Boolean networks, see Thieffry and Romero 1999).

During the development of metazoans, cells become gradually restricted in their developmental potential. As a result, these cells become irreversibly committed to developing along specific pathways, giving rise to the different cell types of the adult (cellular determination). Each determination state involves a particular combination of genes specifically activated as a response to an external signal (morphogen). According to the concept of positional information (Wolpert 1969), the pattern of cell types that form the individual results from the different responses (positional values) of cells to a morphogen, which shows an asymmetric distribution (positional information). It has been shown that positive feedback circuits behave as devices that fulfill two related cellular functions during the development of multicellular organisms. First, positive circuits transform positional information into discrete positional values; that is, into discrete genetic signals made up of particular combinations of active genes, which, in turn, drive the cells into specific developmental pathways. In addition, positive feedback circuits can ensure the maintenance of the developmental choice triggered by the morphogen (Lewis et al. 1977; Meinhardt 1978, Sánchez et al. 1997; Sánchez and Thieffry 2001).

This prominent role of short, positive transcriptional circuits in metazoan developmental networks has to be contrasted with what has been observed in prokaryotes. Indeed, a systematic analysis of existing cross-regulations among transcriptional factors in *Escherichia coli* revealed a wide occurrence of negative autoregulations, few autoactivations, few cascades, and no circuit involving more than one regulatory gene (Thieffry et al. 1998). At this stage, these puzzling results suggest a very loose structure for the transcriptional regulatory network of *Escherichia coli,* though it is probable that many more regulations remain to be uncovered. As autoinhibitory circuits are predominant, the bacterium appears to behave much like a homeostatic machine. However, more extensive studies would be needed before drawing definite conclusions regarding the structure of prokaryote versus eukaryote transcriptional networks.

The Notion of Cross-Regulatory Modules

Another general outcome of our analysis of several gene networks consists in a precise definition of a specific class of regulatory modules. Indeed, on the basis of the notions of positive and negative "feedback circuits" and of the understanding of their biological and dynamical roles (Thomas et al. 1995), our model analyses illustrate the need to consider simultaneously sets of intertwined circuits. But at the same time, we indicate how large networks can be disentangled into sets of intertwined circuits in order to simplify the formal analysis.

To some extent, functional circuits can be considered dynamical building blocks, since proper interconnections between simple circuits can lead to a combination of the corresponding dynamical properties. When several circuits are intertwined, however, it becomes necessary to consider the whole set of circuit elements with their interconnections to predict the corresponding dynamical properties. Input and output elements (not involved in these intertwined feedback circuits) can then be formally treated as input or output variables, as illustrated in the examples discussed above. As a consequence, we can derive a simple, rational method for the decomposition of complex genomic networks into well-defined and relatively autonomous dynamical modules or "*cross*-regulatory modules." More precisely, a "cross-regulatory module" can be defined as the set of genes taking part in a whole series of intertwined circuits (i.e., the circuits sharing genes with each other, or, in other words, the circuits interconnecting through circuits; this definition thus corresponds to that of "strongly connected component" in graph theory).

Note that the term "module" is often used to refer to a set of cis-acting regulatory sequences cooperating in the control of the expression of a given gene. To avoid potential confusions, we propose to apply the more specific term "cis-regulatory module" in such situations. Thus, whereas the term "cross-regulatory module" (or alternatively "trans-regulatory module") refers to a set of cross-regulating genes, the term "cis-regulatory module" would refer to the (eventually complex) organization of the cis-regulatory region of a single gene (see, e.g., Arnone and Davidson 1997).

The expression of several cross-regulatory modules can be coordinated by common upstream factors, as in the case of the two modules found in the gene network involved in the dorsoventral patterning of *Drosophila melanogaster* (the morphogen Dorsal controls *decapentaplegic,* as well as *twist* and *snail*). Alternatively, cross-regulatory modules may feed each other and thereby form regulatory cascades, as in the case of the gene network involved in flower morphogenesis in *Arabidopsis.* Indeed, taking part in a four-element module, *LEAFY* also controls the expression of the two genes forming a second, two-element module. The expression of *LEAFY* thus behaves as an entry variable for the small module. Similarly, in the case of the gene network controlling the anteroposterior patterning in *Drosophila,* the genes forming the "gap" module contribute to the control of the expression of the "pair-rule" genes, themselves forming a module involved in the control of the expression of the segment polarity genes.

The notion of "cross-regulatory module" should also be useful in the context of evolutionary analyses. Indeed, because of their crucial dynamical roles and their collective behavior, we expect these modules to show particular evolutionary stability. Such modules would form a sort of second-order component, which could be used and reused during evolution, in the context of what François Jacob calls "molecular tinkering" (see Jacob 1977; Duboule and Wilkins [1998] provide an interesting update on Jacob's idea). Some of these modules are also likely to play a crucial role in the control of the natural genetic engineering systems that are believed to modulate the very plasticity of the genome (see Shapiro 1999; Radman et al. 1999).

The notion of "cross-regulatory module" proposed here, as well as those of "hierarchical regulatory gene organization" (Britten and Davidson 1969), "syntagm" (García-Bellido 1982), and "gene nexus" (Zuckerkandl 1994), has its roots in the classical notion of "genetic system." This is formally defined as a set of genes that interact to constitute a functional unit. What distinguishes these different notions is how the gene interactions are organized. The last three emphasize the occurrence of hierarchical relationships among regulatory genes. From a formal point of view these interactions are essentially viewed as chains of interactions. In a recent update of García-Bellido's notion, however, Huang (1998) has extended the idea of "syntagms" to sets of interacting genes defining developmental building blocks, and hence forming the basic material for evolution. Our definition of "cross-regulatory module" stresses the organization of the gene interactions into feedback circuits or intertwined networks. This definition is thus more restrictive and implies a clear distinction between the roles of regulatory cascades and feedback circuits. In particular, every time a developmental decision is implemented into a feedback circuit or into a set of intertwined feedback circuits (i.e., implemented into a cross-regulatory module), it becomes impossible to consistently speak of hierarchical relationships among the genes involved. Indeed, any mutation affecting one of the regulatory interactions involved in a circuit will also affect the expression of all the other genes belonging to the module. In contrast, a mutation in a gene belonging to a plain cascade will affect only some of the downstream genes.

This notion of the cross-regulatory module proved to be very powerful in combination with the generalized logical formalism used in our model analyses, allowing the analysis of complex regulatory systems by disentangling their intricate organization into well-defined subsystems. The basic dynamical components (cross-regulatory modules) can then be analyzed individually, while the possibility of "reconstructing" the behavior of the whole system is preserved.

At this point, however, three characteristics of the cross-regulatory modules used during the development of multicellular organisms must be mentioned. First, one has to consider the transient activity of some modules. Indeed, responding to transitory inputs, key regulatory factors are often expressed at specific times (and locations), and the corresponding module can then operate only at that particular developmental stage. This is clearly the case, for example, with the "gap" module, which is involved in segmentation of the *Drosophila* embryo.

The second characteristic is that genes can participate in different developmental pathways. This is the case of gene *Krüppel,* for example, which has been shown to be required for Malpighian tubule development (Gaul and Weigel 1991), as well as for proper formation of the Bolwig larval photoreceptor organ (Schmucker et al. 1992), in addition to being a component of the segmentation gap module that controls the formation of segments in the *Drosophila* embryo. The pleiotropic character of various genetic mutations may reflect such involvement in different developmental pathways.

Finally, the third characteristic consists in the fact that a given module can be repeatedly used in different developmental pathways, as it is in the case of gene modules involved in intercellular signaling (Freeman 2000) and those controlling cell cycle division.

Taken together, these characteristics stress the necessity of considering time (e.g., developmental stage) and space (e.g., cellular location) constraints when disentangling complex genetic networks into cross-regulatory modules (in this respect, see also Niehrs 2004).

Acknowledgments

This work was supported by grant PB98-0466 to L. Sánchez from the D.G.I.C.Y.T., Ministerio de Ciencia y Tecnología of Spain. D. Thieffry acknowledges the financial support of the Programme inter-EPST Bioinformatique CNRS, INSERM, INRA, INRIA, Ministère de la Recherche, France.

References

Arnone, M. I., and E. H. Davidson 1997. The hardwiring of development: organization and function of genomic regulatory systems. *Development* 124:1851–1864.

Aukerman, M. J., I. Lee, D. Weigel, and R. M. Amasino. 1999. The *Arabidopsis* flowering-time gene LUMINIDEPENDENS is expressed primarily in regions of cell proliferation and encodes a nuclear protein that regulates LEAFY expression. *Plant J.* 18: 195–203.

Blazquez, M. A., and D. Weigel. 2000. Integration of floral inductive signal in *Arabidopsis. Nature* 404:889–892.

Bodnar, J. W. 1997. Programming the *Drosophila* embryo. *J. Theor. Biol.* 188:391–445.

Britten, R. J., and E. H. Davidson. 1969. Gene regulation for higher cells: a theory. *Science* 165:342–349.

Burstein, Z. 1995. A network model of developmental gene hierarchy. *J. Theor. Biol.* 174:1–11.

Busch, M. A., K. Bomblies, and D. Weigel. 1999. Activation of a floral homeotic gene in *Arabidopsis. Science* 285:585–587.

Chasan, R., and K. V. Anderson. 1993. Maternal control of dorsal-ventral polarity and pattern in the embryo. In *The development of Drosophila melanogaster*, ed. M. Bates and A. Martínez-Arias, 387–424. Cold Spring Harbor, N.Y.: Cold Spring Harbor Laboratory Press.

Coen, E. S., and E. M. Meyerowitz. 1991. The war of the whorls: genetic interactions controlling flower development. *Nature* 353:31–37.

Duboule, D., and A. S. Wilkins. 1998. The evolution of "bricolage." *Trends Genet.* 14: 54–59.

Freeman, M. 2000. Feedback control of intercellular signalling in development. *Nature* 408:313–319.

García-Bellido, A. 1982. The bithorax syntagma. In *Advances in genetics, development, and evolution of Drosophila*, ed. S. Lakovaara, 135–148. New-York: Plenum Press.

Gaul, U., and D. Weigel. 1991. Regulation of Kr expression in the anlage of the Malpighian tubules in the *Drosophila* embryo. *Mech. Dev.* 33:57–68.

Gouzé, J. L. 1998. Positive and negative circuits in dynamical systems. *J. Biol. Syst.* 6: 11–15.

Granjeaud, S., F. Bertucci, and B. R. Jordan. 1999. Expression profiling: DNA arrays in many guises. *BioEssays* 21:781–790.

Huang, F. 1998. Syntagms in development and evolution. *Int. J. Dev. Biol.* 42:487–494.

Hunding, A., S. A. Kauffman , and B. C. Goodwin. 1990. *Drosophila* segmentation: supercomputer simulation of prepattern hierarchy. *J. Theor. Biol.* 145:369–384.

Jacob, F. 1977. Evolution and thinkering. *Science* 196:1161–1166.

Lacalli, T. C. 1990. Modelling the *Drosophila*-rule pattern by reaction-diffusion: gap input and pattern control in a 4-morphogen system. *J. Theor. Biol.* 144:171–194.

Lawrence, P. A. 1992. *The making of a fly.* Oxford: Blackwell Scientific Publications.

Lewis, J., J. M. Slack, and L. Wolpert. 1977. Thresholds in development. *J. Theor. Biol.* 65:579–590.

Ma, H. 1998. To be, or not to be, a flower: control of floral meristem identity. *Trends Genet.* 14:26–32.

Mauhin, V., Y. Lutz, C. Dennefeld, and A. Alberga. 1993. Definition of the DNA-binding site repertoire for the *Drosophila* transcription factor SNAIL. *Nucleic Acids Res.* 21: 3951–3957.

Meinhardt, H. 1978. Space-dependent cell determination under the control of morphogen gradient. *J. Theor. Biol.* 74:307–321.

Meinhardt, H. 1986. Hierarchical inductions of cell states: a model for segmentation in *Drosophila. J. Cell Sci. Suppl.* 4:357–381.

Mendoza, L., D. Thieffry, and E. R. Alvarez-Buylla. 1999. Genetic control of flower morphogenesis in *Arabidopsis thaliana:* a logical analysis. *Bioinformatics* 15:593–606.

Misener, S., and S. A. Krawetz, eds. 2000. *Methods in molecular biology.* Vol. 132, *Bioinformatics methods and protocols.* Totowa, N.J.: Humana Press.

Monod, J., and F. Jacob. 1961. General conclusions: teleonomic mechanisms in cellular metabolism, growth, and differentiation. *Cold Spring Harbor Symp. Quant. Biol.* 26:389–401.

Niehrs, C. 2004. Synexpression groups: genetic modules and embryonic development. In *Modularity in development and evolution,* ed. G. Schlosser and G. P. Wagner. Chicago: University of Chicago Press.

Nüsslein-Volhardt, C., and S. Roth. 1989. Axis determination in insect embryos. In *Cellular basis of morphogenesis,* ed. D. Evered and J. Marsh, 37–55. Ciba Foundation Symposium 144. New-York: John Wiley and Sons.

Nüsslein-Volhardt, C., and E. Wieschaus. 1980. Mutations affecting segment number and polarity in *Drosophila. Nature* 287:795–801.

Parcy, F., O. Nilsson, M. A. Busch, I. Lee, and D. Weigel. 1998. A genetic framework for floral patterning. *Nature* 395:561–566.

Plahte, E., T. Mestl, and S. W. Omholt. 1995. Feedback loops, stability and multistationarity in dynamical systems. *J. Biol. Syst.* 3:409–413.

Podos, S. D., and E. L. Ferguson. 1999. Morphogen gradients: new insights from DPP. *Trends Genet.* 15:396–402.

Radman, M., I. Matic, and F. Taddei. 1999. Evolution of evolvability. *Ann. N.Y. Acad. Sci.* 870:146–155.

Reinitz, J., D. Kosman, C. E. Vanario-Alonso, and D. H. Sharp. 1998. Stripe forming architecture of the gap gene system. *Dev. Genet.* 23:11–27.

Rivera-Pomar, R., and H. Jäckle. 1996. From gradients to stripes in *Drosophila* embryogenesis: filling in the gaps. *Trends Genet.* 12:478–483.

Sánchez, L., J. van Helden, and D. Thieffry. 1997. Establishment of the dorso-ventral pattern during embryonic development of *Drosophila melanogaster:* a logical analysis. *J. Theor. Biol.* 189:377–389.

Sánchez, L., and D. Thieffry. 2001. A logical analysis of the *Drosophila* gap gene system. *J. Theor. Biol.* 211:115–141.

Schmucker, D., H. Taubert, and H. Jäckle. 1992. Formation of the *Drosophila* larval photoreceptor organ and its neuronal differentiation requires continuous Krüppel gene activity. *Neuron* 9:1–20.

Shapiro, J. A. 1999. Genome system architecture and natural genetic engineering in evolution. *Ann. N.Y. Acad. Sci.* 870:23–35.

Snoussi, E. H. 1998. Necessary conditions for multistationarity and stable periodicity. *J. Biol. Syst.* 6:3–9.

Thieffry, D., A. M. Huerta, E. Pérez-Rueda, and J. Collado-Vides. 1998. From specific gene regulation to global regulatory networks: a characterisation of *Escherichia coli* transcriptional network. *BioEssays* 20:433–440.

Thieffry, D., and D. Romero. 1999. The modularity of biological regulatory networks. *Biosystems* 50:49–59.

Thomas, R. 1978. Logical analysis of systems comprising feedback loops. *J. Theor. Biol.* 73:631–656.

Thomas, R. 1991. Regulatory networks seen as asynchronous automata: a logical description. *J. Theor. Biol.* 153:1–23.
Thomas, R., and R. D'Ari. 1990. *Biological feedback.* Boca Raton, Fla.: CRC Press.
Thomas, R., D. Thieffry, and M. Kaufman. 1995. Dynamical behaviour of biological regulatory networks. 1. Biological role of feedback loops and practical use of the concept of the loop-characteristic state. *Bull. Math. Biol.* 57:247–276.
von Dassow, G., E. Meir, E. M. Munro, and G. M. Odell. 2000. The segment polarity network is a robust developmental module. *Nature* 406:188–192.
Wagner, D., R. W. Sablowski, and E. M. Meyerowitz. 1999. Transcriptional activation of APETALA1 by LEAFY. *Science* 285:582–584.
Wolpert, L. 1969. Positional information and the spatial pattern of cellular differentiation. *J. Theor. Biol.* 25:1–47.
Zuckerkandl, E. 1994. Molecular pathways to parallel evolution. 1. Gene nexuses and their morphological correlates. *J. Mol. Evol.* 39:661–678.

12 Exploring Modularity with Dynamical Models of Gene Networks

GEORGE VON DASSOW AND ELI MEIR

Introduction

If one is seeking a biochemical understanding of development then the language of dynamical systems theory seems a natural one to use.

If we are serious about attempting to understand the hierarchy of developmental decisions in molecular terms then we do not just need to identify the relevant genes and gene products but also to understand their dynamical behaviour. In the past this has proved to be necessary for understanding such things as the mechanism of nerve conduction or aggregation in slime moulds. In the future it seems probable that it will be through the mathematics of dynamical systems theory that embryological and molecular results can meaningfully be brought together.
—J. M. W. Slack, *From Egg to Embryo*

No one doubts the contribution of mathematical models to evolutionary theory, or the necessity of simulations and statistical modeling to ecology, or the role of kinetic models in enzymology, and yet the application of models to developmental biology seems always under question. The anticipations recorded by Jonathon Slack remain unfulfilled. For one thing, molecular biology has only recently begun to provide the kinds of facts from which empirically grounded models could be formulated. On the other hand, the reality of epigenetics is far more complex than envisioned by most earlier workers, although some, like Slack (1983) certainly appreciated the scope of the problem. Thus, the kinds of readily understood dynamical systems models reviewed by Slack for the most part fail to capture the complexity that lab-bench biologists confront. Consider a recent expression of the situation:

> 1980, the year that Christiane Nüsslein-Volhard and Eric Wieschaus embarked on their Nobel Prize–winning screen for embryonic lethal mu-

tants in *Drosophila,* in some ways marked the end of the Age of Beautiful Theories in biology, and the dawn of the Age of Ugly Facts. . . . If Watson and Crick's double-helical model of the structure of DNA showed that imagination (with a sprinkling of data) could triumph over Nature, Nüsslein-Volhard and Wieschaus's saturation mutagenesis showed that evolution can produce biological mechanisms of such unimaginable complexity that it would be useless, if not laughable, to try to intuit them a priori. Nature's imagination, it showed, usually far outstrips that of the human brain . . . the baroque and counterintuitive biological mechanisms that evolution has produced so often mock the human imagination. (Anderson and Walter 1999, 557–558.)

Aside from implying that complex mechanisms are ugly, this passage highlights the fact that molecular biology has finally inverted the habit of biological inquiry. Instead of using phenomenology and perturbation experiments to deduce some mechanism, and then uncovering facts one by one to support that hypothesis, modern biologists increasingly turn to large-scale exploration (e.g., DNA microarrays, genome sequencing) to generate a mass of facts whose relevance is eventually established by phenomenology and from which mechanistic understanding might hopefully emerge.

How to accomplish that last step of making the mechanistic understanding emerge from the sum of the parts? When things get too complicated for human intuition and language, scientists turn to math and models. Our work on the segment polarity and neurogenic networks, reviewed below, is a preliminary exploration of how biologists might use dynamical models to come to grips with their ever-growing maps of epigenetic interactions. Elsewhere we have described our approach and the results of our first case studies (von Dassow et al. 2000; Meir et al. 2002b; von Dassow and Odell 2002; Meir et al. 2002a). To us, modularity is a working assumption: we are trying to build up some network that exhibits some lifelike behavior from parts that do not, by themselves, fully explain that behavior. This is the opposite of starting with a large-scale map, seeking to break it down into more-readily-understood bits. The two approaches will surely lead to different, but complementary, results. Here we address in general terms what we think are the prospects for our approach. We discuss several intertwined issues:

Plausibility: The most basic limit, presently, to making sense of the parts catalog of molecular biology is our own inability to tell in words whether or not a particular conspiracy of molecules *actually does* what we think it might do. When confronted with systems too complex to argue out in words, we need more rigorous methods than human language to sort out plausibility. Computer models can tell us *whether it is plausible* that some phenomenon can be explained by some set of relevant facts.

Hole filling and inference: A converse of the plausibility issue is that the known facts are usually inadequate. A particular model's deficiencies often reflect gaps between the facts, as long as the model's assumptions cannot be trivially questioned. However, efforts to use models for inference will forever suffer from the inability of human imagination, as lamented by Anderson and Walter, to match the creativity of the evolutionary process.

Evolvability and variational tendencies: Assuming one constructs a realistic model that exhibits some lifelike behavior, the dependence of the model's behavior on its parameters constitutes a set of hypotheses about the evolutionary potential of the modeled mechanism. This will become an important use of dynamical models, since it is often difficult to deduce experimentally the variational tendencies of developmental processes.

Functional design: Models allow us to explore whether the particular topology of an epigenetic process is merely contingent, that is, nature assembling mechanisms out of the junk heap of the genetic heritage, or whether in a particular case nature has hit upon a genuinely good way to solve a design problem. We can ask, How does a particular network achieve some systems-level property of functional value, such as robustness against perturbation, or modularity, and are there common mechanistic themes to such properties? Recalling once again Anderson and Walter's lament, are these mechanisms really so baroque?

Using mathematical or computer models to explore ideas about genetic and developmental mechanisms is hardly novel. Pioneers of several major threads include Glass and Kauffman (e.g., Glass and Kauffman 1972, 1973; Kauffman 1993), Turing (1952), and Meinhardt and colleagues (e.g., Gierer and Meinhardt 1972; Meinhardt 1977, 1984), and Waddington and Kacser (1957). These workers developed very different conceptual approaches to the problem of how to capture gene network dynamics in maths. However, until relatively recently most of these efforts have been abstract and phenomenological, rather than grounded in empirical facts, because the puzzle pieces have been mostly missing. Kauffman and his followers and (independently) Thomas and colleagues avoid the issue of missing pieces, while still confronting complex systems, by using randomly connected or reality-inspired networks of Boolean or thresholded interactors to explore the generic properties of complex networks (extensively reviewed by Kauffman 1993; examples in Thieffry et al. 1998; Thieffry and Romero 1999; Thomas et al. 1995). This approach is often intentionally divorced from the constraints of reality in order to get at what Kauffman calls the "statistical mechanics" of complex networks. Exploring Boolean models led Kauffman to a variety of conclusions, especially about the dependence of the existence of steady states on the density of connectivity in the model. Thus, Kauffman's and Thomas's schools have shown that analysis of these models in the ensemble provides insights into the general features to be expected of complex genetic circuits.

Meanwhile Meinhardt and his followers (among others) developed a variety of candidate models for hypothetical developmental mechanisms based on reaction-diffusion processes (Gierer and Meinhardt 1972; Meinhardt 1977, 1984). This approach was based on Turing's insight that coupled systems of diffusible reactants could, under certain conditions, elicit regular spatial patterns, and that developing embryos could employ such processes to differentiate initially homogenous cells in a tissue (Turing 1952). Indeed, Meinhardt (1984) anticipated many of the essential features of the segment polarity network before it was molecularly deduced. Despite widespread (and rather undeserved) contempt among modern molecular biologists for this approach, the Turing-style models deserve credit for showing that simple chemical processes could produce complex spatial patterns. The derision of working biologists comes from the fact that these models have typically been products of the modeler's skill, not derived from facts about the molecular processes causally involved in the phenomenon which the model proposes to explain.[1] In addition, simple reaction-diffusion models exhibit a variety of biologically unrealistic tendencies; Slack (1983) provides an excellent overview of the results and criticisms of reaction-diffusion models.

A third major thread, the use of dynamical systems models, is the direct lineage of our efforts. In his influential book, Slack (1983) justified the use of dynamical systems theory as the natural language for modeling developmental pattern formation and other epigenetic processes. He argued that one could readily capture measurable, and general, properties of biochemical reactions, and furthermore that the phenomenology of development parallels that of dynamical systems. Notably, Slack discussed the intimate connection between the stability of cell states and the attractors of dynamical systems, he pointed out the parallel between progressive determination and the time evolution of a dynamical system toward a steady state, and he highlighted the initial-condition dependence of cell fate specification and the choice of attractors by dynamical systems near the boundaries between basins of attraction. Slack acknowledged the inherent difficulty of working with nonlinear dynamics but recognized that this is a necessary cost of improved realism.

A handful of recent attempts use continuous nonlinear models to

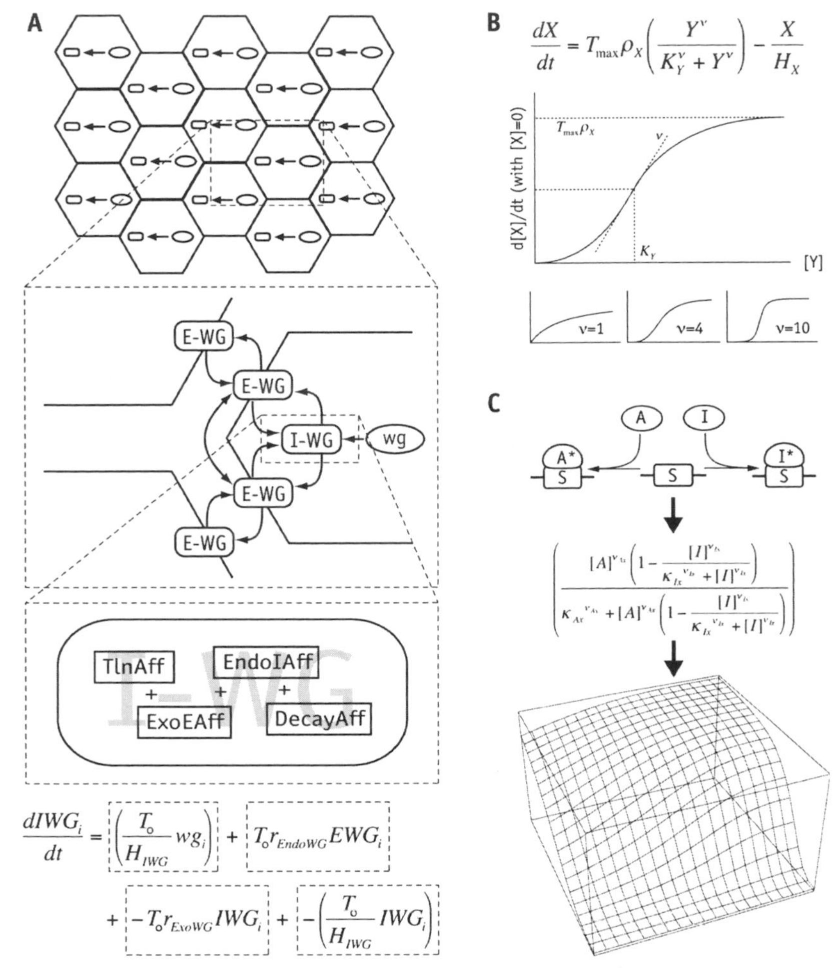

Fig. 12.1.—The Ingeneue modeling framework. *A,* Ingeneue does everything (in its current version) in a grid of hexagonal cells. The network topology is stamped out into every cell in a user-specified grid. The topology consists of nodes that interact according to formulas chosen by the user. Nodes may be intracellular or membrane bound; in the latter case Ingeneue tracks the concentration on each cell face. The equations governing interactions among nodes are built from "affectors" that encapsulate formulas for various dynamical processes; in the example illustrated, the node representing the intracellular form of the Wg protein, I-WG, is translated in proportion to the abundance of wg mRNA (governed by the TlnAff object), it experiences endo- and exocytosis (the affectors ExoEAff and EndoIAff govern equilibration with the extracellular form of Wg, E-WG), and it undergoes first-order decay (the DecayAff object). The actual formulas, in dimensional form, are shown at the bottom; dashed boxes correspond to the individual affector objects. At each time point, Ingeneue invites each node to compute its own time derivative simply by adding together its stable of affectors. This architecture makes it trivial to modify the network topology throughout the entire field; if we wanted to add a new interaction for I-WG, we simply add the appropriate tags to the input script, and Ingeneue handles sorting out all the neighbor relations within the cell grid. *B,* Most regulatory relationships in Ingeneue are represented by sigmoid dose-response curves. Shown here is a simple equation (in dimensional form) in which the first term endows transcriptional activation of X by Y, and the second confers first-order decay. The virtue of this approach is that it enforces several biologically realistic parameters: a saturation level (i.e., maximum transcriptional activity), a half-maximal level of regulator, and a shape parameter, which is equivalent to the Hill coefficient. Modulating the Hill co-

explore real, well-understood epigenetic processes. Edgar and Odell (Edgar et al. 1989) developed one of the earliest realistic, nonlinear models of developmental pattern formation, showing that a subset of the *Drosophila* pair-rule genes can account, through mutual cross-repression, for how pair-rule gene products sharpen each other's expression boundaries. A variety of recent efforts similarly attempt to capture the behavior of entire (if as yet small) genetic circuits, deduced empirically, using continuous nonlinear models, the most emblematic of which is Barkai and Leibler's (1997) model of the core control circuit for bacterial chemotaxis. These authors not only used their model to predict that this mechanism would tolerate variation in the levels of gene expression, but also showed that the real biological circuit has this property as well (Alon et al. 1999).

Most such efforts borrow heavily from the well-developed body of formulations describing enzyme and binding kinetics, which has been under development for over a century and is deeply integrated with lab practice (Gutfreund 1995; Wyman and Gill 1990). We follow the same prescription because many interactions between gene products literally *are* binding reactions, enzyme-catalyzed transformation, or other straightforward chemical processes, so formulas for first- and second-order chemical reactions and so forth can get us pretty far as long as we assume that cells are well-stirred reaction vessels, and as long as we assume that molecular species are abundant enough in cells to use the continuum approximation. As described elsewhere (von Dassow et al. 2000; Meir et al. 2002a) we use a stereotyped formulation for dose-response relationships between regulators and targets, largely inspired by classical treatments of enzymatic processes and allosteric binding phenomena. Because many of the networks we are interested in mediate pattern formation in fields of cells, and because these networks are expressed by systems of differential equations too complex to be wielded comfortably by mere humans, we developed a gene network simulator program (Ingeneue) that weaves the equations together from a library of formulaic building blocks, guided by a text description of the network, and instantiates indexed copies of the network in each cell in a user-specified grid (see fig. 12.1). This program makes it easy to "rewire" network models, testing consequences within a common

---

efficient ν changes the steepness of the dose-response curve. It is this parameter that we call "cooperativity," by analogy to classic allosteric systems; with $\nu = 1$ we say the response is noncooperative, and increasing cooperativity leads to a more and more steplike function. *C*, Individual nodes often must integrate multiple inputs. For example, an activator A and inhibitor I might compete for binding to an enhancer sequence S; this relationship can be captured by nesting dose-response curves to come up with an appropriate behavior, as judged by the graph. For this formula, as appropriate, the inhibitor can squelch the response to low concentrations of activator, but increasing activator concentration overwhelms any particular level of inhibitor.

framework, and makes it feasible to compare the results from models of entirely different circuits.

Lessons from the Segment Polarity Network

Our first task was to synthesize the known facts about the mechanism of segmentation and ask, simply, Do we know enough about this system to make a model that accounts for some aspect of the behavior and function of the real biological entity, and if not, what do we need to know better? The segment polarity network is the last tier in a cascade of ever-finer-scale patterning processes that start with maternally transcribed mRNAs localized to each end of the egg and end with a nearly cell-row-by-cell-row specification of positional information along the anterior-posterior axis of the embryo at about the time of cellularization (summarized in fig. 12.2, *A*, and reviewed by Martinez-Arias 1993; Pankratz and Jäckle 1993). The segment polarity network stabilizes and maintains the boundary between parasegments (the metameric units that patterning genes map out). The tier immediately above them in the segmentation cascade, the pair-rule genes, are expressed just long enough to map out the segments and activate patterned expression of segment polarity genes like *engrailed* (*en*), *wingless* (*wg*), *sloppy-paired* (*slp*). Thereafter the segment polarity genes have to "hold on to" the pattern imprinted by the pair-rule genes. This is accomplished, according to the canonical view, because the segment polarity genes mediate a codependence between cell states on either side of the parasegment boundary. Persistence of the *wg*-expressing cell state in the cells anterior to the compartment boundary depends on signaling by the product of the *hedgehog* (*hh*) gene under the control of En. In turn, the persistence of the *en*-expressing cell state *posterior* to the boundary depends on Wg signaling (fig. 12.2, *B*).[2]

Most of the core segment polarity genes are components of the Hh and Wg signal transduction pathways. Hh acts through the products

Fig. 12.2.—The segmentation cascade and the core segment polarity network. *A*, Cartoon-and-arrows summary of the segmentation cascade. This figure is meant to convey the flavor of the process, not every feature. Maternally expressed gene products such as *bicoid* (*bcd*) and *nanos* (*nos*) are localized within the oocyte; during early development localized synthesis leads to long-range gradient formation; gap genes, including *hunchback* (*hb*), *giant* (*gt*), *knirps* (*kni*) and *Krüppel* (*Kr*) respond to local Bicoid concentration and/or to each other, forming broad bands of expression; they in turn shape the emerging expression patterns of pair-rule genes, including *hairy* (*h*), *even-skipped* (*eve*), *runt*, *paired* (*prd*), and *fushi tarazu* (*ftz*) into finer-scale stripes; and the pair-rule genes shape the initial expression of the segment polarity genes, especially *wingless* (*wg*) and *engrailed* (*en*) and *sloppy-paired* (*slp*) *B*, A common textbook summary of the segment polarity cascade. *wg*-expressing cells on the anterior side of the boundary depend on *en*/*hh*-expressing cells on the posterior side, and vice versa. *C*, The wiring diagram for the simplest version of our segment polarity model. The diagram here was rationalized and analyzed in von Dassow et al. (2000); dashed links were added after an initial model without them failed to exhibit lifelike behavior. CID, Cubitus interruptus; CN, repressor fragment of Cubitus interruptus; PH, Patched-Hedgehog complex; PTC, Patched.

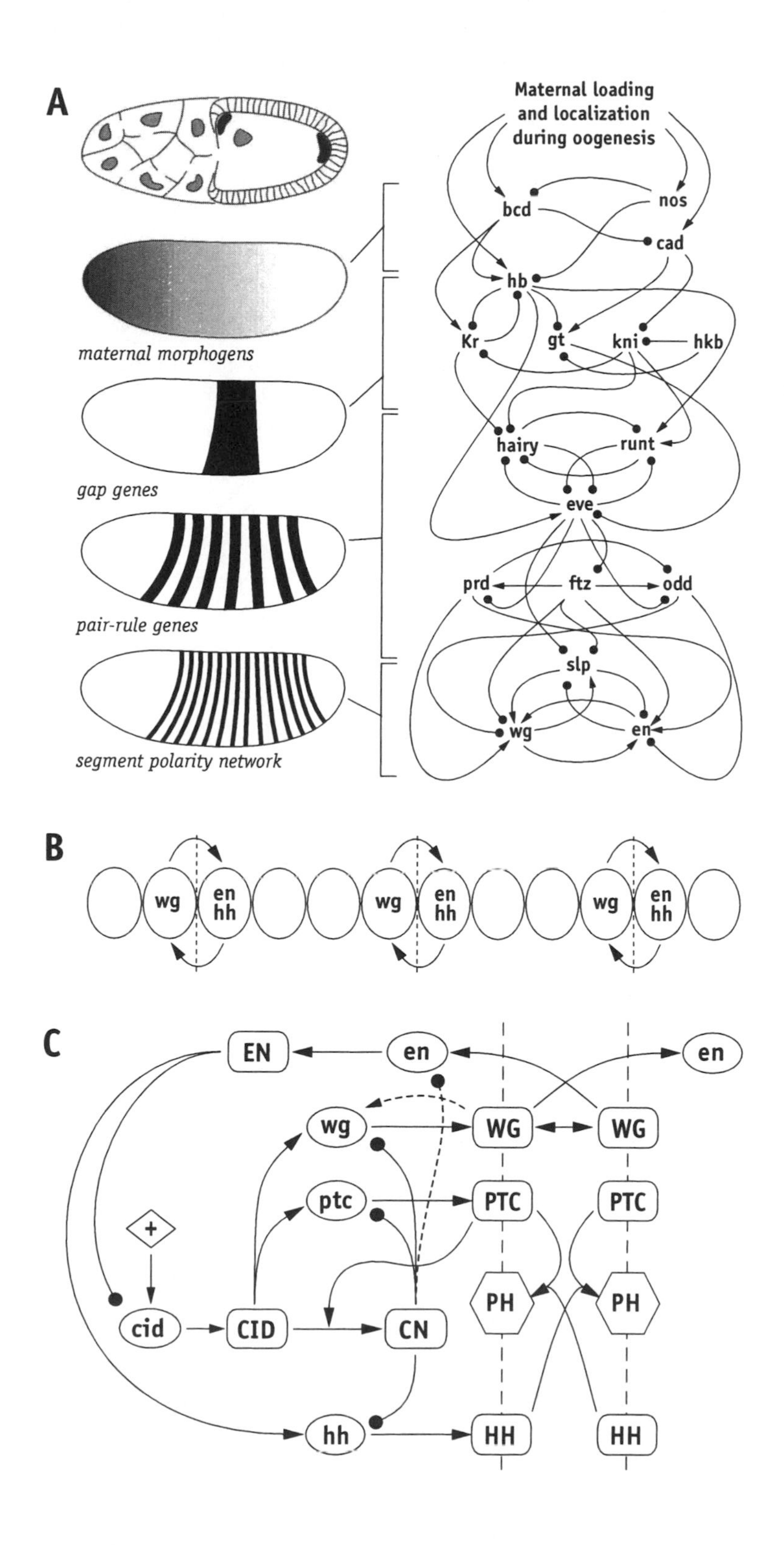
A
Maternal loading
and localization
during oogenesis
maternal morphogens
gap genes
pair-rule genes
segment polarity network
bcd
nos
cad
hb
Kr
gt
kni
hkb
hairy
runt
eve
prd
ftz
odd
slp
wg
en
B
wg
en
hh
C
EN
en
wg
WG
ptc
PTC
PH
cid
CID
CN
hh
HH
+

of the genes *patched* (*ptc*) (Hooper and Scott 1989; Marigo et al. 1996; Stone et al. 1996) and *smoothened* (*smo*) (Alcedo et al. 1996; van den Heuvel and Ingham 1996) and the transcriptional switch encoded by *cubitus interruptus* (*ci*) (Alexandre et al. 1996; Dominguez et al. 1996; Hepker et al. 1997; Von Ohlen et al. 1997). Ptc is the Hh-binding component of a complex that includes Smo, and in the absence of Hh, Ptc prevents Smo from sending an as-yet-poorly-understood signal (Chen and Struhl 1996; Alcedo and Noll 1997; Chen and Struhl 1998). The result of this signal (or signals), whatever it is, is to liberate Ci protein, the full-length form of which is a transcriptional activator, from a complex that includes the products of the genes *fused, Suppressor of fused* (*Su*(*fu*)), and *costal;* this complex both keeps Ci in the cytoplasm and directs it to be proteolyzed to yield a truncated protein (CN) that behaves as a repressor (Aza-Blanc et al. 1997; Ohlmeyer and Kalderon 1998; Wang and Holmgren 1999). Thus, in the absence of Hh, Ci is converted to a repressor that keeps Hh target genes off (including *ptc* and *wg*), and in the presence of Hh, due to removal of Ptc, Ci remains intact, mysteriously passes through various activation steps and enters the nucleus, and activates Hh target genes.

Wg signal transduction begins with products of *frizzled*-family genes (Bhanot et al. 1996; Bhat 1998; Bhanot et al. 1999). The details of Frizzled signaling remain mysterious, but, analogous to Hh signaling, the crucial switch involves a cytoplasmic complex that restrains a transcriptional regulator. In this case it is the product of the *armadillo* gene that is targeted for proteolysis; Wg signaling leads to the release of Arm from a cytoplasmic complex that includes the kinase encoded by *shaggy* and a *Drosophila* homologue of the oncogene APC (a very complex literature is reviewed by Cadigan and Nusse 1997). Free Arm binds to the product of *pangolin,* and Arm and Pan together act as a transcriptional activator (Brunner et al. 1997; van de Wetering et al. 1997).

There are many other segment polarity genes that participate in the process in flies, and that, when mutated, yield various phenotypes. We believed that the basic codependence outlined above would be sufficient to account for the basic function of the segment polarity network, namely, to maintain an asymmetric boundary with *wg* expressed on one side and *en* expressed on the other. Our goal was to start with the simplest dynamical representation of the network and build up piece by piece. Thus, to start with we abbreviated signal transduction pathways and refrained from including apparently redundant or "extra" components. We thought that perhaps the core network, the simplest network we could get to do the job required, would probably be relatively fragile, and that all the other segment polarity gene products—for instance, the transcription factor encoded by *gooseberry* (Li and Noll 1993), or the Wg-inducible signaling inhibitor encoded by *naked*

(Zeng et al. 2000)—might be required to make the network robust to various kinds of perturbations. Thus, we were not dismayed in the least when it turned out that our initial attempt to concoct a model was very hard to get to behave properly.

In fact it slowly dawned on us that it was completely impossible for our first attempt to work under any conditions. We had to add two specific links, one of which was more or less well demonstrated but ignored, and the other of which was, at the time, more or less a guess (fig. 12.2, C). Completely to our surprise, however, it turned out that with those two links in place, the core network was fabulously robust to variation in both governing parameter and initial conditions. As described in von Dassow et al. (2000), random sampling for parameter values throughout an enormous, high-dimensional parameter space allowed us to find "working" sets of values with unbelievably high frequency. Further explorations (von Dassow and Odell 2002) showed that the core network's boundary-maintaining function is also robust to architectural variations. In other words, once the right links are in place, there is no one single way to make the network function; once all the pieces are hooked up right, the lifelike behavior we sought to reproduce *in silico* became intrinsic to the topology of the network, rather than to any particular tuning of the connections and components within it.

This is a very satisfying finding given a certain evolutionary hypothesis that had originally been in our minds when we started the work. It appears that the upstream aspects of the segmentation cascade are not conserved among insects (see, for examples, Patel et al. 1992; Dawes et al. 1994; Dearden and Akam 1999), the furthest-upstream components not even beyond Diptera, but the segment polarity network *might* be involved in segmentation in everything from flies to beetles to grasshoppers and beyond (Patel et al. 1989; Nagy and Carroll 1994; Patel 1994). Although there is not as much evidence supporting this hypothesis as one might like, it remains appealing; the suggestion is that the upstream mechanism that lays out segment boundaries in other insects must be very different, despite the homology of all insect segments and the conservation of the gene network assigned to stabilizing those boundaries throughout development. Our results say that this hypothesis is plausible: intrinsic to the topology of this network is the ability to do the thing it does in embryogenesis, absent any extrinsic guidance, and if we could make an animal with only these genes, then practically any bias on their expression among the cells of that animal would result in at least one segmental boundary!

To summarize, our initial modeling effort resulted in at least six specific, empirically testable predictions:

1. Our model explicitly highlights a need for a repressor of *engrailed*

in the anterior compartment and suggests that the N-terminal fragment of Ci could fill this role. Another candidate, explored in later models (von Dassow and Odell 2002), is Sloppy-paired (Cadigan et al. 1994b; Grossniklaus et al. 1992); the model merely focuses attention on the missing link; obviously, it is an empirical problem to figure out what that link might be.

2. *wingless* autoactivation is functionally important and probably conserved, and must follow certain guidelines (described in von Dassow and Odell 2002) to fulfill its role. This phenomenon has received very little attention in the literature, and our model explains its importance.

3. Interactions among segment polarity genes should exhibit moderate to high cooperativity, except for interactions mediating negative feedback between *ci* and *ptc* (Meir et al. 2002a; von Dassow and Odell 2002).

4. Our model "prefers" the Wg diffusion rate to be low, suggesting that rapid diffusion makes pattern formation by this mechanism more difficult. Indeed, several findings show that Wg cell-cell transport is under fairly specific control (Dierick and Bejsovec 1998; Moline et al. 1999).

5. Since the model tolerates a variety of initial prepatterns, we would predict that the specific inputs to the segment polarity network from the pair-rule and gap genes will *not* be rigorously conserved even within long-germ insects.

6. We predict that the segment boundary maintenance mechanism is robust to quantitative variation in gene function. We are currently trying to test whether the real segment polarity network exhibits the same degree of robustness as our model.

Points 1 and 2 directly illustrate plausibility and inference applications for dynamical models. Although our initial model expressed a more detailed summary of the network topology than was at the time typical even of workers studying the segment polarity network empirically, we found that there were two specific defects that could not be overcome even by choosing kinetic parameters carefully. One of these defects was cured with the documented, but little-attended, phenomenon of *wg* autoregulation (Hooper 1994; Vincent and Lawrence 1994; Manoukian et al. 1995; Yoffe et al. 1995). The mechanism for this remains poorly understood to date, but our model showed that the *wg* autoregulation mechanism, whatever it is, may be central to the function of the segment polarity network even though it had heretofore figured barely at all in discussions of how segmentation works. The second defect also concerned a lack of attention by the community of experimental biologists to a detail of the mechanism, in this case to the regulation of *en*. Almost all the attention has gone to either the specification of the initial *en* expression pattern by pair-rule genes (DiNardo et al. 1988; Ingham et al. 1988), the subsequent dependence of *en* on Wg or En itself (the classic account is Heemskerk et al. 1991), or to the stabilization of the *en* activation state late in embryogenesis under the control of the Polycomb—and Trithorax—group genes (e.g., Moazed and O'Farrell 1992). The model forced us to notice what should have been obvious in the first place: something has to shut *en* off in the anterior compartment.[3]

In experimentally tractable model organisms like *Drosophila,* biologists are quite efficient enough to fill in these sorts of details sooner or later with or without the help of models like ours. Maybe work like ours can accelerate the process. Our approach is of far greater potential value if applied to less willing organisms where making transgenes and knockouts and the like is either a major technical challenge or otherwise out of the question. We have little intention of focusing our own efforts in such areas ourselves, but the point we want to underscore is the time has come that realistic enough computer models can make plausible suggestions about how to fill in the holes; computer models, unlike us, cannot be fooled by an arrow diagram backed by a rhetorically compelling word salad.

Points 3 and 4 represent inferences from the models about how the real mechanism might work, but also bespeak functional design. The segment polarity network model can be thought of as a set of spatially entrained switches in which the various stable states for each switch are mutually exclusive within an individual cell, but the network is structured such as to make these switches entrain each other to alternate states in neighboring cells. The switches are based on nonlinear responses, which could be due to cooperative binding effects; higher cooperativity increases the likelihood of choosing parameter values for which both switched states will be stable. In addition the negative feedback loop between *ci* and *ptc* keeps cells in the "ground" state responsive to Hh signaling; low cooperativity within this loop makes it more likely that it behaves as a homeostat, rather than generating oscillations. The mutual entrainment of neighboring cell states depends on signals' making it to neighboring cells but not much further; hence it is harder (though not fatally so) to tune up parameters to achieve the desired pattern the more rapidly the intercellular signals are allowed to diffuse. This contrasts with Gierer-Meinhardt-style reaction-diffusion models, in which the diffusion rates of intercellular signals determine the periodicity of the patterns they can make (Slack 1983 provides a critical review of this family of models).

Point 5 has some implications both for our understanding of functional design, and also for the evolvability of the segmentation mechanism. In describing the segment polarity network as a series of switches, it is the initial conditions that determine which switch gets thrown in which cells. Again unlike the Turing-style models that have so often been suggested to explain pattern formation, the segment polarity network does not make patterns out of small perturbations in an undifferentiated field. Rather, the segment polarity network stably maintains (and can sharpen) a prepattern conferred upon it by anything which biases the initial conditions toward one or another switched state on a cell-by-cell basis. Whether or not the network can stably "make" a particular pattern thus depends not just on the kinetic parameters but also on the initial conditions. In the case of the target pattern we tried to get the simple, core model to make (von Dassow et al. 2000), the outcome is based on a race between *en* and *ci,* on the one hand, and on the other hand on making sure that *wg* gets a quick enough assist (from Ci) to keep itself on.[4] Thus, the model's demands on the initial conditions can be crudely stated like this: for any pattern of initial biases that swings these races in the right direction, there can be found some set of parameters that allows the model to lock on and hold that pattern. In other words, the blind watchmaker can fool around with the upstream regulators as long as certain guidelines are not violated, and as long as the kinetics can be tuned up at the same time.[5]

Point 6 surprised us most, and our lab is testing this prediction empirically. We tend to think of robustness as a design feature, and as something difficult to achieve. Certainly human-engineered devices do not exhibit the degree of insensitivity to control parameters that we found in the case of the segment polarity network; does nature need to evolve robust designs, or is this kind of property generic to genetic networks? Moreover it is not yet obvious to us *why* evolution should have made this mechanism so astonishingly robust, if indeed it is in reality. Even less transparent is *how* this module came to be (although we have an idea, discussed in Meir et al. 2002a, and touched on below); is its present state and employment in flies a highly derived, finely honed design, or a lasting legacy of a lucky co-option early in the evolutionary history of animal life? Only comparative data could answer this, and despite the misleading impression given by some authors (e.g., von Dassow et al. 2000), we know very little about whether the segment polarity module is evolutionarily static or whether details of its architecture adapt to different developmental modes, even within fruit flies.

Whence Robustness?

As recounted at the end of our first report (von Dassow et al. 2000), we originally hoped to explore the mechanistic origins of robustness in developmental mechanisms through *in silico* reconstitution. To reiterate, we expected that the simplest (but still realistic) models would require us to carefully select parameters (by intuition or optimization strategies) to make it work, and that only carefully chosen initial conditions would lead to the desired behavior. Our hope was that by making progressively more complex models based on known interactions not incorporated into the simplest model we would reveal which design principles evolution had hit upon to make the process in question robust. There are a variety of flavors of robustness, such as tolerance of parameter variation, stochastic perturbations, or initial conditions, and it seems reasonable to expect that embryos, and cells everywhere, need special circuitry to tolerate all these insults. Much of the complexity we see in biological mechanisms might exist for the purpose of endowing some core process with robustness.

This remains an intuitively appealing general hypothesis, but with the segment polarity network it turned out that the core model is hard to improve upon with respect to the basic tests we can subject it to. Not only does it tolerate the kinds of variation enumerated above, but it also tolerates numerous different wiring choices, including whether or not certain secreted proteins diffuse, whether or not reactions are reversible, whether or not particular links and components are present, and so on.

In a forthcoming report (von Dassow and Odell 2002) we describe a test for stripe sharpening, in which the wiring really makes a difference in the performance of the network. The *wingless* and *engrailed* stripes are both reported to narrow as cells rearrange during germ-band extension; cells that move away from the parasegment boundary lose *wg* and *en* expression as they stray beyond the range of sustaining signals (see, e.g., Vincent and O'Farrell 1992; review in Martinez-Arias 1993). There are specific requirements for the model to mimic these behaviors: for *wg*, autoactivation must synergize with activation by Ci; for *en*, there must be stoichiometric balancing of certain components of the Wg signaling pathway. However, while those tests seem legitimate (especially in *Drosophila*), they are almost certainly not general to all uses of this network. The test in which we ask whether the network can restrict *en* stripes to one cell width in the face of cell rearrangement is probably irrelevant to segmentation in short-germ insects: in both grasshoppers and crayfish the En stripes widen as the segments develop (Patel 1994). Similarly, although we tested the ability of network variants to develop the target pattern from a very crude prepattern, in all cases we are aware of, En first appears in crisp stripes. Thus, we focus the discussion below on the robustness of the boundary-maintenance function of the segment polarity network. This may not be the only biologically relevant behavior, but it is the one we have the best handle on.

We consistently found three determinants of robustness. First, the higher the "cooperativity"[6] of most connections, the more variation the network tolerates (Meir et al. 2002a). Second, the right mix of positive feedback with both positive and negative cross-talk is essential to confer broad domains of parameter space and initial conditions in which the model functions. Third, intermediate steps tend to damp out temporal oscillations. While the second issue is a design concern specific to this network, the others are generic. The role of intermediate steps seems to be a byproduct of cooperative interactions. Because it seems to be such an important generic way to make gene networks robust, we discuss how cooperativity confers robustness.

Consider, for illustration, a trivial signal transduction cascade in which a signal activates a responder (say, a transcriptional activator), which then activates transcription of a secondary target gene. If all the responses are linear-saturating curves, then the output tracks the input: end of story. The more positively cooperative the response (we stretch the notion of cooperativity a little), the more a small change in the input *around the threshold* will result in a large change in output. Since the responses must saturate in the physical world, the higher the cooperativity at each step, the closer the whole chain will be to an all-or-none switch. This may not immediately strike one as the basis for robust behavior, since to call something robust roughly means it behaves the same for a variety of conditions. However, all-or-none responses mean that the behavior of a complex system becomes *less* sensitive to the exact value of off-to-on thresholds. In this pedagogical example, if the signal is moderately above the threshold, then we get a full response from the responder, which means that, as long as "full on" for this gene is also above its activity threshold, we get a full response from the target. This also explains (partly) why intermediate steps increase the robustness of the model; oscillations in the level of some regulator are damped by sharply thresholded dose-response curves as long as the regulator concentration never gets too close to the threshold level. As an aside, a corollary is that introducing delays would not be likely to improve robustness, and indeed might promote oscillatory behavior, in a network composed of more or less linear interactions.[7]

This is fine as long as the behavior we are interested in is one that can be described solely in terms of whether genes and enzymes are "on" or "off." So far, that is all we have demanded of the segment polarity network. But surely there are downstream effects of these genes that are differentially sensitive to quantitative levels of segment polarity gene expression. In imaginal discs Strigini demonstrated the expression patterns of various Hh and Wg targets are sensitive to local differences in the availability of these signals (Strigini and Cohen 1997, 2000). We have not yet explored how these phenomena could work, but these targets may simply be tuned to respond at threshold signal concentrations near the maximum level that that signal (or its effectors) could achieve.

If we modify the signal transduction example slightly, it reveals that cooperativity does not beget *all* kinds of robustness. What if the responder has two targets, the output and an inhibitor? Imagine that this inhibitor both negatively feeds back on the responder and also feed-forward inhibits the output (fig. 12.3, *A*). It turns out that this device has a variety of behaviors if high cooperativity is allowed in all the intermediate connections. The output can have a threshold response to a signal, as before (fig. 12.3, *B*). However, it can also exhibit an upper threshold above which it is inactivated, much like the response of gap genes to maternal morphogens in *Drosophila* (fig. 12.3, C). There are a variety of oscillatory regimes, including ones in which the period is tuned by the signal. However, if connections are all constrained to have low cooperativity (<2) then an entirely different kind of robust behavior emerges: the output can respond at intermediate levels over an enormous range of signal concentrations (fig. 12.3, *D*); in other words, this simple system buffers the input. Thus, cooperativity makes all-or-none switch-based mechanisms more robust but makes it difficult to obtain buffered responses.[8]

Furthermore, we have found that the segment polarity model suggests that some interactions, specifically those between *ci* and *ptc,* should have low cooperativity. Our core model tends to exhibit strong oscillations in the levels of full-length Ci and its derivative, CN. This behavior does not necessarily prevent the model from adopting stable patterns for other components, presumably (again) because cooperativity provides buffering as long as the input well exceeds its activity threshold. Nevertheless, forcing high cooperativity for Ci-*ptc* interactions leads to apparently inaccurate predictions about the relative strengths of Ci and CN and leads to oscillatory behaviors that seem unrealistic. Why? Ci and *ptc* form the only strictly negative feedback loop in the model. Full-length Ci activates *ptc,* but free Ptc causes Ci to be cleaved into CN, which represses *ptc.* If these interactions are governed by linear-saturating curves, then they can easily find a steady state, whose position in state space is tuned by the availability of Hedgehog. Equilibration depends on steady responses to changes in concentration of each component. However, sharp thresholds mean targets in effect fail to respond to changes in regulator levels in a certain range and then respond abruptly near the threshold. As in the toy model above, the result is oscillations.

Slack (1983) and Edgar and colleagues (Edgar et al. 1989) both demonstrated the requirement for threshold responses in the mechanism of cell state switches, so our findings merely pin that architectural principle to another specific case. Furthermore, again reminiscent of Edgar and Odell's model of a subset of the pair-rule genes, in our model the spatial regime of mutually entrained cell states depends on negative cross-talk among the active genes in each state. On the basis of experiments

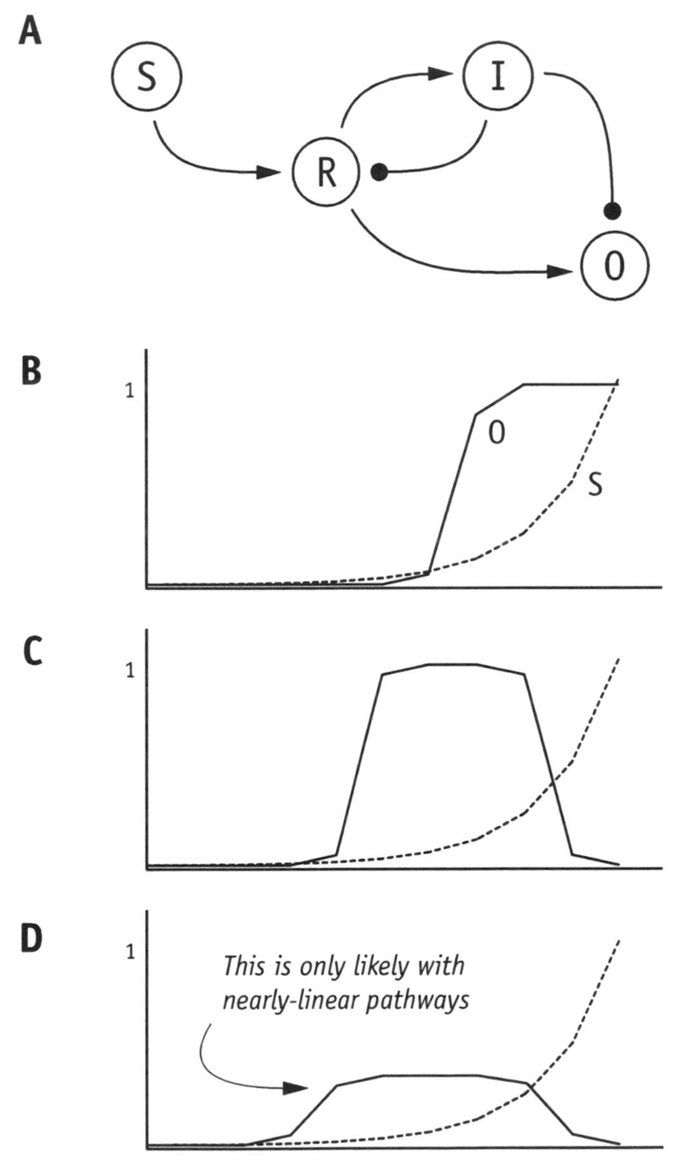

Fig. 12.3.—A potentially homeostatic signaling pathway. *A,* A signal, S, promotes either the activity or the synthesis of a responder, R, which in turn promotes the activity of an output, O. Simultaneously, high levels of R increase the activity of an inhibitor, I, which suppresses both the output and the responder. Each of the links represents a potentially cooperative regulatory effect. *B–D,* Charts showing steady-state output level (*solid lines*) in response to a gradation of input signal (*dashed lines*). When most links are cooperative, the most common responses are a simple threshold response (*B*) and separate thresholds for activation and inhibition (*C*); in both cases the output is either full on or full off. High cooperativity also fosters various oscillatory behaviors (not shown). If most interactions are linear or nearly so, then it becomes possible to find conditions under which the output responds at a fixed level over a large range of input signal strengths (*D*).

with protein synthesis inhibitors, Edgar and Odell modeled the pair-rule genes as if they were basally active. In the case of our segment polarity model, only *ci* is basally activated, so the cell state regime depends also on positive feedback both within some of the cell states and between the different cell states *in neighboring cells.* To repeat, the mechanism is a race between *en* and *ci;* En, via *hh,* enlists the help of *wg* expressed by neighboring cells to keep *en* active, and Wg maintains itself through a still-vague mechanism that probably involves both *slp* and *ci.* Steep thresholds help ensure that any edge in the races pushes the leader toward the attractor in the dynamical system's phase space, the stable cell state, characterized by the leader's expression. Meanwhile, thresholds help make sure that one cell state entrains its neighbors to adopt a different one, and vice versa, thus reinforcing the original choice. This argument at least partially explains the robustness of the segment polarity model's boundary-maintenance function to parameter variation. Further, any spatially varying biases in the initial conditions swing the cell state choice in one or the other direction, and since any such bias will do for some choices of parameters, this explains the robustness to initial conditions.

A final note: in the context of the segment polarity model, there is a direct relationship between (crudely speaking) the average cooperativity of all interactions, and the degree to which the network tolerates variation in either parameter values or initial conditions or architecture. In other words the model predicts that *canalization of this gene network is a direct effect of nonlinear, threshold dose-response functions,* as anticipated by Gibson (1996). The question often arises, How could canalization evolve? Teleologically, it seems that of course canalization should be selectively advantageous in certain circumstances. However, on the basis of population genetic models Gibson and Wagner (2000) suggest that it is actually rather difficult to find conditions under which canalization will arise through positive selection on some "canalizing" allele. These authors express the concern that such results may reflect only the inadequacy of canonical population genetic models.

We admit to such ignorance of population genetics that we could not even begin to agree or disagree, but our models *do* make two interesting suggestions: first, that in the segment polarity network (and the neurogenic network; Meir et al. 2002b), canalizing mutations might arise readily. For example, we found that there are mild to strong variational biases[9] on parameters as diverse as the Wg diffusion rate, the avidity with which En represses *ci,* and the maximum cleavage rate of Ci protein (von Dassow and Odell 2002). These things should be baby steps by mutation, point mutations adjusting the match of enhancer site to regulator, affinity of ligand and receptor, and so on. In other words, the robustness of the model can be "tuned" via quantitative changes in the kinetics of intrinsic components, without the presumably more involved evolutionary step of changing the topology of the network (i.e., by adding new links or components).

Second, on the basis of the arguments above, it is clear that the higher the cooperativity embodied within the model's many positive feedback loops, the more it could tolerate (1) variability in individual parameters (i.e., mimicking genetic mutation), (2) coordinate, simultaneous changes in many parameters (e.g., to mimic variation in temperature or oxygen supply), or (3) stochastic fluctuations over the time course of pattern formation (due to inherent noisiness of gene expression or cell division or whatnot). Thus, for networks that work like these (i.e., coupled cell state switches) canalization against several sources of variability (mutation, environmental perturbation, developmental noise) may be coordinated. Ancel and Fontana (2000) point out that the reduction of phenotypic plasticity (which we think equates with canalization against either environmental variability or developmental noise) "requires a genotype-phenotype map in which plasticity mirrors variability" with respect to genetic mutation. They call this situation "plastogenetic congruence" and show that it is a generic feature of RNA folding, and that therefore "genetic canalization will ensue as a byproduct of selection for environmental canalization." We think something similar holds for gene networks, and we expect that by comparing the level of developmental noise in wild-type versus sensitized mutants of the segment polarity and neurogenic pathways we will be able to test whether such a congruence exists in reality.

Hierarchical Structure of Genetic Modules

We have often wondered how we can define the boundaries of genetic modules. What criteria define a module, as opposed to just another tangle in the genetic web? No one doubts that life as we know it involves gene networks with intrinsic behaviors; the genome of any organism is such a network, as is the genome of any virus. Similarly, no one doubts that such things are organized into modules; genes themselves, after all, are modules of a certain kind, as are genomes, at a very different level of organization. The question is to what extent genomes break down into, or genes and their products conspire to form, *logically separable guilds of the metabolic milieu*—that is, intermediate entities, made of genes and parts of genomes, that do something we can comprehend in isolation. There are a few cases that seem intuitively obvious: the *lac* operon, the yeast mating-type switch, bacteriophage, the cell cycle clock, and the segment polarity network. What do they all have in common? What exactly is it that our intuition tells us about these mechanisms? What criteria can we extract from our intuition that we can generalize? We do not have an answer yet because it turns out that none of the straightforward criteria (like connectivity) or simple analogies (e.g., to object-oriented programming; see below) seem useful.

It happens that the way biologists investigate genetics disposes the discovery process to reveal knots of locally relevant genes whose products all participate in some way in the production of a certain phenotype or characteristic. Perhaps genomics will change this, but presently it is the case that one finds such local tangles and has no way of knowing whether the membership in the tangle represents just the extent of exploration to date or the core membership of a genuine subunit of the genome. One can make a credible argument that developmental genetics is only possible to the extent that such local subunits are realistic; after all, pleiotropic genes are more difficult for geneticists to analyze than are those genes that specifically regulate particular characters (think of the *Drosophila ras* homologue, which seems to be involved in practically everything, versus the *bicoid* gene, which has a fairly specific function).

This hints at some kind of a criterion based on connectivity. It is tempting to suggest that what distinguishes module from not-module is a degree of interconnectivity, or in the strengths of connections. The suggestion is commonplace that modules are composed of "strong" (or "dense") connections, but have only "weak" (or "sparse") connections to other things. Such a notion turns out, for our purposes, to be largely fruitless. Consider an example: Ras has a starring role in EGF signaling, but only among its other roles; EGF signaling, in turn, one might say, is a module unto itself, a part of diverse morphogenetic control processes, appearing in various developmental mechanisms, and not just in a cameo, either. Do we say that EGF signaling does not count as a module unto itself because Ras participates as an essential step in other pathways? Do we say that in *Drosophila* ventral ectoderm patterning and dorsal eggshell patterning are logically inseparable because they share the EGF pathway? We think not. Consider an analogy: The futures markets for various agricultural products are each governed by various causal factors, some unique but many not. No one would claim that the dynamics of the market in pork bellies was inseparable from that in soybeans simply because they share some causal factors (like the weather). And no one would claim that the various futures markets were not separable from treasuries and stocks simply because they are all influenced, and strongly, by the price of crude oil. Instead, we think that the crucial thing our intuition tells us is to look for *things which have their own intrinsic dynamics.*

Furthermore, this suggests that rather than a how-to-break-it-down problem, we really have a how-to-build-it-up issue. And gene networks, like economics, are richly hierarchical. Thus, we see the problem of defining a module in terms of the following thought experiment: Given some behavior of interest, which known facts account for that behavior? And, if there are more known facts than we need to account for the behavior of interest, what do they contribute? Hence, we have often described our approach to modularity by analogy to test-tube biochemical reconstitution, in which the procedure is to add purified components until some desired complex function emerges from the conspiracy of core parts and then try to add more purified components to see how they affect the performance or other aspects of the system of interest. No one would claim that cellular life can proceed without translation of mRNA to proteins, or that translation of the cell's protein complement can proceed without the rest of the cell's activities. No single molecular species can do it. Rather, we consider translation a unified phenomenon among the cell's activities because *biochemists can reconstitute that function from purified extracts.* So, our approach is to use computers to do the thought experiment "What if we had an animal with only these genes?" we add gene products and molecular interactions to a simulation until we get some behavior that seems lifelike.

Figure 12.4 illustrates the hierarchical nature of gene networks using the neurogenic network. At the heart of this network is a bistable switch consisting of the proneural genes, which encode basic helix-loop-helix (bHLH) transcriptional regulators, here represented by *achaete* (*ac*) and *scute* (*sc*). The products of these genes not only feed back positively on their own production; they also cross-activate each other (Martinez and Modolell 1991; Skeath and Carroll 1991; Van Doren et al. 1992). It is better to say the proneural circuit *could* make a bistable switch: there exist sets of parameter values such that there is a stable "off" state in which none of these genes is expressed, and a stable "on" state in which both are. If one pushes the system toward one or the other steady state, beyond some threshold determined by the governing parameters (most significantly those governing the potency of Ac and Sc proteins as regulators), the system will evolve toward and remain at that state until perturbed across the threshold again. For example, a sufficient pulse of *ac* transcription might, under certain conditions, be sufficient to flip the switch on. Thus, we claim this little circuit is a switch module; that is its intrinsic behavior (although it must be kept in mind that the behavior depends on parameter values, etc.).

As it happens, among the direct regulators of the proneural genes are some of the bHLH proteins encoded by genes of the *Enhancer-of-split* complex (E(spl)-C) genes (Oellers et al. 1994; hereafter we discuss these genes for simplicity as if there were a single one, say, *E(spl)*-m8

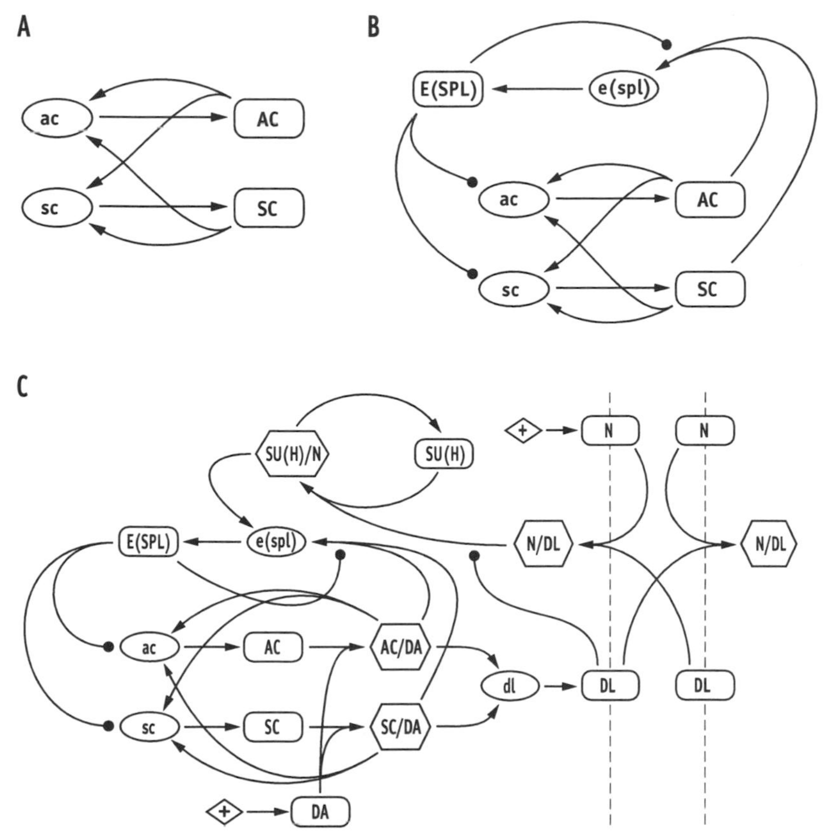

Fig. 12.4.—A hierarchy of modules. The kernel of the neurogenic network is a switch packaged into a homeostat. *A,* The proneural genes *achaete* and *scute* encode transcription factors that stimulate their own and each other's production. As long as these interactions are cooperative, this positive feedback loop can be a bistable switch. If the synthesis rate of Ac and Sc is sufficient to overwhelm degradation, their concentration increases until the synthesis rates saturate, degradation catches up, and the system remains at a stable "on" state; otherwise, degradation turns the switch off. *B, Enhancer of split* is a direct target of Ac and Sc activation, but its product shuts them down. At the same time it interferes with its own activation by the Ac and Sc. This circuit can still function as a switch but can also hold an intermediate steady state or oscillate around some middle expression level. *C,* The Delta/Notch signaling pathway couples the proneural/E(spl) homeostat in one cell to the same circuit in neighboring cells, because high-level Ac and Sc (as the proneural switch heads toward "on") activates *Dl,* and activated N leads to activation of *E(spl),* which shuts the proneural switch off. DA, Daughterless; SU(H), Suppressor of Hairless.

itself). *E(spl)* not only represses the proneural genes, but is also a direct target of them; Ac and Sc activate *E(spl)* transcription (Kramatschek and Campos-Ortega 1994; Singson et al. 1994). Thus, layered around the proneural switch is a negative feedback loop. In addition, E(spl) interferes with its own activation by Ac and Sc. The larger circuit retains the ability to make a bistable switch, albeit the volume fraction of parameter space in which it does so is, while quite large, still much smaller than the analogous fraction for the proneural switch without *E(spl)*. The negative feedback loop adds interesting new behaviors: under some conditions the circuit oscillates; under other conditions it achieves a stable intermediate state, neither on nor off. In either case it is obvious that if E(spl) were suddenly unplugged, we would be left with the proneural switch; that is, the new loop enables the circuit to sit still or wobble around between on and off, undecided, until some extrinsic influence comes along and defeats E(spl).

These clever switches-within-homeostats, one in every cell of some field, are coupled to each other through cell-cell signaling via *Delta* (*Dl*) and *Notch* (*N*). The proneural genes promote *Dl* expression (Kunisch et al. 1994); *Dl* encodes the ligand for a receptor encoded by *N* (Fehon et al. 1990); Notch, upon binding Dl, gets cleaved, and the intracellular portion forms a complex with the transcription factor encoded by *Suppressor of Hairless* (*Su(H)*); together they activate *E(spl)*, which represses the proneural genes (Bailey and Posakony 1995; Lecourtois and Schweisguth 1995). In a cluster of equipotent cells, in which some influence has gotten the proneural switch started (perhaps to the intermediate "deciding" state), the idea is that, because of stochastic differences or initial prepatterns or even specific localized signals, one cell might get a little bit ahead of the others in Dl production, or behind in N activity, such that it experiences less N signaling than the others, and thus flips the proneural switch on, consequently entraining neighbors to switch the same switch *off* (reviewed by Simpson 1997). We have shown that this mechanism is plausible; that is, a model encompassing the facts diagrammed in figure 12.4, *C*, succeeds in picking out a lucky neuroblast and shutting its neighbors off, if there is some initial difference to go on (Meir et al. 2002a).

Now, when we contrast parts *A*, *B*, and *C* of figure 12.4, which one is the module? We say all of them. The proneural switch is no less a switch because of the presence of *E(spl)*, even if *E(spl)*, when unmolested by extrinsic factors, completely abolishes switching. Similarly, the mutual entrainment of switches in a cluster of cells by Dl-N signaling in no way negates the fact that the *Ac-Sc-E(spl)* circuit has certain intrinsic behaviors. Modularity criteria based on connection density or strength would have a hard time putting a pair of scissors into figure 12.4, *B*, or even 12.4, C. That is why we prefer to think of genetic modularity in terms of the intrinsic functional behavior of some network.

It was only recently that we fully appreciated the hierarchical organization of the segment polarity network. That network, if we consider a version that incorporates *sloppy-paired* (fig 12.5, *A*), consists of two subnetworks: one based on the organization of the Hh signaling pathway, and the other, we think, based on interactions between targets of Wg signaling. The former (fig. 12.5, C) makes center-surround patterns (a Hh-producing center surrounded by cells expressing genes

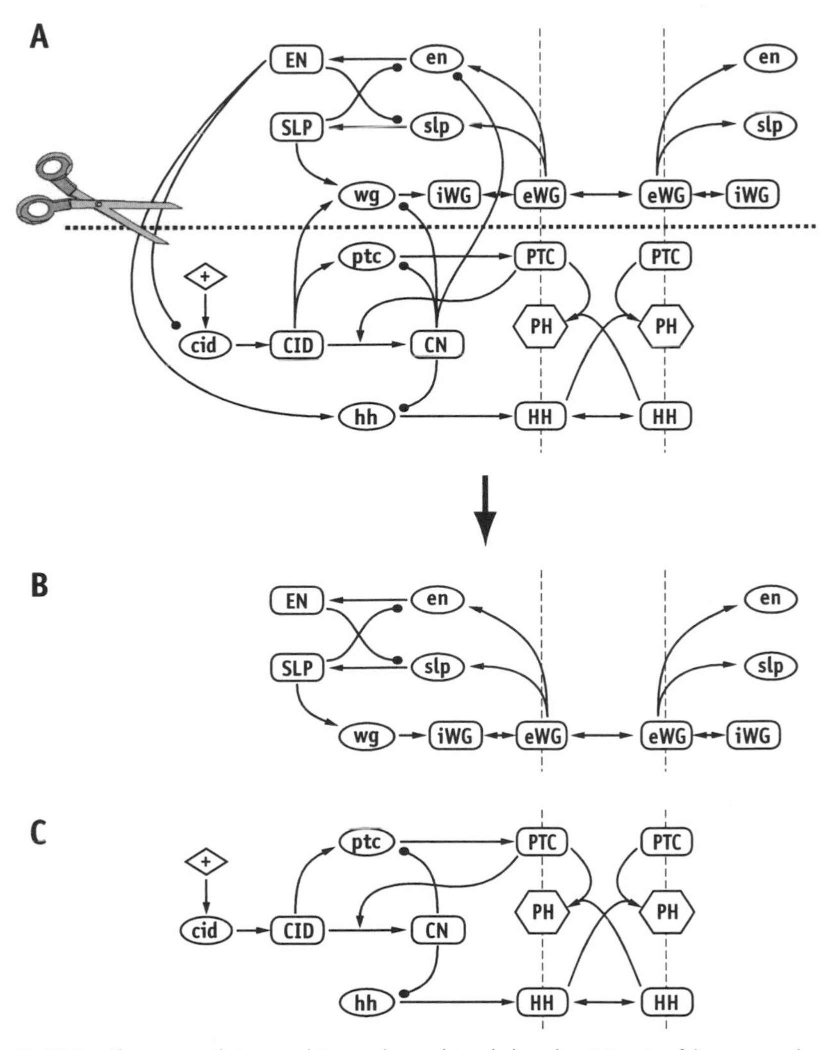

Fig. 12.5.—The segment polarity network is two subnetworks patched together. *A*, A version of the segment polarity network that shows how *sloppy-paired* might mediate *wingless* autoregulation. This network could be snipped in the middle to yield two subnets that each make center-surround patterns on their own: a coupled positive and negative feedback loop consisting of *wg, en,* and *slp* (*B*), and the Hh-Ptc-Ci signaling pathway (*C*).

regulated by Ci, the transcription factor that mediates the response to Hh); the Hh-binding component of the Hh receptor is encoded by the gene *ptc,* which is also activated by full-length Ci and repressed by the N-terminal fragment of Ci. Hh signaling inhibits cleavage of Ci to form the repressor. Thus, Hh signaling leads to increased Ptc expression at the cell surface, which sequesters Hh and limits the range of signaling. This subnetwork is unable to make asymmetric boundaries, the way the complete segment polarity network does.

The other subnetwork may consist of *en*, *wg*, and *slp*. *slp* encodes a transcriptional regulator that represses *en*, activates *wg*, and is activated by Wg and repressed by En (fig. 12.5, *B*) (Bhat et al. 2000; Cadigan et al. 1994b; Grossniklaus et al. 1992; Lee and Frasch 2000). This circuit, too, makes center-surround patterns, but recently we realized that under certain conditions this little subnetwork is able to do the same task as the whole segment polarity network. However, it does that task *much less robustly* than the complete network. Thus, by patching together two center-surround makers, one of which could have been the ancestral asymmetric-boundary module, we get a larger network that does the task of maintaining an asymmetric boundary very robustly. In no sense does the larger network invalidate the existence of the building blocks it is made out of; indeed, it is the *hh-ptc-ci* circuit that seems to have been co-opted most readily over the course of evolution for new roles (see, e.g., Goodrich et al. 1996).

Gene Networks and the Adaptive Landscape

Sewall Wright's metaphor of the "adaptive landscape" (see Futuyma 1998) conceives of a high-dimensional topography in which each phenotype (or genotype) is a point in the space of character states and associated with some fitness value (the independent variable that is the "height"). Thus, fitness is a function of phenotype (or genotype, if one prefers to think in terms of fitness as a function of continuously varying genetic traits), and the surface defined by *that* function is the landscape which evolving organisms populate, driven across it by mutation and winnowed by selection. The topography of the adaptive landscape constrains the evolvability of the traits that determine the landscape itself. Figure 12.6 diagrams some simplistic stereotypes in which fitness is a function of a single quantitatively varying trait. Intuitively, some of these possibilities will be easier to navigate using mutation, selection, and recombination. Should a population find itself on the sloping, low hill in figure 12.6, *A*, there is a trivial path by mutations of small effect (assuming that mutations can quantitatively affect this trait along the entire axis shown) that leads through selectively favored intermediates to the top. Not so for figure 12.6, *E*, although it is hard to imagine such a function for a simple quantitative trait. Even so, the point is that there may be no path (or even one that is easy to find by random mutation) that allows a population to move from local, but suboptimal, peaks to regions of the trait space associated with higher fitness values. Ruggedness in the fitness function may thus seriously constrain the rate of adaptation. Different shapes of the fitness function lead one to expect different levels of within-population diversity; the plateau of

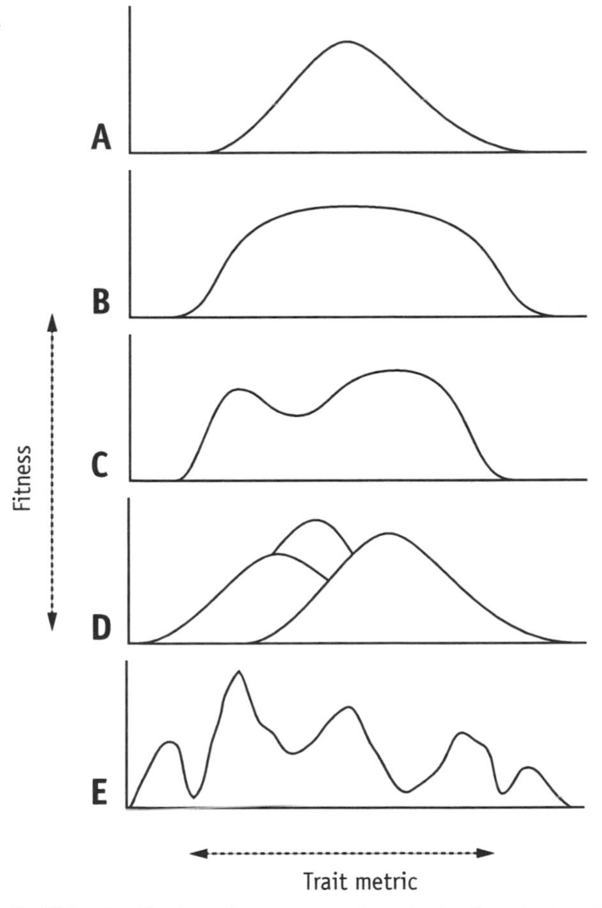

Fig. 12.6.—Possible adaptive landscapes in one dimension. An adaptive landscape is the surface defined by fitness as a function of traits, trait metrics, or character states. For simplicity this figure assumes that the traits of interest are continuous variables of which fitness is a continuous function. Traits of interest might be either phenotypic characters, such as the length of a limb, or genotypic or physiological characters, such as the affinity of an enzyme for its substrate. Mutation causes populations to diffuse across the adaptive landscape; selection causes populations to climb. In *A* there is a Fuji-like smooth-sided peak. If selection is strong enough, then the entire population should eventually cluster around the peak, because starting at any trait value there is a monotonic path to the trait value with the highest fitness. *B* shows a mesa. Again, monotonic paths lead to the top, but in this case a wide range of trait values are virtually indistinguishable. In *C* two peaks are divided by an alpine valley; both peaks are nearly equivalent in fitness, but unless mutation is very strong relative to selection, a population will probably not travel between peaks and instead will cluster around whichever peak it arrives at first. In *D* there is a mountain range, illustrating the possibility that the most fit value for a particular trait might depend on the environment, the genetic background, or even the makeup of the population. Finally, *E* illustrates that it is conceivable that adaptive landscapes could be very rugged indeed. A population trying to navigate *E* faces an adaptive Catch-22: if mutation is weak relative to selection, the population will become trapped on local, but seriously suboptimal, peaks, but if mutation is strong, then the population is very unlikely to be able to remain on any peak that it does find. If ruggedness were very common, we might expect to find that mutation rates, the rate of recombination, and the degree to which the effects of mutations are buffered from phenotypic effects are facultatively variable properties of individual organisms.

figure 12.6, *B*, would allow mutation to disperse populations across the most fit domain with a spectrum of neutral phenotypic variation.

By analogy with the adaptive landscape, consider the surface mapped out by the goodness-of-fit function we employ to evaluate the behavior of gene network models. If we were to pretend that only a single behavior of the gene network was functional, or rather that our function captures everything significant about that network's behavior, independent of ecology, then the surface determined by the objective function would be a *proxy* for the adaptive landscape. We can ask how easy this landscape would be to navigate by a local search in parameter space (analogous to what evolution accomplishes by selecting upon heritable variation in populations). We can ask how structural or architectural features of the network determine the topography of this landscape (for which we do not have a catchy name). We might be able to ask what kinds of mutations are likely to be neutral, and which might result in what kinds of phenotypic variation. Obviously no one can yet answer these questions, but the ever-improving knowledge of how limbs, fins, eyes, eyespots, teeth, toes, and so on are actually made during development is surely opening up this line of exploration. If biologists can develop a picture of the genetic module underlying any of these phenotypic modules, perhaps it will become possible to draw parallels between evolutionary trends manifest in nature and the behavioral repertoires of the genetic networks that shape development.

We have used several rudimentary approaches to come up with a caricature of the terrain in which the segment polarity and neurogenic networks live (von Dassow et al. 2000; von Dassow and Odell 2002; Meir et al. 2002b). Again, the parameter space combined with the goodness-of-fit function we use to judge the pattern produced by the model maps out a topography analogous to the adaptive landscape,[10] but we invert it for mathematical convenience. This function monitors the model's dynamics for measurable qualities that could possibly correspond to adaptive qualities of the segmentation mechanism (see supplement to von Dassow et al. 2000 for details). For example, in the case of the segment polarity network, the model gets better scores the earlier it achieves the desired pattern, the more stable that pattern is, and the sharper the definition of the pattern is; maybe there is evolutionary pressure to develop faster, perhaps oscillations lead to unreliability in quantitative control of downstream modules, and maybe sharply differentiated cell states are more stable than poorly differentiated ones. We can use sampling strategies, or various nonlinear optimization methods, to ask how easily we can navigate the parameter space of the model. The most straightforward approach is to cut transects across the parameter space: starting from a point at which the model gets a good score, we can hold all but one parameter fixed and then

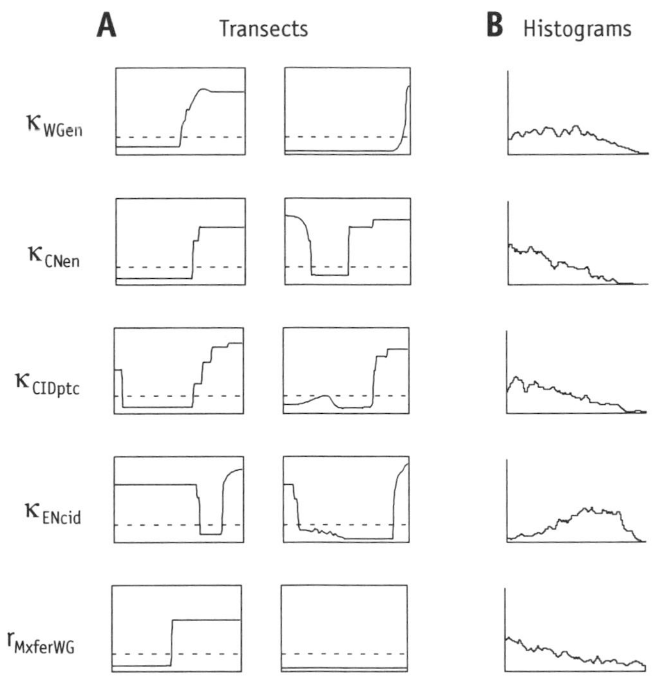

Fig. 12.7.—Profiles of the segment polarity model's parameter space. *A*, Transects along each named parameter's entire allowed range. Each box represents a case in which we started with a parameter set that worked, then varied a single parameter while holding the others fixed, and monitored the behavior of the model. The horizontal axis is the range of variation, which is three orders of magnitude, on a log scale. The vertical axis is the score the model received at that point in parameter space. The dashed line represents the cutoff *above* which we judge the model to have failed. Although the scoring function is designed to respond linearly around this region, note the predominance of sharp thresholds rather than slopes. Two cases are shown for each of five parameters which, from top to bottom, govern how potently Wg activates *en*, how effective CN is at repressing *en*, how effective Ci is at activating *ptc*, how potently En represses *ci*, and how fast Wg equilibrates between opposite faces of neighboring cells (the Wg "diffusion" rate). For the first four, the left side represents potent regulation, the right side weak regulation; for the Wg diffusion rate, the left side represents slow transport, the right side fast. The narrow gap in the fourth row of the first column represents approximately fourfold variation in that particular parameter. At the other extreme, the fifth row of the second column is a case in which the model is completely insensitive to this parameter. *B*, Histograms showing the frequency of working parameter sets as a function of the same parameters as in *A*. The horizontal axis is the same as in *A*. Approximately 1,200 working parameter sets, found in a random sample, are represented here. The interpretation is that working parameter sets are most dense wherever there are peaks in the distribution. For example, the fourth row means that the model is most likely to work when En is a moderately weak, but not *too* weak, repressor of *ci*. The plots in *A* are taken from von Dassow et al. 2000, and those in *B* are taken from von Dassow and Odell 2002.

vary the remaining parameter over several orders of magnitude while tracing the goodness-of-fit function. Several such transects are depicted in figure 12.7, *A*, and they are quite typical: a wide, flat-bottomed canyon, bounded by steep-walled cliffs (compare to fig. 12.6).

One can look for regions of the parameter space in which working sets are especially frequent. Given enough randomly sampled working parameter sets, we can do this by looking at histograms showing the distribution of values for each parameter among all the working sets found within defined boundaries. Some parameter distributions exhibit biases toward some neighborhood within their allowed range, and others do not (fig. 12.7, *B*). We can bracket the peaks and thus narrow the boundaries of the parameter space. For the segment polarity network, we found (von Dassow and Odell 2002) that bracketing the modes with a tenfold range, instead of the thousandfold range in the original sampling, yielded a hit rate of 4 in 5, rather than the 1 in 200 reported for the original search. Thus, for this model there is a vast central canyon in which it is hard to find parameter sets that do *not* work.

In addition, we have tried optimization strategies to test whether one can get from outlying regions of parameter space into the central basin, and it seems that one can. However, the difficulty is that the typical sample point lands either above the canyon rim or on a flat, or at best gently sloping, canyon floor (see transects in fig. 12.7, *A*), and most nonlinear optimization strategies cannot tell where to go from either starting point. One can perturb the parameter set, trying to get the optimizer to ride down the ridgelines instead, and often this enables the optimizer to stumble its way into the central basin (G. von Dassow, unpublished observations). However, so far the most useful strategy has been to mimic what populations do: mutate and recombine (described in Meir et al. 2002b). It turns out that both models have a Grand Canyon in the middle of parameter space, within which the network tolerates essentially neutral variation in every parameter, and many tributaries feed into this canyon from its edges.

The discussion above pretends that the segment polarity network has a single functional behavior, and that we know exactly how to characterize it. This is because we have focused on its role in segmentation, which is relatively well understood, and on the question of whether it is plausible that this circuit could be dissociated from upstream developmental pattern-forming processes. Because of our original motivations in making this model, we have so far explored much less about how the same network could itself generate phenotypic variation, but it is certain that the real segment polarity network has been a major player in the evolution of the insects. In *Drosophila* and other insects, the segment polarity network or a variant thereof provides the basic plan for all the appendage primordia; it is involved in the patterning of the *Drosophila* gut; it lays out the pattern of cuticle structures in the larva; and so on. This module has found re-use in a wide variety of contexts in *Drosophila* alone, and in each case it is used to do something slightly different. Along the anterior-posterior compartment boundary in the imaginal discs, this module establishes a system of morphogens with complex responses by neurogenic patterning circuits (Mullor et al. 1997), vein-producing mechanisms in the wing (Gomez-

Skarmeta and Modolell 1996), the proximo-distal patterning process in the leg (Diaz-Benjumea et al. 1994), and more. In the butterfly wing the segment polarity genes are re-deployed to position the eyespot (Keys et al. 1999). In each case one might hypothesize that "developmental context," whatever that is, selects among the various behaviors in the repertoire of the segment polarity module.

However, the modularity notion cannot go too far: it is not at all clear, in each of these cases, that the "module" is really the same. For instance, in imaginal discs it is impossible to imagine that *engrailed* depends on Wingless signaling since *engrailed* is expressed throughout the entire posterior compartment and Wingless only in a narrow stripe along the anterior-posterior boundary in leg discs, and in an even less suggestive pattern in wing discs (Baker 1988). However, in the context of embryonic segment specification, both in reality (Heemskerk et al. 1991) and in the more detailed versions of our model, Wg signaling is required only to get *engrailed* through an initial phase. Perhaps thereafter *en* expression is clonally inherited (or, in effect, autoactivated), even in all the posterior compartment cells of the disc. Realistically, the picture is somewhat more complicated, because in discs (but perhaps not in embryos) ectopic Hh expression can induce *en* and establish a novel posterior compartment (Gibson and Schubiger 1999; Guillen et al. 1995). Surely, aside from nice, pat ideas about re-deployment of modules and so forth, the evolutionary process will adapt every instance in which the segment polarity gene network is used according to the particular pressures on the trait in question and according to the opportunities the network affords for modifying its behavior.

In our future work we will attempt to account for how this network has been re-deployed in so many contexts, and what is different about the way it works in each one. In which cases has the network been restructured to perform different tasks (i.e., to make variant modules), and in which cases do extrinsic factors (such as the initial prepattern) merely select among the behaviors of the module? We have some preliminary ideas about what the module *could do* given different initial conditions or parameters or even topologies. For example, figure 12.8 shows a few alternate patterns produced by the simplest version of the model. This repertoire (which is incomplete, consisting of just a few patterns that are easy to find and are stable) emerges from variation in parameters but identical initial conditions (referred to as the "crisp" stripe prepattern in von Dassow et al. 2000). Parameter variation is analogous to quantitative changes in gene function but also could represent modulation by extrinsic control. Thus, figure 12.8 shows that the same topology could "do something else" if either mutations or extrinsic factors could tune it up appropriately. For a hypothetical example, some transcription factor, extrinsic to the segment polarity mod-

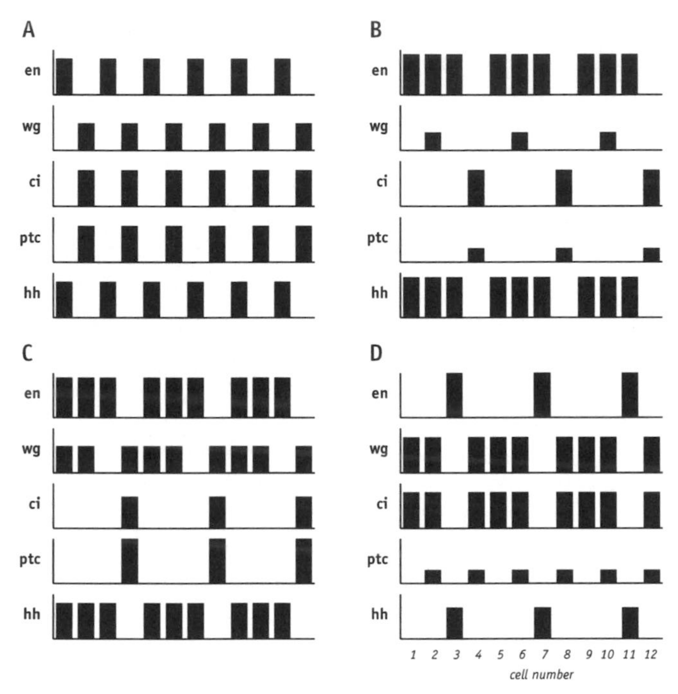

Fig. 12.8.—Near-neighbor patterns made by the segment polarity network. Each pattern results with one in every few hundred or thousand random parameter sets. *A,* A degenerate pattern, the only striped pattern the model could make without *wg* autoregulation and repression of *en* by CN. *B,* A pattern resulting from weak or no repression of *en. C,* Broad, overlapping stripes of *wg* and *en. D,* The pattern resulting if *wg* is strongly activated by Ci but weakly repressed by CN. For all of these, parameter sets that make these patterns are easy to find in the neighborhood of parameter sets that make the standard pattern.

ule, could "tune" the module's behavior by modulating the effect of, say, CN on *engrailed,* thus causing the model to make a different pattern (such as fig. 12.8, *B*) in some developmental context where that extrinsic factor is expressed. This scenario corresponds to the notion of selector genes which locally modulate developmental processes; the Hox complex genes, *engrailed,* and *vestigial* are just a few of the genes known to behave as selectors with respect to processes as diverse as denticle patterns, neuroblast formation, and adult appendage development. Another repertoire (not shown) results from variation in the initial conditions. Both sets represent a sample of the near neighbors in a pattern morphospace that can be explored by tuning the control of the segment polarity network. In other words, they're what's just up the canyon rim. While we cannot yet pin any one of them to a specific, real instance in a living organism, surely this "tuning" analogy captures something analogous to the way in which the evolutionary process tinkers with its tools.

Analogies for Genetic Architecture

He his fabric of the heav'ns hath left to their disputes, perhaps to move his laughter at their quaint opinions wide hereafter, when they come to model heaven and calculate the stars . . . how build, unbuild, contrive, to save appearances.
—Raphael to Adam, in John Milton's *Paradise Lost*

It seems tempting (to us and others) to compare genetic networks to other more familiar or man-made networks: electronic circuits, neural networks (the modeler's kind or the real thing), computer code, the internet, and so on. Analogies are at least as useful for the distinctions as for the similarities. Here we want to critique a common analogy with computer programming code.

One could possibly argue that the prevalence of the "developmental pathway," the "genetic program," or related metaphors owes something, historically, to the development of serial-instruction-chain computing machines. It would be interesting to know in a scholarly way, but our impression is that the embryology literature prior to the invention of digital, instruction-chain computers does not emphasize the notion of a chain of instructions, with the possible exception of the literature on induction phenomena. Even in the case of induction, most of the discussion took place in the context of the dynamical notion of the "morphogenetic field." Since the elaboration of the Central Dogma of Molecular Biology, pathway and program metaphors seem much more common, and developmental biologists do not seem to have used the notion of a morphogenetic field as if it had, any longer, the same explanatory power that it had been invested with in an earlier era.

So it is tempting to draw a parallel between the hypothesis of genetic modularity and object-oriented computer programming. Object-oriented programming means dividing up a computer program into building blocks that each encapsulate certain procedures, functions, and data. Each object is an instance of a "class" definition; the class defines a set of behaviors for all objects of that type and defines which data those objects store. The class also defines the "interface" of objects of that type: what messages they know how to interpret, what data they allow other objects access to, and so on. Once the programmer has defined all the classes that make an entire program, running the program means populating the computer with instances of those classes (the objects), letting them communicate and do whatever they need to do to interact with inputs (like the user) or generate outputs. One of the chief advantages of object-oriented programming is that, if done right, the building blocks not only can be used to build other structures, but also can be swapped with new versions as long as the interface remains the same. Our gene network simulation program, Ingeneue, was written in the object-oriented language Java. It is highly modular: everything from the user interface to the numerical routines to the terms in the differential equations is a class definition that, at runtime, gets instantiated into a bunch of objects as needed. When we need a new piece of a certain type, we need only make sure that it has the right interface. If we need to improve, say, the numerical integration routine, the rest of the program is none the wiser, because we simply replace the internal methods of the integrator module, without changing the interface.

Analogously, it is tempting to think of gene networks as building blocks for the larger programs of development. The segment polarity network has its own behaviors, states, and inputs (the pair-rule genes) and outputs (signals like Wingless and Hedgehog, transcription factors like Sloppy-paired, Gooseberry, and Engrailed), and a lot of "internal" machinery (the mediators of Wg and Hh signaling) that seem, when we look at a network diagram, neither input nor output. We like to think of the evolutionary process co-opting gene networks to use them in new contexts. Objects, like gene networks, are hierarchical; larger-scale building blocks contain smaller ones, and so on. But this analogy is deceptive in several ways: first of all, in the case of gene networks, the distinction between inputs, outputs, and internal methods is an artifact of how we choose to think about things, whereas in the case of software objects, it is a fact of life enforced by syntax.

Probably the most important of all is the issue of data hiding. Good object-oriented code requires data hiding, meaning that objects cannot interfere with the methods and data stored by other objects, unless granted a specific right to do so through defined methods for access. Nature cannot do data hiding. The closest thing to data hiding in gene networks is cellular compartmentalization. At the level of the networks we work with, data hiding is almost a meaningless notion: *nothing whatsoever* prevents some other network from fooling around with the "internal" workings of the neurogenic or segment polarity networks.[11] In fact, this is one of the things about genetic architecture that makes it so creative a substrate for evolution: nature can rewire it in all sorts of ways. For example, the neurogenic network may, in some cases, choose the winning neuroblast because local expression of Wingless, which just happens to bind to Notch and prevent it from functioning as a Delta receptor, reduces the amount of Notch available for signaling (see Wesley 1999; Wesley and Saez 2000 for evidence that Wg interacts with N; we do not know of direct evidence that Wg biases proneural cluster selection by such a mechanism). The closest that object-oriented code can come to *that* sort of thing is through the mechanism of inheritance, in which the programmer can define a subclass of a class she wants to modify, and add methods to it.

Another analogy, which we also find instructive, is to think of gene networks the same way that biochemists think about protein folding. Each amino acid has certain properties, such as physical size, and polarity. Within each small stretch of peptide chain, those properties determine the tendency of that stretch to adopt a particular secondary structure. The tertiary structure emerges from the secondary, and the quaternary from the tertiary. As with gene networks, each of the fundamental components is at once a potential input and a potential output (as long as it ends up on the surface of something such that it can interact with something else, even another domain of the protein, as when buried in the center). But what we find most appealing about this analogy is that in both cases, gene networks and protein folding, the emergence of a coherent higher-level structure subsumes the role of the lower-level building blocks as units of function or selection. In most cases we do not think of amino acid residues as units of function unless we are talking about how secondary structure arises; increasingly, we talk about the function of most individual genes only in the context of how they participate in networks. There are exceptions: the three crucial residues, positioned just so, in a serine protease, the crucial phosphorylatable tyrosines in a signaling protein, or those genes like superoxide dismutase whose function is their own apart from any network. But for most genes, they are not the units of interest once we get to the network level: it is the whole conspiracy we care about.

Limitations

WE DON'T RENT PIGS

—Augustus McCrae's sign, in Larry McMurtry's *Lonesome Dove*

To close we highlight four considerations that limit the usefulness of the approach we have developed over the past several years. Other issues specific to mathematical formulation are treated elsewhere (Meir et al. 2002b).

First, the models we have constructed required a wealth of information about the detailed circuitry of gene networks that is unlikely to be available in more than a few paradigm cases for the near future. This immediately raises certain questions about whether our conclusions should generalize to gene networks in the abstract. Both the segment polarity and neurogenic networks were (and still are being) worked out by armies of diligent molecular biologists, over the course of perhaps as much as a dozen man-millennia for each case, and it is highly unlikely that most developmental mechanisms will ever receive quite as much effort. We do not know how typical these mechanisms really are. R. Strathmann (personal communication) has suggested that developmental mechanisms and developmental biologists coevolve: the nice, tidy modular mechanisms are more easily understood, leading to more fame and money for the biologists who choose to study them, in turn attracting more talent to the now-paradigmatic cases, and so on. Thus, we must keep in mind that our findings may document interesting properties only of the particular networks we have worked with, and we need further work (possibly involving randomly wired network models) to know whether our conclusions have general import.

A second major limitation is a methodological one: our focus on the parameter space as the central problem in gene networks imposes both a computational and conceptual burden. This burden may be so great as to prevent our methods from being useful for problems much more complex than the models we have treated here, especially if robustness turns out *not* to be a general feature of gene networks. While the mathematical formulation we use may be applicable to most gene network problems, it may be impossible to confront most problems using such a blunt tool as a random search in a wide rectangle in parameter space. We continue to research more sophisticated strategies to extend our methods, but from our experience the major challenges appear to be versatile methods for pattern recognition, rational approaches to defining the "reasonable" ranges of parameters, and developing a comprehensive library of formulas to deal with diverse and complex regulatory relationships between macromolecules. We expect to solve these problems as we develop experience with more and more case studies. Meanwhile, we expect to lessen the computational burden through numerical techniques such as the integration recipe described by Meir et al. (2002a), and a related fixed-point iteration method developed by E. Munro (unpublished), but there may be nothing for it in the case of a truly complex, yet not terribly robust, network.

Third is a tactical limit: we doubt seriously that biologists will be able to use models like ours to infer network topologies from DNA microarray (or similar) data on correlated patterns of gene expression. Even if it becomes feasible to obtain large-scale gene expression data from embryos, we fear it will be very difficult to deduce the wiring of such tightly looped networks as we have dealt with from such data sets. This may not be true in every case; "networks" that consist only of simple cascades with little feedback and cross-talk will surely be straightforward to deduce. But how do we know what we are dealing with a priori? Imagine if we tried to understand the segment polarity network from DNA microarray data alone, using models as a deductive aid. Say we gathered data on wild-type versus *engrailed* embryos and discovered that the expression of *patched* depends on *en* function; of course it does, because En causes *hedgehog* to be expressed, and *hedgehog*'s product ultimately causes *ptc* to be transcribed. But En itself represses *ptc* (Hooper and Scott 1989; Hidalgo and Ingham 1990). To be sure, we might notice this if we also tried to measure responses to *en* overexpression, but we would be hard pressed to guess which effects were indirect and which direct and how each worked for even this simple fragment of the network. Perhaps one would try to compose a series of possible topologies that might account for the observed data and then choose specific experiments to decide among them. Ultimately, the combinatorics required to use network models to distinguish the possibilities would be overwhelming, and a sensible biologist would turn to experiments to figure things out. Thus, we worry that attempts to use compute-intensive nonlinear models to deduce network topologies from microarray data are doomed.[12]

Finally, the biggest problem is ascribing functions to gene networks. How, given some conspiracy of genes, are we to know what dynamical behavior they are "meant" to do in the living organism? For the segment polarity network, we used the notion, present in the literature for over a decade, that this network's job is to maintain parasegmental boundaries. For the neurogenic network, we used the notion, also present in the literature for over a decade, that its task is to mediate lateral inhibition within a cluster of equipotent cells. These ideas come *not* from the great mass of molecular data; they come *not* from synthesizing those data into mathematical models; they come, prior to the molecular facts, from careful perturbation experiments and developmental genetics—experiment, not the parts list, reveals the nature of the mechanism. For most networks, notions of function still end with statements like "genes X, Y, and Z are necessary for such-and-such an aspect of the phenotype," or "gene R is a master control gene for such-and-such a process, since activating it ectopically leads to the expression of X, Y, and Z." Sensible models demand a shift from assigning function based on phenotypic effects to assigning function in terms of intrinsic behavior. We suspect this is a matter of waiting for someone clever to do the right set of experiments and have that one critical intuition about what the process is all about. Genomics, microarrays, and all the other avatars of the Age of Ugly Facts hopefully make that process easier (rather than simply overwhelming it).

Lawrence and Sampedro (1993) opened a critique of early efforts to concoct a molecular "explanation" of the segment polarity network with a quote that bears repeating, Banquo's reaction to Macbeth's prophetic witches:

> The instruments of darkness tell us truths;
> Win us with honest trifles, to betray's
> In deepest consequence.—
> (Banquo, in William Shakespeare's *Macbeth,* act 1, scene 3)

Whether Lawrence and Sampedro had in mind the molecular biologists or the theorists in the role of the witches, or both, we do not presume to guess. The point is that the crucial step is an insight into the nature of the mechanism and the bounds of the device, and the Parts Catalog of Life will not tell anyone what all those parts are supposed to add up to. In every historical case we can think of, it has been perturbation experiments (physical, molecular, or genetic), analysis of phenotypes, and comparative studies that have brought about that crucial insight, whether the parts list follows or not.

Acknowledgments

We thank Garry Odell and Richard Strathmann for comments on an early draft of this paper, Gerhard Schlosser and Günter Wagner for constructive and stimulating critique as well as the opportunity to write it, and Virginia Rich for proofreading the manuscript. Our work on gene network models is supported by NSF grants MCB-9732702, MCB-9817081, and MCB-0090835 to G. M. Odell.

Notes

1. A little historical perspective is in order when evaluating the role of earlier theoretical efforts. Turing, Meinhardt, and their followers were trying to imagine how development could work at a time when very little was known about how development does work. The same is true of other theoretical attempts of the same era. All those efforts, given the history of ideas in developmental biology, ought to be seen as remarkable successes: they managed to show that developmental pattern formation could be due to relatively straightforward chemical processes, a conclusion that many classical embryologists, most famously Hans Driesch, had trouble accepting. Thus, such models gave direct encouragement to nascent attempts at developmental genetics: if simple chemical processes could explain development, then there was real reason to hope to discover the developmental control genes and understand their function. Is it possible that *Drosophila* developmental genetics, and the Age of Ugly Facts, was actually inspired by the Beautiful Theories? We suspect that what those theories did is convince a critical mass of people that is was worth looking for the molecular basis of morphogenesis because it might be a simple thing after all. At the same time we suspect that if developmental geneticists of the 1960s and 1970s had stuck with the sentiments of classical embryologists, they might have given up hope. A passing investigation shows that a significant fraction of the embryological literature of the first half of the 20th century reads like a hymn to vital forces; even the nonvitalists among the embryological community were unlikely to be committed reductionists (see the excellent volume edited by Scott Gilbert [Gilbert 1994], and also his recent review [Gilbert et al. 1996]). A relatively small number of workers, especially Waddington, Weiss, and Needham, appreciated the potential of genetics and biochemistry to reveal mechanistic explanations of complex developmental phenomena (Needham 1942; Waddington and Kacser 1957; Weiss 1968). Gierer and Meinhardt, Wolpert, Slack, Kauffman, and other contributors to theoretical developmental biology of the 1970s, constitute an important part of the intellectual weave that connects the morass of classical embryological phenomenology to the equally bewildering morass of modern molecular developmental genetics.

2. This codependence lasts, literally, for a couple of hours; it is a bridge between the transient input provided by the pair-rule genes and longer-term mechanisms for stabilizing cell states within the segment. By the end of germ-band extension, *en* expression no longer depends on *wg* (Heemskerk et al. 1991). Perhaps this is because En represses factors that would otherwise turn En off, and because perhaps, after the initial phase, there is enough free Arm in the absence of Wg signaling to allow *en* transcription in the absence of repressors. Most versions of the segment polarity model, except the simplest, have the potential for this kind of "En autoregulation."

3. To our knowledge, only Cadigan and Grossniklaus and colleagues (Cadigan et al. 1994a; Cadigan et al. 1994b; Grossniklaus et al. 1992), who characterized *sloppy-paired,* highlighted the need for something to keep *en* off in the anterior compartment.

4. These remarks apply to the version of the segment polarity network shown in figure 1. Other versions that incorporate additional components and interactions (von Dassow and Odell 2002) alleviate some of the problems described in this scenario. For example, versions that include *sloppy-paired* do not depend so much on Ci providing an early assist to *wg* expression, because *slp* is expressed early enough to fulfill that role. Nevertheless, the overall description, of a race between mutually exclusive but codependent cell states, remains valid.

5. The "blind watchmaker" is Dawkins's metaphor for the evolutionary process, in which natural design is the outcome of mutation and selection. The metaphor originates with William Paley's famous argument for the existence of divinity, since to Paley a complex device like a watch implies a watchmaker.

6. What we call "cooperativity" may be due to true allosteric cooperativity, or may not; we use it as a convenient, evocative term for the steepness of nonlinear, sigmoid dose-response curves.

7. Intermediate steps provide opportunities to damp out oscillations just by sluggishness. Given a chain of responders, if one step responds on a longer time scale than the step preceding (i.e., a protein with a longer half-life than its mRNA), it will respond slowly to variations in its inputs, thus converting high-amplitude oscillations to lower-amplitude oscillations of the same frequency.

8. This passage should also serve as a reminder that "robustness" is not a unified phenomenon; instead of saying, "this device is robust," we need to say, "such-and-such a functional behavior of this device is robust to such-and-such perturbations." The first statement makes no sense without a context to specify the behavior and perturbation in question.

9. By this term we mean the tendency for randomly sampled "working" parameter sets to include values, for a particular parameter, that cluster in some neighborhood, even when there is no absolute restriction. For example, working parameter sets are about three times as dense in the "slow" third of the range we allow for the Wg diffusion rate as they are in the range as a whole.

10. We emphasize that we are taking this as a proxy for an adaptive landscape. We do not mean to conflate the two. The adaptive landscape refers to a fitness function on a manifold whose axes are either phenotypic traits or genotypic characteristics. We haven't a clue how to relate the dynamical behavior of a gene network directly to survival and reproduction; all we can do is characterize how the dynamics of the model match observed gene expression patterns, which are but a manifestation of some machinery that creates the phenotype, whose fitness is determined by the ecology the organism finds itself in. For the sake of the metaphor we are pretending that the network has to do more or less what it does in a wild-type animal, and that the better it mimics the wild-type gene expression regime, the higher its fitness.

11. But we note that at higher levels of organization, there are barriers that to some extent isolate different developmental modules from one another. For example, the spa-

tial layout and morphogenesis of embryos can either prevent or enforce cross talk between different morphogenetic fields.

12. A more suitable approach to this challenge might be the neural-network-inspired method developed by Reinitz and colleagues, in which the weights governing all possible connections within the network are tuned to achieve the best possible fit to real data (Reinitz et al. 1998; Reinitz and Sharp 1995).

References

Alcedo, J., M. Ayzenzon, T. Von Ohlen, M. Noll, and J. E. Hooper. 1996. The *Drosophila* smoothened gene encodes a seven-pass membrane protein, a putative receptor for the hedgehog signal. *Cell* 86:221–232.

Alcedo, J., and M. Noll. 1997. Hedgehog and its patched-smoothened receptor complex: a novel signalling mechanism at the cell surface. *Biol. Chem.* 378:583–590.

Alexandre, C., A. Jacinto, and P. W. Ingham. 1996. Transcriptional activation of hedgehog target genes in *Drosophila* is mediated directly by the cubitus interruptus protein, a member of the GLI family of zinc finger DNA-binding proteins. *Genes Dev.* 10:2003–2013.

Alon, U., M. G. Surette, N. Barkai, and S. Leibler. 1999. Robustness in bacterial chemotaxis. *Nature* 397:168–171.

Ancel, L., and W. Fontana. 2000. Plasticity, evolvability, and modularity in RNA. *J. Exp. Zool. (Mol. Dev. Evol.)* 288:242–283.

Anderson, and Walter. 1999. Blobel's Nobel. *Cell* 99:557–558.

Aza-Blanc, P., F. A. Ramirez-Weber, M. P. Laget, C. Schwartz, and T. B. Kornberg. 1997. Proteolysis that is inhibited by hedgehog targets Cubitus interruptus protein to the nucleus and converts it to a repressor. *Cell* 89:1043–1053.

Bailey, A. M., and J. W. Posakony. 1995. Suppressor of hairless directly activates transcription of enhancer of split complex genes in response to Notch receptor activity. *Genes Dev.* 9:2609–2622.

Baker, N. E. 1988. Transcription of the segment-polarity gene wingless in the imaginal discs of *Drosophila,* and the phenotype of a pupal-lethal wg mutation. *Development* 102:489–497.

Barkai, N., and S. Leibler. 1997. Robustness in simple biochemical networks. *Nature* 387:913–917.

Bhanot, P., M. Brink, C. H. Samos, J. C. Hsieh, Y. Wang, J. P. Macke, D. Andrew, J. Nathans, and R. Nusse. 1996. A new member of the frizzled family from *Drosophila* functions as a Wingless receptor. *Nature* 382:225–230.

Bhanot, P., M. Fish, J. A. Jemison, R. Nusse, J. Nathans, and K. M. Cadigan. 1999. Frizzled and DFrizzled-2 function as redundant receptors for Wingless during *Drosophila* embryonic development. *Development* 126:4175–4186.

Bhat, K. M. 1998. frizzled and frizzled 2 play a partially redundant role in wingless signaling and have similar requirements to wingless in neurogenesis. *Cell* 95:1027–1036.

Bhat, K. M., E. H. van Beers, and P. Bhat. 2000. Sloppy paired acts as the downstream target of wingless in the *Drosophila* CNS and interaction between sloppy paired and gooseberry inhibits sloppy paired during neurogenesis. *Development* 127:655–665.

Brunner, E., O. Peter, L. Schweizer, and K. Basler. 1997. pangolin encodes a Lef-1 homologue that acts downstream of Armadillo to transduce the Wingless signal in *Drosophila. Nature* 385:829–833.

Cadigan, K. M., U. Grossniklaus, and W. J. Gehring. 1994a. Functional redundancy: the respective roles of the two sloppy paired genes in *Drosophila* segmentation. *Proc. Natl. Acad. Sci. U.S.A.* 91:6324–6328.

Cadigan, K. M., U. Grossniklaus, and W. J. Gehring. 1994b. Localized expression of sloppy paired protein maintains the polarity of *Drosophila* parasegments. *Genes Dev.* 8:899–913.
Cadigan, K. M., and R. Nusse. 1997. Wnt signaling: a common theme in animal development. *Genes Dev.* 11:3286–3305.
Chen, Y., and G. Struhl. 1996. Dual roles for patched in sequestering and transducing Hedgehog. *Cell* 87:553–563.
Chen, Y., and G. Struhl. 1998. In vivo evidence that Patched and Smoothened constitute distinct binding and transducing components of a Hedgehog receptor complex. *Development* 125:4943–4948.
Dawes, R., I. Dawson, F. Falciani, G. Tear, and M. Akam. 1994. Dax, a locust Hox gene related to fushi-tarazu but showing no pair-rule expression. *Development* 120: 1561–1572.
Dearden, P., and M. Akam. 1999. Developmental evolution: axial patterning in insects. *Curr. Biol.* 9:R591–*R*594.
Diaz-Benjumea, F. J., B. Cohen, and S. M. Cohen. 1994. Cell interaction between compartments establishes the proximal-distal axis of *Drosophila* legs. *Nature* 372: 175–179.
Dierick, H. A., and A. Bejsovec. 1998. Functional analysis of Wingless reveals a link between intercellular ligand transport and dorsal-cell-specific signaling. *Development* 125:4729–4738.
DiNardo, S., E. Sher, J. Heemskerk-Jongens, J. A. Kassis, and P. H. O'Farrell. 1988. Two-tiered regulation of spatially patterned engrailed gene expression during *Drosophila* embryogenesis. *Nature* 332:604–609.
Dominguez, M., M. Brunner, E. Hafen, and K. Basler. 1996. Sending and receiving the hedgehog signal: control by the *Drosophila* Gli protein Cubitus interruptus. *Science* 272:1621–1625.
Edgar, B. A., G. M. Odell, and G. Schubiger. 1989. A genetic switch, based on negative regulation, sharpens stripes in *Drosophila* embryos. *Dev. Genet.* 10:124–142.
Fehon, R. G., P. J. Kooh, I. Rebay, C. L. Regan, T. Xu, M. A. Muskavitch, and S. Artavanis-Tsakonas. 1990. Molecular interactions between the protein products of the neurogenic loci Notch and Delta, two EGF-homologous genes in *Drosophila. Cell* 61:523–534.
Futuyma, D. J. 1998. Evolutionary biology. Sunderland, Mass.: Sinauer Associates.
Gibson, G. 1996. Epistasis and pleiotropy as natural properties of transcriptional regulation. *Theor. Popul. Biol.* 49:58–89.
Gibson, G., and G. Wagner. 2000. Canalization in evolutionary genetics: a stabilizing theory? *BioEssays* 22:372–380.
Gibson, M. C., and G. Schubiger. 1999. Hedgehog is required for activation of engrailed during regeneration of fragmented *Drosophila* imaginal discs. *Development* 126: 1591–1599.
Gierer, A., and H. Meinhardt. 1972. A theory of biological pattern formation. *Kybernetik* 12:30–39.
Gilbert, S. F. 1994. A Conceptual history of modern embryology. Baltimore: Johns Hopkins University Press.
Gilbert, S. F., J. M. Opitz, and R. A. Raff. 1996. Resynthesizing evolutionary and developmental biology. *Dev. Biol.* 173:357–372.
Glass, L., and S. A. Kauffman. 1972. Co-operative components, spatial localization and oscillatory cellular dynamics. *J. Theor. Biol.* 34:219–237.
Glass, L., and S. A. Kauffman. 1973. The logical analysis of continuous, non-linear biochemical control networks. *J. Theor. Biol.* 39:103–129.
Gomez-Skarmeta, J. L., and J. Modolell. 1996. araucan and caupolican provide a link

between compartment subdivisions and patterning of sensory organs and veins in the *Drosophila* wing. *Genes Dev.* 10:2935–2945.

Goodrich, L. V., R. L. Johnson, L. Milenkovic, J. A. McMahon, and M. P. Scott. 1996. Conservation of the hedgehog/patched signaling pathway from flies to mice: induction of a mouse patched gene by Hedgehog. *Genes Dev.* 10:301–312.

Grossniklaus, U., R. K. Pearson, and W. J. Gehring. 1992. The *Drosophila* sloppy paired locus encodes two proteins involved in segmentation that show homology to mammalian transcription factors. *Genes Dev.* 6:1030–1051.

Guillen, I., J. L. Mullor, J. Capdevila, E. Sanchez-Herrero, G. Morata, and I. Guerrero. 1995. The function of engrailed and the specification of *Drosophila* wing pattern. *Development* 121:3447–3456.

Gutfreund, H. 1995. Kinetics for the life sciences: receptors, transmitters, and catalysts. Cambridge: Cambridge University Press.

Heemskerk, J., S. DiNardo, R. Kostriken, and P. H. O'Farrell. 1991. Multiple modes of engrailed regulation in the progression towards cell fate determination. *Nature* 352: 404–410.

Hepker, J., Q. T. Wang, C. K. Motzny, R. Holmgren, and T. V. Orenic. 1997. *Drosophila* cubitus interruptus forms a negative feedback loop with patched and regulates expression of Hedgehog target genes. *Development* 124:549–558.

Hidalgo, A., and P. Ingham. 1990. Cell patterning in the *Drosophila* segment: spatial regulation of the segment polarity gene patched. *Development* 110:291–301.

Hooper, J. E. 1994. Distinct pathways for autocrine and paracrine Wingless signalling in *Drosophila* embryos. *Nature* 372:461–464.

Hooper, J. E., and M. P. Scott. 1989. The *Drosophila* patched gene encodes a putative membrane protein required for segmental patterning. *Cell* 59:751–765.

Ingham, P. W., N. E. Baker, and A. Martinez-Arias. 1988. Regulation of segment polarity genes in the *Drosophila* blastoderm by fushi tarazu and even skipped. *Nature* 331:73–75.

Kauffman, S. A. 1993. The origins of order: self organization and selection in evolution. New York: Oxford University Press.

Keys, D. N., D. L. Lewis, J. E. Selegue, B. J. Pearson, L. V. Goodrich, R. L. Johnson, J. Gates, M. P. Scott, and S. B. Carroll. 1999. Recruitment of a hedgehog regulatory circuit in butterfly eyespot evolution. *Science* 283:532–534.

Kramatschek, B., and J. A. Campos-Ortega. 1994. Neuroectodermal transcription of the *Drosophila* neurogenic genes E(spl) and HLH-m5 is regulated by proneural genes. *Development* 120:815–826.

Kunisch, M., M. Haenlin, and J. A. Campos-Ortega. 1994. Lateral inhibition mediated by the *Drosophila* neurogenic gene delta is enhanced by proneural proteins. *Proc. Natl. Acad. Sci. U.S.A.* 91:10139–10143.

Lawrence, P. A., and J. Sampedro. 1993. *Drosophila* segmentation: after the first three hours. *Development* 119:971–976.

Lecourtois, M., and F. Schweisguth. 1995. The neurogenic suppressor of hairless DNA-binding protein mediates the transcriptional activation of the enhancer of split complex genes triggered by Notch signaling. *Genes Dev.* 9:2598–2608.

Lee, H. H., and M. Frasch. 2000. Wingless effects mesoderm patterning and ectoderm segmentation events via induction of its downstream target sloppy paired. *Development* 127:5497–5508.

Li, X., and M. Noll. 1993. Role of the gooseberry gene in *Drosophila* embryos: maintenance of wingless expression by a wingless-gooseberry autoregulatory loop. *EMBO J.* 12:4499–4509.

Manoukian, A. S., K. B. Yoffe, E. L. Wilder, and N. Perrimon. 1995. The porcupine gene is required for wingless autoregulation in *Drosophila*. *Development* 121:4037–4044.

Marigo, V., R. A. Davey, Y. Zuo, J. M. Cunningham, and C. J. Tabin. 1996. Biochemical evidence that patched is the Hedgehog receptor. *Nature* 384:176–179.

Martinez, C., and J. Modolell. 1991. Cross-regulatory interactions between the proneural achaete and scute genes of *Drosophila. Science* 251:1485–1487.

Martinez-Arias, A. 1993. Development and patterning of the larval epidermis of *Drosophila*. In *The development of Drosophila melanogaster,* ed. M. Bate and A. Martinez-Arias, 517–608. Cold Spring Harbor, N.Y.: Cold Spring Harbor Laboratory Press.

Meinhardt, H. 1977. A model of pattern formation in insect embryogenesis. *J. Cell Sci.* 23:117–139.

Meinhardt, H. 1984. Models for positional signalling, the threefold subdivision of segments and the pigmentation pattern of molluscs. *J. Embryol. Exp. Morphol.* 83 (suppl): 289–311.

Meir, E., E. M. Munro, G. M. Odell, and G. von Dassow. 2002a. Ingeneue: a versatile tool for reconstituting genetic networks, with examples from the segment polarity network. *J. Exp. Zool.* 294:216–251.

Meir, E., G. von Dassow, E. Munro E, and G. M. Odell. 2002b. Robustness, flexibility, and the role of lateral inhibition in the neurogenic network. *Curr. Biol.* 12: 778–786.

Moazed, D., and P. H. O'Farrell. 1992. Maintenance of the engrailed expression pattern by Polycomb group genes in *Drosophila. Development* 116:805–810.

Moline, M. M., C. Southern, and A. Bejsovec. 1999. Directionality of wingless protein transport influences epidermal patterning in the *Drosophila* embryo. *Development* 126:4375–4384.

Mullor, J. L., M. Calleja, J. Capdevila, and I. Guerrero. 1997. Hedgehog activity, independent of decapentaplegic, participates in wing disc patterning. *Development* 124: 1227–1237.

Nagy, L. M., and S. Carroll. 1994. Conservation of wingless patterning functions in the short-germ embryos of *Tribolium castaneum. Nature* 367:460–463.

Needham, J. 1942. Biochemistry and morphogenesis. Cambridge: Cambridge University Press.

Oellers, N., M. Dehio, and E. Knust. 1994. bHLH proteins encoded by the Enhancer of split complex of *Drosophila* negatively interfere with transcriptional activation mediated by proneural genes. *Mol. Gen. Genet.* 244:465–473.

Ohlmeyer, J. T., and D. Kalderon. 1998. Hedgehog stimulates maturation of Cubitus interruptus into a labile transcriptional activator. *Nature* 396:749–753.

Pankratz, M. J., and H. Jäckle. 1993. Blastoderm segmentation. In *The development of Drosophila melanogaster,* ed. M. Bate and A. Martinez-Arias, 467–516. Cold Spring Harbor, N.Y.: Cold Spring Harbor Laboratory Press.

Patel, N. H. 1994. The evolution of arthropod segmentation: insights from comparisons of gene expression patterns. *Development Suppl,* pp. 201–207.

Patel, N. H., E. E. Ball, and C. S. Goodman. 1992. Changing role of even-skipped during the evolution of insect pattern formation. *Nature* 357:339–342.

Patel, N. H., E. Martin-Blanco, K. G. Coleman, S. J. Poole, M. C. Ellis, T. B. Kornberg, and C. S. Goodman. 1989. Expression of engrailed proteins in arthropods, annelids, and chordates. *Cell* 58:955–968.

Reinitz J., D. Kosman C. E. Vanario-Alonso and D. H. Sharp. 1998. Stripe forming architecture of the gap gene system. *Dev. Genet.* 23:11–27.

Reinitz J., and D. H. Sharp. 1995. Mechanism of eve stripe formation. *Mech. Dev.* 49: 133–158.

Simpson, P. 1997. Notch signalling in development: on equivalence groups and asymmetric developmental potential. *Curr. Opin. Genet. Dev.* 7:537–542.

Singson, A., M. W. Leviten, A. G. Bang, X. H. Hua, and J. W. Posakony. 1994. Direct downstream targets of proneural activators in the imaginal disc include genes involved in lateral inhibitory signaling. *Genes Dev.* 8:2058–2071.

Skeath, J. B., and S. B. Carroll. 1991. Regulation of achaete-scute gene expression and sensory organ pattern formation in the *Drosophila* wing. *Genes Dev.* 5:984–995.

Slack, J. M. W. 1983. From egg to embryo: determinative events in early development. New York: Cambridge University Press.

Stone, D. M., M. Hynes, M. Armanini, T. A. Swanson, Q. Gu, R. L. Johnson, M. P. Scott, D. Pennica, A. Goddard, H. Phillips, M. Noll, J. E. Hooper, F. de Sauvage, and A. Rosenthal. 1996. The tumour-suppressor gene patched encodes a candidate receptor for Sonic hedgehog. *Nature* 384:129–134.

Strigini, M., and S. M. Cohen. 1997. A Hedgehog activity gradient contributes to AP axial patterning of the *Drosophila* wing. *Development* 124:4697–4705.

Strigini, M., and S. M. Cohen. 2000. Wingless gradient formation in the *Drosophila* wing. *Curr. Biol.* 10:293–300.

Thieffry, D., A. M. Huerta, E. Perez-Rueda, and J. Collado-Vides. 1998. From specific gene regulation to genomic networks: a global analysis of transcriptional regulation in *Escherichia coli. BioEssays* 20:433–440.

Thieffry, D., and D. Romero. 1999. The modularity of biological regulatory networks. *Biosystems* 50:49–59.

Thomas, R., D. Thieffry, and M. Kaufman. 1995. Dynamical behaviour of biological regulatory networks. 1. Biological role of feedback loops and practical use of the concept of the loop-characteristic state. *Bull. Math. Biol.* 57:247–276.

Turing, A. M. 1952. The chemical basis of morphogenesis. *Philos. Trans. R. Soc. B Biol. Sci.* 237:37–72.

van den Heuvel, M., and P. W. Ingham. 1996. smoothened encodes a receptor-like serpentine protein required for hedgehog signalling. *Nature* 382:547–551.

van de Wetering, M., R. Cavallo, D. Dooijes, M. van Beest, J. van Es, J. Loureiro, A. Ypma, D. Hursh, T. Jones, A. Bejsovec, M. Peifer, M. Mortin, and H. Clevers. 1997. Armadillo coactivates transcription driven by the product of the *Drosophila* segment polarity gene dTCF. *Cell* 88:789–799.

Van Doren, M., P. A. Powell, D. Pasternak, A. Singson, and J. W. Posakony. 1992. Spatial regulation of proneural gene activity: auto- and cross-activation of achaete is antagonized by extramacrochaetae. *Genes Dev.* 6:2592–2605.

Vincent, J. P., and P. A. Lawrence. 1994. *Drosophila* wingless sustains engrailed expression only in adjoining cells: evidence from mosaic embryos. *Cell* 77:909–915.

Vincent, J. P., and P. H. O'Farrell. 1992. The state of engrailed expression is not clonally transmitted during early *Drosophila* development. *Cell* 68:923–931.

von Dassow, G., E. Meir, E. M. Munro, and G. M. Odell. 2000. The segment polarity network is a robust developmental module. *Nature* 406:188–192.

von Dassow, G., and G. M. Odell GM. 2002. Design and constraints of the *Drosophila* segment polarity module: robust spatial patterning emerges from intertwined cell state switches. *J. Exp. Zool.* 294:179–215.

Von Ohlen, T., D. Lessing, R. Nusse, and J. E. Hooper. 1997. Hedgehog signaling regulates transcription through cubitus interruptus, a sequence-specific DNA binding protein. *Proc. Natl. Acad. Sci. U.S.A.* 94:2404–2409.

Waddington, C. H., and H. Kacser. 1957. The strategy of the genes; a discussion of some aspects of theoretical biology. London: Allen and Unwin.

Wang, Q. T., and R. A. Holmgren. 1999. The subcellular localization and activity of *Drosophila* cubitus interruptus are regulated at multiple levels. *Development* 126: 5097–5106.

Weiss, P. A. 1968. Dynamics of development: experiments and inferences: selected papers on developmental biology. New York: Academic Press.

Wesley, C. S. 1999. Notch and wingless regulate expression of cuticle patterning genes. *Mol. Cell. Biol.* 19:5743–5758.

Wesley, C. S., and L. Saez. 2000. Notch responds differently to Delta and Wingless in cultured *Drosophila* cells. *J. Biol. Chem.* 275:9099–9101.

Wyman, J., and S. J. Gill. 1990. Binding and linkage: functional chemistry of biological macromolecules. Mill Valley, Calif.: University Science Books.

Yoffe, K. B., A. S. Manoukian, E. L. Wilder, A. H. Brand, and N. Perrimon. 1995. Evidence for engrailed-independent wingless autoregulation in *Drosophila. Dev. Biol.* 170:636–650.

Zeng, W., K. A. Wharton, Jr., J. A. Mack, K. Wang, M. Gadbaw, K. Suyama, P. S. Klein, and M. P. Scott. 2000. naked cuticle encodes an inducible antagonist of Wnt signalling. *Nature* 403:789–795.

13 Basins of Attraction in Network Dynamics: A Conceptual Framework for Biomolecular Networks

ANDREW WUENSCHE

Introduction

Some of the outstanding questions in genetics, evolution, evolvability, and development, including notions of modularity, will involve unraveling and comprehending networks of interacting elements.

In differentiation and gene regulation, feedback makes the one-way signaling pathway paradigm inadequate. It has been superseded by a dynamical network approach. The dynamics manifests itself in two ways: first, by the changing pattern of activation *on* network elements (genes, or neurons in a neural network) and, second, by the dynamics *of* the network—how the network architecture itself changes and how its elements connect, driven by evolution, development, and learning.

Another issue is the quality of different network architectures in terms of connectivity, and also the updating logic. Of the possible types of network architectures and logical schemes, has nature selected some tiny subset? Are there characteristic biases which affect dynamics? Probably. This kind of ensemble approach to biological networks might complement the search for the detailed biomolecular interactions.

Recent studies have shown that network topologies in real cell signaling and metabolic networks share some universal features (Albert et al. 2000; Vogelstein et al. 2000; Jeong et al. 2001). The connectivity appears to be "scale free," where link frequency (the number of connections at each node) roughly follows a power law distribution. Most elements have few connections, and a few are highly connected. The "small world" average distance between pairs of elements depends on those few highly connect nodes, which if disconnected can break the network into separate components. Similar topologies have been found in many different contexts, both natural and artificial, ranging from

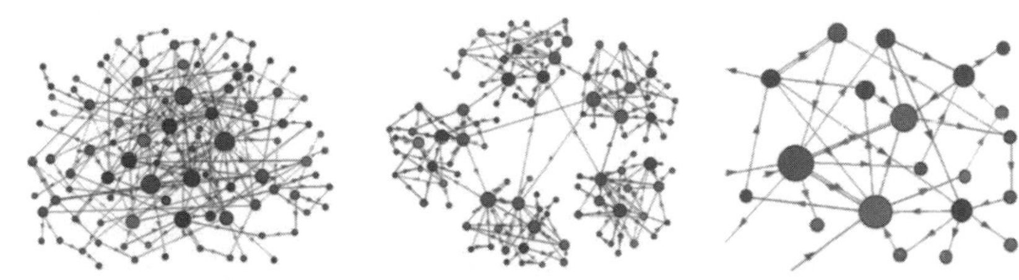

Fig. 13.1.—Hypothetical networks of interacting elements (size $n = 100$) with an approximate power law distribution of connections, both inputs (k) and outputs, which are represented by directed links (with arrows). Nodes are scaled according to k, and average $k \simeq 2.2$. *Left,* a fully connected network. *Center,* a network made up of five weakly interlinked $n = 20$ subnetworks or modules. *Right,* a detail of the top right subnetwork. These are examples of random Boolean networks defined in the section "Random Boolean Networks."

Fig. 13.2.—Histograms of link frequency (Y axis) against link size (X axis), for inputs + outputs, in the networks in figure 13.1. The fully connected network (*left*) and modular network (*right*) have a similar link frequency profile. However, their dynamics are very different, as described in "Attractors in Fully Connected and Modular Networks."

ecosystems to the World Wide Web. Are gene regulatory networks of this type? Or are they broken into modules: small, semi-independent subnetworks responsible for useful adaptations and functions which are conserved over long stretches of the evolutionary tree (Relaix and Buckingham 1999). Note that each module itself could also have a scale-free topology. This question goes to the heart of modularity and evolvability. Figure 13.1 illustrates these two types of hypothetical network. Although their link frequency profiles are similar (fig. 13.2), their dynamics turn out to be very different. Results described in "Attractors in Fully Connected and Modular Networks" below indicate that the modular network has more basins of attraction, which are relatively more stable, than the fully connected network.

Dynamical networks in biology are found wherever one cares to look, from the brain to ecology. In the context of modularity in evolution and development the networks in question are genetic regulatory networks in cell differentiation, protein networks in cell signaling, networks of cells in tissues and organs, and of organs in the body. These networks overlap, making supernetworks, and break down into subnetworks through many levels. The finer details of signaling pathways and network fragments are being discovered, both by painstaking experiments on *Drosophila* and other organisms (Wolpert 1998) and by the new microarray methods providing floods of data on the dynamics of gene expression patterns in development. There appears to be an urgent need for theoretical approximations and concepts to keep pace with the data (Solé et al. 2000).

Can methods from complex systems, network theory, and discrete dynamics contribute? Cellular automata have provided models of pattern information in biological organisms (Ball 1999). Boolean networks have provided models of genetic networks underlying cell differentiation based on basins of attraction (Kauffman 1969; Glass and Kauffman 1973; Somogyi and Sniegoski 1996; Wuensche 1998). Basins of attraction can be computed and portrayed (Wuensche and Lesser 1992; Wuensche 1994, 2001), revealing the global dynamics on small networks. Could this provide a conceptual framework for networks in biology?

Discrete dynamical networks, as abstract systems, manifest ubiquitous emergent properties which transcend any particular context, studied for their own sake, not just as models of something else. This chapter will outline some of these properties for cellular automata and for random Boolean networks, which are more general and probably have more biological relevance. The key emergent property is that the dynamics on the networks converge and thus fall into a number of "basins of attraction," which hierarchically categorize patterns of activation, or "state-space," creating memory as a function of the network architecture. High convergence implies order; low convergence implies disorder or chaos. The most interesting phenomena occur at the transition, sometimes called the "edge of chaos" (Langton 1990).

Basins of attraction represent the network's global dynamics and are analogous to Poincaré's "phase portrait," which provided powerful insights in continuous dynamics. For small systems, it is possible to compute and draw basins of attraction and measure their convergence and stability to perturbation. Analogous basins of attraction can be imagined in biological networks. That these networks are very simple might be a strength rather than a weakness when extrapolating the ideas to real systems, by the argument that if the simple network has particular emergent properties, a real biological network, infinitely more subtle and complex, should be capable of that *at least*. Also, a biologically faithful model might become so elaborate as to mask the underlying phenomena that one seeks to capture, which might be clear in a simple model.

Basins of Attraction

Figure 13.3 provides a summary of the idea of state-space and basins of attraction in idealized networks of interacting elements, some-

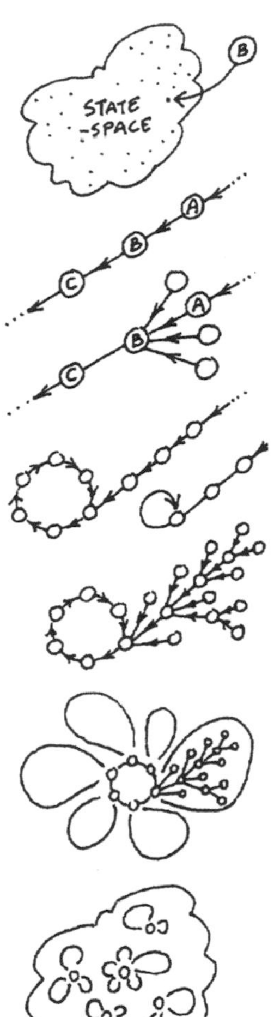

For a network of size *n*, an example of one of its states *B* might be 1010 . . . 0110. *State-space* is made up of all 2^n states, the space of all possible bit-strings or patterns.

Part of a *trajectory* in state-space, where *C* is a successor of *B*, and *A* is a *pre-image* of *B*, according to the dynamics on the network.

The state *B* may have other pre-images besides *A*; the total is the *in-degree*. The pre-image states may have their own pre-images or none. States without pre-images are known as *garden-of-Eden* states.

Any trajectory must sooner or later encounter a state that occurred previously—it has entered an attractor cycle. The trajectory leading to the attractor is a *transient*. The period of the attractor is the number of states in its cycle, which may be just one—in which case it is a point attractor.

Take a state on the attractor, find its pre-images (excluding the pre-image on the attractor). Now find the pre-images of each pre-image, and so on, until all garden-of-Eden states are reached. The graph of linked states is a *transient tree* rooted on the attractor state. Part of the transient tree is a subtree defined by its root.

Construct each transient tree (if any) from each attractor state. The complete graph is the *basin of attraction*. Some basins of attraction have no transient trees, just the bare "attractor."

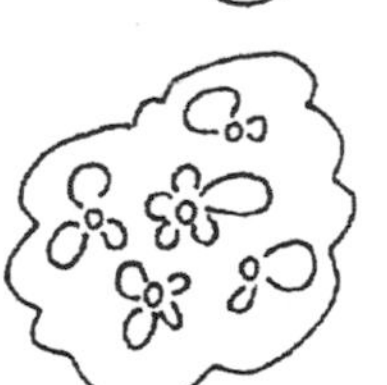

Now find every attractor cycle in state-space and construct its basin of attraction. This is the *basin of attraction field* containing all 2^n states in state-space but now linked according to the dynamics on the network. Each discrete dynamical network imposes a particular basin of attraction field on state-space.

Fig. 13.3.—State-space and basins of attraction.

times called "decision-making" networks. The dynamics depends on the connections and update logic of each element, which "decides" its next value on the basis of the values of the few elements that provide its inputs, which might include self-input. The result is a complex web of feedback making the dynamics difficult to treat analytically, despite the extreme simplicity of the underlying network. In fact, although the dynamics are deterministic, the future is in general unpredictable. Understanding these systems relies chiefly on computer simulation.

A more precise definition of the network architecture is given in later sections. For the moment we will define just some essential concepts and properties, and also some terminology. At a given moment, each element in the network has a value, for simplicity 0 or 1 in a binary network, though the same arguments apply to a multistate network. The pattern of 0s and 1s across the whole network is the network's configuration, pattern of activation, or "state" at that moment in time, which can be represented as a bit-string. Time proceeds in discrete steps, "time-steps." At each time-step, all signals transmitted by network links are processed simultaneously, that is, synchronously, in parallel. This transforms a state at time-step t to another state at time-step $t + 1$, then another state at $t + 2$, and so on; the system is iterated. This process is deterministic: there is just one successor to each state, and the iteration continues indefinitely, while the network architecture itself (its wiring scheme and rule scheme) stays the same. That is, the source of the inputs (the "wiring") and the logical function (the "rule") that each element executes on its inputs to update its value do not change. The sequence of states is called a "trajectory." A "space-time" pattern representation of the dynamics is made by placing one-dimensional bit-strings, representing successive states, one below the other (time proceeds down), as in figure 13.4, where 0s and 1s are shown as white and black dots. This demonstrates how the pattern evolves from an initial state. It might stabilize, become periodic, become chaotic, or show some interesting pattern formation.

Because the network is finite, we can define a "state-space" as the space of all possible bit-strings or states. There are 2^n unique states for a binary network of size n. We can say that a trajectory started from some initial state moves through its state-space. Because the state-space is finite, sooner or later the trajectory must encounter a state that occurred before. When this happens, because the system is deterministic, the trajectory must become trapped in a perpetual cycle of repeating states—a state cycle, or "attractor." The number of time-steps between the repeats of a state is the attractor period, which could be just one, in the case of a "point attractor," where the space-time pattern has completely stabilized. Conversely, a chaotic space-time pattern might have a very large attractor period.

Each state has just one successor, but what about a state's immediate predecessors? It turns out that a state can have any number of these, called "pre-images," including none. The number of pre-images a state has is its "in-degree." The existence of in-degrees other than exactly one and the existence of states outside the attractor are conditions that must imply each other; otherwise, the basin would be just a bare attractor cycle. States outside an attractor lie on trajectories that flow to the attractor, known as "transients." A state with an in-degree of zero is known as a "garden-of-Eden" state, and except for highly chaotic

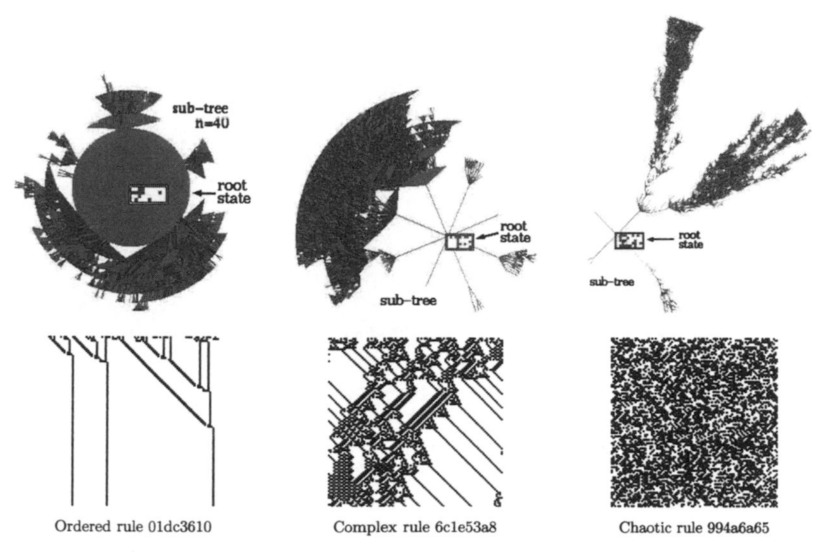

Fig. 13.4.—Ordered, complex and chaotic dynamics of one-dimensional cellular automata are illustrated by the space-time patterns and subtrees of three typical $k = 5$ rules (shown in hex; see "Random Boolean Networks"). The bottom row shows the space-time patterns from the same random initial state. The bit-strings $n = 100$ of successive time-steps (represented by white and black dots) are shown horizontally one below the other; that is, time proceeds down. Above each space-time pattern is a typical subtree for the same rule. In this case $n = 40$ for the ordered rule, and $n = 50$ for the complex and chaotic rules. The root states were reached by first iterating the system forward by a few steps from a random initial state and then tracing the subtree backward. Note that the convergence in the subtrees, their branchiness or typical in-degree, relates to order-chaos in space-time patterns, where order has high and chaos low convergence.

systems, most states—almost all, for large networks—turn out to be garden-of-Eden states.

Some time ago I invented algorithms for finding the pre-images of a state directly, without having to exhaustively test the entire state-space (Wuensche and Lesser 1992; Wuensche 1994). This allowed the efficient backward tracing and reconstruction of the branching transients that flow into attractors, the "transient tree," where the "leaves" are garden-of-Eden states. Conversely, the flow toward attractors is convergent, like a river. High convergence implies order in space-time patterns: short, bushy, highly branching transient trees, with many leaves and small attractor periods. Conversely, low convergence implies chaos in space-time patterns: long, sparsely branching transient trees, with few leaves and long attractor periods. Figure 13.4 illustrates this, showing a transient subtree for three representative cellular automata rules, for order, complexity and chaos, and their corresponding space-time patterns. A simple measure of convergence, taken on a basin of attraction field, a single basin, a subtree, or just part of a subtree for larger

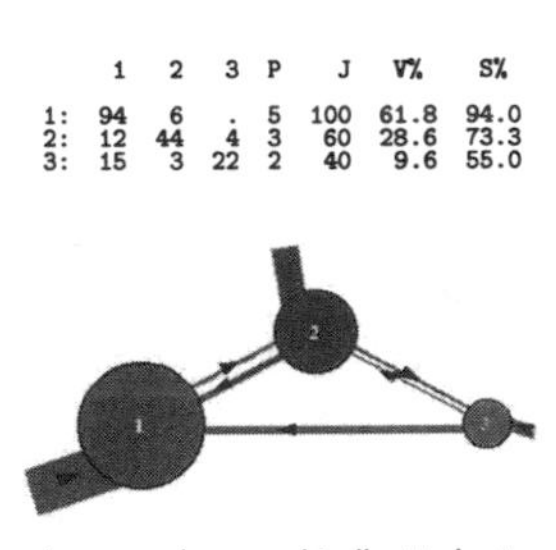

	1	2	3	P	J	V%	S%
1:	94	6	.	5	100	61.8	94.0
2:	12	44	4	3	60	28.6	73.3
3:	15	3	22	2	40	9.6	55.0

Fig. 13.5.—The basin of attraction field of the $n = 20$ subnetwork shown in detail in figure 13.1 (*right*). The rules (input logic) were assigned at random. State-space (size $2^{20} \simeq 1.05$ million) is partitioned into three basins of attraction. The attractor states are shown as 5×4 bit patterns. The table and diagram on the right show the probability of jumping between basins due to one-bit perturbations of their attractor states. P, attractor period; J, possible jumps ($P \times n$); V%, the basin "volume" as a percentage of state-space; S%, the percentage of self-jumps for each basin. For example, in basin 1, $P = 5$, $J = 100$ possible jumps, with 6 of these jumps to basin 2, none to basin 3, and 94 back to itself. All three basins are relatively stable because $S > V$. The diagram below the table, the "metagraph" (see "Jumping between Basins Due to Attractor Perturbations"), shows the same data graphically. Node size reflects basin volume; link thickness, percentage jumps; arrows, the direction; and the short stubs, self-jumps. The fraction of garden-of-Eden states in all three basins is 0.999+, indicating high convergence and order.

systems, is the density of garden-of-Eden states, and its rate of increase with n. A more comprehensive measure is the in-degree frequency (Wuensche 1997, 1999).

The set of transient trees rooted on an attractor cycle is a "basin of attraction." The dynamics of the network connects state-space into a number of distinct basins, the "basin of attraction field," representing the system's global dynamics, as in figures 13.5, 13.6, and 13.7. State-space has now been partitioned, or categorized, by the dynamics of the network into a number of separate basins of attraction. In addition, the root of each subtree within a transient tree forms a subcategory. The precise way that states are linked into a basin of attraction field depends on the details of the network architecture.

Subtrees and basins of attraction are portrayed as state transition graphs, where vertices or nodes representing states are connected by directed edges. The direction of edges (i.e., time) flows inward from garden-of-Eden states to the attractor and then clockwise around the attractor cycle, as indicated in figures 13.6 and 13.7. According to the graphic convention (Wuensche and Lesser 1992; Wuensche 2001), the length of the edges decreases with distance from the attractor, and the diameter of the attractor cycle approaches an upper limit with increasing period.

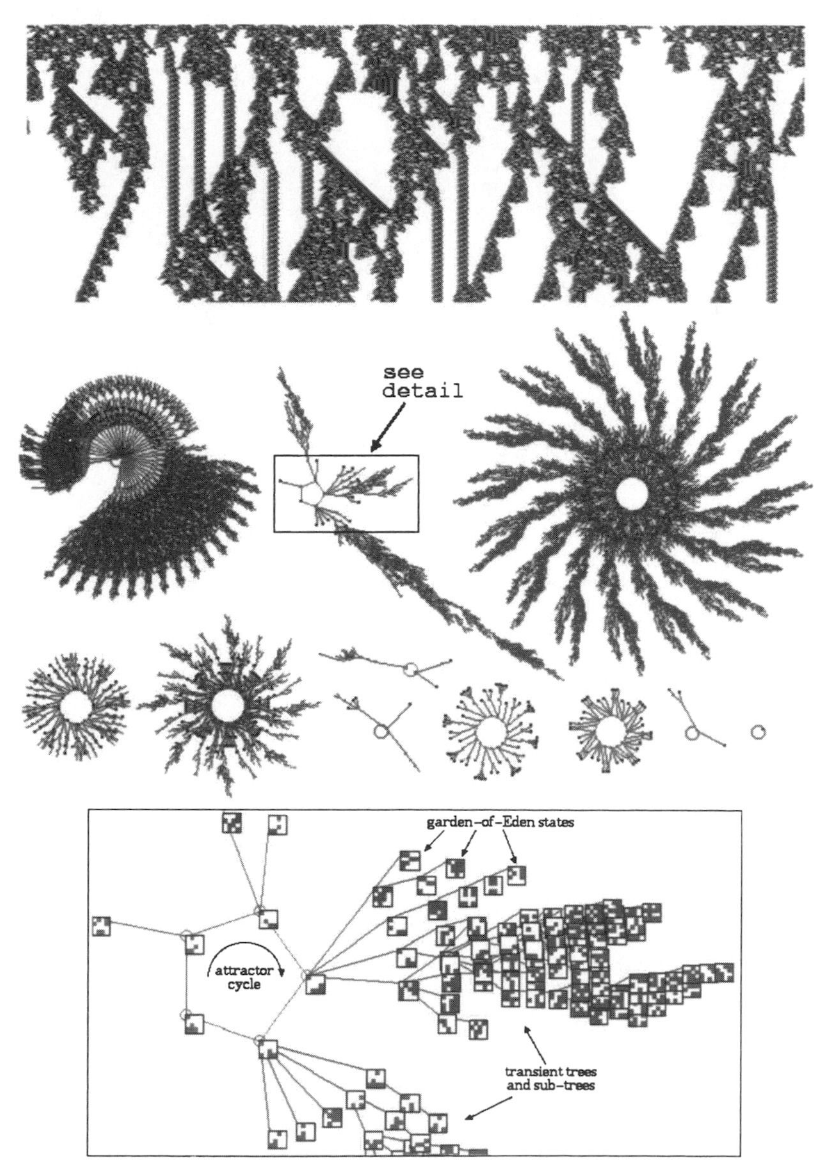

Fig. 13.6.—*Top,* the space-time pattern of a one-dimensional complex cellular automaton with interacting gliders (Wuensche 1999), $n = 700$, $k = 7$, showing 308 time-steps from a random initial state. *Center,* the basin of attraction field for the same rule, $n = 16$. The 2^{16} states in state-space are connected into 89 basins of attraction, but only the 11 nonequivalent basins are shown, with symmetries characteristic of cellular automata. *Bottom,* a detail of the second basin in the basin of attraction field, where states are shown as 4×4 bit patterns.

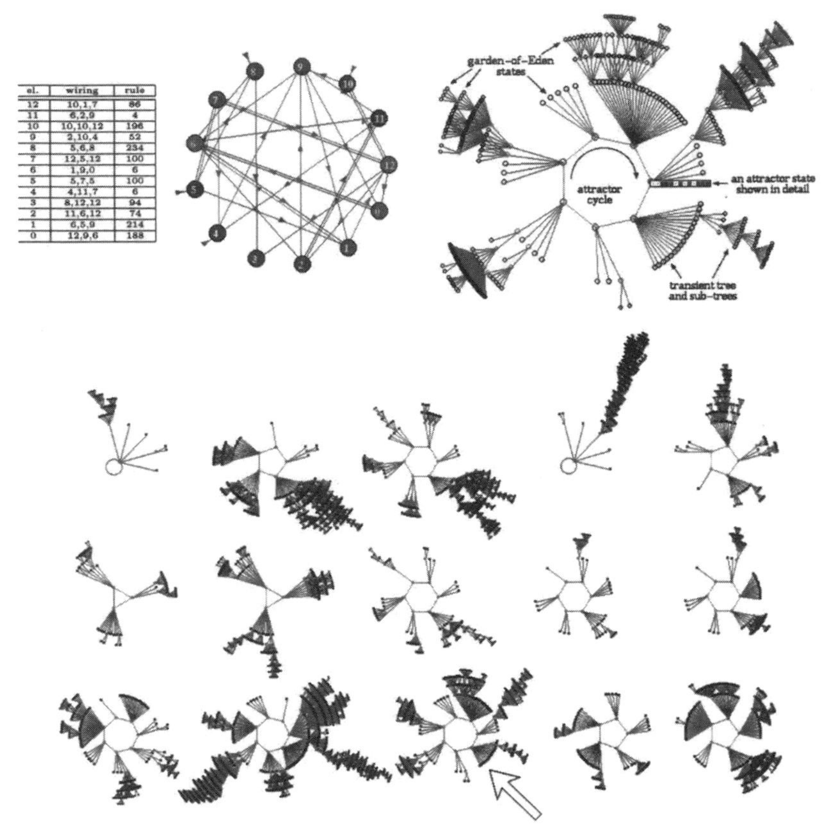

el.	wiring	rule
12	10,1,7	86
11	6,2,9	4
10	10,10,12	196
9	2,10,4	52
8	5,6,8	234
7	12,5,12	100
6	1,9,0	6
5	5,7,5	100
4	4,11,7	6
3	8,12,12	94
2	11,6,12	74
1	6,5,9	214
0	12,9,6	188

Fig. 13.7.—A small random Boolean network, $n = 13$, with homogeneous $k = 3$ wiring, though some elements have more than one input from the same element. *Top left,* the wiring/rule scheme, which fully defines the network. *Top center,* the network shown as a directed graph with numbered nodes. Self-links are short arrows sticking into nodes. *Top right,* one of the basins of attraction. The basin links 604 states, of which 523 are garden-of-Eden states. The attractor period is 7, and one of the attractor states is shown in detail as a bit pattern. The direction of time is inward from garden-of-Eden states to the attractor, then clockwise. *Bottom,* The basin of attraction field. The $2^{13} = 8{,}192$ states in state-space are organized into 15 basins, with attractor periods ranging between 1 and 7, and basin volume between 68 and 2,724. The arrow points to the basin shown in more detail.

Cellular Automata

Cellular automata are networks whose elements form a regular lattice, possibly in one, two, or three dimensions, with inputs from nearest neighbors (and possibly next nearest, etc.) and homogeneous logic. In fact, the lattice is a consequence of the input connections. Cellular automata can be seen as a special case, a subset, of the more general system, random Boolean networks, and there are a variety of hybrid systems between the two. Cellular automata have provided models in computational and physical systems, in biological systems, such as pattern formation (for example, stripes on mammalian coats and patterns on nautilus shells; Ball 1999), and in ecology (for example, modeling forest fires). Large-scale surface pattern can emerge from just local interactions, by reaction-diffusion, as in figure 13.8 (*left*) and other mechanisms (Wolfram 1986).

Another type of pattern formation which provides a striking example of self-organization is the emergence of coherent interacting structures, sometimes known as "particles" or "gliders" (Conway 1982), as in figures 13.9 and 13.6 (*top*). Only a small fraction of cellular automaton rules generate gliders. They are classified as complex, in contrast to ordered or chaotic (Wolfram 1984). Gliders can be embedded within a uniform or periodic background and propagate at various velocities up to the system's "speed of light," set by the neighborhood diameter. Colliding gliders can self-destruct, pass through each other, or produce new gliders. Compound gliders may emerge made up of subgliders recolliding periodically, which can combine into yet higher-order structures, and the process could unfold without limit in large enough systems. Once gliders have emerged, cellular automata dynamics can be described at a higher level, by glider collision rules as opposed to the underlying cellular automata rules (Wuensche 1999).

Figure 13.8 shows examples of emergent patterns in two- and three-dimensional cellular automata. Figure 13.6 (*top*) shows a large space-time pattern of a one-dimensional complex cellular automaton, and also the basin of attraction field of the same rule for a small system. Note that the basin symmetries can be explained by the regularity of cellular automata architecture (Wuensche and Lesser 1992). In random Boolean networks these symmetries, and also gliders, are absent.

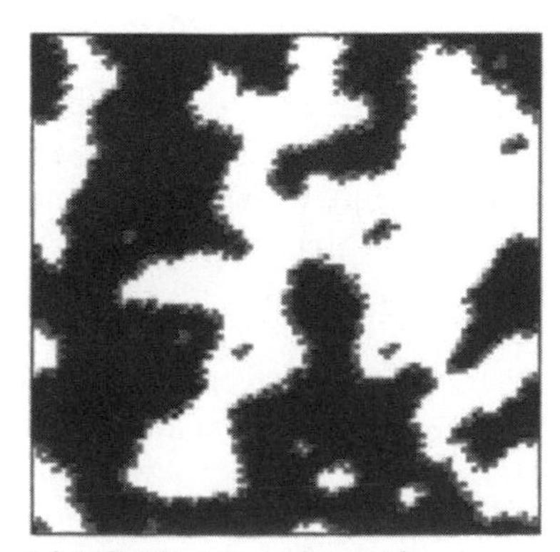

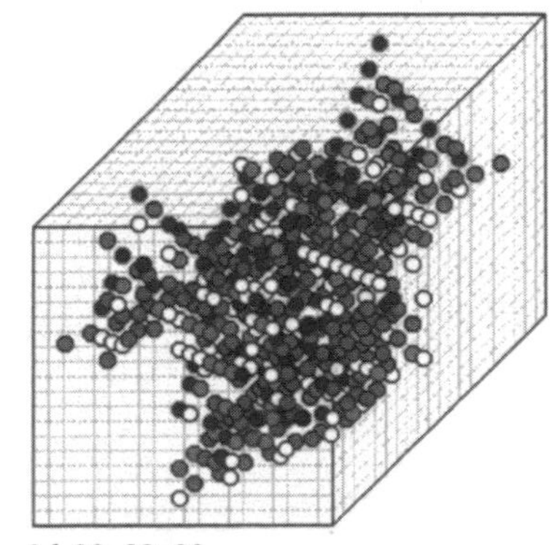

Fig. 13.8.—Examples of emergent patterns in two-dimensional and three-dimensional cellular automata. *Left,* an evolved time-step of a two-dimensional cellular automaton on a $k = 7$ triangular lattice with a reaction-diffusion rule. *Right,* a time-step of a three-dimensional nearest-neighbor ($k = 7$) cellular automaton with a randomly selected rule, starting from a single central 1. View this as if looking up into a transparent box.

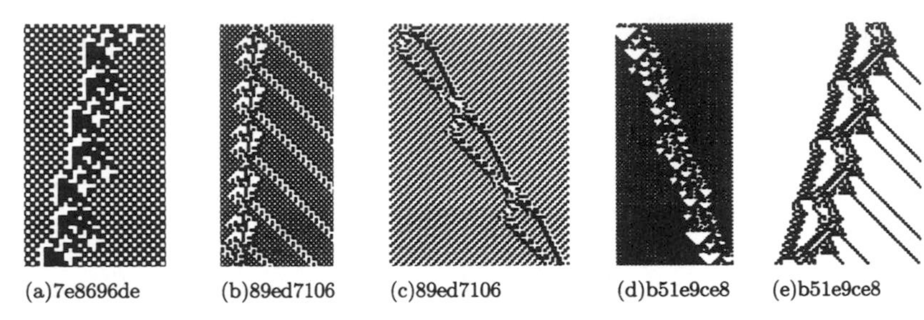

Fig. 13.9.—Gliders, "glider guns" (which generate subgliders), and compound gliders in $k = 5$ one-dimensional cellular automata. *c* is a compound glider made up of two independent gliders locked into a cycle of repeating collisions. *d* is a glider with a period of 106 time-steps. *e* is a compound glider-gun.

Glider dynamics is said to occur at a phase transition in rule-space between order and chaos, relative to static rule parameters (Langton 1990; Wuensche 1999). Measures on space-time patterns allow rule-space to be classified automatically, and complex rules to be identified (Wuensche 1999). There is a sense here of modularity. In cases where a biocellular or biomolecular substrate approximates a regular geometric lattice with local interactions, cellular automata might provide useful models.

Random Boolean Networks

A random Boolean network is a more general system than a cellular automaton. The connections, or inputs, are nonlocal, and both inputs and logic can be different for each network element. Such a network seems more biological than a cellular automaton, for example, in relation to neural networks, or genetic regulatory networks. The nonlocality and heterogeneity prevent the sort of pattern formation possible with cellular automata, though one is of course free to introduce any measure of locality and homogeneity if appropriate, like taking inputs randomly from a local patch, or restricting, and thus biasing, the range of logic.

The network consists of a set of elements, which could be model genes or neurons, taking inputs (0 or 1) from each other and changing their value (0 or 1) according to some logical function on these inputs (the rule). The connectivity is usually sparse. The number of network elements, n, is usually much bigger that the typical number of inputs to each element k. The updating is synchronous, in discrete time-steps. The pattern of 0s and 1s across the network is a global "state," changing over discrete time. The network architecture itself does not change. Although there are various interesting possibilities for elaborating this model, such as having a larger range of values (or alphabet) than just 0 or 1, and various schemes for asynchronous updating, this discussion stays with the simple model.

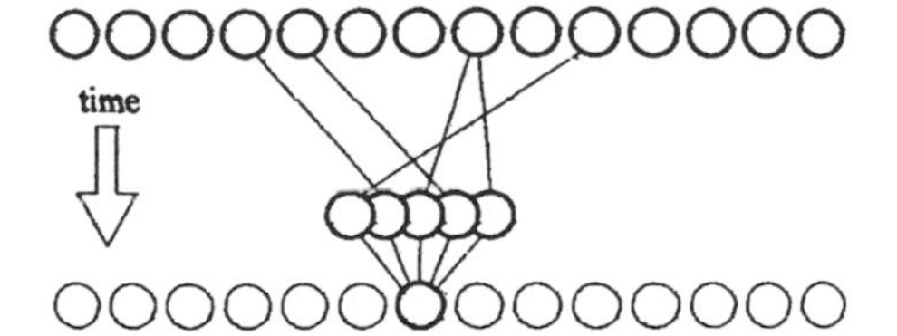

Fig. 13.10.—Random Boolean network architecture. Each element in the network synchronously updates its value according to the values in a pseudo-neighborhood, set by inputs (wiring) from anywhere in the network. Each network element may have a different number of input wires *k*, different wiring, and a different rule. The system is iterated.

Figure 13.10 shows one element's inputs "wired" between two consecutive time-steps, where the wires connect to a "pseudo-neighborhood," equivalent to a cellular automaton's real neighborhood. The number of possible input patterns on a pseudo-neighborhood of size k is 2^k. The most general expression of the Boolean function or rule is a look-up table (the rule-table) with 2^k entries, giving 2^{2^k} possible rules. Subcategories of rules can also be expressed as simple algorithms, concise logical statements, totalistic rules (Wolfram 1984), or threshold functions.

By convention (Wolfram 1984) the rule-table, as in cellular automata, is arranged in descending order of the values of neighborhoods as binary numbers, and the resulting bit-string converts to a decimal or hexadecimal (hex) rule number. For example, the $k = 3$ neighborhoods are set out below in the conventional order, and the bit-string formed by the output of each neighborhood is the rule-table, or rule, in this case rule 30 in decimal, or rule "1e" in hex.

7	6	5	4	3	2	1	0	. . . neighborhoods, decimal
111	110	101	100	011	010	001	000	. . . all possible neighborhoods, binary
0	0	0	1	1	1	1	0	. . . outputs, the binary rule-table, equals 30 in decimal

The possible alternative network architectures, and thus the behavior space of a random Boolean network, is very large, taking into account all possible wiring schemes and rule schemes, but there are also equivalence classes which reduce the number somewhat. In general, the number of effectively different random Boolean networks of size n, and thus basin of attraction fields, cannot exceed $(2^n)^{(2^n)}$.

The examples in figure 13.7 illustrate a small random Boolean network, its network architecture, its basin of attraction field, and the stability of the attractors to perturbation. Note that because of the heterogeneity of wiring and rules, the basin topology lacks the symmetries characteristic of cellular automata.

Jumping between Basins Due to Attractor Perturbations

In a basin of attraction field, perturbations to network states will reset the dynamics, which may then jump to another basin, or to a different position in the same basin. Stability requires a high probability of returning to the same basin, whereas adaptability or differentiation requires appropriate jumps to other basins in response to specific signals. Perturbations or external signals are most likely to affect attractor states, because that is where the dynamics spends the most time.

Taking single bit-flips, or perturbations, to attractor states as the simplest case, the metagraph of the basin of attraction field represents the probabilities of jumping between basins and gives some insight into the stability and mobility between the basins of attraction. Figure 13.11 shows the metagraph for the basin of attraction field in figure 13.7. Link thickness represents jump probability, and self-jumps are shown

Fig. 13.11.—The metagraph of the basin of attraction field in figure 13.7, showing the probability of jumping between basins due to single bit-flips to attractor states. Nodes representing basins (shown inside each node) are scaled according to the number of states in the basin (basin volume). Links are scaled according to both basin volume and the jump probability. Arrows indicate the direction of jumps. Short stubs are self-jumps.

as short stubs. The largest basin receives the most jumps and sends jumps to other basins as well as back to itself. Although the network was set up at random, the proportion of self-jumps indicates an unexpected degree of stability.

Stability of Basins Due to Mutations in the Network Architecture

Very fine mutations can be made to the wiring/rule scheme of a random Boolean network, such as moving one wiring input or flipping one bit in the rule of just one element. In general, the basin of attraction field is stable to such small mutations, which result in slight changes to the overall topology. This is especially true for larger networks, where single mutations become relatively less significant. However, even a single mutation can sometimes have a drastic effect, such as breaking an attractor cycle. The consequences of moving a connection wire are usually greater than a one-bit mutation in a rule (Wuensche 1994).

In the software DDLab (Wuensche 2001) there are "learning" algorithms that allow pre-images to be a automatically attached to (learned) or detached from (forgotten) a selected target state by mutations, moving wires, or flipping bits in rules (Wuensche 1994). The aim is to "sculpt" a basin of attraction field to correspond to some desirable structure. The method is useful for fine-tuning a network but not for building a network from scratch, because of side effects—unplanned changes in other parts of the basin of attraction field.

One way to see the effect of network mutations is to highlight a set of states and track them as the basin of attraction field changes for successive mutations. Figure 13.12 shows one such mutation for a very small network so that the pattern at each node can be seen. Larger networks are affected in analogous ways.

Figure 13.13 shows the effects of all possible one-bit mutations to a cellular automaton rule, where all states fall into one very regular basin. As a cellular automaton has just one rule, effectively, the same bit in each element's rule was mutated, so these mutations are more drastic than in figure 13.12. If the rule is seen as the genotype and the basin of attraction as the phenotype, the relationship of the mutants to the original is striking. This is not atypical of mutations of cellular automata (Wuensche and Lesser 1992). The phenotype is more sensitive to the mutation of some bits than others.

Order-Chaos Measures for Large Random Boolean Networks

In contrast to cellular automata, glider dynamics and coherent pattern formation in general cannot occur in random Boolean networks be-

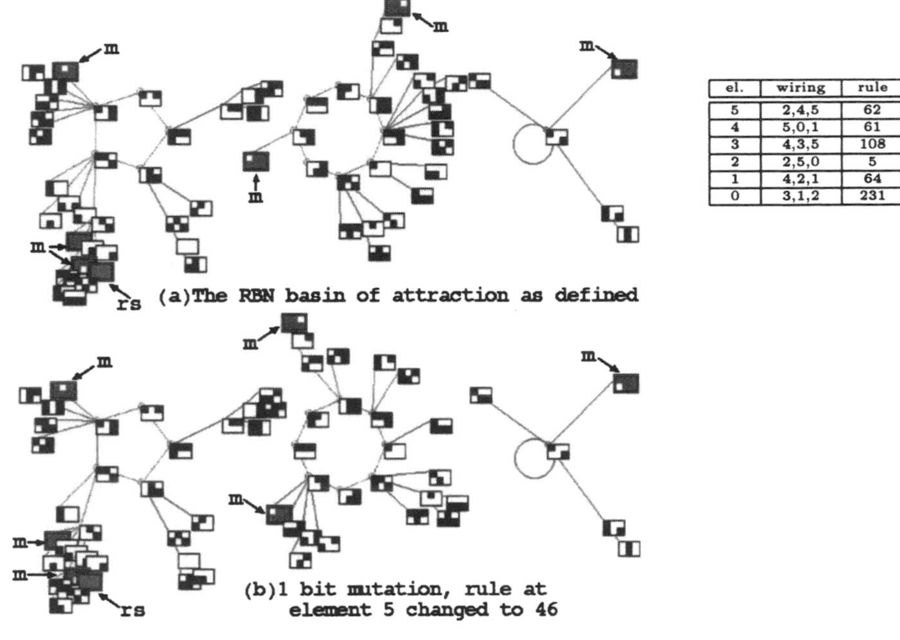

el.	wiring	rule
5	2,4,5	62
4	5,0,1	61
3	4,3,5	108
2	2,5,0	5
1	4,2,1	64
0	3,1,2	231

Fig. 13.12.—The basin of attraction field of (*a*) the random Boolean network ($n = 6$, $k = 3$) as defined in the table above, and (*b*) the network following a one-bit mutation to one of its rules. The attractors are stable to this mutation, though some differences in subtrees are evident. The all-1s reference state (rs) and the six states that differ from it by one bit (m) are indicated in the two fields.

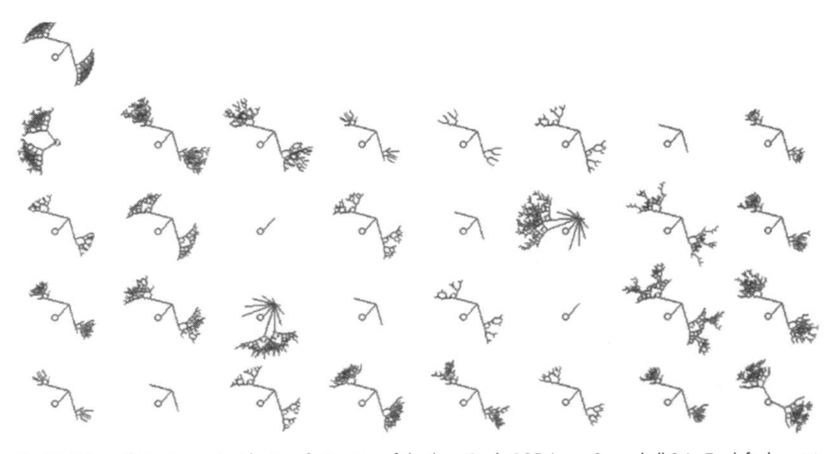

Fig. 13.13.—Thirty-two mutant basins of attraction of the $k = 3$ rule 195 ($n = 8$, seed all 0s). *Top left,* the original rule, where all states fall into just one very regular basin. The rule was first transformed to its equivalent $k=5$ rule (f00ff00f in hex), with 32 bits in its rule-table. All 32 one-bit mutant basins are shown. If the rule is the genotype, the basin of attraction can be seen as the phenotype.

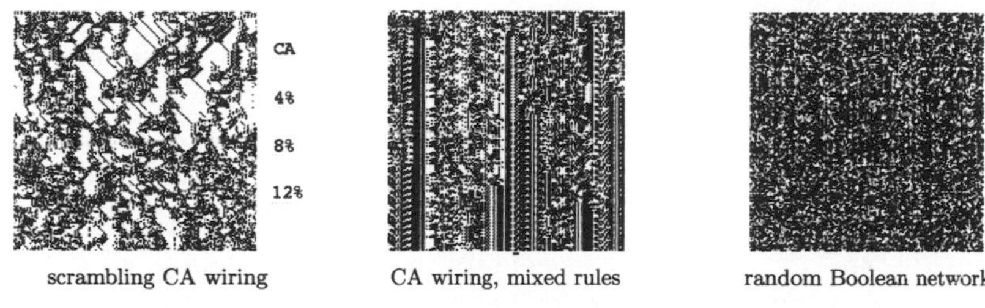

Fig. 13.14.—Space-time patterns for intermediate one-dimensional architecture, from cellular automata (CA) to random Boolean networks. $n = 150$, $k = 5$, 150 time-steps from a random initial state. *Left,* starting off as a complex cellular automaton (as in fig. 13.4), 4% (30 of 750) of available wires are randomized at 30-time-step intervals. The coherent pattern is progressively degraded. *Center,* a network with cellular automata wiring but mixed rules. Vertical features are evident. *Right,* a random Boolean network, with no bias, shows extremely chaotic dynamics.

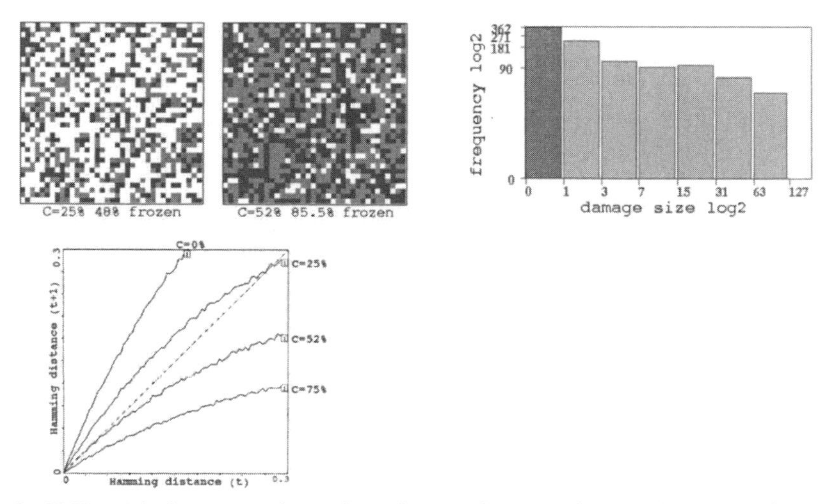

Fig. 13.15.—Order-chaos measures for a random Boolean network 36×36, $k = 5$. C, the percentage of canalizing inputs in the randomly biased network. *Top left,* frozen elements (that have stabilized for 20 time-steps) are shown in gray, for $C = 25\%$ and 52%. *Top right,* the log-log "damage spread" histogram for $C = 52\%$, with sample size about 1,000. *Left,* the Derrida plot for $C = 0\%$, 25%, 52%, and 75%.

cause of their irregular architecture. Figure 13.14 (*left*) shows glider dynamics degrading as local wiring is progressively scrambled.

One order-chaos measure for random Boolean network space-time patterns is the balance between "frozen," stabilized regions and changing regions (Kauffman 1993), as in figure 13.15 (*top left*). Stable regions are characteristic of networks with low connectivity, $k \leq 3$, because rules which induce stability are relatively frequent in these rule-spaces. To induce stability (i.e. order) for $k \geq 4$, where chaotic rules become overwhelmingly predominant, biases on rules must be imposed. One way is to skew the proportion of 0s and 1s in a rule's look-up table away from an even distribution. Another is to set a high proportion of "canalizing" inputs, *C*. Although these biases overlap to a certain extent, a strong case has been made (Harris et al. 2002) that the latter type of bias is the most significant in control rules governing the activity of eukaryotic genes.

In a rule's look-up table, an input wire is said to be canalizing if a particular input value (0 or 1) determines the output irrespective of the other inputs. A rule's degree of canalization can range from 0 to k (for the same output). For the network as a whole, the percentage of all inputs that are canalizing, *C*, can be randomly set. A random Boolean network's order-chaos characteristics for varying *C*, applied especially for large networks, are captured by the measures illustrated in figure 13.15 and described below (Derrida and Stauffer 1986; Harris et al. 2002; Wuensche 2001).

The "Derrida plot," analogous to the Liapunov exponent in continuous dynamics, measures the divergence of trajectories based on normalized Hamming distance H_t, the fraction of bits that differ between two patterns. Pairs of random states separated by H_t are independently iterated forward by one time-step. For a sample of random pairs, the average H_{t+1} is plotted against H_t, and the plot is repeated for increasing H_t. A curve above the main diagonal indicates divergent trajectories and chaos; a curve below, convergence and order. A curve tangential to the main diagonal indicates balanced dynamics.

A related measure is the distribution of "damage" resulting from a single bit-flip at a random position in a random state, for a sample of random states. The bit-flip might cause an avalanche of difference, or damage, relative to the original space-time pattern. The size of the resulting damage is measured once it has stabilized. A histogram of damage size against the frequency of sizes is plotted. Its shape indicates the degree of order or chaos in the network, where a balance between order and chaos approximates a power law distribution. Results according to these measures for $k = 5$ indicate a balance at $C \approx 52\%$.

These and other methods are applied in the context of random Boolean network models of genetic regulatory networks. The conjecture is that evolution maintains genetic regulatory networks marginally on the ordered side of the order-chaos boundary to achieve stability and adaptability in the pattern of gene expression which defines the cell type (Harris et al. 2002).

Attractors in Fully Connected and Modular Networks

For large networks too big to generate the basin of attraction field, information on the field, the attractors, and basins of attraction can still be found by statistical methods (Walker and Ashby 1966), though for excessively chaotic networks, transients and attractor periods can be

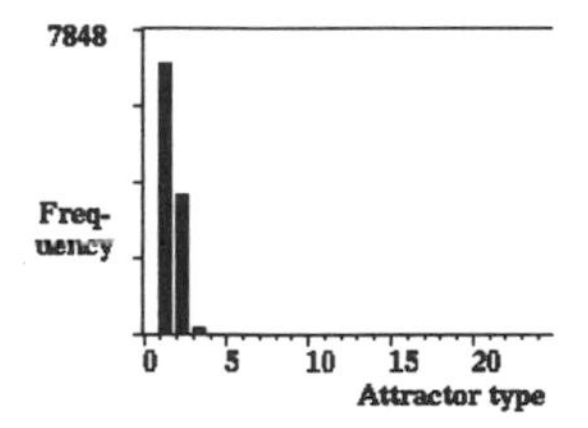

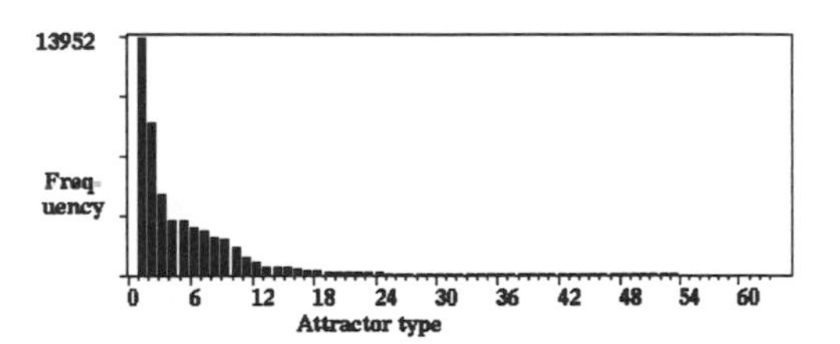

Fig. 13.16.—Attractor frequency histograms, showing the frequency (Y axis; scale indicated at top left) of falling into different attractors (X axis, sorted by basin volume) from a large sample of random initial states. These examples are for the "scale-free" networks in figure 13.1, where $n = 100$. The frequency of each attractor is a statistical measure of basin volume, the fraction of state-space occupied by each basin. *Left,* the fully connected network. The number of attractor types stabilized at just three after 10,000+ runs. Their periods are 2,046, 553, and 380, with average transient length 605, 673, and 97. *Right,* the modular network. The number of attractor types stabilized at 53 after 50,000+ runs, though about two-thirds of these represent very small basins. The three most frequent basins have periods 30, 14, and 2, with average transient length 54, 46, and 47.

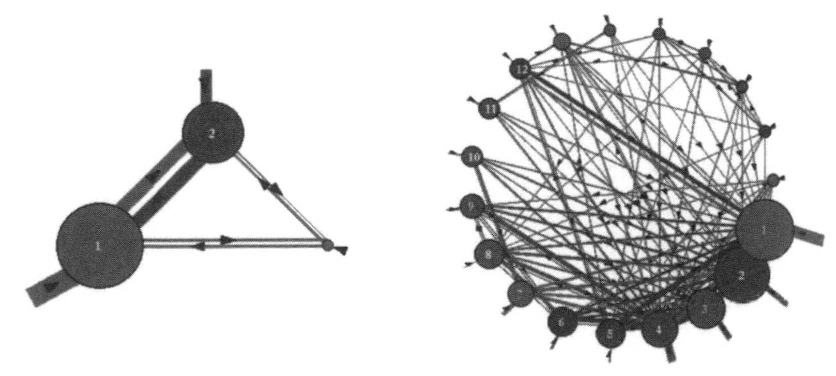

Fig. 13.17.—The metagraphs of the attractor frequency histograms in figure 13.16, showing the probability of jumping between basins due to single bit-flips to attractor states. Nodes representing basins are scaled according to basin volume. Links are scaled according to both basin volume and the jump probability. Arrows indicate the direction of jumps. Short stubs are self-jumps. *Left,* the fully connected network. The percentages of self-jumps in the three basins are 60%, 36%, and 6%. *Right,* the modular network: (for the 19 largest basins only, out of 53). The percentage of self-jumps is 41%, 20%, and 31% for the three largest basins. Eleven of the smallest basins (not shown) are unreachable.

too long for the method to be practical. The network is run forward from many random initial states looking for state repeats to identify the different attractors, noting the periods and attractor states. Although this does not reveal subtree topology, the frequency of falling into different attractors is a statistical measure of basin volume, the fraction of state-space occupied by each basin, so all but the smallest basins are likely to be found in a large sample of initial states.

Figure 13.16 shows the attractor frequency histograms of the scale-free networks in figure 13.1, one fully connected, and the other modular, but with similar link frequency and random rules. Figure 13.17 shows the metagraphs of the attractor frequency histograms, showing the probability of jumping between basins due to single bit-flips to attractor states. The metagraph for the modular network shows the 19 largest basins only.

The attractor frequency histograms and data indicate that the modular network has markedly more basins, with smaller attractor periods and shorter transients, than the fully connected network. Other experiments on similar networks confirm the same tendency. The jump probabilities indicate that basins in the modular network are more stable than in the fully connected network, because the fraction of self-jumps for a typical basin is significantly greater than the basin volume. This is not the case for the fully connected network, where self-jumps for the two biggest basins correlate closely with basin volume.

These results point to significant differences between the basin of attraction fields of modular and fully connected network architectures, where both are scale-free. Breaking a network into weakly linked modules increases both the number and stability of basins. Conversely, adding more links between the modules reduces both the number and stability of basins. The modules in the modular network are behaving like discrete coupled oscillators, perturbing each other between their alternative subattractors (see fig. 13.5). This is where an explanation of the differences might be found.

Typical transient length and attractor period usually give some indication of the order-chaos characteristics of a network. The smaller attractor periods, and shorter transients, found in the modular network would usually be a sign of a greater degree of "order." However, the order-chaos measures in "Order-Chaos Measures for Large Random Boolean Networks" do not show a significant difference. These are preliminary results, based on a limited sample. Further research is needed to better understand the behavior of these types of scale-free, modular, random Boolean networks.

Genetic Regulatory Networks

The cells of multicellular organisms differentiate within the developing embryo into the various cell types that form tissues by a process that is regulated at the molecular level by DNA sequences, encoding genes that produce proteins that regulate other genes. All eukaryotic cells in an organism carry essentially the same set of genes, some of which are expressed while others are not. A cell type depends on the particular subset of active genes, where the gene expression pattern needs to be stable but also adaptable. What, then, is the mechanism that maintains the many alternative patterns of gene expression of the various cell types making up a multicellular organism?

Genes regulate each other's activity by coding for transcription factors, which may enhance or repress the expression of other genes by binding, often in combination, at particular sites. A given gene may directly regulate just a small set of other genes; those genes regulate other genes in turn, so a gene may indirectly influence the activity of many genes downstream. Conversely, a gene is indirectly influenced by many genes upstream. A gene may directly or indirectly contribute to regulating itself. The result is a genetic regulatory network, a complex feedback web of genes turning each other on and off.

This has been modeled as an idealized dynamical system, Kauffman's random Boolean network, where model genes are connected by directed links (transcription factors), updating their (on-off) state in parallel, according to the combinatorial logic of their inputs (Kauffman 1969, 1993; Somogyi and Sniegoski 1996). Depending on the network's current state, and biases in the connections and logic, the dynamics on such a network, as has been shown, can settle onto a number of attractors, which are interpreted in this context as cell types. This provides a mechanism for explaining the apparent paradox that the same genome can create and maintain a variety of cell types and subtypes, such as skin, muscle, liver, and brain. In stem cells, trajectories leading to attractors have been interpreted as pathways of differentiation because small perturbations can nudge the dynamics toward different attractors.

The model has been criticized for being too simple in that the parallel (synchronous) updating and the on-off characterization of genes are implausible idealizations when applied to real genetic networks, given that transcription is asynchronous and driven at different rates. However, gene activity at the molecular scale consists of discrete events occurring concurrently. Variable protein concentrations can be accounted for by genes being on for some fraction of a given time span. More elaborate models have been proposed but are harder to interpret. The random Boolean network idealization is arguably the simplest possible abstract model that is computationally tractable, well understood in fields outside biology, yet still reflecting the essential qualities of the dynamics of genetic networks.

In a cell type's gene expression pattern over a span of time (i.e., its space-time pattern), a particular gene may, broadly speaking, be on, off, or changing. If a large proportion of the genes are changing (chaotic dynamics), the cell will be unstable. On the other hand, dynamics that settles to a pattern where a large proportion of the genes are permanently on or off (frozen) may be too inflexible for adaptive behavior. Cells constantly need to adapt their gene expression pattern in response to a variety of hormone and growth/differentiation factors from nearby cells. A cell type is probably a set of closely related gene expression patterns, not just on the attractors, but shifting around within the basin of attraction, allowing an essential measure of flexibility in behavior. Too much flexibility might allow a perturbation to flip the dynamics too readily into a different basin of attraction, to a different cell type such as a cancer cell, or from a bone cell to a fat cell. The conjecture is that the appropriate dynamical regime has evolved to find a delicate balance between stable on-off regions and dynamically changing regions (Harris et al. 2002).

The basins of attraction in random Boolean networks are idealized models for the stability of cell types against mutations, and also for their response to perturbations of the current state of gene activation. Jumping between basins due to these effects was described in the sections "Jumping between Basins Due to Attractor Perturbations" and "Stability of Basins Due to Mutations in the Network Architecture." If a particular pattern of gene expression undergoes a perturbation, the dynamics may jump to a different subtree in the same basin, which would temporarily adapt the pattern, or it may be flipped to another basin, that is, a different cell type. In this case the basin of attraction field remains unchanged. Alternatively, the network itself may undergo a mutation, resulting in an altered basin of attraction field. This model provides a mechanism for both the stability and adaptability of gene expression.

Memory

Attractors classify state-space into broad categories; it can be thought of as the network's content addressable "memory" (Hopfield 1982). Furthermore, state-space is categorized along transients, by the root of each subtree forming a hierarchy of subcategories. This notion of memory far from the equilibrium condition of attractors greatly extends the classical concept of memory by attractors alone (Wuensche 1994, 1996).

It can be argued that in biological networks such as neural networks in the brain or networks of genes regulating the differentiation and adaptive behavior of cells, basins of attraction and subtrees must be just right for effective categorization. The dynamics need to be sufficiently versatile for adaptive behavior but short of chaotic to ensure reliable behavior, and this in turn implies a balance between order and chaos in the network.

A current research topic, known as the "inverse problem," is to find ways to deduce network architecture from usually incomplete data on transitions, such as a trajectory. This is significant because it could help us to infer the genetic regulatory network, modeled as a random Boolean network, from data on successive patterns of gene expression in the developing embryo (Somogyi et al. 1997). In pattern recognition and similar applications in the area of artificial neural networks, solutions to the inverse problem would provide "learning" methods for random Boolean networks to make useful categories (Wuensche 1994, 1996).

Conclusions

In the context of random Boolean network models of genetic regulatory networks, basins of attraction represent a kind of modular functionality, in that they allow alternative patterns of gene expression in the same genome, providing a mechanism for cell differentiation, stability, and adaptability.

Although the entire genome can be viewed as a vast genetic network, the notion of modularity suggests that the network is in fact built from functionally semi-independent, weakly interlinked subnetworks. Each subnetwork may itself consist of further subnetworks in a sort of nested hierarchy. Figure 13.1 gave a hypothetical example. "Attractors in Fully Connected and Modular Networks" gave some preliminary results indicating that the dynamics on scale-free modular networks are markedly different from the dynamics on their fully connected cousins, having more basins with smaller attractors and transients, which are relatively more stable.

There is evidence that genetic regulatory subnetworks, highly adapted to useful functions, are conserved by evolution, and that examples of essentially the same subnetwork are redeployed across different species, genera, and even families (Relaix and Buckingham 1999). Examples of subnetworks, such as self-contained transcriptional circuits in regulated eukaryotic genes, are being found by experiment, together with the biases on their connections and logic (Harris et al. 2002).

Each subnetwork module could behave semi-independently within its own basin of attraction field, but cross-talk—that is, perturbations—between subnetworks, and intercellular signals, would allow adaptive jumps between basins of attraction, or to transients within the same basin, shifting the dynamics away from the equilibrium at the attractor and constantly adapting to the metabolic environment.

There are many open questions stemming from this discussion. From the network theory perspective, a better understanding is needed of the dynamics on modular random Boolean networks and related systems, their basins of attraction, the stability of basins, and jump probabilities between basins. How does the dynamics change as more links are added between interacting subnetworks? What biases on connectivity and logic are required for network models to reflect data coming from biology?

From the biological perspective, data is accumulating on the makeup of genetic and other biological networks. What kind of biases might exist in general, in connectivity, modularity, and logic. How does the biological network strike the right balance between stability and adaptability. Are genetic regulatory networks "scale-free"? And if modular, to what extent are the modules interlinked? In short, what are the general principles that describe the quality of these networks. The hope is that the ideas, methods, and tools for studying network dynamics on idealized networks can provide a conceptual framework for comprehending networks in biology.

Acknowledgments

Thanks to Ricard Solé, Steve Harris, Gerhard Schlosser, and Günter Wagner for comments and suggestions. Also to Pietro Speroni di Fenizio for suggesting the idea of analyzing jumps between basins. Thanks to the Santa Fe Institute, where a significant part of the research was done. The simulations and figures in this chapter were made with DD-Lab, Discrete Dynamics Lab (Wuensche 2001). The software and documentation is available at www.santafe.edu/~wuensch/ddlab.html and www.ddlab.com.

References

Albert, R., H. Jeong, and A.-L. Barabási. 2000. Error and attack tolerance in complex networks. *Nature* 406:378–381.

Ball, P. 1999. *The self-made tapestry: pattern formation in nature.* New York: Oxford University Press.

Conway, J. H. 1982. What is life? In *Winning ways for your mathematical plays,* ed. E. Berlekamp, J. H. Conway, and R. Guy, vol. 2, chap. 25. Academic Press, New York.

Derrida, B., and D. Stauffer. 1986. Phase transitions in two-dimensional Kauffman random network automata. *Europhys. Lett.* 2:739.

Glass, L., and S. A. Kauffman. 1973. The logical analysis of continuous, non-linear biochemical control networks. *J. Theor. Biol.* 39:103.

Harris, E. S., B. K. Sawhill, A. Wuensche, and S. Kauffman. 2002. A model of transcriptional regulatory networks based on biases in the observed regulatory rules. *Complexity* 7 (4): 23–41.

Hopfield, J. J. 1982. Neural networks and physical systems with emergent collective computational abilities. *Proc. Natl. Acad. Sci. U.S.A.* 79:2554–2558.

Jeong, H., S. P. Mason, and A.-L. Basabási. 2001. Lethality and centrality in protein networks. *Nature* 441:41–42.

Kauffman, S. A. 1969. Metabolic stability and epigenisis in randomly constructed genetic nets. *J. Theor. Biol.* 22:437–467.

Kauffman, S. A. 1993. *The origins of order.* New York: Oxford University Press.

Langton, C. G. 1990. Computation at the edge of chaos. *Physica D (Nonlinear Phenomena)* 42:12–37.

Relaix, F., and M. Buckingham. 1999. From insect eye to vertebrate muscle: redeployment of a regulatory network. *Genes Dev.* 13:3171–3178.

Solé, R. V., I. Salazar-Cuidad, and S. A. Newman. 2000. Gene network dynamics and the evolution of development. *Trends Ecol. Evol.* 15:479–480.

Somogyi, R., S. Fuhrman, M. Askenazi, and A. Wuensche. 1997. The gene expression matrix. In *Proceedings of the World Congress of Non-linear Analysis* 30 (3): 1815–1824.

Somogyi, R., and C. Sniegoski. 1996. Modeling the complexity of genetic networks. *Complexity* 1 (6): 45–63.

Vogelstein, B., D. Lane, and A. J. Levine. 2000. Surfing the p53 network. *Nature* 408: 304–310.

Walker, C. C., and W. R. Ashby. 1966. On the temporal characteristics of behavior in certain complex systems. *Kybernetik* 3:100–108.

Wolfram, S. 1984. Universality and complexity in cellular automata. *Physica D (Nonlinear Phenomena)* 10:1–35.

Wolfram, S., ed. 1986. *Theory and application of cellular automata.* Singapore: World Scientific.

Wolpert, L. 1998. *Principles of development.* Oxford: Oxford University Press.

Wuensche, A. 1994. The ghost in the machine. In *Artificial life III,* ed C. G. Langton, Santa Fe Institute Studies in the Sciences of Complexity. Reading, Mass.: Addison-Wesley.

Wuensche, A. 1996. The emergence of memory. In *Towards a science of consciousness,* ed. S. R. Hameroff, A. W. Kaszniak, and A. C. Scott. Cambridge, Mass.: MIT Press.

Wuensche, A. 1997. Attractor basins of discrete networks. CSRP, no. 461. D.Phil. thesis, University of Sussex.

Wuensche, A. 1998. Genomic regulation modeled as a network with basins of attraction. In *Pacific Symposium on Biocomputing '98.* Singapore: World Scientific.

Wuensche, A. 1999. Classifying cellular automata automatically. *Complexity* 4 (3): 47–66.

Wuensche, A. 2001. *The DDLab manual* and *Discrete Dynamics Lab* software. Online at www.santafe.edu/~wuensch/ddlab.html and www.ddlab.com.

Wuensche, A., and M. J. Lesser. 1992. *The global dynamics of cellular automata.* Santa Fe Institute Studies in the Sciences of Complexity. Reading, Mass.: Addison-Wesley.

Part 3

The Evolutionary Dynamics and Origin of Modules

Perhaps the biggest challenge in the theory of modularity is to understand the variety of roles modularity plays in evolution. This difficulty has many facets. On the one hand there is the fact that modularity may assume different meanings in different contexts. For instance, something that is a module in development, in accordance with the operational criteria from developmental biology, may not act as a module of evolutionary change. Also, the role of modularity in influencing the evolvability of a lineage is far from easy to understand, since evolvability results from an interplay between function, variation, and transmission, each process with its own structure and with a rich array of possible conflicts. Finally and most frustratingly, the evolutionary origin of developmental and variational modularity has escaped a unitary and robust explanation. In this part we bring together chapters that contribute to various combinations of these open questions, starting off with two chapters with different but perhaps complementary views about the evolutionary origin of modules.

Allan G. Force, William A. Cresko, and F. Bryan Pickett start with the idea that organismal modularity of developmental and anatomical units may ultimately derive from the parcellation of genotypic modules. They propose a model in which organismal modularity results from a two-step process. In the first step modular genetic subfunctions evolve, and in a second step these are resolved among duplicated genes. This model extends their recent influential model for the subfunctionalization of duplicated genes based entirely on nonadaptive evolutionary mechanisms, the so-called degeneration-duplication-complementation model. This approach thus contrasts in philosophy with the next chapter, which takes a more traditional selectionist point of view.

In their chapter, Günter P. Wagner and Jason G. Mezey review various (failed) attempts to understand the origin of variational modularity as a direct result of natural selection. Using results from the evolution of RNA secondary structure, they then argue that some underlying variational differences between gene effects are necessary for natural selection to create phenotypic modules.

William C. Wimsatt and Jeffrey C. Schank explore an apparent paradox between generative entrenchment and modularity. Simulation results suggests that parts of an organism have a tendency to either get lost or become increasingly entrenched and thus less variable. This trend seems to be in conflict with the notion of quasi independence and modularity.

Urs Schmidt-Ott and Ernst A. Wimmer discuss one of the most baffling facts about the comparative developmental biology of insects—namely, the fact that a fundamental patterning mechanism found in *Drosophila melanogaster,* anterior identity determination by *bicoid,* is found only in a taxonomically very restricted range of insects. They suggest that *bicoid* is not part of a module but provides an input into the modular segmentation cascade that can be replaced by other inputs.

In his contribution, Ralf J. Sommer continues with the topic of variable developmental mechanisms, in his case vulva development in nematodes. He provides exciting new examples of the changing developmental roles of homologous cells in the evolution of nematode development. The examples show how developmental modules are changeable in their developmental role.

Neil H. Shubin and Marcus C. Davis review recent evidence about the evolution of paired appendages in sarcopterygian fish and the origin of the tetrapod limb. One underlying theme is the interaction during evolution of the dermal and endoskeletal modules of the paired fin. They argue that while autopodium-like endoskeletal structures have evolved convergently in sarcopterygians, the autopodium could become an autonomous module in tetrapods only after the dermal module was lost.

14 Informational Accretion, Gene Duplication, and the Mechanisms of Genetic Module Parcellation

ALLAN G. FORCE, WILLIAM A. CRESKO, AND F. BRYAN PICKETT

Introduction

Interactions in biological systems are highly clustered, leading to the emergence of modules (Nijhout 1997; Hartwell et al 1999; Michod 2000). A module comprises numerous interacting parts that are functionally integrated and are largely independent of components of other modules (Bolker 2000). Many extant organisms appear to be modular at a variety of levels, from the structures of individual genes (Kirchhamer and Davidson 1996; Yuh and Davidson 1996; Yuh et al. 1996, 1998), to the topology of genetic pathways (Wagner 1999, 2000, 2001), to distinct morphological characters contributing to the phenotype (Wagner 1996, Cheverud 1996, Wagner et al. 1997). Modules at higher levels of the biological hierarchy are likely to be organized by modules functioning at lower levels (Wolf et al. 2001). Thus, phenotypic modularity is probably connected to genotypic modularity, but the manner of this connection is still an open question for biologists.

How has phenotypic modularity changed over time, and what are the processes that govern its evolution? Two processes, parcellation and integration, have been implicated in the evolution of increased phenotypic modularity (Wagner and Misof 1993; Cheverud 1996; Wagner 1996; Wagner and Altenberg 1996; Bolker 2000). Parcellation produces modularity from an integrated whole through the differential suppression of interactions among components between modules, while integration increases modularity by strengthening interactions within modules. Wagner and Altenberg (1996) suggested that increases in phenotypic modularity are primarily due to the process of parcellation, because the fossil record indicates that taxa, over evolutionary time, have developed increasingly specialized structures that cover a larger range of morphospace, within relatively conserved body plans (Arthur

1997; Shubin et al. 1997; Carroll 2000; Thomas et al. 2000). For instance, few highly specialized arthropods with homogeneity of segments and appendage types appear in the Cambrian, while in later epochs there is an increase of forms having greater diversification of appendages (Carroll et al. 1995; Knoll and Carroll 1999; Carroll 2000). However, the basic body plan of arthropods has remained relatively unchanged (Averof and Patel 1997).

In this chapter, we explore the mechanisms behind the parcellation of genotypic modules, which may ultimately lead to the parcellation of phenotypic modules over evolutionary time. We suggest that genotypic parcellation may be brought about by a two-step process, involving, first, the origin of new gene subfunctions and, second, the resolution of these subfunctions between gene duplicates. Subfunction formation occurs through processes we have termed "informational accretion," where potential functional DNA segments such as regulatory elements or functional domains in coding regions are added or evolve de novo within a gene. After exploring subfunction formation through informational accretion, we turn our attention to the subdivision of this information among genes and fit our new findings into the existing gene duplication framework as presented in the duplication-degeneration-complementation (DDC) model. We find that informational accretion may initiate the parcellation of developmental modules, and that subfunction resolution after gene duplication can complete genotypic parcellation. These processes together result in the parcellation of developmental pathways. We find that subfunction origin and resolution phases may occur under directional selection but, more important, may occur under stabilizing selection by near-neutral or even slightly deleterious population mechanisms. These findings lead to the conclusion there is a general process, driven by mutation and drift, acting on developmental genes and pathways that may result in increases in genotypic modularity over evolutionary time. Therefore, genotypic parcellation may be a direct consequence of the actions of mutation and genetic drift acting on finite metazoan populations. Such an increase in genotypic modularity may eventually be realized as an increase in phenotypic modularity, and adaptive mechanisms need not be invoked as an explanation.

Subfunction Definition and the Modularity of Genes

Admittedly, modularity is a nebulous concept, and the term has been used liberally in recent years to describe a variety of biological entities, to the point that to some it has lost any meaning (Lewontin 1970; Hull 1980; Streelman and Kocher 2000). However, when defined explicitly, modularity provides a useful shorthand for describing patterns seen in the evolution of life and therefore provides a starting point for the exploration of the mechanisms that produce these patterns (Wagner 1996; von Dassow and Munro 1999; Bolker 2000). In this section, we present first an explicit definition of both modularity and increase of modularity over evolutionary time. Next, we explore what modularity at the level of the gene looks like, and we posit that subfunctions are the basic units of genetic networks. We then explore the components of subfunctions, and how they might arise. If we would ultimately like to understand the origin of subfunctions, we must first understand the origin of building blocks that make up subfunctions. Therefore, in this section we examine pathways for the origins of subfunction components so that in the next section we can begin to explore the mechanisms by which these building blocks become aggregated into distinct subfunctions.

Modules of biological phenomena at all levels share the common properties of comprising numerous interacting parts that are (1) functionally integrated, (2) largely independent of other modules, and (3) capable of following evolutionary trajectories that are independent of other modules at the same level in the biological hierarchy (Wagner 1996; von Dassow and Munro 1999; Bolker 2000). Therefore, interactions within and among modules are highly skewed, such that interactions among components within a module are much more likely to occur than interactions between components of different modules. One of the most striking examples of biological modularity is the development of organismal structures during ontogeny, where different developmental pathways are called into action at particular times to produce discrete structures (Hartwell et al. 1999; Stern 2000; von Dassow et al. 2000; Guss et al. 2001). Differences in organismal modularity can therefore be defined in terms of differences in the numbers of modules, and increases in modularity through evolution are the product of addition of new modules. Because different components of biological systems can be members of different modules, increases in modularity are not necessarily the product of increases in components, but may merely be the consequence of increases in, and partitioning of, connections among existing components (Wagner and Altenberg 1996; Müller and Wagner 1996; Hartwell et al. 1999). Thus, parcellation of interactions allows the evolution of functionally independent modules, since changes in the components of one may have little effect on components of other modules.

Although modules at the same level may be independent of one another, modules are of course not evolutionarily independent of the developmental and organismal systems in which they reside. Biological modules often inhabit nested hierarchies, where the interacting parts of a more inclusive module are themselves modules (Rice 2000). For example, most traits of interest to evolutionary biologists are actually mosaics of developmental modules, with each module contributing to the expression of the trait as well as covariance between traits (Lande 1976; Lande and Arnold 1983). Further embedded in the hierarchy, these developmental modules are composed of component parts themselves. This regression through nested levels of modularity, in which modules at one level are components of modules at another level, can proceed for numerous steps and is the hallmark of multicellular life (Michod 2000; Maynard Smith and Szathmary 1995). Although the reductionist approach of examining one system while keeping all others constant can be very fruitful when examining biological modularity, it must be remembered that in reality all the modules within an organism must interact coherently to produce an adult capable of reproduction. Evolutionary independence among modules is therefore a relative term, since any changes in one module will probably affect other systems. However, the effects of perturbations, either environmental or mutational, will have the largest effects on the module in which the perturbed components reside, whereas other modules will be largely insulated from these effects. This modular structure allows the rapid adaptive evolution of some characters while maintaining the integrity of other systems and therefore the entire organism. In sum, even though aspects of the phenotype that are under the control of entirely different genetic networks interact to produce the fitness of a particular organism, the modularity of organisms primarily exists in the genotype-to-phenotype map. This genetic network modularity allows developmental modules to be evolutionarily independent of one another (see Schlosser 2004).

How is one genetic network module largely insulated from changes in others? To answer this question we must examine the structure and logic of genetic networks. Most models of genetic networks conceptualize genes as being the components (e.g., Omholt et al. 2000; Becksei and Serrano 2000). However, we suggest that the basic unit of developmental gene networks is subfunctions. We have previously defined a gene subfunction to be an independently mutable function of a gene which falls into a distinct complementation class (Force et al. 1999). For instance, a gene that carries out multiple functions in several tissues and has distinct DNA sequences that are required for those functions which can be removed by mutations in regulatory regions or protein coding regions is a subfunction. Thus, a subfunction includes both regulatory elements and coding sequences (Force et al. 1999). Therefore, gene subfunctions fulfill the design criteria we established above for a biological module, comprising interacting components that are largely independent of other components and are therefore capable of independent evolutionary trajectories. Subfunctions are therefore modules within genes and are the basic units of developmental modules.

What are the components of a subfunction module? In order to simplify our discussion, we have coined the phrase "molecular information building blocks" (MIBBs) to describe a small DNA sequence that has its own indivisible function and may lie either in the regulatory or in the coding region of a gene. Examples of MIBBs include codons, splice sites, a transcription factor binding site, promoter sequences, and polyadenylation signal sequences among others (fig. 14.1; Arnone and Davidson 1997). Even though MIBBs themselves have central roles to play as components of a gene subfunction, their individual functions do not equate to a gene subfunction. A set of MIBBs act collectively to form one or more subfunctions in a gene. For example, figure 14.1 shows the structure of a multifunctional gene (14.1, *A*) and the two subfunctions into which it may be decomposed (14.1, *B* and *C*). The multifunctional gene is composed of multiple MIBBs, some of which are labeled in figure 14.1, including transcription factor binding sites (1, 2, and 3), the transcription start site (4), a codon in the first exon (5), a donor splice site (6), and a polyadenylation signal (7). Independent and overlapping sets of MIBBs of the multifunctional gene function together to produce two distinct subfunctions. For instance, the transcription start site and polyadenylation signal are shared by both subfunctions, whereas transcription factor binding site 1 and part of the third exon, and transcription factor binding site 3 and the second exon, are unique to subfunctions 1 and 2, respectively. In this case mutations in the independent MIBBs can reduce or eliminate a specific subfunction at the locus. Thus, the critical characteristic defining a subfunction is not the number or type of MIBBs that it contains, but their integrated operation in performing a function that is mutationally independent of other suites of MIBBS functionally integrated into other subfunctions.

Ultimately, we would like to understand the mutational and population-level mechanisms by which subfunctions arise. In order for a subfunction to form, a subset of MIBBs at a gene must evolve interactions in order for them to take on the role of a module. Therefore, understanding the evolution of MIBB interactions is crucial to understanding the origin of subfunctions. First, however, we must understand where MIBBs come from in the first place. There are a number of mechanisms which may produce new gene function in regulatory and coding regions of DNA (Long and Langley 1993; Chen et al. 1997; Piatigorsky and Wistow 1991). For our discussion we will ignore ad-

A) Multifunctional Gene

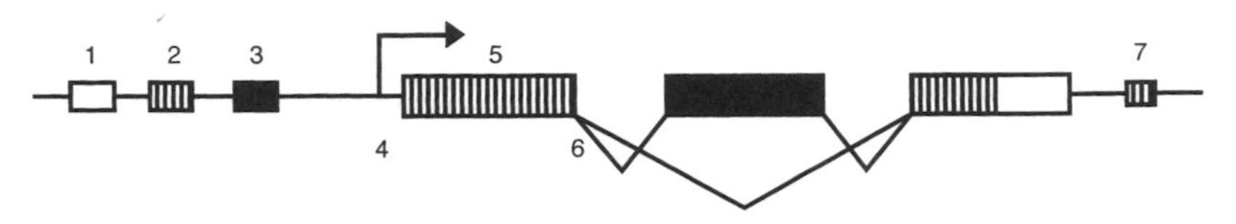

B) Subfunction Module 1

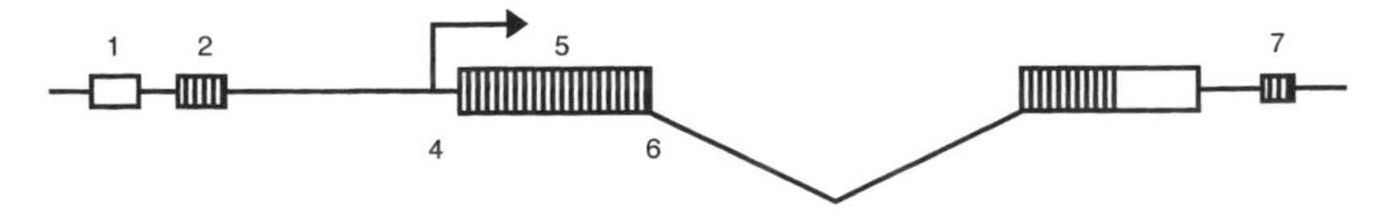

C) Subfunction Module 2

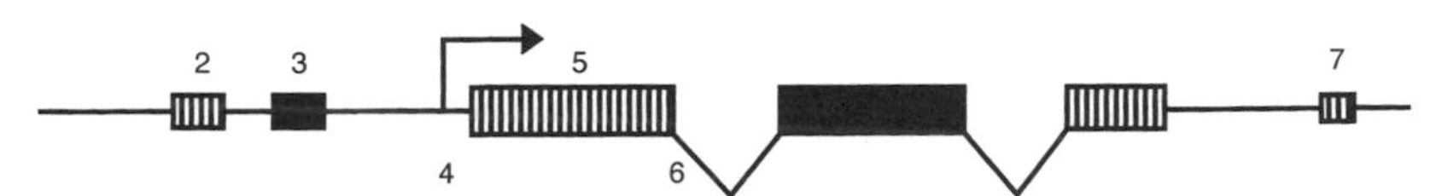

Fig. 14.1.—A mulitfunctional gene (*A*) comprising several molecular information building blocks (MIBBs) that are aggregated into two distinct subfunctions (*B, C*). MIBBs 1, 2, and 3 are transcription factor binding sites. MIBB 4 is the transcription start, MIBB 6 is the first splice donor site, MIBB 7 is the polyadenylation signal, and MIBB 5 is a codon within the first exon. Gray regions are shared embedded functional regions between the subfunctions. The colors black and white denote independent regions of subfunctions 1 and 2 which are required for the developmental stage or tissue-specific function of the gene. The subfunctions are composed of multiple MIBBs, where some are shared (*hatched*) and some independent (*black or white*). A subfunction is a module within a gene composed of multiple interacting MIBBs.

ditions to coding regions and focus our discussion on the mechanisms of formation of new regulatory modules, but the same concepts apply to each. Putative functional transcription factor binding sites occur naturally with high frequency in random sequences of DNA. For instance, the ETS recognition sequence (TCTCCT) would occur on average once every 2,048 times in a randomly derived sequence if we assumed that the transcription factor binding site was functional in either the sense or antisense strand. Point mutation, slippage replication, and small insertions and deletions all have the capacity to produce a new functional binding site. Furthermore, many random DNA sequences when placed into an appropriate context have been shown to possess regulatory function. This surprising finding is amply demonstrated by the work of Edelman et al. (2000). They constructed a randomized library of 18-mers inserted in front of a minimal promoter in a eukaryotic cell line and, amazingly, found that out of 160 random clones sampled, more than 6% exhibited activity greater than twice the basal level of expression and 1 out of 160 (0.6%), exhibited expression 5.5 times as high as the minimal promoter alone! Therefore, a large number of random DNA sequences could contain the precursors to new functional activity. The library was subjected to two rounds of selection for increased activity; approximately 25% of the selected constructs had activity 4–50-fold greater than the minimal promoter by itself, while only approximately 1% of the constructs from the unselected population showed activity greater than fourfold. In addition, different combinations of transcription factor binding sites produced similar levels of expression, demonstrating that there are multiple (convergent) pathways for producing a similar output level of expression. Between random formation of DNA binding sites through mutation and selection for activity it would appear relatively simple to evolve expression from small, novel DNA sequences. However, many transcription factor binding sites found in genomic DNA of organisms are not functionally active in vivo, so their presence is not sufficient for their activity.

In addition to de novo mutation of regulatory elements, new sequences might be incorporated into a gene by the movement of preexisting regulatory or coding sequences from other locations in the genome. The movement of sequences may be mediated by chromosomal inversions, nonhomologous recombination events, translocations, and transposable element insertion events (Zhang et al. 2000). This mechanism is well illustrated by the recent, extensive, and preferential insertion of members of the miniature inverted-repeat transposable element family *Heartbreaker* into genic regions of maize (Zhang et al. 2000). In some cases the transposition process moves known regulatory DNA sequences into proximity with genic regions (Oosumi et al. 1995). Extensive evidence has recently been accumulated demonstrating that transposable elements have become incorporated as functional parts of genes, including alternative splice sites, coding exons, promoters, polyadenylation signals, mRNA stability elements, transcriptional enhancers, and transcriptional silencers (see Brosius 1999; Kidwell and Lisch 2001, for excellent reviews).

In summary, genes are composed of one or more subfunctions, which are modules at the level of the gene. The components of subfunction modules are MIBBs, which include codons, splice sites, transcription factor binding sites, promoter sequences, and polyadenylation signal sequences. Thus, the problem of the evolutionary increase of modularity at a gene through the origin of a new subfunction is a problem of how sets of MIBBs gain functional independence that did not previously exist. We explore these processes in the next section.

The Evolution of Gene Subfunctions by Informational Accretion

In order for a new subfunction to form, a new regulatory element may arise for every new expression domain and function of a gene. Alternatively, partitioning of existing global patterns of regulation may occur, in essence forming new genetic modules through the splitting of existing modules. These two processes can be termed subfunction co-option and subfunction fission, respectively. Subfunction fission posits that the ancestral functions of a gene are not independently mutable because the same regulatory and protein sequences carry out all functions of the gene. Therefore, small changes in regulatory elements and/or protein coding domains can lead to the formation of incipient subfunctions, which become further modified into distinct subfunctions. On the other hand, subfunction co-option occurs when a functional DNA sequence becomes inserted into a gene or arises de novo and confers at least the incipient phase of a totally new function, not carried out by the ancestral gene. Below we present theoretical discussions for each mechanism.

Subfunction Fission by Informational Accretion

In order for new subfunctions to form through informational accretion and subsequent fission two conditions must hold. First, the set of MIBBs which comprise a subfunction need to carry out more than one function, and all of the MIBBs must be required for each function. Therefore, the set of MIBBs belonging to a particular subfunction have an embedded functional structure. Second, a mutational pathway must exist for the replacement of embedded subfunction components by MIBB accretion. Consider the specific case of transcriptional regulation, which is illustrative of the more general case of accretion processes affecting different types of MIBBs at a gene. Regulatory elements of a single subfunction may undergo fission into two subfunctions when there are tissue-specific transcription factors expressed in the first tissue and a different set of transcription factors expressed in the second tissue, which currently do not have transcription factor binding sites within the subfunction. These conditions set up the possibility of replacement of the nonspecific factor with specific factors through a series of transcription factor binding site substitution events. In the following sections, we trace the MIBB accretion and substitution processes in more detail.

Subfunction Fission by Neutral Redundant Informational Accretion under Stabilizing Selection

Through this process, a general positive regulator within a subfunction can be replaced sequentially by specific, redundant positive regulators (fig. 14.2, *A*). For example, a subfunction composed of the embedded regulatory elements (positive transcription factor binding site A1 and negative transcription factor binding site R1) drives expression in two domains, D1 and D2, by A1-positive regulation, while R1 represses expression in D3. Positive transcription factors TFA2 and TFA3 are expressed within D1 and D2, respectively. Although they may be expressed in other domains, TFA2 cannot be expressed in D2 and likewise TFA3 cannot be expressed in D1. A transcription factor binding site arises by a mutation accretion process for TFA2 near the regulatory elements A1 and R1, and this allele drifts to fixation. Following this event, a second transcription factor binding site arises for TFA3 and drifts to fixation. Finally, degenerative mutations arise within A1, and this allele drifts to fixation. At this point, the subfunction has been divided by fission into two subfunctions, beginning with an embedded regulatory arrangement and moving to an independent regulatory arrangement.

Subfunction Fission by Nonredundant Informational Accretion under Stabilizing Selection and Directional Selection

In this case, a general positive regulator within a subfunction could be replaced sequentially by specific nonredundant positive regulators (fig. 14.2, *B*) under positive Darwinian selection. For example, embedded regulatory elements (positive transcription factor binding site A1 and negative transcription factor binding site R1) from a single subfunction drive expression in two domains, D1 and D2, and actively repress expression in D3. In this case the expression pattern of the positive regulator associated with A1 is similar to that seen by the transcription factors TFA2 and TFA3, which are expressed within D1 and D2, respectively, although they play no role in the regulation of this gene prior to establishment of new subfunctions by fission.

On the basis of its dependence on a single subfunction, it is possible that A1-R1-driven regulation of this gene is suboptimal in D1 and D2. One possibility is that increases in the level of expression of the gene in D1 or D2 would provide a selective advantage. Incorporation of A2 and A3 binding sites adjacent to the existing subfunction components would permit an increase in expression if this change led to increased fitness. Under this scenario, A1 may be either retained or lost depending on the extent of overlap of regulation between the A1, A2, and A3

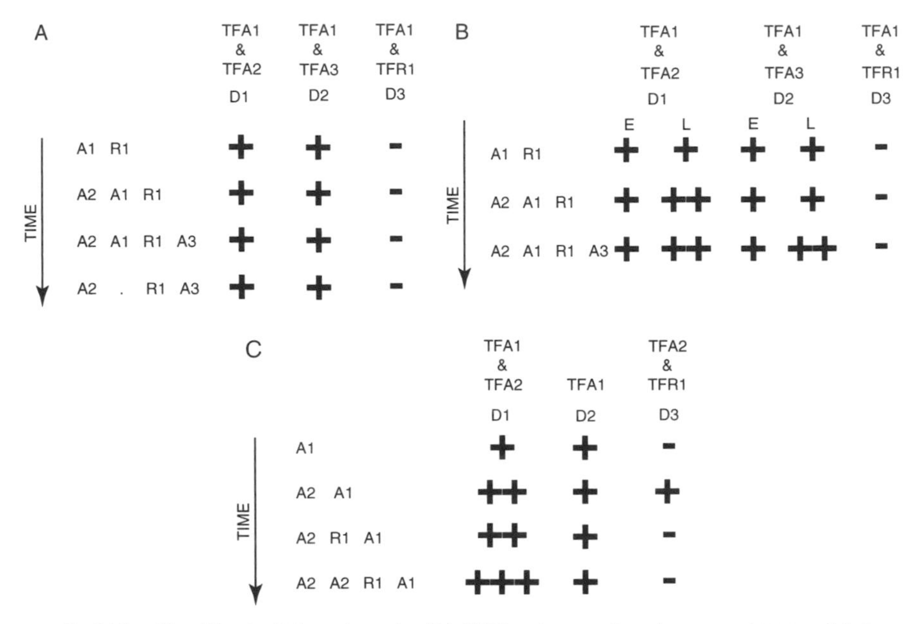

Fig. 14.2.—The subfunction fission pathways by which MIBBS can become coherently aggregated into two distinct subfunctions through neutral (*A*) or selective (*B, C*) mechanisms. Each one of these tables is an evolutionary time series that should be read from top to bottom. *A,* Subfunction fission under stabilizing selection. The ancestral expression domains, D1 and D2, are driven by the same transcription factor, TFA1, and repressed by the expression of transcription factor TFR1 in expression domain D3. The plus sign indicates the level of expression and the minus sign indicates the absence of expression. Before the fission process begins, transcription factor TFA2 is expressed in D1, and TFA3 in D2. Two new transcription factor binding sites, A2 and A3, for transcription factors TFA2 and TFA3, arise by mutation and sequentially replace the ancestral transcription factor binding site, A1 for transcription factor TFA1. The process occurs under near-neutral population mechanisms and results in specific transcription factors for expression domains D1 and D2. Thus, a general ancestral subfunction is split into two descendant subfunctions. *B,* Subfunction fission under directional selection. E stands for early expression, and L stands for late expression domains. An ancestral subfunction becomes differentiated into two distinct subfunctions by the addition of new transcription factor binding sites, A2 and A3, that increase expression specifically in the late stage. *C,* Subfunction fission under directional selection and stabilizing selection. In this case there is strong selection for increased expression in D1 through a new transcription factor binding site, A2, which simultaneously results in ectopic expression in D3 that is deleterious. The pleiotropic effects of A2 result in selection for transcription factor TFR1, which reduces expression in D3 and reduces the deleterious effect. A2 and R1 sites may then increase in number, resulting in higher expression in D1 and no expression in D3, resulting in the formation of an independent subfunction.

expression domains. In either case, two independent subfunctions now exist, one associated with D1 activity driven by the A2 transcription factor binding site and one associated with D2 acting through the A3 transcription factor binding site.

An alternative path exists when higher levels of gene expression are beneficial within one existing domain but are deleterious in another domain (stabilizing selection). For instance, in figure 14.2, C, the expression of the TFA2 factor within D1 and D3 causes expression in both domains, and more A2 sites cause increasing expression in both

D1 and D3. However, while increased expression is beneficial in D1, it is deleterious in D3. In this case, pleiotropy creates an amenable situation for coevolution of activators and repressors that act together to increase expression in D1 while reducing expression in D3. The action of both directional selection and stabilizing selection in this example leads to the formation of a D1-specific regulatory module which, again, creates two distinct subfunctions from one ancestral subfunction.

Subfunction Co-option by Informational Accretion

Subfunction formation through informational accretion and co-option requires the existence of a pool of tissue-specific transcription factors that could drive novel domains of expression. In contrast to subfunction fission, a gene acquires a completely new subfunction that uses MIBBs that may or may not presently be part of an existing subfunction. The stable inclusion of accreted MIBBs in a lineage and the associated ectopic expression of a particular gene or pathway is an example of subfunction formation through co-option. Below we outline subfunction co-option in more detail.

Subfunction Co-option by Neutral Informational Accretion under Stabilizing Selection

Ectopic expression of developmental genes may be deleterious in certain circumstances but not in others; thus, it is possible that some genes can sample a variety of novel expression patterns without placing an undue burden on the fitness of individuals that posses these types of alleles. Figure 14.3, *A*, illustrates how a new transcription factor binding site (A2) regulating a transcription factor (TFX) may produce low levels of expression in a novel domain (D3) with little deleterious effect, permitting this allele to drift to fixation. Following this event a transcription factor binding site for TFX, X, arises in another gene (TFY) which, in turn, undergoes subfunction fission such as in (fig. 14.2, *A*) above. X becomes required for TFY expression in D3 if the subfunction it interacts with undergoes fission and incorporates the X transcription factor binding site. Therefore, a once neutral expression domain acquisition event has led to the emergence of a new subfunction through entirely neutral mechanisms under stabilizing selection. The second step leading to incorporation of the X transcription factor binding site may also occur under positive selection.

Subfunction Co-option by Nonredundant Informational Accretion under Directional Selection

A transcription factor binding site arising within a gene that produces a new expression domain which is immediately beneficial may cause a

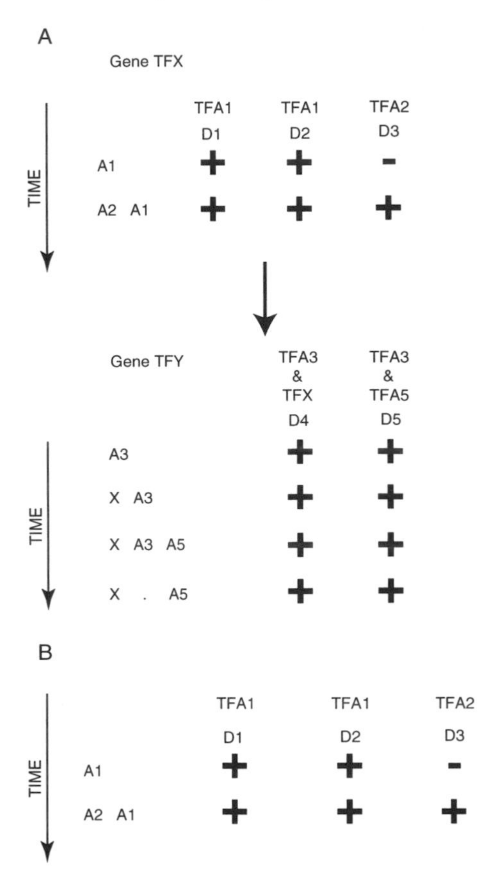

Fig. 14.3.—The co-option pathways by which MIBBs can become coherently aggregated into new subfunctions through neutral (*A*) or selective (*B*) mechanisms. *A,* Subfunction co-option under stabilizing selection. The top panel is the regulation of gene TFX, and the bottom panel is the regulation of gene TFY. There are two steps to neutral subfunction co-option. In the first step a new DNA binding site, A2, for transcription factor A2, TFA2, becomes incorporated into a gene. This gene encodes transcription factor X, which causes new expression in D3. Secondarily, gene TFX becomes required in D3 by being functionally incorporated into the regulation of a different gene, TFY, by subfunction fission. Expression domains D3 and D4 partially or entirely overlap, so that TFX can become required for TFY expression in D4 under subfunction fission. *B,* Subfunction co-option under directional selection. A new DNA binding site, A2, for transcription factor A2, TFA2, becomes incorporated into a gene, which causes new expression in D3. This new expression domain is immediately beneficial and driven to fixation by positive selection.

new subfunction to arise by placing it under immediate, positive Darwinian selection (fig. 14.3, *B*). A new factor, TFA2, causes expression in an ectopic expression domain, through the A2 transcription factor binding site, which is then driven to fixation through positive Darwinian selection. The incorporation and/or deletion of additional MIBBs leads to the refinement of the new subfunction. This type of co-option may be tied to a single gene with direct effects in the domain but with no cascading downstream effects. Or, at the other extreme, a master regulator activating an entire developmental pathway with extreme cascading effects in the new tissue might arise, producing homeotic changes.

It appears that there are multiple pathways, both involving directional selection and/or stabilizing selection that could lead to the evolution of new subfunctions at the transcriptional level in genes and therefore a corresponding increase in the level of gene modularity. However, at this time we do not have a firm grasp on the relative probabilities of occurrence of the alternative mutational pathways leading to the origin of new subfunctions. Similar alternative pathways may also be envisaged for coding region changes, incorporation of new introns, and incorporation of alternative promoters. Clarification of the origin of subfunctions is a central issue in the evolution of modularity and also the preservation of gene duplicates, which we will explore in more detail in the next section.

The Duplication-Degeneration-Complementation-Model and Parcellation of Developmental Modules

In the previous section we discussed the mechanisms for the origin of new subfunctions and concomitant increases in modularity at the level of the gene. Here we discuss the resolution of subfunctions within multifunctional genes following gene duplication, and how these may produce changes in the modularity of the developmental modules of which they are a part. Several models of gene duplication have attempted to incorporate the multifunctional nature of genes within the past 10 years (Piatigorsky and Wistow 1991; Hughes 1994; Force et al. 1999; Lynch and Force 2000). Each of these models can be placed under a general framework for duplicate gene evolution, the DDC model based on the aforementioned observations concerning gene structure, the high level of duplicate preservation, and extensive partitioning of duplicate gene expression patterns and functions (Ferris and Whitt 1977, 1979; Force et al. 1999; Westin and Lardelli 1997; Normes et al. 1998; De Martino et al. 2000; Lister and Raible 2001). Under the DDC model, initially redundant duplicate genes may experience three fates (fig. 14.4). The first is nonfunctionalization (Nei and Roychoudhury 1973; Takahata and Maruyama 1979; Li 1980; Watterson 1983), where one duplicate copy is silenced by degenerative mutations and genetic drift. The second fate is neofunctionalization, whereby beneficial mutations conferring a new or improved subfunction occur in either copy and are fixed by positive Darwinian selection (Ohno 1970; Hughes 1994; Cooke et al. 1997; Sidow 1996; Walsh 1995). The third fate is subfunctionalization, whereby both genes accumulate complementary degenerative mutations that result in their joint requirement for the maintenance of the ancestral functions (Force et al. 1999; Stolzfus 1999; Lynch and Force 2000). Under subfunctionalization, an allele bearing a subfunction mutation drifts to fixation at one of the duplicate loci. Subsequently, a complementary subfunction mutation at the second locus also drifts to fixation. Under subfunctionalization or neofunctionalization, the remaining redundant subfunctions are resolved.

The process of subfunctionalization provides a plausible mechanism to explain why so many duplicate-gene pairs exhibit a partitioning of ancestral expression patterns and protein functional domains. One ex-

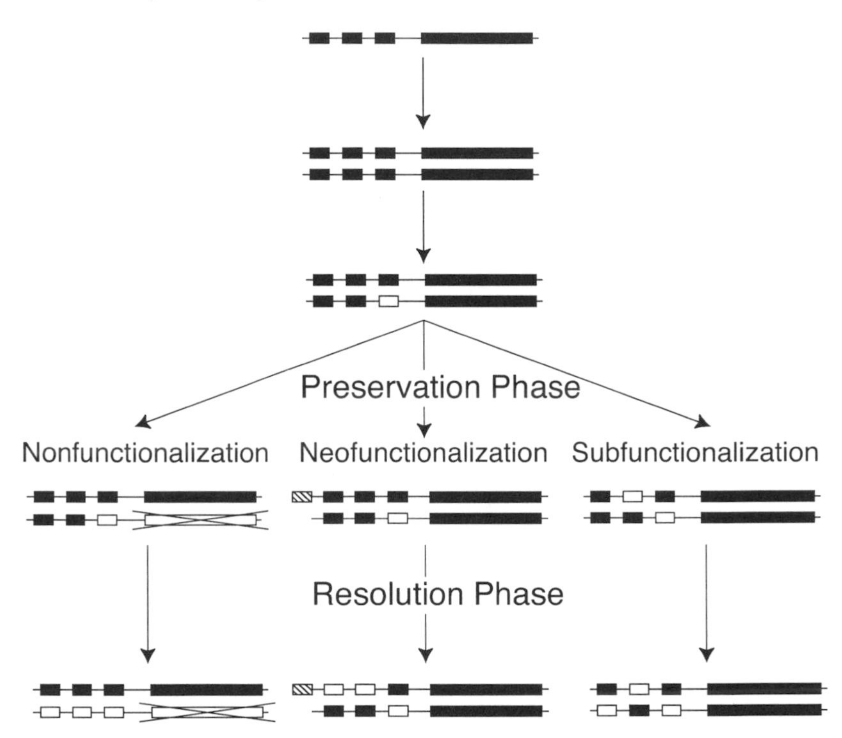

Fig. 14.4.—The duplication-degeneration-complementation (DDC) process for the preservation of duplicate genes and resolution of subfunctions between them. The small boxes represent independently mutable subfunctions that eliminate only specific functions of a gene, whereas the large boxes represent the target area for null mutations that eliminate all functions of a gene. An ancestral multifunctional gene is duplicated and undergoes the preservation and resolution phases. During the preservation phase three fates are possible for initially redundant duplicate genes. The first fate is nonfunctionalization, where one duplicate becomes silenced by the accumulation of null mutations. The second fate is neofunctionalization, where one duplicate acquires a new function (*hatched box*) through beneficial mutation and positive selection, while the second copy loses an ancestral subfunction by degenerative mutation and drift. The last fate is subfunctionalization, where each duplicate loses complementary subfunctions by degenerative mutation and drift, resulting in their joint preservation. During the resolution phase, remaining redundant subfunctions are resolved by degenerative mutation and drift between duplicates preserved by subfunctionalization or neofunctionalization.

ample that is consistent with the subfunctionalization hypothesis is the zebrafish engrailed duplicates (*eng1* and *eng1b*), which originated after the divergence of ray-finned fishes and tetrapods (Force et al. 1999). The *eng1* gene is expressed in the pectoral appendage bud, while *eng1b* is expressed in hindbrain and spinal cord interneurons. In tetrapods, however, *En1* is expressed and is functional in both regions. A number of studies clearly indicate that many duplicate genes typically contain subsets of the functions of the ancestral genes (Westin and Lardelli 1997; Normes et al. 1998; Force et al. 1999; De Martino et al. 2000; Lister and Raible 2001). A similar pattern of duplication of *engrailed* genes and resolution of ancestral expression domains has also occurred independently in multiple arthropod lineages (Marie and Bacon 2000; Peterson et al. 1998; Abzhanov and Kaufman 2000).

The resolution of subfunctions between gene duplicates completes the process of developmental module parcellation that began with the formation of new subfunctions. The resolution of subfunctions occurs under both the subfunctionalization and neofunctionalization mechanisms for the preservation of gene duplicates. Thus, the exact mechanism of duplicate-gene preservation, whether near-neutral under subfunctionalization or beneficial under neofunctionalization, does not change the outcome that the preservation of duplicates must inevitably complete the parcellation for some subfunctions within multifunctional genes. The parcellation process can be entirely near-neutral for some subfunctions under both preservation mechanisms. This is because under subfunctionalization the process is completely near-neutral by definition, and under neofunctionalization some of the subfunctions of a gene are not under positive selection and thereby resolve themselves under near-neutrality. In both cases, connections are pruned and the number of subfunctions per gene decreases, leading to an increase in compartmentalization of the developmental genetic network.

Informational accretion events increase the modularity within genes, since the formation of a new subfunction results in the addition of a new gene module. Simultaneously, the formation of a new gene function can begin the process by which genetic networks become more modular. If the nodes of gene networks are subfunctions, the addition of a new subfunction increases the nodes available to be partitioned between networks. However, if two subfunctions of a single gene are nodes in two different genetic networks, those networks are not as mutationally independent of one another as they would be if the subfunctions were resolved among different duplicates. While subfunctions of a single gene are mutationally independent with respect to one another, as defined previously, both their physical proximity and sharing of minimal components of the gene intertwine the networks in which they reside. The DDC process, by breaking this physical linkage and sharing of minimal components of the gene, increases modularity at the level of the developmental pathway by severing the residual pleiotropy between developmental genes in related modules. Interestingly, while the modularity of the genetic network increases at this point through parcellation, the modularity of the duplicate genes actually decreases, since each loses a subfunction. Modularity therefore does not appear to change in a coordinated manner in all levels of a hierarchy at all times. Instead, fluctuations in the level of modularity at the level of the gene apparently can be transferred to a more inclusive level, that of the developmental genetic network. Thus, although modularity of the global system is increasing throughout the entire process,

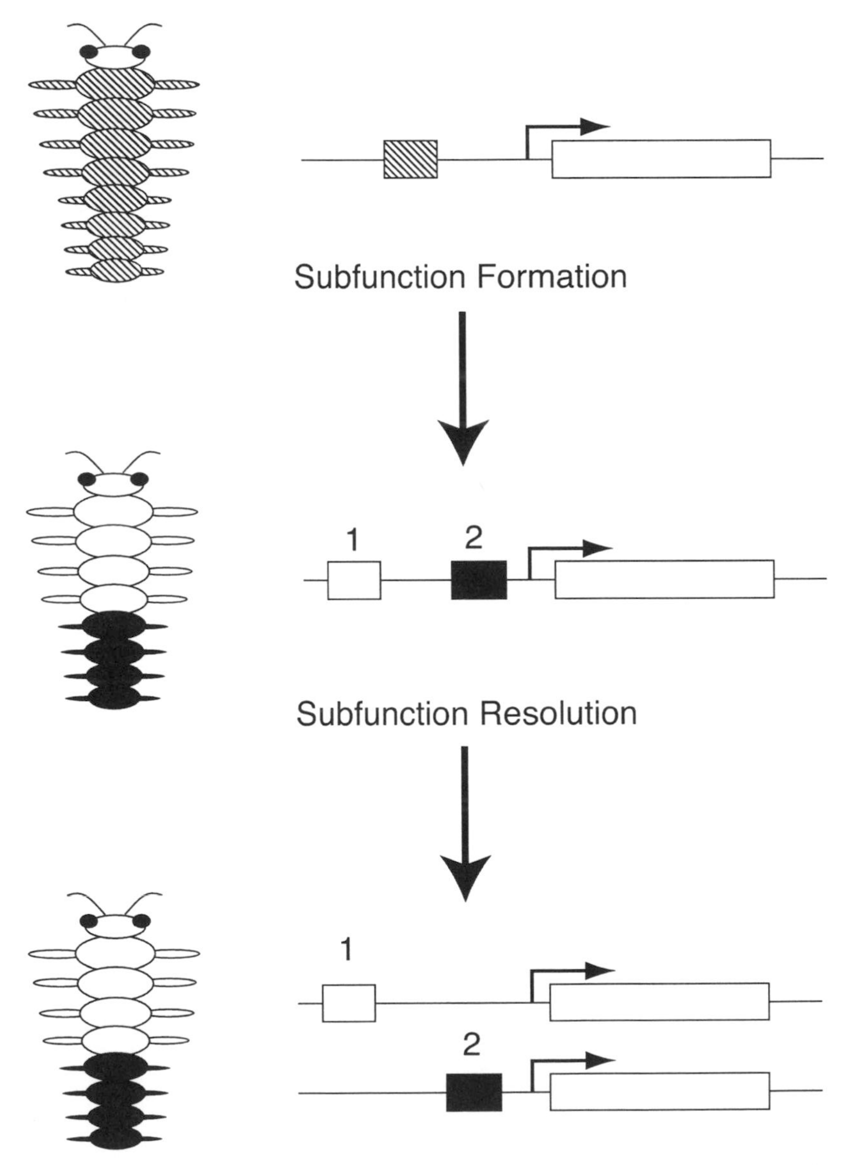

Fig. 14.5.—A simplified representation of how subfunction formation and resolution may allow genotypic modules of organisms to become parcellated over evolutionary time and lead to an increase in phenotypic modularity. An ancestral gene, which is involved in morphological patterning of segments, may have ubiquitous expression along the anterior-posterior body axis of the hypothetical organism associated with a single subfunction (*hatched box*). Informational accretion leads to subfunction fission, forming subfunctions 1 and 2 (*white and black boxes*), which allows morphological divergence of anterior and posterior segments. Following gene duplication and preservation, two independent genes, each with one of the ancestral subfunctions, are formed, severing the remaining ancestral pleiotropy between the two subfunctions. The dual processes of subfunction formation and resolution give rise to increased genotypic modularity.

decreasing modularity at the level of the gene may be integral for completing the parcellation process of genetic networks. Understanding the origin of phenotypic modularity requires understanding the mechanisms affecting developmental genetic network organization, which appears to be strongly affected by processes affecting modularity at the level of the gene (fig. 14.5).

Nonadaptive Processes for the Evolution of Genotypic and Phenotypic Modularity

What are the mechanisms responsible for changes in the epistatic and pleiotropic relationships in the genotype-phenotype map leading to parcellation? Adaptive mechanisms, at the level of selection either among individuals (Cheverud 1996; Wagner 1996; Bolker 2000) or among clades (Gerhart and Kirschner 1997), have been proposed for the process of parcellation and compartmentation producing the modularity of the phenotype. One adaptive hypothesis suggests that variants in a population which exhibit modularity could have a higher fitness and therefore increase in frequency by selection in the population. An alternative, higher-level adaptive hypothesis suggests that clades exhibiting increased levels of modularity are more successful over evolutionary time (Hartwell et al. 1999). While each of these models may be true in specific cases, what other mechanisms exist for the evolution of genotypic and phenotypic modularity?

An alternative view suggests that phenotypic modularity emerges after major restructuring changes occur in genotypic modularity, where the changes in genotypic modularity may be due to nonadaptive mutation and drift-driven processes. As discussed above, near-neutral mechanisms exist for both the origin and resolution of gene subfunctions. The origin of new subfunctions by MIBB fission or co-option processes may occur either under near-neutrality or under directional selection. Therefore, subfunction origin can occur by adaptive mechanisms but may also occur by nonadaptive mechanisms. Extensive theory and simulation work has not been completed for the subfunction fission and co-option models presented above. However, extensive work has been completed on the subfunctionalization model (Force et al. 1999; Lynch and Force 2000; Lynch et al. 2001). Key insights regarding changes in genotypic modularity can be obtained by further analysis of the origin of subfunctions and the subfunctionalization mechanism for duplicate gene retention.

Genotypic modularity may be increased by subfunctionalization which involves small decreases in fitness. Subfunctionalization leads to the formation of genes that are only required for a specific subfunction or subfunctions, thus allowing distinct subfunctions to evolve independently. Therefore, DDC processes can lead to an increase in developmental modularity. However, different models for duplicate preservation, such as subfunctionalization versus neofunctionalization, lead to different predictions about changes in fitness associated with an increase in modularity. Is an increase in modularity always adaptive? Under the subfunctionalization model, partitioning of subfunctions and preservation of gene duplicates is a consequence of degenerative mutations and genetic drift, coupled to purifying selection. Theory and computer simulations show that subfunctional genotypes are at a small selective disadvantage relative to null genotypes because differences in the null mutation rate for each genotype lead to the formation of nonviable individuals (Lynch and Force 2000). Whether or not these small mutationally induced selective differences affect the probability of subfunctionalization depends on population size. The selective disadvantage of a subfunctional genotype versus a nonfunctional genotype is approximately two to four times the null mutation rate for unlinked duplicate loci (Lynch and Force 2000) and equal to the null mutation rate for linked duplicate loci. For example, if the population size is about 1,000, then selection has little effect on the probability of subfunctionalization, yet at a population size of 100,000, the probability of subfunctionalization is close to zero.

Now let us consider a concrete example by asking what the fitness difference is between an ancestral organism with a total of 10,000 genes and a descendant genome with 14,000 genes, where 40% of the genes were preserved by subfunctionalization following a genome duplication event. Here, we assume multiplicative fitness effects across loci as the simplest case, but other, more complicated models involving synergistic epistasis may be explored in the future (Hansen and Wagner 2001). The fitness (W) of the descendant genome (DG) relative to the ancestral genome (AG) is approximately

$$W(DG \mid AG) = \exp-(2\mu_n \Pr(Sub))$$
$$W(DG \mid AG) = .923$$

where n is the number of genes in the genome ($n = 10{,}000$), the Pr(Sub) is equal to the fraction of duplicates preserved by subfunctionalization (Pr(Sub) = .4), and μ is the average null mutation rate per locus ($\mu = .00001$). A large number of genes now carry out smaller subsets of the ancestral gene's subfunctions, and there has been a significant increase in genotypic modularity or potential independence of developmental modules. However, the fitness of the descendant genome is .923 of the ancestral genome. Thus, subfunctionalized duplicates now have entirely independent functions and have increased modularity at the expense of decreased fitness of the descendant genome relative to the ancestral genome. Similar calculations for the neofunctionalization model would show an increase in fitness of the descendant genome versus the ancestral genome, if the Pr(Neo) is assumed to be .4. However when we assume both preservational processes are operating and the Pr(Neo) is small relative to the probability of subfunctionalization, there is a substantial probability there will be either no change in fitness or an actual decrease in fitness associated with increased modularity. How can the fitness of the descendant genome become reduced to such an extent? Maybe it cannot. However, each fixation event proceeds independently of all others, reducing overall fitness by a slow ratchet process that prevents exposing genotypes with extreme differences to direct competition. Only variants with small fitness differences are ever compared by selection. Therefore, although modularity could potentially be adaptive within clades over the long term, modularity could also increase by a maladaptive ratchet mechanism, that is, the successive fixation of slightly deleterious duplicate genes under subfunctionalization (Lynch and Force 2000). Thus, understanding long-term trends in metazoan evolution should be examined in terms of the population genetic mechanisms of informational accretion processes and the DDC process at the level of the gene.

Conclusion

In this chapter, we discussed potential mechanisms behind the parcellation of genotypic modules which may ultimately lead to the parcellation of phenotypic modules over evolutionary time. We proposed that genotypic parcellation is brought about by a two-step process. The first step is the origin of new gene subfunctions by informational accretion mechanisms. Informational accretion processes lead to the formation of new gene subfunctions by either subfunction fission or subfunction co-option. The observation that increases in phenotypic modularity exhibited since the Cambrian occur mostly through the subdivision and specialization of existing phenotypic modules (Knoll and Carroll 1999) suggests that a parallel process may be occurring at the level of the gene. Therefore, subfunction fission may be more important for the evolution of genetic modularity than co-option, but this is an empirical question that remains to be answered. The second step is the resolution of these subfunctions between duplicates following gene duplication by the DDC process. These processes together may result in a global increase of modularity and the parcellation of developmental pathways, whereas the modularity of genes actually decreases during the resolution phase. Subfunction origin and resolution phases may occur under directional selection but may also occur under stabilizing selection by near-neutral or even slightly deleterious population mechanisms. These findings lead to the conclusion that there is a general process, driven by mutation, acting on developmental genes and pathways that may result in increases in genotypic modularity over evolutionary time. Therefore, genotypic parcellation is a direct consequence of the actions of mutation and genetic drift acting on finite metazoan populations. Such an increase in genotypic modularity may eventually be realized as an increase in phenotypic modularity, and adaptive mechanisms need not be invoked as an explanation.

References

Abzhanov, A., and T. C. Kaufman. 2000. Evolution of distinct expression patterns for engrailed paralogues in higher crustaceans (Malacostraca). *Dev. Genes Evol.* 210: 493–506.

Arnone, M. I., and E. H. Davidson. 1997. The hardwiring of development: organization and function of genomic regulatory sequences. *Development* 124:1851–1864.

Arthur. 1997. *The origin of animal body plans: a study in evolutionary developmental biology.* Cambridge: Cambridge University Press.

Averof, M., and N. H. Patel. 1997. Crustacean appendage evolution associated with changes in Hox gene expression. *Nature* 388:682–686.

Becksei, A., and L. Serrano. 2000. Engineering stability in gene networks by autoregulation. *Nature* 405:590–593.

Bolker, J. A. 2000. Modularity in development and why it matters to evo-devo. *Am. Zool.* 40:770–776.

Brosius, J. 1999. Genomes were forged by massive bombardments with retroelements and retrosequences. *Genetica* 107:209–238.

Carroll, S. B. 2000. Chance and necessity: the evolution of morphological complexity and diversity. *Nature* 409:1102–1109.

Carroll, S. B., S. D. Weatherbee, and J. A. Langeland. 1995. Homeotic genes and the regulation and evolution of insect wing number. *Nature* 375:58–61.

Chen, K. S., P. Manian, T. Koeuth, L. Potocki, Q. Zhao, et al. 1997. Homologous recombination of a flanking repeat gene cluster is a mechanism for a common contiguous gene deletion syndrome. *Nat. Genet.* 17:154.

Cheverud, J. M. 1996. Developmental integration and the evolution of pleiotropy. *Am. Zool.* 36:44–50.

Cooke, J. M., M. A. Nowak, M. Boerlijst, and J. Maynard Smith. 1997. Evolutionary origins and maintenance of redundant gene expression during metazoan development. *Trends Genet.* 13:360–364.

De Martino, S., Y.-L. Yan, T. Jowett, J. H. Postlethwait, Z. M. Varga, et al. 2000. Expression of sox11 gene duplicates in zebrafish suggests the reciprocal loss of ancestral gene expression patterns in development. *Dev. Dyn.* 217:279–292.

Edelman, G. M., R. Meech, G. C. Owens, and F. S. Jones. 2000. Synthetic promoter elements obtained by nucleotide sequence variation and selection for activity. *Proc. Natl. Acad. Sci. U.S.A.* 97:3038–3043.

Ferris, S. D., and G. S. Whitt. 1977. Loss of duplicate gene expression after polyploidization. *Nature* 265:258–260.

Ferris, S. D., and G. S. Whitt. 1979. Evolution of the differential regulation of duplicate genes after polyploidization. *J. Mol. Evol.* 12:267–317.

Force, A., M. Lynch, F. B. Pickett, A. Amores, Y. Yan, et al. 1999. Preservation of duplicate genes by complementary, degenerative mutations. *Genetics* 151:1531–1545.

Gerhart, J., and M. Kirschner. 1997. Cells, embryos, and evolution. Malden, Mass.: Blackwell Science.

Guss, K. A., C. E. Nelson, A. Hudson, M. E. Kraus, and S. B. Carroll. 2001. Control of a genetic regulatory network by a selector gene. *Science* 292:1164–1167.

Hansen, T. F., and G. P. Wagner. 2001. Epistasis and the mutation load: a measurement theoretical approach. *Genetics* 158:477–485.

Hartwell, L. H., J. J. Hopfield, S. Leibler, and A. W. Murray. 1999. From molecular to modular cell biology. *Nature* 402:C47–C52.

Hughes, A. L. 1994. The evolution of functionally novel proteins after gene duplication. *Proc. R. Soc. Lond. B Biol. Sci.* 256:119–124.

Hull, D. L. 1980. Individuality and selection. *Annu. Rev. Ecol. Syst.* 11:311–332.

Kidwell, M. G., and D. R. Lisch. 2001. Perspective: transposable elements, parasitic DNA, and genome evolution. *Evolution* 55:1–24.

Kirchhamer, C. V., and E. H. Davidson. 1996. Spatial and temporal information processing in the sea urchin embryo: modular and intramodular organization of the CYIIIa gene cis-regulatory system. *Development* 122:333–348.

Knoll, A. H., and S. B. Carroll. 1999. Early animal evolution: emerging views from comparative biology and geology. *Science* 284:2129–2147.

Lande, R. 1976. Natural selection and random genetic drift in phenotyic evolution. *Evolution* 30:314–334.

Lande, R., and S. J. Arnold. 1983. The measurement of selection on correlated characters. *Evolution* 37:1210–1226.

Lewontin, R. C. 1970. The units of selection. *Annu. Rev. Ecol. Syst.* 1:1–18.

Li, W.-H. 1980. Rate of gene silencing at duplicate loci: a theoretical study and interpretation of data from tetraploid fishes. *Genetics* 95:237–258.

Lister, J. A., J. Close, and D. W. Raible. 2001. Duplicate mitf genes in zebrafish: complementary expression and conservation of melanogenic potential. *Dev. Biol.* 237: 333–344.

Long, M., and C. H. Langely. 1993. Natural selection and the origin of jingwei, a chimeric processed functional gene in *Drosophila*. *Science* 260:91–95.

Lynch, M., and A. Force. 2000. The probability of duplicate gene preservation by subfunctionalization. *Genetics* 154:459–473.

Lynch, M., M. O'Hely, B. Walsh, and A. Force. 2001. The probability of preservation of a newly arisen gene duplicate. *Genetics* 159:1789–1804.

Marie, B., and J. P. Bacon. 2000. Two engrailed-related genes in the cockroach: cloning, phylogenetic analysis expression and isolation of splice variants. *Dev. Genes Evol.* 210:436–448.

Maynard Smith, J., and E. Szathmary. 1995. *The major transitions in evolution.* Oxford: Oxford University Press.

Michod, R. 2000. *Darwinian dynamics: evolutionary transitions in fitness and individuality.* Princeton, N.J.: Princeton University Press.

Müller, G. B., and G. P. Wagner. 1996. Homology, Hox genes and developmental integration. *Am. Zool.* 36:4–13.

Nei, M., and A. K. Roychoudhury. 1973. Probability of fixation of nonfunctional genes at duplicate loci. *Am. Nat.* 107:362–372.

Nijhout, H. F., and S. M. Paulsen. 1997. Developmental models and polygenic characters. *Am. Nat.* 149:394–405.

Normes, S., M. Clarkson, I. Mikkola, M. Pedersen, A. Bardsley, J. P. Martinez, S. Krauss, and T. Johansen. 1998. Zebrafish contains two Pax6 genes involved in eye development. *Mech. Dev.* 77:185–196.

Ohno, S. 1970. *Evolution by gene duplication.* Berlin: Springer-Verlag.

Omholt, S. W., E. Plahte, L. Oyehaug, and K. Xiang. 2000. Gene regulatory networks generating the phenomena of additivity, dominance and epistasis. *Genetics* 155: 969–980.

Oosumi, T., B. Garlick, and W. R. Belknap. 1995. Identification and characterization of putative transposable DNA elements in solanaceous plants and *Caenorhabditis elegans*. *Proc. Natl. Acad. Sci. U.S.A.* 92:8886–8890.

Peterson, M. D., A. Popadic, and T. C. Kaufman. 1998. The expression of two engrailed-related genes in an apterygote insect and a phylogenetic analysis of insect engrailed-related genes. *Dev. Genes Evol.* 208:547–557.

Piatigorsky, J., and G. Wistow. 1991. The recruitment of crystallins: new functions precede gene duplications. *Science* 252:1078–1079.

Rice, S. 2000. The evolution of developmental interactions: epistasis, canalization and integration. In *Epistasis and the evolutionary process,* ed. J. B. Wolf, E. D. Brodie III, and M. J. Wade, 82–98. Oxford: Oxford University Press.

Schlosser, G. 2004. The role of modules in development and evolution. In *Modularity in development and evolution,* ed. G. Schlosser and G. P. Wagner. Chicago: University of Chicago Press.

Shubin, N., C. Tabin, and S. Carroll. 1997. Fossils, genes and the evolution of animal limbs. *Nature* 388:639–648.

Sidow, A. 1996. Gen(om)e duplications in the evolution of early vertebrates. *Curr. Opin. Genet. Dev.* 6:715–722.

Stern, D. L. 2000. Evolutionary biology: the problem of variation. *Nature* 408: 529–531.

Stoltzfus, A. 1999. On the possibility of constructive neutral evolution. *J. Mol. Evol.* 49:0169–0181.

Streelman, J. T., and T. D. Kocher. 2000. From phenotype to genotype. *Evol. Dev.* 2: 166–173.

Takahata, N., and T. Maruyama. 1979. Polymorphism and loss of duplicate gene expression: a theoretical study with application to tetraploid fish. *Proc. Natl. Acad. Sci. U.S.A.* 76:4521–4525.

Thomas, R. D. K., R. M. Shearman, and G. W. Stewart. 2000. Evolutionary exploitation of design options by the first animals with hard skeletons. *Science* 288:1239–1242.

von Dassow, G., E. Meir, M. Munro, and G. M. Odell. 2000. The segment polarity network is a robust developmental module. *Nature* 406:188–192.

von Dassow, G., and E. Munro. 1999. Modularity in animal development and evolution: elements of a conceptual framework for evo-devo. *J. Exp. Zool.* 285:307–325.

Wagner, A. 1999. Redundant gene functions and natural selection. *J. Evol. Biol.* 12: 1–16.

Wagner, A. 2000. Robustness against mutations in genetic networks of yeast. *Nat. Gen.* 24:355–361.

Wagner, A. 2001. The yeast protein interaction network evolves rapidly and contains few redundant duplicate genes. *Mol. Biol. Evol.* 18:1283–1292.

Wagner, G. P. 1996. Homologues, natural kinds and the evolution of modularity. *Am. Zool.* 36:36–53.

Wagner, G. P., and L. Altenberg. 1996. Complex adaptations and the evolution of evolvability. *Evolution* 50:967–976.

Wagner, G. P., G. Booth, and H. Bagheri-Chaichian. 1997. A population genetic theory of canalization. *Evolution* 51:329–347.

Wagner, G. P., and B. Y. Misof. 1993. How can a character be developmentally constrained despite variation in developmental pathways? *J. Evol. Biol.* 6:449–455.

Walsh, J. B. 1995. How often do duplicated genes evolve new functions? *Genetics* 110: 345–364.

Watterson, G. A. 1983. On the time for gene silencing at duplicate loci. *Genetics* 105: 745–766.

Westin, J., and M. Lardelli. 1997. Three novel notch genes in zebrafish: implications for vertebrate Notch gene evolution and function. *Dev. Genes Evol.* 207:51–63.

Wolf, J. B., W. A. Frankino, A. F. Agrawal, E. D. Brodie III, and A. J. Moore. 2001. Developmental interactions and the constituents of quantitative variation. *Evolution* 55 (2): 232–245.

Yuh, C.-H., H. Bolouri, and E. H. Davidson. 1998. Genomic cis-regulatory logic: experimental and computational analysis of a sea urchin gene. *Science* 279:1896–1902.

Yuh, C.-H., and E. H. Davidson. 1996. Modular cis-regulatory organization of Endo16, a gut-specific gene of the sea urchin embryo. *Development* 122:1069–1082.

Yuh, C.-H., J. G. Moore, and E. H. Davidson. 1996. Quantitative functional interrelations within the cis-regulatory system of the *S. purpuratus* Endo16 gene. *Development* 122:4045–4056.

Zhang, Q., J. Arbuckle, and S. R. Wessler. 2000. Recent, extensive, and preferential insertion of members of the miniature inverted-repeat transposable element family Heartbreaker into genic regions of maize. *Proc. Natl. Acad. Sci. U.S.A.* 97:1160–1165.

15 The Role of Genetic Architecture Constraints in the Origin of Variational Modularity

GÜNTER P. WAGNER AND JASON G. MEZEY

Introduction

In recent years the concept of modularity has emerged as a crystallization point for ideas about the evolution and organization of development (Bolker 2000; Callebaut and Rasskin-Gutman in press; von Dassow and Munro 1999; Leroi 2000; Raff 1996; Schlosser 2002; Wagner and Altenberg 1996; Winther 2001). A reason for the popularity of this concept seems to be that it captures a level of integration important in developing and evolving organisms that is outside the purview of other recognized units in biology, such as genes, cells, and populations. Each science has to define the units that play a causal role in the processes that one seeks to understand (Wagner 1996; Wagner 2001). For instance, the units recognized in chemistry are atoms, molecules, and electrons, as well as less well defined entities like complexes among molecules. In biology four fundamental units have been recognized as playing crucial roles in life processes in addition to those that are chemical entities: genes, cells (including their parts, like mitochondria, which are derived from cells), populations, and species. Comparative developmental biology as well as morphological evolution describes additional units that are more inclusive than individual genes or cells and less inclusive than organisms and populations. Examples are gene networks, developmental fields, and characters, to name a few. The modularity concept serves to fix the idea that there are these units or levels of integration which are not captured by the conceptual inventory of classical molecular biology. In fact, as discussed below, the modularity concept is a cluster of at least three independent ideas that relate to each other in subtle ways.

Most of the efforts in developing the modularity concept go toward documenting the existence of modules and their impact on morphological evolution (Brandon 1999; Raff and Sly 2000; Yang 2001). Much less attention has been dedicated to the question of how modules have originated in evolution. The purpose of this chapter is to discuss the problem of whether natural selection alone can account for the origin of variational modules (see below for a definition). An overview of various attempts to explain the evolutionary origin of modules has been published elsewhere (Wagner et al. in press). In this chapter we argue that variational modularity most likely originates from an interaction between natural selection and variational biases in the underlying genetic architecture.

Kinds of Modularity

In recent biological thought the modularity concept has three largely independent roots. These are experimental developmental biology, comparative developmental genetics, and population genetic theory. All three of these concepts were proposed in work published in the mid-1990s which argues for their independent origins (Gerhart and Kirschner 1997; Raff 1996; Shubin et al. 1997; Wagner and Altenberg 1996; Zuckerkandl 1994). Each of them gave rise to a different notion of modularity, and these notions play distinct but related roles in developmental evolution and evolutionary theory. In this section we want to discuss the meaning of these concepts and their possible relationships to other concepts.

Developmental Modules

The concept of developmental modules was introduced by Rudy Raff in his overview of developmental evolution published in 1996. The concept is related to a variety of older notions, most notably that of morphogenetic fields (Gilbert et al. 1996) and developmental dissociability (Gould 1977; Needham 1933). It essentially refers to any developmentally autonomous part of the embryo that can develop all or most of its structure outside its normal context. The operational criterion for recognizing a developmental module is therefore the ability to experimentally induce the development of the module at another location or outside the body (i.e., tissue culture). The paradigm for a developmental module is the limb bud. Once initiated, the development of the limb bud follows an autonomous path. This was known for a long time to experimental developmental biologists (Hinchliffe and Johnson 1980). More recently it was shown that the insect compound eye is also highly autonomous, as shown by its induction at various places in the fly body by ectopic *Pax-6* expression (Halder et al. 1995). The evolutionary importance of developmental modularity derives from the fact that the developmental autonomy of modules allows two major forms of evolutionary variations. On the one hand, developmental autonomy allows for temporal dissociation between the development of various body parts and thus is required for heterochrony (Gould 1977). On the other hand, developmental autonomy also allows the deployment of the developmental modules in different parts of the body and thus allows so-called heterotopy, that is, the development of homologous parts in different parts of the body (Sattler 1984).

Genetic Process Modules

Perhaps the most important empirical generalization that have emerged from developmental genetics so far is the realization that very similar developmental genes are involved in the development of quite disparate body plans, as in the notorious fly-mouse comparisons (Carroll et al. 2001). Furthermore, it turned out that the same genes are also used in the development of very different body parts within the same organism. For instance, in mammals *Hoxa-11* is used in axial and limb development as well as in kidney and female urogenital organ development. Even more surprising, not only individual genes are used over and over again but in fact whole networks of interacting genes (Gerhart and Kirschner 1997; Zuckerkandl 1994; Shubin et al. 1997). These networks tend to serve "abstract" functions, like making or maintaining an asymmetric boundary or picking out a single cell among a group of developmentally competent cells (see Borycki 2004; de Celis 2004; and Kardon et al. 2004). Many of these networks have been characterized, so that it is difficult to point to a particular paradigm, but the Notch/Delta network is certainly a good exemplar (de Celis 2004). Thus, the operational criterion for recognizing a genetic process module is finding the same network used in different developmental contexts. The evolutionary importance of genetic process modules derives from the fact that they apparently can be deployed in toto in new developmental contexts by a few mutations. This is a new level of phenotypic variation, where whole molecular functions can be deployed and redeployed with relative ease.

There is a certain affinity between developmental and genetic process modules. Both recognize the context insensitivity of certain functional units in the developmental process, but there are also important differences that warrant conceptually distinguishing between them. Developmental modules are spatially bounded parts of the embryo that usually give rise to definite morphological body parts. For instance, the feather bud leads to a feather and the limb bud to extremities. Developmental modules are also usually multicellular, while genetic process modules are mostly subcellular molecular machines, which are spatially not bounded. They are thus only functionally or dynamically individualized (see also Schlosser 2004; and von Dassow and Meir 2004).

Variational Modules

While developmental and genetic process modules are primarily defined through their proximate mechanistic integration and individuality, variational modules are based on the statistical notion of the genotype-phenotype map. A genotype-phenotype map is the relationship between genetic variation and phenotypic variation; that is, it specifies which genetic differences give rise to which phenotypic differences (Mezey et al. 2000; Wagner and Altenberg 1996). A variational module is a collection of phenotypic traits that are variationally integrated through the pleiotropic effects of genes and independent of other such clusters because of the relative lack of pleiotropic effects among them (fig. 15.1). The concept is very closely related to the notion of morphological and genetic integration. Operationally modules are recognized through the correlation patterns of quantitative characters (Chernoff and Magwene 1999; Cheverud 1982; Cheverud 1984; Cheverud 1996; Cowley and Atchley 1990; Olson and Miller 1958; Riska 1985) or by the pattern of effects of QTLs[1] on phenotypic traits (Cheverud et al. 1997; Mezey et al. 2000; and see also Cheverud 2004). The paradigm of variational modules is the two main parts of the mouse mandible, the ascending ramus and the alveolar region (Cheverud et al. 1997; Mezey et al. 2000) (fig. 15.2). Thc mouse mandible is a single structural unit—that is, it is a single bone that arises from several developmental modules (Atchley and Hall 1991). The evolutionary importance of variational modules consists in their relation to patterns of natural selection (Magwene 2001) and their response to selection. Because of the high integration among the parts of a variational module, these parts or traits react to natural selection as a unit, both at the level of the selection differential as well in terms of the selection response (Magwene 2001). In addition, their variational independence also assures that they act as independent units of phenotypic evolution.

The relationship between developmental and variational modules is not entirely straightforward. It is clear that not every developmental module can also be a variational module. For instance, all limb buds are developmental modules based on their developmental autonomy, but none of them individually gives rise to a variational module. The left and right forelimb buds give rise to two highly correlated structures, the forelimbs, because they basically express the same genetic information. Not every developmental module is necessarily also a variational module. Conversely, it is not clear whether each variational

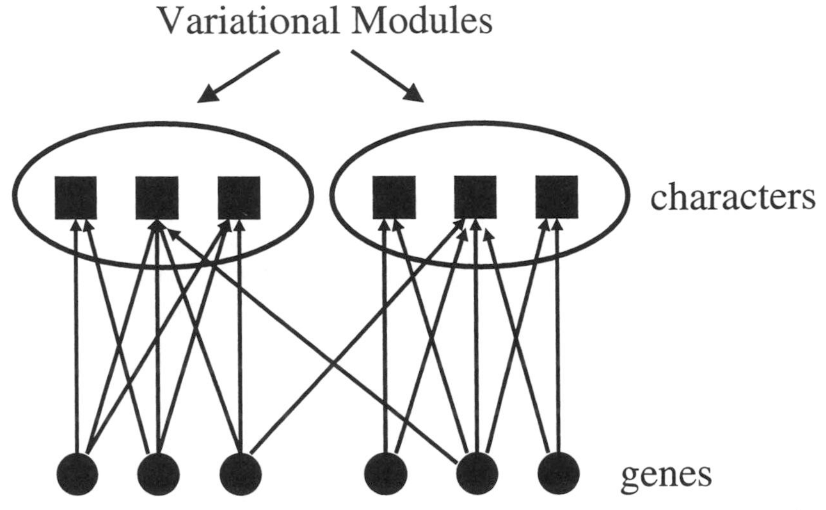

Fig. 15.1.—Schematic representation of variational modules. A variational module is a set of phenotypic traits that are genetically correlated through the pleiotropic effects of the genes affecting them but are relatively independent of other traits due to the relative lack of pleiotropic effects shared among them.

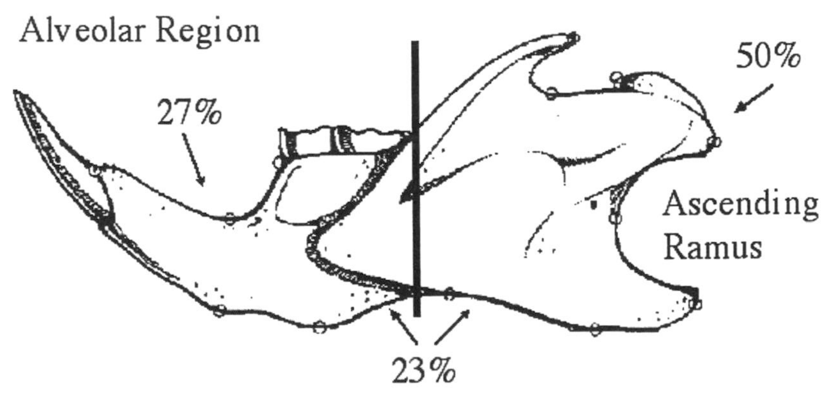

Fig. 15.2.—Example of variational modules in the mouse mandible. Only 23% of QTLs (see note 1) identified affect traits in both the alveolar and the ascending ramus, while 27% and 50% of QTLs have effects restricted to either region, respectively (data from Cheverud et al. 1997). (Figure courtesy of J. M. Cheverud, modified by the authors.)

module consists of one of more developmental modules, since too few examples of variational modules have been studied to provide a basis for generalizations.

The Origin of Variational Modularity

In the following we will discuss the problem of how variational modules may have originated in evolution. A variety of models have been proposed to explain the origin of variational modules in evolution. An overview of these models (Wagner et al. in press) led to the conclusion that the proposals are mechanistically highly heterogeneous, and only a few of the models actually work if modeled and are biologically plausible. In this chapter we do not want to recount these models but, rather, to focus on two specific questions. The first is whether natural selection alone can account for the origin of variational modularity. The answer to this question will be negative. The second question is why, during the evolution of phenotypic robustness, variational modularity can arise spontaneously. From this we will seek to generalize the reasons found in the specific examples and apply them to a wider range of characters.

Natural Selection Cannot Do It Alone

It is surprisingly difficult to come up with a reasonably plausible and general population genetic model for the origin of variational modularity. Similar difficulties have been found in attempts to understand the evolution of other variational properties[2] like mutation rate, recombination rate, dominance, canalization, and sex (Feldman et al. 1997; Gibson and Wagner 2000; Maynard-Smith 1978; Mayo and Bürger 1997). But the difficulty of explaining variational modularity is even greater. Any of the classical models for the evolution of variational properties like canalization or mutation rate seek to understand the evolution of only a single variable, for instance, mutation rate or mutational variance. In contrast, the origin of variational modules requires the evolution of two opposing tendencies. On the one hand, it is expected that the correlation or "integration" among the parts of a module will increase or at least be maintained while the correlation between variational modules decreases. It requires integration as well as parcellation among traits at the same time.

One of us has proposed that perhaps alternating patterns of directional selection could lead to variational modularity (Wagner 1996). The proposal was based on the finding that directional selection on one character or a linear combination of quantitative characters induces quite strong selection against the pleiotropic effects on other characters that are simultaneously under stabilizing selection (fig. 15.3). These results were based on simple modifier models where one gene had no direct effect but modified the mutational distribution of many genes. These results were further extended, showing the effectiveness of this type of selection in models of gene interaction where there is no distinction between primary and modifier genes (Mezey 2000). These results encouraged the idea that any two sets of characters that never simultaneously experience directional selection may be sorted into variational

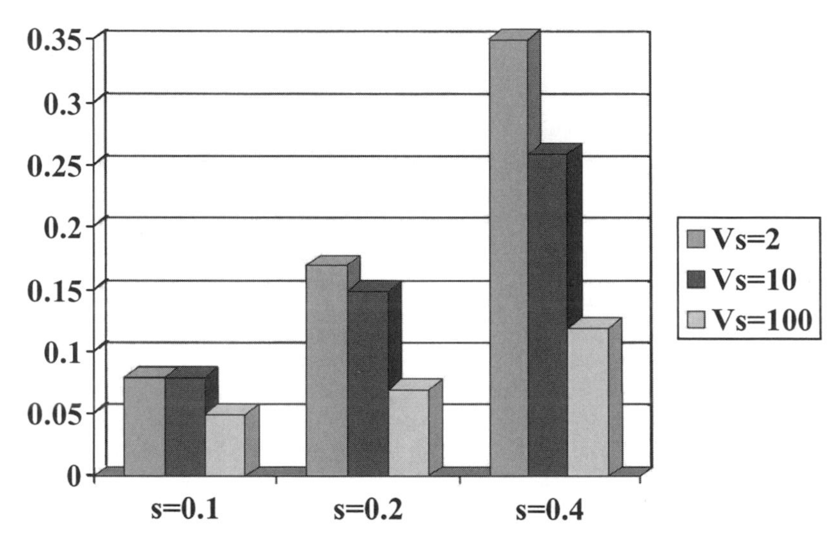

Fig. 15.3.—Selection coefficient, S(M), of a modifier gene that suppresses the pleiotropic effects of 50 genes. The phenotype was represented by two quantitative characters z_1 and z_2, one under directional selection and the other under stabilizing selection. Initially each locus had equally strong effects on both characters. The modifier suppressed half of the effects on the character under stabilizing selection in the heterozygous state, and completely when homozygous for the derived allele. The selection coefficient was estimated in individual-based simulations from the mean time to fixation. The fitness function was $w(z_1,z_2) = \exp\{sz_1 - z_2^2/V_s\}$, where s gives the strength of directional selection and $1/V_s$ the strength of stabilizing selection. Note that the selection coefficient for modification of pleiotropic effects can be quite high, especially under strong directional selection. The reason for this is that sustained directional selection maintains large amounts of genetic variation in the population, and thus, reducing the effects on a character under stabilizing selection is beneficial.

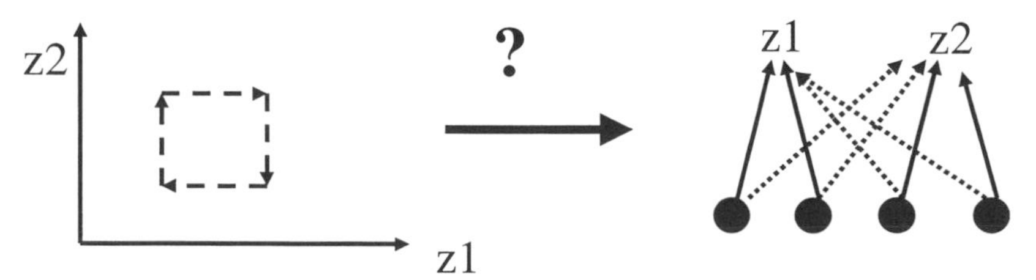

Fig. 15.4.—A model for the evolution of variational modularity after Wagner 1996. The idea is that two characters, z_1 and z_2, which are never coselected may recruit a nonoverlapping set of genes and suppress the pleiotropic effects among them. The stippled arrows show the direction of evolution of these characters. Note that only one character is under directional selection at any one time, and thus, this selection scenario may lead to modularity among the traits. This idea, however, was not supported by the results; see figure 15.5.

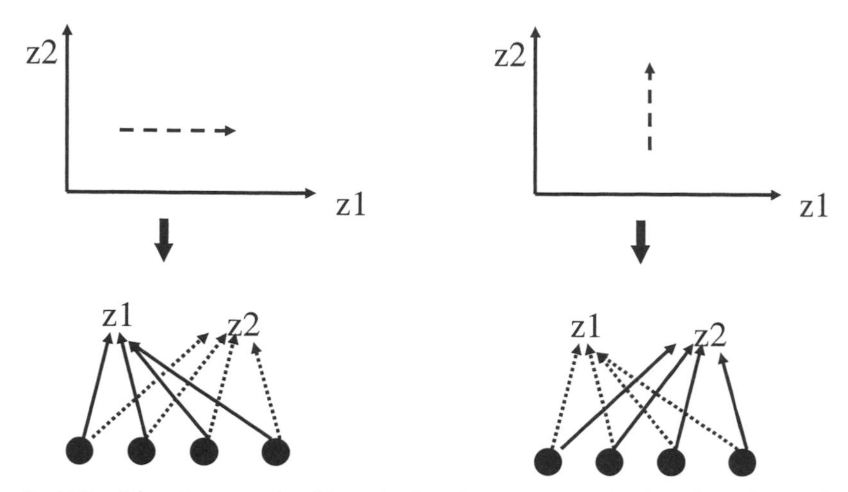

Fig. 15.5.—Schematic representation of the results of a study attempting to simulate the evolution of variational modularity. Each character was subjected to directional selection independently, with the expectation that each character would recruit a subset of genes. The results, however, showed that in this model variational modularity was structurally unstable. Whenever a character was under directional selection, all genes were recruited to affect this character. A separation of genes in two independent groups was not possible. The model of gene interaction used in this study is the one analyzed in Wagner and Mezey 2000.

modules. Each set would experience selection against pleiotropic effects, and all that would be required is a kind of a threshold effect where the rate of recruitment depends strongly on the variational bias, that is, what the relative effects are on each character (fig. 15.4). This effect did not materialize. Any episode of directional selection recruited genes into that direction, and there was no separation into different variational modules in our models (fig.15.5). We now understand these models much better and think we know why this is the case in this class of models (Wagner and Mezey 2000).

While this result was certainly disappointing, it is conceptually interesting. The class of models we analyzed assumed no or only very weak structure in the production of genetic interaction effects. Everything should be possible, like a continuum of genetic possibilities. It is Fisher's dream world: no limitations to what mutations can produce by small incremental steps. The question we asked was whether variational structure can arise from such a continuum only due to the forces of natural selection. Our results suggest that this is impossible, mostly because the nonlinearities in the model are too weak to allow pattern formation. This and the results discussed below imply that the evolution of variational modularity may result from the interaction between natural selection and the intrinsic genetic architecture of characters.

RNA Secondary Structure and Robustness: The Spontaneous Origin of Variational Modularity

In this section we want to summarize a surprising result that was obtained by Lauren Ancel and Walter Fontana in a simulation study of RNA secondary structure evolution (Ancel and Fontana 2000). It showed the spontaneous evolution of variational modularity of RNA secondary structure components (for an explanation of the term "RNA secondary structure component" see fig. 15.6). In their study Ancel and Fontana used powerful folding algorithms for RNA (Hofacker et al. 1994–1998) to simulate the evolution of a simple phenotype, RNA secondary structure. Evolution proceeds through base substitutions and selection based on the phenotypic distance to a "target" structure, the phenotypic optimum. A critical feature in their model is that they used not only the lowest free energy configuration to evalu-

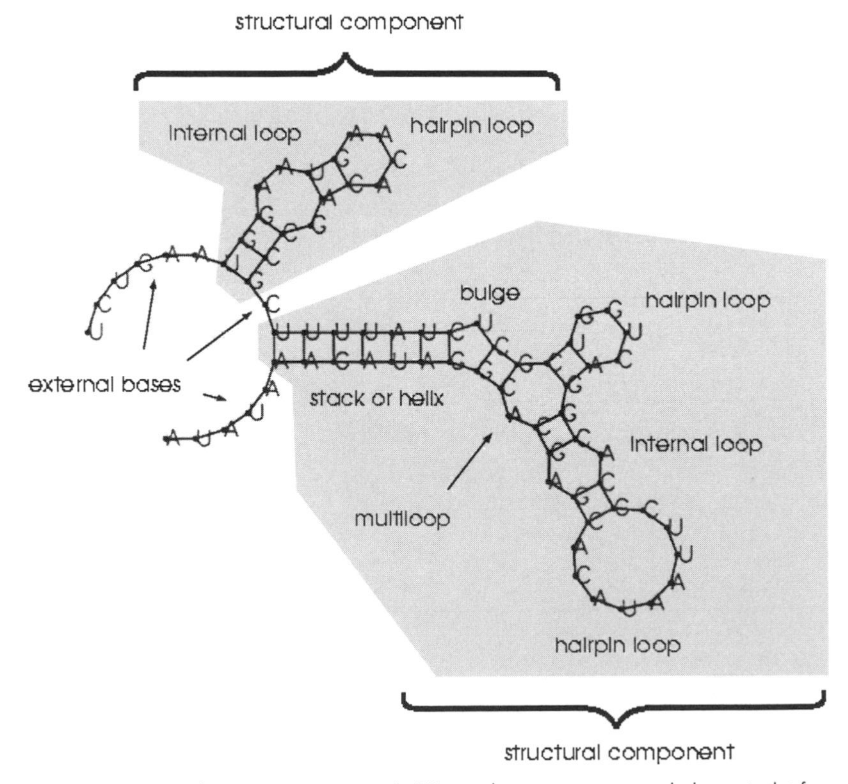

Fig. 15.6.—RNA secondary structure components. An RNA secondary structure represents the base pairs that form during the folding of a single-stranded RNA. A structure component is a set of contiguous stems and loops that are separated from other such components by single-stranded RNA. (Figure courtesy of W. Fontana.)

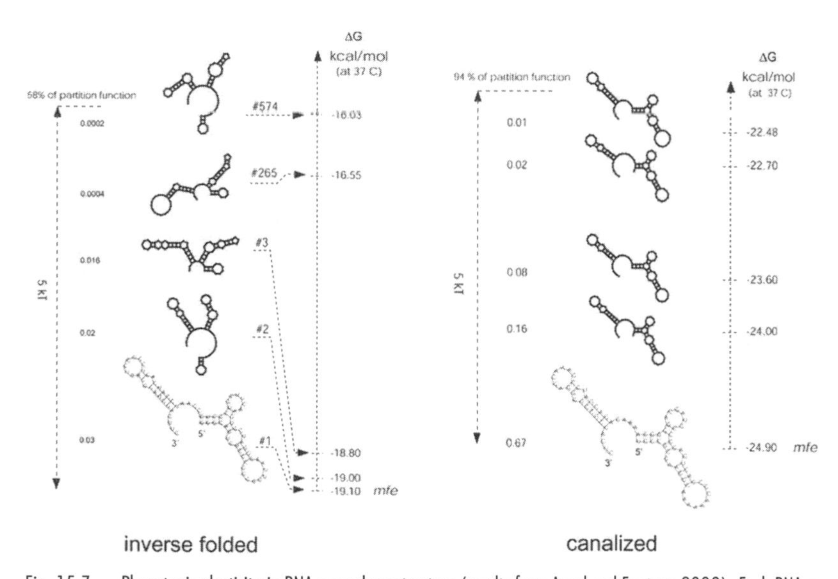

Fig. 15.7.—Phenotypic plasticity in RNA secondary structure (results from Ancel and Fontana 2000). Each RNA sequence can fold in a variety of secondary structures, but with different free energies. If one randomly selects a RNA sequence that can fold into a given structure, the lowest energy structure often is a minority in the ensemble of structures. If the sequence has evolved phenotypic robustness, fewer secondary structures are realized by the sequence, a larger fraction of molecules fold in the lowest energy configuration, and the energy differences between the structures are larger. (Figure courtesy of W. Fontana.)

ate the fitness of an RNA sequence. Instead they used all secondary structures into which the sequence can fold within a 5 kT energy band (fig. 15.7). This is a form of "phenotypic plasticity," or "developmental noise." To include developmental noise is important, since in many cases the lowest free energy structure represents a statistical minority among the secondary structures realizable by a sequence. The fitness of an RNA sequence then is the weighted average of the fitness of all the phenotypes it folds into.

Several unexpected results were obtained. First it turned out that the populations did not always reach the optimum when phenotypic plasticity was taken into account, while the optimum was often reached without phenotypic plasticity. In pursuing why this happens Ancel and Fontana found that the sequences developed strong phenotypic robustness; that is, the phenotypes became resistant to mutations. This is a form of genetic canalization. Furthermore, it turned out that genetic canalization evolved as a correlated response to the evolution of environmental canalization. In other words, the lowest free energy structures became more stable and more frequent in the ensemble. Ancel and Fontana found that there is a correlation between phenotypic and genetic robustness, and thus, genetic canalization evolves as a consequence of environmental canalization. Most surprisingly, though,

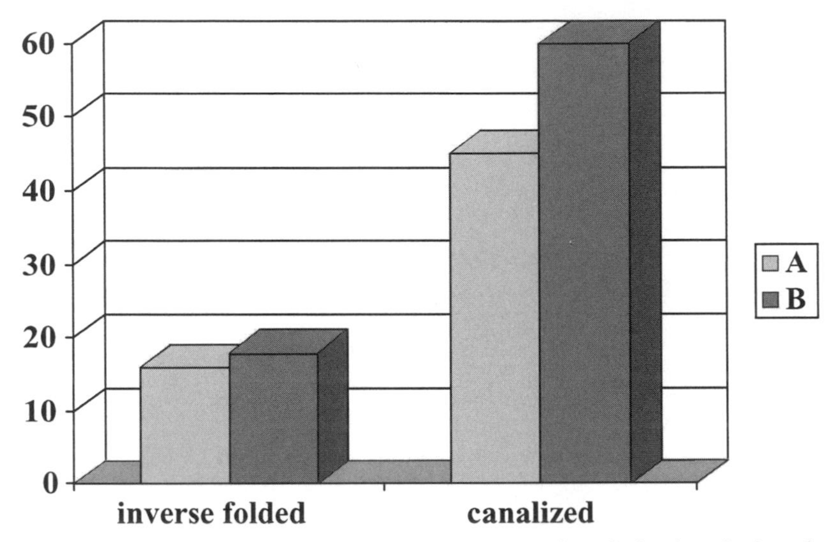

Fig. 15.8.—Variational modularity of RNA secondary structure components. The graphs show the results of a test for the context sensitivity of secondary structure components. The test consists of taking a sequence that folds into a certain secondary structure and cutting out the part of the sequence that makes up a certain structure component. This part of the sequence is then put into a random sequence so that the total length is the same as that of the original sequence. Then we ask whether the secondary structure of the new hybrid sequence also reproduces the structure component. The Y-axis of these graphs gives the percentage of times that the structure component was found in the hybrid sequence. A and B are two different secondary structures tested. Two structure components are tests, once from a random sequence folding into given secondary structure (inverse folded) and once after the sequence has evolved phenotypic robustness (canalized). Note that random sequences maintain a structure component in a random context in only 10%–20% of cases. In contrast, a canalized sequence reproduced the same structure component in about 50% of random contexts. Hence, the evolution of phenotypic robustness leads to variational modularity of secondary structure components (data from Ancel and Fontana 2000).

and most relevant to the present chapter, was the discovery that the canalized sequences were also modular. Modularity here refers to the variational behavior of secondary structure components like stem-loop regions.

Modularity in canalized secondary structures is revealed in two ways: in the melting behavior of the structure components and their context insensitivity. If one raises the temperature, the lowest free energy structure will change until an unfolded RNA strand is the only stable form. If one takes a random sequence that folds in a given secondary structure, the melting behavior will lead to many new secondary structures in which most of the structure components are not conserved among the different states. In contrast, if one melts a canalized secondary structure, each of the structure components exhibits its own melting behavior, independent of the rest of the molecule. The structure components have attained individuality with their own melting point. In addition, it can be shown that canalized secondary structure components are less context sensitive than their noncanalized counterparts (fig, 15.8). This has been shown by taking a structure component and splicing it into a random sequence and asking how likely this component is to be part of the secondary structure of this new hybrid sequence. In other words, the question is how sensitive the secondary structure is to the nucleotide sequence outside the structure component itself. It was found that structure components of canalized secondary structures are less sensitive than noncanalized ones. This result shows that canalized secondary structure components are legitimate exemplars of variational modules because nucleotide changes outside the structure component itself become less important—that is, there is a suppression of pleiotropic effects.

It is important to note that in this study natural selection was not set up to specifically produce modularity. The model is not even set up to study the evolution of phenotypic stability. All that was asked of the model was to evolve toward a given target secondary structure. Rather, the chain of causality runs like this: Phenotypic variation is selected against because the average fitness of an RNA sequence decreases with the number of different structures into which it can fold. This leads to phenotypic robustness or canalization. Genetic robustness results from the fact that there is a strong correlation between genetic and phenotypic robustness, also called plastogenetic congruence by Ancel and Fontana. In turn, variational modularity is correlated with genetic robustness. Why is genetic robustness correlated with variational modularity?

We think the key is found in the question of how genetic changes can influence the resulting phenotype (fig. 15.9). A base substitution can change the phenotype in two ways. It can either affect the secondary structure component of which it is a physical part or affect another part of the secondary structure. We will call the former "focal" effects and the latter "indirect" effects. Indirect effects exist for RNA secondary structure components because the existence of a certain secondary structure component depends on the sequence context in which it occurs. We think it is intuitively pretty obvious that focal effects are harder to eliminate than indirect effects. It is almost impossible to evolve a sequence where mutations in a secondary structure have no effect, in particular if the affected nucleotide is part of a base pair. On the other hand, the indirect effects on a particular structure component can be minimized by increasing the thermodynamic stability of the structure component. A more stable structure component is less sensitive to the rest of the sequence. Thus, base substitutions outside the structure component are less likely to affect it, as shown in the simulation experiments cited above. In genetic terms, with a more stable structure component the mutations outside the structure itself have weaker pleiotro-

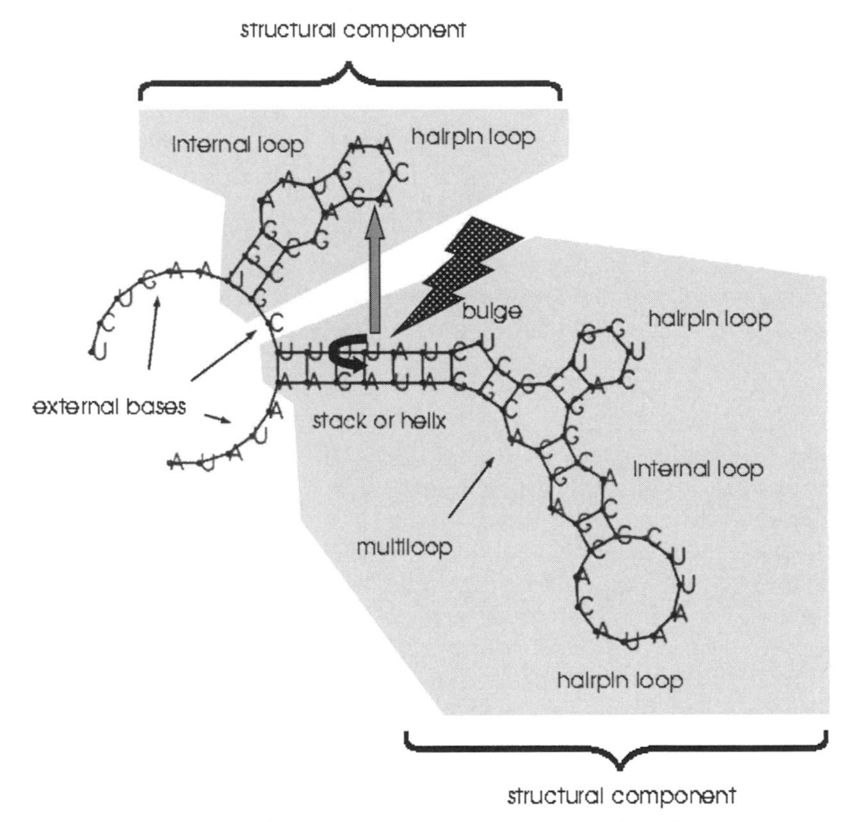

Fig. 15.9.—Focal and indirect effects of mutations in RNA secondary structures. A base substitution in a structure component has two types of effects. Focal effects are on the structure component in which the nucleotide participates. These effects are nearly unavoidable and are thus "hard" effects. On the other hand, a base substitution can also affect the stability of another structure component in which it is not directly participating. This is the case because the existence of a structure component in the folding process depends on the sequence context of the component (see fig. 15.8). These effects are indirect, but they can be mitigated if the other structure component has evolved greater thermodynamic stability and thus lower context sensitivity. These indirect effects are thus "soft" in the sense that their magnitude can be modulated quite easily. (Modified from a figure courtesy of W. Fontana.)

pic effects on the structure component. There is an effective suppression of pleiotropic effects. On the other hand, the focal effects are not so easily suppressed or not at all. This implies that indirect effects are preferentially removed by natural selection if the structure evolves phenotypic stability and the result is variational modularity.

Note that this mode of origin of variational modules critically depends on the "genetic architecture" of the phenotype. In particular, differences in the ability to modify different classes of effects are critical for the outcome. The assumption is that indirect effects are more easily mitigated than focal effects. In the next section we discuss whether this reasoning can be extended to other biological characters.

Hard and Soft Mutation Effects in Organisms

In the previous section it was argued that in RNA secondary structure variational modularity arises through the preferential elimination of indirect pleiotropic effects during the evolution of phenotypic robustness. The question then is whether this model can be generalized to organismal traits. In other words, is it possible to argue that certain effects of genes are more difficult to eliminate than other effects?

Before we consider a possible analogy between organismal traits and the models of RNA secondary structure, a brief note on terminology is in order. For the RNA example we made the distinction between focal effects and indirect effects. This distinction was possible because many bases in an RNA secondary structure are physically involved in the structure component and because each base can be part of only one structure component. This means that for each base there is at most one character for which it can be focal. All other effects are indirect.

In transmission genetics it is usual to talk about pleiotropic effects, that is, mutations which affect more than one character, but there is no transmission genetic way of distinguishing which of the various effect of a gene are primary and which are indirect. With molecular genetic techniques, however, it is possible to distinguish between focal effects and epigenetic effects in the following way: On the one hand, a mutation can and in many cases will affect the development of the cells in which the gene is expressed. These effects can be seen as analogous to focal effects as defined for RNA secondary structure with the modification that a gene can be focally involved in the development of more than one character. This is the case because the same gene can be expressed during the development of various characters. On the other hand, a mutation can also affect a character in which the gene is never expressed. These effects can be mediated through hormonal, biomechanical, or other indirect ways. These epigenetic effects cannot be distinguished from focal effect with transmission genetic methods (Cowley and Atchley 1992) but may differ with respect to their variational properties. We propose that these epigenetic effects play the same role as indirect effects in RNA secondary structure evolution. To explain this idea we start by asking how any two phenotypic traits can be genetically coupled, that is, affected by the same mutation.

If two characters are affected by the same mutation, this can be due to a combination of focal and epigenetic effects. First, the same gene can be expressed in the development of both characters, and it is thus possible (but not necessary; see Chiu and Hamrick 2002; Stern 2000; and Force et al. 2004) that a particular mutation affects both characters. On the other hand, it is possible that one character is also the focal character of the gene but that the effect on the second character is due to epigenetic effects. Finally, it is possible that both characters are affected by epigenetic effects and that the gene is expressed in yet another character or characters. Pleiotropy can thus be due to shared genes, and thus shared focal effects, or due to any combination of focal and epigenetic effects. While this distinction is irrelevant to transmission genetic predictions, it may be very relevant if we consider models for the evolution of genetic architecture, that is, the question which of these effects can be modified or suppressed by natural selection. In other words, the question is whether focal and epigenetic effects differ in their likelihood of being modified by a mutation at some other locus.

Epigenetic effects have to be mediated through processes depending on other genes, such as hormones or other extracellular signals and all the factors that influence the release, transport, and reception of these signals. These effects are mediated through cell-cell interaction and complex cascades of molecular interactions. It is thus plausible that the effect of a mutation mediated through epigenetic processes depends on the genotype of the other genes responsible for mediating this process. Most obviously, the production of a receptor and all the intracellular signaling molecules is required for a signal to influence a target cell. Similarly, the reaction norm of, say, skeletal tissues to stress depends on the genes expressed in the bone itself, and thus biomechanical effects on bone structure do not depend only on muscle strength and activity but also on the genes expressed in the reacting tissue. Epigenetic effects result from cell-cell communication and thus depend on the combination of genes expressed in the interacting tissues. These effects are thus highly modifiable by mutations in genes expressed in both participating cell populations.

For a gene that is focally involved in the development of a cell it is hard to see how mutation effects can be effectively eliminated as long as the gene product serves a necessary function. There is only one way in which this can happen, namely, genetic redundancy (Tautz 1992), perhaps due to gene or genome duplication (Wilkins 1997). While this is certainly true, genetic redundancy is evolutionarily unstable.[3] The reason is that genetic redundancy influences not only the focal effects of the gene but all gene effects and thus leads to weak or no selection on maintaining gene function. Loss-of-function mutations thus will accumulate in the population at the rate of the (loss-of-function) mutation rate and eliminate the genetic redundancy.

Genetic canalization of epigenetic effects does not lead to a complete loss of selection on the genes and thus will not lead to complete neutrality of the affected genes. The reason is that there will still be selection due to the focal effects, and thus, the resulting genetic architecture can be evolutionarily stable. Hence, it is conceivable that epigenetic effects are "soft" in the sense that their magnitude can be more easily decreased than the average size of focal effects, which are thus "hard" genetic effects.

The Erosion Model for the Evolution of Modularity

If in fact epigenetic effects in organisms are "soft" in the same sense as indirect effects on RNA secondary structure components are, and if focal effects are "hard" in the same sense as they are in RNA, then one can imagine a kind of "erosion" model for the origin of variational modularity (fig. 15.10). Whatever the genetic architecture of the ancestor is, there is a tendency to preferentially eliminate epigenetic effects whenever selection favors either a decrease of overall genetic effects or simple phenotypic robustness. The result is that focal effects remain, basically reflecting the boundaries of developmental modules, such that developmental and variational modules tend to coincide. Accepting this assumption leads to the question of which kinds of selection forces can drive the erosion process.

Certainly, selection for phenotypic robustness is a possible mechanism, as it is in the RNA example (Ancel and Fontana 2000). Current mathematical models suggest that selection for phenotypic robustness

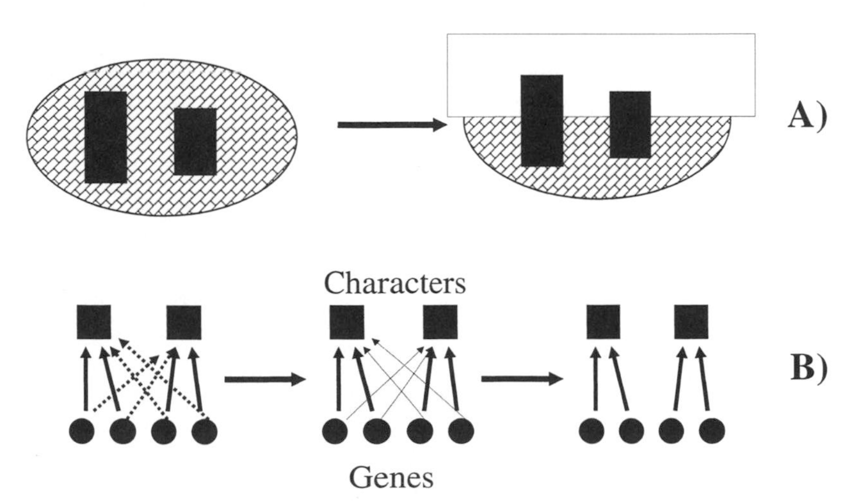

Fig. 15.10.—The "erosion model" for the evolution of variational modularity. *A,* A structure can arise if a "hard" material is embedded in a softer material. Erosion preferentially eliminates the soft material, revealing the hidden structure of the material. *B,* Just as physical structure can arise from differences in hardness, variational modularity can arise if gene effects have different degrees of modifiability. If focal effects (*solid arrows*) are less likely to be mitigated than epigenetic effects (*stippled arrows*), then any selection pressure against the overall mutational effects will preferentially eliminate epigenetic (soft) effects, leading to variational modularity, if the hard effects show a tendency to cluster.

directly selects for stability against environmental variation and that genetic robustness, mainly caused by eliminating epigenetic effects, can arise as a correlated response (for a discussion of this point see Gibson and Wagner 2000). As in the RNA example, this mechanism requires a correlation between genetic and environmental robustness. Whether organisms do show this kind of correlation between environmental and genetic variability is an open empirical question that it would be important to answer.

Direct selection for genetic robustness can happen only if the genetic variation in the population is high and largely independent of the strength of stabilizing selection. These conditions can be met in situations where there is strong spatial heterogeneity and migration (J. Hermisson, unpublished results). On the other hand, directional and stabilizing selection can also conspire to create strong selection to eliminate pleiotropic effects. This was shown by analytical as well as simulation studies (Mezey 2000; G. P. Wagner, unpublished results). Hence, there are a number of selective scenarios that have the potential to select against the magnitude of mutational effects, either directly or indirectly. In this situation the exact selective scenario seems less important as long as it creates strong enough selection against pleiotropic effects in one way or another.

An erosion model for the origin of variational modularity makes specific predictions about the genetic and developmental architecture of the resulting modules. QTLs affecting a module are expected to be also expressed in the module during development. Hence, when it becomes possible to identify the actual genes that correspond to a QTL, it is predicted that these genes will also be focal genes for that module (see also Schlosser 2002).

The erosion model also makes predictions about the relationship between developmental and variational modules. While it is clear that not every developmental module can also be a variational module (see "Kinds of Modularity" above), the erosion model predicts that a variational module will consist of a set of developmental modules that share a significant number of genes. In other words, according to the erosion model variational modules consist of developmental modules that are integrated through shared focal genes. There is some preliminary evidence that correlated skeletal characters develop from cells sharing focal genes, genes that are expressed in the developmental rudiments (Nemeschkal 1999).

Conclusion

The proposed solution to the problem of how variational modularity arises in evolution is based on an interaction between the developmental genetic architecture, that is, the distinction between focal and epigenetic effects of genes, and natural selection. Hence, the explanatory force (Amundson 1989) rests neither with population genetics nor with developmental mechanisms. The result is essentially an interaction effect. Thus, the origin of variational modularity may be another example where developmental genetic information is an essential part of an evolutionary explanation, like evolutionary novelties (Wagner et al. 2000) or the frequency of homoplastic character states (Donoghue and Ree 2000; Wake 1991).

Notes

1. A QTL (quantitative trait locus) is a region of a chromosome that has been shown to influence a quantitative trait by association between the segregation of molecular markers and trait expression (Lynch and Walsh 1998).

2. The variational properties of a genotype describe its propensity to create variation. These are the rate of mutations, the recombination rates, and the distribution of mutational effects on phenotypic characters.

3. It is possible that redundancy is actively selected for, but this is effective only if the mutation rate is high (Nowak et al. 1997; Wagner 1999).

References

Amundson, R. 1989. The trials and tribulations of selectionist explanations. In *Issues in evolutionary epistemology*, ed. K. Hahlweg and C. A. Hooker, 413–432. New York: State University of New York Press.

Ancel, L. W., and W. Fontana. 2000. Plasticity, evolvability and modularity in RNA. *J. Exp. Zool. (Mol. Dev. Evol.)* 288:242–283.

Atchley, W. R., and B. K. Hall. 1991. A model for development and evolution of complex morphological structures. *Biol. Rev.* 66:101–157.

Bolker, J. A. 2000. Modularity in development and why it matters in evo-devo. *Am. Zool.* 40:770–776.

Borycki, A.-G. 2004. Sonic hedgehog and Wnt signaling pathways during development and evolution. In *Modularity in development and evolution,* ed. G. Schlosser and G. P. Wagner. Chicago: University of Chicago Press.

Brandon, R. N. 1999. The units of selection revisited: the modules of selection. *Biol. Philos.* 14:167–180.

Callebaut, W., and D. Rasskin-Gutman, eds. In press. *Modularity: understanding the development and evolution of natural complex systems.* Cambridge, Mass.: MIT Press.

Carroll, S. B., J. K. Grenier, and S. D. Weatherbee. 2001. *From DNA to diversity.* Malden, Mass.: Blackwell Science.

Chernoff, B., and P. M. Magwene. 1999. Morphological integration: forty years later. In *Morphological integration,* ed. E. C. Olson and R. L. Miller, 319–348. Chicago: University of Chicago Press.

Cheverud, J. M. 1982. Phenotypic, genetic, and environmental morphological integration in the cranium. *Evolution* 36:499–516.

Cheverud, J. M. 1984. Quantitative genetics and developmental constraints on evolution by selection. *J. Theor. Biol.* 101:155–171.

Cheverud, J. M. 1996. Quantitative genetic analysis of cranial morphology in the cotton-top (*Saguinus oedipus*) and saddle-back (*S. fuscicollis*) tamarins. *J. Evol. Biol.* 9: 5–42.

Cheverud, J. M. 2004. Modular pleiotropic effects of quantitative trait loci on morphological traits. In *Modularity in development and evolution,* ed. G. Schlosser and G. P. Wagner. Chicago: University of Chicago Press.

Cheverud, J.M., E. J. Routman, and D. K. Irschick. 1997. Pleiotropic effects of individual gene loci on mandibular morphology. *Evolution* 51:2004–2014.

Chiu, C.-H., and M. W. Hamrick. 2002. Evolution and development of the primate limb skeleton. *Evol. Anthropol.* 11:94–107.

Cowley, D. E., and W. R. Atchley. 1990. Development and quantitative genetics of correlation structure among body parts of *Drosophila melanogaster. Am. Nat.* 135: 242–268.

Cowley, D. E., and W. R. Atchley. 1992. Quantitative genetic models for development, epigenetic selection and phenotypic evolution. *Evolution* 46:495–518.

de Celis, J. F. 2004. The Notch signaling module. In *Modularity in development and evolution,* ed. G. Schlosser and G. P. Wagner. Chicago: University of Chicago Press.

Donoghue, M. J., and R. H. Ree. 2000. Homoplasy and developmental constraint: a model and an example from plants. *Am. Zool.* 40:759–769.

Feldman, M. W., S. P. Otto, and F. B. Christiansen. 1997. Population genetic perspectives on the evolution of recombination. *Annu. Rev. Genet.* 30:261–295.

Force, A. G., W. A. Cresko, and F. B. Pickett. 2004. Informational accretion, gene duplication, and the mechanisms of genetic module prcellation In *Modularity in development and evolution,* ed. G. Schlosser and G. P. Wagner. Chicago: University of Chicago Press.

Gerhart, J., and M. Kirschner. 1997. *Cells, embryos, and evolution.* Malden, Mass.: Blackwell Science.

Gibson, G., and G. P. Wagner. 2000. Canalization in evolutionary genetics: a stabilizing theory? *BioEssays* 22:372–380.

Gilbert, S. F., J. Opitz, and R. A. Raff. 1996. Resynthesizing evolutionary and developmental biology. *Dev. Biol.* 173:357–372.

Gould, S. J. 1977. *Ontogeny and phylogeny.* Cambridge, Mass.: Harvard University Press.

Halder, G., P. Callerts, and W. J. Gehring. 1995. Induction of ectopic eyes by targeted expression of the eyeless gene in *Drosophila. Science* 267:1788–1792.

Hinchliffe, J. R., and D. R. Johnson. 1980. *The development of the vertebrate limb.* New York: Oxford University Press.

Hofacker, I. L., W. Fontana, P. F. Stadler, and P. Schuster. 1994–1998 *Vienna RNA Package.* Vienna: University of Vienna, Department of Theoretical Chemistry.

Kardon, G., T. A. Heanue, and C. J. Tabin. 2004. The *Pax/Six/Eya/Dach* network in development and evolution. In *Modularity in development and evolution,* ed. G. Schlosser and G. P. Wagner. Chicago: University of Chicago Press.

Leroi, A. M. 2000. The scale independence of evolution. *Evol. Dev.* 2:67–77.

Lynch, M., and B. Walsh. 1998. *Genetics and analysis of quantitative traits.* Sunderland, Mass.: Sinauer Associates.

Magwene, P. M. 2001. New tools for studying integration and modularity. *Evolution* 55:1734–1745.

Maynard-Smith, J. 1978. The evolution of sex. Cambridge: Cambridge University Press.

Mayo, O., and R. Bürger. 1997. The evolution of dominance: a theory whose time has passed? *Biol. Rev.* 72:97–110.

Mezey, J. 2000. Pattern and evolution of pleiotropic effects: analysis of QTL data and an epistatic model. Ph.D. diss., Yale University, New Haven, Conn.

Mezey, J. G., J. M. Cheverud, and G. P. Wagner. 2000. Is the genotype-phenotype map modular? a statistical approach using mouse quantitative trait loci data. *Genetics* 156:305–311.

Needham, J. 1933. On the dissociability of the fundamental processes in ontogenesis. *Biol. Rev.* 8:180–223.

Nemeschkal, H. L. 1999. Morphometric correlation patterns of adult birds (Fringillidae: Passeriformes and Columbiformes) mirror the expression of developmental control genes. *Evolution* 53:899–918.

Nowak, M. A., M. C. Boerlijst, J. Cooke, and J. Maynard-Smith. 1997. Evolution of genetic redundancy. *Nature* 38:167–171.

Olson, E. C., and R. L. Miller. 1958. *Morphological integration.* Chicago: University of Chicago Press.

Raff, R.-A. 1996. *The shape of life.* Chicago: University of Chicago Press.

Raff, R. A., and B. J. Sly. 2000. Modularity and dissociation in the evolution of gene expression territories in development. *Evol. Dev.* 2:102–113.

Riska, B. 1985. Group size factors and geographic variation of morphological correlation. *Evolution* 39:792–803.

Sattler, R. 1984. Homology: a continuing challenge. *Syst. Bot.* 9:382–394.

Schlosser, G. 2002. Modularity and the units of evolution. *Theory Biosci.* 121:1–80.

Schlosser, G. 2004. The role of modules in development and evolution. In *Modularity in development and evolution,* ed. G. Schlosser and G. P. Wagner. Chicago: University of Chicago Press.

Shubin, N., C. Tabin, and S. Carroll. 1997. Fossils, genes and the evolution of animal limbs. *Nature* 388:639–648.

Stern, D. L. 2000. Evolutionary developmental biology and the problem of variation. *Evolution* 54:1079–1091.

Tautz, D. 1992. Redundancies, development and the flow of information. *BioEssays* 14: 263–266.

von Dassow, G., and E. Meir. 2004. Exploring modularity with dynamical models of gene networks. In *Modularity in development and evolution,* ed. G. Schlosser and G. P. Wagner. Chicago: University of Chicago Press.

von Dassow, G., and E. Munro. 1999. Modularity in animal development and evolution: elements of a conceptual framework for evodevo. *J. Exp. Zool. (Mol. Dev. Evol.)* 285:307–325.

Wagner, A. 1999. Redundant gene functions and natural selection. *J. Evol. Biol.* 12: 1–16.

Wagner, G. P. 1996. Homologues, natural kinds and the evolution of modularity. *Am. Zool.* 36:36–43.

Wagner, G. P., ed. 2001. *The character concept in evolutionary biology.* San Diego: Academic Press.

Wagner, G. P., and L. Altenberg. 1996. Complex adaptations and the evolution of evolvability. *Evolution* 50:967–976.

Wagner, G. P., C.-H. Chiu, and M. Laubichler. 2000. Developmental evolution as a mechanistic science: the inference from developmental mechanisms to evolutionary processes. *Am. Zool.* 40:819–831.

Wagner, G. P., and J. Mezey. 2000. Modeling the evolution of genetic architecture: a continuum of alleles model with pairwise AxA epistasis. *J. Theor. Biol.* 203:163–175.

Wagner, G. P., J. Mezey, and R. Calabretta. In press. Natural selection and the origin of modules. In *Modularity: understanding the development and evolution of natural complex systems,* ed. W. Callebaut and D. Rasskin-Gutman. Cambridge, Mass.: MIT Press.

Wake, D. B. 1991. Homoplasy: the result of natural selection, or evidence of design limitations? *Am. Nat.* 138:543–567.

Wilkins, A. S. 1997. Canalization: a molecular genetic perspective. *BioEssays* 19: 257–262.

Winther, R. G. 2001. Varieties of modules: kinds, levels, origins and behaviors. *J. Exp. Zool. (Mol. Dev. Evol.)* 291:116–129.

Yang, A. S. 2001. Modularity, evolvability, and adaptive radiations: a comparison of the hemi- and holometabolous insects. *Evol. Dev.* 3:59–72.

Zuckerkandl, E. 1994. Molecular pathways to parallel evolution. 1. Gene nexuses and their morphological correlates. *J. Mol. Evol.* 39:661–678.

16 Generative Entrenchment, Modularity, and Evolvability: When Genic Selection Meets the Whole Organism

WILLIAM C. WIMSATT AND
JEFFERY C. SCHANK

Modularity and the Problem of Generative Entrenchment

There are in this book almost as many concepts of modularity as arguments for it. Anatomical elements in serial homology, chromosomes, supergenes, neural circuits, and traits have all been treated as modules. Now we have new kinds of things: HOX clusters, morphogenetic fields, and undefined quasi-independent traits that must be nearly independently selectable if evolution is to work. There are also the various detailed developmental-genetic dynamical structures and the hypothesized adaptive complexes they generate. Winther (2001, in prcss) provides a taxonomy. We prefer the term *evolutionarily significant modules* for the units we discuss, because evolvability appears to require such modules, and because our argument concerning them derives from considering the nature of selection processes.

We begin with a class of models for the evolution of developmental structures which include an important and general dynamical structural feature: *generative entrenchment,* or GE (Wimsatt 1981, 1986; Rasmussen 1987; Schank and Wimsatt 1988).[1] To summarize where we will go, in 1988, we first described a new and puzzling phenomenon in some multilocus selection models designed originally to test and modify models by Kauffman (1985, 1993). These systems generated interactions among alleles at different loci in cases where one might suppose they should not occur. This phenomenon had an interesting explanation involving heterogeneous fitness classes and a new kind of frequency-dependent selection, but we were unsure then how to generalize it (Wimsatt and Schank 1988). We present and discuss more striking demonstrations of it here, whose analysis suggest that it is robust and should apply quite broadly. Prima facie, this phenomenon should make significant evolutionary changes, including modular ones, more difficult. We are convinced that dealing with the processes described here must be a part of the solution to how modularity is possible. We would like to know what conditions allow or prevent the evolution of modular architectures.

Generative Entrenchment

Consider the phenotype as a generative structure: a smaller number of elements interact with each other and with inputs from the environment, producing a growing structure of interacting elements having a broadly adaptive organization. Circuits in the genome and their products fit this description (as elaborated and reviewed by Davidson [2001]), as do particular genes (e.g., *bicoid;* Schmidt-Ott and Wimmer 2004). Different elements of the phenotype from individual genes, through circuits, to developing morphological elements and behaviors should all fit this broad description. Environmental inputs, maintained conditions, and periodic or other patterns of change (e.g., daily or seasonal variations) can also have significant generative roles in the developing phenotype (Greene 1989).[2]

The generative structure of the system (including the organism plus relevant aspects of its environment) has a characteristic set of causal interactions which could be variously represented. One of the simplest representations is a directed graph, where nodes are parts, processes, or events, and arrows are consequences of the presence or operation of nodes on other nodes. For each node, consider how many other nodes can be reached from it by following the arrows. This indicates how much of the phenotype is downstream of, causally dependent upon, or affected by a given node. We define the *generative entrenchment* of a node as the magnitude of this downstream dependence. (This is left deliberately vague—it could be operationalized in various ways.) See figure 16.1. GE is not limited to classical developmental processes. It spans the "extended phenotype" or the "recurrently assembled" resources reaching beyond traditional organismal boundaries as posited by developmental systems theory (Wimsatt 2001). Downstream consequences considered should be limited to characteristic effects, common and systematic enough that they could be consequences of selection.

Generative structures are found at various levels of organization and the interactions between them. Consider generative structures produced in self-organizing systems such as social insect colonies. Several species of army ants go through emergent reproductive cycles of over a month, in which they enter statary and nomadic phases (Schnierla 1971). Nomadic phases are driven in part by the development of larvae (nodes) and emergence of new workers (nodes) both chemically

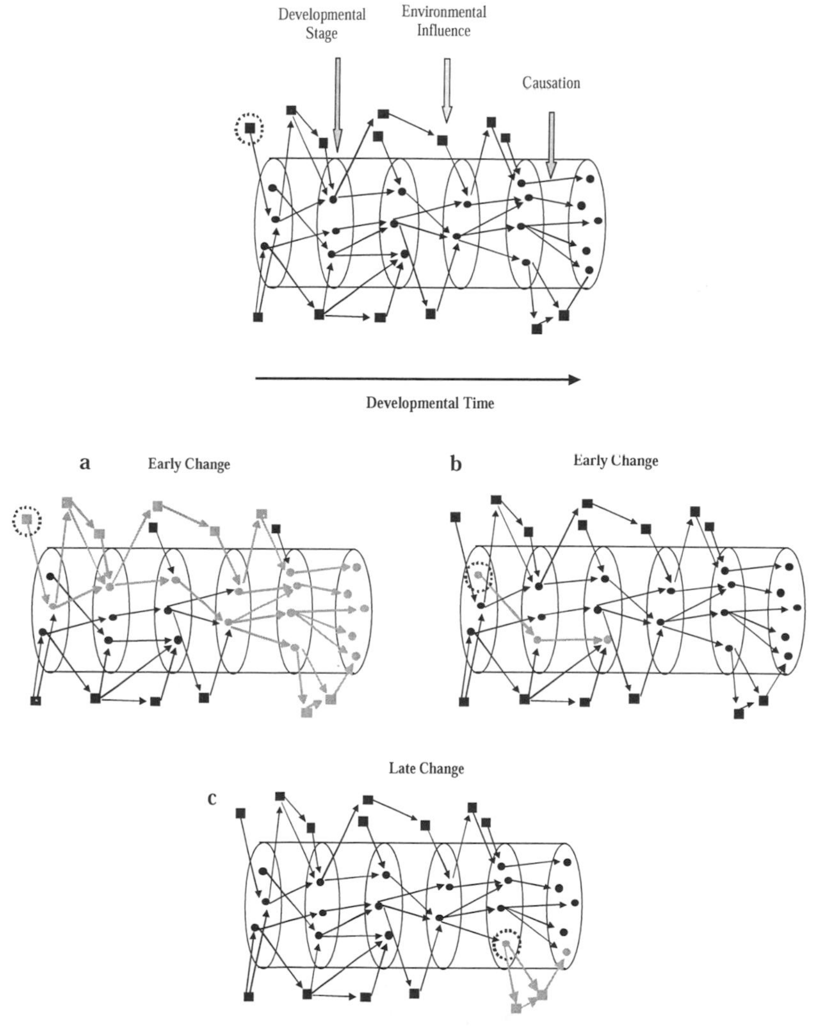

Fig. 16.1.—Representations of the life cycle, with developmental stages and generative entrenchment (*a*). Circles are features internal to the phenotype, squares are features of the environment, both characteristic of normal development and interacting causally through a succession of developmental stages. Nodes in dotted circles are assumed to be changed or perturbed, with downstream consequences propagated along arrows to other more lightly shaded nodes. Activation or change of features early in development characteristically has more pervasive effects, as in *b*, than later changes, as in *d*, but some early changes have relatively small effects, as in *c*. Some may have no effects (as with a mutation in an inactive pseudogene). And many effects exit and reenter across organism-environment boundary.

and tactilely interacting (connections) with adult workers (downstream nodes), "agitating" them and initiating the nomadic phase with colony-level patterns of behavior and spatial relationships (colony-level phenotypic patterns of behavior). Here the rules of individual interactions (between nodes, larvae, newly emerging workers, and workers) become entrenched, as do mating rituals, more generally, in animal systems of reproduction.

All other things being equal, nodes with more downstream connections should be more evolutionarily conservative, because there are more consequences of their activity, which can be disrupted if connections are changed arbitrarily. Thus, there is a much greater chance that something will go wrong if they are changed. In the simplest (multiplicative) selection models, the probability that a modification exhibits improvement over a variant or variants already present declines exponentially with increasing number of downstream nodes. The probability of *strongly* deleterious or lethal consequences also increases. In simple models, perturbations earlier in development should be more likely to have widespread downstream effects. This conceptual model strongly suggests a probabilistic analogue of von Baer's law that differentiation proceeds from the general to particular, with taxonomically more general things expressed earlier in development. This also suggests a cone of increasing variation with progress through development. Greater generative entrenchment yields greater evolutionary conservatism, which leads to greater taxonomic generality (Riedl 1978; Arthur 1984; Wimsatt 1981, 1986; Rasmussen 1987, Schank and Wimsatt 1988). This is confirmed, with some qualifications, in all of our studies, including the simulations reported below.

Various critics have proposed a pattern different from the "cone of variation": developing from a single-celled zygote to the variously called "neck" of an "hourglass." Elinson (1987) compared early stages to a "funnel," with a now famous figure adding the diverse earlier stages left out by Haeckel to his widely copied drawing of diverging later stages of diverse vertebrates. The waist of the hourglass has been called variously the "phylotypic stage" (Sander 1983), the "Bauplan," and the "zootypic stage" (Slack et al. 1993; Duboule 1994). These terms have different theoretical associations, and different mechanisms are proposed for the "neck," but each is taken to correspond to a specific and significant reduction in variation at an intermediate developmental stage across specific animal groups and possibly more generally across taxa of intermediate generality. Raff (1996, 208) refers to this pattern as the "developmental hourglass." This pattern has recently become a center of debate over how rigidly constrained development is at various earlier stages in ontogeny, and the possible evolutionary consequences of different patterns of constraints. And how, above all, is the early developmental variability consistent with the downstream dependence we expect? (All concerned appear to accept GE as acting and explaining patterns of variability downstream of the neck.)

None of the models above deal adequately with complications arising from gene (or other) duplications, functional redundancy, changing degrees of modularity, or canalization at different stages of development, at least not if these structural organizational parameters are free to vary independently. (They probably cannot vary independently, but this cannot be assessed until we have models incorporating them.) Modularity, duplication, and functional redundancy should each decrease entrenchment by reducing interdependence between (for modularity) or dependence upon specific system components (for duplication or redundancy), but assessing how all of these factors interact demands both more empirical and theoretical work. Nonetheless, GE does suggest a prima facie constraint acting in development, and developmental geneticists now commonly infer from evolutionary conservatism or phylogenetic breadth to greater causal centrality, and conversely. It remains a good default inference for later development in spite of various still unexplained kinds of exceptions showing greater divergence in earlier developmental stages of amphibians, ascidians, and echinoids (Raff 1996). Some or all of these exceptions may be due to the above lacunae in the models. If so, the problem lies not with GE per se, but with overly simple models or inaccurate conceptualizations for evaluating GE in complex organizational structures.

Darwinian processes should almost inevitably give rise to generative structures (Wimsatt 2001).[3] However, we are still left with two perplexing questions: How can complex adaptive systems evolve and continue to evolve in any other than a predominantly accretionary way if their generative elements become increasingly entrenched with increasing complexity (Schank and Wimsatt 2000)? How does this permit continued modular evolvability?[4] It is no surprise, therefore, that a fundamental research focus of the evolutionary sciences is to figure out how complex systems can continue to evolve when evolutionary processes generically give rise to entrenched structures. We will call this the GE paradox.

We suspect that elements of early stages are quite entrenched within a lineage—that is, experimental manipulation of pre-Bauplan stages causes problems, but there are many ways of getting to highly similar phylotypic Bauplan stages in different related lineages. On this hypothesis, the Bauplan stage is in effect canalized. (1) Generative elements early in development are more likely entrenched in complex adaptive systems, (2) they are more likely more entrenched, (3) more entrenched elements have lower probability of adaptive modification when changed, and (4) greater GE increases the probability of *major* dysfunction when changed. These are *only* probabilities because of variations in network structure, in their adaptive design, and in detailed conditions that modulate consequences. Some entrenchments reflect differences in adaptive design at the species level or higher, some, genetic variation within species, and some, developmental noise and environmental variation. Specific solutions to the GE paradox may provide a starting point for a deeper understanding of how complex adaptive systems evolve despite expectations that GE grows nonlinearly with complexity. That early developmental changes are relatively common in amphibians indicates something systematic if they do not have a common origin: we want to know why so many exceptions to the GE paradox should occur there. Are these exceptions piecemeal fixes or do they indicate general principles or both? Much work clearly remains.

Simulating the Effects of Generative Entrenchment

Our simulations were originally designed to test hypotheses by Stuart Kauffman (1985, 1993) on the evolution of gene control networks. Kauffman argued that selection could not maintain more than a small fraction of the connections in a gene control network of realistic size in the face of mutation. His model networks were directed graphs, with genes as nodes and arrows (connections) between them representing influences of one gene on another (fig. 16.2). Mutations involved random reassignment of the head or tail of an arrow, changing gene interactions, and altering the connectivity of the network. Each gene thus had two ways to mutate, so n connections had $2n$ mutable "sites." The ratio of genes to connections determined the connectivity of the network and, through this, many of its generic (overwhelmingly likely, and thus robust) topological properties (Kauffman 1985, 1993). Generic properties are (commonly statistical) properties which are widely distributed among members of an ensemble of systems constructed subject to certain constraints. Generic properties of networks of M nodes with N connections per node might include the average number of connections reachable from nodes, the number of closed cycles in the network, and the mean cycle length, expressed as functions of M and N.[5] We considered the robustness of his conclusions in more realistic simulations which included the effects of generative entrenchment. GE is also a generic property of such structures, but one which Kauffman failed to model (Schank and Wimsatt 1988). At least some results from these models are more general, affecting any phenotypes whose genes or generative elements are subject to a diversity of selection forces of different magnitudes. This diversity may be produced by differential entrenchment but need not be. Since having a nonhomogeneous array of selection pressures is a robust property of machines and evolving systems generally (Wimsatt and Schank 1988), this should apply widely to virtually all phenotypes.

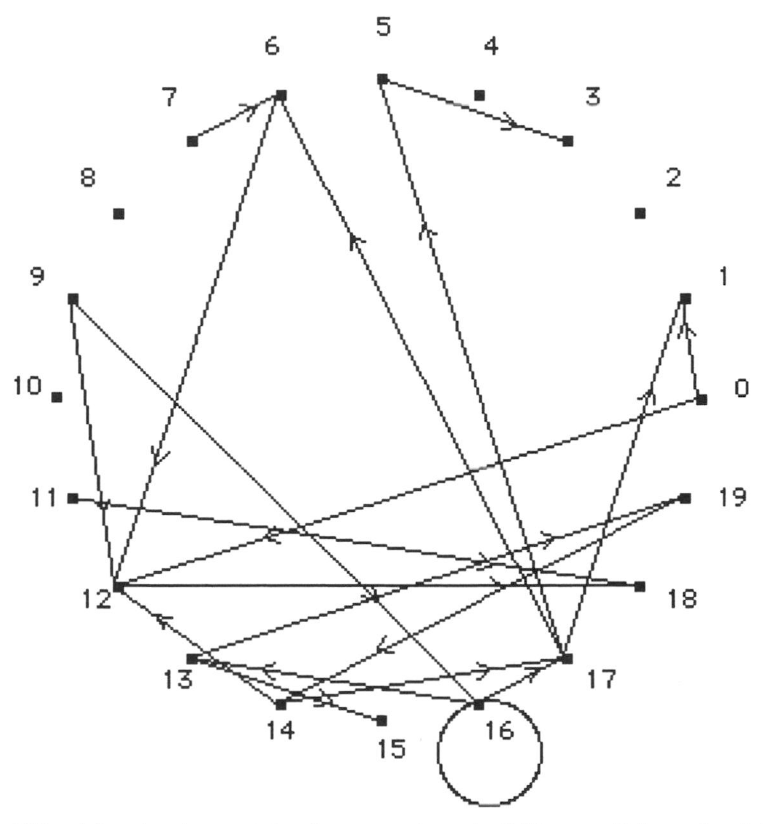

Fig. 16.2.—A directed graph representation of a gene control network with 20 genes and 20 connections. Nodes are genes, and a directed arrow indicates action of the gene at the tail on the expression of the gene at the head of the arrow. Mutations act on connections, and may randomly reassign the gene at the head or the tail of the arrow. With 20 connections, it thus has 40 mutable sites. This particular graph (data output of the computer program) was produced by 100 mutation events acting on a closed loop model gene system of 20 genes and 20 connections. It is indistinguishable in generic properties from one constructed at random.

Simulation Procedure

In Kauffman's simulations, populations of "ideal" networks with an arbitrarily chosen "wiring diagram" were subjected to mutation and allowed to reproduce with fitness determined by similarity with an "ideal" network. (The similarity measure was the number of shared connections). This was a relatively simple "selection-mutation balance" model, though with many loci rather than just a few. Populations had 100 circuits (or "organisms"), each with 20 genes and 20 connections (so 40 mutable sites), a mutation rate of .005 per site per generation, and selection coefficients of .05 per connection. (Kauffman's selection coefficients were the same for all connections, a crucial assumption.) After 1,000 generations of selection, approximately 45% of the "good" connections in a circuit were randomly replaced by others. With his assumptions, this yields a reduction in fitness of 45% and, more important, the loss of any structure involving more than a few connections.[6]

These changes displayed the relevant properties of generic networks (fig. 16.2). (A generic property is in effect a "high entropy" property—possessed by almost every member of the ensemble of structures possible for the parameter values used.) If mutations randomly sample the state-space of possible connections, mutations to structures already in generic states will likely remain generic while structures in nongeneric states will probably change to more generic states. It should be harder to maintain structures in nongeneric states the less generic they are. From the easy transition to generic states observed in randomly mutated circuits, Kauffman argued that selection could not maintain specific wiring diagrams in large circuits and thus that almost all relevant properties of large gene control networks would be generic and not explained by selection. For generic outcomes, any circuit would have a specific structure, but the specificity was irrelevant: what mattered was that the same outcomes would result also from a large variety of alternative circuit structures. It is good if important things can be "generic" (making them highly robust), but we were bothered by the strong limitations on selection suggested by his model, and particularly by some of the assumptions he made to get them.

Kauffman's assumption of equal fitness decrements for the loss of any connection was simple but problematic. In virtually any circuit (random or not), different nodes or connections influence different numbers of other nodes. In figure 16.2, the connection from 5 to 3 has no further consequences (no arrows leave 3), but the connection from 16 to 13 has many. From node 16, we can travel 16 → 13 → 19 → 14 → 17 → 5 → 3, with other divergent paths along the way. It is rare—essentially impossible for robust reasons—to find interesting networks in which all nodes have equal influence (Wimsatt and Schank 1988). The loss of connections through which many nodes are reached should cause more disruption than the loss of those leading to only a few. So larger selection coefficients should be assigned to nodes with more nodes and connections downstream. Our simulations paralleled and extended Kauffman's, with this crucial difference: in our models, fitness decrements for the loss of a "good" connection were always at least a monotonically increasing function of the number of the nodes downstream of it.[7]

Initial Results and Discussion

We simulated various-sized populations of circuits with different numbers of nodes and connections, connection densities, mutation rates, and fitness functions for different measures of GE. With similar parameters, we observed a decline in the proportion of good connections to equilibriums of about 50%, as noted by Kauffman (1985, 1993). Identical selection coefficients caused random loss of good connections in Kauffman's simulations, but our similar losses occurred almost entirely among connections with lower GE, with consequences elaborated elsewhere (Schank and Wimsatt 1988; Wimsatt and Schank 1988) and further below. Differential preservation of generatively entrenched connections occurred even with selection coefficients not much stronger than those lost: the effect was strongly nonlinear.

These simulations were designed to assess the effects of differential generative entrenchment of phenotypic traits, but this assumption was used only to assign different selection coefficients. Thus, the results are more general. Even relatively small differences among selection coefficients of genes (whether due to differential generative entrenchment or differential adaptive importance) were sufficient to bias the preservation of nodes toward the more strongly selected ones, even with quite strong losses due to drift.

A More Sophisticated Model

These results were confirmed and elaborated with a more realistic and adjustable model (Wimsatt and Schank 1988) allowing simulations with much larger circuits, and more revealing and useful graphic output. The two most significant changes were:

1. Fitness decrements were *assigned* to connections, rather than *calculated* from their generative entrenchment. This was usually unproblematic, because the model dynamics were affected only by the selection coefficients.[8] This throws away information, but the ability to choose distributions of selection coefficients from scratch allowed crisper tests and demonstrations that were crucial below. Plausible distributions of selection intensities for randomly constructed networks could still be determined and then plugged into the model (as we did), but one could also test distributions of selection coefficients that might not occur in nature to determine their effects.

2. Most critically, Kauffman's (1985, 1993) assumption that all fitness contributions summed to 1 was relaxed. This assumption seems initially plausible for relative fitness: we commonly let fitnesses of genotypes range from 0 to 1. Kauffman's model assumed something that looked equivalent, but is not: that the sum of fitness *decrements* for deleterious mutations in all genes summed to 1. This version is not realistic either for populations or for individuals. (So if there were a lethal mutation, then all other mutations would have to be neutral!) In our model, the maximum sum of fitness decrements for an individual

(which we call the *exposure*) can be greater than 1. The value of this sum is a key parameter. Fitness decrements were calculated for each circuit in the population, and circuits with total fitness decrements (*realized exposure*) greater than or equal to 1 were assigned fitness 0. So on this model fitnesses still range from 0 to 1, but the sum of fitness decrements do not. (When you're dead, you're dead, but there are many ways to kill a cat, sometimes simultaneously.) All organisms and any complex machines (those with more than one part!) characteristically exhibit this kind of potential overkill.

This failure mode deserves to be a principle of metaengineering. You are not even in the game unless most ways in which you could be dead will not happen and none of the others have happened yet. Some things you just can't mess with—those are protected before we ever start talking about *differential* selection. As we will see below, this fact preserves more genes, gene combinations, and adaptations than would be possible if Kauffman's picture were right. In effect, Kauffman's model is too optimistic: with his fitness assignments, in the worst possible case, you're dead—but just barely. But a "squashed bug" or "road kill" is not just barely dead. Truncation selection (in which any organism with phenotypic values on the wrong side of critical "thresholds" fails to reproduce) probably characterizes all real cases of selection.

In this model, phenotypic properties are characterized only in terms of their contribution to fitness—which may itself be thought of as a very abstract phenotypic property. The model here is a modified truncation selection model (with a threshold at fitness 0, and reproduction with probabilities proportional to fitness if it is greater than 0). It is more realistic than Kauffman's models (1985, 1993), and also more realistic than truncation selection models with selection as an all-or-nothing affair with 0 fitness below threshold, and 1 above. It allows lethals and a (kind of) conditional lethal. (A lethal is a mutation with a fitness decrement of 1. It is treated as unconditional. So even the fittest genotype [fitness 1] modified by adding a lethal mutation then has fitness 0. By extension, one kind of conditional lethal for a genotype [there are others][9] is one whose assigned decrement makes the fitness of *that* genotype less than or equal to 0. It is conditional because there are other [fitter] genotypes for which that decrement would not be lethal. This kind of conditional lethal is also captured by the model). This model is also convenient for the simulation of genetic load phenomena: the genetic load of a population is $1 - W$ (where W is the mean Darwinian fitness of the population) if the maximum fitness of a genotype is set equal to 1. The effects of further simplifications in these models will be assessed in the discussion.

Simulation Procedures, Results, and Discussion

Figure 16.3 illustrates a model with these new assumptions (Wimsatt and Schank 1988), showing how connection frequencies change over time in 10 averaged runs ("replications") for 2,000 generations with 100 haploid asexual circuits each having 100 genes with 100 connections assigned equally (20 each) to five fitness classes of .02, .03, .04, .05, and .06. The total exposure of 4 is contributed proportionally (.4, .6, .8, 1.0, and 1.2) by these different classes. (This is high enough to show the effects of higher exposures, but probably still much lower than found in nature. Higher exposure should allow maintenance of more genes by selection if the reproductive potential is high enough to handle the greater expected mortality.[10] This is a variant of the genetic load problem.) A mutation rate of .005 per connection end per generation yields a (plausible) average of 1 mutation per organism per generation, and a wait for back mutations comparable to the length of the simulation. Mean population fitness (plotted every 500 generations) declines to around 10%–11%. Therefore, a reproductive potential of 10 or more is required for the population to maintain itself.

The top graph is a continuous stacked bar graph showing the cumulative frequency of good connections in all classes over time (incremented every 50 generations). (Separate data for different fitness classes were unavailable for graphs in Schank and Wimsatt 1988.) The cumulative frequency stabilized at about 70%, considerably higher than Kauffman's 50%–55%. With circuits five times larger, six to seven times as many connections were preserved as in the original Kauffman simulations. With an exposure of 4 rather than 1, circuits that lost even a few more connections were lethals. So Kauffman's (1985, 1993) low connection equilibrium was at least partially an artifact of not using a truncation-selection model.

Two particularly salient results emerged from these simulations:

1. As with simulations for smaller circuits, a strong differential bias favors retention of more strongly selected alleles. This bias is visible in both the top and bottom graphs of figure 16.3.

2. A new phenomenon emerged in the bottom graph of figure 16.3—a bifurcation between the fates of connections with lower and higher fitness decrements. The two lowest fitness classes (.02 and .03) decreased in frequency throughout the simulation. The two top fitness classes (.05 and .06) also decreased initially but reversed direction after generation 50 (the plotting interval) and went nearly to fixation. The middle fitness class (.04) seemed right on the fence: after generation 100 it fluctuated indefinitely at 76%–80%. So there were very different fates for connections with relatively small differences in selection coefficients.

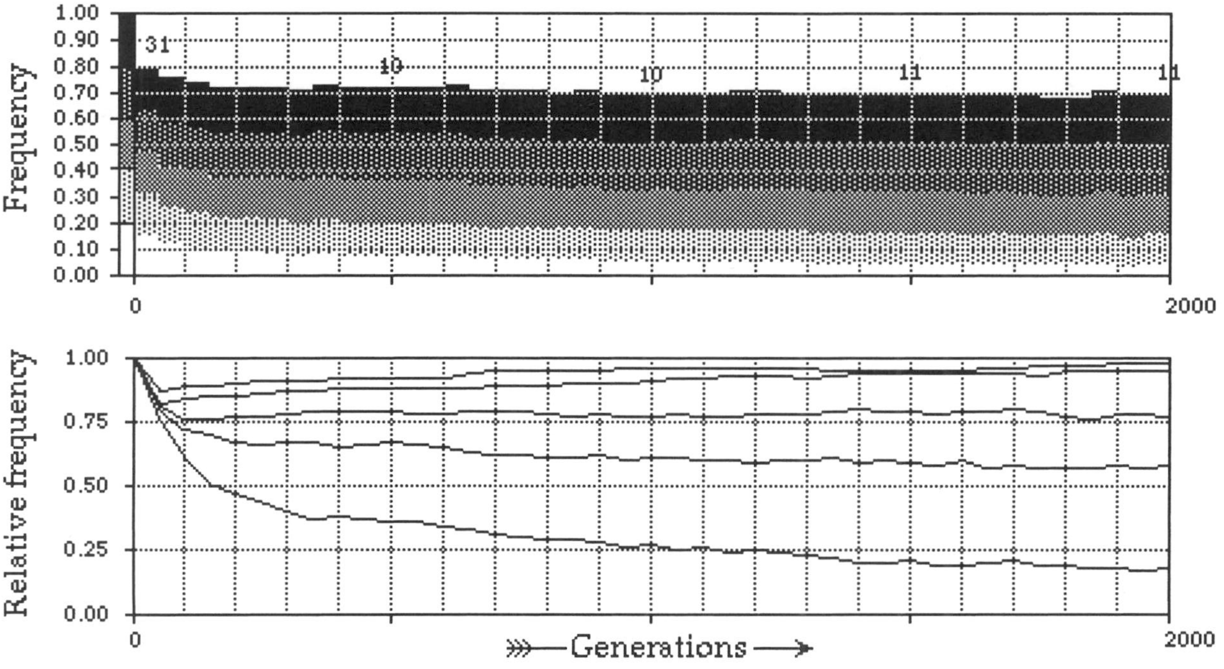

Fig. 16.3.—Evolution of connection frequencies (of good connections) in five fitness classes in gene control networks having 100 genes with 100 connections, so each network has 200 mutable sites. Population size is fixed at 100. The mutation rate is .005 per connection end (thus averaging 1 mutation per network per generation). The five fitness classes have 20 connections each of fitness decrements of .06, .05, .04, .03, and .02. Total exposure, the sum of fitness decrements is thus 4. Frequencies are averages for 10 runs. Mean population fitness (in % of start) is indicated every 500 generations in the top graph. Shading gives the relative selection intensity of different classes, with black for the greatest. *Top graph,* a stacked bar of frequencies of connection types as a fraction of the total at the beginning. *Bottom graph,* class frequencies as proportions of their initial values. Their average frequencies through time follow the same order as their assigned selection coefficients. Note frequency reversals in the top two classes, discussed in the text.

Why did this bifurcation of trajectories occur? Our explanation has broad implications for selection models. There are three contributing factors. The first is a kind of fitness rescaling which applies even in Kauffman's original simulations. Also necessary (but missing from Kauffman's models) is the existence of different fitness classes, which with the fitness rescaling produced a dynamics which can yield trajectory reversals. The third factor (also missing from Kauffman's models) is truncation selection, which amplifies effects of the first two.

Consider the fitness rescaling: start as Kauffman (1985, 1993) did with a population of identical organisms each with an ideal circuit having 20 connections, a fitness of 1, and a fitness loss of .05 per connection for mutated connections—a 5% loss in relative fitness. Now suppose after some generations with drift and random mutation, the fittest circuit has but 10 good connections and a fitness of .50 (on the original scale). But the loss of a connection with.05 in fitness now represents a 10% loss in *relative* fitness. Thus, as the mean Darwinian fitness of the population declines, the loss of an allele with a constant absolute fitness contribution yields a successively larger loss in relative fitness.

This is a new kind of frequency-dependent variation in selection in what was supposedly a strictly additive model with constant fitnesses. A multiplicative model would scale appropriately to produce no changes in relative fitness if fitnesses asymptoted to zero, but not if the model were adjusted (as below) to allow for truncation selection. So if truncation selection is general, so is this mechanism. The additive versus multiplicative assumptions and their effects in population genetics are discussed in Wade et. al 2001.

A trajectory reversal (causing a bifurcation) requires also more than one fitness class. Any continuous distribution of fitnesses would do, but discrete classes make the phenomena clearly visible.[11] Kauffman's original simulations in which all connections have equal fitnesses cannot do it. With multiple fitness classes, mutations and drift cause initially highly fit organisms to lose alleles, stochastically, at equal rates. Losses occur until those loci reach their respective selection-mutation equilibriums. Classes with smaller fitness contributions lose proportionately more alleles because those alleles have lower equilibriums with mutation. They all started at fixation, so allele frequencies of lower fitness alleles have further to go and take longer to get there.

Alleles need not start at fixation in natural populations, but larger changes in selection coefficients will still lead to more rapid changes in the affected allele frequencies. Less strongly selected alleles cannot "keep up." A vector of changes in selection coefficients of different sizes acting on different alleles should have an effect similar to those demonstrated here. The processes discussed here are not limited to the particular initial conditions of these simulation experiments, and also not limited to selection for a single optimum circuit.

In these simulations, as the population's mean absolute fitness decreases over time, *relative* selection coefficients of alleles in all fitness classes increase by the same scalar factor ($1/W_{max}$) until the fitter ones stop decreasing, reaching their (temporary) equilibrium first. But alleles in the lower fitness classes have not yet reached theirs, continue to be lost, and in doing so further inflate all relative selection coefficients. Alleles in the higher fitness classes now reverse (in order of their selection coefficients) and increase in frequency. In effect, the population reached a rough fitness equilibrium (in fig. 16.3, from roughly generation 500 on), after which evolution proceeded by the substitution of fitter alleles for less fit ones at other loci.[12]

Without these fitness differences, all classes would have reached equilibrium at the same time, with no selection reversals. With ubiquitous fitness differences at different loci in nature, occurrences like these must be common: this mechanism should be actuated by any environmental shocks or arbitrary perturbations from equilibrium. Indeed, this phenomenon raises new questions about whether it ever makes sense to expect frequency-independent selection—broadening a conclusion Lewontin reached by another route in 1955.[13]

Further Validation of These Proposed Mechanisms

The simulations of figure 16.3 had equal numbers of alleles in the different fitness classes, an unrealistic assumption. Evaluation of generative entrenchment for connections or nodes in circuits generated at random (fig. 16.2) suggested a roughly exponential distribution, with geometrically increasing numbers of connections having lower fitness decrements. Cellular descent trees or branching cascades in gene control mechanisms in which all cells or nodes downstream of a given event are affected (or even just a constant proportion of them) suggest similar relations (though redundancy, canalization, and differential modularity could affect these). We have investigated models with a variety of different exponential distributions of connection numbers (ratios of 2, 3, 4, and 5 of connections in neighboring fitness classes, giving various "rates of expansion" for the exponential). They all showed similar kinds of interactions.[14]

The model circuits with the particular parameter choices discussed below had exponential distributions, and also elegant symmetry properties, which facilitated detection of patterns of interclass compensation on several levels (fig. 16.4 and subsequent figures). These circuits had five fitness classes contributing to a total exposure of 4, as above, but different fitness distributions and a larger number of connections.

Crucial for symmetry properties, the connection-class fitness decrements were halved as the number of connections doubled, so that each class made the same total fitness contribution (of .8). The largest such circuit type simulated had 8 connections with a fitness contribution or decrement of .10, and the other fitness classes, 16 with a contribution of .05, 32 with a contribution of .025, 64 with a contribution of .0125, and 128 with a contribution of .00625, for a total of 248 connections with a 16-fold range of selection intensities ranging from very strong to quite weak.[15]

Is there any way to further test the form of fitness interactions suggested in the discussion of figure 16.3? Suppose that trajectories were products of equilibrating processes of selection and mutation operating both within and across fitness classes. (Standard mutation-selection balance models would cover equilibration within a single fitness class, but there is no reason to suppose that equilibration processes are compartmentalized in this way.) Such interactions were assumed in our explanation for the trajectory reversals above. The fact that two mechanisms, truncation selection and rescaling of relative fitness for alleles with constant "absolute" fitness contributions, both produce this effect should make it especially robust.

The proposed mechanism should be at least partially compensating as well: if one class fluctuated higher than its expected value while all others were at theirs, the higher mean Darwinian fitness would depress the realized *relative* fitnesses, relaxing all classes (stochastically) to lower levels under mutation pressure, and similarly for deviations in the other direction. This suggests interclass compensation for frequency deviations. Testing specifically for it would further validate this proposed mechanism. We want a situation in which if one fitness class deviates from its equilibrium trajectory, causing a net fitness deviation from equilibrium of a given amount, we can determine whether and how closely the sum of frequency deviations in the other classes produces an equal and opposite fitness deviation. In the model of figures 16.4–16.7 we produce this in a situation which also has a more realistic distribution of selection coefficients. We look for returns to equilibrium after the system is perturbed from it by "natural" fluctuations, produced by sampling processes.

In prior research we had focused on trajectory averages, accumulating individual replications to improve estimates. Smaller populations (only 10 rather than the 100 used earlier) showed much larger fluctuations in class frequencies and led to a crucial reassessment of the data—a figure-ground reversal. These fluctuations were remarkable, with nearly simultaneous visibly countervailing fluctuations in different classes. But instead of being noisier and more problematic for calculating averages, the character of these fluctuations became the targets of interest. If present, an equilibration process should be at work in each replication, with individual noise as "natural experiments" yielding different sets of perturbations on it. Averages were no longer data goals, but became reference standards to calibrate perturbations in individual replications.[16] Fluctuations far from the equilibrium trajectory in one class should be opposed by fluctuations in one or more of the other classes. If evenly balanced, fluctuations in any class ought to be compensated by the sum of deviations in the other four.

Even with populations of 10, very few simulations had large fluctuations permitting easy analysis.[17] The most striking, replication 50, had massive fluctuations in an unusual and very useful pattern (fig. 16.4): two out of eight connections in the largest fitness class were lost by drift very early (by generation 100). These both subsequently back mutated, going rapidly to fixation—the first between generations 400 and 450, and the second between generations 1,300 and 1,350. This gave three of the largest possible perturbations: the double loss at the beginning, and two back mutations restoring connections to populations, which (by hypothesis) had adapted to their absence. (Back-mutated large connections [classes 1–3] go from 0 to fixation within the 50-generation sampling period of the data). The double loss in the largest fitness class was remarkable, and its presence in the first class allowed a fortuitous form of data aggregation by pairing the other four classes to show patterns not visible otherwise.

The analysis began with graphs output by the simulation: two each of the 50-run averages and the trajectories of replication 50 (fig. 16.4). Pairing average and replication 50 trajectories (the graphs second and fourth from the top in figure 16.4) together, we could extract the deviations (fig. 16.5). The deviations revealed clearly countervailing trajectories, with the deviations in classes 2–5 above the x-axis while the massive perturbation in class 1 was below. Classes 2–5 then recentered their (still oppositely directed) fluctuations twice, matching their compensation as the deviation in class 1 relaxed twice precipitously back to the zero line.

We then plotted the sum of deviations in classes 2–5 with the primary deviation. The lumped sum was a noisy mirror image. Reversing the sign of the primary deviation (fig. 16.6) permitted direct comparison—and a remarkably good fit: the scale of the noise was small compared to the large excursions in the primary deviation that it followed so closely. Summing all five deviations to see their net effect deviation from the equilibrium trajectory revealed a large transient excursion downward made and reversed in the first 200 generations as mean Darwinian fitness decreased and the relative selection pressures increased, forcing it back into line. This was followed by relatively close tracking

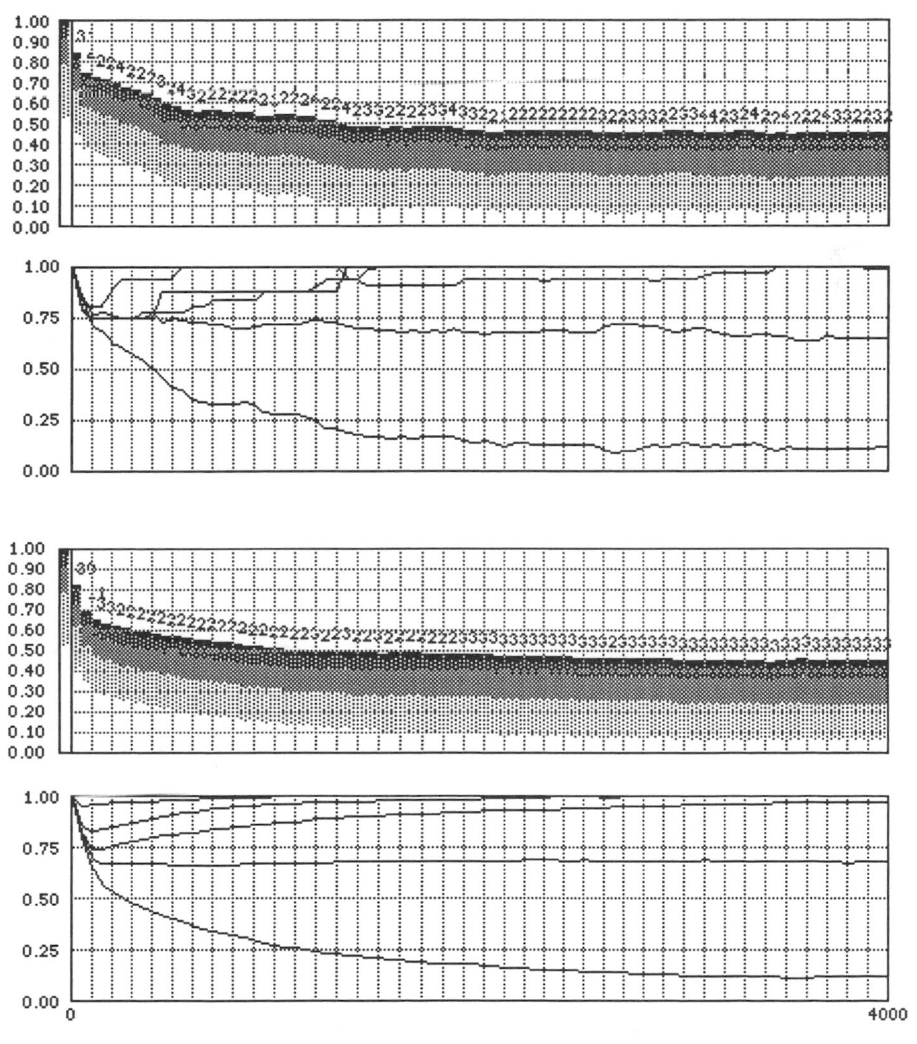

Fitness Types: 1: -0.1000 2: -0.0500 3: -0.0250 4: -0.0125 5: -0.0063
Genes = 100 Connections = 248 Mut. Rate = 0.0025000 Pop. N = 10

Fig. 16.4.—Gene frequency trajectories for connections in five fitness classes in a 248-locus haploid model, with 8, 16, 32, 64, and 128 connections each having fitness contributions, respectively, of .10, .05, .025, .0125, and .00625. Thus, each fitness class makes the same fitness contribution, of .8. Summed over all five classes this gives a total exposure of 4. The mutation rate is .0025 per connection end per generation. *Top two graphs,* run 50. *Bottom two graphs,* average trajectories for runs 1–50. The top graph in each pair is a cumulative stacked bar graph of frequencies in all classes. The bottom in each pair gives the relative frequency within each type by class. Note stochastic fluctuations in the top two graphs from average trajectories in the bottom two produced by the use of small ($N = 10$) populations. Here there are anomalous trajectory reversals with initial decreases and later increases in the top three fitness classes.

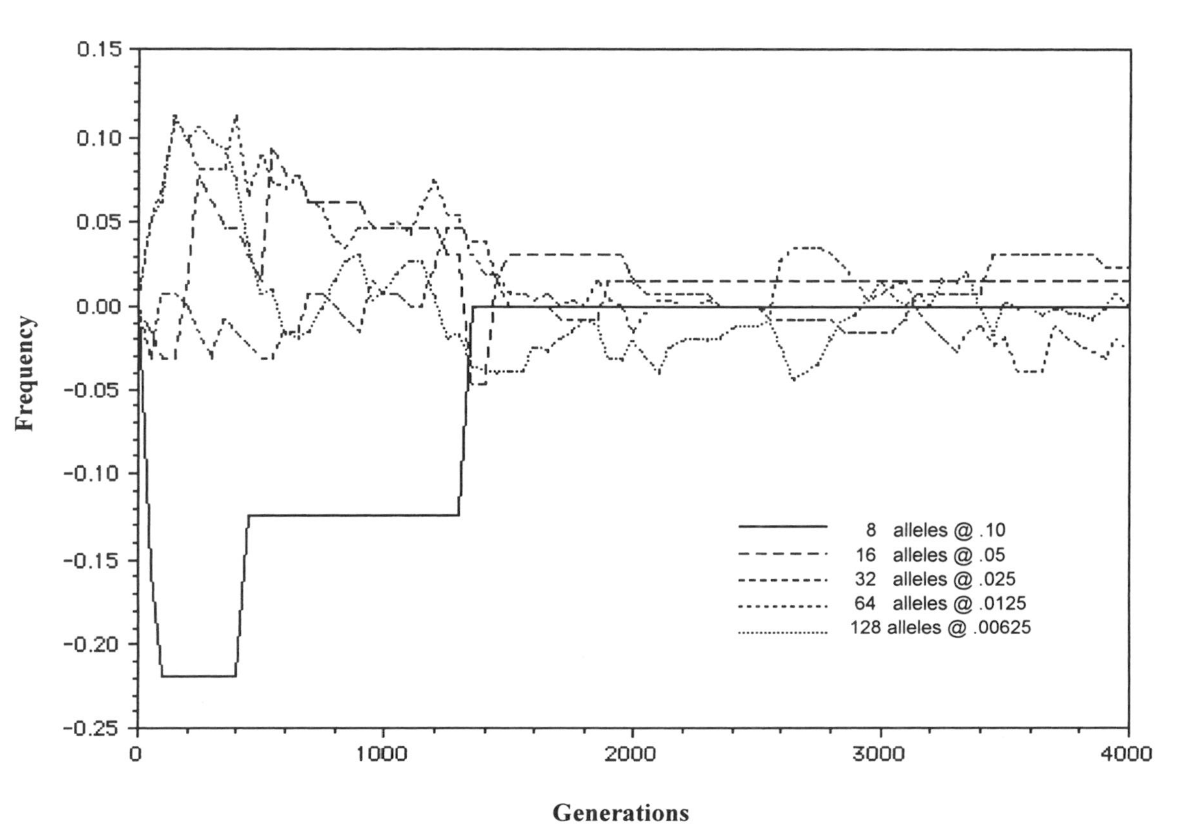

Fig. 16.5.—Combination of the data from the second and fourth graphs from the top of the preceding figure to show deviations from 50-run averages of frequencies of different fitness classes of connections for replication 50, having an unusually large fluctuation in the first fitness class (loss by drift of alleles at two out of eight loci). Larger fitness class frequencies (.1, .05, and .025) seem quantized after the first 200 generations (suggesting that back-mutated loci go to fixation within the same period they reappear, i.e., within 50 generations). Positive deviations in four smaller classes approximately offset negative deviations in the first class.

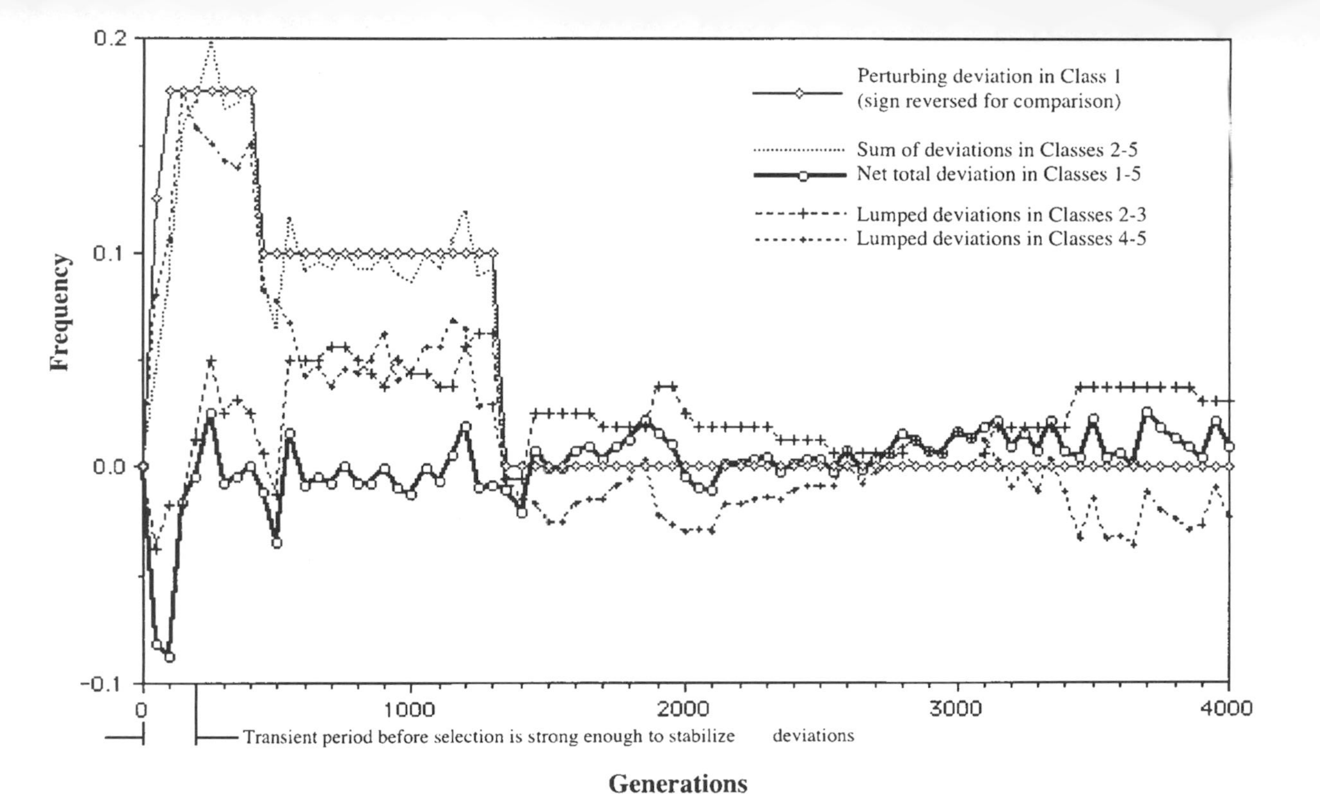

Fig. 16.6.—Equilibrating deviations in smaller fitness classes following a large perturbation with loss thru sampling error of two out of eightalleles in the highest fitness class in the early transient stage of the first 200 generations of run 50. These alleles were reestablished by back mutation after generations 400 and 1,300 producing secondary perturbations. All three perturbations induced countervailing fluctuations in all other fitness classes. The 50-run average is used to estimate equilibrium tendencies of the system. The initial perturbing fluctuations in class 1 are shown here with negative sign (contrast the preceding figure) to allow direct comparison with the sum of fluctuations 2–5, yielding a net deviation for all classes (sum 1–5) which is less than deviations in any single class. Classes 2 and 3 and classes 4 and 5 are lumped together. Original losses occur within a 200-generation transient period, while decreases in mean fitness inflate relative fitnesses of all alleles, producing trajectory reversals for alleles in the two highest fitness classes. At loci in five fitness classes (.10, .05, .025, .0125, and .00625) having 8, 16, 32, 64, and 128 alleles, 248 alleles are subjected to mutation and approach selection-mutation equilibrium frequencies determined by their fitnesses, the mutation rate, and population size.

of the zero point with excursions usually smaller than those in any of the contributing five classes for the rest of the run (figs. 16.5, 16.6).

So not only did equilibration of classes 2–5 occur with the massive deviations in class 1, but there also appeared to be equilibration among classes, as one would expect: no class should be privileged. How can we visualize this? Because each class made the same fitness contribution, relative frequencies of the different classes could be meaningfully graphed at the same scale, signifying fitness deviations. To understand what is happening, note the following:

1. The number of connections affects loss rates in a given fitness class at two points: larger classes of alleles (in the lower fitness connections) yields more expected mutations (and losses), though the proportional loss rate should be the same. But the class size affects the number of possible "correct" hits for back mutations restoring missing "good" connections.[18] Thus, in class 1, the first reversion (with two possible targets) occurred within 350 generations, but the second (with only one left) took another 900. For lower fitness classes with many more lost connections, "good" back mutations were virtually continuous.

2. Since alleles in the more numerous classes had corresponding smaller fitness contributions, their finer-scale adjustments in genotypic fitnesses were more likely to remain at intermediate frequencies in mutation-selection balance (not driven immediately to fixation if present), providing potential buffers for changes at other loci. Stepwise changes in frequencies in the top three classes across sampling intervals suggest movements from absence to fixation in one 50-generation interval, reflecting strong selection (fig. 16.5). The bottom two classes wandered more continuously up and down, reflecting smaller changes in total frequency if one went from absence to fixation, and many alleles at intermediate frequencies in fluctuating equilibriums. This is expected, given the larger number of alleles in those classes (with more back mutation) and the weaker selection on them, allowing rapid (but weak and noisy) small-scale response.

Moreover, the fact that the perturbing fluctuations were in class 1, and the compensations occurred in four contiguous classes showing doubling of connections and halving of fitness contributions in successive classes allowed another data transformation:

For more clarity we lumped neighboring classes *2 + 3* and *4 + 5* to form two heterogeneous classes with larger differences in numbers and in fitness contributions,[19] making scale-dependent phenomena more visible with the two new composite classes than in four unlumped neighboring classes. The two compound classes now provided all possible sources for change counteracting deviations in the first class or each other, allowing direct visualization of opposing compensation in the deviations.

The three levels of primary perturbation in effect generated three experimental treatments for the other two classes at different stages in the trajectory. Several things now became visible in figure 16.6:

a) The primary perturbation is initially offset almost completely by the more frequent and smaller-effect alleles in class *4 + 5*. Class *2 + 3* did not compensate significantly until after the first back mutation in class 1 (at generation 500). With expected mutations at four times the rate in *4 + 5*, and fourfold smaller fitness contributions, fewer alleles would be fixed. With more alleles at intermediate frequencies, selective response is faster both because frequencies change more rapidly in the middle range and because there is no wait for a mutation. This is crisper confirmation of the explanation in 2 above, as are the other observations below.

b) From generation 550 to 1,300 (the second back mutation in class 1), classes *2 + 3* and *4 + 5* compensated roughly equally, suggesting compensation proportional to their *exposure*. Even fluctuations in *2 + 3* seemed balanced by oppositely directed fluctuations in *4 + 5*.

c) From 1,350 to 4,000, the primary deviation had disappeared. Classes *2 + 3* and *4 + 5* compensated for each other alone, with oppositely directed fluctuations about 0.

d) Class *2 + 3* is significantly "grainier" in responses than *4 + 5*, as expected. It moves less often, with visibly quantified "hops." With its fixations, and longer waits for rarer back mutations, it acts as a secondary driver to *4 + 5*.[20]

We thus see remarkably fine-structured adjustment in and between different fitness classes for deviations in any one of them, modulated by significant differences in the stochastic character of responses in different classes because of their sizes. The summed fit of classes 2–5 to class 1 deviations was so good (and tracked back mutations so well with changes on different scales) as to suggest a stochastic analogue to a Fourier series approximation to a square wave function. Stochastic fluctuations are endemic, but each large fluctuation was answered by countervailing ones in other fitness classes. With more connections and the smallest fitness scale, class 5 (or *4 + 5*) was normally first to respond to a fluctuation elsewhere, and with the finest-scaled adjustments, with the others roughly in order as their increasing scale.

A Methodological Note

Why spend so much time discussing a single rare case? Isn't this a Monte Carlo simulation? And don't simulations establish credibility

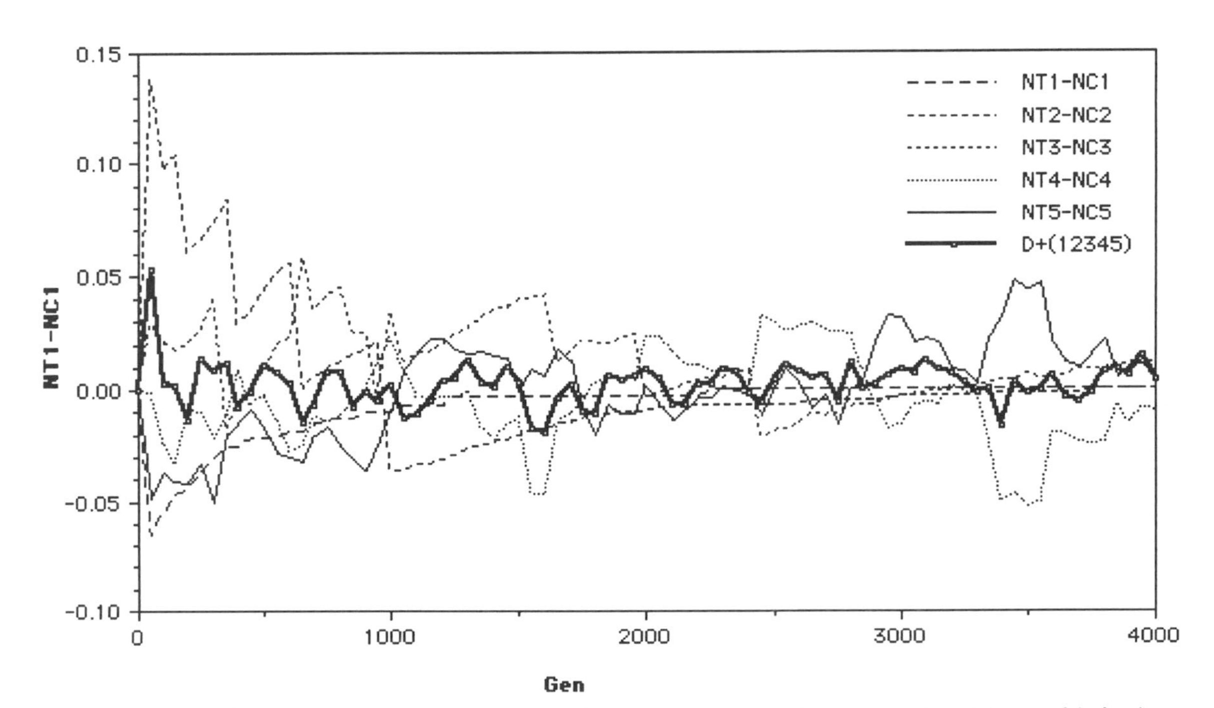

Fig. 16.7.—Equilibrating frequency deviations in fitness classes 1–5 from the 100-run average for run 100, graphed with the total deviation of the five classes added together, showing compensation in classes 1, 3, 4, and 5 for drift-induced perturbations in class 2. This is otherwise analogous to figure 16.6. The "sawtooth" trajectory of class 2 is a joint product of a sequence of back mutations followed by quantal fixations (within the same 50-generation interval) occurring in generations 50, 150, 400, 650, and 950 and the exponential approach of the equilibrating average trajectory for class 2 (visible as the second trajectory from the top in the bottom graph of fig. 16.3). Note particularly the countervailing deviations in 3, 4, and 5 after about generation 1,200, when class 1 has gone to equilibrium fixation, and class 2 is asymptoting there. As a result, the sum of the five classes remains much more tightly bounded than any of remaining three fluctuating classes.

by showing that an outcome is very frequent in an ensemble as a whole, or at least under a desired array of parameter values? This objection would be misplaced. It was appropriate for us to do many runs to get good estimates for the average trajectories. And interclass compensation is also a phenomenon which we think occurs generally. But no matter how common these phenomena are, cases meeting conditions which allow clear demonstration are rare. Any simulations with deviations large enough to see above the normal noise in the process exhibited apparent compensatory behavior. Seeing it requires larger fluctuations, and to analyze it, the fluctuations need to meet still more conditions, but for those which did, we saw compensation on all scales. Very few of our 100 replications were "natural experiments" which made "good" cases. Statistical analysis of the compensation would be a complex multivariate problem and not worth the effort, given how clearly it is demonstrated in replications 50 and 100.

For a clear demonstration we sought trajectories which (1) deviated substantially from their averages—larger deviations provide larger compensations, easier to detect and analyze; (2) had most of the deviations concentrated in a single class, so its sign and magnitude could be compared with deviations in the other classes; (3) were in one of the top three classes, which characteristically produced "quantized" transitions—frequency changes large and sudden enough to generate noticeable responses in other classes; and (4) deviated substantially in the top class, so the other classes could be paired to better see scale-dependent effects and reduce the number of trajectories.

With all of these conditions, a replication like 50 is like a rare and useful mutation. Only three others involved loss and back mutation of even a single allele in the top fitness class. The next best case was run 100 (fig. 16.7) with the loss of 5 out of 16 alleles in class 2, and their successive back mutations over the next 950 generations. Figure 16.7 shows ("by inspection") nice and fine-structured qualitative compensation involving all five classes. But location of the perturbation in class 2 prevented the class lumping which was so revealing in replication 50. Just eyeballing this case shows why it would be hard to analyze further.

Discussion and Conclusions

Dilemmas of Evolutionary Modularity

Modular adaptive systems would seem to have great advantages over comparable systems with less modularity.[21] Herbert Simon ([1962] 1996) pioneered arguments that "nearly decomposable" systems modularly organized in "stable subassemblies" should evolve more rapidly than things that are not. Lewontin urges related principles of quasi independence and continuity: with any variance for modularity, there should be selection for more of it (Lewontin 1978; Wimsatt 1981; Brandon 1999; Schank and Wimsatt 2000). Wagner and Laublicher (2000) have recently offered a formal basis for this notion. Perhaps over the long run, selection or reduplication groups the architecture of gene expression and developmental interaction into units which can recombine together efficiently to address the combinatorial array of special problems organisms face in different environments. But Wagner and Altenberg (1996) consider such a mechanism and, on further analysis, raise doubts over its efficacy. The simulations here may raise further doubts. Thus, the aims of this discussion are first to set some of the context and correlative complications, and second to see how within the context of our results modular changes can most likely be established in evolution. Given the structure of our simulations, our results bear on the fate of changes in parts or modules insofar as they affect the fitness of the whole organism. They make no predictions directly of the relative strengths, numbers, or fates of intra- versus intermodular interactions.

Modules must also function integratively (Wimsatt 1997; Schank and Wimsatt 2000). Here a tension emerges: Developmental biologists speak of anatomical or developmental modules, but the evolutionary modularity of the parts of a system—the ability to substitute and rearrange them without perilous fitness consequences—might seem at odds with the functional integration necessary to make something an evolutionary unit. What is the mapping between genetic organization and phenotypic organization such that evolutionary rearrangements or modifications of the genetic architecture does not scramble the functional organization of the phenotype?

To gene selectionists Dawkins's (1976) fable of evaluating rowers by interchanging them among crews so they perform against different backgrounds and supposed parallels of this process with genetic recombination seem to promise that it is possible. Our simulations suggest that this may be misleading. Beyond issues of direct causal functional integration, an even deeper process binds even apparently unrelated elements that combine to build the phenotype, and results in a kind of unanticipated interaction spanning the whole organism. If the very processes that make evolution possible—variation and natural selection—poise organisms at the competitive edge of extinction, then our simulations suggest that evolutionarily significant modules become tightly coupled. But do they stay so? If they do, the answer may lie in entrenchment processes. We reify modules at various levels, but do we yet understand how such modularity is evolutionarily possible? We suggest below that such evolutionary modularity may be transient and rare and made possible by the relaxation of selection from either external or internal sources. But even so, this may suffice to generate significant evidence of what we call modularity, on a geological time scale.

What is a modular change? We assume it should be a change that can be made without affecting numerous other things in the phenotype as a whole. (We have seen that this may not often be possible for fitness, though that says nothing definite about the frequency of modular change thus conceived.) Reversibility is also evidence for relative modularity, because it suggests independent modifiability, and consequently, less fitness epistasis of that element with others in the phenotype. But also, many things that look modular and internally integrated may have been products of rapid elaboration of smaller pivotal changes which led to a bifurcation of divergent trajectories. If this involves changes in localized (but not decoupled) components of phenotype, we would have apparent modular change in at least one of two corresponding highly integrated subsystems of divergent species.

Summary of Conditions under Which Modular Replacements Could Occur

All of our simulations of generative entrenchment showed an almost overwhelming bias favoring retention of strongly selected or of entrenched connections. Yet in the simulation just analyzed, we had the loss of two major connections. How is this possible? We believe these simulations show conditions relevant to the transitory appearance and disappearance of modularity. Under what conditions can such major losses occur? There are four, all important and none surprising. But their conjunction permits conclusions which seem both surprising and important.

First, *as important as the connections were, their losses were not lethal.* They were conditional lethals, but lost early in the simulation when those conditions were not met—before other losses which would have made them lethal. (Their competitors were likely in comparable states.) Many deeply GE'd features would be lethals under broader or even all conditions, so *changes in evolutionary modules should not disrupt deeply entrenched features.* Modularity tends to decrease the effect and scope of entrenchments (Schank and Wimsatt 2000).

Second, *events like those in replication 50 were rare,* even though we manipulated conditions to make them much more common.[22] But how common must modular changes be on what are very long time scales for population geneticists (say 10^5–10^8 generations) to appear moderately common in the paleontological record? The answer, we suggest, is "not very." (Such changes do not commonly flip back and forth, for example, though reversibility would strengthen the case for modularity, because it suggests independent modifiability.) The lovely example provided by Wimmer (Wimmer etal. 2000; Schmidt-Ott and Wimmer 2004) of how *bicoid* could entrain and integrate existing control circuits provides a plausible way of making changes deep in development while preserving the downstream consequences whose disruption would be lethal. But this change became irreversible through the loss of once useful but now redundant control linkages—another expression of generative entrenchment. And even if most such deeper changes employed similar scenarios, they would not be common in absolute terms.

Third, *population sizes were very small:* only then could we expect to lose important adaptations or make major changes through drift. In addition, the more we learn about evolutionary history, the more such bottlenecks in population size (or local ones, leading to local differentiation of populations) seem common.

Finally, you have a better chance of losing a crucial adaptation, or surviving a seriously deleterious mutation (or a potentially beneficial but poorly integrated one) if conditions are very good—presumably both genetically and in the environment, as they were early in the simulation. Then traditional "hill-climbing" tinkering can improve the fit. The pivotal importance of this condition (first argued by Arthur 1984, 1997) emerges much more strongly here than before. Why? Because only with the excess slack cut by very good conditions do we have room for quasi independence to work on modules of any appreciable "size." We return to this below.

Could Such a Simple Model Possibly Have Anything to Tell Us?

This model might have seemed quite unrealistic for its positing a single "best" genotype and the consequences of mutational loss from that point. But there are important and generalizable things to learn from it. Properly construed, this is not a single-peaked optimizing argument, but a satisficing one (Simon [1962] 1996) about behavior just above the rising flood. The prime function of starting the population as composed of a focal adaptive phenotype is not to provide a realistic model of variability and focal adaptation in nature—it is not—but to show the consequences of mutation-selection balancing processes under conditions where a population starts with reserve reproductive capacity and degrades to just making it. And the loss of smaller-contributing elements documented here and their effects suggests that degradation. With the occurrence of the major adaptive back mutations, these simulations also show the consequences of sudden relaxations in selection intensity. And the mutation event may be to the domain of attraction of a quite different adaptive peak: what is scored as a back mutation to an original state is characterized only in terms of its fitness and could just as well be another important mutation to an alternative state, and the release of variability with the relaxation of selection an exploration of new modifiers to the new state. Virtually every empirical population genetic study tracking stasis or small adaptational reversals or local genetic differentiation confirms this as the common picture. Losses of weakly selected alleles and anchoring of more strongly selected alleles says things about maintenance, but also about evolution, which can pivot around the stronger alleles. The effects arising through back mutation to restore stronger alleles can also just as well stand for the establishment of new fitter alleles, which then permit further changes, both through the relaxation of selection (like following exploitation of a new niche) but also through the accumulation of newly possible specific modifiers to improve performance or remove deleterious consequences of these new mutations.

The stratification of allowable replacements near a truncation selection threshold (in which deleterious mutations in only a narrow window of selection coefficients can be tolerated) suggests the remarkable picture of "neutral percolation" through a high-dimensional adaptive topography for RNA configurations presented in Huynen, Stadler, and Fontana 1996. A neutral mutation is one that is (selectively) neutral relative to the allele it replaces, not one that is silent or does nothing. Then substitution of one allele for another in the same fitness class is, for these purposes, a neutral substitution. So their qualitative picture of unconstrained motion from one attractor to another through neutral percolation should apply also in the context of our model. And addition of small ranges in selection coefficients as "nearly neutral" broadens the scope of their picture.

Indeed, consider this model as a partial description of a much more complex system, rather than as a complete model of a simple system. If we treat the 248-locus model as a lower-dimensional slice—a sampling from the loci and alleles in a much larger system—the chance of back mutations for any of the alleles is effectively zero, and with losses and back mutations we are doing a random walk through a very high-dimensional genotypic space. Indeed, this is a far more realistic interpretation of this model as applied to nature. From the point of view of these selection models, it does not matter if the allele that "reappears" (as defined in the model) is a genuine reappearance or just an appearance of another allele with the same selection coefficient as the one lost.[23] It does not even matter whether what appears is a gene or a change in fitness reflecting an environmental change. Fitness is a relational property between organism and environment, so conclusions drawn in our models about the response of selection on genotypes to fluctuations in fitness apply equally whether those fluctuations are genetically or environmentally induced.[24]

Organismal Selection Revivified?

Most strikingly, these simulations suggest that the evolutionary outcome for an adaptive element in a phenotype commonly depends on the fates of other phenotypic elements which may be quite developmentally and functionally disconnected from each other save for just being parts of the same organism. This kind of interdependence reveals a new arena in which discussions of modularity in development and in evolution must take place. Selection is always selection of whole organisms: We have rediscovered what might be called the "last straw" principle. For an evolutionary unit close enough to the threshold of viable fitness, loss or compromising of almost any functional component (no matter how functionally decoupled from other components) can drag it under, and the more fitness "buoyancy" a unit loses the less it has to play with. And genetic load and "Red Queen" (Van Valen 1973) arguments each independently suggest that organisms spend most of their time close to that threshold.

The interdependence among phenotypic elements emerging from these simulations seems remarkably nonspecific. Though phenotypic and environmental details clearly impinge in determining fitness, what ultimately determines whether an evolutionary unit can lose a phenotypic component is the magnitude of its fitness contribution, and how close the evolutionary unit is to the critical fitness threshold. It should be emphasized that this is not necessarily loss of *modules* of different sizes, but of any parts whatsoever, modular or not, including ones which may be quite distributed in location and quite pleiotropic in their effects. Indeed, *what we appear to have demonstrated unusually clearly in these models is organismal selection,* for here organismal fitness still appears to matter enormously even in a case where we have removed all obvious sources of gene interaction. But that is ultimately illusory, for *the existence of truncation selection itself creates a maximally global form of epistasis. And truncation selection is ubiquitous.*

A Strength of Weak Ties?

Another surprising result from these simulations is the role played by genes of small fitness effect. Things that are "nearly neutral" are readily lost. With lots of them (as we suppose there are), allowing populations to approach more closely to the critical fitness threshold and inflating the relative fitness contributions of all, their loss serves to better anchor the larger contributing elements more securely. Larger-scale mutations act as constraints on smaller-scale mutations because of their rarer occurrence on longer time scales, but they are in turn cemented in (and made more constraint-like—in their presence or absence, which must be adapted to) by the loss of smaller-scale mutations with negligible individual importance. Paradoxically, phenotypes with more homogeneity in the importance of their components (if they could exist) would have their parts less secure than comparably important components of phenotypes, which had lots of "small change" adaptations.[25]

This shows up in our simulations. Compare figure 16.3 (100 loci, fitness range from .02 to .06), with figure 16.4 (248 loci, fitness range from .00625 to .10). The simulations depicted in these figures both had an exposure of 4, and the larger circuit's lower mutation rate offsets its larger size. The main difference between them is their equilibrium mean Darwinian fitnesses—smaller by a factor of 3 or more for the larger circuits (about .10 vs. about .03). The effect is clear and to be expected for the greater rescaling of relative fitness in the latter case (fig. 16.4). Connections with assigned fitness of .02 and .03 in the first simulation go to *lower* frequencies than connections with assigned fitness of only .0125 in the second simulation, and the top three classes exhibit similarly divergent effects. This is due to the presence of more weakly selected alleles (connections) in the second simulation.

The Consequences of Breathing Space

Gaining (or regaining) a larger-contributing element gives more "breathing space" and deflates relative fitness contributions. It should thus increase loss of smaller-contributing elements, *so the net effect is to generate a substitution of fitter alleles at some loci for less fit ones at other loci.* This is the lesson of the trajectory reversals in the upper fitness classes accompanying the losses in the lower fitness ones (fig. 16.4). Even without epistasis as normally conceived, there can be significant global interlocus interaction of the whole phenotypically embodied genome in the coproduction of fitness.

But this "relaxation" phenomenon has another important consequence: either a significant relaxation of environmental conditions, improving fitness (and facilitating Arthur's [1984, 1997] "n-selection"), or a mutation doing the same thing should lead to a wave of further possible (usually smaller) innovations, and that for reasons apparently "internal" to the fitness architecture of the genome and design of the phenotype. Looking now to modes of phenotypic architecture which can yield these benefits, redundancy seems a natural, giving reserve room for experiment through simple duplication of function. Gene duplication provides a "bottom-up" example. Or phenotypic architectures may provide highly context-sensitive and conditional arrangements, which work only contingently and transiently. An emerging and important example of this last is provided by the heat-shock proteins, whose dosage and activity can interact with environmental stress to allow developmental expression of an extremely diverse range of morphological variation (Rutherford and Lindquist 1998).[26] But canalization, maternal effect, symbiosis, social support, and all of the modes of behavioral plasticity, as well as the relaxation of competition provided by the extinction of competitors or windfall fluctuations of environmental conditions, can provide rich opportunities of the second sort, and all of them can open up new opportunities for elaboration or transformation of the niche. We need to look to the ecological dimensions of the problem.

If we look to the now best-documented cases of selection in the literature, the extended work on Darwin's finches by the Grants (e.g., Grant and Grant 1989) and their students (reviewed in Weiner 1994), we see lots of opportunities for the kinds of fluctuations in fitness suggested as important in our simulations. A difference of a few percent in beak depth in a bad year make the difference between life and death, in the next year between being mated or not, and in those and other years, in food preferences and niche differentiation. Yet in good years, with excess food, such marginal differences may have no impact on mating and survival. Even larger variants could survive and, with the right conditions, spread. Intense selection, with frequent reversals in direction of favored variants provide excellent opportunities for the importance of conditions suggested by our simulations in allowing relatively major and even modular changes, confirming at least part of Wagner and Altenberg's (1996) conclusions.

The Significance of Context Independence

Finally, what kinds of adaptations make the relatively context-independent contributions which make their relative fitness scale up when they approach the threshold of a truncation selection regime?[27] Our models tell us that these kinds of adaptations are important. They are adaptive elements whose contribution does not scale multiplicatively, or proportionately to the fitness of the organism, but tend to make a fitness contribution whose effect is at least partially independent of the other factors producing differential fitness. Any time we have threshold effects, or things whose losses are not readily compensated for, we violate multiplicative scaling and can produce the conditions for these models to apply. And threshold effects are ubiquitous in organic systems.

Paradoxically, the things whose fitness effects are most insensitive to context are large and far-reaching enough in effect that their loss is fatal, and so they never can be gained or lost independently. But this is not modularity: the context insensitivity is one-way: loss is unconditionally lethal. For modularity, we need a context insensitivity at the other end of the fitness scale: interchangeability that does not disturb relatively high fitness too much—fitness of a well-functioning system. Only with design for modularity and for portability (as with plasmid bodies, or commonly commensal parasites) can we imagine true modularity—involving frequent gain or loss, or exchange of coordinated clusters of traits.[28]

Quasi independence seems theoretically and observationally robust. This appears to be in direct conflict with generative entrenchment, unless we keep the scale of events in mind. Frequent events on a paleontological scale may be quite rare to a population geneticist. So perhaps our constraints do not prevent modularity on a macroevolutionary scale. If we suppose that quasi independence is highly context sensitive—suggesting that if there are modules, they are shifting arrangements which show relatively high context sensitivity or, while modular, are requirements for almost any living thing of that type—then such modules can, nevertheless, still be deeply generatively entrenched. So if we are right, modular changes may happen, but rarely, and require near "garden-of-eden" conditions to support these experiments.

Acknowledgments

W. C. Wimsatt thanks the National Humanities Center for time, library assistance, and pleasant surroundings and company, David Hull for bibliographic advice, and the Research Triangle Philosophy of Biology group (Duke, the University of North Carolina, North Carolina State University) for comments at its first trial. Simulations were done (over a decade ago) with help from Apple Computer and the National Science Foundation (grant nos. SES-8709586, SES-8807869). J. C. Schank thanks the National Institutes of Health (through a subcontract with Indiana University) for support on this project. Gerhard Schlosser, Günter Wagner, and Kim Sterelny all made pointed, full, and productive comments on an earlier draft which were enormously helpful.

Notes

1. Similar detailed arguments were first made by Riedl (1978). Arthur has been elaborating a similar account since 1982 (Arthur 1984, 1997). It is in most respects now the fullest available, though each approach has its own twists.

2. Our research biases (Wimsatt 1980) are such that we should find proportionately more of these as our knowledge of developmental genetics progresses (Oyama et. al. 2001).

3. Kim Sterelny (in conversation) urges qualification. Darwinian processes give rise to GE globally, but this does not mean that in any given case it will favor more exposed dependence over, say, canalization. For a start on relations between generative entrenchment and canalization, see Wimsatt 1999, though much more is required.

4. No simple answers to this question meet all necessary constraints. Some things early in development escape entrenchment, such as pseudogenes and code synonymy (Wimsatt 1986). Riedl (1978) considers other proposals. But pruning unentrenched branches from a tree still leaves a tree (albeit one growing with a slower exponent). So an hourglass of variation (Duboule 1994; Raff 1996) requires additional mechanisms, such as canalization at the "neck." These mechanisms and others probably all play hybrid roles in different specific cases, since phenotypes are assembled opportunistically by evolution.

5. What properties are generic and their generic values may (but need not) be quite sensitive to the exact specification of ensemble properties. Thus, circuits with three connections per node (and, more surprisingly, those with an *average* of two connections rather than exactly two connections per node) may have quite different generic properties. Thus, the structural specification of the ensemble is at least as important as the determination of its generic properties. The claims we use about differential generative entrenchment are much less sensitive than many of Kauffman's to exact ensemble specifications (Schank and Wimsatt 1988; Wimsatt and Schank 1988; Gelfand and Walker 1984).

6. Kauffman used high mutation rates to give observable effects with smaller populations of smaller circuits in shorter times. He makes arguments for how these effects should scale up to larger circuits in larger populations with smaller mutation rates. Our results (which point in unanticipated directions) raise questions about their relevance.

7. The simplest such measure is the sum of downstream nodes. This treats all causal interactions as equally important, but this is not required. We have tried various measures of entrenchment, and various circuit architectures, some more conservative, with similar qualitative results (Schank and Wimsatt 1988). This robustness is surely due to the strong nonlinearity of the results reported here.

8. Detailed circuit connectivities affect how GEs and thus selection coefficients change with the loss of connections. We treated selection coefficients as constant for the duration of a replication. More realistically, they should be vectors with different components realized against different genetic and environmental backgrounds, which thus change during the course of the simulation. This idealization underestimates epistasis in the network. Since we argue that unanticipated interactions are (already) causing troubles, underestimating interactive effects is conservative and should not undercut our conclusions. But keeping selection coefficients constant when they are changing as a result of changing connectivities forces us to forgo any conclusions about trajectories of change in such networks. For various reasons, the analyses given here are of only local validity, but for that they appear to be quite robust.

9. This generates a simple "additive" kind of context dependence for lethality which could be thought of as a kind of cumulative overload ("the straw that broke the camel's back"). It would not cover many important kinds of conditional lethality, for example, of the kind suggested by polyploidy or other forms of redundancy, where the fitness decrement of a mutation would be 0 or much less if there were a functional allele at the other locus.

10. Expected mortality from this source can be reduced by reducing the mutation rate, or by diploidy, tandem duplication or other modes of increasing redundancy, or canalization or other modes of regulation of effect (Wimsatt and Schank 1988).

11. Surprisingly, classical additive quantitative genetic models assumed that all alleles contributed the same fitness increment, and so they missed the phenomenon which is so important here.

12. This mechanism make these systems striking demonstrations of some of Simon's ([1962] 1996) ideas of near decomposability. They have the same association of force and relaxation times (stronger forces go more rapidly to equilibrium), but also the unusual and counterintuitive feature (which Simon's lacked) that *the long range behavior of the weak interactions dominates ultimate outcomes for the strong interactions.*

13. Lewontin 1955 is the experimental basis and Lewontin 1958 the theoretical treatment. The idea used here is actually a broader concept than frequency-dependent selection: rather than the frequency of single allele types as individuated by sequences and linkage relations, it involves the frequency of genes in a given fitness class, which should be the same for purposes of selection. Perhaps it deserves a different kind of name because the mechanism producing it is so different.

14. The interactions could be detected most easily by the presence of trajectory reversals in the higher fitness classes but should also affect the equilibrium ratios of fitness class frequencies.

15. Initial simulations with population sizes of 100 showed similar but less extreme trajectory reversals than those discussed below. Kauffman was puzzled by the reversals and proposed smaller population sizes to test whether they were a kind of sampling artifact. We were convinced they would not go away, but did not anticipate how revealing this would be. Also, computation time was limiting. Single replications of $N = 100$ for 4,000 generations took 26 hours (in 1989!) and could not be scheduled often. With populations of 10, we did three or four per night.

16. This then became an (unplanned) nearly perfect example of a common model-building strategy found often in classical genetics: building a model which generates a pattern which can be compared with the data, not to confirm the pattern, but to see the structure of deviations from it. The deviations are often more interesting than the original pattern (Wimsatt 1987).

17. Desirable fluctuations had unusually large losses in just one class. The quantized character of replacements in the higher fitness classes produces crisper shocks on the system. Run 100 was thus also interesting (fig. 16.7): no losses in the first class but 5 (out of 16) in the second. These back mutated over the next 950 generations (in generations 50, 150, 400, 650, and 950), producing a "sawtooth" pattern of deviation, with adjustment in the other classes and a smaller mean deviation for all five classes together than in run 50. But with the perturbation in the second class rather than the first, one could not lump classes *2* + *3* and *4* + *5* to see the clear scale effects we saw there.

18. In this model, genotype is just a list of pairs of numbers, for connections between pairs of genes, the second number the target of gene action and the first the actor. Random mutations would (stochastically) generate good connections in a class in proportion to the number of missing good connections.

19. Each larger class had identical fitness contributions, with two subclasses in a 1:2 connection ratio with 2:1 fitness ratio. Though heterogeneous, they were identical in structure and differed only in size. With the nonlinearity of interactions, these strong similarities were crucial to clear demonstrations of the effects. Class *2* + *3* had 16 + 32 = 48 connections, and class *4* + *5* had 64 + 128 = 192 connections. These heterogeneous classes had three times as many connections in them as classes 2 and 4, respectively, improving sample size. Numbers and fitnesses in the new heterogeneous classes now differed by a factor of 4 rather than 2, yielding much more pronounced scale effects.

20. After generation 550, when equilibration is well established, class *2* + *3* stayed in place in successive sampling periods for 42 out of 69 possible transitions, for a transition probability of 39%, whereas class *4* + *5* did so only 8 times—a transition probability of 88%. Class *2* + *3* had 8 longer runs of 9, 6, 6, 6, 5, 4, 3, and 3 constant periods, while class *4* + *5* had no constant periods longer than 2.

21. This supposes that the ability to adapt rapidly to environmental change is advantageous (and that such rapid environmental changes are relatively common over evolutionary and ecological time). Spatial and temporal patchiness has been accepted as common since the population biology of the 1970s, and Van Valen's (1973) "Red Queen" hypothesis assumes and provides evidence for constant environmental degradation for a species as a result of escalation by its competitors, predators, and parasites. In

a long-static environment, high integration could have bigger advantages, but not in all cases: modularity (e.g,, the ability to lose and regenerate limbs) could well be beneficial for some evolutionarily stable aspects of the environment.

22. Thus, we used very small populations and artificially high mutation rates, but the absence in our model of recombination and population structure in a patchy environment would act in the other direction. But such changes do not have to be common, even on paleontological time scales.

23. Though it should be (exponentially) more difficult to find allele replacement in the same fitness class with increasing size of the change.

24. The conclusions of this model do depend upon (1) the distribution of selection coefficients, (2) the closeness of the fitnesses of an organism and its deme to a truncation selection threshold, and the dynamics of (3) gain, (4) loss, and (5) fitness changes of alleles in genotypes near this boundary. Our conclusions are fairly robust under plausible variations in 1. Anything that could cause sufficiently large uncontrolled or unpredictable changes in 2–5 during the course of the simulation (i.e., on a scale comparable or faster than the relaxation times of the equilibrating processes modeled here) could affect outcomes of these simple models. Three features of endogenetic architecture require further analysis. Diploidy (and polyploidy), linkage, and sex, if present in a system seen only through this model, could appear to affect mutation rates and stability of selection coefficients in ways not predicted by this model. Whether (or when) these effects are significantly large needs analysis. Aspects of population structure (exogenetics) and environmental patterning could also be important.

25. This is a thought experiment with a result that is more surprising than threatening: all real organisms should have lots of "small change" adaptations.

26. The mechanisms they have uncovered would tend to be activated not in "relaxed" environments, but in various kinds of environmental stress (particularly but not only heat shock), presumably like those described by Grant and Grant (1989). However, new morphological variations produced by heat-shock proteins may boost fitness and thus generate what is for those variants a "relaxed" environment in the sense used above.

27. To scale in this way, an adaptation must contribute a fitness component which is relatively context independent, and thus contributing, say, on average .1 offspring, rather than increasing reproductive output by 10%. The former contribution yields fitness of 1.1 and 2.1 for an organism to which it is added having fitnesses of 1 and 2. The latter would yield fitnesses of 1.1 and 2.2.

28. Kim Sterelny's skepticism about an earlier incompletely and erroneously formulated version of this point was crucial here.

References

Arthur, W. 1984. *Mechanisms of morphological evolution: a combined genetic, developmental and ecological approach.* New York: John Wiley and Sons.

Arthur, W. 1997. *The origin of animal body plans.* New York: Cambridge University Press.

Brandon, R. 1999. The units of selection revisited: the modules of selection. *Biol. Philos.* 14:167–199.

Dawkins, R. 1976. *The selfish gene.* Oxford: Oxford University Press.

Davidson, E. H. 2001. *Genomic regulatory systems: development and evolution.* San Diego: Academic Press.

Duboule, D. 1994. Temporal colinearity and the phylogenetic progression: a basis for the stability of a vertebrate bauplan and the evolution of morphologies through heterochrony. *Development* (suppl), pp. 135–142.

Elinson, R. P. 1987. Changes in developmental patterns: embryos of amphibians with

large eggs. In *Development as an evolutionary process,* ed. R. A. Raff and E. C. Raff, 1–21. New York: Liss.
Gelfand, A. E., and C. C. Walker. 1984. *Ensemble modelling: inference from small-scale properties to large-scale systems.* New York: Marcel Dekker.
Grant, B. R., and P. Grant. 1989. *Evolutionary dynamics of a natural population.* Chicago: University of Chicago Press.
Greene, E. 1989. A diet-induced developmental polymorphism in a caterpillar. *Science* 243:643–646.
Huynen, M., P. F. Stadler, and W. Fontana. 1996. Smoothness within ruggedness: the role of neutrality within adaptation. *Proc. Natl. Acad. Sci. U.S.A.* 93:397–401.
Kauffman, S. A. 1985. Self-organization, selective adaptation and its limits: a new pattern of inference in evolution and development. In *Evolution at a crossroads: the new biology and the new philosophy of science,* ed. David J. Depew and Bruce H. Weber. Cambridge, Mass.: MIT Press.
Kauffman, S. A. 1993. *The origins of order.* New York: Oxford University Press.
Lewontin, R. C. 1955. The effects of population density and composition on viability in *Drosophila melanogaster. Evolution* 9:27–41.
Lewontin, R. C. 1958. A general method for investigating the equilibrium of gene frequency in a population. *Genetics* 43:419–434.
Lewontin, R. C. 1978. Adaptation. *Sci. Am.* 239 (3, September): 212–230.
Oyama, S., P. Griffiths, and R. Gray, eds. 2001. *Cycles of contingency: developmental systems and evolution.* Cambridge, Mass.: MIT Press.
Raff, R. 1996. *The shape of life: genes, development, and the evolution of animal form.* Chicago: University of Chicago Press.
Rasmussen, N. 1987. A new model of developmental constraints as applied to the *Drosophila* system, *J. Theor. Biol.* 127 (3): 271–301.
Riedl, R. 1978. *Order in living organisms: a systems analysis of evolution.,* New York: J. H. Wiley.
Rutherford, S. L., and S. Lindquist. 1998. Hsp90 as a capacitor for morphological evolution. *Nature* 396 (26 November) :336–342.
Sander, K. 1983. The evolution of a patterning mechanisms: gleanings from insect embryogenesis and spermatogenesis. In *Development and evolution: the sixth symposium of the British Society for Developmental Biology,* ed. B. C. Goodwin, N. Holder, and C. C. Wylie, 137–159. Cambridge: Cambridge University Press.
Schank, J. C., and W. C. Wimsatt. 1988. Generative entrenchment and evolution. In *PSA-1986,* ed. A. Fine and P. K. Machamer, 2:33–60. East Lansing, Mich.: Philosophy of Science Association.
Schank, J. C., and W. C. Wimsatt. 2000. Modularity and generative entrenchment. In *Thinking about evolution: historical, philosophical, and political perspectives,* ed. R. S. Singh, C. B. Krimbas, D. B. Paul, and J. Beatty. Cambridge: Cambridge University Press.
Schmidt-Ott, U., and E. A. Wimmer. 2004. Starting the segmentation gene cascade in insects. In *Modularity in development and evolution,* ed. G. Schlosser and G. P. Wagner. Chicago: University of Chicago Press.
Schneirla, T. C. 1971. *Army ants: a study in social organization.* Edited by H. R. Topoff. San Francisco: Freeman.
Simon, H. A. [1962] 1996. The architecture of complexity. In *The sciences of the artificial,* 3d ed., chap. 7. Cambridge, Mass.: MIT Press.
Slack, J. M. W., P. W. H. Holland, and C. F. Graham. 1993. The zootype and the phylotypic stage. *Nature* 361 (6412, 11 February): 490–492.
Van Valen, L. 1973. A new evolutionary law, *Evol. Theory* 1:1–30.
Wade, M., R. Winther, A. Agrawal, and C. Goodnight. 2001. Alternative definitions of epistasis: dependence and interaction. *Trends Ecol. Evol.* 16 (9): 498–504.

Wagner, G. P., and L. Altenberg. 1996. Complex adaptations and the evolution of evolvability. *Evolution* 50:967–976.

Wagner, G. P., and M. D. Laublicher. 2000. Character identification: the role of the organism. *Theory Biosci.* 119:20–40.

Weiner, J. 1994. *The beak of the finch*. New York: Knopf.

Wimmer, E., A. Carleton, P. Harjes, T. Turner, and C. Desplan. 2001. *bicoid*-independent formation of thoracic segments in *Drosophila. Science* 287 (March 31): 2476–2479.

Wimsatt, W. C. 1980. Reductionistic research strategies and their biases in the units of selection controversy. In *Scientific discovery*, vol. 2, *Case studies*, ed. T. Nickles, 213–259. Dordrecht: Reidel.

Wimsatt, W. C. 1981. Units of selection and the structure of the multi-level genome. In *PSA-1980*, ed. P. D. Asquith and R. N. Giere, 2:122–183. East Lansing, Mich: Philosophy of Science Association.

Wimsatt, W. C. 1986. Developmental constraints, generative entrenchment, and the innate-acquired distinction. In P. W. Bechtel. ed. *Integrating scientific disciplines.* Dordrecht: Martinus-Nijhoff. pp. 185–208.

Wimsatt, W. C. 1987. False models as means to truer theories. In *Neutral models in biology*, ed. M. Nitecki and A. Hoffman, 23–55. London: Oxford University Press.

Wimsatt, W. C. 1997. Functional organization, functional analogy, and functional inference. *Evol. Cogn.* 3 (2): 2–32.

Wimsatt, W. C. 1999. Generativity, entrenchment, evolution, and innateness. In *Biology meets psychology: philosophical essays*, ed. V. Hardcastle, 139–179. Cambridge, Mass.: MIT Press.

Wimsatt, W. C. 2000. Generative entrenchment and the developmental systems approach to evolutionary processes. In *Cycles of contingency: developmental systems and evolution*, ed. S. Oyama, R. Gray, and P. Griffiths, 219–237. Cambridge, Mass.: MIT Press.

Wimsatt, W. C., and J. C. Schank. 1988. Two constraints on the evolution of complex adaptations and the means for their avoidance. In *Evolutionary progress*, ed. M. Nitecki, 231–273. Chicago: University of Chicago Press.

Winther, R. 2001. Varieties of modules: kinds, levels, origins, and behaviors. *J. Exp. Zool. (Mol. Dev. Evol.)* 291:116–129.

Winther, R. In press. Evolutionary developmental biology meets levels of selection: modular integration or competition, or both? In *Modularity: understanding the development and evolution of natural complex systems*, ed. W. Callebaut and D. Rasskin-Gutman. Cambridge, Mass.: MIT Press.

17 Starting the Segmentation Gene Cascade in Insects

URS SCHMIDT-OTT AND ERNST A. WIMMER

Introduction

Development is based on a series of modules that can be dissociated, duplicated, modified, and recombined. A key feature of such modules is their robustness, as their "relative autonomy" requires insensitivity to contextual perturbations (see Schlosser 2004). In computer simulations, a genetic circuitry active in the segmentation process of the *Drosophila* embryo has shown such robustness, since its function and output are very insensitive to the exact nature of the input stimuli (von Dassow et al. 2000). These results suggest that a segmentation module involving Hedgehog and Wingless signal transduction is relatively independent of the genetic context from which it appears. Thus, the specificities of how the segmentation gene cascade is started seem of minor importance, and high plasticity in early developmental events is expected. Moreover, evolutionary processes should be able to vary the inputs into the module relatively easily.

For two reasons, insects are very suitable for testing the hypothesis of early developmental plasticity and evolvability at the molecular level. First, the systematics and evolution of the insect clade have been intensely studied and provide a good frame for comparative work (Grimaldi 2000; Kristensen 1991). Second, the fruit fly *Drosophila* provides an excellent reference system for molecular developmental processes (Martinez-Arias and Bate 1993). Here we focus on studies in the field of developmental genetics, addressing the plasticity of factors that are required for starting the segmentation gene cascade. Two approaches have been used to study the plasticity of input stimuli for segmentation in insects. One is based on the comparison of different species. The second involves testing the consequence of experimentally changing the genetic input required for starting the segmentation gene cascade in *Drosophila*. Both approaches indicate that initial steps of the segmentation process can differ substantially without impeding normal segmentation and morphological differentiation, thereby demonstrating relative context insensitivity of the segmentation module active during *Drosophila* embryonic development.

The *Drosophila* Model for Early Development

Before reviewing facts and speculations about the evolution of early acting developmental genes, we will give a synopsis of their interactions in *Drosophila*. Anterior-posterior polarity (to which we will restrict the discussion) relies on localized mRNAs of maternal origin and on signaling between the embryo and somatic cells at the poles of the egg (Rivera-Pomar and Jäckle 1996; St Johnston and Nüsslein-Volhard 1992). In the freshly laid *Drosophila* egg, the mRNA of *bicoid* is localized at the anterior pole, while the mRNA of *nanos* (and the germ cell determinants) are concentrated at the posterior pole. These mRNAs are translated after egg activation and their protein products, Bicoid and Nanos, build up as shallow gradients in opposite directions. At this stage, the embryo consists of a syncytium, rich in yolk with some 6,000 nuclei (Foe et al. 1993). The role of Nanos in segmentation is to restrict the translation of ubiquitously distributed maternal *hunchback* transcripts to the anterior half of the embryo (Lipshitz and Smibert 2000). Hunchback is one of several transcription factors involved in anterior gene activation (Pankratz and Jäckle 1993; Rivera-Pomar and Jäckle 1996), and the posterior repression of early *hunchback* activity is essential to allow abdominal segment formation (St Johnston 1993). In contrast to Nanos, the Bicoid gradient regulates zygotic gene expression directly. It has a predominant function in anterior development (Driever 1993). Loss of *bicoid* activity results in embryos without head and thorax (which are replaced by duplicated posterior terminal structures), and in variable deletions and fusions of abdominal segments. At the molecular level Bicoid acts on the one hand as a transcription factor by activating zygotic segmentation genes like the gap genes *orthodenticle, hunchback,* or *Krüppel* (Driever and Nüsslein-Volhard 1989; Gao and Finkelstein 1998; Hoch et al. 1992). On the other hand Bicoid acts as a translational repressor of the maternal *caudal* mRNA, which is distributed throughout the early embryo, thereby forming a maternal posterior-to-anterior Caudal protein gradient (Dubnau and Struhl 1996; Rivera-Pomar et al. 1996). In addition, terminal signaling at the poles causes the expression of terminal gap genes like *tailless* (Jiménez et al. 2000; Paroush et al. 1997). In summary, Bicoid, Caudal, Hunchback, and transcription factors under the control of terminal signaling are responsible for asymmetric zygotic gene activation in

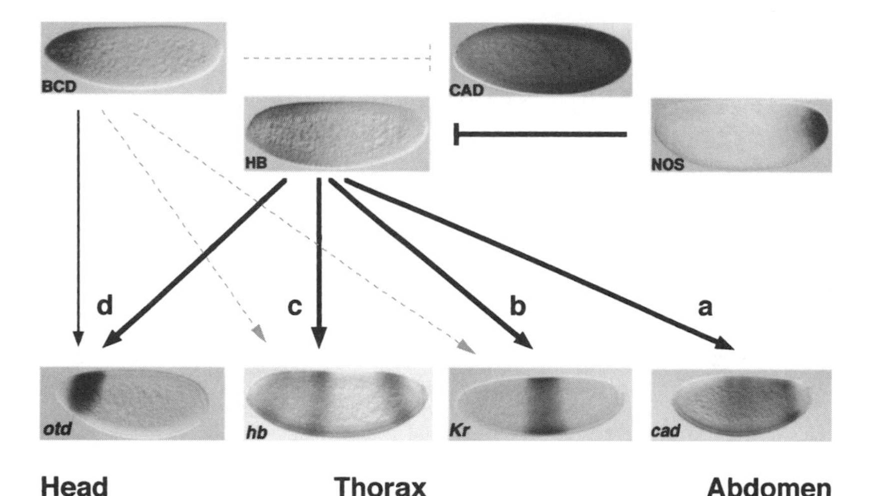

Fig. 17.1.—Several interactions involved in starting the segmentation gene cascade in *Drosophila.* Bicoid (BCD) protein distributed in an anterior-to-posterior gradient represses the translation of maternal *caudal* mRNA, thereby inducing a Caudal (CAD) protein gradient in the oposite direction. Nanos (NOS) protein is distributed in a posterior-to-anterior gradient and represses the translation of maternal *hunchback* mRNA, thus causing an early anterior-to-posterior Hunchback (HB) protein gradient. BCD, CAD, and HB act as transcription factors and together with the transcription factors under the control of the terminal system (not shown) provide the polarity in zygotic gene activation. The lowercase letters (a, b, c, d) refer to redundancies between Bicoid-dependent and Hunchback-dependent gap gene activation, which allow for part of the plasticity in early developmental processes described in the text. Note that the gene expression patterns are dynamic. Some change considerably during pregastrular development.

the early embryo (fig. 17.1). Their first target genes—the gap genes (Nüsslein-Volhard and Wieschaus 1980)—encode other transcription factors, which further subdivide the embryo in spatially restricted domains of gene activity. In the trunk region, the gap genes, together with the maternal determinants, activate pair rule genes, which are expressed in alternate segment equivalents, reflecting for the first time during embryogenesis the metameric organization of the insect body plan (Pankratz and Jäckle 1993). Finally, these genes define expression patterns of segment polarity genes in subregions of each segment (Martinez-Arias 1993). In addition, the gap genes together with the pair rule genes also define spatial domains of *Hox* gene expression. The *Hox* genes are required for differential segment identities (Castelli-Gair 1998; Kaufman et al. 1990; Lewis 1998; McGinnis and Krumlauf 1992).

Starting Segmentation in Other Insects

How conserved are genes involved in starting the segmentation gene cascade in other insect species? Expression studies with homologues to the *Drosophila* genes have been the predominant approach to this question. Clear-cut homologues of *orthodenticle, hunchback, Krüppel, caudal, nanos,* and *tailless* are known from different insects (and other animal phyla) and exhibit expression patterns similar to those observed in flies (Curtis et al. 1995; Li et al. 1996; Schröder et al. 2000; Schulz et al. 1998; Sommer and Tautz 1991; Wolff et al. 1995; Xu et al. 1994). However, a noteworthy exception is *bicoid.* Direct *bicoid* homologues are confined to cyclorrhaphan flies, a derived subgroup of the insect order Diptera (Schröder and Sander 1993; Sommer and Tautz 1991; Stauber et al. 1999).

Comparative functional data are still very scarce. An interesting example of functional differences among early segmentation gene homologues is provided by the characterization of mutants in the parasitic wasp *Nasonia* (Hymenoptera) (Pultz et al. 1999). Two zygotic phenotypes obtained resemble those of *hunchback* and *caudal,* although more segments are missing in each of the *Nasonia* mutants. The observation still awaits confirmation of the mutated genes but is suggestive of an increased zygotic function of *hunchback* and *caudal* in *Nasonia.* A second example is given by the function of *bicoid* in different flies (Shaw et al. 2001; Stauber et al. 2000). In the phorid *Megaselia, bicoid* appears to be required in a larger region of the embryo than in *Drosophila* (Stauber et al. 2000), and in the housefly *Musca domestica* it might not be necessary for thorax development (Shaw et al. 2001).

Genes encoding primarily localized axis determinants such as *bicoid* and *nanos* in insect orders other than Diptera still await identification. Therefore, we will briefly review three types of experiments that support such localized determinants (for a detailed review of the older literature see Sander 1976). First, transversal ligation of intravitelline cleavage stage embryos of beetles, cicadas, and flies result in the formation of anterior segments in the anterior half and posterior segments in the posterior half, but these segments together do not add up to a complete set of segments. This "gap phenomenon" does not result simply from defects caused by the ligation, as demonstrated in the leafhopper *Euscelis* (Hemiptera) and in *Drosophila,* where displacement of cytoplasm can rescue the ligation-induced gap in the segmental pattern (Sander 1960; Schubiger et al. 1977). Therefore, it was postulated that the development of a central portion of the insect embryo requires an interaction of anterior and posterior localized cytoplasmic factors. Second, in chironomid midges ("lower" Diptera) ablation of head and thorax was obtained by 285 nm UV light irradiation of the anterior tip of early embryos by Kalthoff (1979). When these irradiations were immediately followed by irradiation at 310–460 nm wavelength, normal development was rescued. Together with other findings, these results suggest localized RNA as the ablation target. The identity of this RNA remains unknown. For the red flour beetle *Tribolium* (Coleoptera) a Bicoid-like molecule has been proposed, because the direct Bicoid target genes *caudal* and *hunchback* of this species are regulated in a Bicoid-dependent manner when introduced into *Drosophila* embryos (Wolff et al. 1998).

Thus, some experimental data suggest localized factors involved in anterior-posterior patterning in many insects. However, comparative embryology also suggests important differences. Insects of the order Diptera develop as long-germ-band insects. That is, the body segments form more or less simultaneously, and their anlagen are established during the blastoderm stage, before the onset of gastrulation (Anderson 1966). *Drosophila* is a typical example. However, this mode of development is derived, and it is found only in some holometabolous taxa (Schwalm 1987). In other insects, such as the locust *Schistocerca* (Orthoptera), or in *Tribolium,* most segments are generated sequentially from a cellular posterior growth zone (short or intermediate germ-band development). Even more strikingly, some insects, such as the parasitic wasp *Copidosoma* (Hymenoptera), develop by total cleavage (Grbic et al. 1996). In *Copidosoma* a morula-like stage is generated prior to gastrulation, which consists of an inner cell mass and a syncytial extraembryonic layer. The inner cell mass splits into thousands of arbitrarily oriented embryos, which develop into typical germ bands. It is difficult to imagine how the maternally provided and localized determinants known from *Drosophila* could establish the segmental blueprint in these species.

Origin of the *bicoid* Gene

Sequence analysis of a *bicoid* homologue from the phorid fly *Megaselia* provides direct support for a sister gene relationship with *zerknüllt,* a derived *Hox* class 3 gene of insects (Stauber et al. 1999). The genes are located next to each other in the *Hox* gene complex (*Hox*-C) in all species where linkage between both genes has been analyzed (*Drosophila, Lucilia, Calliphora*) (fig. 17.2) (Brown et al. 2001; Kaufman et al. 1990; Randazzo et al. 1993; Terol et al. 1995). The recognition of *bicoid* and *zerknüllt* as a gene pair indicates that the position of *bicoid* in the *Hox*-C is ancestral. This information is important for establishing the phylogenetic age of *bicoid,* because mapping the gene origin on the phylogenetic tree requires that the absence of *bicoid* orthologues be demonstrated for out-group taxa. In the case of very diverged *bicoid* orthologues (corresponding genes in different organisms) even a complete genome sequence might not settle the question, because of the possible failure to recognize the orthologue. Yet the absence

a

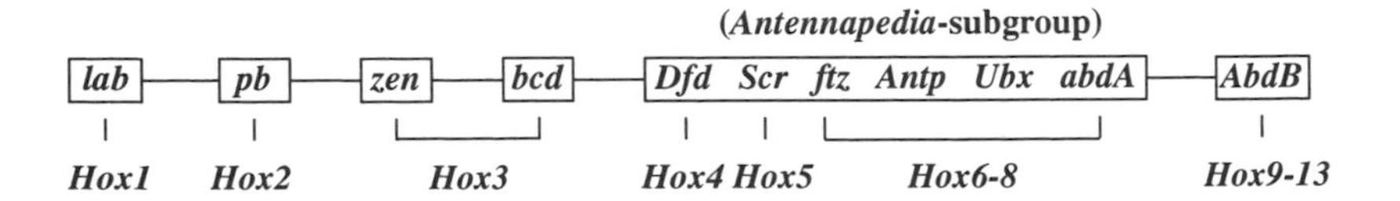

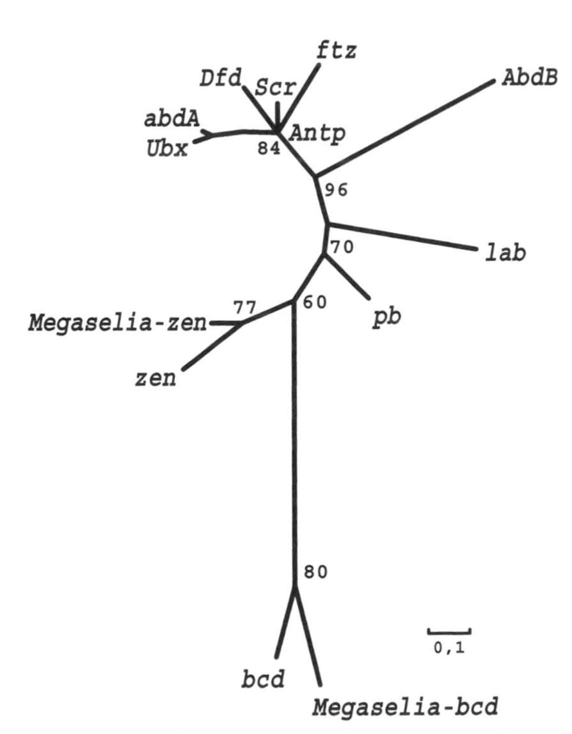

Fig. 17.2.—*bicoid* and *zerknüllt* are sister genes (Stauber et al. 1999). *a,* The *Hox* gene complex of insects (Beeman 1987; Brown et al. 2001; Falciani et al. 1996; Kaufman et al. 1990; Lewis 1998; Randazzo et al. 1993; von Allmen et al. 1996) and homologous genes in vertebrates (*Hox-1—13*) (McGinnis and Krumlauf 1992). *Dfd* (*Deformed*), *Scr* (*Sex combs reduced*), *ftz* (*fushi tarazu*), *Antp* (*Antennapedia*), *Ubx* (*Ultrabithorax*), and *abdA* (*Abdominal A*) form a subgroup of *Antennapedia*-related genes. *lab, labial; pb, proboscipedia; zen, zerknüllt; bcd, bicoid; AbdB, Abdominal B. b,* Homeodomain tree generated with the Quartett Puzzling maximum likelihood method using 1,000 puzzling steps (maximum reliability = 100) (Strimmer and von Haesseler 1996). The branch length indicates the number of amino acid exchanges (0.1 amino acid exchanges per scale). Note that the *bicoid* and *zerknüllt* sequences form a clade. For details see Stauber et al. 1999.

of any candidate gene at a predicted site in the genome provides a reliable indication, if the same result is obtained in several independent (paraphyletic) out-group taxa. The *Hox-C* provides an ideal frame, because linkage of *Hox* genes is extremely well conserved over long periods of time (Falciani et al. 1996). By applying the approach to *bicoid*, it should be possible to ultimately pinpoint the origin of *bicoid* on the phylogenetic tree with a high degree of confidence. Currently, only the relevant *Hox-C* portion of *Tribolium* has been sequenced in insects, and no *bicoid*- but two *zerknüllt*-like genes were found between the *Hox-2* and *Hox-4* interval of *Tribolium* (Brown et al. 2001). It is unlikely that one of these *zerknüllt*-like genes is actually a *bicoid* orthologue that failed to diverge as much as *bicoid* in the fly lineage. Sequence similarity between the two *Tribolium-zerknüllt*-like genes rather suggests a recent gene duplication specific to the *Tribolium* lineage. Therefore, the findings in *Tribolium* argue that *bicoid* originated after the basal radiation of holometabolous insects.

The identification of *zerknüllt* as an evolutionary sister gene of *bicoid* also provides an entry point for reconstructing the functional evolution of these genes. *bicoid* is involved in embryonic patterning throughout cyclorrhaphan flies, while *zerknüllt* is required for establishing extraembryonic tissues. Within Diptera, Cyclorrhapha differ from more basal taxa in the morphogenetic process of head involution (Hennig 1968) and with respect to extraembryonic tissue formation (Anderson 1966; Schmidt-Ott 2000). In noncyclorrhaphan Diptera and many other insect taxa extraembryonic tissue consists of two epithelia, the amnion and the serosa. The serosa abuts the egg-shell on the inner side. The amnion covers the embryo ventrally and generates an amniotic cavity between itself and the germ band. Importantly, in noncyclorrhaphan Diptera and other insects the extraembryonic anlage is established in an area which includes the anterior portion of the early embryo (fig. 17.3) (Rohr et al. 1999). In cyclorrhaphan flies, where *bicoid* is active in the anterior portion of the early embryo, extraembryonic tissue is at all stages restricted to the dorsal side of the embryo and no amniotic cavity is formed. *zerknüllt* homologues from insects as diverse as locusts, beetles, and flies are zygotically expressed in the anlage of extraembryonic tissue, suggesting that the progenitor of *bicoid* and *zerknüllt* was *zerknüllt*-like in that it had a role in establishing extraembryonic tissue. More recently, it has been shown that *zerknüllt* homologues of several noncyclorrhaphan Diptera (Stauber et al. 2002) and of the locust *Schistocerca* (Orthoptera) (Dearden et al. 2000) are, like *bicoid*, expressed maternally. However, in contrast to anterior localized *bicoid*, their mRNAs are equally distributed throughout the oocyte and early embryo. These observations led to the speculation

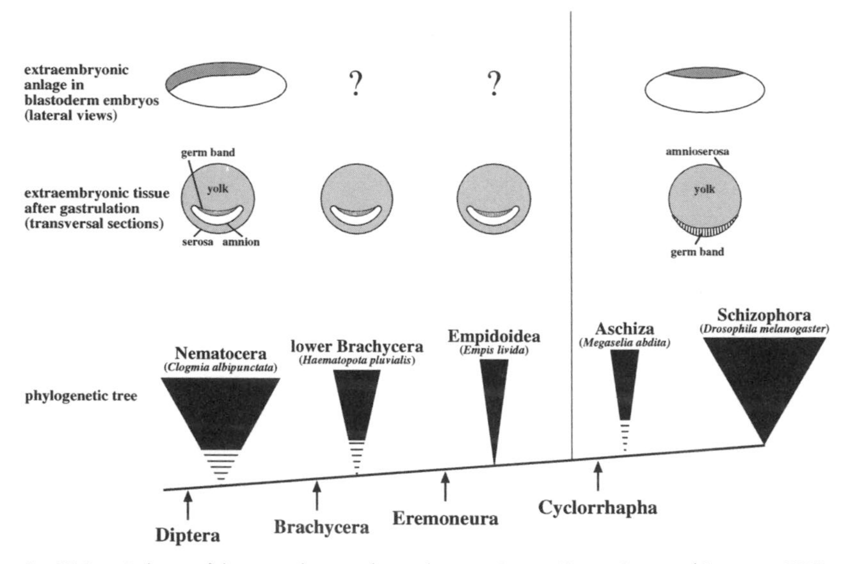

Fig. 17.3.—Reduction of the extraembryonic anlage and tissue in Diptera (Campos-Ortega and Hartenstein 1997; Rohr et al. 1999; Schmidt-Ott 2000; Stauber et al. 1999). The width of the triangles in the phylogenetic tree of Diptera (Yeates and Wiegmann 1999) indicates relative species abundance in each taxon. Dorsal is up and anterior (*upper row*) is to the left.

that the *bicoid/zerknüllt* progenitor had a maternal role in embryonic as well as a zygotic role in extraembryonic development, which were distributed to the two daughter genes separately. Nevertheless, the data do not support a morphogenetic role for maternal gene expression of the *bicoid/zerknüllt* progenitor.

Plasticity of Early Developmental Mechanisms

The results discussed so far suggest that the morphogenetic gradient of Bicoid was added to the factors involved in starting the segmentation gene cascade of insects during the evolution of holometabolous insects, possibly only in the stem lineage of cyclorrhaphan flies. The question of whether *bicoid* inherited functions from its *Hox-3* progenitor is currently unsolved. An alternative but not necessarily exclusive scenario is that *bicoid* replaced unrelated activators either completely or partially. Under the assumption that noncyclorrhaphan Diptera do not have a *bicoid* gene, evidence for a tightly localized anterior determinant in chironomid midges, that is, noncyclorrhaphan Diptera, might be taken as an indication of complete replacement of a yet unknown ancestral anterior determinant. An alternative although again not mutually exclusive view is that Bicoid gradually replaced ancestral activators, which might still function in cyclorrhaphan flies, albeit with less influence on anterior patterning. Viewed in this way, partial redundancies might reflect the evolutionary history of the genetic circuitry. This line of thinking led to a number of experiments aiming to replace *bicoid* activity by potentiating evolutionarily conserved genes that are believed to play important roles in the embryogenesis of other organisms. We will review these conditions below and indicate factors that substitute for specific Bicoid functions. Finally, we will discuss the finding that enhanced Bicoid activity can substitute for the anterior terminal signaling system.

Translational Repression of *caudal*

In the developing embryo of different insects, Caudal is expressed in a posterior-to-anterior gradient. In *Drosophila,* this gradient is formed during the syncytial blastoderm stage (Macdonald and Struhl 1986), when Bicoid represses the translation of the uniformly distributed maternal *cad* mRNA in the prospective head region (Mlodzik and Gehring 1987). In the flour beetle *Tribolium castaneum* a similar Caudal gradient forms, but, initially, translation of Caudal protein occurs throughout the embryo. Since the *Tribolium* germ rudiment develops in a posterior portion of the egg, Caudal expression actually covers the prospective head region (Schulz et al. 1998). In the silk moth *Bombyx mori* Caudal protein is distributed evenly at blastoderm stage. A gradient forms only when gastrulation is already underway and is most likely of zygotic origin (Xu et al. 1994). Anterior repression of *caudal* homologues occurs in other animal phyla as well. In the nematode *Caenorhabditis elegans* the mRNA of the *caudal* homologue *pal-1* is translationally repressed by a KH-domain protein, Mex-3 (Draper et al. 1996). In the frog *Xenopus,* it has been proposed that a mutually antagonistic relationship between the *orthodenticle* homologue *Xotx2* and the *caudal* homologue *Xcad3* is important for the transcriptional gene activation required for embryonic patterning (Isaacs et al. 1999). In spite of the conservation of Caudal gradients in different species, "when" and "how" this gradient is established vary. Moreover, in *Drosophila,* the lack of maternal *caudal* function can be compensated for by early zygotic *caudal* expression (Macdonald and Struhl 1986), which is actually regulated by *hunchback* (fig. 17.1, a) (Schulz and Tautz 1995). Furthermore, if Caudal is expressed ectopically in the prospective head region of the early *Drosophila* embryo, development can proceed normally at low temperatures (Niessing et al. 1999; Niessing et al. 2000). In this case Hunchback seems to block Caudal function by interfering posttranslationally (Wimmer et al. 2000).

Therefore, the translational repression of *caudal* mRNA by Bicoid appears not to be absolutely required for starting the segmentation gene cascade.

Transcriptional Activation of Abdominal Genes

In the abdomen, loss of Bicoid has variable consequences, depending on the temperature (Schulz and Tautz 1994) and presumably the genetic background. In some cases the abdomen appears normal, but more often loss of the second and fourth abdominal segments is observed. Bicoid contributes to the activation of genes required for abdomen formation, such as *Krüppel* and *knirps*. However, *knirps* is also activated by the posterior-to-anterior gradient of Caudal, and only in the absence of both Caudal and Bicoid is *knirps* expression completely suppressed (Rivera-Pomar and Jäckle 1996). Similarly, *Krüppel* expression is completely repressed only if Bicoid and Hunchback are removed together (Hülskamp et al. 1990; Schulz and Tautz 1994; Struhl et al. 1992). An enhanced anterior-to-posterior gradient of Hunchback can compensate for the absence of Bicoid in the formation of abdominal segments despite the fact that these genes are unrelated (fig. 17.1, b) (Schulz and Tautz 1994; Struhl et al. 1992).

Transcriptional Activation of Thoracic Genes

For proper thorax development, high levels of *hunchback* expression are crucial. Both maternal and zygotic *hunchback* contribute to anterior patterning of the embryo. The zygotic, Bicoid-dependent expression of *hunchback* (Tautz 1988) causes an intrinsic problem in studying the specific roles of the Bicoid and Hunchback gradients, because whenever *bicoid* activity is altered, *hunchback* activity is changed as well (Driever and Nüsslein-Volhard 1989; Struhl et al. 1989). Thus, in principle, many of the effects attributed to Bicoid might be indirect and eventually caused by Hunchback. To overcome this problem, an in vivo system has been developed to allow the study of the two gradients independently of each other (Wimmer et al. 2000). This system replaces the Bicoid-dependent, zygotic expression of *hunchback* with an additional contribution of maternal Hunchback, which allowed a small percentage of embryos to develop all larval segments. This finding demonstrates that Bicoid-dependent *hunchback* expression is in principle dispensable for segmentation and can be compensated for by maternal *hunchback* (fig. 17.1, c). However, in embryos without Bicoid an artificial anterior-to-posterior Hunchback gradient can rescue only the meta- and the mesothorax, and not the prothorax or the head (Wimmer et al. 2000). Thus, maternal Hunchback can substitute for

Bicoid in determining thoracic structures but cannot replace *bicoid* activity required for head development.

Transcriptional Activation of Head Genes

Wild-type levels of *bicoid* activity are sufficient to pattern the anterior embryonic head including pregnathic and the first gnathic segments in the absence of any *hunchback* activity (Lehmann and Nüsslein-Volhard 1987). Both Bicoid and Hunchback independently suppress the formation of posterior terminal structures at the anterior end, but only Bicoid is able to induce the development of the anterior head segments (Wimmer et al. 2000). However, Hunchback becomes absolutely essential for patterning the anterior head when the levels of *bicoid* activity are reduced. In fact, the same amount of Bicoid can essentially result in the development of either anterior head or be unable to even repress ectopic posterior abdominal structures, depending on whether Hunchback is present or absent (E. A. Wimmer and C. Desplan, unpublished results). Bicoid and Hunchback are therefore able to synergize with each other in order to organize anterior development (fig. 17.1, *d*) (Simpson-Brose et al. 1994). It is important to note that Bicoid and Hunchback code for transcription factors with different DNA-binding specificity (Driever and Nüsslein-Volhard 1989; Treisman and Desplan 1989). Therefore, it is likely that all genes controlled by Bicoid also contain cis-regulatory elements with functional Hunchback binding sites.

Whereas Hunchback without Bicoid fails to rescue head structures, another Bicoid target—the homeodomain encoding the transcription factor *orthodenticle,* which, like *hunchback,* is a Bicoid-target—can rescue expression of some Bicoid target genes and head structures of still undefined anterior origin when provided independently of Bicoid (J. Reischl and E. A. Wimmer, unpublished results). This finding is interesting, because the evolutionarily conserved Orthodenticle molecule shares DNA-binding specificity with Bicoid based on a lysine in position 50 of the homeodomain. A lysine in this position is found only in a few homeodomains and constitutes an important difference from all other homeodomain genes in the Hox-C (Bürglin 1995). Therefore, Orthodenticle-mediated rescue of *bicoid* mutants raises the question of whether acquisition by Bicoid of an Orthodenticle-like DNA-binding specificity has partially released *orthodenticle* from its role in patterning the head.

Bicoid Can Replace the Anterior Terminal System

In *Drosophila,* the poles of the embryo are patterned by the maternal terminal system (Casanova and Struhl 1993; Klingler et al. 1988;

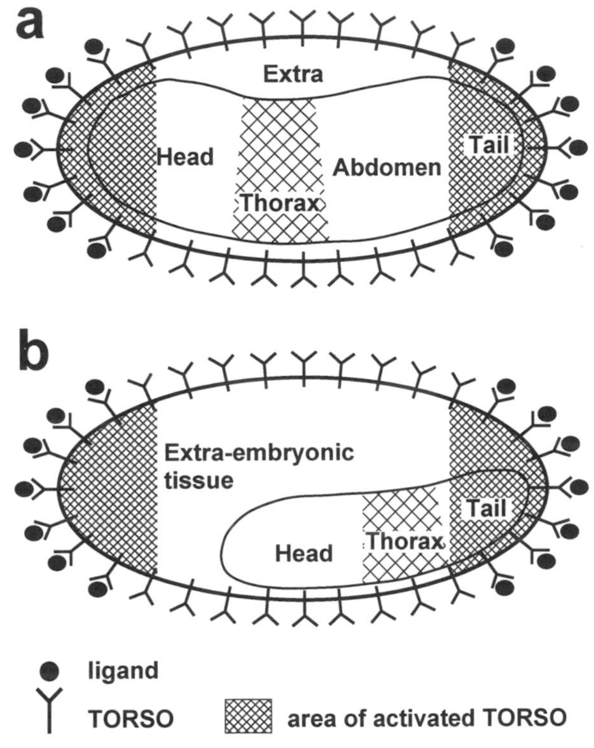

Fig. 17.4.—Activity of the terminal maternal system in long- and short- or intermediate-germ-band insects. The terminal maternal system is activated at the egg poles after a ligand has bound to the receptor tyrosine kinase Torso. *a*, In the long-germ-band insect *Drosophila* Torso-mediated signaling participates in anterior embryonic development. *b*, In short- or intermediate-germ-band insects, such as *Tribolium*, anterior *torso* activity is restricted to the extraembryonic anlage (Schröder et al. 2000).

Sprenger and Nüsslein-Volhard 1992). This system includes the generation of a localized extracellular ligand that activates the receptor tyrosine kinase Torso. *Drosophila* embryos deficient in the terminal system fail to develop the anteriormost head structures. *torso* activity contributes to the expression of several Bicoid-dependent head gap genes, and Torso-responsive elements are expected in most natural regulatory regions of Bicoid target genes (Gao et al. 1996; Grossniklaus et al. 1994; Schaeffer et al. 1999; Wimmer et al. 1995). In wild-type embryos, the Torso pathway functions by derepression of terminal gap genes like *tailless* (Jiménez et al. 2000; Paroush et al. 1997), and it seems likely that the effects on Bicoid target genes are caused by a derepression mechanism as well. Consistent with the fact that the Torso signaling pathway and Bicoid act independently on their common targets, very high levels of Bicoid can compensate for the lack of the anterior terminal system (Schaeffer et al. 2000). Again, this observation illustrates partial redundancies among the various factors involved in starting head development. Apparently, independent regulatory inputs can compensate for each other, as long as the remaining input is appropriately enhanced. However, there is currently no evidence that an enhanced terminal system can also compensate for the loss of Bicoid. Does directionality in the rescue imply that the anterior terminal system has been taking over functions that were previously assigned to *bicoid?* Comparative embryology supports this idea. The early embryo of most insects is restricted to the posterior portion of the egg (fig. 17.4). In the intermediate-germ-band insect *Tribolium,* activity of the terminal system at the anterior pole of the egg occurs in a region devoid of embryonic anlagen (Schröder et al. 2000). Therefore, the involvement of terminal signaling in the long-germ-band insect *Drosophila,* where all of the anterior blastoderm is specified as embryonic tissue, might be derived.

Concluding Remarks

Segmentation in *Drosophila* relies on several transcription factors which together start the segmentation process. However, none of the individual genes functions as a "master regulator" for some part of the embryo. Rather, the zygotic segmentation genes integrate several independent inputs. The independent inputs can to some extent compensate for each other and provide a very plastic early developmental system. It is currently unclear whether compensation occurs preferentially in one direction, reflecting the evolutionary history of the system, or works both ways. In the latter case, heterogenetic rescue experiments might not be suitable for revealing the order of recruitment in the evolution of the individual factors. Nevertheless, it is now clear that developmental plasticity allows for substantial reorganizations in the gene network underlying early development, and an increasing body of comparative data strongly suggests that considerable molecular reorganizations indeed took place during insect evolution. The most conspicuous example is *bicoid.* The origin of this molecule must have had a profound effect on the entire patterning system, and only on this system, because *bicoid* is not required during later stages of the life cycle. There is good evidence that *bicoid* is a diverged *Hox-3* gene and that an ancestral progenitor was *zerknüllt*-like in sequence and in—at least part of—its functions. In the future, a major challenge to understanding the evolution of *bicoid* and the developmental process of segmentation associated with it will be to bridge the functional gap between *zerknüllt* and *bicoid* genes. Which *bicoid* functions were inherited from the common progenitor? Which functions were gained independently of the progenitor? Which unrelated genes—such as *hunchback* or *orthodenticle*—did the emerging Bicoid molecule replace? Mapping each event individually on the phylogenetic tree will be a necessary prerequisite for addressing the important question of how developmental change at an early stage affects the evolution of the clade.

The early developmental processes starting the segmentation gene cascade do not appear to be modular in nature. The activity gradients of genes such as *bicoid* and other maternally expressed factors provide a highly interactive but partially redundant and decomposable genetic system, which puts an upstream limit on a proposed segmentation module. Thus, *bicoid* does not seem to be part of a developmental module, but provides one of different potential inputs into an executing segmentation module. The robustness of this module, however, might allow the plasticity and evolvability of the early developmental inputs. The case of *bicoid* seems to fulfill criteria for the creation of functional evolutionary novelties by a two-step process: gene duplication and independent evolution of ancestral subfunctions (Force et al. 1999; Lynch and Force 2000; Stauber et al. 2002; Schlosser 2004). A maternally and zygotically expressed *Hox-3* gene duplicates and generates maternal *bicoid* and zygotic *zerknüllt*. *zerknüllt* continues to fulfill an original zygotic function in extraembryonic tissue formation and is not compromised when maternal *bicoid* changes DNA-binding specificity and presumably acquires its role in anterior patterning.

Acknowledgments

We thank Claude Desplan (New York) for very helpful comments on an earlier draft of the manuscript and Ralf Pflanz (Göttingen) for the picture of the Caudal gradient shown in figure 1.

References

Anderson, D. T. 1966. The comparative embryology of the Diptera. *Annu. Rev. Entomol.* 11:23–64.

Bate, M., and A. Martinez-Arias, eds. 1993. *The development of Drosophila melanogaster.* Cold Spring Harbor, N.Y.: Cold Spring Harbor Laboratory Press.

Beeman, R. W. 1987. A homeotic gene cluster in the red flour beetle. *Nature* 327: 247–249.

Brown, S., J. Fellers, T. Shippy, R. Denell, M. Stauber, and U. Schmidt-Ott. 2001. A strategy for mapping *bicoid* on the phylogenetic tree. *Curr. Biol.* 11:R43–R44.

Bürglin, T. R. 1995. The evolution of homeobox genes. In *Biodiversity and evolution,* R. Arai, M. Kato, and Y. Dai, 665–710. Tokyo: National Science Museum Foundation.

Campos-Ortega, J. A., and V. Hartenstein. 1997. *The embryonic development of Drosophila melanogaster.* Berlin: Springer-Verlag.

Casanova, J., and G. Struhl. 1993. The *torso* receptor localizes as well as transduces the spatial signal specifying terminal body pattern ion *Drosophila. Nature* 362:152–155.

Castelli-Gair, J. 1998. Implications of the spatial and temporal regulation of *Hox* genes on development and evolution. *Int. J. Dev. Biol.* 42:437–444.

Curtis, D., J. Apfeld, and R. Lehmann. 1995. *nanos* is an evolutionarily conserved organizer of anterior-posterior polarity. *Development* 121:1899–1910.

Dearden, P., M. Grbic, F. Falciani, and M. Akam. 2000. Maternal expression and early zygotic regulation of the *Hox3/zen* gene in the grasshopper *Schistocerca gregaria. Evol. Dev.* 2:261–270.

Draper, B. W., C. C. Mello, B. Bowerman, J. Hardin, and J. R. Priess. 1996. MEX-3 is a KH domain protein that regulates blastomere identity in early *C. elegans* embryos. *Cell* 87:205–216.

Driever, W. 1993. Maternal control of anterior development in the *Drosophila* embryo. In *The development of Drosophila melanogaster,* ed. M. Bate and A. Martinez-Arias, 301–324. Cold Spring Harbor, N.Y.: Cold Spring Harbor Laboratory Press.

Driever, W., and C. Nüsslein-Volhard. 1989. The bicoid protein is a positive regulator of *hunchback* transcription in the early *Drospohila* embryo. *Nature* 337:138–143.

Dubnau, J., and G. Struhl. 1996. RNA recognition and translational regulation by a homeodomain protein. *Nature* 379:694–699.

Falciani, F., B. Hausdorf, R. Schröder, M. Akam, D. Tautz, R. Denell, and S. Brown. 1996. Class 3 Hox genes in insects and the origin of *zen. Proc. Natl. Acad. Sci. U.S.A.* 93:8479–8484.

Foe, V. E., G. M. Odell, and B. A. Edgar. 1993. Mitosis and morphogenesis in the *Drosophila* embryo: point and counterpoint. In *The development of Drosophila melanogaster,* ed. M. Bate and A. Martinez-Arias, 149–300. Cold Spring Harbor, N.Y.: Cold Spring Harbor Laboratory Press.

Force, A., M. Lynch, F. B. Pickett, A. Amores, Y.-L. Yan, and J. Postlethwait. 1999. Preservation of duplicate genes by complementary, degenerative mutations. *Genetics* 151:1531–1545.

Gao, Q., and R. Finkelstein. 1998. Targeting gene expression to the head: the *Drosophila orthodenticle* gene is a direct target of the Bicoid morphogen. *Development* 125: 4185–4193.

Gao, Q., Y. Wang, and R. Finkelstein. 1996. *Orthodenticle* regulation during embryonic head development in *Drosophila. Mech. Dev.* 56:3–15.

Grbic, M., L. M. Nagy, S. B. Carroll, and M. Strand. 1996. Polyembryonic development: insect pattern formation in a cellularized environment. *Development* 122: 795–804.

Grimaldi, D. 2000. Mesozoic radiations of the insects and origins of the modern fauna. In *Proceedings of the XXI International Congress of Entomology,* Iguassu, Brazil, 1: xix–xxvii.

Grossniklaus, U., K. M. Cadigan, and W. Gehring. 1994. Three maternal coordinate systems cooperate in the pattering of the *Drosophila* head. *Development* 120:3155–3171.

Hennig, W. 1968. *Die Larvenformen der Dipteren.* Berlin: Akademie-Verlag.

Hoch, M., N. Gerwin, H. Taubert, and H. Jäckle. 1992. Competition for overlapping sites in the regulatory region of the *Drosophila* gene *Krüppel. Sience* 256:94–97.

Hülskamp, M., C. Pfeifle, and D. Tautz. 1990. A morphogenetic gradient of Hunchback protein organizes the expression of the gap genes *Krüppel* and *knirps* in the early *Drosophila* embryo. *Nature* 346:577–580.

Isaacs, H. V., M. Andreazzoli, and J. M. W. Slack. 1999. Anteroposterior patterning by mutual repression of orthodenticle and caudal-type transcription factors. *Evol. Dev.* 1:143–152.

Jiménez, G., A. Guichet, A. Ephrussi, and J. Casanova. 2000. Relief of gene repression by Torso RTK signaling: role of *capicua* in *Drosophila* terminal and dorsoventral patterning. *Genes Dev.* 14:224–231.

Kalthoff, K. 1979. Analysis of a morphogenetic determinant in an insect embryo (*Smit-*

tia Spec., Chironomidae, Diptera). In *Determinants of spatial organization,* ed. S. Subtelny and I. R. Konigsberg, 79–126. New York: Academic Press.

Kaufman, T. C., M. A. Seeger, and G. Olsen. 1990. Molecular and genetic organization of the *Antennapedia* gene complex of *Drosophila melanogaster. Adv. Genet.* 27: 309–362.

Klingler, M., M. Erdelyi, J. Szabad, and C. Nüsslein-Volhard. 1988. Function of *torso* in determining the terminal anlagen of the *Drosophila* embryo. *Nature* 335:275–277.

Kristensen, N. P. 1991. Phylogeny of extant hexapods. In *Insects of Australia,* ed. CSIRO, Division of Entomology, 125–140. Melbourne: Melbourne University Press.

Lehmann, R., and C. Nüsslein-Volhard. 1987. *hunchback,* a gene required for segmentation of an anterior and posterior region of the *Drosophila* embryo. *Dev. Biol.* 119: 402–417.

Lewis, E. 1998. The *bithorax* complex: the first fifty years. *Int. J. Dev. Biol.* 42: 403–415.

Li, Y., S. Brown, B. Hazsdorf, D. Tautz, R. E. Denell, and R. Finkelstein. 1996. Two *orthodenticle*-related genes in the short-germ beetle *Tribolium castaneum. Dev. Genes Evol.* 206:35–45.

Lipshitz, H. D., and C. A. Smibert. 2000. Mechanisms of RNA localization and translational regulation. *Curr. Opin. Genet. Dev.* 10:476–488.

Lynch, M., and A. Force. 2000. The probability of duplicate gene preservation by subfunctionalization. *Genetics* 154:459–473.

Macdonald, P. M., and G. Struhl. 1986. A molecular gradient in early *Drosophila* embryos and its role in specifying the body pattern. *Nature* 324:537–545.

Martinez-Arias, A. 1993. Development and patterning of the larval epidermis of *Drosophila.* In *The development of Drosophila melanogaster,* ed. M. Bate and A. Martinez-Arias, 517–608. Cold Spring Harbor, N.Y.: Cold Spring Harbor Laboratory Press.

McGinnis, W., and R. Krumlauf. 1992. Homeobox genes and axial patterning. *Cell* 68: 283–302.

Mlodzik, M., and W. J. Gehring. 1987. Expression of the *caudal* gene in the germ line of *Drosophila:* formation of an RNA and protein gradient during early embryogenesis. *Cell* 48:465–478.

Niessing, D., N. Dostatni, H. Jäckle, and R. Rivera-Pomar. 1999. Sequence interval within the PEST domain of Bicoid is important for translational repression of *caudal* mRNA in the anterior region of the *Drosophila* embryo. *EMBO J.* 18:1966–1973.

Niessing, D., W. Driever, F. Sprenger, H. Taubert, H. Jäckle, and R. Rivera-Pomar. 2000. Homeodomain position 54 specifies transcriptional versus translational control by Bicoid. *Mol. Cell* 5:395–401.

Nüsslein-Volhard, C., and E. Wieschaus. 1980. Mutations affecting segment number and polarity in *Drosophila. Nature* 287:795–801.

Pankratz, M. J. and H. Jäckle. 1993. Blastoderm segmentation. In *The development of Drosophila melanogaster,* ed. M. Bate and A. Martinez-Arias, 467–516. Cold Spring Harbor, N.Y.: Cold Spring Harbor Laboratory Press.

Paroush, Z., S. M. Wainwright, and D. Ish-Horowicz. 1997. Torso signaling regulates terminal patterning in *Drosophila* by antagonizing Groucho-mediated repression. *Development* 124:3827–3834.

Pultz, M. A., J. N. Pitt, and N. M. Alto. 1999. Extensive zygotic control of the anteroposterior axis in the wasp *Nasonia vitripennis. Development* 126:701–710.

Randazzo, F. M., M. A. Seeger, C. A. Huss, M. A. Sweeney, J. K. Cecil, and T. C. Kaufman. 1993. Structural changes in the *Antennapedia* complex of *Drosophila pseudoobscura. Genetics* 133:319–330.

Rivera-Pomar, R., and H. Jäckle. 1996. From gradients to stripes in *Drosophila* embryogenesis: filling in the gaps. *Trends Genet.* 12:478–483.

Rivera-Pomar, R., D. Niessing, O. U. Schmidt, W. J. Gehring, and H. Jaeckle. 1996. RNA binding and translational suppression by *bicoid. Nature* 379:746–749.

Rohr, K. B., D. Tautz, and K. Sander. 1999. Segmentation gene expression in the moth-midge *Clogmia albipunctata* (Diptera, Pschodidae) and other primitive dipterans. *Dev. Genes Evol.* 209:145–154.

Sander, K. 1960. Analyse des ooplasmatischen Reaktionssystems in *Euscelis plebejjus* Fall (Cicadina) durch Isolieren und Kombinieren von Keimteilen. *Roux's Arch. Entwicklungsmech.* 151:660–707.

Sander, K. 1976. Specification of the basic body pattern in insect embryogenesis. *Adv. Insect Physiol.* 12:125–238.

Schaeffer, V., F. Janody, C. Loss, C. Desplan, and E. A. Wimmer. 1999. Bicoid functions without its TATA-binding protein-associated factor interaction domains. *Proc. Natl. Acad. Sci. U.S.A.* 96:4461–4466.

Schaeffer, V., D. Killian, C. Desplan, and E. A. Wimmer. 2000. High Bicoid levels render the terminal system dispensable for *Drosophila* head development. *Development* 127:3993–3999.

Schlosser, G. 2004. The role of modules in development and evolution. In *Modularity in development and evolution,* ed. G. Schlosser and G. P. Wagner. Chicago: University of Chicago Press.

Schmidt-Ott, U. 2000. The amnioserosa is an apomorphic character of cyclorrhaphan flies. *Dev. Genes Evol.* 210:373–376.

Schröder, R., C. Eckert, C. Wolff, and D. Tautz. 2000. Conserved and divergent aspects of terminal patterning in the beetle *Tribolium castaneum. Proc. Natl. Acad. Sci. U.S.A.* 97:6591–6596.

Schröder, R., and K. Sander. 1993. A comparison of transplantable *bicoid* activity and partial *bicoid* homeobox seqences in several *Drosophila* and blowfly species (Calliphoridae). *Roux's Arch. Dev. Biol.* 203:34–43.

Schubiger, G., R. C. Moseley, and W. J. Wood. 1977. Interaction of different egg parts in determination of various body regions in *Drosophila melanogaster. Proc. Natl. Acad. Sci. U.S.A.* 74:2050–2053.

Schulz, C., R. Schröder, B. Hausdorf, C. Wolff, and D. Tautz. 1998. A *caudal* homologue in the short germ band beetle *Tribolium* shows similarities to both, the *Drosophila* and the vertebrate *caudal* expression patterns. *Dev. Genes Evol.* 208:283–289.

Schulz, C., and D. Tautz. 1994. Autonomous concentration-dependent activation and repression of *Krueppel* by *hunchback* in the *Drosophila* embryo. *Development* 120: 3043–3049.

Schulz, C., and D. Tautz. 1995. Zygotic *caudal* regulation by *hunchback* and its role in abdominal segment formation of the *Drosophila* embryo. *Development* 121: 1023–1028.

Schwalm, F. 1987. *Insect morphogenesis.* Basel: S. Karger.

Shaw, P. J., A. Salameh, A. P. McGregor, S. Bala, and G. A. Dover. 2001. Divergent structure and function of the *bicoid* gene in Muscoidea fly species. *Evol. Dev.* 3:251–262.

Simpson-Brose, M., J. Treisman, and C. Desplan. 1994. Synergy between the *hunchback* and *bicoid* morphogens is required for anterior patterning in *Drosophila. Cell* 78: 855–865.

Sommer, R., and D. Tautz. 1991. Segmentation gene expression in the housefly *Musca domestica. Development* 113:419–430.

Sprenger, F., and C. Nüsslein-Volhard. 1992. Torso receptor activity is regulated by a diffusible ligand produced at the extracellular terminal regions of the *Drosophila* egg. *Cell* 71:987–1001.

Stauber, M., H. Jäckle, and U. Schmidt-Ott. 1999. The anterior determinant *bicoid* of *Drosophila* is a derived Hox class 3 gene. *Proc. Natl. Acad. Sci. U.S.A.* 96: 3786–3789.

Stauber, M., A. Prell, and U. Schmidt-Ott. 2002. A single Hox3 gene with composite *bicoid* and *zerknüllt* expression characteristics in non-cyclorrhaphan flies. *Proc. Natl. Acad. Sci. U.S.A.* 99:274–279.

Stauber, M., H. Taubert, and U. Schmidt-Ott. 2000. Function of *bicoid* and *hunchback* homologs in the basal cyclorrhaphan fly *Megaselia* (Phoridae). *Proc. Natl. Acad. Sci. U.S.A.* 97:10844–10849.

St Johnston, D. 1993. Pole plasm and the posterior group genes. In *The development of Drosophila melanogaster,* ed. M. Bate and A. Martinez-Arias, 325–363. Cold Spring Harbor, N.Y.: Cold Spring Harbor Laboratory Press.

St Johnston, D., and C. Nüsslein-Volhard. 1992. The origin of pattern and polarity in the *Drosophila* embryo. *Cell* 68:201–220.

Strimmer, K., and A. von Haesseler. 1996. Quartet puzzling: a quartet maximum-likelihood method for reconstructing tree topologies. *Mol. Biol. Evol.* 13:964–969.

Struhl, G., P. Johnston, and P. A. Lawrence. 1992. Control of *Drosophila* body pattern by the *hunchback* morphogen gradient. *Cell* 69:237–249.

Struhl, G., K. Struhl, and P. M. Macdonald. 1989. The gradient morphogen bicoid is a concentration-dependent transcriptional activator. *Cell* 57:1259–1273.

Tautz, D. 1988. Regulation of the *Drosophila* segmentation gene *hunchback* by two maternal morphogenetic centers. *Nature* 332:281–284.

Terol, J., M. Perez-Alonso, and R. de Frutos. 1995. Molecular characterization of the *zerknüllt* region of the Antennapedia complex of *D. subobscura. Chromosoma* 103: 613–624.

Treisman, J. E., and C. Desplan. 1989. The products of the *Drosophila* gap genes *hunchback* and *Krüppel* bind to the *hunchback* promoters. *Nature* 341:335–337.

von Allmen, G., I. Hogga, A. Spierer, F. Karch, W. Bender, H. Gyurkovics, and E. Lewis. 1996. Splits in fruitfly *Hox* gene complexes. *Nature* 380:116.

von Dassow, G., E. Meir, E. M. Munro, and G. M. Odell. 2000. The segment polarity network is a robust developmental module. *Nature* 406:188–192.

Wimmer, E. A., A. Carleton, P. Harjes, T. Turner, and C. Desplan. 2000. *bicoid*-independent formation of thoracic segments in *Drosophila. Science* 287:2476–2479.

Wimmer, E. A., M. Simpson-Brose, S. M. Cohen, C. Desplan, and H. Jäckle. 1995. Trans- and cis-acting requirements for blastodermal expression of the head gap gene *buttonhead. Mech. Dev.* 53:235–245.

Wolff, C., R. Schröder, C. Schulz, D. Tautz, and M. Klingler. 1998. Regulation of the *Tribolium* homologues of *caudal* and *hunchback* in *Drosophila:* evidence for maternal gradient systems in a short germ embryo. *Development* 125:3645–3654.

Wolff, C., R. Sommer, R. Schröder, G. Glaser, and D. Tautz. 1995. Conserved and divergent expression aspects of the *Drosophila* segmentation gene *hunchback* in the short germ band embryo of the flour beetle *Tribolium. Development* 121:4227–4236.

Xu, X., P. X. Xu, and Y. Suzuki. 1994. A maternal homeobox gene, *Bombyx caudal,* forms both mRNA and protein concentration gradients spanning anteroposterior axis during gastrulation. *Development* 120:277–285.

Yeates, D. K., and B. M. Wiegmann. 1999. Congruence and controversy: toward a higher-level phylogeny of Diptera. *Annu. Rev. Entomol.* 44:397–428.

18 The Evolution of Nematode Development: How Cells and Genes Change Their Function

RALF J. SOMMER

Introduction

One of the most important features of developmental systems is their hierarchical and modular organization. Organs, tissues, cells, genetic networks, and signaling cascades build a hierarchical system, which is organized in a modular fashion generating networks of interacting units within the organism (Riedl 1975). During evolution, morphological novelty can result from two types of modifications. Existing modules can change without completely altering the rest of a network. Alternatively, new developmental modules can be constructed. Now that developmental processes have been studied in great detail using integrated genetic and molecular approaches in selected model systems, one can start to analyze how developmental systems evolve and how developmental modules are modified on an evolutionary timescale (Raff 1996; Gerhard and Kirschner 1997). Evolutionary developmental biology thereby provides the unique opportunity to combine experimental approaches to developmental biology with the comparative research tradition of evolutionary biology.

Nematode Developmental Biology: Cells as Modules in Development

Understanding the relationship between evolution and development relies on case studies. Several such case studies have been carried out in recent years, mostly starting from the detailed knowledge of developmental processes in a limited number of model organisms. The free-living soil nematode *Caenorhabditis elegans* is one of the important model systems in developmental biology. Free-living nematodes have invariant cell lineages, and the adult organisms consists only of around 1,000 somatic cells (Sulston and Horvitz 1977). In *C. elegans*. for example, self-fertilizing hermaphrodites consist of 959 somatic cells, whereas males consist of 1,031 cells (Wood 1988; Riddle et al. 1997). Given the invariability of the cell lineage, developmental processes can be studied at the level of individual cells, and the interactions among cells can be analyzed. When the results are complemented with a genetic and molecular analysis, an integrated understanding of developmental processes at the cellular, genetic, and molecular levels in nematodes can be achieved.

When organisms are considered as networks of interacting units (modules) in a hierarchical system, the example of nematodes allows us to study how cellular and genetic networks work during development and evolution. In this context, the concept of a module can be applied to organs, cells, genes, and even domains within a gene. Thus, a module can be a unit at any hierarchical level and as such can interact with many similar units in a network to regulate pattern formation at the next higher level of organization. In what follows, I will use nematode vulva development as a case study and argue that the vulva itself, but also the cells forming this organ and the genes acting in these cells to regulate vulva formation, can be considered modules within more complex systems (networks). During evolution the vulva module is stable, while other modules, such as cells and genes, can change their function without affecting higher modules, like the vulva itself. Thus, the comparative analysis of vulva development in nematodes indicates how developmental modules and networks evolve.

Nematode Vulva Development: A Case Study

A developmental process that has been studied in great detail in *C. elegans* is the formation of the vulva, the egg-laying structure and copulatory organ of the hermaphrodite. The vulva is generated by three precursor cells, which divide three times within five hours to give rise to the 22 cells eventually forming the complete organ. The ease with which this cellular system can be analyzed attracted researchers soon after the postembryonic lineage was originally described by Sulston and Horvitz (1977). Over the years, studies of *C. elegans* vulva development evolved from experimental studies on the cell-cell interactions during cell fate specification to genetic and molecular approaches addressing the mechanisms underlying this pattern formation process.

The vulva is a derivative of the ventral epidermis, which consists of 12 precursor cells, called P1.p to P12.p. They form a linear array of cells between pharynx and rectum and differentiate in a position-specific manner (fig. 18.1) (Sulston and Horvitz 1977). P1.p, P2.p, and P9.p–P11.p remain epidermal and fuse with the hypodermal syncytium (fig. 18.1, *B*). In contrast, P3.p–P8.p in the central body region remain

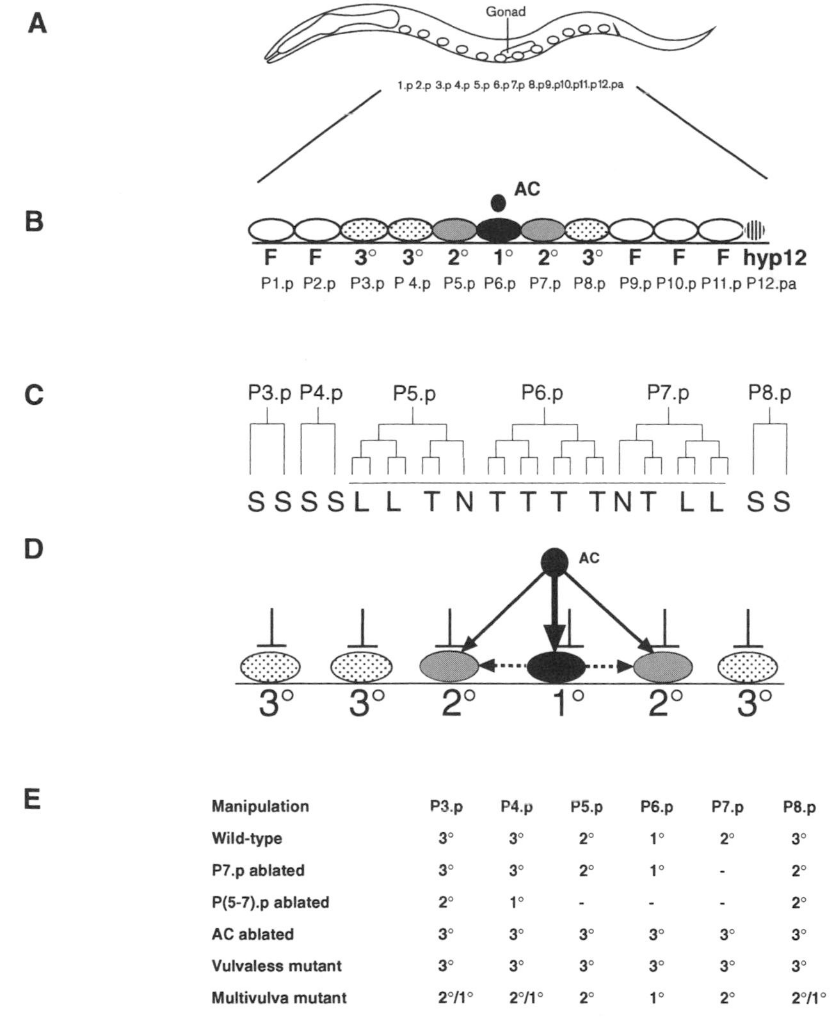

Manipulation	P3.p	P4.p	P5.p	P6.p	P7.p	P8.p
Wild-type	3°	3°	2°	1°	2°	3°
P7.p ablated	3°	3°	2°	1°	-	2°
P(5-7).p ablated	2°	1°	-	-	-	2°
AC ablated	3°	3°	3°	3°	3°	3°
Vulvaless mutant	3°	3°	3°	3°	3°	3°
Multivulva mutant	2°/1°	2°/1°	2°	1°	2°	2°/1°

Fig. 18.1.—Schematic summary of vulva formation in *C. elegans. A,* During the L1 stage, the 12 ventral epidermal cells, P1.p–P12.p, are equally distributed between pharynx and rectum. *B,* P1.p, P2.p, and P9.p–P11.p fuse with the hypodermal syncytium *hyp7* (F; *white ovals*). P3.p–P8.p form the vulva equivalence group and adopt one of three alternative cell fates. P6.p has a 1° fate (*black oval*), and P5.p and P7.p have a 2° fate (*grey ovals*). P3.p, P4.p, and P8.p have a 3° fate and remain epidermal (*dotted ovals*). The anchor cell (AC; *black circle*) provides an inductive signal for vulva formation. *C,* Cell lineage pattern of the vulval precursor cells. P3.p, P4.p, and P8.p divide once and then fuse with *hyp7* (S). P5.p and P7.p generate seven progeny each. The first two cell divisions occur along the anteroposterior axis; the third division can be longitudinal (L), transversal (T), or absent (N). P6.p generates eight progeny. *D,* Schematic summary of signaling interactions during vulva formation in *C. elegans.* An inductive EGF-like signal originates from the AC (*black arrows*). P6.p signals its neighbors to adopt a 2° fate via "lateral signaling" (*dotted arrows*). Negative signaling (*bars*) prevents inappropriate vulva differentiation. See text for details. *E,* Summary of cell ablation experiments. After ablation of P7.p, P8.p adopts a 2° fate and forms part of the vulva. After ablation of P5.p–P7.p, P3.p, P4.p, and P8.p can form a functional vulva. After ablation of the AC, all precursor cells adopt a 3° fate. Vulvaless mutants have a similar phenotype as AC-ablated animals. Multivulva mutants show a phenotype opposite to vulvaless mutants, namely, the ectopic proliferation of P3.p, P4.p, and P8.p.

unfused and form the so-called vulva equivalence group (fig. 18.1, *B*). In principle, all of these six cells can participate in vulva formation. However, in a wild-type animal only the three central cells, P5.p–P7.p, divide several times and form vulval tissue, whereas P3.p, P4.p, and P8.p remain epidermal. The latter cell fate has been designated the 3° fate (Sulston and White 1980; Sternberg and Horvitz 1986). P5.p and P7.p generate seven progeny, which form the outer part of the vulva (fig. 18.1, *C*). P6.p generates eight progeny, forming the central part of the vulva which connects the uterus to the outside environment (fig. 18.1, *C*). The fate of P5.p and P7.p has been designated the 2° fate, and that of P6.p, the 1° fate (Sulston and White 1980; Sternberg and Horvitz 1986). These hierarchical cell fate designations resulted from combinatorial cell ablation experiments, which indicated that cells with a lower fate (i.e., 3° cells) can replace cells with a higher fate (2° or 1° cells) (fig. 18.1, *E*).

Additional cell ablation experiments revealed that vulva formation is induced by a signal from the gonadal anchor cell (AC) (fig. 18.1, *D, E*) (Kimble 1981). The AC is morphologically distinct from the surrounding cells of the somatic gonad and eventually makes contact with the progeny of P6.p. When the AC was ablated, P5.p–P7.p had an epidermal fate, like P3.p, P4.p, and P8.p, and no vulva was formed (fig. 18.1, *E*).

To understand the cell replacement capacities among the six vulval precursor cells (VPCs) and the inductive interaction with the AC at the molecular level, intensive genetic studies have been carried out, and many vulva-defective mutants have been isolated (Ferguson et al. 1987). Most mutants can be divided into two phenotypic classes. In "vulvaless" mutants, no vulva is formed, and in many such mutants, all VPCs have a 3° cell fate, resembling AC-ablated animals (fig. 18.1, *E*). In contrast, "multivulva" mutants show ectopic vulva differentiation by P3.p, P4.p, and P8.p (fig. 18.1, *E*).

Molecular studies, most of which have been carried out during the past 10–15 years, have revealed that at least four different signaling systems have to interact with one another during vulva formation. The inductive signal for vulva development is an EGF-like molecule encoded by the gene *lin-3* and is specifically expressed in the AC (Hill and Sternberg 1992). This signal is transmitted by an EGFR-RAS/MAPK pathway within the VPCs (for review see Sternberg and Han 1998). This inductive signal acts in a redundant fashion with a lateral signal, involving the Notch-like molecule *lin-12* of *C. elegans* (for review see Greenwald 1997). Both pathways act together to induce the proper 2°-1°-2° pattern (fig. 18.1, *D*).

Negative signaling prevents inappropriate and precocious vulva differentiation and consists itself of two redundant signaling systems (Ferguson and Horvitz 1989). Recent work has indicated that *lin-35* and *lin-53* encode for an Rb-like molecule and its binding protein RbAp48 (Lu and Horvitz 1998). In addition, the canonical Wnt signaling was also shown to play a role in the regulation of the transcription factor *lin-39,* an important regulator of vulval cell fate specification (Eisenmann et al. 1998). In summary, intensive genetic and molecular studies over the past 20 years have provided a detailed mechanistic understanding of vulva formation in *C. elegans,* providing a platform for the evolutionary analysis of this process.

Cells as Modules: How Cells Change Their Function

Free-living nematodes can be isolated from soil samples around the world, and many strains can be cultured in the laboratory on lawns of *E. coli* as a food source (Sommer et al. 1994). In recent years, collections of free-living nematodes, spanning species of many different families, have been established in several laboratories, mainly in Northern America and Europe. Comparative studies on vulva development were initiated by cell lineage and cell ablation studies in different species of the Rhabditidae, the family to which *C. elegans* belongs (Sommer and Sternberg 1994, 1995). By now, these studies have been extended to the Diplogastridae (Sommer and Sternberg 1996; Sommer 1997) and members of the Cephalobidae, Panagrolaimidae, and Myolaimidae (Félix and Sternberg 1997, 1998; Félix et al. 2000). In the following section, the major cellular changes during nematode vulva evolution are reviewed.

Vulva Formation by Three versus Four Precursor Cells

All nematodes analyzed so far have 12 ventral epidermal cells that can be homologized based on their position along the anteroposterior body axis. However, the cell fate specification and developmental competence of individual cells differ among nematode species. The most prominent difference is seen between species of the Rhabditidae and Diplogastridae, such as *C. elegans* and *Pristionchus pacificus,* on the one hand, and members of the Cephalobidae, Panagrolaimidae, and Myolaimidae, on the other hand (fig. 18.2). In the former, the vulva is formed by three VPCs, P5.p–P7.p, which adopt the 2°-1°-2° fate pattern, as described for *C. elegans* (figs. 18.1, 18.2). In contrast, in the latter, the vulva is formed by four precursor cells, P5.p–P8.p, which adopt a 2°-1°-1°-2° fate pattern (fig. 18.2). Associated with a change in the number of VPCs forming vulval tissues is an altered position of the AC relative to the VPCs. In species with a three-celled vulva, the AC is located above P6.p and makes contact only with the progeny of

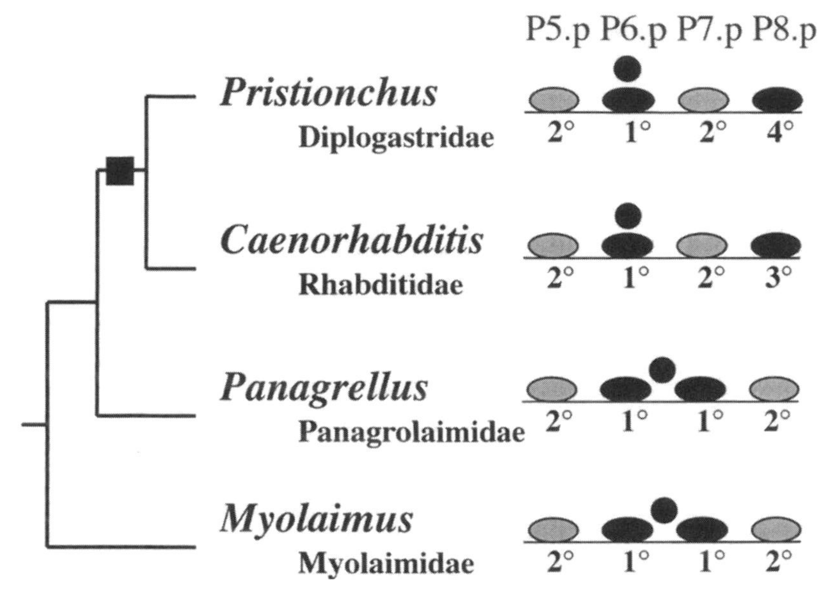

Fig. 18.2.—Evolutionary variation in the number of precursor cells participating in vulva formation. In species of the Rhabditidae and Diplogastridae, such as *C. elegans* and *P. pacificus,* respectively, P5.p—P7.p form vulva tissue with the 2°-1°-2° pattern, as indicated in figure 18.1. In species of the Panagrolaimidae or Myolaimidae, such as *Panagrellus redivivus* or *Myolaimus* sp., the vulva is formed by the progeny of four precursor cells. P5.p—P8.p adopt a 2°-1°-1°-2° pattern. The fate of P8.p in *C. elegans* and *P. pacificus* is described in more detail in figures 18.1 and 18.4, respectively.

P6.p to form the connection between the uterus and the hypodermis (fig. 18.1, *B, D;* fig. 18.2). In contrast, in species with a four-celled vulva, the AC is located between P6.p and P7.p and establishes a contact with some of the progeny of both cells, resulting in a completely different spatial arrangement of vulval cells.

Phylogenetic analysis suggest that the four-celled vulva represents an ancestral character, whereas vulva formation by three VPCs is a derived character (fig. 18.2) (Félix et al. 2000). The transition from a four-celled to a three-celled vulva seems to have been a one-time evolutionary event, as no other alterations of this aspect of patterning have been observed in species of the families analyzed so far.

Evolution of Vulva Position

The position of the vulva within the animal represents another evolving character. However, in contrast to vulva formation by three versus four precursor cells, vulva position evolved several times independently. Whereas most species form their vulvae in the central body region, others form the vulva in a species-specific position in the posterior body region. In extreme cases, such as in members of the genus *Teratorhabditis,* the vulva forms at 95% body length, just anterior to the rectum (Sommer and Sternberg 1994; Sommer et al. 1994). Cell lineage analysis in members of the genera *Mesorhabditis, Teratorhabditis, Cruznema* (Sommer and Sternberg 1994), *Brevibucca* (Félix et al. 2000), and *Diplogastrellus* (I. Carmi and R. J. Sommer, unpublished observation) revealed that in all analyzed species with a posterior vulva a similar mechanism has been used: the centrally born cells migrate toward the posterior, stop in a species-specific position, and undergo vulva differentiation (fig. 18.3, *A*).

In all cases except for *Cruznema,* vulva formation occurs in a gonad-independent way; that is, a proper vulva is formed even after the ablation of the somatic gonad, including the AC (fig. 18.3, *C*) (Sommer and Sternberg 1994; Félix et al. 2000; I. Carmi and R. J. Sommer, unpublished observation). Thus, posterior vulva formation involves two important deviations from the *C. elegans* pattern: First, in all species the central cells constitute the vulva equivalence group and the VPCs migrate toward the posterior prior to differentiation. Second, a shift in vulva position is associated with a different mode of vulval patterning, which no longer relies on gonadal signaling. Projecting these vulval character states onto a phylogenetic tree of nematodes indicates that posterior vulva formation represents a derived character that evolved several times independently (Sommer 2000). Thus, posterior vulva formation represents an example of convergent evolution (Sommer 1999).

As a whole, the cellular analysis of vulva formation in representative nematodes of several different families indicates that the vulva itself is a stable developmental module. The vulva is a homologous organ (module) and is formed by homologous precursor cells. Nonetheless, at the next lower level of organization, the cells forming the vulva change their function and behavior during evolution: vulva formation in the posterior rather than the central body region represents an example of how modifications at one level of organization—here, the introduction of cell migration—can result in evolutionary novelty. This example also shows that cells are both evolutionarily stable and changeable developmental modules. They are stable developmental modules because the group of cells forming vulval tissue is basically conserved between nematode species. They are changeable modules because their behavior, interactions, and functions do change during evolution.

The comprehensive analysis of patterning alterations, however, also indicates the presence of important constraints which limit the types of modifications observed during evolution. For example, posterior vulva formation by the posterior ventral epidermal cells, like P9.p–P11.p,

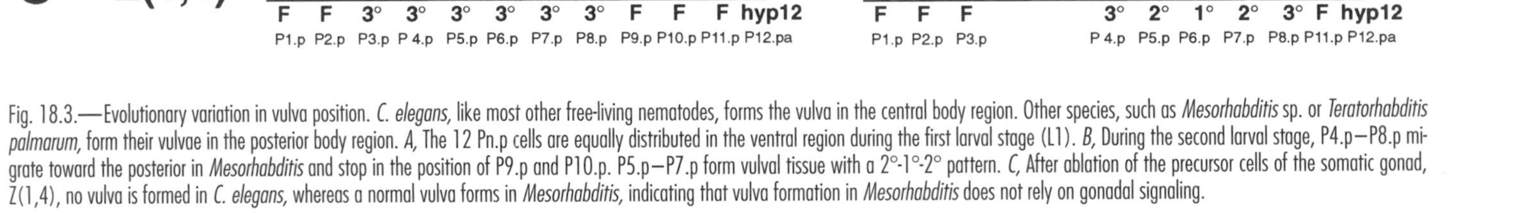

Fig. 18.3.—Evolutionary variation in vulva position. *C. elegans,* like most other free-living nematodes, forms the vulva in the central body region. Other species, such as *Mesorhabditis* sp. or *Teratorhabditis palmarum,* form their vulvae in the posterior body region. *A,* The 12 Pn.p cells are equally distributed in the ventral region during the first larval stage (L1). *B,* During the second larval stage, P4.p–P8.p migrate toward the posterior in *Mesorhabditis* and stop in the position of P9.p and P10.p. P5.p–P7.p form vulval tissue with a 2°-1°-2° pattern. *C,* After ablation of the precursor cells of the somatic gonad, Z(1,4), no vulva is formed in *C. elegans,* whereas a normal vulva forms in *Mesorhabditis,* indicating that vulva formation in *Mesorhabditis* does not rely on gonadal signaling.

has not been observed in any species, suggesting that a developmental constraint restricts the type of possible modifications of developmental modules (Sommer and Sternberg 1994; Sommer 1999).

Genes as Modules: Homologous Genes with Different Functions

To understand how evolutionary modifications at the cellular level are achieved, we have to extend comparative studies to the genetic and molecular levels. The diplogastrid species *Pristionchus pacificus* has been selected as a satellite species for the investigation of vulva formation using genetic and molecular tools (Sommer et al. 1996; Eizinger et al. 1999; Sommer 2000). *P. pacificus* has a four-day life cycle and is a hermaphroditic species, like *C. elegans,* and fulfills a number of additional technical requirements (Sommer et al. 1996).

Cell fate specification of the ventral epidermal cells in *P. pacificus* differs from specification in *C. elegans,* and many of the characters seen in *P. pacificus* are unique to this species or its close relatives. First, seven of the 12 ventral epidermal cells die of apoptosis, shortly after these cells are born during late embryogenesis (fig. 18.4, *A*) (Sommer and Sternberg 1996). The vulva itself is made from P5.p–P7.p with a cell lineage pattern very similar to the one in *C. elegans* (fig. 18.4, *B*). Second, vulva induction is a continuous process involving several cells of the somatic gonad rather than the single AC, as in *C. elegans* (Sigrist and Sommer 1999) (fig. 18.4, *C*). Third, of the four surviving cells in the central body region, P8.p, the only cell not participating in wild-type vulva formation, represents a new cell type with characteristics unknown in VPCs in any other analyzed nematode species (Jungblut and Sommer 2000) (fig. 18.4, *C*). P8.p provides a lateral inhibitory signal that influences the cell fate decision of P5.p and P7.p but not P6.p upon gonadal induction (fig. 18.4, *D*). This new type of lateral inhibition also requires the mesoblast M, the precursor of all postembryonically derived mesodermal tissue (fig. 18.4, *C*). In addition, P8.p also provides a negative signaling, the molecular nature of which remains unknown.

Given all of these differences between *P. pacificus* and *C. elegans,* large-scale mutagenesis screens have been carried out in order to identify important genes for *P. pacificus* vulva formation (Eizinger et al. 1999). Many vulva-defective mutants have been isolated, and the molecular characterization of several of them provides an insight into the molecular mechanisms of vulva formation in *P. pacificus.* Comparing the function of homologous genes between *P. pacificus* and *C. elegans* allows the investigation of the evolvability of developmental modules in vulva formation to be taken one step further, to the level of gene function.

A

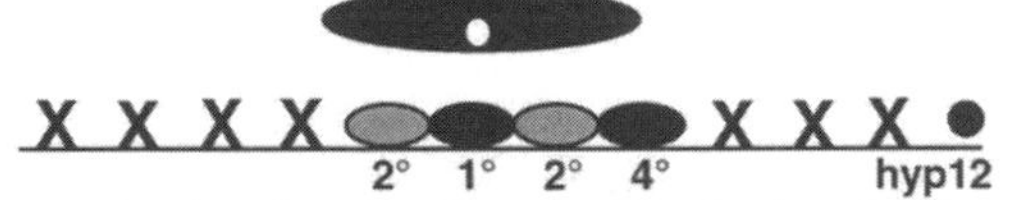

B

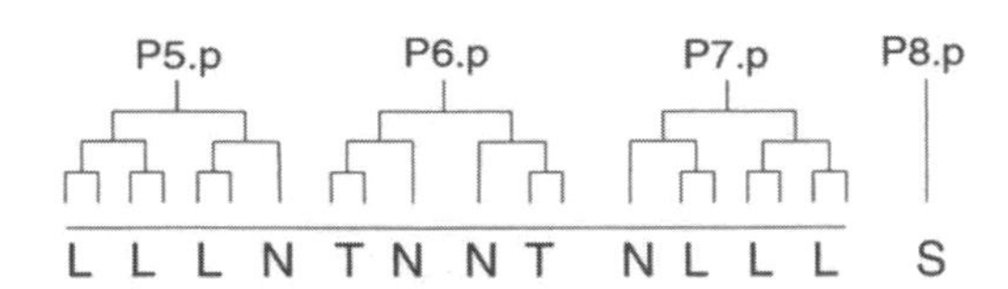

C

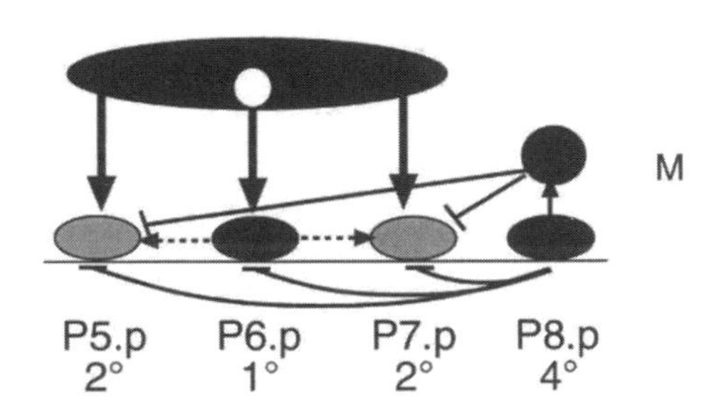

D

Manipulation	P5.p	P6.p	P7.p	P8.p
Wild-type	2°	1°	2°	4°
P(6,7).p ablated	2°	-	-	4°
P(6-8).p ablated	1°	-	-	-

E

Species/genet. Const.	P3.p	P4.p	P5.p	P6.p	P7.p	P8.p
Cel Wild-type	3°	3°	2°	1°	2°	3°
Cel-lin-39	F	F	F	F	F	F
Cel-lin-39, early hs	3°	3°	3°	3°	3°	3°
Ppa Wild-type	X	X	2°	1°	2°	4°
Ppa-lin-39	X	X	X	X	X	X
Ppa-lin-39, Ppa-ced-3	F	F	2°	1°	2°	4°
Cel-mab-5	3°	3°	2°	1°	2°	3°
Ppa-mab-5	X	X	2°	1°	2°	D

Evolution of *lin-39* Function

The first gene to be studied in detail was *lin-39*, a homeotic gene that specifies the vulva equivalence group in *C. elegans* early in development. *lin-39* shows its highest sequence similarity to the *Drosophila* gene *Deformed*. *Cel-lin-39* mutants have a generation-vulvaless phenotype; that is, the vulva equivalence group is not formed, and P3.p–P8.p fuse with the surrounding hypodermis, like their anterior and posterior counterparts (fig. 18.4, *E*) (Wang et al. 1993; Clark et al. 1993). Using a temperature-sensitive allele of *Cel-lin-39*, Maloof and Kenyon (1998) overcame the requirement for *lin-39* in the VPCs early in development, and they were able to show that *Cel-lin-39* has a second important function during vulva induction (fig. 18.4 *E*). LIN-39 acts as a transcription factor downstream of the EGF/RAS/MAPK signaling pathway, providing specificity to vulval signaling (Maloof and Kenyon 1998). If *Cel-lin-39* is not provided during vulva induction, P5.p–P7.p have a 3° fate. Thus, *Cel-lin-39* has two distinct functions during *C. elegans* vulva formation.

In *P. pacificus*, several generation-vulvaless mutants have been isolated in which P5.p–P8.p die of programmed cell death, like their anterior and posterior counterparts. Molecular analysis indicates that mutations in *Ppa-lin-39* cause this phenotype (fig. 18.4, *E*) (Eizinger

Fig. 18.4.—Schematic summary of vulva formation in *P. pacificus*. *A*, Cell fate specification of the 12 ventral epidermal cells. P1.p–P4.p and P9.p–P11.p die of programmed cell death during late embryogenesis. P5.p–P7.p form the vulva with a 2°-1°-2° pattern. P8.p (*black oval*) has a special fate designated as 4°. *B*, Cell lineage pattern of the vulval precursor cells. P5.p and P7.p generate seven progeny each, whereas P6.p generates six progeny. P8.p does not divide and finally fuses with *hyp7*. *C*, Model for cell-cell interactions during vulva development in *P. pacificus*. P8.p provides a lateral inhibition to P5.p and P7.p, mediated by the mesoblast M (*black circle*). Lateral inhibition influences the 1° versus 2° cell fate decision of P5.p and P7.p. P8.p also provides a negative signal (*black bars*), which influences the vulval versus nonvulval cell fate decision. For clarity, negative signaling is shown here as interactions between P8.p and P5.p, between P8.p and P6.p, and between P8.p and P7.p. It is possible that indirect interactions involving other cells could exist. Inductive signaling from the somatic gonad is a continuous process (*black arrows*). Lateral signaling occurs between P6.p and P8.p (*not indicated*) and perhaps also between P6.p and P5.p and between P6.p and P7.p (*dotted arrows*). *D*, Summary of cell ablation experiments. After ablation of P6.p and P7.p, P5.p adopts a 2° fate in the presence of P8.p. After ablation of P6.p–P8.p, P5.p predominantly has a 1° fate, indicating that the presence of P8.p influences the cell fate decision of P5.p. *E*, Comparison of the function of the homeotic genes *lin-39* and *mab-5* between *C. elegans* and *P. pacificus*. In *C. elegans lin-39* mutant animals, the central body region shows a homeotic transformation and P3.p–P8.p fuse with the surrounding hypodermis, like their anterior and posterior lineage homologues. *P. pacificus lin-39* mutant animals also show a homeotic transformation, and P5.p–P8.p die of programmed cell death. If the first function of *Cel-lin-39* is rescued by providing *lin-39* under the control of a heat-shock promoter, P3.p–P8.p have a 3° fate because *Cel-lin-39* is required during vulva induction. The first function of *Ppa-lin-39* can be overcome by generating a *Ppa-lin-39, Ppa-ced-3* double mutant. Ppa-CED-3 is a general regulator of programmed cell death, and mutations in *Ppa-ced-3* result in animals unable to undergo apoptosis. Such double mutants form a normal vulva, indicating that in contrast to *Cel-lin-39*, *Ppa-lin-39* is not required during vulva induction. *C. elegans mab-5* mutant animals show no vulval patterning defects. *P. pacificus mab-5* mutant animals show a homeotic transformation, and P8.p forms an ectopic vulva-like structure. X, programmed cell death; F, cell fusion; D, ectopic vulva differentiation.

and Sommer 1997). Thus, *Cel-lin-39* and *Ppa-lin-39* have similar functions early in development in setting up the vulva equivalence group. *Cel-lin-39* prevents the fusion of P3.p–P8.p, whereas *Ppa-lin-39* prevents the programmed cell death of P5.p–P8.p.

Is *Ppa-lin-39* also reused later in development, like *Cel-lin-39?* To study the potential second role of *Ppa-lin-39* during vulva formation, *Ppa-lin-39, Ppa-ced-3* double mutants have been generated (Sommer et al. 1998). *Ppa-ced-3* encodes for an ICE-related caspase, which is one of the key regulators of programmed cell death in *C. elegans* and *P. pacificus.* In *Ppa-ced-3* mutants, programmed cell death is not executed, resulting in animals with 12 Pn.p cells in the ventral epidermis. In *Ppa-lin-39, Ppa-ced-3* double mutants, the early function of *Ppa-lin-39*—the inhibition of programmed cell death of P5.p–P8.p—is no longer necessary because cell death cannot be executed at all (fig. 18.4, *E*) (Sommer et al. 1998). Surprisingly, such double mutants form a normal vulva by P5.p–P7.p indicating that *Ppa-lin-39* has no second role during vulva induction, as *Cel-lin-39* does.

Taken together, the comparison of *Ppa-lin-39* and *Cel-lin-39* during vulva formation shows both conservation and change of gene function: the first function is conserved between both nematodes, whereas the second function is present only in *C. elegans.*

Evolution of *mab-5* Function

Besides *Ppa-lin-39, Ppa-mab-5* is a second homeotic gene that has been shown to have an important role during cell fate specification of P5.p–P8.p. *mab-5* is the *Antennapedia* homologue of nematodes and was originally described on the basis of its abnormalities in *C. elegans* males (Kenyon 1986). *Ppa-mab-5* was isolated on the basis of the ectopic differentiation of P8.p in the hermaphrodite (Jungblut and Sommer 1998). In most *Ppa-mab-5* mutant animals, P5.p–P7.p form a normal vulva, but P8.p also differentiates and forms a small pseudovulva (fig. 18.4, *E*). The differentiation of P8.p is not dependent on gonadal induction but rather depends on a signal from the lineage of the mesoblast M. In *Ppa-mab-5* mutants, the M lineage undergoes overproliferation, resulting in an egg-laying-defective phenotype and an inappropriate signaling to P8.p. If the M cell is ablated in *Ppa-mab-5* mutant animals, the ectopic differentiation can be significantly reduced, indicating a new type of cell-cell interaction (Jungblut and Sommer 2000).

Although Cel-MAB-5 plays a role in the specification of P7.p and P8.p, *Cel-mab-5* mutants have no obvious vulva phenotype and no ectopic proliferation can be seen, as in *Ppa-mab-5* (fig. 18.4, *E*) (Clandinin et al. 1997). Thus, in contrast to *lin-39, mab-5* plays a more important role during vulva formation in *P. pacificus.*

As a whole, the genetic and molecular analysis of vulva formation between *C. elegans* and *P. pacificus* indicates that genes—like cells—are stable developmental modules. However, while some functions of genes are conserved, other functions clearly differ between species. It is this interplay between conservation and change that can ensure the continuity of a homologous developmental system, on the one hand, and the evolvability of the structure, on the other hand. When genes are considered modules that interact in complex networks within a cell, a striking similarity to the consideration of cells is observed: at both levels, modules can be simultaneously stable and changeable.

Modules within Genes: Regulatory versus Coding Evolution

If the function of homologous genes can change during evolution, the comparative approach ultimately has to be taken to the mechanistic level. The most obvious question is whether changes of the regulatory and/or the coding regions of developmental control genes, like *lin-39* or *mab-5,* can account for the observed functional differences. This type of question can be addressed using *C. elegans* mutants as a test tube: one can generate hybrid molecules between *Cel-lin-39* and *Ppa-lin-39* and test such transgenes for the rescue of *Cel-lin-39* mutants.

As *C. elegans* and *P. pacificus* have been separated for at least 100–200 million years (Eizinger and Sommer 1997), it is very likely that the regulatory control regions of homologous genes differ between these two nematodes. Therefore, the real question of the mechanism of *lin-39* evolution is whether, in addition to changes in the regulatory region, changes in the coding region of *lin-39* can contribute to the observed functional differences. If the evolution of the regulatory regions alone can account for the observed functional difference between *C. elegans* and *P. pacificus,* a *Ppa-lin-39* cDNA, when expressed under the control of a *Cel-lin-39* promoter, should be able to rescue the *Cel-lin-39* mutant phenotype. In contrast, if there are important differences in the coding region as well, such a construct should be unable to rescue the *Cel-lin-39* mutant phenotype. Preliminary studies indicate that in the case of *lin-39* it is only the regulatory region, but not the coding region, that has substantially evolved between *C. elegans* and *P. pacificus* (Grandien and Sommer 2001). Thus, using *C. elegans* as a test tube, one can extend the comparative analysis of vulva development from the cellular and genetic levels to the final molecular mechanistic level.

Conclusion

In free-living nematodes, developmental processes can be studied at the cellular, genetic, and molecular levels. Comparative studies on vulva development indicate that cells are stable developmental modules. Nonetheless, they can change their function during evolution, generating novelty, for example, in the position of vulva formation. Thus, cells are both evolutionarily stable and changeable developmental modules. Besides *C. elegans,* the diplogastrid species *P. pacificus* is amenable to genetic analysis. The comparison of the functions of the homeotic genes *lin-39* and *mab-5* indicates that genes, just like cells, are evolutionarily stable and changeable developmental modules. It is the integrative comparative approach at the cellular, the genetic, the molecular, and the mechanistic levels that makes the analysis of nematode vulva development a fruitful case study, indicating how developmental modules evolve.

Acknowledgments

I thank J. Srinivasan for critically reading the manuscript and members of the lab and my nematode colleagues for discussion.

References

Clandinin, T. R., W. S. Katz, and P. W. Sternberg. 1997. *Caenorhabditis elegans* HOM-C genes regulate the response of vulval precursor cells to inductive signal. *Dev. Biol.* 182:150–161.

Clark, S. G., A. D. Chisholm, and H. R. Horvitz. 1993. Control of cell fates in the central body region of *C. elegans* by the homeobox gene *lin-39. Cell* 74:43–55.

Eisenmann, D. M., J. N. Maloof, J. S. Simske, C. Kenyon, and S. K. Kim. 1998. The β-catenin homolog BAR-1 and LET-60 Ras coordinately regulate the Hox gene *lin-39* during *Caenorhabditis elegans* vulva development. *Development* 125:3667–3680.

Eizinger, A., B. Jungblut, and R. J. Sommer. 1999. Evolutionary change in the functional specificity of genes. *Trends Genet.* 15:191–196.

Eizinger, A., and R. J. Sommer. 1997. The homeotic gene *lin-39* and the evolution of nematode epidermal cell fates. *Science* 278:452–455.

Félix, M. A., P. De Ley, R. J. Sommer, L. Frisse, S. A. Nadler, K. Thomas, J. Vanfleteren, and P. W. Sternberg. 2000. Evolution of vulva development in the Cephalobina (Nematoda). *Dev. Biol.* 221:68–86.

Félix, M.-A., and P. W. Sternberg. 1997. Two nested gonadal inductions of the vulva in nematodes. *Development* 124:253–259.

Félix, M.-A., and P. W. Sternberg. 1998. A gonad-derived survival signal for vulval precursor cells in two nematode species. *Curr. Biol.* 8:287–290.

Ferguson, E. L., and H. R. Horvitz. 1989. The multivulva phenotype of certain *C. elegans* mutants results from defects in two functionally redundant pathways. *Genetics* 123:109–121.

Ferguson, E. L., P. W. Sternberg, and H. R. Horvitz. 1987. A genetic pathway for the specification of the vulval cell lineages of *Caenorhabditis elegans. Nature* 326:259–267.

Gerhard, J., and M. Kirschner. 1997. *Cells, embryos and evolution.* Oxford: Blackwell Science.

Grandien, K., and R. J. Sommer. 2001. Functional comparison of the nematode Hox *gene lin-39* in *C. elegans* and *P. pacificus* reveals evolutionary conservation of protein function despite divergence of primary sequences. *Genes Dev.* 15:2161–2172.

Greenwald, I. 1997. Development of the vulva. In *C. elegans II*, ed. D. L. Riddle, T. Blumenthal, B. J. Meyer, and J. R. Priess, 519–542. Cold Spring Harbor, N.Y.: Cold Spring Harbor Laboratory Press.

Hill, R. J., and P. W. Sternberg. 1992. The gene *lin-3* encodes an inductive signal for vulval development in *C. elegans. Nature* 358:470–476.

Jungblut, B., and R. J. Sommer. 1998. The *Pristionchus pacificus mab-5* gene is involved in the regulation of ventral epidermal cell fates. *Curr. Biol.* 8:775–778.

Jungblut, B., and R. J. Sommer. 2000. Novel cell-cell interactions during vulva development in *Pristionchus pacificus. Development* 127:3295–3303.

Kenyon, C. 1986. A gene involved in the development of the posterior body region of *C. elegans. Cell* 46:477–487.

Kimble, J. 1981. Lineage alterations after ablation of cells of the somatic gonad of *Caenorhabditis elegans. Dev. Biol.* 87:286–300.

Lu, X., and H. R. Horvitz. 1998. *lin-35* and *lin-53,* two genes that antagonize a *C. elegans* Ras pathway, encode proteins similar to Rb and its binding protein RbAp48. *Cell* 95:981–991.

Maloof, J. N., and C. Kenyon. 1998. The HOX gene *lin-39* is required during *C. elegans* vulval induction to select the outcome of Ras signaling. *Development* 125:181–190.

Raff, R. A. 1996. *The shape of life.* Chicago: University of Chicago Press.

Riddle, D. L., T. Blumenthal, B. J. Meyer, and J. R. Priess, eds. 1997. *C. elegans II*. Cold Spring Harbor, N.Y.: Cold Spring Harbor Laboratory Press.

Riedl, R. 1975. *Die Ordnung des Lebendigen.* Hamburg: Paul Parey Verlag.

Sigrist, C. B., and R. J. Sommer. 1999. Vulva formation in *Pristionchus pacificus* relies on continuous gonadal induction. *Dev. Genes Evol.* 209:451–459.

Sommer, R. J. 1997. Evolutionary change of developmental mechanisms in the absence of cell lineage alterations during vulva formation in the Diplogastridae. *Development* 124:243–251.

Sommer, R. J. 1999. Convergence and the interplay of evolution and development. *Evol. Dev.* 1:8–10.

Sommer, R. J. 2000. Evolution in worms. *Curr. Opin. Genet. Dev.* 10:443–448.

Sommer, R. J., L. K. Carta, S. Y. Kim, and P. W. Sternberg. 1996. Morphological, genetic and molecular description of *Pristionchus pacificus* sp. n. (Nematoda: Neodiplogastridae). *Fundam. Appl. Nematol.* 19:511–521.

Sommer, R. J., L. K. Carta, and P. W. Sternberg. 1994. The evolution of cell lineage in nematodes. *Development 1994* (suppl.), pp. 85–95.

Sommer, R. J., A. Eizinger, K. Z. Lee, B. Jungblut, A. Bubeck, and I. Schlak. 1998. The *Pristionchus* Hox gene *Ppa-lin-39* inhibits programmed cell death to specifiy the vulva equivalence group and is not required during vulval induction. *Development* 125:3865–3873.

Sommer, R. J., and P. W. Sternberg. 1994. Changes of induction and competence during the evolution of vulva development in nematodes. *Science* 265:114–118.

Sommer, R. J., and P. W. Sternberg. 1995. Evolution of cell lineage and pattern formation in the vulval equivalence group of rhabditid nematodes. *Dev. Biol.* 167:61–74.

Sommer, R. J., and P. W. Sternberg. 1996. Apoptosis and change of competence limit the size of the vulva equivalence group in *Pristionchus pacificus:* a genetic analysis. *Curr. Biol.* 6:52–59.

Sulston, J. E., and H. R. Horvitz. 1977. Postembryonic cell lineages of the nematode *Caenorhabditis elegans. Dev. Biol.* 56:110–156.

Sulston, J. E., and J. G. White. 1980. Regulation and cell autonomy during postembryonic development in *Caenorhabditis elegans. Dev. Biol.* 78:577–597.

Sternberg, P. W., and M. Han. 1998. Genetics of RAS signaling in *C. elegans. Trends Genet.* 14:466–472.

Sternberg, P. W., and H. R. Horvitz. 1986. Pattern formation during vulval development in *C. elegans. Cell* 44:761–772.

Wang, B. B., M. M. Müller-Immerglück, J. Austin, N. T. Robinson, A. Chisholm, and C. Kenyon. 1993. A homeotic gene cluster patterns the anteroposterior body axis of *C. elegans. Cell* 74:29–42.

Wood, W. 1988. *The nematode C. elegans.* Cold Spring Harbor, N.Y.: Cold Spring Harbor Laboratory Press.

19 Modularity in the Evolution of Vertebrate Appendages

NEIL H. SHUBIN AND MARCUS C. DAVIS

Introduction

Much of the adaptive evolution of vertebrates and arthropods has involved modification of appendages. The evolution of diverse locomotor strategies, from burrowing to flying, often involves changes in the number, pattern, or size of limbs, wings, legs, or fins. The number and placement of appendages has important consequences for the behavior, ecology, and performance of the organism. Indeed, much of this adaptive evolution involves redeploying ancient genetic and developmental processes in new ways. The remarkable similarities in the genetic circuits that pattern the axes of metazoan appendages suggest that modular evolution is involved in the generation of both homologous and analogous organs (Shubin et al. 1997).

The modular construction of vertebrate appendages has been a major factor in their adaptive evolution. One can think of modules as existing at several levels in limbs: from the whole organ to combinations of its constituent skeletal and developmental parts. Limbs are compartments that project from the body wall and develop relatively independently of other structures. As such, much of the growth and morphogenesis of the limb skeleton happens in an environment that is physically separated from the main axis of the body. In addition, the cascade of signals that is involved in limb patterning can be initiated ectopically (e.g., Cohn and Tickle 1996; Sessions et al. 1999). Whether the developmental perturbation is chemical, parasitic, or the result of tissue or gene manipulation, a normal limb can be produced in a new location. This modular construction has played a major role in the evolution of the vertebrate body plan. For example, different types of bony fish have evolved novel locomotor and feeding strategies by changes in the placement, size, and number of appendages (fig. 19.1).

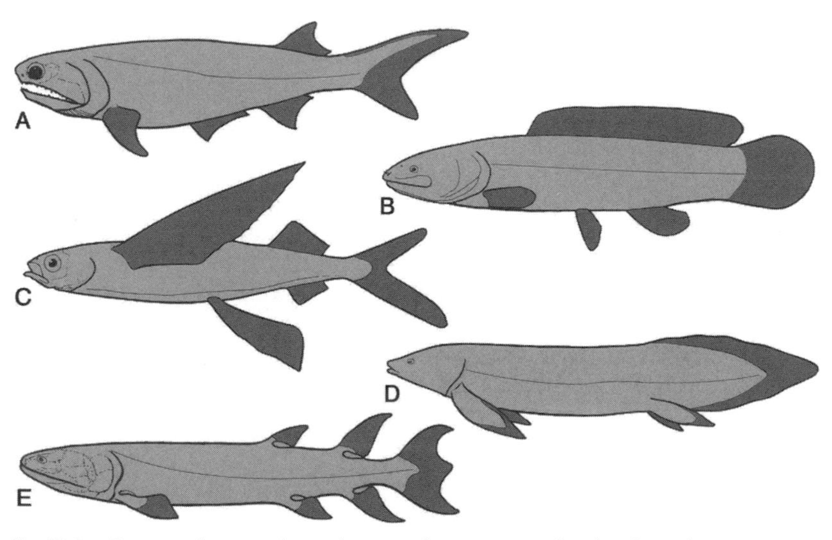

Fig. 19.1.—Various gnathostomes, showing diversity in placement, size, and number of appendages. *A,* An extinct basal actinopterygian, the paleoniscid *Cheirolepis; B,* an extant basal actinopterygian, the bowfin *Amia; C,* an extant derived actinopterygian, the flying fish *Cypselurus; D,* an extant sarcopterygian, the lungfish *Neoceratodus; E,* an extinct sarcopterygian, the osteolepiform *Eusthenopteron.* Bony fish can vary in the number of dorsal fins: for example, there are two in *E,* one in *A–C,* and none in *D.* In addition, paired fins can vary in size and position: note the ventrally positioned fins in *A* and *B* and the enlarged and dorsally positioned in fins in *C.* Actinopterygians and sarcopterygians also differ in the relative contribution of dermal (*dark gray*) and endochondral (*light gray*) bone in the paired fins. Compare *A–C* to *D* and *E.* Fish are not to scale. (Images modified from Nelson 1976.)

Appendage skeletons can also vary in a modular manner. Vertebrate appendages contain two different types of skeleton, each of which is formed by a distinct developmental process (Patterson 1977). These components can be thought of as developmental modules, because their patterns emerge at different times and they appear to be independently regulated developmentally (e.g., Wood 1982; Laforest et al. 1998). The dermal skeleton forms by the direct ossification of noncartilaginous connective tissue. Typically, the dermal skeleton of bony fish consists of a series of rays (lepidotrichia) that compose the bulk of the surface area of the fin. The other skeletal component is endochondral and is formed by ossification of a cartilage model. These skeletal units often lie at the proximal margin of the fin and can articulate directly with the shoulder girdle. In addition, endochondral bones can have a nodular or elongate shape (Shubin 1995). In a typical appendage, the endochondral elements and dermal rays lie in series, with mobile joints lying between dermal ray and endochondral bone. Consequently, the function of the fin as a whole is a product of motions of both types of skeleton.

Major clades of bony fish differ in the relative amounts of dermal and endochondral skeletons in their fins. Derived actinopterygians, such as teleosts, have a fin that is composed mostly of dermal radials. In these forms, the endochondral skeleton is typically small and restricted to the extreme proximal margin of the fin. The dermal radials are richly jointed and can flex throughout their entire length. Derived sarcopterygians have a more robust and distally extensive endoskeleton than actinopterygians. The fin of the Late Devonian sarcopterygian *Eusthenopteron* serves as an example (fig. 19.2). Here, the endochondral skeleton forms a branched axis and consists of rod-shaped elements of different sizes. The dermal skeleton envelops the endochondral one and extends to the distal margin of the fin.

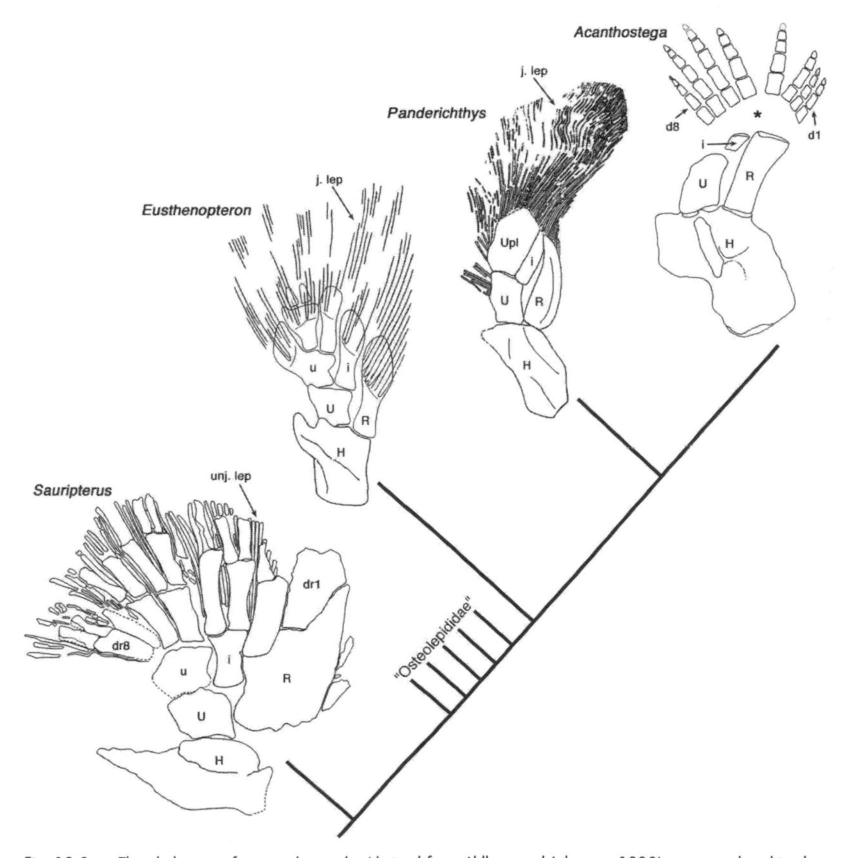

Fig. 19.2.—The phylogeny of tetrapodomorphs (derived from Ahlberg and Johanson 1998) suggests that rhizodontids such as *Sauripterus* occupy a basal position in phylogeny. All fins are depicted in ventral view with preaxial to the right. The branches marked "Osteolepididae" denote a paraphyletic grade of taxa having the same basic biserial endochondral fin skeleton as the more derived *Eusthenopteron*. Fins are not to scale. In *Acanthostega* the asterisk marks the location of presumably unossified carpals. d1, digit 1; d8, digit 8; dr1, distal radial 1; dr8, distal radial 8; j. lep, jointed lepidotrichia; H, humerus; i, intermedium; R, radius; U, ulna; u, ulnare; unj. lep, unjointed lepidotrichia; Unp, Ulnar plate. (*Eusthenopteron* modified from Andrews and Westoll 1970; *Panderichthys* modified from Vorobyeva 1992; *Acanthostega* modified from Coates 1996.)

One of the key features of appendage evolution is that basic skeletal, developmental, and genetic units are established relatively early in the history of the vertebrates; subsequent evolution largely consists of redeployment and modification of these components (Shubin et al. 1997). Differences between the appendages of actinopterygians and sarcopterygians largely derive from changes in the relative size, shape, and position of the dermal and endochondral skeletons. Indeed, tetrapods represent the most extreme condition, having lost the dermal rays altogether. Thorogood (1991) proposed that a simple heterochronic shift in the development of these units was behind the shift between ray-finned and lobe-finned designs (fig. 19.3). The relative amounts of dermal skeleton and endoskeleton in the fin are hypothesized to relate to the timing of the shift between the apical ectodermal ridge (AER) and apical ectodermal fold (AEF). In this model, ray-finned fish, such as teleosts, would have an early shift from ridge to fold, causing a greater proportion of the skeleton to be of dermal origin. Tetrapods would be derived in the loss of the AEF, with the ridge never transforming into a fold.

The origin of the tetrapod limb can be considered as much a shift in modular design as it is in the origin of novel structures. The origin of tetrapods involved the loss of the dermal skeleton of the limbs, the reduction of the dermal skeleton in the girdles, and the enhancement of the endochondral skeleton, particularly in the distal region of the appendage and pectoral girdle (Ahlberg and Milner 1994; Coates 1996). Tetrapod limbs are derived in several features of the endoskeleton (Coates 1996). The first of these is digits. Digits are rodlike endochondral bones that articulate as a series of phalanges. Digits primitively terminate at a common proximodistal level. Tetrapods also have a mesopodium—wrist or ankle bones that are mostly nodular and carry both proximodistal and mediolateral articulations with other bones of the hand or foot. For instance, the ulnare of tetrapods articulates proximally with the ulna, distally with digits or distal carpals, and medially with another mesopodial element, the intermedium or centrale. These elements can have variable shapes but typically lack the expanded postaxial processes seen in equivalent osteolepiform fins (Daeschler and Shubin 1998).

Endochondral bones provide the primary structural support for the tetrapod limb. This situation contrasts with that of fish, where the function of the fin is a product of the dermal and endochondral skeletons together. The origin of the limb, then, represents a situation where shifts in modular construction during development translate into changes of functional modules—the appendage transforms from a combined organ into one whose skeletal function relies on a single

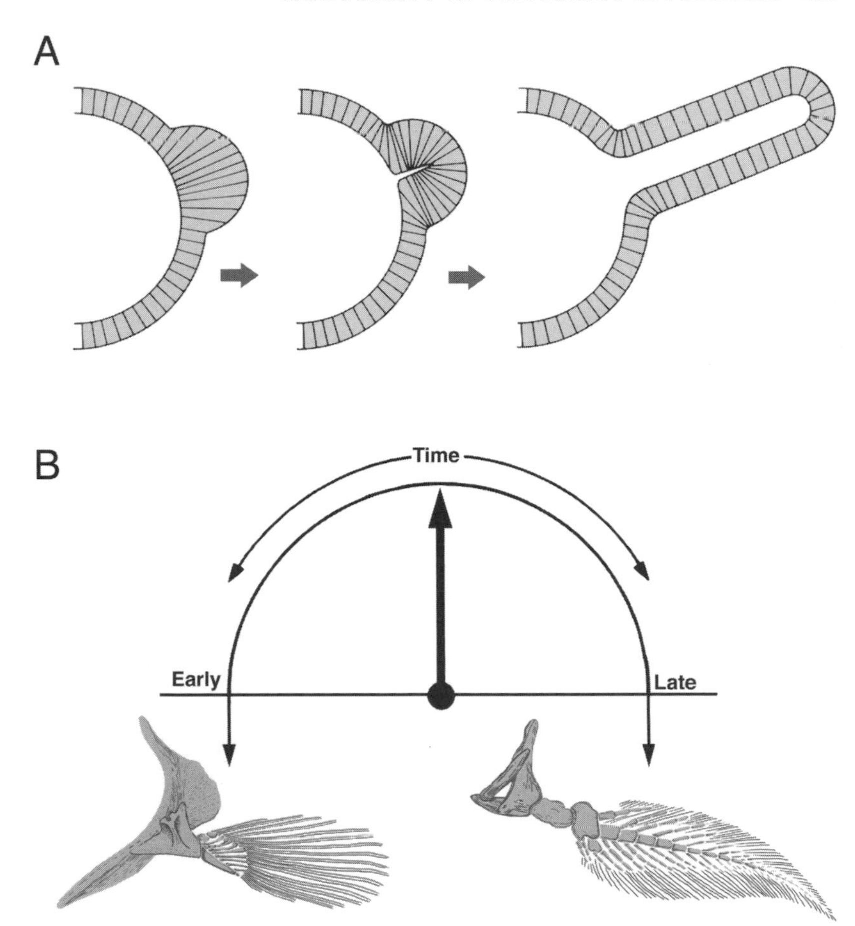

Fig. 19.3.—*A,* Transition from apical ectodermal ridge (*left*) to apical ectodermal fold (*right*) during early fin development as shown in a cross section along the proximodistal axis. *B,* "Clock model" of heterochrony and phenotype. A developmentally early transition from ridge to fold would result in an "actinopterygian" phenotype (here represented by the pectoral fin of *Amia*). A developmentally later transition would result in a "sarcopterygian" phenotype (here represented by *Neoceratodus*). (*A* and *B* are modified from Thorogood 1991; fins are modified from Jarvik 1980.)

module. The implication is that digits interact more directly with the environment than do endochondral bones of other gnathostomes.

The differences between actinopterygian and sarcopterygian designs, so dramatically revealed in comparisons of teleosts and tetrapods, blur in basal members of each group (Shubin 1995; Mabee 2000; Davis et al. 2001). Basal actinopterygians such as *Polypterus, Polyodon,* and *Acipenser* all have both extensive endochondral and dermal skeletons. In terms of the relative amounts of each type of skeleton in the fin, these basal taxa are intermediate. New fossils blur this distinction even further and provide an understanding of the adaptive radiation of gnathostomes into shallow freshwater environments, as described below.

The Rhizodontid *Sauripterus:* Phylogeny, Development, and Function in the Evolution of Modular Design

The Late Devonian (370 mya) contains the elaboration of highly productive nonmarine ecosystems and the early radiation of jawed vertebrates and land plants. By the Middle to Late Devonian, tetrapod and nontetrapod sarcopterygians have evolved a range of derived novelties that appear intermediate for life on land.

The Late Devonian Catskill Formation contains fossil localities that are uniquely diverse. During this time period, the Catskill Delta extended from the Acadian Highlands of the Old Red Continent (Euramerica) to the epicontinental Catskill Sea, which lay to the west (Woodrow 1985). The most productive fossiliferous horizons have been interpreted as representing channel margin and overbank deposits of a wide river flowing across a low-gradient floodplain under a subtropical climate (Woodrow et al. 1995). These sites have yielded a diverse assemblage of freshwater vertebrates, terrestrial plants, and invertebrates. Among the vertebrates are at least two early tetrapod taxa, at least three taxa of osteolepiform sarcopterygians, an early actinopterygian, groenlandaspidid and phyllolepid placoderms, gyracanthid acanthodians, and chondrichthyians. This vertebrate fauna is associated with progymnosperm and lycopsid plants as well as palynomorphs. In addition, trigonotarbid arthropods are preserved both as body fossils and traces.

We have recovered new specimens of an important taxon, *Sauripterus,* originally discovered in 1840 (Daeschler and Shubin 1998; Davis et al. in press). The specimens are of several types. The adults of this species are known only from isolated pectoral fins; these fins are well preserved and articulated (fig. 19.4, *C*). Immature specimens are also known (fig. 19.4, *A, B;* fig. 19.5). These specimens reflect different stages of growth, and the skeleton is more completely preserved than in the adults (Davis et al. 2001).

The adult pectoral fin of this creature is a broad, flat paddle that contains a uniquely flattened radius. It is identified as a rhizodontid fish because of the distinctive overlap between its clavicle and cleithrum. This pattern of reversed overlap is seen only in rhizodontids (Andrews 1985). The skeletal pattern and structure of the new fin presents obvious similarities to tetrapods. Most of these similarities lie in the distal region of the fin, where there are eight jointed radials, all of which face distally. Six of these radials articulate at a common proximodistal level; the other two articulate with the radius. None of these characters

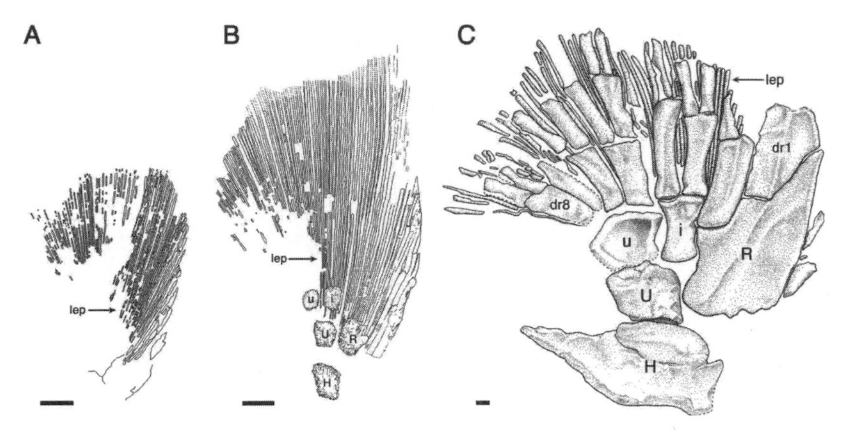

Fig. 19.4.—An ontogenetic series of the pectoral fins of the rhizodontid *Sauripterus*. All fins are depicted in ventral view with preaxial to the right. In *A* and *B*, blank areas reflect regions where lepidotrichia are obscured by matrix or were damaged when part and counterpart slabs were separated. In C, lepidotrichia are incompletely preserved distally. *A* and *B* are to the same scale. All scale bars = 5 mm. dr1, distal radial 1; dr8, distal radial 8; lep, lepidotrichia; H, humerus; i, intermedium; R, radius; U, ulna; u, ulnare.

are seen in the more derived tetrapodomorph fish such as *Eusthenopteron* and *Panderichthys*, and several are uniquely shared between the six to eight digits of tetrapods and the eight radials of the new taxon (fig. 19.2). Similar comparisons can be made between parts of the tetrapod carpus and more proximal regions of the new fin. The intermedium is uniquely similar to that of tetrapods in that it terminates at the same proximodistal segment as the ulnare, articulates laterally with the ulnare, and carries several radials. In addition, the homologue of the ulnare does not contain a postaxial process (a character that is seen in all osteolepiforms; e.g., in *Eusthenopteron* but not in *Panderichthys*).

The paradox of *Sauripterus* is that it has an exceedingly limblike endochondral pattern set in an entirely different structural, functional, and phylogenetic context from that of tetrapods (Daeschler and Shubin 1998; Davis et al. 2001; Davis et al. in press). Structural equivalents of both mesopodials and digits are present. Yet this richly branched endoskeleton lies within an extensive exoskeleton. A series of elongate and unjointed dermal rays cover the dorsal and ventral surfaces of the endoskeleton. These rays extend proximally all the way to the "elbow" joint between the humerus and radius (fig. 19.4, *B*, *C*). The fact that these structures are unjointed is paradoxical: why have an extensive endoskeleton with mobile joints between several of the major bones if the endoskeleton does not interact directly with the environment? Answers to this question come from understanding the ontogeny, the muscle attachments, and the phylogenetic setting in which this fin evolved.

Current phylogenetic analyses support the hypothesis that rhizo-

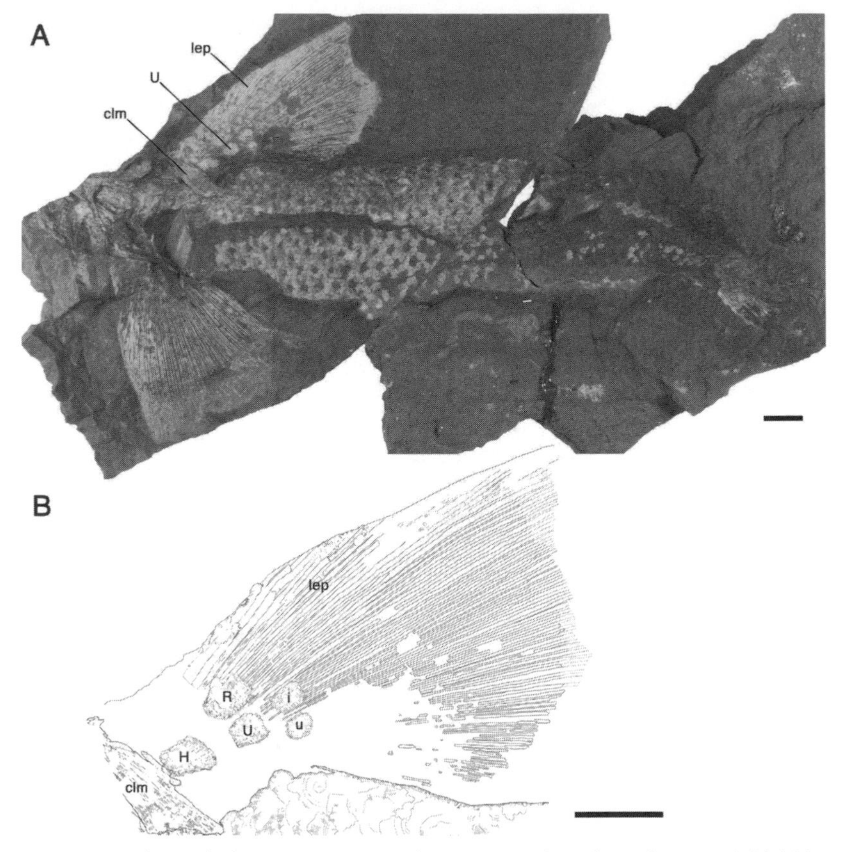

Fig. 19.5.—*A,* Photograph of an immature specimen of *Sauripterus* (Academy of Natural Sciences of Philadelphia 20980). Anterior is to the left. *B,* Illustration of the right pectoral fin shown in *A.* Scale bars = 10 mm. Abbreviations are as listed in the caption for figure 19.4 except clm, cleithrum.

dontids such as *Sauripterus* are not the sister group of tetrapods (Ahlberg and Johanson 1998). Indeed, there appear to be several stem taxa, each lacking autopodial equivalents, separating tetrapods and *Panderichthys* from rhizodontids. This phylogenetic distribution suggests that virtually all of the autopodial specializations of *Sauripterus* are independently derived. Most of the stem taxa that separate *Sauripterus* from tetrapods have a very simple biserial fin design (fig. 19.2). This implies that the similarities seen in the ulnare, the elongate preaxial radials, and the termination of most of the preaxial radials at a common proximodistal level evolved in parallel. Unfortunately, complete skeletal material for adult rhizodontids has not been discovered: adult *Sauripterus* is known from isolated fins. Rhizodontid crania appear to disarticulate easily and, hence, are often poorly preserved. We cannot exclude the notion that the phylogenetic situation may change with the discovery and analysis of new material.

Regardless of the specific phylogenetic hypothesis used, *Sauripterus* reveals that autopodial structures are not unique to tetrapods. The key differences between *Sauripterus* and tetrapods may relate to the role that the dermal skeleton plays in locomotion and support. In non-tetrapod sarcopterygians, the unit that interacts with the substrate or water column is a combined element consisting of both dermal and endochondral skeletons. The function of the endochondral skeleton in locomotion is mediated by the dermal skeleton: the appendage is a higher-order unit composed of two types of skeletal tissues. In tetrapods, this is not the case: the loss of the dermal skeleton implies that the endochondral skeleton interacts more directly with the substrate or water column. The autopodium of tetrapods has important differences from the autopodium of rhizodontids, and these relate to the loss of a dermal module. The elongate unjointed radials that sandwich the endochondral skeleton would inhibit flexion and extension. Importantly, the most distal endochondral joint that allows flexion and extension is the elbow—at the most proximal limit of the lepidotrichia. The evolution of a mesopodium with complex rotatory motions (including components of flexion and extension) would not be possible in this system. The loss of the dermal module would, then, be correlated to the evolution of a functional mesopodium. This constraint would operate differently in creatures like *Panderichthys,* and *Eusthenopteron.* In these taxa, the dermal skeleton is richly jointed throughout its length—flexion and extension would be possible distally. Interestingly, osteolepiforms also appear to have evolved the ability to flex and extend endochondral bones distal to the elbow. Evidence of this comes from the elaborate postaxial processes on the ulnare.

These discoveries suggest that some of the patterning mechanisms involved with the autopod are not unique to tetrapods. Together with paleoenvironmental data, we are led to the hypothesis that Late Devonian freshwater ecosystems were the source of major novelties in skeletal and appendage evolution. Either there were multiple experiments in the origin of autopodial designs or the autopodium was originally established in large, freshwater fish such as rhizodontids. The former notion is supported by the bulk of phylogenetic data. Parallel evolution of similar novelties would be the expected condition in taxa that share common developmental mechanisms and evolve within the same novel ecosystems. In this case, the origin of novel freshwater habitats in the Devonian would be linked to the establishment of major regularities in the origin of novelties.

The juvenile specimens offer a glimpse into the patterns of ossification and growth in basal sarcopterygians (Davis et al. 2001). This series contains several different-sized individuals, and there seems to be a direct correlation between the size of the animal and the extent to which it is ossified (fig. 19.4). The rhizodontid affinities of these specimens are supported by a variety of characters of the dermal fin skeleton and shoulder girdle. Indeed, they uniquely share features with *Sauripterus,* such as the expanded and flattened radius. The smallest of these animals have no endochondral elements in either the fins or the axial skeleton. The entire surface of the fin is formed by elongate and unjointed lepidotrichia. Slightly larger animals possess a set of small and weakly ossified homologues of the humerus, radius, ulna, ulnare, and intermedium. Accordingly, immature *Sauripterus* are largely dermal animals. The support for the head, axial skeleton, and appendages is largely dermal at these stages of growth.

The relative amount and position of the dermal and endochondral skeletons in *Sauripterus* is an unexpected condition, given our current knowledge of fin skeletal development in living taxa (e.g., Sordino et al. 1995; Grandel and Schulte-Merker 1998). Contrary to the predictions of the Thorogood model, *Sauripterus* possesses an elaborate dermal and endochondral skeleton. Furthermore, the high degree of overlap between the two fin skeletons is unlike the conditions seen in derived taxa: teleosts exhibit little or no overlap between the dermal and endochondral skeletons, and tetrapods lack the dermal component altogether. These characters suggest that basal taxa may confound a developmental hypothesis based solely on derived taxa.

Limb Modules and Evolvability

One of the most fundamental questions of evolutionary biology centers on the origin of morphological variation—how does new variation arise in populations? The modular design of limbs offers insights into the large-scale patterns of skeletal evolution seen in their origin and early evolution. The distal endochondral bones in sarcopterygian fish and tetrapods may share similar aspects of pattern, that is, their connectivity (sensu Saint Hilaire 1818). Despite this similarity, their patterns of variation are largely different. One reason for this may be the loss of the dermal module in tetrapod limbs. Because digits interact more directly with the environment than do the distal elements of sarcopterygian fins, the autopodial region becomes an independently functioning unit. Digits are free to vary in different ways than do the endochondral bones of other sarcopterygians, largely because the functional context is different. Selection acting on digital function is likely to result in different kinds of variation from those brought about by selection acting on the function of individual endochondral radials in the distal fin of *Sauripterus*. In this case, evolution of developmental modules has entailed functional shifts that change the ways in which features evolve. The loss of the dermal module may have shifted the constraints acting on appendage endochondral bones—these bones are now free to adapt in ways different from those taken by more basal forms. In this sense, the origin of the variation that led to bird wings, whale flippers, mole forelimbs, and bat wings was, in part, made possible by the loss of the fin dermal skeleton in the Devonian.

References

Ahlberg, P. E., and Z. Johanson. 1998. Osteolepiforms and the ancestry of tetrapods. *Nature* 395:792–794.

Ahlberg, P. E., and A. R. Milner. 1994. The origin and early diversification of tetrapods. *Nature* 368:507–514.

Andrews, S. M. 1985. Rhizodont crossopterygian fish from the Dinantian of Foulden, Berwickshire, Scotland, with a re-evaluation of this group. *Trans. R. Soc. Edinb.* 76: 67–95.

Andrews, S. M., and T. S. Westoll. 1970. The postcranial skeleton of *Eusthenopteron foordi* Whiteaves. *Trans. R. Soc. Edinb.* 68:207–329.

Coates, M. I. 1996. The Devonian tetrapod *Acanthostega gunnari* Jarvik: postcranial anatomy, basal tetrapod interrelationships and patterns of skeletal evolution. *Trans. R. Soc. Edinb.* 87:363–421.

Cohn, M. J., and C. Tickle. 1996. Limbs: a model pattern formation within the vertebrate body plan. *Trends Genet.* 12:253–257.

Daeschler, E. B., and N. H. Shubin. 1998. Fish with fingers? *Nature* 391:133.

Davis, M. C., N. H. Shubin, and E. B. Daeschler. 2001. Immature rhizodontids from the Devonian of North America. *Bull. Mus. Comp. Zool.* 156 (1): 171–187.

Davis, M. C., N. H. Shubin, and E. B. Daeschler. In press. A new specimen of *Sauripterus taylori* (Sarcopterygii, Osteichthyes) from the Famennian Catskill Formation of North America. *J. Vertebr. Paleontol.*

Grandel, H., and S. Schulte-Merker. 1998. The development of the paired fins in the zebrafish (*Danio rerio*). *Mech. Dev.* 79:99–120.

Jarvik, E. 1980. *Basic structure and evolution of vertebrates.* 2 vols. London: Academic Press.

Laforest, L., C. W. Brown, G. Poleo, J. Geraudie, M. Tada, M. Ekker, and M.-A. Akimenko. 1998. Involvement of the Sonic Hedgehog, patched 1 and bmp2 genes in patterning of the zebrafish dermal fin rays. *Development* 125:4175–4184.

Mabee, P.M. 2000. Developmental data and phylogenetic systematics: evolution of the vertebrate limb. *Am. Zool.* 40 (5): 789–800.

Nelson, J.S. 1976. *Fishes of the world.* New York: John Wiley and Sons.

Patterson, C. 1977. Cartilage bones, dermal bones and membrane bones, or the exoskeleton versus the endoskeleton. In *Problems in vertebrate evolution,* London Linnean Society Symposium no. 4, ed. S. M. Andrews, R. S. Miles, and A. D. Walker, 77–122. London: Academic Press.

Saint-Hilaire, E. G. 1818. Philosophie anatomique des organes respiratoires sous le rapport de la détermination et de identité de leurs pieces osseus. Paris: Méquignon-Marvis.

Sessions, S. K., R. A. Franssen, and V. L. Horner. 1999. Morphological clues from multilegged frogs: are retinoids to blame? *Science* 284:800–802.

Shubin, N., C. Tabin, and S. Carroll. 1997. Fossils, genes and the evolution of animal limbs. *Nature* 388:639–648.

Shubin, N. H. 1995. The evolution of paired fins and the origin of tetrapod limbs. *Evol. Biol. (N.Y.)* 28:39–85.

Sordino, P., F. van der Hoeven, and D. Duboule. 1995. Hox gene expression in teleost fins and the origin of vertebrate digits. *Nature* 375:678–681.

Thorogood, P. 1991. The development of the teleost fin and implications for our understanding of tetrapod limb evolution. In *Developmental patterning of the vertebrate limb,* ed. J. R. Hinchliffe, J. M. Hurle, and D. Summerbell, 347–354. New York: Plenum Press.

Vorobyeva, E. I. 1992. The role of development and function in formation of tetrapod-like pectoral fins. *Zh. Obshch. Biol.* 53 (2): 149–158.

Wood, A. 1982. Early pectoral fin development and morphogenesis of the apical ectodermal ridge in the killifish *Aphysemion scheeli. Anat. Rec.* 204:349–356.

Woodrow, D. L. 1985. Paleogeography, paleoclimate, and sedimentary processes of the Late Devonian Catskill Delta. In *The Catskill Delta,* Special Paper 201, ed. D. L. Woodrow and W. D. Sevon, 51–63. Boulder, Colo.: Geological Society of America.

Woodrow, D. L., R. A. J. Robinson, A. R. Prave, A. Traverse, E. B. Daeschler, N. D. Rowe, and N. A. Delaney. 1995. Stratigraphic, sedimentologic, and temporal framework of Red Hill (Upper Devonian Catskill Formation) near Hyner, Clinton County, Pennsylvania: site of the oldest amphibian known from North America. In *1995 field trip guide,* ed. J. Way, 1–8. Lock Haven, Pa.: 60th Annual Field Conference of Pennsylvania Geologists.

Part 4

Individuals as Modules in Higher-Level Units

Most modules treated in this book (e.g., signaling cascades, gene regulatory networks, and organs) have never lived a life of their own. Nevertheless, they are important units during the development of organisms as well as during their evolutionary transformation. However, individuals themselves can also act as modules, and one way to generate modularity in evolution is the association of independently reproducing individuals (such as cells) into a higher-level individual (a multicellular organism). While this may happen only rarely even on a macroevolutionary scale, such transitions in individuality are important, because they may have quite profound implications for the subsequent evolvability of a lineage. This scenario is represented in this section in three different versions.

Brad Davidson, Molly W. Jacobs, and Billie J. Swalla write about the evolution of colonies where multicellular individuals are organized into higher order entities. The interesting aspect of this transition is that it provides the colony with ready-made physiological and developmental modules which in turn are the units of differentiation of colonies. Conversely, an ancestral colonial lifestyle may leave deep marks in the life history of species that adopt a solitary lifestyle secondarily. The authors review and discuss the developmental and life history modifications that are associated with transitions between solitary lifestyles and coloniality, with a special emphasis on ascidians.

In the chapter by Aurora M. Nedelcu and Richard E. Michod the authors investigate the three-way relationship between the origin of multicellularity, the modularity of the genotype-phenotype map, and evolvability. They argue that the way the transition to multicellularity is achieved is critical for the genotype-phenotype map of the resulting multicellular lineage. Evidence from the evolution and development of volvocalean green algae is used to argue that this group arrived at a genotype-phenotype map particularly inconducive to its evolvability. This finding may explain the low phenotypic disparity achieved by this group.

Kim Sterelny writes about the role of symbiosis in making one organism a physical and informational module of another organisms. He then goes on to argue that the transmission of these symbiotic modules constitute an inheritance system that complements the genome. In particular, Sterelny focuses on the question of how this sample-based inheritance system and the genetic inheritance system differ in their contribution to the evolvability of the lineage.

20 The Individual as a Module: Solitary-to-Colonial Transitions in Metazoan Evolution and Development

BRAD DAVIDSON, MOLLY W. JACOBS,
AND BILLIE J. SWALLA

Introduction

Modularity, as explored in this volume, can encompass units ranging from molecules to entire biological individuals. In this chapter, we examine the role of coloniality in metazoan evolution and discuss the potential for individuals to function as developmental modules within colonial organisms. Colonial organisms are defined as groups of physiologically interconnected, asexually produced zooids or ramets (Boardman and Cheetham 1973; Mackie 1986). Colonial organisms are essentially modular in their development, as each individual zooid arises from a semiautonomously patterned bud. Selection can therefore act independently on individual zooids, creating the potential to explore innovative body plans as variations in zooid form.

Although the ability to reproduce asexually is widespread in most metazoan clades, true coloniality is limited to a relatively small number of metazoan phyla (fig. 20.1, *A*). It has generally been assumed that these colonial forms are derived from solitary ancestors within each phylum. Alternatively, coloniality may be ancestral for metazoans as a whole, or for one or more of the metazoan phyla. Consideration of individuals as developmental modules within colonial forms has different evolutionary implications, depending on the direction of the solitary-colonial transition. If coloniality is a derived state, it is important to understand what developmental traits may have predisposed solitary ancestral forms to transform into modular colonies. On the other hand, if coloniality is ancestral, the potential for each zooid of the colony to act as a developmental module may have been crucial in the evolution of innovative body plans (Dewel 2000). It is possible that major developmental reorganizations could occur within the individuals of an ancestral colony that could be exploited by future solitary

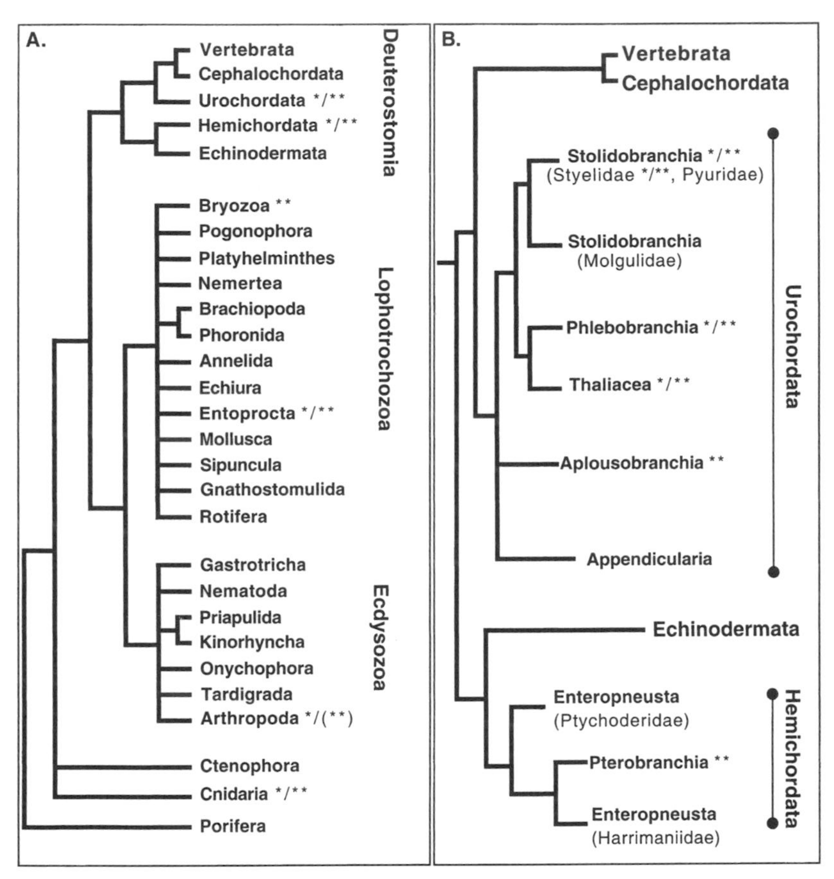

Fig. 20.1.—Colonial species within the metazoans. *A*, Metazoan phylogeny (adapted from Adoutte et al. 2000) showing the occurrence of colonial forms. *B*, Deuterostome phylogeny (adapted from Swalla et al. 2000; and Cameron et al. 2000) showing the occurrence of colonial forms. **, entirely colonial clade; */**, solitary and colonial forms both present. The (**) sign for the arthropods is in parentheses to indicate that the occurrence of coloniality in this clade is very limited and almost certainly derived (see text).

descendants. However, without a solid developmental and molecular framework, such evolutionary hypotheses remain speculative.

Presently, there is very little known about the molecular mechanisms that underlie evolutionary transitions between solitary and colonial life histories. However, transitions between solitary and colonial forms are frequently correlated with transformations in reproductive and developmental modes that are amenable to molecular analyses. These colonial developmental and reproductive traits include a significant reduction in body size, increased capacity for regeneration, the ability to reproduce asexually, and larval brooding. Comparative analyses of molecular mechanisms that underlie distinctive colonial reproductive and developmental traits may provide a necessary framework for understanding the evolutionary basis of colonial-solitary transitions.

Synopsis

We begin with a phylogenetic survey of colonial life history modes within the metazoans. We propose that further developmental investigations are necessary to establish the nature and direction of shifts between solitary and colonial animals. In the next section we focus on the deuterostomes, as they are uniquely suited for developmental comparisons between relatively closely related solitary and colonial extant species. Within the deuterostomes, ascidians contain many colonial species including several instances of evolutionarily recent solitary-colonial shifts. Additionally, previous work on ascidian development has established molecular techniques for analyzing gene expression and function.

We propose that ascidians represent an excellent system for exploring

1. Molecular mechanisms underlying solitary-colonial transitions
2. A developmental framework to test evolutionary hypotheses concerning solitary-colonial transitions within the deuterostomes

We then explore the developmental and life history modes of solitary versus colonial ascidians, emphasizing a suite of developmental traits that correlate with coloniality within the deuterostomes. We hypothesize that solitary-colonial transitions in ascidians may be linked to a dissociation of distinct larval and adult developmental modules. Next we review recent progress that has been made in investigating ascidian adult differentiation, including our preliminary data on the role of thyroid hormones in coordinating ascidian metamorphosis. Finally, we discuss evolutionary hypotheses concerning solitary-colonial life history transformations and how we plan to assess them through developmental comparisons.

Individuals as Modules: The Evolution of Metazoan Coloniality

Sponges and Cnidarians

Sponges, the basal metazoans (Willmer 1990; Cavalier-Smith et al. 1996; Borchiellini et al. 1998), have a unique mixture of colonial and solitary characteristics. Many ecologists place them in the same functional group as other sessile colonial invertebrates (Jackson 1977) because of their indefinite growth and the way they occupy ecological space. However, the resemblance is mostly superficial, and individual "modules" cannot be detected in most adult sponges. For the purposes of this review, we consider them solitary animals with amorphous growth patterns (fig. 20.1, *A*).

Coloniality takes many different forms in the Cnidaria. Anthozoans have solitary pelagic larvae that transform into sessile adults that may be colonial, solitary and clonal, or solitary and aclonal (Bridge et al. 1995). Many hydrozoans also display the opposite pattern: solitary pelagic adults (medusae) with colonial sessile larvae (polyps), a strategy not seen elsewhere in the metazoans (Boero and Bouillon 1987). Asexual reproduction is common throughout the group, and coloniality may have evolved multiple times within the cnidarians (Bridge et al. 1995).

Lophotrochozoans

In the Lophotrochozoans, a large and varied clade (Halanych et al. 1995), asexual reproduction is extremely common but true coloniality is found only in bryozoans (ectoprocts) and kamptozoans (entoprocts) (fig. 20.1, *A*). Bryozoans are entirely colonial, and polymorphism (the specialization of individual zooids for functions such as reproduction or defense) is common. See Harvell (1994) for a cogent review of polymorphism in colonial invertebrates and its possible significance in the organization and evolution of modular forms. Kamptozoans may be either solitary or colonial, and it is unclear which is the derived state in this phylum (Nielsen 1995). Relationships between most groups of lophotrochozoans remain unresolved (fig. 20.1, *A*) (Adoutte et al. 2000), so it is unknown whether the ancestral lophotrochozoan was more likely to be solitary or colonial.

Ecdysozoans

The Ecdysozoans, a large clade of bilaterians (Halanych 1996b; Aguinaldo et al. 1997), are composed almost entirely of solitary species (Strathmann 1987; Gullan and Cranston 2000) (fig. 20.1, *A*). When asexual reproduction is present, it takes place by parthenogenesis rather than budding, fragmentation, or fission (Ruppert and Barnes 1994; Gullan and Cranston 2000). The only exception to this rule is found in a specialized group of parasitic barnacles, the rhizocephalans. Several groups within this clade have been described as colonial, having multiple externae derived from a single cyprid infection (Hoeg and Lutzen 1995). It is clear, however, that coloniality is among a number of derived traits associated with the rhizocephalans' unusual lifestyle; the ancestral ecdysozoan was almost certainly solitary.

Deuterostomes

There are currently three described phyla within the deuterostomes: the echinoderms, hemichordates, and chordates, although it has been suggested that the chordate subphylum Urochordata should be considered a separate phylum (Cameron et al. 2000). Asexual reproduction is present in all of the deuterostome phyla, and coloniality is found in the urochordates and hemichordates (Strathmann 1987; Cameron et al. 2000) (fig. 20.1, *A*). If coloniality is derived, it has evolved more than once independently in the deuterostomes, at least once in the hemichordates (Cameron et al. 2000), and several times in the urochordates (Wada et al. 1992; Swalla et al. 2000; Swalla 2001) (fig. 20.1, *B*). The evolution of coloniality within the deuterostomes is discussed in detail in the following sections.

Summary

We can conclude from this brief review that the ancestral metazoan probably had the capacity to reproduce asexually, but it is unclear whether or not that ancestor was truly colonial. For years, the general assumption has been that of a solitary ancestor (Ruppert and Barnes 1994; Willmer 1990). However, several recent authors have challenged this assumption. Reiger (1994) hypothesized that the ancestral metazoan had a biphasic life history with a microscopic solitary larva and a macroscopic clonal adult, as in the cnidarians. Dewel (2000) has suggested that the ancestral bilaterian may have been a cnidarian-like colony that underwent a process of "individuation" to produce a stem (solitary) bilaterian. Blackstone and Ellison (2000) take this analysis one step further, suggesting that the phenomena of set-aside cells and maximal indirect development (Davidson et al. 1995) may have evolved as a result of the need to minimize levels-of-selection conflicts and maximize size in a colonial bilaterian ancestor.

Based on a strict parsimony analysis of the presence or absence of coloniality in the major extant metazoan clades, we can make a safe assumption only regarding the ecdysozoans, which were almost certainly derived from a solitary ancestor. Better resolution is needed, particularly within the cnidaria and the lophotrochozoa, to determine whether the colonial groups in those clades are basal or derived. The deuterostome clade is better resolved phylogenetically and may provide us with the best opportunity to tease out patterns of colonial evolution.

Individuals as Modules: The Evolution of Coloniality in Deuterostomes

Echinoderms and Hemichordates

Recent molecular phylogenies of the deuterostomes have shown that the echinoderms and hemichordates are sister groups (Halanych 1995; Bromham and Degnan 1999; Cameron et al. 2000). Within the echinoderms, there are instances of asexual reproduction but no colonial forms. The hemichordates include both solitary, exclusively sexual enteropneust worms and colonial pterobranchs that can undergo asexual and sexual reproduction (Hyman 1959; Barrington 1965; Cameron et al. 2000) (fig 20.1, *B*). One clade of enteropneusts includes the ptychoderid and balanoglossoid families that have indirect-developing, feeding larvae. The other enteropneust clade, the harrimaniid worms, has direct-developing, nonfeeding larvae (Hyman 1959; Barrington 1965; Cameron et al. 2000).

The colonial pterobranch hemichordates vary dramatically in morphology and life history from the solitary enteropneust worms (Hyman 1959; Barrington 1965; Cameron et al. 2000). Individual body size is greatly reduced in pterobranchs, and reproduction is primarily asexual. During sexual reproduction, the pterobranchs brood a small number of directly developing embryos. Thus, in hemichordates, colonial life history is linked to a shift in developmental and reproductive mode. In colonial hemichordates larvae are brooded until they are ready for settlement, while solitary enteropneusts are all free spawners. Colonial hemichordate larvae develop into miniaturized adults, capable of asexual reproduction and extensive regeneration. This correlation between colonial life history and developmental and reproductive mode also occurs in urochordates (see below).

Until recently, the colonial pterobranchs were believed to represent the "ancestral" hemichordate on the basis of the presumed homology of the tentacular feeding structures of pterobranchs and lophophorates (Romer 1967; Nielsen 1995). However, present-day molecular phylogenies suggest that the lophophorates are protostomes (Halanych et al. 1995; Adoutte et al. 2000). Therefore, the lophophore-like feeding structures of the pterobranchs and lophophorates probably evolved convergently (Halanych 1996a). Additionally, recent 18S rDNA molecular phylogenies suggest that the pterobranchs may have arisen from within the lineage of the solitary enteropneust worms (Halanych 1995; Cameron et al. 2000). Thus, there are strong indications that the ancestral hemichordate may not have resembled a colonial pterobranch. Even though phylogenies of hemichordates that combine 18S and 28S rDNA give conflicting results (Winchell et al. 2002), the similarities of the enteropneust and echinoderm feeding larvae suggest that the *Ptychodera*/*Balanoglossus* clade is likely to be ancestral in hemichordates (Peterson and Eernisse 2001).

The Chordates: Urochordates, Cephalochordates, and Vertebrates

Recent molecular phylogenies indicate that the chordates should be divided into two phyla, the monophyletic urochordates (Swalla et al.

2000) and the true chordates, which include the relatively closely related cephalochordates and vertebrates (Cameron et al. 2000; Winchell et al. 2002). Within the cephalochordates and vertebrates there are no instances of either coloniality or asexual reproduction, excepting rare instances of parthenogenesis among the vertebrates (fig. 20.1, *B*). Within urochordates, the planktonic appendicularians are exclusively solitary and sexual (Bone 1998; Wada 1998; Swalla et al. 2000). However, coloniality is pervasive in the remaining urochordate group, the ascidians (fig. 20.1, *B*).

Coloniality is found in three of the four ascidian clades (fig. 20.1, *B*) (Wada et al. 1992; Swalla et al. 2000; Swalla 2001), including the styelid stolidobranchs (Wada et al. 1992; Cohen et al. 1998; Swalla et al. 2000), the aplousobranchs (Christen and Braconnot 1998), and the phlebobranchs (Wada et al. 1992; Swalla et al. 2000). Coloniality also occurs within the pelagic thaliaceans (Bone 1998) that appear to be related to phlebobranch ascidians (Wada 1998; Swalla et al. 2000). The only group of ascidians that does not contain colonial species is the molgulid stolidobranchs (Huber et al. 2000; Swalla et al. 2000). Colonial life histories within the urochordates and hemichordates are accompanied by the ability to reproduce asexually, larval brooding, miniaturization of the body plan, and an increased capacity for regeneration (Van Name 1945; Berrill 1950; Wada et al. 1992; Cohen et al. 1998; Swalla et al. 2000).

Summary

Within the deuterostomes, colonial species are found only within the hemichordates and urochordates. All colonial deuterostomes share a distinctive suite of developmental and reproductive traits. The correlation between coloniality and these developmental and reproductive traits may be explained in several ways. It may represent a convergent phenomenon related to selective pressures specific to colonial forms. For example, because of their small body size, larval brooding may be the optimal reproductive strategy for colonial zooids. Alternatively, these traits may reflect fundamental developmental mechanisms of colony formation. For instance, colonies expand through asexual budding, and the ability to regenerate and to brood may be consequences of the developmental potential inherent in budding. Lastly, these linked traits may represent homologies indicating a common deuterostome colonial ancestor. Comparisons between adult differentiation in closely related solitary and colonial forms might uncover the developmental basis for key colonial traits such as miniaturization, budding, and brooding. In turn, this developmental framework will direct further experimentation related to questions of coloniality and evolution. Ascidians are an optimal clade for pursuing this approach due to the prevalence of colonial forms and established access to molecular techniques necessary to analyze developmental mechanisms. Therefore, we are concentrating on investigating the molecular basis of ascidian metamorphosis and adult differentiation as a first step in investigating the evolution of coloniality within the deuterostomes.

Adult Differentiation and Life History Transitions in Ascidians

The transition between solitary and colonial life histories within the ascidians is accompanied by a shift in the timing and nature of adult organ differentiation (Berrill 1935,1936; Jeffery and Swalla 1992, 1997) (fig. 20.2). In most solitary species, adult differentiation begins after settlement of the larvae and resorption of larval structures, while in colonial species differentiation of the adult begins precociously within brooded larvae (Berrill 1936). Solitary species generally spawn small eggs that develop over 7–48 hours into simple larval forms before hatching. These simple larvae possess only rudimentary adult structures within the head and trunk region (Cloney 1978; Jeffery and Swalla 1992, 1997; Satoh 1994) (fig. 20.2, *A*, *C*). During a fixed period after hatching, swimming larvae gain competence to respond to external settlement cues. Upon settlement, larval structures such as the tail and sensory organs are rapidly resorbed. After settlement, the differentiation of functional adult organs occurs over a period of 3–10 days (fig. 20.2, *A*). Colonial larvae begin as large eggs that are almost exclusively brooded or viviparous (Cloney 1978). Colonial free-spawning forms of the genus *Diozona* are the exception to this rule, and they possess a limited form of coloniality without true propagative budding (Satoh 1994). Colonial larvae have a prolonged larval brooding period lasting from days to weeks as adult structures differentiate to varying degrees within the head and trunk region of the larvae (fig. 20.2, *B*, *C*). Larval release, followed by settlement and tail resorption, does not begin until substantial adult differentiation has already occurred (Berrill 1935; Cloney 1978) (fig. 20.2, *B*). Asexual budding is often observed in the larvae as well (fig. 20.2, *C*). After settlement, the larval head and trunk rapidly complete differentiation into a functional oozoid that can only reproduce asexually. Sexual reproduction occurs solely in budded zooids (Satoh 1994).

The radical shift in the developmental timing of adult differentiation between solitary and colonial ascidians may depend upon the dissociation of two distinct developmental modules (Berrill 1935). One module consists of the transitory larval organs, the tail, notochord, larval central nervous system (CNS), and sensory organs. In both solitary and colonial ascidians, these purely larval structures are similar in morphol-

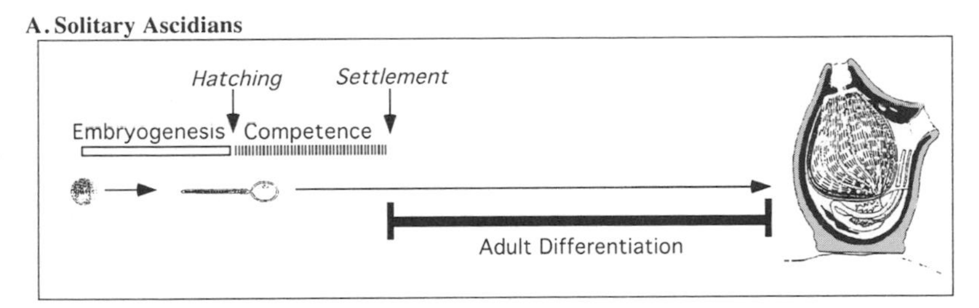

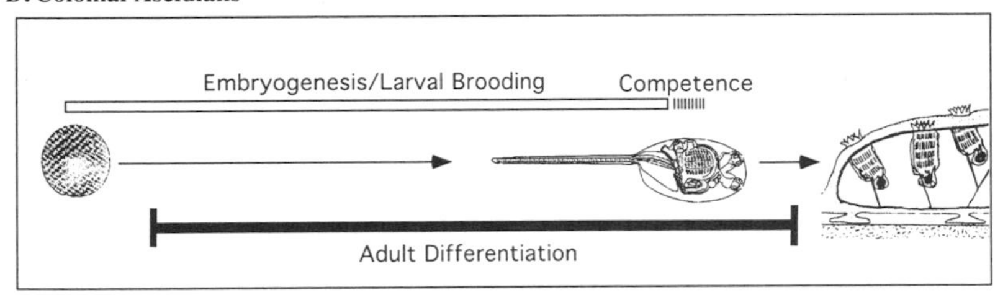

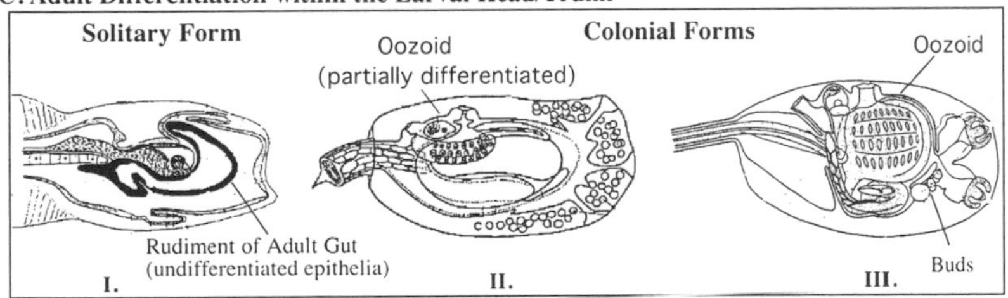

Fig. 20.2.—Ascidian development in solitary versus colonial forms. *A,* Solitary ascidian development (the image on the far right represents an adult). Note that adult differentiation does not begin until after settlement. *B,* Summary of colonial ascidian development (the image on the far right represents a young colony in which three zooids are embedded in a common tunic; modified from Lahille 1890). Note that adult differentiation begins during larval brooding in colonial ascidians. *C,* Lack of adult differentiation within the larval head and trunk region in solitary ascidians (*I*) and examples of variation in the degree of adult differentiation within the larval head and trunk region of colonial ascidians: partial differentiation of oozoid in larvae of *Amaroucium constellatum* (*II*) (modified from Scott 1945), and full differentiation of oozoid and beginning of budding in larval *Distaplia occidentalis* (*III*) (modified from Cloney 1978).

ogy and are likely to have conserved determinative developmental patterning (Cloney 1978; Satoh 1994). The other module consists of the adult organs such as the gut (including the pharynx and gill slits), siphons, body wall musculature, heart, and adult nervous system. In both solitary and colonial ascidians, these adult structures are similar in morphology and may have a conserved regulatory mode of development (Satoh 1994) (see below). The independence of these two developmental modules is further emphasized by frequently observed abnormalities in ascidian metamorphosis. Such abnormalities include the retention of a functional larval tail in settled colonial juveniles (Cloney 1978) and the spontaneous differentiation of adult structures in the simple larva of solitary species before settlement (Berrill 1935). Hinman and Degnan (2001) point out that this ascidian dissociation between the development of larval "somatic" (notochord and neural) structures and adult "visceral" (gut) structures parallels observations by Romer (1972) on the modularity of vertebrate development.

Although much is known about embryogenesis of solitary ascidian larvae, research on the differentiation of adult organs in colonial larvae and solitary juveniles has just begun. Once ascidian adult development has been better characterized, we should be able to make developmental comparisons between solitary and colonial forms that will clarify the evolution of this life history transition.

Molecular Characterization of Ascidian Adult Differentiation

Epidermal Growth Factor Pathways during Ascidian Metamorphosis

Recent studies are uncovering conserved signaling pathways which are used to trigger and coordinate adult differentiation in solitary ascidians. A number of studies have focused on differential gene transcription during metamorphosis (Eri et al. 1999; Davidson and Swalla 2001, 2002; Nakayama et al. 2001). Uridine incorporation experiments reveal a dramatic increase in transcription shortly after settlement in the solitary ascidian *Herdmania curvata* (Green et al. 2002). Treatment with actinomycin D has demonstrated that transcription is necessary for the acquisition of larval competence as well as settlement and adult differentiation in the solitary ascidians *Boltenia villosa* and *Herdmania curvata* (Davidson and Swalla 2001; Green et al. 2002). Therefore, the use of differential display and subtractive hybridization to reveal which genes are being transcribed during settlement and metamorphosis is a promising pathway for investigating the signals that trigger and coordinate adult differentiation.

Arnold et al. (1997) used differential display to isolate genes expressed in larvae of *Herdmania curvata*. This technique led to the isolation of Hemps, a protein with epidermal growth factor (EGF) motifs (Arnold et al. 1997). The EGF family includes a diverse assemblage of secreted proteins with a wide range of developmental roles involving signaling between cells (Schweitzer and Shilo 1997). Degnan et al. (1997) previously demonstrated that a signal for metamorphosis was secreted at the anterior end of *Herdmania* larva. This anterior papillary region is precisely where Hemps is detected. Furthermore, there is strong evidence that Hemps is both necessary and sufficient to trigger metamorphosis in competent *Herdmania* larva (Eri et al. 1999). Nakayama et al. (2001) have also used differential display to isolate settlement genes from *Ciona intestinalis,* leading to the isolation of the EGF-like transcript *Ci-meta1*.

We used subtractive hybridization to isolate genes that are differentially transcribed during larval competence, settlement, and adult differentiation in *Boltenia villosa* (Davidson and Swalla 2001, 2002). These screens led to the isolation of the *Boltenia* homologue of the gene *cornichon* (Bv-*Cni*) (Davidson and Swalla 2001). Our research indicates that Bv-*Cni* expression is up-regulated as larvae become competent to respond to settlement cues (Davidson and Swalla 2001). *Cornichon* was first isolated in *Drosophila,* where it has a role in regulating the EGF signaling involved in establishing the dorsal-ventral and anterior-posterior axes during oogenesis (Roth et al. 1995). Recent work in yeast has established a role for *cornichon* in the directed secretion of targeted vesicles (Powers and Barlowe 1998). Our hypothesis is that expression of the ascidian *cornichon* acts to potentiate the secretion of EGF-like proteins, such as Hemps, in response to settlement cues. Taken together, these investigations point toward a significant role for EGF signaling in competence and settlement of solitary ascidian larvae and the subsequent initiation of adult differentiation.

Nuclear Receptor Hormones and Ascidian Metamorphosis

Recent work suggests that nuclear receptor hormones, including retinoic acid (RA) and thyroid hormones (THs), may play a role in ascidian adult differentiation. Hinman and Degnan (1998, 2000, 2001) have conducted a series of elegant experiments examining the effect of ectopic RA on *Herdmania curvata* development. These experiments demonstrate that RA differentially effects larval versus adult patterning and gene expression (Hinman and Degnan 2001). In vertebrates and cephalochordates, RA plays a central role in the embryonic patterning of the CNS and endoderm by coordinating the expression of transcription factors such as *Hox1, Otx2* and *Pax2* (Holland and Holland 1996; Ross et al. 2000). The application of ectopic RA in cephalochordates and vertebrate embryos disrupts the expression of these patterning genes causing posteriorization of the CNS and endoderm. In the solitary ascidian *Halocynthia roretzi,* embryonic RA treatment resulted in a loss of anterior larval structures and expansion of *Hox1* gene expression. These results were interpreted as indicating a conserved role for RA in patterning the embryonic CNS in vertebrates and ascidians (Katsuyama et al. 1995). However, in *Herdmania curvata,* Hinman and Degnan (1998, 2000, 2001) found that ectopic RA did not perturb the embryonic expression of the CNS patterning genes *Hec-Otx* and *Hec-Pax2/5/8*. Furthermore, their examination of the treated embryos indicated that the posteriorizing effects of RA were limited to adult primordial structures instead of the larval CNS. In contrast, treatment of juvenile *Herdmania* led to a dramatic posteriorization of the endoderm and a significant down-regulation in *Hec-Otx*, which more closely parallels the effect of RA treatment on vertebrate embryos. Therefore, they conclude that RA-mediated patterning of the embryonic CNS is not conserved between vertebrates and ascidians. Additionally, they hypothesize that RA-mediated patterning of the vertebrate embryonic endoderm has been heterochronically shifted in ascidians to pattern the adult endoderm after settlement. This work represents the first molecular insight into the modularity of ascidian larval and adult development. In particular, RA treatments on ascidians delineate the contrast between a determinant/RA insensitive, ascidian-specific developmental patterning of the larval CNS versus a more regulative/RA sensitive, vertebrate-like patterning of adult endoderm.

Work on asexual reproduction in colonial ascidians has demonstrated a potential role for retinoic acid in bud patterning (Kawamura et al. 1993; Kamimura et al. 2000). RA treatment of buds in the colonial ascidian *Polyandrocarpa misakiensis* can induce ectopic gut formation or posteriorization of the developing bud (Kawamura et al. 1993). A retinoic acid receptor, *PmRAR*, and its dimerization partner, *PmRXR*, have been cloned from *Polyandrocarpa* and demonstrated to be functional mediators of RA signaling (Kamimura et al. 2000). It is hypothesized that mesenchymal cells express both *PmRAR* and *PmRXR*, thereby mediating bud polarity and pattern generation by reacting to the differential synthesis of RA at the proximal end of the bud (Kamimura et al. 2000). Thus, studies of retinoic acid in ascidian development have established a molecular link between the regulative patterning of adult endoderm in solitary ascidian metamorphosis and the regulative patterning of the endoderm central to colonial bud morphogenesis.

There has been a long-standing effort to determine if thyroid hormones (THs) have a role in ascidian metamorphosis similar to the well-documented role in amphibians. The ascidian endostyle is homologous to the endostyle of larval lamprey (which develops into the thyroid gland after lamprey metamorphosis) (Manzon et al. 2001). Additionally, the ascidian endostyle is able to bind iodine and synthesize thyroid hormones (Eales 1997). Furthermore, ascidian homologues of the thyroid peroxidase gene (*TPO*) have been cloned from the solitary ascidians *Ciona intestinalis* and *Halocynthia roretzi* (Ogasawara and Satoh 1998). In vertebrates, *TPO* is specific to the thyroid gland and is involved in thyroid hormone synthesis. Expression of *TPO* in ascidians is also specific to the TH synthesis region of the endostyle (Ogasawara et al. 1999).

Early experimental work on THs in ascidian metamorphosis focused exclusively on settlement. At the time, ascidian settlement was conceived as possibly homologous to amphibian metamorphosis because of the recurrence of tail resorption. Not surprisingly, tests of this hypothesis gave ambiguous results. Application of thyroid hormones and goitrogens (TH synthesis blockers) to solitary and colonial ascidian embryos caused a variety of contradictory effects on settlement and metamorphosis (Barrington 1968; Patricolo et al. 1981).

The demonstration that echinoderm metamorphosis is coordinated by the thyroid hormone thyroxine (Chino et al. 1994) has reinvigorated the hypothesis that THs, particularly thyroxine (T4), may be involved in ascidian metamorphosis. In echinoderms and vertebrates, THs coordinate the transformation of larval structures and adult rudiments into differentiated adult structures. If ascidians have a similar role for one of the THs, one would expect it to coordinate adult differentiation instead of settlement. In order to test this idea we treated larvae of the solitary ascidian *Boltenia villosa* with the TH synthesis inhibitors thiourea (50 mM) and potassium thiocyanate (10 mM) and observed the effect of these inhibitors on settlement and subsequent adult differentiation. The TH inhibitors had no effect on *Boltenia* settlement, but they had a dramatic effect on the subsequent differentiation of the adult (fig. 20.3). Treated *Boltenia* larva settled normally, extended ampullae, and began differentiation of the adult test, but the differentiation of adult rudiments was completely arrested at approximately 2–4 days after settlement (fig. 20.3, *D*). However, this effect could not be rescued through the application of exogenous THs, and the addition of ectopic THs did not noticeably accelerate adult differentiation (fig. 20.3).

Preliminary radioimmunoassay (RIA) measurements of TH levels in *Boltenia* show an intriguing rise in T4 production 2–4 days after settlement, just as juvenile differentiation is initiated (fig. 20.4). The timeline of T4 production corroborates the TH blocker results. T4 synthesis seems to be initiated at the same stage that TH inhibitors cause developmental arrest (between 2–4 days after settlement). However, these preliminary results are far from conclusive. The effect of the TH blockers may be nonspecific, and the RIA measurements are at the low end of detectability for this assay. More extensive RIA studies involving larger amounts of tissue and greater sample sizes are required. Additionally, the presence of TH needs to be confirmed through HPLC fractionation. Nevertheless, these results suggest that small amounts of T4 may be synthesized by newly settled juveniles and used to coordinate adult differentiation.

Recent work by Patricolo et al. (2001) further substantiates our findings. They have measured a significant level of thyroxine ($\leq$ 300 pg/mg) in *Ciona intestinalis* larvae in HPLC-fractionated samples. They also report that thiourea blocks settlement and further differentiation.

Superficially, this effect of thiourea on settlement in *Ciona* contradicts our finding that thiourea has no effect on *Boltenia* settlement. However, *Ciona* larvae prior to settlement show greater differentiation of adult rudiments than *Boltenia* larvae. Subsequently, settled *Ciona* larvae complete their morphogenesis into feeding juveniles much more rapidly than *Boltenia* larvae. Therefore, it is probable that thiourea is disrupting a crucial juncture in adult differentiation in both species, but this juncture is reached earlier (prior to settlement) in *Ciona*. To test this hypothesis, a better time course of juvenile differentiation and TH levels in both *Boltenia villosa* and *Ciona intestinalis* is required.

The potential involvement of THs in ascidian metamorphosis is particularly exciting because research in amphibians has demonstrated that a shift in the TH axis may underlie some aspects of the transitions from indirect to direct development in anurans (Jennings and Hanken 1998) as well as neoteny in urodeles (Rosenkilde and Ussing 1996; Safi

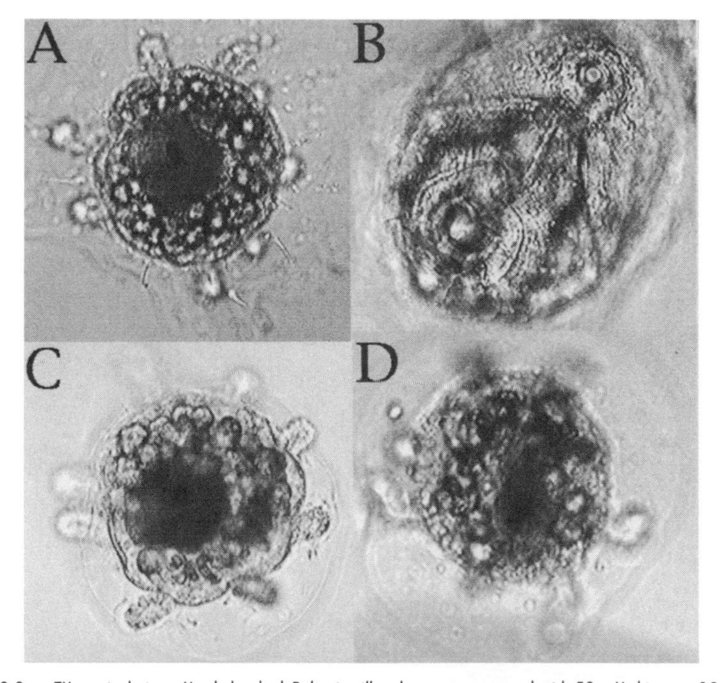

Fig. 20.3.—TH manipulations. Newly hatched *Boltenia villosa* larvae were treated with 50 mM thiourea, 10 mM K thiocyanate, 100 nM/1 μM/5 μM T3 or T4, or a combination of 50 mM thiourea + 1 μM T3 + 1 μM T4. They were then induced to undergo settlement 10 hours after hatching through the addition of 50 mM KCl into the seawater. *A, B,* Typical *B. villosa* juveniles 2 days (*A*) and 8 days (*B*) after settlement. TH treated *Boltenia* showed no significant variation from these typical morphologies ($n = 25-36$ per treatment). *C, D,* Typical thiourea-treated juveniles 2 days (*C*) and 8 days (*D*) after settlement. Results were identical for K thiocyanate–treated juveniles. TH-inhibited juveniles showed no significant variation from these typical arrested morphologies ($n > 100$ per treatment). Application of exogenous THs along with thiourea had no effect on this arrested morphology ($n > 100$).

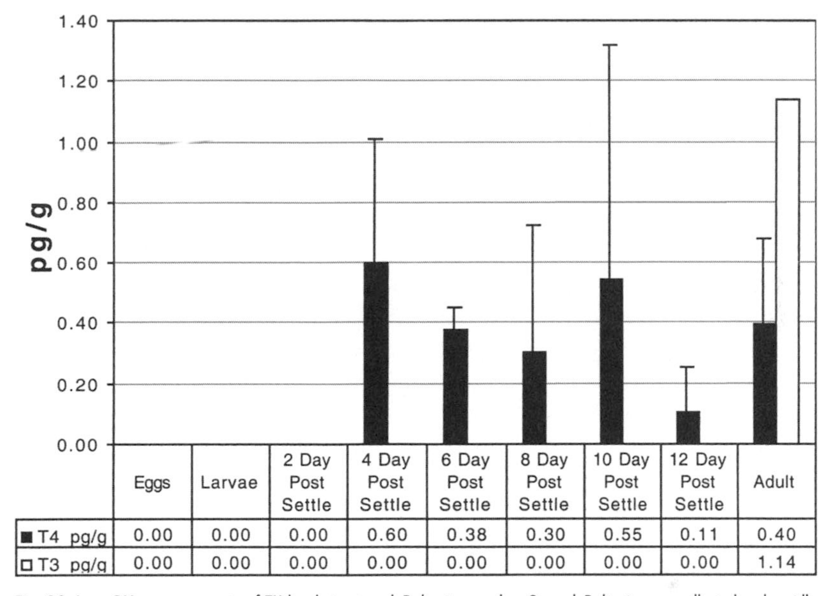

Fig. 20.4.—RIA measurements of TH levels in staged *Boltenia* samples. Staged *Boltenia* were collected and rapidly frozen in liquid nitrogen. THs were extracted according to the method of Gordon et al. (1982). Competitive binding RIA was performed using the Amerlex-M T3 and T4 Kits (Amersham International). Two separate samples of each stage were tested for T4, while only one set of samples was used for T3 measurements. The error bars represent one standard deviation.

et al. 1997). Thus, ascidian life history transformations may also involve such a shift in hormonal timing. Recently the first ascidian TH receptor family member was cloned from the ascidian *Ciona intestinalis* (Carosa et al. 1998). Structural and functional analysis showed this to be a non-TH binding form. It is likely that another, functional TH receptor is also present in the ascidians. Any further research into the role of TH in ascidian metamorphosis should therefore focus on obtaining a clone of this TH receptor and analyzing its developmental expression and function.

Conclusions and Future Directions

In a survey of bilaterian life history patterns there are surprisingly few groups in which coloniality occurs. It has generally been assumed that colonial forms are highly derived within their clades. However, it is also possible that there was a sessile colonial ancestor for the Bilateria as a whole or for one or more of the bilaterian clades (Van Name 1921; Blackstone and Ellison 2000; Dewel 2000). In this scenario, positive selection for a more active, mobile life style may have led to multiple lines of derived solitary forms. The evolutionary potency of the individual as a module within an ancestral colony could have been instrumental in the evolution of new body plans among derived clades. As described by Harvell (1994), colonial anthozoans and bryozoans have abundant potential for morphological innovation, as evidenced by the extreme variation among zooids in polymorphic colonies. A number of factors contribute to this plasticity, including the lability of signal transduction inherent in asexual budding, the low selective cost of losing zooids to failed innovations, and the ability of the colony as a whole to support partially successful "hopeful monsters" along the path to novel functional morphologies (Harvell 1994). Thus, the evolutionary potential of a colonial ancestor to produce divergent morphologies is closely tied to the developmental modularity of individual colony members.

In order to assess these evolutionary hypotheses, a developmental understanding of solitary-colonial transitions is necessary. Developmental comparisons between closely related solitary and colonial forms can be used to analyze specific molecular mechanisms required for asexual budding of new individuals, the key innovation required to form a colony. Furthermore, comparisons of colonial developmental patterns among a wide range of taxa could be used to assess the extent to which these patterns have been conserved. A highly conserved colonial developmental pattern would suggest a colonial ancestor, and vice versa. Ascidian biologists in the middle of the twentieth century attempted to use this approach to determine whether the ancestral ascidian was solitary or colonial. Berrill (1955) analyzed modes of asexual replication in ascidians and concluded that they were too divergent to have evolved in a common lineage. Millar (1966) disagreed with Berrill's analysis, suggesting that similarities can indeed be traced in modes of budding in ascidians. Comparative analyses of molecular mechanisms of development across different colonial clades of ascidians may resolve this debate.

The only major metazoan clades with the admixture of colonial and solitary forms necessary for meaningful developmental comparisons are cnidarians and deuterostomes. Cnidarians, along with ctenophores (all solitary), are consistent out-groups to the bilaterian metazoans in most modern phylogenetic analyses (Collins 1998; Adoutte et al. 2000). Consequently, close examination of patterns of coloniality in cnidarians may give clues about the nature of the ancestral bilaterian. However, the poor phylogenetic resolution of the cnidarians will complicate attempts to make evolutionary hypotheses based on developmental comparisons within this group. As we have discussed above, the deuterostomes, particularly the ascidians, represent the optimal clade for such developmental comparisons due to good phylogenetic resolution, the availability of molecular techniques for assessing developmental patterns, and the widespread occurrence of colonial and solitary forms within the ascidians.

Colonial ascidians display a suite of distinctive developmental and reproductive traits including miniaturization, enhanced regeneration, asexual reproduction, and larval brooding. These traits may reflect fundamental developmental pathways central to colony formation. In particular, acquiring the developmental potential to bud and/or brood may have been critical in the transition from solitary to colonial forms. Conversely, the loss of this developmental potential may have played a significant role in colonial-to-solitary transitions. Budding is a highly regulative process requiring the transdifferentiation of multiple tissue types. The developmental potential for such a process in ascidians is clearly restricted to the more regulative adult module of their development as opposed to the more determinative larval developmental module. The interplay between these two distinctive developmental modules may have had a key role in ascidian solitary-colonial transitions. For instance, the predominance of a highly regulative developmental program originally sequestered to the viscera of a solitary ascidian ancestor may have created the potential for budding. On the other hand, the predominance of a highly determinative developmental program originally relegated to the larvae of a colonial ancestor may have precluded the potential for budding. Therefore, in order to gain insight into the developmental basis of ascidian coloniality it is critical to investigate adult differentiation and the presumed modularity of adult versus larval development.

We have surveyed a number of strong candidates for conserved developmental pathways that may coordinate ascidian metamorphosis and the differentiation of adult structures, including EGF, retinoic acid, and thyroid hormones. Ongoing research will clarify the role of EGF-like signaling in solitary ascidian development and how it may interact with other signaling pathways to trigger the events of settlement and the subsequent differentiation of the adult rudiment. Once these signals are better described in solitary species, it will be possible to test their involvement in colonial ascidian metamorphosis. These comparisons between solitary and colonial species may provide valuable insights into the nature of the presumed adult and larval developmental modules and how the coordination between them has been manipulated during ascidian evolution.

The RA-induced posteriorization of endoderm in both solitary and colonial ascidians as well as in cephalochordates and vertebrates (Hinman and Degnan 2001; Holland and Holland 1996; Kamimura et al. 2000) indicates a role for RA in the patterning of adult endoderm in the chordate ancestor. However, the refractoriness of the ascidian larval CNS to RA treatment indicates that RA patterning of the CNS was either derived within higher chordates or lost in the solitary ascidians. In order to understand the ancestral role of RA in the chordates and the deuterostomes, further developmental studies involving retinoic acid must be conducted, particularly on colonial ascidian larvae, hemichordates, and echinoderms. There is also research on RA in cnidarians that implies that RA effects on developmental patterning may be plesiomorphic for the metazoans (Müller 1984; Kostrouch et al. 1998).

The role for THs in coordinating temporal patterning during metamorphosis in lampreys, echinoderms, teleosts, amphibians, and possibly ascidians (see "Nuclear Receptor Hormones and Ascidian Metamorphosis" above) may indicate a conserved role for THs in the development of the ancestral deuterostome (Chino et al. 1994; Eales 1997; Manzon et al. 2001). It is unlikely that metamorphosis within these disparate deuterostome clades represents a homologous process. Instead, we hypothesize that THs are part of an ancestral deuterostome adult developmental module that has been used convergently to coordinate adult developmental timing independently of larval timing. Along with working to confirm a role for THs in solitary ascidian metamorphosis, it is critical to investigate a role for THs in colonial ascidians and hemichordates. Additionally, in vertebrates TH and RA signaling pathways can overlap, particularly through the shared dimerization with RXR (Shi et al. 1998), a phenomenon which may have important implications for analyzing RA and TH effects within the deuterostomes.

As our knowledge of the developmental basis for metamorphosis and adult differentiation within the ascidians and other nonvertebrate deuterostomes becomes more refined, careful comparisons of the extent to which these developmental pathways are conserved within the deuterostomes could help resolve vital questions about the ancestral life history pattern of the urochordates, the chordates, and the deuterostomes as a whole. For instance, as mentioned above, recent phylogenies have seriously challenged the proposed homology between the feeding tentacles of pterobranch hemichordates and the lophophores of phoronids, brachiopods, and bryozoans. In turn, theories that placed colonial pterobranchs as representatives of the deuterostome stem group seem unlikely. However, if developmental comparisons between colonial pterobranchs and ascidians demonstrated a conserved basis for coloniality, this would be strong evidence that the pterobranchs' colonial life history pattern is plesiomorphic for the deuterostomes. Developmental comparisons may thus eventually indicate that one or more of the metazoan clades had a colonial ancestor. Such a finding would represent an exciting opportunity to explore how the ability of an individual to act as a module within an ancestral colony influenced the evolution of novel morphologies among its descendants.

Acknowledgments

The authors would like to thank Gerhard Schlosser and Günter Wagner for organizing the Delmenhorst meeting on modularity and providing insight and encouragement in the preparation of the manuscript. We thank Richard Strathmann for his comments on the manuscript and his unique insights into life history evolution. B. Davidson received support from a National Institutes of Health Developmental Biology Training Grant, and an National Science Foundation Graduate Fellowship supports M. W. Jacobs. This manuscript was funded by a University of Washington Royalty Research Grant and National Science Foundation grant IBN-0096266 to B. J. Swalla.

References

Adoutte, A., G. Balavoine, N. Lartillot, O. Lespinet, B. Prud'homme, and R. de Rosa. 2000. The new animal phylogeny: reliability and implications. *Proc. Natl. Acad. Sci. U.S.A.* 97:4453–4456.

Aguinaldo, A. M. A., J. M. Turbeville, L. S. Linford, M. C. Rivera, J. R. Garey, R. A. Raff, and J. A. Lake. 1997. Evidence for a clade of nematodes, arthropods, and other moulting animals. *Nature* 387:489–493.

Arnold, J. M., R. Eri, B. M. Degnan, and M. F. Lavin. 1997. A novel gene containing multiple EGF-like motifs transiently expressed in the papillae of the ascidian tadpole larvae. *Dev. Dyn.* 210:264–273.

Barrington, E. 1965. *The biology of hemichordata and protochordata.* San Francisco: W. H. Freeman.

Barrington, E. J. W. 1968. Metamorphosis in lower chordates. In *Metamorphosis: a problem in developmental biology,* ed. W. Etkin and L. I. Gilbert, 223–270. New York: Appleton-Century-Crofts.

Berrill, N. J. 1935. Studies in tunicate development. 3. Differential retardation and acceleration. *Philos. Trans. R. Soc. Lond. B Biol. Sci.* 225:255–326.

Berrill, N. J. 1936. Studies in tunicate development. 5. The evolution and classification of ascidians. *Philos. Trans. R. Soc. Lond. B Biol. Sci.* 226:43–70.

Berrill, N. J. 1950. *The tunicata with an account of the British species.* London: Adlard and Son.

Berrill, N.J. 1955. *The origin of vertebrates.* London: Bernard Quaritch.

Blackstone, N. W., and A. M. Ellison. 2000. Maximal indirect development, set-aside cells, and levels of selection. *J. Exp. Zool. (Mol. Dev. Evol.)* 288:99–104.

Boardman, R. S., and A. H. Cheetham, eds. 1973. *Animal colonies.* Stroudsburg: Dowden, Hutchinson and Ross.

Boero, F., and J. Bouillon. 1987. Inconsistent evolution and paedomorphosis among the hydroids and medusae of the athecate/anthomedusae and the thecate/leptomedusae (cnidaria, hydrozoa). In *Modern trends in the systematics, ecology, and evolution of hydroids and hydromedusae,* ed. J. Bouillon et al. Oxford: Clarendon Press.

Bone, Q. 1998. *The biology of pelagic tunicates.* Oxford: Oxford University Press.

Borchiellini, C., N. Boury-Esnault, J. Vacelet, and Y. Le Parco. 1998. Phylogenetic analysis of the HSP 70 sequences reveals the monophyly of metazoa and specific phylogenetic relationships between animals and fungi. *Mol. Biol. Evol.* 15:647–655.

Bridge, D., C. W. Cunningham, R. DeSalle, and L. Buss. 1995. Class-level relationships in the phylum cnidaria: molecular and morphological evidence. *Mol. Biol. Evol.* 12: 679–689.

Bromham, L. D., and B. M. Degnan. 1999. Hemichordate and deuterostome evolution: robust molecular phylogenetic support for a hemichordate + echinoderm clade. *Evol. Dev.* 1:166–171.

Cameron, C. B., J. R. Garey, and B. J. Swalla. 2000. Evolution of the chordate body plan: new insights from phylogenetic analyses of deuterostome phyla. *Proc. Natl. Acad. Sci. U.S.A.* 97:4469–4474.

Carosa, E., A. Fanelli, S. Ulisse, R. Di Lauro, J. E. Rall, and E. A. Jannini. 1998. *Ciona intestinalis* nuclear receptor. 1. A member of steroid/thyroid hormone receptor family. *Proc. Natl. Acad. Sci. U.S.A.* 95:11152–11157.

Cavalier-Smith, T., M. T. Allsop, E. E. Chao, N. Boury-Esnault, and J. Vacelet. 1996. Sponge phylogeny, animal monophyly, and the origin of the nervous system: 18s rRNA evidence. *Can. J. Zool.* 74:2031–2045.

Chino, Y., M. Saito, K. Yamasu, T. Suyemitsu, and K. Ishihara. 1994. Formation of the adult rudiment of sea urchins is influenced by thyroid hormones. *Dev. Biol.* 161: 1–11.

Christen, R., and J. C. Braconnot. 1998. Molecular phylogeny of tunicates: a preliminary study using 28s ribosomal RNA partial sequences: implications in terms of evolution and ecology. In *The biology of pelagic tunicates,* ed. Q. Bone, 265–271. Oxford: Oxford University Press.

Cloney, R. A. 1978. Ascidian metamorphosis: review and analysis. In *Settlement and metamorphosis of marine invertebrate larvae,* ed. F. S. Chia and M. E. Rice, New York: Elsevier.

Cohen, C. S., Y. Saito, and I. L. Weissman. 1998. Evolution of allorecognition in botryllid ascidians inferred from a molecular phylogeny. *Evolution* 52:746–756.

Collins, A. G. 1998. Evaluating multiple alternative hypotheses for the origin of bilateria: an analysis of 18s rRNA molecular evidence. *Proc. Natl. Acad. Sci. U.S.A.* 95: 15458–15463.

Davidson, B. J., and B. J. Swalla. 2001. Isolation of genes involved in ascidian metamorphosis: EGF signaling and metamorphic competence. *Dev. Genes Evol.* 4:190–194.

Davidson, B. J., and B. J. Swalla. 2002. A molecular analysis of ascidian metamorphosis reveals elements of an innate immune response. *Development* 129:4739–4751.

Davidson, E. H., K. J. Peterson, and R. A. Cameron. 1995. Origin of bilaterian body plans: evolution of developmental regulatory mechanisms. *Science* 270:1319–1325.

Degnan, B. M., D. Souter, S. M. Degnan, and S. C. Long. 1997. Induction of metamorphosis with potassium ions requires development of competence and an anterior signaling centre in the ascidian *Herdmania momus. Dev. Genes Evol.* 206: 370–376.

Dewel, R. A. 2000. Colonial origin for eumetazoa: major morphological transitions and the origin of bilaterian complexity. *J. Morphol.* 243:35–74.

Eales, J. G. 1997. Iodine metabolism and thyroid-related functions in organisms lacking thyroid follicles: are thyroid hormones also vitamins? *Proc. Soc. Exp. Biol. Med.* 214:302–317.

Eri, R., J. M. Arnold, V. F. Hinman, K. M. Green, M. K. Jones, B. M. Degnan, and M. F. Lavin. 1999. Hemps, a novel EGF-like protein, plays a central role in ascidian metamorphosis. *Development* 126:5809–5818.

Gordon, J. T., F. L. Crutchfield, A. S. Jennings, and M. B. Dratman. 1982. Preparation of lipid-free tissue extracts for chromatographic determination of thyroid hormones and metabolites. *Arch. Biochem. Biophys.* 216:407–415.

Green, K. M., B. D. Russel, R. J. Clark, M. K. Jones, M. J. Garson, G. A. Skilleter, and B. M. Degnan. 2002. A sponge allelochemical induces ascidian settlement but inhibits metamorphosis. *Mar. Biol.* 140:355–364.

Gullan, P. J., and P. S. Cranston. 2000. *The insects: an outline of entomology.* Oxford: Blackwell Science.

Halanych, K. M. 1995. The phylogenetic position of the pterobranch hemichordates based on 18S rDNA sequence data. *Mol. Phylogenet. Evol.* 4:72–76.

Halanych, K. M. 1996a. Convergence in the feeding apparatuses of lophophorates and pterobranch hemichordates revealed by 18S rDNA: an interpretation. *Biol. Bull.* 190:1–5.

Halanych, K. M. 1996b. Testing hypotheses of chaetognath origins: long branches revealed by 18S rDNA. *Syst. Biol.* 45:223–246.

Halanych, K. M., J. D. Bacheller, A. M. Aguinaldo, S. M. Liva, D. M. Hillis, and J. A. Lake. 1995. Evidence from 18S ribosomal DNA that the lophophorates are protostome animals. *Science* 267:1641–1643.

Harvell, C. D. 1994. The evolution of polymorphism in colonial invertebrates and social insects. *Q. Rev. Biol.* 69:155–445.

Hinman, V. F., and B. M. Degnan. 1998. Retinoic acid disrupts anterior ectodermal and endodermal development in ascidian larvae and postlarvae. *Dev. Genes Evol.* 208:336–345.

Hinman, V. F., and B. M. Degnan. 2000. Retinoic acid perturbs Otx gene expression in the ascidian pharynx. *Dev. Genes Evol.* 210:129–139.

Hinman, V. F., and B. M. Degnan. 2001. Homeobox genes, retinoic acid and the development and evolution of dual body plans in the ascidian *Herdmania curvata. Am. Zool.* 41:664–675.

Hoeg, J. T., and J. Lutzen. 1995. Life cycle and reproduction in the cirripedia rhizocephala. *Oceanogr. Mar. Biol. Annu. Rev.* 33:427–485.

Holland, L. Z., and N. D. Holland. 1996. Expression of *Amphihox-1* and *Amphipax-1* in amphioxus embryos treated with retinoic acid: insights into evolution and patterning of the chordate nerve cord and pharynx. *Development* 122:1829–1838.

Huber, J. L., K. B. da Silva, W. R. Bates, and B. J. Swalla. 2000. The evolution of anural larvae in molgulid ascidians. *Semin. Cell Dev. Biol.* 11:419–426.

Hyman, L. H. 1959. The enterocoelous coelomates: phylum Hemichordata. Chap. 17 of *The invertebrates: smaller coelomate groups,* ed. E. J. Boell, 5:72–207. New York: McGraw-Hill.

Jackson, J. B. C. 1977. Competition on marine hard substrata: the adaptive significance of solitary and colonial strategies. *Am. Nat.* 111:743–767.

Jeffery, W. R., and B. J. Swalla. 1992. Evolution of alternate modes of development in ascidians. *BioEssays* 14 (4): 219–226.

Jeffery, W. R., and B. J. Swalla. 1997. Embryology of the tunicates. In *Embryology: constructing the organism,* ed. S. Gilbert, 331–364. Sunderland: Sinauer.

Jennings, D. H., and J. Hanken. 1998. Mechanistic basis of life history evolution in anuran amphibians: thyroid gland development in the direct-developing frog, *Eleutherodactylus coqui. Gen. Comp. Endocrinol.* 111:225–232.

Kamimura, M., S. Fujiwara, K. Kawamura, and T. Yubisui. 2000. Functional retinoid receptors in budding ascidians. *Dev. Growth Differ.* 42:1–8.

Katsuyama, Y., S. Wada, S. Yasugi, and H. Saiga. 1995. Expression of the labial group hox gene *Hrhox-1* and its alteration induced by retinoic acid in development of the ascidian *Halocynthia roretzi. Development* 121:3197–3205.

Kawamura, K., K. Hara, and S. Fujiwara. 1993. Developmental role of endogenous retinoids in the determination of morphallactic fields in budding tunicates. *Development* 117:835–845.

Kostrouch, Z., M. Kostrouchova, W. Love, E. Jannini, J. Piatigorsky, and J. E. Rall. 1998. Retinoic acid x receptor in the diploblast, *Tripedalia cystophora*. *Proc. Natl. Acad. Sci. U.S.A.* 95:13442–13447.

Lahille, F. 1890. Recherches sur les tuniciers des côtes de France. Toulouse: Lagarde & Sebille.

Mackie, G. 1986. From aggregates to integrates: physiological aspects of modularity in colonial organisms. *Philos. Trans. R. Soc. Lond. B Biol. Sci.* 313:175–176.

Manzon, R. G., J. A. Holmes, and J. H. Youson. 2001. Variable effects of goitrogens in inducing precocious metamorphosis in sea lampreys (*Petromyzon marinus*). *J. Exp. Zool.* 289:290–303.

Millar, R. H. 1966. Evolution in ascidians. In *Some contemporary studies in marine sciences,* ed. H. Barnes, 519–534. London: George Allen and Unwin.

Müller, W. A. 1984. Retinoids and pattern formation in a hydroid. *J. Embryol. Exp. Morphol.* 81:253–271.

Nakayama, A., Y. Satou, and N. Satoh. 2001. Isolation and characterization of genes that are expressed during *Ciona intestinalis* metamorphosis. *Dev. Genes Evol.* 211: 184–189.

Nielsen, C. 1995. *Animal evolution: interrelationships of the living phyla.* Oxford: Oxford University Press.

Ogasawara, M., and N. Satoh. 1998. Isolation and characterization of endostyle-specific genes in the ascidian *Ciona intestinalis. Biol. Bull.* 195:60–69.

Ogasawara, M., R. Di Lauro, and N. Satoh. 1999. Ascidian homologs of mammalian thyroid peroxidase genes are expressed in the thyroid-equivalent region of the endostyle. *J. Exp. Zool.* 285:158–169.

Patricolo, E., G. Ortolani, and A. Cascio. 1981. The effect of thyroxine on the metamorphosis of *Ascidia malaca. Cell Tissue Res.* 214:289–301.

Patricolo, E., M. Cammarata, and P. D'Agati. 2001. Presence of thyroid hormones in ascidian larvae and their involvement in metamorphosis. *J. Exp. Zool.* 290:426–430.

Peterson, K. J., and D. J. Eernisse. 2001. Animal phylogeny and the ancestry of bilaterians: inferences from morphology and 18S rDNA gene sequences. *Evol. Dev.* 3: 170–205.

Powers, J., and C. Barlowe. 1998. Transport of ax12p depends on erv14p, an ER-vesicle protein related to the *Drosophila cornichon* gene product. *J. Cell Biol.* 142:1209–1222.

Reiger, R. H. 1994. The biphasic life cycle—a central theme of metazoan evolution. *Am. Zool.* 34:484–491.

Romer, A. S. 1967. Major steps in vertebrate evolution. *Science* 158:1629–1637.

Romer, A. S. 1972. The vertebrates as a dual animal—somatic and visceral. *Evol. Biol. (N.Y.)* 6:121–156.

Rosenkilde, P., and A. P. Ussing. 1996. What mechanisms control neoteny and regulate induced metamorphosis in urodeles? *Int. J. Dev. Biol.* 40:665–673.

Ross, S. A., P. J. McCaffery, U. C. Drager, and L. M. De Luca. 2000. Retinoids in embryonal development. *Physiol. Rev.* 80:1021–1054.

Roth, S., F. S. Neuman-Silberberg, G. Barcelo, and T. Schupbach. 1995. *Cornichon* and the EGF receptor signaling process are necessary for both anterior-posterior and dorsal-ventral pattern formation in *Drosophila. Cell* 81:967–978.

Ruppert, E. E., and R. D. Barnes. 1994. *Invertebrate zoology.* Orlando, Fla.: Saunders College.

Safi, R., A. Begue, C. Hanni, D. Stehelin, J. R. Tata, and V. Laudet. 1997. Thyroid hormone receptor genes of neotenic amphibians. *J. Mol. Evol.* 44:595–604

Satoh, N. 1994. *Developmental biology of ascidians.* New York: Cambridge University Press.

Schweitzer, R., and B. Z. Shilo. 1997. A thousand and one roles for the *Drosophila* EGF receptor. *Trends Genet.* 13:191–196.

Scott, S. F. M. 1945. The developmental history of *Amaroecium constellatum.* 1. Early embryonic development. *Biol. Bull.* 88:126–138.

Shi, Y. B., Y. Su, Q. Li, and S. Damjanovski. 1998. Auto-regulation of thyroid hormone receptor genes during metamorphosis: roles in apoptosis and cell proliferation. *Int. J. Dev. Biol.* 42:107–116.

Strathmann, M. 1987. *Reproduction and development of marine invertebrates of the northern pacific coast.* Seattle: University of Washington Press.

Swalla, B. J. 2001. Phylogeny of the urochordates: implications for chordate evolution. In *Biology of ascidians (proceedings of the First International Symposium on the Biology of Ascidians),* ed. H. Sawada and C. Lambert, 219–224. Tokyo: Springer Verlag.

Swalla, B. J., C. B. Cameron, L. S. Corley, and J. R. Garey. 2000. Urochordates are monophyletic within the deuterostomes. *Syst. Biol.* 49:52–64.

Van Name, W. G. 1921. Budding in the compound ascidians and other invertebrates, and its bearing on the question of the early ancestry of the vertebrates. *Bull. Am. Mus. Nat. Hist.* 44, article 15:275–282.

Van Name, W. G. 1945. The North and South American ascidians. Bulletin of the American Museum of Natural History 84. New York.

Wada, H. 1998. Evolutionary history of free-swimming and sessile lifestyles in urochordates as deduced from 18s rDNA molecular phylogeny. *Mol. Biol. Evol.* 15: 1189–1194.

Wada, H., K. W. Makabe, M. Nakauchi, and N. Satoh. 1992. Phylogenetic relationships between solitary and colonial ascidians, as inferred from the sequence of the central region of their respective 18s rDNAs. *Biol. Bull.* 183:448–455.

Willmer, P. 1990. *Invertebrate relationships: patterns in animal evolution.* New York: Cambridge University Press.

Winchell, C. J., J. Sullivan, C. B. Cameron, B. J. Swalla, and J. Mallatt. 2002. Evaluating competing theories of deuterostome evolution with new LSU and SSU ribosomal DNA phylogenies. *Mol. Biol. Evol.* 19:762–776.

21 Evolvability, Modularity, and Individuality during the Transition to Multicellularity in Volvocalean Green Algae

AURORA M. NEDELCU AND
RICHARD E. MICHOD

Overview

Evolvability, viewed as the capacity of a lineage to generate heritable, selectable phenotypic variation (Altenberg 1995; Kirschner and Gerhart 1998), is a general feature of biological life. Evolvability is thought to depend critically on the way genetic variation maps onto phenotypic variation (the representation problem) such that improvement becomes possible through mutation and selection (Wagner and Altenberg 1996). It is not known how the genotype-phenotype maps are formed or how they are able to change in evolution and what the selective forces are (Wagner and Altenberg 1996). Properties that reduce constraints on change and allow the accumulation of nonlethal variation are thought to confer evolvability on a system (Kirschner and Gerhart 1998). One example of such a variational property is modularity (Wagner and Altenberg 1996). When defined as a genotype-phenotype map in which there are few pleiotropic effects among characters serving different functions (with pleiotropic effects falling mainly among characters that are part of a single functional complex), modularity is expected to improve evolvability by limiting the interference between the adaptations of different functions (Wagner and Altenberg 1996). Modules can be relatively easy to dissociate, recombine, or redeploy in new contexts; some modules are, nevertheless, resistant to dissociation and can lead to covariation and developmental constraints. Modular evolution may integrate previously separate functions, or create new separate modules from a formerly integrated one. How modules interact and evolve during transitions in units of evolution, or whether these interactions affect the evolutionary potential of a lineage, is not yet understood.

The current hierarchical organization of life reflects a series of transitions in the units of evolution, such as from genes to chromosomes, from prokaryotic to eukaryotic cells, from unicellular to multicellular individuals, and from multicellular organisms to societies. During these evolutionary transitions, new levels of biological organization are created (Buss 1987; Maynard Smith and Szathmáry 1995); moreover, individuality and new levels of heritable fitness variation have to emerge at the higher level (Michod 1999). We argue here that the emergence of individuality during the unicellular-multicellular transition requires the reorganization at the higher level of certain basic life properties (such as immortality, totipotency, growth, and reproduction). We think that the way in which this is achieved not only is instrumental for the emergence of individuality at the higher level but can also affect the potential for evolution, that is, evolvability, of the newly emerged higher-level unit.

We suggest that during evolutionary transitions in individuality, a new genotype-phenotype map must be created to reflect the emergence of the new higher-level unit. Furthermore, the way in which the lower-level genotype-phenotype maps are reorganized at the higher level can influence the potential for evolution of the newly emerged multilevel system. To this end, we use the volvocalean green algal group to argue that (i) during transitions in individuality some processes have to be dissociated at the lower level and recombined or redeployed at the higher level; (ii) the way in which certain complex sets of traits (and the genotype-phenotype maps associated with them) are reorganized during the transition affects the flexibility and robustness of the new genotype-phenotype map at the higher level and can interfere with the potential for further evolution of the lineage; and (iii) although modularity is generally expected to improve evolvability, during transitions in individuality this expectation is complicated and sometimes compromised by constraints at the lower level.

The Volvocalean Green Algal Group: A Case Study in the Transition to Multicellularity

Few groups of organisms hold such a fascination for evolutionary biologists as the Volvocales. It is almost as if these algae were designed to exemplify the process of evolution
—G. Bell, "The Origin and Early Evolution of Germ Cells as Illustrated by the Volvocales"

Rationale

Our reasons for choosing the volvocalean green algal group to investigate the transition to multicellularity and individuality are threefold. First, volvocalean green algae comprise both unicellular (*Chlamydomonas*-like) algae and colonial forms in different stages of organiza-

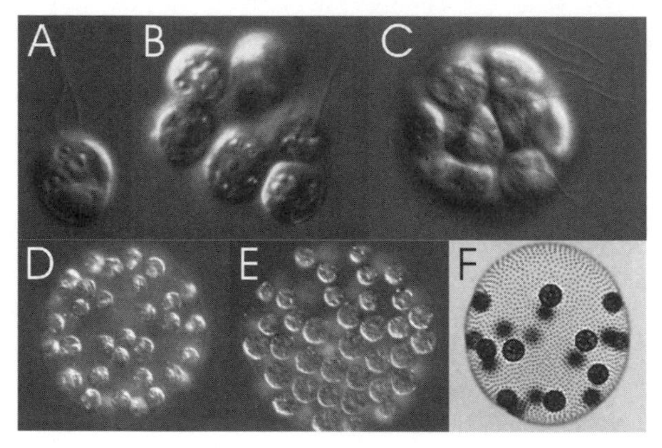

Fig. 21.1.—The "volvocine lineage": a subset of colonial volvocalean green algae that show a progressive increase in cell number, volume of extracellular matrix per cell, division of labor between somatic and reproductive cells, and proportion of vegetative cells. *A*, *Chlamydomonas reinhardtii*; *B*, *Gonium pectorale*; *C*, *Pandorina morum*; *D*, *Eudorina elegans*; *E*, *Pleodorina californica*; *F*, *Volvox carteri*. Where two cell types are present, the smaller cells are the vegetative or somatic cells, whereas the larger cells are the reproductive cells (gonidia). (Images kindly provided by David L. Kirk.)

tional and developmental complexity. The so-called volvocine lineage contains the genus *Chlamydomonas* as well as a subset of colonial volvocalean genera that show a progressive increase in cell number, volume of extracellular matrix per cell, division of labor between somatic and reproductive (gonidia) cells (i.e., germ-soma separation), and proportion of vegetative cells (Larson et al. 1992) (fig. 21.1). Second, multicellularity and individuality evolved multiple times in this group; the different levels of organizational and developmental complexity are thought to "represent alternative stable states, among which evolutionary transitions have occurred several times during the phylogenetic history of the group" (Larson et al. 1992), rather than a monophyletic progression in organizational and developmental complexity. Third, despite the multiple and independent acquisitions of the multicellular state and germ-soma separation in this group, none of these multicellular lineages attained high levels of complexity and/or phenotypic variability (as did other green algal lineages, especially the ancestors of land plants, the charophytes). We believe that understanding the reasons for this apparent limited spurt of diversification and complexity in this lineage will provide insight into how transitions in individuality can affect the evolvability of a lineage.

Complexity

Many traits are known to be rather diverse in this green algal group. The observed morphological and developmental diversity among volvocalean algae appears to result from the interaction of conflicting structural and functional constraints and strong selective pressures.

All volvocalean algae share the so-called flagellation constraint (Koufopanou 1994), which has a different structural basis from the one invoked in the origin of metazoans (Margulis 1981; Buss 1987). In most green flagellates, during cell division the flagellar basal bodies remain attached to the plasma membrane and flagella and behave like centrioles (which is not possible in other protists); however, in volvocalean algae, due to a coherent rigid cell wall the position of flagella is fixed, and thus, the basal bodies cannot move laterally and take the position expected for centrioles during cell division while remaining attached to the flagella (as they do in other green flagellates). Therefore, cell division and motility can take place simultaneously only for as long as flagella can beat without having the basal bodies attached (i.e., only up to five cell divisions).

The presence of a coherent cell wall is coupled with the second conserved feature among volvocalean algae, which is their unique way of cell division. The volvocalean cells do not double in size and then undergo binary fission. Rather, each cell grows about 2^n-fold in volume, and then a rapid, synchronous series of n divisions (under the mother cell wall) is initiated; this type of cell division is referred as to multiple fission and palintomy (i.e., the process during which a giant parental cell undergoes a rapid sequence of repeated divisions, without intervening growth, to produce numerous small cells). Because clusters, rather than individual cells, are produced in this way, it is suggested that this type of division has been an important precondition facilitating the formation of volvocacean colonies (Kirk 1998). In *Chlamydomonas,* the cells (2^2–2^4 cells) separate from each other after division. However, in many species, the cluster of 2^n cells does not disintegrate, and coenobial forms (i.e., a type of multicellular organization in which "the number of cells is determined by the number of cleavage divisions that went into its initial formation, and in which cell number is not augmented by accretionary cell divisions"; Kirk 1998) are produced. In *Gonium,* the resulting cells (2^2–2^5) stay together and form a convex discoidal colony. In *Eudorina* and *Pleodorina* the cells (2^4–2^6, 2^6–2^7, respectively) are separated by a considerable amount of extracellular matrix and form spherical colonies. Finally, in *Volvox,* a high number of cells (2^{15}–2^{16}) form colonies up to 3 mm in size (fig. 21.1).

The two selective pressures that are thought to have contributed to the increase in complexity in all volvocalean lineages are the advantages of a large size (potentially to escape predators, achieve faster motility or homeostasis, or better exploit eutrophic conditions) and the need for motility (e.g., to access to the euphotic or photosynthetic zone)

(Bell 1985). Interestingly, given the background offered by the volvocalean type of organization presented above—namely, the flagellar constraint and the multiple fission type of cell division—it is difficult to achieve the two selective advantages simultaneously. As the colonies increase in size and number of cells, so does the number of cell divisions (up to 16 in some *Volvox* species); consequently, the motility of the colony during the reproductive phase is negatively affected for longer periods of time than are acceptable in terms of the need to access the euphotic zone. This negative impact of the flagellation constraint is overcome by cellular specialization and division of labor: some cells are involved mostly in motility, while the rest of the cells become specialized for reproduction. The proportion of cells that remain motile throughout most or all of the life cycle is directly correlated with the number of cells in a colony, from none in *Chlamydomonas* and *Gonium* to up to one-half in *Pleodorina* and more than 99% in *Volvox* (Larson et al. 1992). In *Volvox,* the division of labor is complete: the motile (somatic) cells are sterile, terminally differentiated, and thought to be genetically programmed to undergo cellular senescence and death once the progeny have been released from the parental colony (Pommerville and Kochert 1981); only the reproductive cells (the gonidia) undergo cleavage to form new colonies (Pommerville and Kochert 1982). The present diversity in morphological and developmental complexity in the volvocalean algae reflects distinct strategies and solutions to the same set of constraints and pressures.

Transition in Individuality during the Transition to Multicellularity in Volvocalean Green Algae

In certain circumstances, a large size can be advantageous. However, cells cannot exceed a particular size, because, as they increase in size, the surface/volume ratio and thus the efficiency of metabolic processes decreases. Consequently, for unicellular organisms to increase in size, the number rather than the size of cells has to increase. Groups of cells can evolve in this way. Nevertheless, the stability of such groups is low, because cells can leave the group and live as free unicellular individuals. As a consequence, individuality at the higher level is difficult to achieve. However, individuality at the higher level evolved in many multicellular lineages. There are several different ways individuality can be defined, based on genetic homogeneity and uniqueness, physiological autonomy and unity, or units of selection (Michod 1999; Santelices 1999). Below, we use the physiological autonomy and unity criterion and define an individual as the smallest unit that is physiologically and reproductively autonomous. The question we are concerned with here is, How can individuality emerge during the unicellular-multicellular transition? What are the constraints that have to be broken in order for a group to become a multicellular individual?

One can approach this question from many perspectives. Below, we present a comparative approach and focus on several general life properties (such as growth and reproduction) and basic life traits (immortality and totipotency). We suggest that for individuality to be created at a higher level certain processes, traits, and functions have to be dissociated at the lower level and reorganized in new ways at the higher level. Moreover, we think that some of the differences among lineages can be explained by the way in which the reorganization of these processes and traits has been achieved during the transition to multicellularity and the emergence of individuality at the higher level. Volvocalean algae exemplify this suggestion well. In this group, the transition to multicellularity embraced unique paths, partly due to the constraints inherited from their unicellular ancestors, mainly the multiple fission type of division. Furthermore, although individuality at the higher level has been achieved in many volvocalean lineages, the way in which this was achieved interfered with the potential for further evolution of these lineages (discussed below in "The Gordian Knot and Evolvability in *Volvox*").

To facilitate the understanding of these issues, we first discuss the concepts used in further discussion. Our goal is to pinpoint the differences in the way that various traits are expressed between unicellular and multicellular forms, and to suggest (in the next section) how they have been reorganized during the transition from unicellular to multicellular individuals. We also apply these concepts to our study case, the volvocalean green algae.

Unicellularity versus Multicellularity: Basic Concepts

General Life Properties and Traits

VEGETATIVE AND REPRODUCTIVE FUNCTIONS

Any biological entity features two main sets of functions, vegetative and reproductive; these basic biological functions are coupled at the level of the individual, as a functional or physiological unit. However, the two sets of functions are realized differently between a unicellular and a multicellular individual (fig. 21.2, *A*). In unicellular forms, the same cell is responsible for both vegetative and reproductive activities (i.e., they are coupled at the cell level). Nevertheless, at the level of the individual, these functions do not take place simultaneously (i.e., they are dissociated in time). In multicellular individuals with germ-soma separation, the two sets of functions are uncoupled at the cell level; some cells do only vegetative functions, whereas other cells are specialized for reproductive functions. Consequently, the two sets of func-

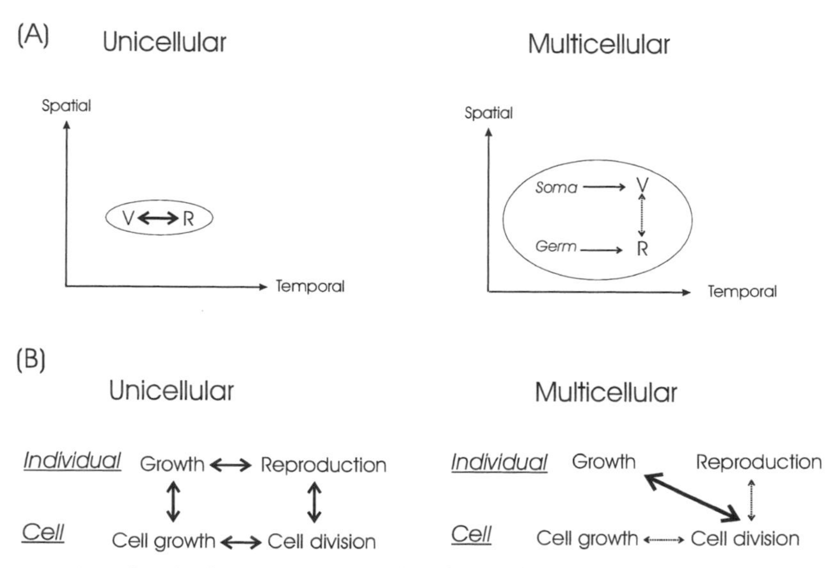

Fig. 21.2.—Relationships between vegetative (V) and reproductive (R) functions, on the one hand, and spatial and temporal contexts, on the other hand (panel *A*), in unicellular versus multicellular individuals, and between processes and properties at the level of the cell and the individual, respectively (panel *B*). Broken arrows denote relationships in which the two components are not necessarily dependent on each another.

tions can take place simultaneously (i.e., they need not to be separated in time anymore).

Growth is an important property of life. Interestingly, growth has different implications in unicellular versus multicellular individuals (fig. 21.2, *B*). In the former, growth is coupled with reproduction; growth to a specific size (cell surface/volume ratio) will generally trigger the reproduction of the individual, and vice versa, reproduction requires achieving a preset size. In multicellular individuals, on the other hand, growth and reproduction of the individual are uncoupled; reproduction is not necessarily dependent on growth, and growth does not necessarily trigger reproduction.

IMMORTALITY AND TOTIPOTENCY

"Immortality" is used here to mean the capacity to divide indefinitely, and "totipotency" is defined as the ability of a cell, such as zygote or spore, to create a new individual. We use the term "pluripotent" to mean the ability of a cell lineage to produce cells that can differentiate into all the cell types (but not into a new functional individual); lastly, "multipotency" refers to the potential of one cell to differentiate into more than one cell type.

Immortality and totipotency are basic life traits. In unicellular forms, they are manifested or expressed in all cells; cells have the potential both to divide indefinitely (i.e., they are potentially immortal) and to create new individuals, either asexually or sexually (i.e., they are totipotent). In unicellular individuals, immortality and totipotency are thus coupled at the cell level. In multicellular individuals, on the other hand, only one or a few cell lineages manifest both immortality and totipotency; most other cell lineages have only certain degrees and combinations of potential for cell division and differentiation. For instance, in groups without an early segregated germ line, the somatic cell lineages are incapable of continuous division or redifferentiation, and thus, they have to be replenished from one or a few pluripotent lineages that remain mitotically active throughout ontogeny and can also differentiate into germ cells (e.g., the interstitial I-cells in *Hydra;* Bode 1996) (fig. 21.3, *A*). In lineages with a germ line that is terminally differentiated in earliest ontogeny, various degrees of mitotic capacity (approaching immortality in some stem cell lineages) and/or potential for differentiation are maintained in the many multipotent somatic stem cells (i.e., secondary somatic differentiation; Buss 1987) (fig. 21.3, *B*).

Cellular Processes and Life Traits

Cell division is an important process in all cellular life forms. The mechanisms underlying cell division are, however, different between unicellular and multicellular individuals (fig. 21.2, *B*). In unicellular

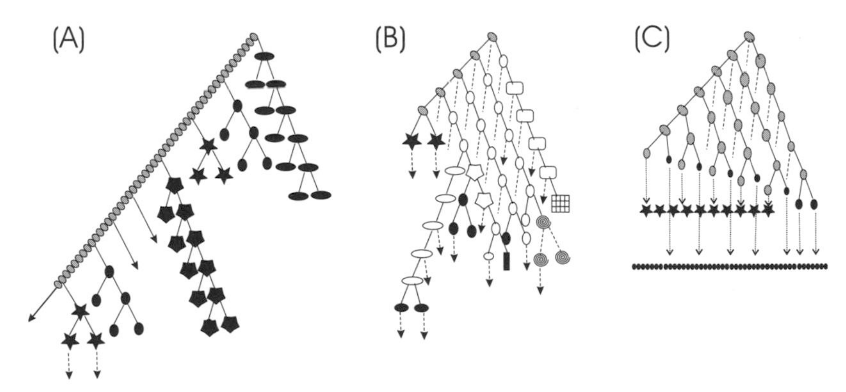

Fig. 21.3.—The reorganization of immortality and totipotency in three types of development. Gray ellipses denote totipotent/pluripotent cell lineages; open ellipses mark multipotent cell lineages; various solid forms indicate different differentiated cell lineages (stars represent the germ cells; the other shapes represent various somatic cell types); solid-headed arrows indicate postembryonic cell divisions in the corresponding cell lineages. *A,* The ancestral mode of development (Buss 1987): a mitotically active and pluripotent lineage gives rise to somatic lineages (which may or may not divide further), as well as to germ cells throughout ontogeny. *B,* The derived mode of development: a totipotent lineage gives rise to multipotent stem cells (which produce various cell types during ontogeny) and then differentiates into germ cells early in the development. *C,* The *V. carteri* mode of development: a totipotent lineage gives rise to a defined and early-segregated germ line as well as to somatic initials (*solid ellipses*) that have limited mitotic potential and produce somatic cells with no mitotic and differentiation potential (note the lack of multipotent stem cells, the presence of only one type of somatic cells, and the lack of postembryonic cell divisions).

individuals, cell division is strictly dependent on cell growth (cells do not divide unless a specific set size is achieved). In many multicellular forms, however, this is not always the case: factors other than cell size (such as intercellular or systemic signals) can trigger cell division. In addition, in unicellular forms cell division is strictly coupled with immortality, whereas in multicellular individuals, cell division has a limited and variable potential in most cell lineages (i.e., they are mortal) and is under the control of the higher-level individual.

Cellular Processes and Higher-Level Functions

Interestingly, cell division and cell growth have different roles at the level of the individual in unicellular compared to multicellular forms (fig. 21.2, *B*). In unicellular forms, every cell division results in the reproduction of the individual (cell division is strictly coupled with reproduction). In multicellular forms, cell division is uncoupled from the reproduction of the individual in most cells (i.e., cell divisions do not necessarily result in the reproduction of the higher level). Also, whereas in unicellular forms cell growth is the main contributor to the growth of the individual (with the exception of extracellular deposits in some lineages), in multicellular forms the growth of the individual is mostly achieved through increasing the number rather than the size of cells (with some exceptions in lineages where there is a significant increase in the volume of extracellular matrix, internal space, or even cell size).

Transition in Individuality

We argue here that the unicellular-multicellular transition and the emergence of individuality at a higher level requires (i) changing the temporal expression of vegetative and reproductive functions into a *spatial context,* (ii) *reorganizing* basic life traits (such as immortality and totipotency) between and within lower levels, (iii) *decoupling* processes from one another at the lower level (e.g., cell division from cell growth), (iv) decoupling certain cellular processes from functions and traits (e.g., cell division from reproduction and immortality), and (v) *co-opting* them for new functions at the higher level (e.g., the co-option of cell division for multicellular growth).

Changing Temporal into Spatial

During the transition to multicellularity with a germ-soma separation, the expression of vegetative and reproductive functions changes from a temporal to a spatial context (fig. 21.2, *A*). For instance, in *Chlamydomonas,* the reproductive phase follows the vegetative and cell growth phase and is paralleled by the loss of some of the vegetative functions, including motility. In *Volvox,* on the other hand, the *spatial disso-*

ciation of reproductive and vegetative functions between gonidia and somatic cells allows the two sets of functions to take place simultaneously; this is very important in these algae, in which the flagellar constraint leads to a strong trade-off between reproduction and vegetative functions.

Reorganizing Immortality and Totipotency

During the transition to multicellularity, and the emergence of individuality at the higher level, immortality and totipotency become restricted to one or a few specific cell lineages, namely, those involved in the reproduction of the higher level. However, many cell lineages maintain various degrees and combinations of mitotic and differentiation potential. This requires the reorganization (i.e., the differential expression) of these traits both among cell lineages and within a cell lineage. As discussed earlier, this reorganization has been achieved differently among the extant multicellular groups (fig. 21.3).

In *Volvox carteri,* immortality and totipotency are restricted to the zygote (if after a sexual cycle; not shown in fig. 21.3) or the asexual spore (i.e., gonidia; *a* in fig. 21.4), the 16 cells following the first four embryonic cell divisions (*b* in fig. 21.4), and the 16 germ line precursors (*c* in fig. 21.4) (Kirk 1994; Kirk et al. 1993). Both traits are lost in one-half of the 32-celled spheroid (*d* in fig. 21.4), as well as in the small cells (i.e., somatic initials) formed during the asymmetric divisions that take place in the germ line precursor lineage (*e* in fig. 21.4). The 16 large cells produced by the first asymmetric division of the germ line precursors (i.e., the germ line blastomeres; *f* in fig. 21.4) go on and divide asymmetrically another two or three times (each time renewing themselves and producing a somatic initial) and arrest mitosis two or three cell division cycles before the somatic blastomeres do. These 16 cells (*g* in fig. 21.4) will differentiate into the germ cells of the next generation. After a total of 11–12 cell divisions, the somatic initials stop dividing and differentiate into somatic cells (*h* in fig. 21.4), which have no mitotic or differentiation potential (they are terminally differentiated).

It is interesting that in *Volvox,* although immortality and totipotency have become fully restricted to the germ line (and reproduction and individuality at the higher level emerged), somatic lineages have no mitotic or differentiation potential. The two traits have been reorganized between germ and soma, but not within somatic cell lineages. The two sets of traits are still very linked in *V. carteri;* they are either both fully expressed (in gonidia) or both suppressed (in somatic cells). Noteworthy, the sequestration of the germ line was achieved without the evolution of secondary somatic differentiation processes (fig. 21.3, C). This is rather surprising, because it has been suggested that the evolution of an early-defined germ line was possible because, due to the

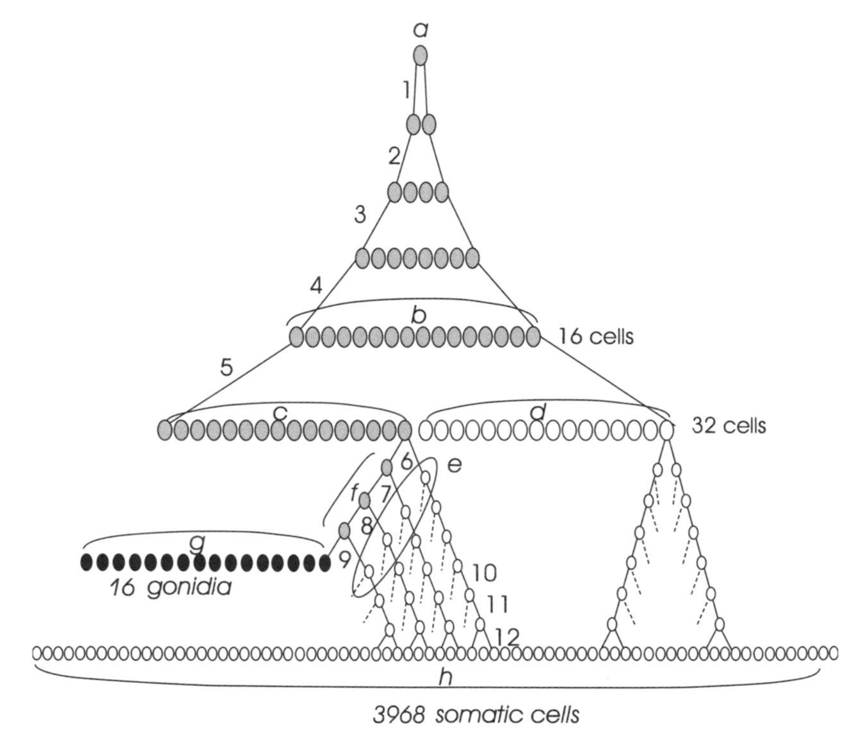

Fig. 21.4.—Schematic representation of the development and germ-soma separation in an asexual *V. carteri.* Gray ellipses denote the totipotent and multipotent cell lineages: the asexual spore, that is, gonidia (*a*), the 16 totipotent blastomeres (*b*), the germ line precursors (*c*), and the germ line blastomeres (*f*); white ellipses indicate unipotent (i.e., the somatic blastomeres and initials, *d* and *e*) and terminally differentiated somatic cells (*h*), and black ellipses indicate terminally differentiated reproductive cells, that is, gonidia (*g*). Note that somatic cells (*h*) have two distinct origins, from germ line blastomeres (*f*) via asymmetric divisions and from somatic blastomeres (*f*) via symmetric divisions. Numbers mark the succession of cell divisions in the embryo. Cells are not represented at scale (*a* is c. 2^9-fold larger than *g,* and there is a one-half reduction in cell size with every symmetric cell division); all divisions take place under the mother cell wall, in a rather rapid fashion without intervening growth (i.e., palintomy and multiple fission).

evolution of the multipotent stem cells and secondary somatic differentiation, the ancestral pluripotent germinative lineage was released from the task of producing the somatic tissues and was able to terminally differentiate into germ cells early in development (Buss 1987).

Decoupling Cell Division from Cell Growth

To ensure the functionality of the soma, factors other than cell size must be used to determine which cells divide, when, and how often. This requirement necessitates decoupling cell division from cell growth; consequently, a better and more finely tuned control on the replicative potential of the lower level can be achieved. However, this has not been accomplished in *V. carteri;* cell division is still strictly dependent on cell growth; reproductive cells have to increase 2^{12}-fold in volume before dividing 12 times to produce the final number of cells in the multicellular individual.

Decoupling Cell Division from Cell Reproduction

To ensure the reproduction of a cell group (and the heritability of the group traits), cell division has to be uncoupled from cell reproduction (i.e., the reproduction of the previously independent unicellular individual) and be co-opted for the reproduction of the higher level (the group). The ability to reproduce the group can be achieved either by all or by only some members of the group.

The case in which all cells have higher-level reproductive capabilities is best exemplified by a reproductive mode called autocolony, in which when the group or colony enters the reproductive phase, each cell within the colony produces a new colony similar to the one to which it belongs; cell division no longer produces unicellular individuals, but multicellular groups. This mode of reproduction characterizes the volvocacean green algae without a germ-soma separation, such as *Gonium* and *Eudorina.*

In *Eudorina,* all cells go through a vegetative and reproductive phase (i.e., divide and each produce a 32-celled embryo). However, cell division does not produce anymore a number of free unicellular individuals (such as in *Chlamydomonas*), but rather a new group; cell division has been thus decoupled from cell reproduction and has been coupled with the reproduction of the group in all members of the group. Nevertheless, cell division is still strictly dependent on cell growth: each cell will start dividing only after a 2^5-fold increase in size has been attained, and once cell divisions are initiated they will continue synchronously until 32 new embryos are formed. Although the stability, heritability, and reproduction of the higher level are ensured in this way, its individuality is not; because every member can be separated from the group, live independently, and create a new group, such a group is not the smallest physiological and reproductive autonomous unit and thus is not a true individual (in the sense used here).

The case in which only some cells have higher-level reproductive capabilities characterizes lineages with a separation between germ and soma. To achieve this, the coupling between cell division and reproduction is broken in most cells, namely, the somatic cells; they reproduce neither themselves (as former free-living unicellular individuals) nor the higher-level unit; cell division is decoupled from the reproduction of both the lower and higher levels. In this way, somatic cells lose their individuality as well as the right to participate in the next generation, but in doing so they contribute not only to the emergence of individuality at the higher level but also to the emergence of a new level of organization, the multicellular soma. Soma is thus the expected consequence of uncoupling cell division from reproduction in order to achieve individuality at the higher level. *V. carteri* follows this pathway; however, the way in which germ-soma separation was achieved is unique among multicellular forms (discussed later).

Co-opting Cell Division for Growth at the Higher Level

When cell division was decoupled from reproduction, this very important process became available for new functions. We suggest that this event was paralleled by the co-optation of cell division for a new function at the higher level, namely, the growth of the multicellular individual. Later, the use of cell division for more than cell multiplication, (i.e., division which "gives rise to more entities of the same kind"; Szathmáry and Maynard Smith 1997) may have provided the multicellular lineages with an additional advantage, namely, cell differentiation; indeed, in many multicellular lineages asymmetric cell divisions are involved in cell differentiation.

In *Chlamydomonas,* as in other unicellular individuals, cell division is coupled with the reproduction of the individual. Interestingly, in *Volvox,* although the coupling of cell division and reproduction has been broken in the somatic cells, cell division was not co-opted for the postembryonic growth of the higher-level individual; rather, cell division was simply turned off in somatic cells. The somatic cells lack the ability to divide postembryonically; all the cell divisions responsible for the final number of cells in the adult take place during embryonic development (the further growth of the young spheroid is accomplished only through small increases in cell size and through a massive deposition of extracellular matrix). The implications of this outcome are multiple and profound. A direct implication is that the soma in *Volvox* differs from the soma of most multicellular organisms. Because somatic cells do not divide, further growth and/or regeneration of the individual is not possible during ontogeny; in addition, because the somatic cells undergo senescence and genetically programmed cell death at the age of 5 days (Pommerville and Kochert 1981, 1982), the life span of the higher-level individual is limited to the life span of the lower-level somatic cell. Furthermore, although asymmetric cell divisions are involved in the differentiation of somatic cells, the way this process is achieved precludes further redifferentiation. The evolutionary implications of this aspect are discussed next.

Evolvability, Genotype-Phenotype Maps and Modularity during the Transition in Individuality in *Volvox carteri*

The transition to multicellularity has occurred multiple times in the evolutionary history of life; in addition, it has generally been followed by an increase in diversity and complexity. Nevertheless, although green algae (the charophytes, in particular) are the closest relatives of the more complex land plants, and 50–75 million years have presumably passed since the divergence of *Volvox* from its unicellular *Chlamydomonas*-like ancestors, none of the *Volvox* lineages appears to have attained high levels of complexity in spite of the multiple events that gave rise to multicellularity and germ-soma separation in this group. What are the reasons for this apparent "slowdown" in the evolutionary potential of this group? Could they be traced back to the events associated with the transition to multicellularity and the emergence of individuality at the higher level? In *Volvox,* the genotype-phenotype map that emerged during the transition to multicellularity reveals some aspects that might be of relevance to the evolvability of the lineage. We suggest that the way in which the lower-level genotype-phenotype maps are reorganized at the higher level can influence the potential for evolution of the newly emerged multilevel system.

Insights from Mutant Forms

Numerous mutant forms have been described in *V. carteri* (see Kirk 1998 for a review); in some forms, features that emerged during the transition to multicellularity, mainly individuality and germ-soma separation at the higher level, are affected. Two types of mutants are of relevance in this context, and they provide us with invaluable insight into how the transition to multicellularity has been achieved in this lineage.

In the somatic regenerator mutants, or Reg mutants, the somatic cells start out as small flagellated cells and then enlarge, lose flagella, and redifferentiate into gonidia. Thirty-nine mutants in four phenotypic classes have been investigated, and all had mutations at the same locus, *regA* (Huskey and Griffin 1979). The gene affected in these mutants has been shown to encode for an active repressor (Kirk et al. 1999) that targets at least 13 nuclear genes whose products are required for chloroplast biogenesis (Choi et al. 1996; Meissner et al. 1999). This finding suggests that the mechanism for the establishment of a stable germ-soma separation in *V. carteri* is based on preventing the somatic cells from growing enough to trigger cell division (by repressing chloroplast biogenesis in these cells; Meissner et al. 1999).

In another class of mutants, the Gls/Reg mutants (Huskey and Griffin 1979), all the cells (though far fewer than in the wild type, i.e., no more than 128 or 256) act first as somatic cells and then redifferentiate into reproductive cells. These mutants are a reversal to the ancestral *Eudorina*-like type of organization and represent a step back in terms of both complexity (there is no germ-soma separation in these forms) and individuality (each cell will produce a new colony; these mutants are "divisible," and thus, they are not true individuals). The Gls mutation has been mapped to a gene, *glsA,* which encodes a protein required for the asymmetric divisions responsible for the segregation of germ line blastomeres and somatic initials (*f* and *e,* respectively, in fig. 4) (Miller and Kirk 1999). Consequently, all cells are equal both in size and in potential for differentiation and undergo the ancestral *Chlamydomonas*-like pathway of acting first as vegetative and then as reproductive cells (Tam and Kirk 1991); it should be mentioned that this mutation is recovered only on a *regA* background that allows the growth and thus differentiation of somatic cells into reproductive cells.

In Reg mutants, both immortality and totipotency are regained by the somatic cells, and these cells "join" the germ line in participating in the next generation. On the other hand, cells in the Gls/Reg mutants never lose either immortality or totipotency. In neither of these mutants are the two traits expressed partially (e.g., limited mitotic capacity or multipotency). Furthermore, it is interesting that somatic mutant cells in which immortality is regained but not totipotency (analogous to the "cancer-like" mutant cells in various other multicellular lineages) are missing in *Volvox,* suggesting that immortality and totipotency are still strongly linked at the lower level in this lineage.

Genotype-Phenotype Maps

During the unicellular-multicellular transition, a new genotype-phenotype map has to be created to reflect the emergence of individuality at the higher level. It is rather intriguing that in *V. carteri* immortality can be regained and individuality can be destroyed by single mutations (such as in *regA* and *glsA*). In other multicellular lineages, such as humans, multiple mutations (each of which requires a minimum of 20–30 cell divisions) are required for immortality (i.e., cancer cells) to be regained (e.g., Wright and Shay 2001). The fact that single mutations have such large effects on individuality traits suggests that in *V. carteri* the genotype-phenotype map at the higher level has been realized through a rather small number of genetic changes. Any attempt to increase the evolvability of these lineages has to first affect the current genotype-phenotype map to allow increased variability of the traits associated with immortality and totipotency (so as to decouple them in the somatic cells) without affecting the individuality of the system (e.g., by evolving mechanisms to control these traits independently, thereby allowing cell replication and/or differentiation in the soma). In other words, the genotype-phenotype map has to at first become more robust (so that small genetic changes will not lead to the recreation of the maps associated with the previously independent lower levels, as is currently the case) but flexible (so as to allow improvement through mutation and selection).

To gain such properties a number of small-effect mutations, in a very precise order (such that the viability of the individual under selection is not affected) is required. However, the way in which cell division, cell growth, immortality, and potency have been reorganized in *Volvox,* as well as the way the genotype-phenotype map has been created at the higher level, makes the evolution of such traits more difficult. For example, the fact that (i) the decoupling of cell division from reproduction in somatic cells was not achieved by inventing new ways to control cell division, but rather by blocking it altogether, and (ii) the suppression of cell division was not achieved through evolving some new mechanisms but rather through inhibiting the growth of the cell strongly limits the evolution of traits that are dependent on these processes. These important complex sets of processes have not been decoupled from one another through their dissociation at the lower level and their co-option for new functions at the higher level, but rather through the suppression of some of the processes at the lower level (see discussion below); in this way, processes such as cell growth, cell division, and differentiation are not represented in the higher-level map and thus cannot contribute to phenotypic variability.

Improvement is expected to come from mutations that, for instance, allow the somatic cells to regain controlled mitotic activity and some degree of differentiation potential during ontogeny. To achieve this, the multiple fission type of division should be replaced by a binary type, such that cell divisions during adulthood do not result in the duplication of the entire organism (as they do in the *V. carteri* mutants in which somatic cells regain mitotic capabilities); in addition, a binary type of cell division would allow a more finely tuned increase in size, via small increments. In this way, more phenotypic variability can be achieved and become available for selection. It should be mentioned that the multiple fission type of division is a derived trait, which is thought to have evolved through the modification of the cell cycle via very conserved type of proteins involved in the key pathway that controls both cell division and differentiation in animal cells, namely, the retinoblastoma (RB) family of tumor suppressors (e.g., Sage et al. 2000). Mutations of this gene in *Chlamydomonas reinhardtii* result in the initiation of the cell cycle at a below-normal size, followed by an increased number of cell divisions (Umen and Goodenough 2001). Such an alteration of the cell cycle might have been involved in the evolution of the multiple fission type of cell division, which is considered a precondition for the origin of multicellularity in *Volvox* (Kirk 1998).

If this is the case, it would argue for another example of achieving an important trait at the higher level (i.e., multicellularity) through a small number of genetic changes, and thus for the potential instability and inflexibility of the higher-level genotype-phenotype map that emerged in this way.

Properties that reduce constraints on change are thought to be very important for the evolvability of a system by conferring flexibility and robustness on processes and consequently increasing nonlethal phenotypic variation and evolvability (Kirschner and Gerhart 1998). Among these, weak linkage contributes greatly to constraint reduction; in *V. carteri,* however, the linkage (i.e., the dependence of one process on another) between some processes is still very strong, which increases the constraint on change, decreases the potential for phenotypic variation unevenly among the organism's activities, and reduces the evolvability of the lineage.

"Divide and Rule": Dissociate and Control

The fact that in unicellular forms cell division does not occur in the absence of cell growth, and growth to a set size unconditionally triggers cell division, suggests that in these lineages the two processes are part of a single functional module, or two strongly linked modules that are dissociated in time during the life cycle of the individual. Likewise, the vegetative and reproductive functions are realized by the same cell (i.e., they are associated in space), and the latter is dependent on the former (i.e., they are coupled); however, they cannot take place simultaneously (they are dissociated in time). The dissociation in time of these processes can be seen as analogous to the dissociation in timing of specific modular interactions during the development of complex multicellular organisms.

In contrast, distinct cell types perform these two main functions in multicellular forms with germ-soma separation; the two sets of functions are thus dissociated in space, among different cell lineages, and they can be realized simultaneously. We suggest that during the transition to multicellularity, the translation of the temporal dissociation of certain processes and functions into a spatial one can be accomplished via the dissociation, recombination, and redeployment of modules or domains associated with these functions in a spatial (rather than temporal) context at the higher level. In other words, during the unicellular-multicellular transition, the ancestral temporal linkages, such as between cell growth and cell division, between cell division and reproduction, and between growth and reproduction, have to be broken and these processes and functions reorganized in a spatial context. For instance, in some cells (i.e., the somatic cells) the domain associated with cell division has been decoupled from cell reproduction and recombined into a new functional module at the higher level, namely, that associated with multicellular growth (in somatic cells, cell division now shares the most pleiotropic interactions with somatic growth; the two domains are linked to each other more closely than they are with other modules). In other cells (e.g., germ line) cell division has been decoupled from cell reproduction but co-opted for reproduction at the higher level. In this way, multicellular growth and reproduction become dissociated in space rather than in time. Likewise, the decoupling of the modules associated with vegetative and reproductive functions between soma and germ allows for the two sets of functions to be dissociated in space and thus to be achieved simultaneously and independently (not successively, as in unicellular forms). The other two integrated domains of the ancestral module, namely, cell growth and cell division, have also been dissociated; in a multicellular organism, cell division and cell growth are not necessarily dependent on one another. Furthermore, cell division (i.e., asymmetric cell division) has been redeployed in the context of a multicellular organism with distinct cell types and has been co-opted for cell differentiation.

The modular domains associated with the two complex sets of traits, immortality and totipotency, appear to have also been dissociated and recombined or redeployed during the transition to multicellularity. The domains associated with immortality and totipotency have themselves been further dissociated into subdomains, such that the potential for cell division and differentiation can be controlled and expressed outside the context of immortality and totipotency, respectively. Somatic cell lineages either can be mitotically active throughout ontogeny (e.g., some stem cell lineages) or can maintain only a reduced (and/or preset) mitotic potential, with or without any potential for differentiation; if they do have such potential, they can differentiate in one or more cell types, depending on the type of cell lineage. Therefore, interactions between modular domains associated with immortality and totipotency have been spatially dissociated in a multicellular organism, both between germ (in which immortality and totipotency are still coupled) and soma (in which the two traits are uncoupled), as well as among the various somatic cell lineages which can enjoy distinct combinations and degrees of replicative and differentiation potential.

The Gordian Knot and Evolvability in *Volvox*

How have the domains and modules associated with cell division, cell growth, immortality, and totipotency become reorganized during the transition to multicellularity in *Volvox*? It is interesting that a single mutation, in the *regA* gene, results in the expression of reproductive traits (both immortality and totipotency) in the somatic cells; thus, *regA* can be seen as a master regulatory gene for reproduction (analogous, for example, to the master control for the complex formation of the eye, *eyeless,* in *Drosophila;* Halder et al. 1995), and the two sets of traits can be seen as part of the same module. Furthermore, it is noteworthy that *regA* manifests its effect on the reproduction of the individual indirectly, by suppressing cell growth, which in turns blocks cell division in somatic cells. Therefore, because a single gene, *regA,* affects both cell growth and division in *V. carteri,* this argues that the two processes are associated with domains of the same genetic module with strong pleiotropic effects within.

Thus, although the switch from a temporal to a spatial dissociation of certain domains has been accomplished in *V. carteri,* and a germ line (with immortality and totipotency) and a soma have evolved, the reorganization of the domains associated with these complex sets of traits at the higher level was achieved in a rather peculiar way. *Volvox* was not able to dissociate and control (i.e., differentially express) the ancestral module associated with immortality and totipotency in somatic cells; instead, in *Volvox,* both domains are entirely suppressed in the somatic cells. Furthermore, the suppression of both domains was achieved by acting on a single process, namely, cell division. Moreover, the way in which *Volvox* suppressed cell division was not by acting directly on the domain associated with the mitotic potential of the cells but, rather, indirectly by acting on a domain that was still very linked to it—that is, that associated with the growth of the cell. By suppressing cell growth in somatic cells, cell division is repressed and the potential for gaining immortality and totipotency is "under control"; however, this type of "ultimate" control later interfered with the potential for evolution in this lineage (discussed below).

The mechanism that is responsible for germ-soma differentiation in *V. carteri* reveals another peculiar way of ensuring the emergence of reproduction at the higher level. Although cell differentiation involves asymmetric cell divisions, they are not involved in the differential segregation of germ line factors (such as the P granules in the nematode *Caenorhabditis elegans;* e.g., Seydoux and Schedl 2001); rather, asymmetric divisions ensure that the gonidia precursors remain large enough that the capacity to grow and further divide is not lost (as it is in the somatic precursor cells, due to the expression of *regA*). Thus, the way in which asymmetric cell division determines the cell fate of the somatic cells is by acting on the ancestral linkage between cell growth and cell division; it does not involve new mechanisms or new pathways of gene regulation.

We suggest that the developmental path observed in *V. carteri* is a consequence of its "inability" to dissociate lower-level modules and recombine or redeploy certain domains into new functional modules at the higher level; as an alternative strategy to ensure and maintain individuality at the higher level, *V. carteri* entirely suppressed domains of some of these modules at the lower level, including cell growth and cell division. The knot that could not be untied was cut: a difficult problem was solved by a quick and decisive action. In this way, the risk of regaining immortality and totipotency at the lower level (as exemplified by the somatic regenerator mutants) was somewhat avoided; but so were other processes, including postembryonic growth and cell differentiation. By completely suppressing the domains associated with cell growth and cell division in the somatic cells, certain sets of processes and traits were not recombined or redeployed in the new context and were not co-opted for new functions at the higher level. Unfortunately, these traits proved to be important for the evolutionary adaptability of a multicellular lineage. Without them, *Volvox* did not and will not easily attain higher levels of complexity. Due to its unique type of soma, *Volvox* is missing more than the ability to grow, regenerate, or live longer (whose lack evidently does not constitute strong disadvantages in the environment to which these algae are adapted, namely, temporal aquatic habitats).

An important evolutionary consequence of modularity is allometry; this occurs when different parts of the body grow at different rates. Allometry can generate evolutionary novelty by small, incremental changes that eventually can cross developmental thresholds; a change in quantity can become a change in quality (e.g., Brylski and Hall 1988). Under the constraint of multiple fission and palintomy, the body parts in *V. carteri* grow at the same rate, so the potential for generating novel traits in this way is not possible in this lineage. In addition, without a mitotically active multipotent stem cell lineage or secondary somatic differentiation there is less potential for cell differentiation and further increases in complexity.

Volvox managed to dissociate the vegetative functions (motility in particular) from the reproduction of the multicellular individual such that both selective advantages, namely, large size and mobility, are achieved. However, although the solution found provides the lineage with the immediate increase in fitness, it affected the potential for modularity to participate in further altering developmental processes to increase the evolutionary adaptability of the lineage. Thus, evolutionary modularity was traded off for functional modularity. The inability to dissociate some of the domains of the lower-level modules might be reflected in the developmental constraints and the low degree of freedom of the phenotype in this lineage, especially with respect to body size (Koufopanou and Bell 1991).

Concluding Remarks

The transition to multicellularity has happened numerous times in the evolutionary history of eukaryotes; of some 23 protist groups, 17 have multicellular representatives (Buss 1987). However, only three major groups, namely, fungi, animals, and plants, have achieved high levels of complexity. In addition, the extant groups appear to vary in their levels of diversity, suggesting distinct potentials for evolutionary adaptability, or evolvability. Various processes, such as modularity, robustness to genetic variability (Conrad 1990), robustness to developmental variation (Kirschner and Gerhart 1998), and heritability of fitness (Michod 1999; Michod et al. 2003), play important roles in evolvability. Here, we suggest that some of these processes gain new dimensions in the context of evolutionary transitions in individuality, and that the potential for further evolutionary adaptability of a lineage might be at some extent influenced by the way that the transition in individuality has been achieved.

A new genotype-phenotype map has to be created at the newly emerged higher level through the reorganization of the genotype-phenotype maps of the previously independent lower levels. Modularity plays a crucial role in this process; *the way in which modules become dissociated at the lower level and recombined or redeployed at the higher level is reflected in the flexibility and robustness of the newly emerged higher-level genotype-phenotype map*. Some modules are more resistant to dissociation than others; if undissociable, their domains might not be represented in the genotype-phenotype of the higher level and thus cannot contribute to phenotypic variability. Moreover, the strong linkage between modules at the lower level can be reflected in developmental constraints at the higher level.

The differential expression of immortality and totipotency traits between cell lineages and among phylogenetic groups is reflected in the various developmental programs in the extant multicellular lineages. The co-option of cell division for growth and cell differentiation at the higher level sets the premises for the evolution of soma and increase in complexity in lineages with a germ-soma separation. Likewise, the decoupling of cell division from cell growth allowed a better control of the replicative potential at the lower level, and thus a better functionality of the higher level. Selfish mutants that occur at the lower level and threaten the individuality of the higher-level might be indicative of the way in which individuality has been achieved in a particular lineage, as well as of the way that modules have been dissociated and certain domains co-opted for new functions at the higher level. Lastly, the diversity in developmental types and complexity levels among multicellular lineages might represent outcomes of distinct strategies to reach "good solutions" to various problems associated with the transition in individuality. The specific paths, however, can interfere with the potential for further evolution of a lineage. *Differences in evolvability among lineages might therefore be traced back to early events associated with the transition in individuality.*

Summary

During evolutionary transitions in the units of evolution, individuality emerges at a new and higher level. Here, we argue that *the transition from unicellular to multicellular organisms requires the reorganization at the higher level of certain basic life properties, such as growth, reproduction, immortality, and totipotency, as well as of the cellular processes associated with them (e.g., cell division and cell growth).* Furthermore, we suggest that the way in which this reorganization is achieved is not only instrumental for the emergence of individuality at a higher level but can also affect the potential for evolution, that is, evolvability, of the newly emerged higher-level unit. We use the volvocalean green algal group to argue that during the unicellular-multicellular transition (i) fundamental processes and functional modules have to be dissociated at the lower level and recombined or redeployed to ensure the emergence of individuality and new functions at the higher level; (ii) although modularity is generally expected to improve evolvability, during transitions in individuality this expectation is complicated and sometimes compromised by constraints at the lower level; and (iii) the way in which complex sets of traits (and the genotype-phenotype maps associated with them) are reorganized during the transition in individuality affects the flexibility and robustness of the new genotype-phenotype map which emerges at the higher level, and can interfere with the potential for further evolution of the lineage. We think that the unique way in which cell division, cell growth, immortality, and totipotency have been reorganized in the multicellular green alga *Volvox carteri,* as well as the way in which a new genotype-phenotype map has been created at the higher level, limited the evolvability of this lineage.

References

Altenberg, L. 1995. The schema theorem and the Prices's theorem. In *Foundations of genetic algorithms 3,* ed. D. Whitley and M. D. Vose, 23–49. Cambridge, Mass.: MIT Press.

Bell, G. 1985. The origin and early evolution of germ cells as illustrated by the Volvocales. In *The origin and evolution of sex,* ed. H. O. Halvorson and A. Monroy, 221–256. New York: Alan R. Liss.

Bode, H. R. 1996. The interstitial cell lineage of *Hydra:* a stem cell system that arose early in evolution. *J.Cell Sci.* 109:1155–1164.

Brylski, P., and B. K. Hall. 1988. Ontogeny of a macroevolutionary phenotype: the external cheek pouches of geomyoid rodents. *Evolution* 42:391–395.
Buss, L. W. 1987. *The evolution of individuality.* Princeton, N.J.: Princeton University Press.
Choi, G., M. Przybyiska, and D. Straus. 1996. Three abundant germ line–specific transcripts in *Volvox carteri* encode photosynthetic proteins. *Curr. Genet.* 30:347–355.
Conrad, M. 1990. The geometry of evolution. *Biosystems* 24:61–81.
Halder, G., P. Callaerts, and W. J. Gehring. 1995. Induction of ectopic eyes by targeted expression of the eyeless gene in *Drosophila. Science* 267:1788–1792.
Huskey, R. J., and B. E. Griffin. 1979. Genetic control of somatic cell differentiation in *Volvox. Dev. Biol.* 72:226–235.
Kirk, D. L. 1994. Germ cell specification in *Volvox carteri.* In *Germline development,* Ciba Symposium 184, ed. J. Marsh and J. Goode, 2–30. Wiley: Chichester.
Kirk, D. L. 1998. Volvox. Molecular genetic origins of multicellularity and cellular differentiation. New York: Cambridge University Press.
Kirk, M., A. Ransick, S. E. McRae, and D. L. Kirk. 1993. The relationship between cell size and cell fate in *Volvox carteri. J. Cell. Biol.* 123:191–208.
Kirk, M., K. Stark, S. Miller, W. Muller, B. Taillon, H. Gruber, R. Schmitt, and D. L. Kirk. 1999. regA, a *Volvox* gene that plays a central role in germ soma differentiation, encodes a novel regulatory protein. *Development* 126:639–647.
Kirschner, M., and J. Gerhart. 1998. Evolvability. *Proc. Natl.Acad.Sci.U.S.A.* 95:8420–8427.
Koufopanou, V. 1994. The evolution of soma in Volvocales. *Am. Nat.* 143:907–931.
Koufopanou, V., and G. Bell. 1991. Developmental mutants of *Volvox:* does mutation recreate the patterns of phylogenetic diversity? *Evolution* 45:1806–1822.
Larson, A., M. Kirk, and D. L. Kirk. 1992. Molecular phylogeny of the volvocine flagellates. *Mol. Biol. Evol.* 9:85–105.
Margulis, L. 1981. Symbiosis in cell evolution. San Francisco: W. H. Freeman.
Maynard Smith, J., and E. Szathmáry. 1995. The major transitions in evolution. San Francisco: W. H. Freeman.
Meissner, M., K. Stark, B. Cresnar, D. L. Kirk, and R. Schmitt. 1999. *Volvox* germline-specific genes that are putative targets of RegA repression encode chloroplast proteins. *Curr. Genet.* 36:363–370.
Michod, R. E. 1999. Darwinian dynamics, evolutionary transitions in fitness and individuality. Princeton, N.J.: Princeton University Press.
Michod, R. E., A. Nedelcu, and D. Roze. 2003. Cooperation and conflict in the evolution of individuality. 4. Conflict mediation and evolvability in *Volvox carteri. Biosystems* 69:95–114.
Miller, S., and D. L. Kirk. 1999. glsA, a *Volvox* gene required for asymmetric division and germ cell specification, encodes a chaperone-like protein. *Development* 126:649–658.
Pommerville, J., and G. Kochert. 1981. Changes in somatic cell structure during senescence of *Volvox carteri. Eur. J. Cell Biol.* 24:236–243.
Pommerville, J., and G. Kochert. 1982. Effects of senescence on somatic cell physiology in the green alga *Volvox carteri. Exp. Cell Res.* 140:39–45.
Sage, J., G. J. Mulligan, L. D. Attardi, A. Miller, S. Chen, B. Williams, E. Theodorou, and T. Jacks. 2000. Targeted disruption of the three Rb-related genes leads to loss of G1 control and immortalization. *Genes Dev.* 14:3037–3050.
Santelices, B. 1999. How many kinds of individual are there? *Trends Ecol. Evol.* 14:152–155.
Seydoux, G., and T. Schedl. 2001. The germline in *C. elegans:* origins, proliferation, and silencing. *Int. Rev. Cytol.* 203:139–185.

Szathmáry, E., and J. Maynard Smith. 1997. From replicators to reproducers: the major transitions leading to life. *J. Theor. Bio.* 187:555–572.
Tam, L. W., and D. L. Kirk. 1991. The program for cellular differentiation in *Volvox carteri* as revealed by molecular analysis of development in a gonidialess/somatic regenerator mutant. *Development* 112:571–580.
Umen, J. G., and U. W. Goodenough. 2001. Control of cell division by a retinoblastoma protein homolog in *Chlamydomonas. Genes Dev.* 15:1652–1661.
Wagner, G. P., and L. Altenberg. 1996. Complex adaptations and the evolution of evolvability. *Evolution* 50:967–976.
Wright, W. E., and J. W. Shay. 2001. Cellular senescence as a tumor-protection mechanism: the essential role of counting. *Curr. Opin. Gen. Dev.* 11:98–103.

22 Symbiosis, Evolvability, and Modularity

KIM STERELNY

Evolvability

Inheritance Systems

This chapter explores the connections between inheritance systems, evolvability, and modularity. I argue that the transmission of symbiotic microorganisms is an inheritance system, and one that is evolutionarily significant because symbionts generate biologically crucial aspects of their hosts' organization through modular developmental pathways. More specifically, I develop and defend five theses.

1. Inheritance is multiple. Any mechanism whereby members of generation N influence their offspring in generation $N + 1$ in ways that tend to make those offspring resemble their parents is an inheritance mechanism. But not all inheritance mechanisms are of equal importance in the evolution of biological diversity and complex adaptation. Genetic inheritance is only one of the systems that supports cross-generation similarity. But it is one of exceptional importance. It is ancient. It is (almost) universal. And it is highly evolvable.

2. The vertical transmission of symbionts, especially of symbiotic microorganisms, is also an evolutionarily important inheritance mechanism.

3. Though symbiotic transmission is an inheritance system, it is one with a character very different from that of the transmission of genes. Genetic inheritance is an information-based system, whereas symbiotic transmission is a sample-based system.

4. Evolution based on sample-based inheritance has evolutionary dynamics different from those based on information. Sample-based inheritance has less variance. But it is more robust and more modular than inheritance based on information.

5. Sample-based inheritance has the potential to support evolutionary innovation. It is highly evolvable, essentially because samples gen-

erate their phenotypic effects in ways that are relatively independent of other developmental resources.

In short, the aim of this chapter is to defend a connection between sample-based inheritance, evolvability, and modularity. I begin by sketching out the reasons for treating genetic inheritance as an information-based system and then link genetic inheritance to evolvability. Inheritance systems that make evolutionary changes in a lineage possible must have a set of important characteristics; they meet evolvability conditions. Genetic inheritance, but not only genetic inheritance, has those characteristics to a high degree. In the final sections of the chapter I develop a case for thinking that symbiotic microorganism transmission often meets them too.

Genes and Information

Though it is very widely assumed that the genome of a fertilized egg consists of a set of instructions for making an adult organism, and that the flow of genes between the generations is the flow of information about how to construct phenotypes, it has proved surprisingly hard to defend that view. The trouble has been that the most unproblematic notion of information is too weak to sustain this view of the role of the genes in development and evolution. According to the simplest conception of information, information is covariation between a signal and a source. If the condition of the peacock's tail covaries with the number, variety, or virulence of that peacock's parasites, then that tail carries information about the peacock's parasite load. The genes in the fertilized egg, the signal, carry information about the developed organization, the source, if aspects of egg genotype predict aspects of the phenotype. This is *predictive information.*

Genes certainly carry predictive information about the organization of the organism. But so do many nongenetic factors. The island on which a seabird hatches predicts where that fledgling will itself lay eggs. The plant on which butterfly eggs hatch carries information about where the developed creature will lay eggs herself. The genes an organism inherits are one of many developmental resources that predict its organization. So though genes do carry predictive information, that does not explain the special role genes are thought to have in evolution and development.

So if the genes in a fertilized egg are a set of instructions, the information in those instructions is not just predictive information. A particular gene complex does not just covary with an aspect of an organism's organization, and hence predict it; it is selected to have those effects. It has those developmental effects by design. Such genes carry *bioinformation* about their selected effects: "gene Q carries bioinformation about trait T = gene Q has the biological function of making T = gene Q has evolved and/or is sustained in the population because of its T-making propensities" (Sterelny et al. 1996).

Many developmental resources carry predictive information without carrying bioinformation. Islands are not adapted to help seabirds return to their own natal sites. However there are nongenetic adapted developmental resources. Chemical gradients in the early developing egg are inherited from the mother and have the function of telling cells where they are. Thus, they control the initial differentiation of the embryo. Moreover, that is their function. Should we say that the gradients, like the genes, carry bioinformation? Maynard Smith (2000) argues that genes do not just have developmental biofunctions; those biofunctions are *arbitrary*. His example is an inducer gene, which switches off a repressor protein made by a regulatory gene. Any protein that would bind to the repressor protein, altering its shape, would do. Many proteins could do that. The same gene products can act as either inhibitors or inducers; for inhibitors can have as their target other inhibitors, so when they bind to their targets they release the inhibition. We cannot read the meaning of a gene from its physical structure.

Moreover, genes, especially structural genes, are part of gene reading systems: there are mechanism (including other genes) which systematically map differences in structural genes into differences in organization. The mouse *eyeless* gene is part of such a system. It does not carry information about how to make a mouse eye. It reads this information from other genes. For when the mouse *eyeless* gene is transplanted to a fruit fly and activated, the result is an ectopic *fruit fly* eye. A fully developed compound eye will form on a fruit fly leg. This suggests that *eyeless* is part of a system capable of reading many messages. In mice, this system executes mouse-eye-making instructions; in fruit flies, it executes compound-eye-making instructions. The arbitrariness of the genetic message, together with the existence of developmental mechanisms which map gene differences onto organizational differences, helps explain why there are so many possible genetic messages. And the genetic channel is of great evolutionary importance in part because it is so rich. Lineages which have evolved inheritance channels through which many instructions can flow are, for that reason, highly evolvable. But the range of possible messages is not the only factor important to evolutionary potential.

Evolvability Conditions

Evolutionary change generated by natural selection depends on variation, fitness differences, and heritability. But though these conditions are necessary, they are not sufficient. In particular, if complex adaptations and a diverse biota are to evolve, there are important conditions on the mechanisms that generate cross-generation similarity. The flow of replicators between the generations should meet three general specifications.

First, the system must somehow *block outlaws*. Complex living systems depend on the cooperation of many components. The division of labour and adaptive specialization seen in, for example, an ant nest depend on the linked reproductive fate of the ants within the nest. If some ants in the nest could reproduce independently of others, very likely circumstances would arise in which the reproductive interests of individuals within the nest would diverge, and defection would undermine cooperative integration. Similarly, if an organism is to be built by a team of replicators cooperating together, those replicators must have a shared evolutionary fate. For otherwise the temptation to defect will undermine cooperative organism building.

Second, the replication system should ensure the *stable transmission* of phenotypes over the generations. Evolutionary innovation depends on cumulative selection. If complex adaptations are to evolve, biological organization must be reliably rebuilt over many generations, not just a few.

Third, selection depends on the *generation of variation*. If evolution is to build a disparate biota, or one characterized by adaptive complexity, the replication system must have the potential to generate a large number of distinct organizations. Given these desiderata, I suggest that a highly evolvable replication system should have the following characteristics.

Antioutlaw Conditions

C1. Replicators should be transmitted vertically. Replicators should flow from parents to offspring, and to them alone.[1]

C2. Replicators should be transmitted simultaneously.

C3. The transmission of the replicator set should not be biased. Either all an organism's replicators are transmitted to each descendant, or each replicator has an equal chance of being transmitted to each descendant.

Stability Conditions

C4. The copy fidelity of the generation of replicators from generation to generation should be high.

C5. The replicator-to-organization map should be robust. To the extent that the causal channel from replicator to organization depends on context, both internal and external, that context should be stable and predictable.

Generation of Variation

C6. The array of possible replicator sets should be very large, possibly even unbounded.

C7. The effect of a replicator on the biological organization of its carrier should normally be well behaved. That is, the replicator-organization map should be smooth. A map is smooth if a small change in the replicator set generates a small change in biological organization; and the smoother the map, the more evolvable the lineage using that inheritance channel.

C8. The generation of biological organization from the replicator set should be modular. The replicators as a whole should not generate the biological organization of the organism as a whole. Rather, replicators, or small sets of replicators, should be designed so that they make a distinctive contribution to the generation of one or a few traits, and relatively little distinctive contribution to others.

I discussed these conditions in some detail in Sterelny 2001. So I shall comment in detail only on C8, since modularity is central to my case for the evolutionary importance of symbiotic inheritance. But first a few words on the others, beginning with antioutlaw criteria. The problem of defection was first noted in the context of group selective explanations of altruism (Sober and Wilson 1998), and it has subsequently been applied to the evolution of multicelled organisms (Buss 1987; Maynard Smith and Szathmary 1995; Michod 1999). To the extent that complex adaptation requires the cooperative integration of elements which have separate evolutionary histories, adaptation can be undercut by defection. The evolution of the genome from independent replicators, the evolution of the eukaryotic cell, the evolution of multicelled organisms, the evolution of cooperative symbiosis, and the evolution of complex social groups all faced potential defection problems. The most evolvable inheritance systems are those that suppress outlaws. Outlaws, in turn, are replicators that go it alone.

Vertical transmission is important for this same reason, since one way a replicator can go it alone is through horizontal or oblique transmission. But though vertical transmission is important, it is not sufficient in itself to police outlaws. The inclusion of a single-celled stage in the life cycle of multicelled organisms may partly be an adaptation against outlaws.[2] For it is a way of preventing the evolution of somatic cell parasites—that is, somatic mutations which induce cell lineages to promote their own prospects of replication at the expense of the fitness of the organism. If organisms routinely reproduced via multicellular propagules, selection could favor mutant cell lineages that competed for access to the germ line. Early segregation of the germ line, and single-celled propagules, minimizes such within-organism conflict. The single-celled stage also exposes other somatic mutations to selection; once more, if organisms reproduced via multicelled buds, malignant cell lineages, mixed with nonmalignant ones, could have access to the next generation (Grosberg and Strathmann 1998).

Steven Frank has emphasized the risk of between-lineage competition in the context of vertical transmission in a series of papers on the evolution of virulence and avirulence and extended it to genetic conflict and genetic inheritance (Frank 1989, 1995, 1996a, 1996b, 1996c). Mitochondria, for example, are vertically transmitted, as are many symbiotic microorganisms. But even vertically transmitted mitochondria can evolve outlawry, if there are different lineages within the one host, and if these can compete for access to the next generation. In such competition, the winner is unlikely to be the lineage that contributes optimally to the organism's fitness. Frank argues that unbiased transmission requires mechanisms which tend to damp down variation between the replicators transmitted from one generation to the next. There are such mechanisms in genetic inheritance: mitochondria are inherited only from the mother, and most cell lineages, with their new somatic mutations, are denied access to the germ line. As we shall see, there seem to be such mechanisms in symbiont transmission, too.

The stability criteria are uncontroversial. C4 demands high-fidelity copying of the replicator set—copying which is faithful enough to satisfy the replicator condition. That condition is satisfied if a replicator miscopied between, say, the F1 and the F2 generations is then faithfully transmitted in its new version to the F3 generation and beyond. Unless inheritance satisfies the replicator condition, cumulative selection, and the gradual evolution of complex adaptive structure, is not possible. For small improvements are not preserved as the basis for further cycles of variation and selection. Genes satisfy this condition, but not all cross-generation influences do. Michael Tomasello has argued that in most species social learning does not. If a macaque discovers a new way of opening hard fruits, that improvement will probably not be faithfully transmitted to the next generation (Tomasello 2000). If Tomasello is right, his insight is important precisely because it bears on the possibility of cumulative cultural evolution.

C5 demands high-fidelity use of the replicator set. Even if replicators were copied with high fidelity, if development were too sensitive to noise, small variations in the replicator set would not reliably correspond to small variations in organization, and cumulative evolution would not be possible. The same set should generate the same organism given the same signals.

Let us turn to the generation of variation. Selection depends on the existence of selectable variation. Hence, we need a rich array of replicator packages (C6). Maynard Smith argues for the centrality of this criterion: he argues that a crucial transition in evolution is the shift from limited systems of replication to unlimited systems. The argument links back to the importance of cumulative selection in adaptive change. Evolution by a series of small steps requires a rich supply of variation, and that requires a rich array of replicator sets, transmitted and used with high fidelity (Maynard Smith and Szathmary 1995; Szathmary and Maynard Smith 1997). But that is not all it requires. We also need to consider the conditions under which replicator packages are mapped onto selectable variation. C7, the requirement for a smooth replicator-organization map, flows from the work of Kauffman (Kauffman 1993, 1995). Selection is ineffective, unless small differences in replicator sets translate into small differences in organization, and then into small differences in fitness.

As we shall see after discussing modularity in the next section, conditions 1–8 specify a highly evolvable inheritance system. Genetic replication does not, of course, meet these conditions perfectly. No real system could. But it does satisfy them to a high degree. Gene transmission is vertical. It is simultaneous and early. It is not quite outlaw-proof, of course. But genetic replication largely solves the problem of preventing individual replicators from going it alone. The fidelity of genetic replication depends in part on how we identify genes, and how we define accuracy. But if we identify genes and measure the accuracy of their replication in any of the reasonable ways, genetic replication is a high-fidelity system. Equally, there is no serious doubt that genetic replication is rich. In concert with much else, it generates a vast range of biological forms. Let me turn now to modularity.

Evolvability and Modularity

Wimsatt, Müller, Wagner, Raff, Lewontin, and Dawkins have all defended the significance of developmental modularity (see Wimsatt 1980; Wagner and Altenberg 1996; Müller and Wagner 1996; Raff 1996; Dawkins 1996; Lewontin 1978). Wimsatt and Lewontin have pointed out that unless development is to some important degree modular, selection will be unable to move a lineage away from its current organization. That organization will become generatively entrenched (Wimsatt and Schank 1988; Wimsatt 2001). If the developmental program of an organism is holistic, then development of any given trait will be connected to that of many others. Hence, that trait cannot change without other changes. But as the number of changes in a organization goes up, so too does the probability that one of those changes will be disastrous. The more modular the developmental network is, the more contained are the consequences of change. Thus, Wimsatt and Lewontin have pointed out that adaptive change requires traits to be "quasi-independent"; there are at least some developmental trajectories that allow one to be changed without affecting others (Wimsatt 1980; Lewontin 1978). In short, the modularity of development is correlated with the evolutionary viscosity of a lineage. If development is holistic, phenotypes will be very viscous. If they are more modular, they will be less viscous.

But modularity plays a second role in evolvability. Modules are often, with modification, reused. Thus, Müller and Wagner claim that "The more we learn about molecular mechanisms of development in widely different organisms, the higher the number of conserved mechanisms that become known. Some of them do indicate homology of morphologically divergent characters. . . . Still others illustrate that highly conserved molecular mechanisms may be used in radically different development contexts, indicating that the machinery of development consists of modular units that become recombined during evolution" (Müller and Wagner 1996, 11). Developmental complexes once discovered can then be co-opted and used for other purposes. The invention of cell types illustrates these themes of flexibility and multiple use. Animals are mostly built from the same types of cell, so different structures can be built from similar cell toolkits. Evolution only had to discover the trick of making a certain type of cell once, and the same theme appears on other scales. The snake lineage did not depend on the invention of new means of vertebrate formation as extra vertebrae were added to their skeleton. But that, of course, depends on developmental modularity. Moreover, there are links between modularity, redundancy, and evolvability. If the replicator-organization map is modular, there is a fair chance that introducing redundancy into the *replicator set* through duplication will result in the duplicated, and hence redundant, *structure in the organism.* These extra structures will usually be deleterious, and selection will dispose of the resultant replicator set. But this will not quite always be true, and then the new structures can form the basis of genuine evolutionary novelty. One possibility is that the different body plans of the Metazoa depend on duplication followed by changes in the activation patterns of Hox genes (Holland and Garcia-Fernandez 1996; Gellon and McGinnis 1998).

There is good reason to believe that the gene-driven development of phenotypic traits exhibits a significant degree of modularity, though its full extent remains to be discovered. Wagner has explored the conditions under which a lineage could evolve toward more modular development through a combination of directional and stabilizing selection. If there is directional selection on one trait, and stabilizing selection on traits developmentally linked to it, there will be selection for mutations reducing those developmental connections (Wagner 1996). So there is no theoretical bar to the evolution of modular developmental systems. And there are empirical reasons to suppose that such systems have evolved. Gilbert, Opitz and Raff have recently argued for the revival of the concept of morphogenetic fields as the key organizing concept of developmental biology, and such fields are developmental modules, with rich internal connections and weak connections to other fields (Gilbert et al. 1996).

Though terminologies differ, such ideas are increasingly widely defended. Kirschner and Gerhart, for example, argue that extant metazoan lineages are highly evolvable because their developmental processes are compartmentalized, are weakly linked, and include exploratory mechanisms; these are all aspects of modularity. Muscle cell tissue development, for example, is compartmentalized. Muscle cell tissues develop via the intermediate form of myoblasts, and myoblast development is triggered by a linked sequence of four genes which probably evolved by duplication of an original single sequence. These genes are both compartmentalized and, as their products are cross-active, they are partially redundant: once any one reaches a threshold, the entire network becomes active and develops into a myoblast without needing further input (Kirschner and Gerhart 1998, 8424). The same is true of the development of the phylotypic stage in *Drosophila*. The development of the basic body axes requires only simple triggering conditions, and once these axes develop, they initiate a cascade that results in the formation of 50–60 spatial compartments that then develop relatively independently of their neighbours. This basic developmental organization depends only on simple and partially redundant genetic signals for its initiation (Kirschner and Gerhart 1998, 8425; Schmidt-Ott and Wimmer 2004).

So, tentatively, it seems reasonable to conclude that there is an unknown but significant degree of modularity in metazoan developmental pathways. Modularity, a little surprisingly, is connected with developmental robustness. The degree to which development is canalized remains to be settled. But modular mechanisms help insulate development against noise; indeed, that may be one of the selective advantages of modularity. This feature of modularity is seen most clearly in Kirschner and Gerhart's "exploratory mechanisms": developmental mechanisms which rely on feedback from the local cellular environment rather than on fine-grained genetic control. Development driven by such mechanisms is both modular and robust, buffering adaptive outcomes against developmental noise. Consider, for example, the genesis of the microtubule cytoskeleton. Among other roles, these structures mediate the segregation of the chromosomes to the spindle poles by hooking onto a specific structure on each chromosome. These tubules have to find chromosomal targets despite the fact that these are scattered in the cell, and despite the various sizes and shapes a cell can have. As so often happens, the trick is turned by a random generation and selective retention mechanism. Microtubules are generated in random orientations, but they are not stable, and they degenerate unless they hook a chromosome. This process is so robust, and so independent of other mechanisms, that it is insensitive both to ordinary developmental perturbations, allowing cell division to proceed normally, and to evolutionary change: "it need not be modified when the cell's morphology is modified by other mutations" (Kirschner and Gerhart 1998, 8422). Somewhat counterintuitively, mechanisms which support the stable replication of phenotypes also support their evolutionary lability.

Symbiosis, Inheritance, and Evolvability

There is no serious doubt that genetic inheritance is highly evolvable. The flow of genes across the generations is a similarity-generating mechanism of especial evolutionary significance. But it is not the only evolutionarily significant inheritance channel. A certain class of symbiotic transmissions is another—namely, specific mutualistic associations where hosts have adaptations to ensure that the next generation is appropriately stocked with its symbiotic associates. When these conditions are met, the host transmission of symbiotic passengers often meets important elements of these evolvability conditions. The transmission of these symbionts is often early. It is typically unbiased. It is often highly evolutionarily stable and mutually obligatory. The symbiotic microorganisms cannot survive alone, and neither can their partners. It is a mechanism with high fidelity. It is reliable, with often delicate adaptations to ensure successful transmission. A specific species associates with a specific species, sometimes so much so that the species branching pattern of the one models that of the associated clade. However, since each member of the partnership retains a good deal of metabolic and developmental integrity, there is reason to believe that development is both modular and robust. Even restricting that domain to obligatory vertical transmission, this category is important. It is not a minor quirk of a few clades.

So I think that the transmission of symbionts between host generations is inheritance. But it is not a flow of information. These are sample-based systems of inheritance; they are preformationist. A leaf cutter queen takes a sample of her fungus on her way to found a new colony. Termites acquire their cellulose digesters from anal secretions of their nest mates. They do *not* acquire bacteria-making instructions from their nest mates. Since such inheritance is sample based, it dispenses with information. Because the leaf cutter ant queen has a sample of the fungus that can be grown, the leaf cutter colony does not need information on how to make the fungus.

I conjecture that the difference between an arbitrary system and a sample-based system leads to two different evolutionary dynamics.[3] In sample-based systems, development is direct and modular. Inheritance is not arbitrary: the inherited resource prefigures its effect. The pathways through which mutualistic relations evolve from exploitative or neutral interactions are still far from clear. No doubt, as Ewald has argued, vertical transmission is important in linking the reproductive interests of the associates (Ewald 1994; Dawkins 1990), and so is the extent to which different lineages compete within the host (Frank 1996b). But whatever the character of the selective landscape, the effect of selection for mutualism is to amplify and entrench preexisting characteristics of the symbiont lineage (Frank 1995, 404). The fungal symbionts of the attine ants are preadapted for their role in the ants' gardens. But they cannot otherwise be incorporated into ant developmental programs. Bivalve chemoautotrophs cannot be recruited to repel echinoderms. For much the same reason, there is probably only a limited range of heritable variation in these sample-based systems: they are high-fidelity but low-variance systems. Finally, because development is both robust and modular, viable macromutations may be more common. Indeed, this is the only inheritance system in the Metazoa in which viable macromutations might occur at a significant frequency. Hence, sample-based inheritance may be one means by which a lineage can cross a fitness trench.

Information-based systems are arbitrary and are potentially extremely rich, with high fidelity and high resolution. But perhaps they are less modular and less robust. After all, there are trade-offs between sensitivity to base pair differences and developmental canalization. Moreover, the more complex and indirect the causal chain from primary replicator action to phenotypic effect, the more ways that effect can be altered. The arbitrariness of replicator function is linked to the fragility of that function; there are more places for the causal train to be derailed.

In this and the next sections I develop and illustrate these ideas about symbiotic inheritance by discussing three core cases. In "Outlaws," I then discuss the problem of outlaws, using *Wolbachia* as my illustration of a symbiont-like outlaw. In "Encoding Your Symbiont," below, I discuss a test case—a symbiont-like system which by my lights is an instance of information-based inheritance, not sample-based inheritance.

My paradigm cases of symbiotic inheritance illustrate a pattern of great evolutionary significance. These associations are important, not just in generating a biologically critical part of their host's organization, but also in enabling the host lineage to diversify. They are examples of high-fidelity transmission over long periods of time. And they show the connection between symbiotic inheritance and developmental modularity. More specifically,

1. In these cases a closely related group of host species is associated with a closely related group of symbiotic species. Sometimes the congruence between host and symbiotic phylogeny is so close that we can reasonably infer that the symbiotic association originated only once and has been inherited from deep in the host tree. Sometimes there is only partial congruence, and so there has been some lateral movement of the symbiotic associate and/or multiple origins of that symbiotic association. But all of these examples show stable replication of the symbiont over evolutionarily significant periods of time.

2. The symbiotic passengers generate a biologically critical part of the host organization. The alliance is obligatory for the host, and sometimes for the symbiont as well.

3. The phenotypic effect of the symbiont on the host organization remains constant as host and symbiont speciate together. The most natural explanation for the resilience of symbiont effect on host organization is that symbionts drive modular developmental pathways in building their distinctive phenotypic effects. That modularity explains their robustness. Even major evolutionary changes in the host lineage have not subverted the phenotypic effect of the symbiont on the host. Modularity, in turn, is surely connected to the fact that the effect of the symbiont on the host is not arbitrary. It is constrained and often predictable from the phenotype of free-living relatives of the symbiotic species. But the example of *Wolbachia,* as we shall see, suggests that the developmental robustness of the symbiont's impact on the host is a coevolutionary construction of some kind, rather than a passive side effect of the symbiont's biology. If (Frank 1995) is right, this coevolutionary construction occurs if the association is fortunate enough to reach a threshold of mutual benefit as a side effect of their preexisting phenotypes.

4. These symbiotic associations are of great evolutionary importance. For they open up ecological opportunity to the host lineage that would be otherwise closed. This is certainly true of the bivalve-chemoautotroph, insect-bacteria, and ant-fungus relationships. Modularity is important in evolution for at least two reasons. To the extent that development is modular, it prevents the developmental gridlock of evolutionary plasticity; that is, it prevents the generative entrenchment of the current phenotypes of the lineage. But also critical adaptations need evolve only once. The enzymes that allow the digestion of lignocellulose do not have to be invented anew in every lineage exploiting woody plants. Bivalves do not have to invent sulphur digestion

for themselves. Symbiosis can play the same role between reproductively isolated lineages that sex plays within them. It allows evolutionary changes with distinct origins to be brought together in a single organism.

I shall illustrate these patterns by discussing bivalve-bacteria and insect-bacteria associations, and also the classic association between the attine ants and their fungal partners.

Symbiotic Paradigms

Bivalve-Chemoautotroph Symbioses

In the late 1970s hydrothermal vent communities were discovered. These communities have a very unusual ecology. They are not based on primary production from photosynthesis. Instead, carbohydrates are produced using energy from sulphur compounds. The primary producers are not plants but chemoautotrophic bacteria living in symbiotic association with a number of metazoan groups. I shall concentrate on the association with bivalves, for this alliance is ancient (perhaps very ancient) and is found outside bivalves in vent communities. Anaerobic environments rich in sulphur compounds exist in a number of other habitats, and symbiont-dependent bivalves are found in them, too. These chemoautotrophs make life possible for their hosts,[4] who in turn have important morphological adaptations for housing them, for the bacteria are found in specialized structures in the gills.

This association exemplifies faithful vertical replication, but also the multiple origins of an association and/or occasional host switching. There seem to be two bacterial lineages involved in these symbiotic associations, though both are drawn from the Proteobacteria. Moreover, the bivalve species involved in symbiotic alliance with their partners are not a clade. There are symbiont-dependent bivalve species on widely separated, long-diverged branches of the bivalve family tree. But there is good evidence that this association is predominantly transmitted vertically, and across speciation events. For though the symbiotic bivalves as a whole do not form a clade, within the bivalves there are a number of symbiotic clades. Of the five bivalve families known to have symbiotic associates, in four, most—perhaps all—species depend on symbionts. Even in the fifth family (the Bathymodiolinae) the symbiotic species form a single lineage. So there are bivalve lineages specializing in chemoautotroph associations. That life history is inherited across speciation events. Moreover, in at least some cases, the symbiotic partners of clades of symbiont-dependent bivalves themselves form a clade. The Vesicomyidae and Mytilidae are both families of bivalves dependent on their symbiotic bacteria, and in each case their symbiotic partners form a clade (Distel 1998, 281).

So though this symbiotic association has multiple origins, and though the biological details of its transmission are not known, it is replicated with high fidelity. Indeed, there is some reason to believe that the association may be very ancient. Bivalve families that specialize in symbiotic association have ancient origins. The Lucinidae, the Solemyidae, and the Mytilidae are symbiotic families, and they are first found, respectively, in the Ordovician, the Silurian, and the Devonian. Moreover, there is paleoecological evidence that suggests that these lineages have always depended on symbiotic associates. The fossil communities of which they were a part are similar to living chemoautotrophic communities: "for example, solemyids, lucinids, and thyasirids form unique burrows whose features are specifically related to their simultaneous need to extract sulphide from deep in anaerobic sediments while obtaining oxygen from the overlying sea water. Studies of trace fossils and life positions indicate that these same adaptations were present in ancient species" (Distel 1998, 284).

The association of bivalves with chemoautotrophs illustrates the macroevolutionary pattern characteristic of symbiotic inheritance systems. This association is very stable. It is ancient. It is replicated with great fidelity. Within particular bivalve lineages, bivalve hosts have had long associations with particular lineages of bacterial passengers. This system of inheritance has made possible the invasion of habitats, and the exploitation of resources, that would otherwise be inaccessible to the bivalves. Bivalve capture of chemoautotrophic bacteria was an adaptive breakthrough that enabled it to invade a set of niches formerly inaccessible to the metazoan lineage.

Furthermore, the association illustrates developmental modularity. For upwards of 400 million years these bacteria have contributed in the same way to the organization of their associates. Four hundred million years of bivalve evolution have made no difference to the symbiotic contribution to their joint phenotype. For the other side of symbiont modularity is conservatism. Many adaptive innovations are further modified to become the basis of new rounds of adaptive change. Once vertebrate limbs were invented, in different lineages they were elaborated in various ways, with their functions being both changed and refined. The same is true of the invention of segmentation in arthropods, and of the invention of sensory and control systems. The development of symbiotic association is an adaptive breakthrough, but one not forming the basis for further elaboration or change of function (perhaps with one small exception: some bivalves host a second class of chemoautotrophs, the methanogens). My conjecture is that symbiont-based host traits are conservative because the inherited materials have an intrinsic rather than an arbitrary connection with their phenotypic upshot.

Insect-Bacteria Symbioses

Insect-bacteria relations are common. There are enormous numbers of insect species, and perhaps as many as 10% form symbiotic liaisons (Moran and Telang 1998, 295). Insect and bacteria show mutual adaptation. The insects house their guests in special cells, bacteriocytes, and often have quite elaborate adaptations for vertical transmission of the bacteria to their eggs. Bacterial lineages also show signs of adaptation to, and for, their hosts. Symbiotic *Buchnera* have gene duplications which massively boost the rate at which they produce specific amino acids for their hosts. In these bacteria, the gene that makes tryptophan is excised from the nucleus and replicated in multiple copies in plasmids instead. These strains of *Buchnera* make far more tryptophan than free-living relatives of the symbiont. These associations are obligatory for both host and symbiotic partner. Finally, these associations are ancient: there is good evidence that this association has been replicated with great fidelity over hundreds of millions of years. So this association, too, illustrates the macroevolutionary pattern characteristic of symbiotic inheritance.

The mechanisms which ensure that bacteria are passed on to the next generation seem to be very effective at ensuring strict vertical transmission. Each major clade of insects has coevolved with a specific clade of bacteria. Aphids, mealybugs, whiteflies, carpenter ants, weevils, tsetse flies, cockroaches, and termites are clades that exist in symbiotic association with specific clades of bacteria. The bacterial associates of, for example, the whitefly lineage themselves form a clade. Furthermore the two clades are congruent. The branching pattern in the whitefly lineage is reflected in the branching pattern of its associates' lineage. The same is true (though to a varying extent) of the other insect-associate lineages. These symbiotic associations were in place at the root of these clades, and they have been transmitted with great reliability through the history of these lineages. Moreover, some of these associations are probably very ancient. The cockroach-bacteria association may be 300 million years old; that of aphids and their partners 200 million years. These facts suggest that the vertical transmission of these associates is very strict indeed. Even over long periods of evolutionary history, host switching events are very rare or nonexistent.

The evolutionary importance of this association for the insect lineages is analogous to the importance of such associations for bivalves. For these insects typically subsist on a narrow and nutritionally unbalanced diet (sap, blood, and the like), and their associates have distinctive adaptations which compensate for that unbalanced diet. Their associates allow them to specialize on these specific resources. For that reason, though in many cases the specific nutrient the associate provides for the host is not known, there is little doubt that insects' symbionts are essential to their lives. Experiments that strip insects of their bacteria (for example, by giving them antibiotics) reduce host fitness (Moran and Telang 1998, 296), and in various lineages where the symbiont is not passed on to every egg, the result is growth suppression (Morgan and Baumann 1994).

Not surprisingly, since these bacteria are so crucial to their lives, hosts have adaptations that ensure that bacteria are transmitted to the next generation. For example, in aphids the bacteriocytes develop in close proximity to the developing egg. As those eggs develop, a passage forms between a bacteriocyte (which opens) and the egg surface through which the bacteria move. As the bacteriocyte cells develop in the egg, bacteria migrate into them. Sucking lice have a similar system. During larval development, symbionts are stored in an organ near the gut, but in adult females some of them (the germ-line equivalents) migrate to a second organ between ovary and oviduct, and each egg is infected before it matures. In other groups, the mechanism is via behavioral rather than morphological adaptation. In some beetles that depend on symbiotic yeasts, the female smears the yeasts on her eggs during oviposition, and the symbiont is transferred after hatching when the young beetle eats its eggshell (Frank 1996b, 341). As we shall see in "Outlaws," many aspects of host transmission mechanisms seem to have the function not just of transmitting symbionts reliably, but also of ensuring that they do not become outlaws.

Attine Ant-Fungus Symbioses

One of the classic examples of symbiosis is the association between attine ants, which cultivate fungus gardens, and their fungal symbiotic partners. The ants are dependent on their fungal partners, for gardening is their only source of food, and the dependence is mutual. So much so, in fact, that until the development of molecular techniques, the phylogeny of the fungal associates remained largely unresolved, for fungus fruiting bodies—their most phylogenetically informative morphological characters—were very rarely produced. Most fungal propagation was clonal, from founder to daughter nests.

This alliance illustrates particularly clearly the importance of specific adaptations for vertical transmission. The attine ants originated around 50 million years ago, and fungus gardening is a primitive character of the 200 or so living attine species (Chapela et al. 1994, 1693). But though fungus gardening has arisen only once, molecular investigation of fungus phylogeny shows that a number of independent fungus lineages are involved in symbiotic alliance with the attines. Taken as a whole, the fungal partners of attine ants do not form a monophyletic lineage, and sometimes closely related ant species have quite distantly related fungal partners (Chapela 1994, 1692).

However, though fungus gardening is a primitive attine trait, adaptations for vertical transfer, where the foundress queen takes a pellet of the fungus in specialized cheek pouches, are derived. Such adaptations are characteristics of the so-called higher attines (Chapela et al. 1994, 1692). And these ant clades have congruent phylogenetic branching patterns with their fungal partners. As ants in these "advanced" genera speciated, their associated fungi diverged from one another (Hinkle et al. 1994, 1696). In the species with adaptations for vertical transmission, partner switching seems to be entirely absent. Moreover the evolutionary depth of this association is significant: the attine clade is reckoned to be about 50 million years old, and the subclades whose phylogenies are congruent with that of their partners are between 25 and 40 million years old (Hinkle et al. 1994, 1696).

This example, too, illustrates my themes: vertical transmission is evolutionarily stable, symbionts play a critical role in generating the host phenotype, the phenotypic significance of the symbiont is modular and conservative, and the acquisition of the symbiont is a breakthrough that allows the invasion of a new adaptive zone.

Outlaws

One of the central problems in understanding the evolution of genetic inheritance is the problem of outlaw genes. How were outlaws sufficiently suppressed when gene collectives—genomes—first evolved? Transmission of microorganisms poses a similar problem. How did this system evolve, and what prevents its subversion by defectors? I have neither the space nor the expertise for a full review of this problem, but I shall discuss some aspects of it here, centring my discussion around *Wolbachia* and vertical transmission.[5]

Steven Frank has shown that even when symbionts are transmitted vertically, competition between different symbiotic lineages in the same host can drive the evolution of symbiotic traits that depress host fitness. For example, bacteria that are not overproducing tryptophan or leucine for the benefit of their host will replicate faster and contribute disproportionately to the sample transmitted to the next host generation. Thus, host transmission mechanisms are often evolutionary compromises between the need to ensure reliable transmission and the need to minimize genetic diversity in the within-host symbiont population; for example, in the sucking louse discussed above, the storage organ between ovary and oviduct contains about 3,000–6,000 bacteria, and each eggs is dosed with about 200 (Frank 1996b, 341);

from a review of these mechanisms, Frank concludes that in general hosts limit transmission to a subset of the symbiont lineages they carry, thus damping down the possibility of within-host competition (Frank 1996a).

Uniparental inheritance itself may be a mechanism with that effect: vertically transmitted symbionts, including of course mitochondria, typically travel in the female line. One of the effects of that inheritance pattern is to reduce lineage mixing in the host, though, as we shall see, this also has costs. But there are other mechanisms as well. Whiteflies have evolved a system that both ensures faithful replication and seems likely to reduce symbiont diversity within the host. "[A]n entire maternal bacteriocyte (or in some species, several bacteriocytes) is transferred into each egg, with the maternal nucleus later degenerating" (Moran and Telang 1998, 297). In those species in which only a single bacteriocyte is transmitted, a small and probably uniform bacterial population is inoculated into the next generation.

Frank's analysis of the importance of within-host competition is important, but even so, received wisdom has been that vertical transmission, by locking the reproductive interest of the symbiont onto that of the host, selects for symbiont traits which promote host fitness. *Wolbachia* seem to show that this view of vertical transmission is too simple, but I shall suggest that this impression is misleading. *Wolbachia* are bacteria transmitted in the female line in many lineages of arthropods. They do not live in specialized cell structures, and the host is not adapted to ensure their faithful replication to the next generation. Instead, they infect the sexual tissues of their host and are transmitted in eggs. That is no surprise, for the association is not a mutualism. *Wolbachia* are outlaws. Though their transmission depends on the successful reproduction of their hosts, they are "reproductive parasites" (Werren and O'Neill 1997, 2).

Vertically transmitted associates can improve their prospects for replication by boosting their host's fitness. But many associates that are transmitted vertically are transmitted only maternally. Those in males are in dead-end lineages. Hence, associates are under selection to bias the sex ratio of their hosts toward females. *Wolbachia* have a number of strategies by which they manipulate hosts to promote their own replication at the cost of host fitness (Boutzis and O'Neill 1998, 288). These strategies fall into two general kinds: direct sex ratio manipulations, and effects which depress the fitness of uninfected host lineages.

1. In some lineages, in particular, isopod crustaceans, they feminize their bearers. That is, they manipulate the sex ratio of the host species by causing infected genetic males to develop a female phenotype.

2. In some lineages of parasitic wasps, they do not just feminize their

hosts, they inaugurate parthenogenetic reproduction. The infected individual hosts in these lineages are not just females; they are females that reproduce without mating or meiosis.[6]

3. In some lineages, they reduce the fitness of uninfected members of the host species by "cytoplasmic incompatibility." If a host has no genetic stake in uninfected members of its own population, the fitness-reducing effect of the host's associates on uninfected organisms increases the relative fitness of infected organisms. That is so even if the average fitness of the population is depressed. Some paramecia have killer endosymbionts that kill other, uninfected, paramecia. To the extent that paramecia reproduce clonally, this is a mutualism.

But in the case of *Wolbachia,* associates fatally sabotage their hosts' potential descendants. *Wolbachia* that drive cytoplasmic incompatibility have evolved a sabotage/repair system. Sperm of infected hosts are sabotaged in ways that make them incapable of fusing with another haploid gamete to produce a functional, developing diploid. But if the egg is also infected with the same strain of *Wolbachia,* it can repair the damage, and the fertilized egg will be viable. These mechanisms are specific to particular strains of *Wolbachia,* both in the sense that some strains induce cytoplasmic incompatibility and others do not and in the sense that the damage to a sperm that one strain inflicts often cannot be repaired by an alternate strain. Thus, in one species of fruit fly, three strains of *Wolbachia* have been identified, each of which is incompatible with the others (Werren 1997, 594). Multiple infections result in sperm disabled in multiple ways, so if the fusion is to lead to a viable embryo, the egg too must be multiply infected. (Werren 1997, 595).

In haplodiploid arthropods, eggs which fail to incorporate male gametes do not fail to develop. Rather, they develop as haploid males. The cytoplasmic incompatibility mechanisms which produce aborted eggs in beetles, flies, mites, and butterflies produce sex ratio distortions in Hymenoptera. *Wolbachia* depress uninfected Hymenoptera females' fitness by causing them to overproduce sons.

These effects are expressed to varying degrees in different lineages and even in different species of the one lineage. Moreover, there are many groups of arthropods which carry *Wolbachia* of unknown phenotypic effect. In a recent survey using PCR techniques to assay for the presence of the bacterium, it was found in 16% of all Neotropical insects sampled (Boutzis and O'Neill 1998, 288). In contrast to, say, attine-fungus relationships or those between bivalves and chemoautotrophs, the phenotypic upshot of *Wolbachia* is neither constant across lineages nor robust within a lineage. This fact suggests that the modularity of symbiont contribution to host organization is an evolutionary achievement, not a passive by-product of their cohabitation. Where outlawry is rife, modularity fails. *Wolbachia* in different hosts have different phenotypic effects, and different *Wolbachia* in the same host produce the same effect by different means.

Wolbachia transmission is vertical in ecological time. The bacteria are transmitted in eggs but not (usually) sperm. But it is not vertical over evolutionary time. *Wolbachia* show clear evidence of frequent horizontal movement. Phylogenetic analyses of *Wolbachia* and their hosts show that sideways movement over evolutionary time scales is quite common. Molecular phylogenies show that very closely related strains of *Wolbachia* are found in hosts as diverse as beetles, flies, Hymenoptera, and butterflies (Werren 1997, 590). Some of these host switching events are quite recent: one strain has moved into a number of new insect hosts in the last few million years (Boutzis and O'Neill 1998, 288–289). The fact that *Wolbachia* and host phylogenies show no signs of concordance, whereas *Buchnera* and its host phylogenies do, suggests that strict vertical transmission—vertical transmission over evolutionary time—imposes good behaviour on associates, perhaps by clade selection (though, equally, there is clade selection on bacteria driving cytoplasmic incompatibility to evolve mechanisms of horizontal transfer; Hurst and McVean 1996). So I read *Wolbachia* as confirming rather than disconfirming the importance of *deep* vertical transmission.

Suppose this suggestion is right. If a symbiont is transferred with very high fidelity (and low internal diversity) over evolutionarily significant periods of time, it does not evolve into an outlaw. At best, this offers only a partial solution to the problem. For we need an explanation of why some lineages are committed to vertical transmission, and others are only predominantly vertical, with occasional sideways transfer. Why are *Buchnera* and similar species so tied to vertical transfer? Moran and Telang suggest that a combination of positive selection on hosts for hospitality and a relaxation of selection on bacteria for the characteristics that make it possible for them to survive independently, ultimately lead to permanent capture. The bacteria becomes incapable of living outside its host.[7] Moreover, there may be phylogenetic constraints on symbionts blocking the evolution of outlaw adaptations. Thus, the bacterial associates of bivalves have a sulphide-using metabolism, and that basic metabolic commitment blocks a whole raft of bacterial lifestyles. More salient, perhaps, is the active role of hosts in policing potential defectors. For example, in those strains of *Buchnera* where the bacteria overproduces the amino acid, recombination mechanisms which would normally edit out the gene repeats responsible for the overproduction seems to be reduced or suppressed (Moran and Telang 1998, 301).

Let us summarize this. *Wolbachia* seem to show that outlaws can be vertically transmitted. But they are transmitted only in the female line, and hence, they can improve their own fitness at the cost of their host through various manipulations of host sex ratios. And while *Wolbachia* are a very striking example of a sex ratio parasite, they are certainly not the only one. In some species of plants, mitochondria are outlaws that manipulate sex ratios in an analogous fashion (Frank 1989; Hanson 1991). But *Wolbachia* are not a clear counterexample to the connection between symbiosis and vertical transmission. For on evolutionary time frames, *Wolbachia* switch lineages, and symbionts do not. That, however, leaves us with another question. What locks in some but not all microorganisms into strict vertical transmission? It seems unlikely that there is a unitary answer to this question. Some may be captured by host policing mechanisms. Others may have been seduced by host provisioning and have lost the ability to live outside the special structures provided by the host. The basic metabolism of some associates may leave them with no temptation to defect.

Encoding Your Symbiont: Wasp–Poly-DNA-Virus Alliances

Many wasps are parasitoids. Their life cycle involves laying eggs in invertebrate hosts that hatch into larvae that consume the host as they develop. Not surprisingly, hosts are not entirely defenseless against parasitic attack. Some species of caterpillar can kill the eggs by an immune reaction. Blood cells of the caterpillar can encapsulate and kill wasp eggs. In two wasp lineages, caterpillar immune defenses are disabled by a symbiotic alliance between the parasitoid wasp and its symbiotic virus (Summers and Dib-Hajj 1995). As the wasp eggs mature, special host cells in the wasp rupture, and the wasp eggs are coated with virions. If these virions are washed off before they are injected into the caterpillar, the caterpillar kills the eggs. If the wasp + virion package is successfully injected into the caterpillar, the virions migrate through the caterpillar tissues, invading the cells. These caterpillar cells express some virion genes, and hence, the gene products of this alliance can be expressed simultaneously throughout the caterpillar (Beckage 1998, 307), disabling its defenses globally. Viral genes are expressed in caterpillar cells, and the expression of those genes prevents encapsulation, either by destroying the hemocytes that encapsulate the wasp egg or by changing their shape so that they cannot adhere to the egg, disabling encapsulation (Beckage 1998, 307–308).

There is no reason to believe that the wasp–poly-DNA-virus system is an ancient association. It is not phylogenetically widespread; the system seems to have evolved independently in two wasp lineages. It is not a major evolutionary innovation, opening up modes of life that would be otherwise blocked to the lineage. There are astonishing numbers of parasitoid wasp species. But this case contrasts most importantly with, say, the ant-fungus association in that the inheritance of symbionts across the wasp generations is *not* an instance of sample-based inheritance. It is informational inheritance. *Viral DNA* rather than virions are transmitted across the generations. The viral DNA of these poly-DNA-viruses is integrated within the genome of the wasps carrying them. These sequences are replicated only in special cells in the female ovaries, as the adult develops in the pupal cocoon. These cells rupture and release their virions onto the eggs—which themselves already contain viral DNA integrated within their own genome. These eggs, virions, and a protein cocktail are injected into the caterpillar. The virions that destroy caterpillar immune systems are evolutionary dead-ends. They will die with the caterpillar.

Thus, if the hypothesis of this chapter is right, and evolutionary patterns based on sample-based inheritance are different from those based on the flow of coded information, then the wasp-virus alliance should show genelike inheritance patterns, not those typical of other host-symbiont systems. In those, functioning symbionts rather than symbiont-making instructions are passed to the next host generation. This system, then, is important to me as a potential test of the ideas on evolvability developed here.

Moreover, the contrast, if it proves to be real, is the key to my response to Maynard Smith. He proposes to see symbiont transmission as the transmission of symbiont genes. Hence, he reduces sample-based inheritance to a slightly nonstandard case of genetic inheritance, along the same lines as mitochondrial inheritance (Maynard Smith 2001, 1496–1497). If my conjecture about the contrast in evolutionary patterns is right, incorporating *Buchnera*-type cases into the mitochondrial paradigm is mistaken. Of course, these symbionts do have genetic inheritance systems, and their genes are copied and transmitted to the new host generation as part of the symbiotic cell. But the genetic inheritance mechanisms of the symbiont are insulated from the host by the continuing (though partial) physical and metabolic autonomy of the symbiont. That will not be true of the wasp–poly-DNA-virus alliance, where the genetic material is integrated into the wasp genome, rather than being segregated by the continuing integrity of the symbiont. The symbionts' systems of genetic inheritance may be segregated in a second sense. The host interacts, both over developmental and evolutionary time, with stabilized, evolutionarily conservative features of the symbiont lineage: with, for example, the basic metabolic profile of the chemoautotrophs. There certainly are evolutionary changes in symbiont lineages. But once the association is stabilized, microevolutionary changes within the symbiont are probably irrelevant to the role of the symbiont in the host lineage. For that role depends only on coarse-grained features of the symbiont lineage.

Conclusion

The view of inheritance defended in this chapter is heterodox. But no one doubts that the transmission of mitochondria is an inheritance mechanism, and mitochondria originated as an independent bacterial lineage that was reengineered as part of the inheritance system of eukaryotic lineages. Moreover, the transmission of symbiotic associates by hosts satisfies any reasonable definition of an inheritance mechanism. It is a systematic and adapted process by which offspring come to resemble their parents, including those aspects of their parents which differ from other organisms in the population.

Most important, we see interesting and important problems more easily from this perspective. It is a productive way of thinking about host-partner associations. For example, once we think of symbiont transmission as an inheritance system the following questions are inescapable:

1. Can we find anti-outlaw mechanisms? These might include simultaneous transfer, host adaptations for blocking all but vertical transmission, host adaptations for limiting the number of individual organisms transferred, and host adaptations for evolutionary capture of symbionts by taking over the provision of critical metabolic resources to the associate.

2. Is there evidence about the range of variation? Does the host phenotype differ, if strains of symbionts are transferred, or if there is more than one clone transmitted? Do genetic differences in the host or associate change the nature of their association?

3. If symbiotic transmission is a highly evolvable inheritance system, how and why could evolution assemble such an inheritance mechanism? Replicators are adapted for their role of ensuring that offspring are like their parents. That is no surprise, for most departures from similarity will be bad news. But a system that ensures accurate replication across a generation is one thing; an evolvable replication system quite another. Might there be lineage-level selection for evolvability (Kirschner and Gerhart 1998)? That is an idea we can test. For example, we can look for evidence of increased evolvability by comparing symbiont-rich clades with symbiont-poor sister clades. Are symbiotically rich clades more species rich, more morphologically disparate, or ecologically diverse than their sisters?

It might be possible to formulate these questions while seeing hosts and their associates as separate but coevolving lineages. But they are obvious and inescapable if we think of symbiont transmission as an inheritance system.

So there are excellent reasons to believe that symbiont transmission is an inheritance mechanism. Hosts copy their associates to their offspring with great reliability. Moreover, as we have seen, it is a mechanism of great evolutionary significance. For symbiotic association enables host lineages to invade new adaptive zones. In turn, that is because the formation of symbiotic associations is probably the only evolutionary process that generates in animals adaptive saltational changes at any appreciable frequency, even on evolutionary time scales (plants, with their more modular organization, are less constrained). Organisms that acquire, for the first time in their lineage's history, a new symbiotic associate may acquire a whole new capacity ready-made, though doubtless one that is subject to coevolutionary fine-tuning. They are hopeful monsters. On human time scales, such events must be vanishingly rare. But mutualisms are fairly common. Over evolutionarily significant periods such associations must be formed quite often. So symbiosis is significant partly because it is one way for lineages to cross fitness trenches and overcome historical constraints. A bivalve shift from, say, filter feeding to sulphide metabolism might well be blocked by historical constraints. No metazoan has evolved for itself these biochemical pathways. Perhaps the only way an animal can invade these sulphide-rich, oxygen-poor environments is by acquiring an appropriate symbiont.

The role of symbiosis in innovation is intimately linked to the fact that symbionts drive the development of modules. Symbionts make an unvarying contribution to host organization, even in host lineages that are deep and species rich. Symbionts make their distinctive contribution to host phenotypes relatively independently of other developmental resources, for their contribution is stable over deep time and considerable phenotypic change in their hosts.

Acknowledgments

Thanks to Gerhard Schlosser, Günter Wagner, Bill Wimsatt, Russell Gray, Steve Downes, and the audience at the Genes and Information II workshop for their feedback on this chapter.

Notes

1. In an earlier discussion of these ideas, I included a fourth antioutlaw condition on evolutionary escape from the role of replicator (Sterelny 2001). But that is just a special version of the condition on vertical transmission; it is *permanent* vertical transmission.

2. Though Dawkins (1982) argues that it is also important for the generation of variation.

3. I present this distinction as if it were sharp. Almost certainly, though, there is a continuum of cases between highly modular, robust, low-variation, nonarbitrary channels and less modular, rich-variation, arbitrary channels. The focus of this chapter is on inheritance channels close to one or the other end of this continuum, and on their evolutionary dynamics.

4. Whether the symbionts are dependent on their hosts is another matter. No symbiotic species has ever been cultured or observed free-living in the environment. But a relatively small proportion of bacteria grow in pure cultures, so in itself that tells us little. Moreover, the mechanism of transfer to the next generation is unknown. Only in a few cases is there evidence of direct vertical transfer via the reproductive tissues of the host (Distel 1998, 280).

5. Vertical transmission is not strictly necessary for mutualism: figs and their fig wasp pollinators, and squid of the genus *Eupryma* and their bioluminescent bacteria, are two examples of symbiotic relationships with horizontal rather than vertical transmission (MacFall-Ngai and Ruby 1998). I shall be discussing the conditions under which it is sufficient.

6. Though, strikingly, in at least one species, parthenogenetic reproduction remains facultative rather than obligatory; such females will mate with males and produce diploid females using the male genome (Werren 1997, 602).

7. Indeed, there may even be positive selection on hosts to take over key housekeeping functions of the bacteria, though Moran and Telang (1998) point out that since bacteria do not live in germ line cells, there is no obvious way key bacterial genes can be transferred to the host germ line—somatic mutations might transfer them to the nucleus of the bacteriocyte cell, but those are dead-end replicators.

References

Beckage, N. 1998. Parasitoids and polydnaviruses. *Bioscience* 48:305–311.

Boutzis, K., and S. O'Neill. 1998. *Wolbachia* infections and arthropod reproduction. *Bioscience* 48:287–293.

Buss, L. 1987. *The evolution of individuality*. Princeton, N.J.: Princeton University Press.

Chapela, I., S. Rehner, T. Schultz, and U. Müller. 1994. Evolutionary history of the symbiosis between fungus-growing ants and their fungi. *Science* 266:1691–1694.

Dawkins, R. 1982. *The extended phenotype*. Oxford: Oxford University Press.

Dawkins, R. 1990. Parasites, desiderata lists and the paradox of the organism. *Parasitology* 100:S63–S73.

Dawkins, R. 1996. *Climbing mount improbable*. New York: W. W. Norton.

Distel, D. 1998. Evolution of chemoautotrophic endosymbioses in bivalves. *Bioscience* 48:277–286.

Ewald, P. W. 1994. *Evolution of infectious disease*. Oxford: Oxford University Press.

Frank, S. A. 1989. The evolutionary dynamics of cytoplasmic male sterility. *Am. Nat.* 133:345–376.

Frank, S. A. 1995. The origin of synergistic symbiosis. *J. Theor. Biol.* 176:403–410.

Frank, S. A. 1996a. Host control of symbiont transmission: the separation of symbionts into germ and soma. *Am. Nat.* 148:1113–1124.

Frank, S. A. 1996b. Host-symbiont conflict over the mixing of symbiotic lineages. *Proc. R. Soc. Lond. B Biol. Sci.* 263:339–344.

Frank, S. A. 1996c. Models of parasite virulence. *Q. Rev. Biol.* 71 (1): 37–78.

Gellon, G., and W. McGinnis 1998. Shaping animal body plans in development and evolution by modulation of Hox expression patterns. *BioEssays* 20:116–125.

Gilbert, S. F., J. M. Opitz, and Raff, R. 1996. Resynthesising evolutionary and developmental biology. *Dev. Biol.* 173:357–372.

Grosberg, R., and R. Strathmann. 1998. One cell, two cells, red cell, blue cell: the persistence of a unicellular stage in multicellular life histories. *Trends Ecol. Evol.* 13 (3): 112–116.

Hanson, M. R. 1991. Plant mitochondrial mutations and male sterility. *Annu. Rev. Genet.* 25:461–486.

Hinkle, G., Wetterer, J Schultz, and T. Sogin. 1994. Phylogeny of the attine ant fungi based on analysis of small subunit ribosomal RNA gene sequences. *Science* 266: 1695–1697.

Holland, P., and J. Garcia-Fernandez. 1996. Hox genes and chordate evolution. *Dev. Biol.* 173:382–395.

Hurst, L., and G. McVean. 1996. Clade selection, reversible evolution and the persistence of selfish elements: the evolutionary dynamics of cytoplasmic incompatibility. *Proc. R. Soc. Lond. B Biol. Sci.* 263:97–104.

Kauffman, S. 1995. *At home in the universe.* New York, Oxford University Press.

Kauffman, S. A. 1993. *The origins of order: self-organisation and selection in evolution.* New York: Oxford University Press.

Kirschner, M., and J. Gerhart. 1998. Evolvability. *Proc. Natl. Acad. Sci. U.S.A.* 95: 8420–8427.

Lewontin, R. C. 1978. Adaptation. *Sci. Am.* 239:156–169.

MacFall-Ngai, M., and E. Ruby. 1998. Sepiolids and vibrios: when first they meet. *Bioscience* 48:257–265.

Maynard Smith, J. 2000. The concept of information in biology. *Philos. Sci.* 67:177–194.

Maynard Smith, J. 2001. Reconciling Marx and Darwin. *Evolution* 55 (7): 1496–1498.

Maynard Smith, J., and E. Szathmary. 1995. *The major transitions in evolution.* New York: Freeman.

Michod, R. E. 1999. *Darwinian dynamics: evolutionary transitions in fitness and individuality.* Princeton, N.J.: Princeton University Press.

Moran, N., and A. Telang. 1998. Bacteriocyte-associated symbionts of insects. *Bioscience* 48:295–304.

Morgan, N., and P. Baumann. 1994. Phylogenetics of cytoplasmically inherited microorganisms of arthropods. *Trends Ecol. Evol.* 9:15–20.

Müller, G. B., and G. P. Wagner. 1996. Homology, Hox genes and developmental integration. *Am. Zool.* 36:4–13.

Raff, R. 1996. *The shape of life: genes, development and the evolution of animal form.* Chicago: Chicago University Press.

Schmidt-Ott, U., and E. A. Wimmer. 2004. Starting the segmentation gene cascade in insects. In *Modularity in development and evolution,* ed. G. Schlosser and G. P. Wagner. Chicago: University of Chicago Press.

Sober, E., and D. S. Wilson. 1998. *Unto others: the evolution of altruism.* Cambridge, Mass.: Harvard University Press.

Sterelny, K. 2001. Niche construction, developmental systems and the extended replicator. In *Cycles of contingency,* ed. S. Oyama, P. E. Griffiths, and R. D. Gray, 333–350. Cambridge, Mass.: MIT Press.

Sterelny, K., K. Smith, and M. Dickison. 1996. The extended replicator. *Biol. Philos.* 11: 377–403.

Summers, M., and S. Dib-Hajj. 1995. Polydnavirus-facilitated endoparasite protection against host immune defenses. *Proc. Natl. Acad. Sci. U.S.A.* 92:29–36.

Szathmary, E., and J. Maynard Smith. 1997. From replicators to reproducers: the first major transitions leading to life. *J. Theor. Biol.* 187:555–571.

Tomasello, M. 2000. Two hypotheses about primate cognition. In *Evolution of cognition,* ed. C. Heyes and L. Huber, 165–184. Cambridge, Mass.: MIT Press.

Wagner, G. P. 1996. Homologues, natural kinds and the evolution of modularity. *Am. Zool.* 36:36–43.

Wagner, G. P., and L. Altenberg. 1996. Complex adaptations and the evolution of evolvability. *Evolution* 50:967–976.

Werren, J. 1997. Biology of *Wolbachia. Annu. Rev. Entomol.* 42:587–609.

Werren, J., and S. O'Neill. 1997. The evolution of heritable symbionts. In *Influential passengers: inherited microrganisms and arthropod reproduction,* ed. S. O'Neill, A. Hoffman and J. Werren, 1–41. Oxford: Oxford University Press.

Wimsatt, W. C. 1980. Units of selection and the structure of the multi-level genome. In *Proceedings of the Philosophy of Science Association,* ed. P. Asquith and T. Nickles, 2:122–183. East Lansing, Mich.: Philosophy of Science Association.

Wimsatt, W. C. 2001. Generative entrenchment and the developmental systems approach to evolutionary processes. In *Cycles of contingency,* ed. S. Oyama, P. E. Griffiths, and R. D. Gray, 219–238. Cambridge, Mass.: MIT Press.

Wimsatt, W. C., and J. C. Schank. 1988. Two constraints on the evolution of complex adaptations and the means of their avoidance. In *Evolutionary progress,* ed. M. H. Nitecki, 231–275. Chicago: University of Chicago Press.

Synthesis

23 The Role of Modules in Development and Evolution

GERHARD SCHLOSSER

Introduction

The past two decades have revolutionized our understanding of the roles of genes in development and of their evolutionary history. However, these advances have left us with two apparent paradoxes, a developmental and an evolutionary one (Wray 1994; Gerhart and Kirschner 1997; Duboule and Wilkins 1998). The developmental paradox results from accumulating evidence that many genes (or entire networks of genes) are active at multiple times or locations in development, in each case governing the production of the same protein (if we disregard the possibility of alternative splicing), which presumably can enter into only a few kind of interactions. How can this be reconciled with the vastly divergent functional roles (e.g., involving the activation of different batteries of downstream genes) that a given gene may play in these different contexts? The analogous evolutionary paradox is posed by observations that orthologous genes in many different lineages often encode proteins that enter into conserved types of interactions with other proteins or genes. How can this be reconciled with the striking phylogenetic differences in morphology, life history, and behavior that finally result from the activity of these genes?

Recently, there has been an astonishing convergence of ideas from many different authors suggesting that the modular organization of organisms may hold the key to resolving both of these apparent paradoxes (see., e.g., Zuckerkandl 1994; Wagner 1995, 1996; Wagner and Altenberg 1996; García-Bellido 1996; Raff 1996; Gilbert et al. 1996; Gerhart and Kirschner 1997; Hartwell et al. 1999, von Dassow and Munro 1999; Gilbert 1998, 2000; Dover 2000; Stern 2000; Bolker 2000; Schlosser and Thieffry 2000; Gilbert and Bolker 2001; Carroll et al. 2001; Winther 2001; Schlosser 2002, in press). Modules are building blocks of interacting elements that operate in an integrated and relatively autonomous manner. Developmental modules may exist at different levels of organization, ranging from gene regulation to networks of interacting genes and proteins to organ primordia involving interactions among many different cells. Because developmental modules are relatively insensitive to contextual perturbations, they may behave relatively invariantly, even when they are multiply realized in different developmental contexts. The diversity of their functions in development may then result simply from the employment of different combinations of modules in each context, suggesting a solution to the first paradox. Likewise, developmental modules may preserve their integrity in evolution despite heritable variations of the context in which they are embedded (or despite replacements of some of their submodules by others). Accordingly, they may form coherent and quasi-autonomous units in evolution (modules of evolution) that are repeatedly recombinable with other such units, indicating a solution to the second paradox.

Modularity, therefore, seems to be center stage in many recent attempts to forge a new synthesis between developmental and evolutionary biology (e.g., Wagner and Altenberg 1996; Gilbert et al. 1996; Raff 1996; Gerhart and Kirschner 1997; Brandon 1999; Schlosser 2002). However, "modularity" is in danger of becoming a buzzword, applicable to everything but lacking conceptual force. Modularity needs to be more precisely defined and operational criteria for its application need to be given if the concept is to play an important theoretical role in a new evolutionary-developmental synthesis. Thus, in the first part of this chapter I will try to answer the question of what a module is. Although some may feel the need for pluralistic concepts of modularity in biology, I will opt for a very general definition. In the second part of the chapter, this definition will be applied to modules of *development*. Examples drawn from many different levels of organization will illustrate the broad applicability of the proposed definition. I will then discuss how modularity contributes to the capacity to develop (what could be termed *developability*) by promoting robustness and flexibility (Gerhart and Kirschner 1997) as well as combinatorial generation of complexity. The third part of this chapter will address the question of whether developmental modules can also act as modules of *evolution* and how such claims may be tested. To answer this question, it is necessary to go beyond individual developmental processes and consider the conditions that determine fitness during the reproduction of such processes in the chain of generations after heritable variations (e.g., Arthur 1997; Wilkins 1998; Hall 2000). Taking these parameters into consideration, developmental modules tend to act as coherent modules of evolution only under certain conditions, which may, however, be greatly favored by the modularity of gene regulation itself. I will present several examples for developmental modules that in fact acted as modules of evolution in particular lineages. Finally, I will discuss how developmental modularity contributes to evolvability by favoring the quasi independence of fitness effects and by allowing the evolution of novelty by recombination.

What Are Modules?

Obviously, "module" always refers to some part of a larger unit. But not all parts are modules. The left half of an animal is certainly a part of the animal, but it is not a module. Modules are special kinds of parts, namely, parts that constitute some kind of integrated and autonomous units. In order to flesh out these intuitions we first have to get rid of some prejudices. We usually tend to think of modules as structures, but we need to move toward a process-oriented perspective before we can specify integration and autonomy in a fruitful way (see also von Dassow and Munro 1999; Gilbert and Bolker 2001; von Dassow and Meir 2004).

A Process-Oriented Definition of Modules

Many processes can be described as more or less complex sequences of interactions between several elements (fig. 23.1). The life cycle of an organism, for example, presents itself as a series of interactions between certain environmental parameters and different organs or cells or genes, depending on how closely we look. Any part or *subprocess* of a certain process will comprise only a subset of all the elements and their interactions that jointly constitute the process. As chunks of a larger scale process, subprocesses have temporal as well as spatial boundaries—that is, they are doubly embedded in the larger-scale process, both upstream (input) and downstream (output). Signaling pathways like the hedgehog pathway may serve as an example for such series of interactions (ligand-receptor binding followed by signal transduction followed by transcriptional activation), which will be activated at a certain time and location in the life cycle (e.g., during early embryonic development of vertebrates in the notochord and floor plate) and will contribute to the initiation of other subprocesses (e.g., the development of spinal motor neurons). *Modules* can be characterized as those types of subprocesses that are integrated and relatively autonomous (fig. 23.1). These concepts of integration and autonomy can be operationally specified from a process-oriented perspective.

A subprocess operates in an *integrated* (nondecomposable) fashion, when it fulfills its specific role in the containing process (e.g., a particular life cycle) only due to particular interactions between its

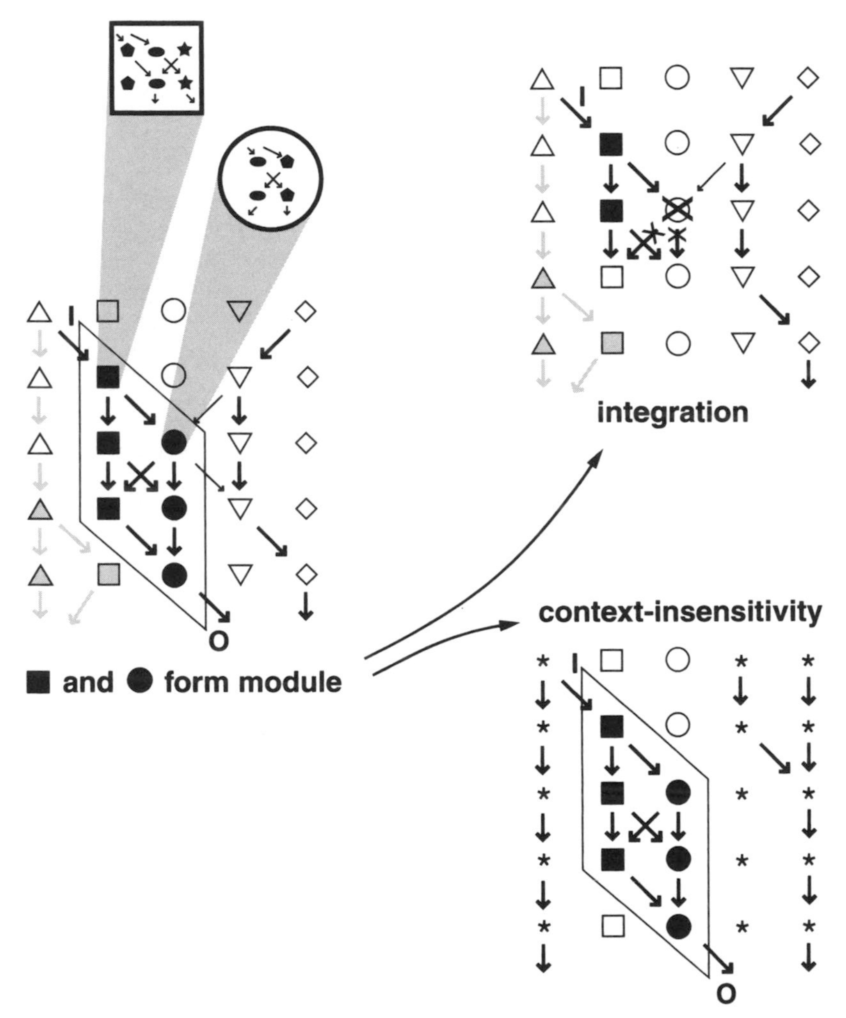

Fig. 23.1.—Modules are subprocesses that make an integrated and context-insensitive contribution to a process. The figure illustrates a sequence of interactions (*arrows*) of various kinds of elements (e.g., cells or molecules; *symbols*) in time. Integration of several interacting elements (*black symbols*) into a module can be demonstrated, when perturbation of one element results in perturbation (or in epistasis; not shown) of other components of the module, so that the relation of input (I) to output (O) is altered. Context insensitivity of a module can be demonstrated when its behavior or input-output relation is unaffected by perturbation of the context (e.g., after transplantation of the module into a different context, indicated by asterisks), for example, because elements of the module are not perturbed (or do not exhibit epistasis; not shown) after perturbation of the context. The components of a module may themselves be modules of smaller-scale elements, as the magnifications of the circular and square component illustrate. Therefore, modules may be embedded in larger-scale modules in a hierarchical fashion. In addition, there may be an overlap of modules, when an element of one module (*black symbols*), such as the square, is also a component of a different module (*grey symbols*).

components, that is, when the input-output relationship (IOR) of the subprocess does not merely represent the additive superposition of the IORs of its components but depends on their particular connectedness. Note that this is a weak sense of integration according to which there can be integration among elements even without reciprocal interactions, for instance, when several elements act in a chain (e.g., in a simple signaling cascade, where ligand-receptor binding provides the input, followed by a sequence of protein-protein interactions, and finally leading to transcriptional activation of target genes as the output) or when one element causally influences or is influenced by all the others.[1] Operationally, the integration of several components of a module can be demonstrated when a perturbation of one component affecting the IOR of the module tends to be associated with perturbation of other components (coperturbation due to common causal dependences) or when the effect of the perturbation of one component on the IOR tends to differ depending on the state of another component (epistasis due to common functional effects).

A subprocess is relatively *autonomous* when its contribution to the larger-scale process (i.e., its IOR) is relatively insensitive to the context (i.e., other subprocesses) in which it operates. Context insensitivity implies that the interactions between the components of the subprocess may proceed relatively "normally," in the sense of guaranteeing a similar input-output relationship of the subprocess, even after perturbation of the context. Operationally, context insensitivity can be demonstrated when, after perturbation of the context, two conditions are met: First, there will be no effects on elements of the module that affect its IOR (i.e., elements of the module will be either not perturbed at all or will be perturbed only in a way that leads to an unaltered IOR due to compensatory changes in other elements of the module). Second, the effects of elements of the module on its IOR will not be epistatically altered (i.e., regardless of whether the context is perturbed or not, the effect of perturbation of an element of the module on its IOR remains the same).

In sum, a subprocess of any process qualifies as a module when its components are likely to interact as an integrated and context-insensitive unit to produce a particular IOR in the face of various perturbations. Of course, integration and autonomy come in degrees. There is no clear-cut dividing line between subprocesses that are modules and those that are not, but certainly some subprocesses qualify to a higher degree than others. The degree of integration and context insensitivity depends on the relative strength of effects of various perturbations on the IOR as well as on the relative frequency (probability) of these various perturbations. Hence, whether something counts as a module or not depends on criteria (e.g., a certain threshold of permitted variation of an IOR, a particular class of permitted perturbations) that have to be defined on a case-by-case basis (1) relative to some specified contribution of a subprocess to a containing process (i.e., some specified IOR) and (2) relative to a specified class of possible perturbations of components. This implies, for instance, that modules of development must fulfill different criteria than modules of evolution. In order to qualify as a module of *development* a unit of elements must (1) make an integrated and context-insensitive contribution to the development of an organism in the face of (2) perturbations that do not need to be heritable (such as environmental perturbations or mutations in genes of somatic cells). However, as will be discussed in more detail below, a unit of elements must fulfill different requirements in order to qualify as a module of *evolution*—that is, it must (1) make an integrated and context-insensitive contribution to its own reproduction in subsequent generations in the face of (2) variations that are heritable (such as mutations in genes of germ line cells).

While this process-oriented concept of modularity is very generally applicable, it makes sense only when applied to subprocesses of a process such as the life cycle. However, structures may be characterized as modules in a derivative (and elliptic) usage of the term, which can be justified by pointing to the integrated and autonomous role a structure plays for some process. Regions of proteins or of DNA are called modules because they jointly enter into particular interactions with other proteins, irrespective of what their neighbors in the sequence do. Organs are modules because they require intimate orchestration of their component cells for their proper function or development, while operating relatively independently from other organs. The degree of structural connectivity between elements in such cases often reflects the degree of integrated and context-insensitive interactions between them (see, e.g., McShea and Venit 2001).

The Architecture of Modules: Connectivities, Hierarchies, and Multiple Instantiations

The simplest kinds of modules may be cascades of interacting elements, where the outputs of one or several elements provide some of the inputs for other element(s). Such cascades count as modules when they are robust enough to display quasi-autonomous behavior, for instance, because the interactions of different proteins in the cascade are relatively insensitive to fluctuations of the molecular composition of the cells in which they act. Often, however, the autonomous behavior of modules will be due to developmental buffering or canalization (e.g., Waddington 1957; Rutherford 2000; Gibson and Wagner 2000) resulting from special types of integration, such as redundant pathways (e.g., Wilkins

1997; Rutherford 2000) and/or some kind of reciprocal (feedback) interactions between their elements (e.g., Thomas 1978, 1991; Thomas et al. 1995; Thieffry and Romero 1999; Bhalla and Iyengar 1999; Freeman 2000; von Dassow et al. 2000). As a corollary of their autonomous behavior, modules can be frequently triggered in a switchlike fashion by a variety of inputs (*contingency*) to which they are only weakly linked (*weak linkage*) and may affect different downstream processes depending on the circumstances (for an detailed discussion of these issues see Gerhart and Kirschner 1997). Despite qualitative differences in inputs and outputs in these cases, the quantitative and spatiotemporal input-output transformation of the module—what could be called its *logical role* (or its intrinsic behavior; see von Dassow and Meir 2004)—stays the same.

It needs to be emphasized that the characterization of modules given here does not rule out that modules may be spatiotemporally embedded in higher-order modules in a hierarchical manner. The apparent contradiction of building a nondecomposable unit from autonomous subunits can be resolved when the behavior of the larger unit depends irreducibly on the particular way its subunits are connected with each other: a lower-order module may be relatively autonomous with respect to its *own* input-output relationship, but it may nonetheless contribute (together with other components) in a nonadditive manner to the realization of a particular input-output relation of a larger-order module (see also von Dassow and Meir 2004). Examples for this, such as the combinatorial employment of the same transcription factors in the regulation of many different genes, will be discussed in more detail below.[2]

Moreover, modules may not be interdependent only by virtue of being spatiotemporally embedded in larger scale modules; they may also overlap by sharing some elements, for example, due to the multiple activity of genes in different developmental modules (fig. 23.1). There may even be multiple instantiation of the "same" type of module in a single process (fig. 23.2). This leads to some particularly tricky problems concerning the individuality and identity of modules, with profound implications for the evolution of modularity. Multiple instantiation is the rule during the development of organisms, because biochemical processes involve parallel interactions of many molecules (as is typical for catalytic interactions, for example, during gene transcription and translation, where interactions of a single template, a small number of polymerase molecules, and many RNA nucleotides and amino acids result in the production of many protein molecules), and because mechanisms for the faithful replication of DNA and of gene activity states generate lineages of cells that behave in a similar manner

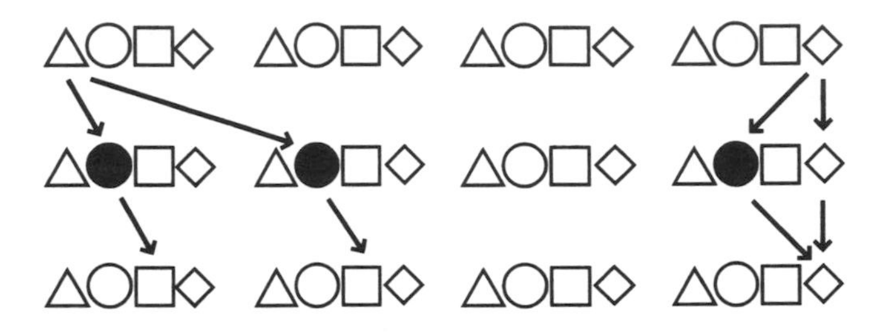

Fig. 23.2.—Modules can be multiply instantiated. Again, the figure illustrates a sequence of interactions (*arrows*) of various kinds of elements in time. However, in this figure the black circle represents an entire module, that is, a context-insensitive and integrated pattern of interactions among elements. Different instantiations of this module belong to the same domain (domain 1) if they are activated by the same upstream factor (*triangle*) but belong to different domains (domains 1 and 2) when they are activated by different upstream factors (*triangle* and *rhombus*, respectively). In a biological interpretation of the figure, each symbol may, for example, represent the activity pattern of a certain gene or gene network, whereas each quartet of different symbols may represent a single cell.

during the development of multicellular organisms. For instance, the hedgehog signaling pathway is active in many different cells during embryonic development of most metazoans, and in each cell it involves interactions among large numbers of identical molecules such as hedgehog and patched (Hammerschmidt et al. 1997; Chuong et al. 2000).

When there is multiple instantiation, an individualized developmental module can be delimited from others of the same type by its independent perturbability during development.[3] In the case in which several individualized modules are initiated by a common input, I will refer to them as a *single domain*. For example, hedgehog signaling is activated in many floor plate cells of the developing vertebrate spinal cord, but they all form a single domain, because they all depend on induction by the notochord (reviewed in Hammerschmidt et al. 1997). As this example illustrates, a single domain is often spatially contiguous, but this is not necessarily the case. Cells that activate the same signaling cascade in response to a particular hormone peak constitute a single domain even when they are sprinkled throughout the body. But individualized modules may also occur in *multiple domains*, namely, when they differ in inputs. Hedgehog signaling is used not only by floor plate cells but also by a population of cells in the zone of polarizing activity in the limb bud, where it is not activated via induction by the notochord (see Hammerschmidt et al. 1997). Furthermore, each domain may have one or several outputs or functions; these may be identical, partially overlapping, or different between different domains. Hedgehog signaling illustrates the latter possibility: Whereas hedgehog signaling in the floor plate is involved in motor neuron induction, in the limb bud it has different functions.

Importantly, however, it makes sense to talk about the *same* type of module being instantiated in different domains only when, regardless of the qualitative differences in its inputs and outputs between different domains, other parameters are preserved, in particular its intramodular patterns of interactions, its spatiotemporal and quantitative input-output relations (its logical role), and, for evolutionarily significant modules of organisms, also its genetic basis.[4]

Modules of Development

Organisms are not static; organisms are life cycles (Bonner 1988; Schlosser 2002). Life cycles proceed via development, activity in adult life, and the production of new propagules that initiate another generation. Such life cycles are modular in all their phases. Here I will concentrate on developmental modules, that is, those subprocesses that make an integrated and context-insensitive contribution to the development of an organism even when challenged with various perturbations such as environmental fluctuations or mutations. I will focus on the embryonic development of vertebrates to illustrate that developmental modules can be recognized at many different levels. Examples will be given for (1) modules of gene regulation, (2) signaling modules, (3) positional modules, (4) cell type modules, (5) organ modules, and (6) systemic modules. In each case, I will proceed by first briefly summarizing evidence for integration of the module and then review evidence for its autonomy during development. Integration can be established by showing that perturbation of one component of a unit tends to be associated with perturbation of other components (or affects them epistatically). Autonomy or context insensitivity is supported when development of the unit proceeds relatively normally in isolation or at ectopic locations, or when it is used in multiple domains. The purpose of these examples is to illustrate the general approach for assessing modularity.

Since knowledge of developmental processes is incomplete, most of the findings reviewed here provide only preliminary evidence for modularity in each case. A new arsenal of methods from functional genomics including cluster analysis of coexpressed genes (Wen et al. 1998; Eisen et al. 1998; Niehrs and Pollet 1999; Lockhart and Winzeler 2000; see also Niehrs 2004; Somogyi et al. 2004) will greatly aid the discovery of context-insensitive units of development in the future. However, because modularity comes in degrees, we should not expect that there will ever be a simple and general test that allows us to draw a sharp dividing line between modules and nonmodules.

Modules in the Development of Vertebrates: Examples from Different Levels

Modules of Gene Regulation

The regulation of gene activity is thought to proceed predominantly by the regulation of transcription, where several levels of modularity have been demonstrated (fig. 23.3). First and foremost, the activation of the transcriptional apparatus clearly forms a module different from the module formed by the sequence-specific assembly of proteins. Transcriptional activation is an integrated process that requires the association of numerous transcription factors with their binding sites in the promoters and enhancers (cis-regulatory regions) of a given gene, followed by direct or indirect (involving cofactors and possibly other cis-regulatory regions; see, e.g., Yuh et al. 1998; Mannervik et al. 1999) interactions with components of the basal transcriptional apparatus (BTA). These interactions are required to stably assemble the latter at the proper location (reviewed in Gerhart and Kirschner 1997; Ptashne and Gann 1997, 1998; Lewin 2000). Enhancer-swapping experiments (e.g., Kermekchiev et al. 1991; Gray et al. 1994; Kirchhamer et al. 1996a) suggest that transcriptional activation can operate relatively autonomously from sequence-specific protein synthesis, because particular promoters or enhancers may work in combination with new coding sequences and then drive the expression of the new protein in domains appropriate for their gene of origin (context insensitivity). This property is also exploited in molecular techniques using reporter genes or enhancer traps (e.g., O'Kane and Gehring 1987), where gene expression is studied by fusion of enhancers with coding sequences for heterologous proteins that can be easily visualized. The modular nature of transcriptional activation is often reflected in the presence of multiple enhancers in the cis-regulatory region of a single gene, each of which controls a particular expression domain of the gene in response to a tissue-specific combination of transcription factors and appears to be often relatively freely combinable with other enhancers (see Gray and Levine 1996; Kirchhamer et al. 1996b; Arnone and Davidson 1997; Davidson 2001; in the latter three publications enhancers or promoters themselves are termed "modules").

There is a second level of modularity in transcriptional regulation, because the activation of the transcriptional apparatus is often itself highly modularized. Promoters or enhancers typically contain many binding sites (fig. 23.3), each of which is around 6–15 nucleotides long and allows specific interaction with the DNA-binding domain of a particular transcription factor (reviewed in Arnone and Davidson 1997; Carroll et al. 2001; Davidson 2001). Another protein domain of transcription factors then interacts with components of the BTA, thereby

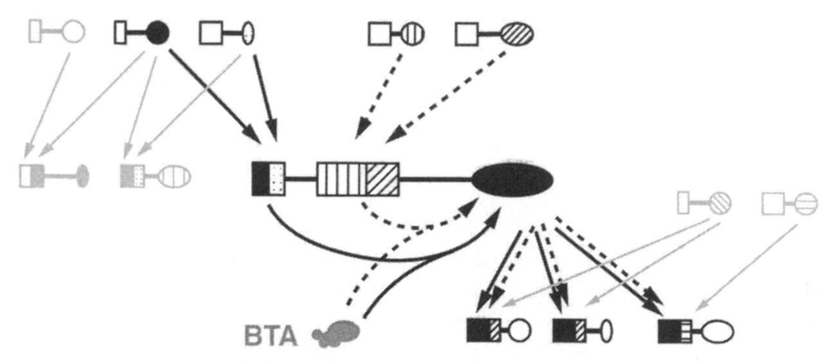

Fig. 23.3.—Schematic diagram showing modularity of gene regulation. Round symbols denote protein coding regions of various genes, rectangles denote binding sites for transcription factors, and solid and broken arrows denote two alternative sequences of causal interactions. Those binding sites that are singly required and jointly sufficient for gene activation forming an enhancer are drawn adjacent to one another without a gap, whereas different enhancers are separated by a gap between them. The gene with the black oval coding region can be activated by recruitment of the basal transcriptional apparatus (BTA) either through a combination of inputs from transcription factors coded by the black and dotted genes (*solid arrows*) or through a combination of inputs from transcription factors coded by the hatched genes (*broken arrows*). The gene will in turn produce a transcription factor that can turn on a number of target genes in combination with different cofactors. This system is highly modular, because the same enhancers (e.g., *black square combined with hatched rectangle*) may work in combination with different coding regions and independent of other enhancers. Moreover, transcriptional activation is itself modularized: The same transcription factor (e.g., encoded by the black oval coding region) exerts similar—for instance, activating—effects on each of the genes containing a binding site (e.g., *black squares*) for it, even when it is combined with different cofactors and binding sites (encoded by the horizontally or obliquely hatched circles and rectangles, respectively).

activating or repressing transcription. The integration of a specific DNA-protein interaction with a particular effect on the BTA is essential for the specificity of transcription factor activity (nonetheless, DNA binding and activation or repression domains may again be swapped between different transcription factors—a third level of modularity; see, e.g., Kessler 1997, Zuber et al. 1999). As noted above, recruitment of the BTA typically requires cooperative interactions of multiple transcription factors, but often these transcription factors do not seem to modify components of the BTA allosterically, but rather interact with them in a local and general, gluelike fashion (reviewed in Ptashne and Gann 1997, 1998). This is supported by experiments demonstrating that different types of activating transcription factors can functionally replace each other, as can different types of repressing transcription factors (e.g., Gray et al. 1994; Gray and Levine 1996; Arnosti et al. 1996; Halder et al. 1998b). In many enhancers there also appears to be some tolerance regarding the spacing between binding sites—it seems to be mainly the number of activators versus repressors that counts. Therefore, different types of DNA binding site–transcription factor–BTA interactions appear to be integrated but context-insensitive modules in that they work in a similar fashion regardless of which other transcription factor activities they are combined with. These features explain the apparent promiscuity of transcription factors and their binding sites, which are multiply employed in different combinations in the cis-regulatory regions of many different genes (see Arnone and Davidson 1997).

Signaling Modules

The number of pathways or networks involved in signal transduction between communicating cells is surprisingly small. At present, only five major pathways (coming in several versions), namely, the hedgehog, TGFβ, Wnt, receptor tyrosine kinase, and Notch pathways, are thought to be important during early embryonic development (Gerhart 1999). These, however, act as relatively autonomous modules that play multiple roles in the development of very different tissues. The evidence for this will be summarized here briefly for the Notch pathway (reviewed in Kimble and Simpson 1997; Beatus and Lendahl 1998; Lewis 1998; Artavanis-Tsakonas et al. 1999; Mumm and Kopan 2000; de Celis 2004).

The integration of Notch, Delta, and other elements into a single pathway is now well established by numerous experiments in vertebrates and invertebrates (reviewed, e.g., in Beatus and Lendahl 1998; Mumm and Kopan 2000). It was first discovered by the common neurogenic phenotype of their mutants in *Drosophila:* these develop supernumerary neurons due to failure of lateral inhibition (reviewed in Campos-Ortega 1993). The Notch pathway is initiated when one of various ligands, for example, the transmembrane protein Delta, binds to the Notch receptor, resulting in the release of the intracellular domain of Notch. The latter is then translocated to the nucleus, where it binds to the transcription factor Suppressor of hairless, which then activates *Enhancer of split*–related genes. These act as transcriptional repressors, which reduce expression of several target genes, including those (e.g., *Neurogenins*) that activate the transcription of *Delta* or other ligands in the same cell (fig. 23.4, *A*). The employment of this pathway in neighboring cells thus creates a positive feedback loop that tends to amplify initial differences in the level of Delta expression: The cell which expresses Delta at slightly higher levels (e.g., due to chance fluctuation) will be able to further upregulate Delta but suppress Delta expression in adjacent cells. This lateral inhibition mechanism creates inequalities between adjacent cells. Subsequently, the Delta expressing cell often adopts a particular fate (e.g., neuron), whereas its Notch-expressing neighbor remains undifferentiated. It should be noted that this is only a simplistic sketch that must suffice here to illustrate a general principle. The real Notch pathway is a rather complex network that involves different types of ligands and a variety of other modulating

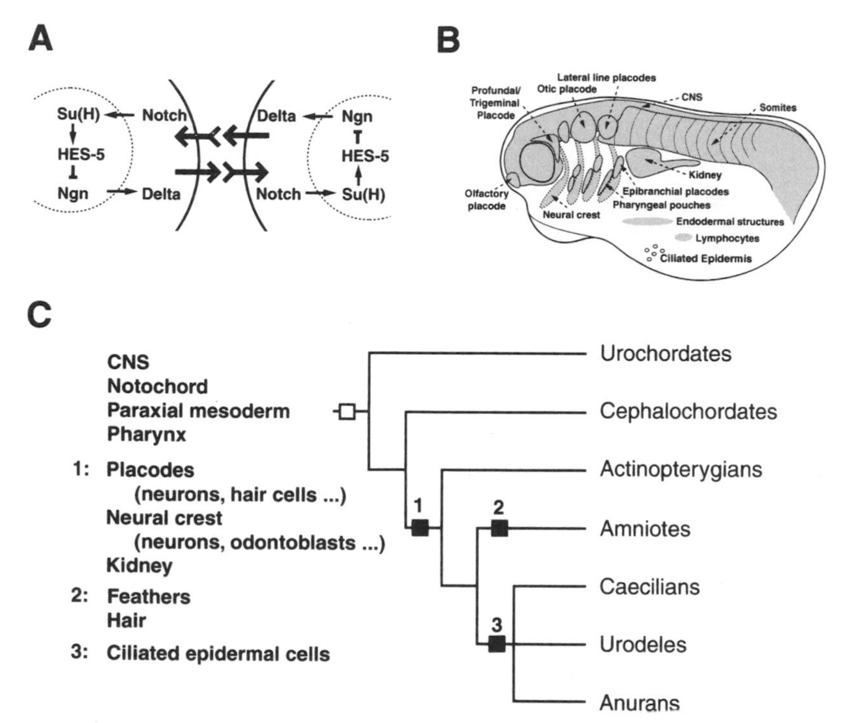

Fig. 23.4.—The vertebrate Notch pathway as a signaling module. *A,* Schematic diagram of interactions between two adjacent cells involving the Notch pathway and resulting in lateral inhibition (based on various sources; see, e.g., Blaschuk and Ffrench-Constant 1998). The activation of the cell-surface receptor Notch by its membrane-bound ligand, for example, Delta, in adjacent cells results in the activation of Suppressor of hairless (Su(H)) in the nucleus (*dotted line*). This transcription factor then activates a hairy/enhancer of split–related gene (e.g., *HES 5*), which represses activators of *Delta,* like *Neurogenins* (Ngn). *B,* Schematic diagram of a vertebrate tailbud stage embryo depicting a subset of all the domains where the Notch pathway is known to be active. This employment of the Notch pathway in multiple domains suggests that it acts as a relatively context-insensitive module in development. (Data are based on the following sources: for central nervous system (CNS) and retina, e.g., Chitnis et al. 1995; Henrique et al. 1995; Dorsky et al. 1995, 1997; review in Beatus and Lendahl 1998; Perron and Harris 2000; for otic placode and derivatives, e.g., Haddon et al. 1998; Riley et al. 1999; Morrison et al. 1999; for other placodal derivatives, e.g., Schlosser and Northcutt 2000; for ciliated epidermal cells, Deblandre et al. 1999; for neural crest derivatives, Mitsiadis et al. 1998; Wakamatsu et al. 2000; for pronephros, McLaughlin et al. 2000; for somites, review in Jiang et al. 2000; for endodermal structures (mainly endocrine cells in gut, pancreas, and lung), Jensen et al. 2000a, 2000b; Post et al. 2000; Lammert et al. 2000; for lymphocytes, review in Deftos and Bevan 2000; Rothenberg 2000.) *C,* Out-group comparison of vertebrates with urochordates (Hori et al. 1997) and cephalochordates (Holland et al. 2001) indicates that the Notch module in ancestral chordates was employed in the CNS, the notochord, paraxial mesoderm, and pharynx. During chordate evolution it was repeatedly redeployed in additional domains as illustrated (feathers: Crowe et al. 1998; hair follicles: Powell et al. 1998; Lin et al. 2000) suggesting that it also acted as a module in chordate evolution.

players. Moreover, while most versions of the Notch pathway promote some sort of cell fate decision, they may do so via lateral induction rater than lateral inhibition, or may amplify intrinsic rather than stochastic differences between adjacent cells, to name only a few recently emerging complexities (see Kimble and Simpson 1997; Artavanis-Tsakonas et al. 1999; Mumm and Kopan 2000; Baker 2000; Frisén and Lendahl 2001; de Celis 2004).

The Notch pathway qualifies as a module because it acts in a relatively autonomous fashion. It is now known to be employed not only during early neurogenesis but in an immense number of tissues derived from all three germ layers (Lewis 1998). In vertebrates these include epidermal derivatives, various parts of the central nervous system, various placodal derivatives, various neural crest derivatives, somites, pronephros, lymphocytes, gut, lung, and pancreas (see fig. 23.4, *B*). In most of these domains, the Notch pathway appears to mediate cell fate decisions among adjacent cells. However, the particular cell fates adopted are domain specific. For example, Delta expressing cells in the early neural plate will differentiate into primary neurons (e.g., Chitnis et al. 1995; Henrique et al. 1995), in the inner ear into hair cells (e.g., Haddon et al. 1998; Riley et al. 1999; Morrison et al. 1999), in the epidermis into ciliated epidermal cells (Deblandre et al. 1999), and in the teeth into ameloblasts and odontoblasts (Mitsiadis et al. 1998). This indicates that the Notch pathway is relatively context insensitive; it can be activated under different conditions and apparently affects different developmental processes in each context.

The notion that signaling pathways are autonomous modules has been challenged by mounting evidence for interactions between different signaling pathways (e.g., Axelrod et al. 1996; Brennan et al. 1999; Nishita et al. 2000; Borycki 2004). We still lack sufficient evidence to evaluate whether this really argues against the modularity of signaling pathways, but two arguments suggest that it does not. First, as parts of a functional whole organism, modules can not be absolutely autonomous, so it would be naive to expect strict compartmentalization barring any cross-talk.[5] Second, it has been argued that at least some of the evidence for interaction does not, in fact, support cross-talk between pathways, but rather a partial sharing of components (Noselli and Perrimon 2000). This suggests that signaling pathways are composed of recombinable submodules but does not preclude their acting as modules themselves.

Positional Modules

In multicellular organisms, particular cell types (e.g., muscle cells, neurons) or organ primordia may develop repeatedly in different regions of the body (see below) but then acquire distinct specializations depending on their position. These are thought to be controlled by position-specific selector genes (García-Bellido 1975; for review see Gerhart and Kirschner 1997; Carroll et al. 2001; Davidson 2001; Nelson 2004) or small networks of genes. Essential for the function of a positional selector is a defined spatiotemporal expression relative to other positional selectors, which may be due to their dependence on a common mechanism of pattern formation and/or cross-regulation of each other's expression. The *Hox* genes, which confer staggered identities along unidimensional axes (such as the anteroposterior body axis) are presumably the most famous example (reviewed, e.g., in Krumlauf 1994; Gellon and McGinnis 1998; Akam 1998). I will focus here, however, on the genes of the *Pax, Six, Eya,* and *Dach* families, which form a regulatory gene network (reviewed in Relaix and Buckingham 1999; Kawakami et al. 2000; Kardon et al. 2004) that appears to work as a multiply employed module affecting proliferation, inductive events, and morphogenetic movements in a position-specific manner depending on which *Pax* paralogue is involved (see, e.g., Mansouri et al. 1996; Dahl et al. 1997; Underhill 2000).

In vertebrate embryos, genes of these four families are coexpressed in many different structures, including the developing eye, kidney, ear, and somites (Heanue et al. 1999; Xu et al. 1999; Wawersik and Maas 2000). In mutants, these structures are often reduced in size or missing (e.g., Dahl et al. 1997; Xu et al. 1999; Kawakami et al. 2000). Moreover, the four genes and their encoded proteins appear to be integrated into a network of cross-regulatory interactions (fig. 23.5, *A*): Eya proteins act as transcriptional coactivators that bind directly to Dach proteins (another transcriptional coactivator) and to Six transcription factors (Heanue et al. 1999; Ohto et al. 1999). Expression of each of these three proteins is stimulated by Pax transcription factors and in turn stimulates *Pax* expression in a positive feedback loop (reviewed in Relaix and Buckingham 1999; Kawakami et al. 2000; Kardon et al. 2004). This simple picture is complicated by the fact that each of the four gene families has several members. Whereas *Eya* and *Dach* are represented by a single gene in invertebrates, four *Eya* paralogues and two *Dach* paralogues exist in vertebrates (Heanue et al. 1999; Xu et al. 1997; Borsani et al. 1999). Moreover, there are four groups of *Pax* genes and three groups of *Six* genes, each with a single member in invertebrates but with several members in vertebrates (Balczarek et al. 1997; Kawakami et al. 2000). *Pax* genes that belong to different groups generally show very different expression patterns and have clearly different functions. *Pax3, Pax2,* and *Pax6,* for example, are essential for the development of neural crest, ears and kidneys, and eyes, respectively.

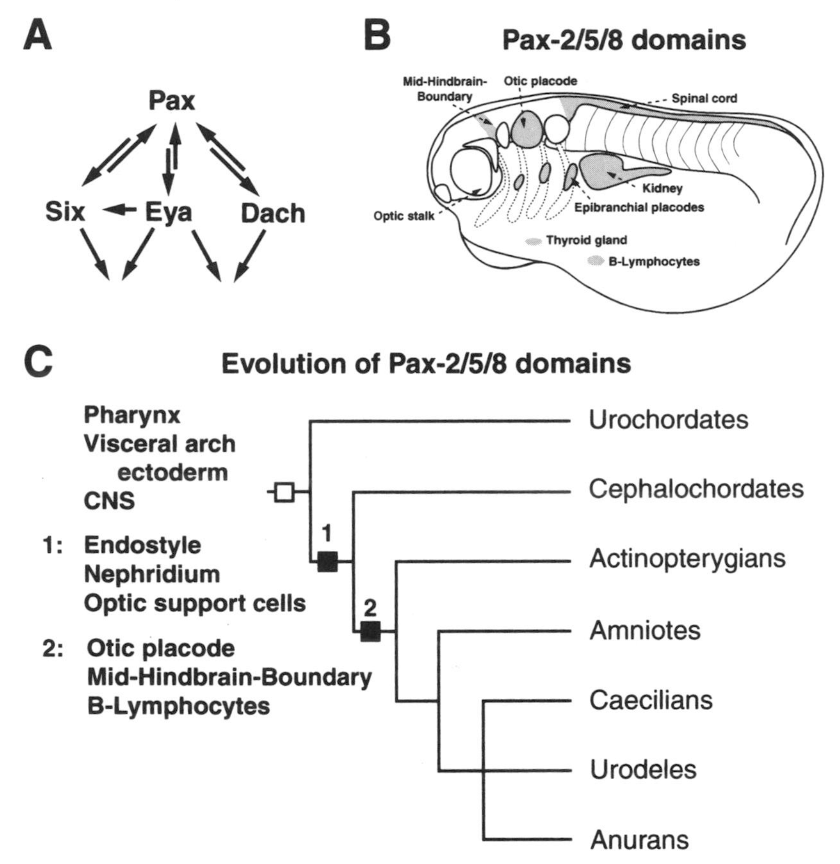

Fig. 23.5.—The vertebrate Pax-Six-Eya-Dach network as a positional module. *A,* Schematic diagram of known regulatory interactions between *Pax, Six, Eya,* and *Dach* genes (based on Relaix and Buckingham 1999; Xu et al. 1999; Loosli et al. 1999; Kawakami et al. 2000; Kardon et al. 2004). *B,* Schematic diagram of a vertebrate tailbud stage embryo depicting a subset of all the domains where vertebrate *Pax2, Pax5,* or *Pax8* genes are expressed (e.g., Adams et al. 1992; Heller and Brändli 1997, 1999; Pfeffer et al. 1998; Morrison et al. 1998), which are all derived from a single ancestral gene (*Pax2/5/8*). Coexpression with *Six* (Oliver et al. 1995; Ohto et al. 1998; Esteve and Bovolenta 1999; Pandur and Moody 2000; Kobayashi et al. 2000; Kawakami et al. 2000), *Eya* (e.g., David et al. 2001; Xu et al. 1997; Sahly et al. 1999), and *Dach* genes (Hammond et al. 1998; Heanue et al. 1999; Kozmik et al. 1999b; Davis et al. 1999, 2001) has been shown for otic placode, kidney, epibranchial placodes, optic stalk, and spinal cord. This suggests that the entire network of *Pax, Six, Eya,* and *Dach* genes is multiply employed and acts as a relatively context-insensitive module in development. *C,* Out-group comparison of vertebrates with urochordates (Wada et al. 1998) and cephalochordates (Kozmik et al. 1999a) suggests that in ancestral chordates *Pax2/5/8* was employed in the pharynx, the visceral arch ectoderm, and the CNS but was later redeployed in additional domains as indicated (alternative evolutionary scenarios are possible, depending on how ascidian expression domains are homologized; see "Developmental Modules as Modules of Evolution: Examples from Different Levels"). This redeployment may have involved the entire network, suggesting that it also acted as a module in chordate evolution. However, confirmation of this scenario awaits the analysis of *Six, Eya,* and *Dach* gene expression in uro- or cephalochordates.

There appears to be, however, some promiscuity regarding which *Six* and *Eya* genes they are coexpressed with, but at present, we have still insufficient data to decide whether these different combinations are functionally equivalent or not.

The Pax-Six-Eya-Dach network acts as a module because it functions in a relatively context-insensitive way. This is supported by two lines of evidence. First, as already noted, these genes are coexpressed and presumably form a similar network in multiple areas that constitute different domains (fig. 23.5, *B*). These domains have distinct spatiotemporal expression profiles and are important for the development of completely unrelated tissues, indicating that at least some inputs and outputs of the network differ between domains. For example, *Pax2, Eya1,* and *Six1* or *Six2* are required for the orchestration of multiple processes in mammalian ear as well as kidney development (Xu et al. 1999). Second, the network is able to operate ectopically, that is, in abnormal cellular environments. Ectopic expression of *Pax* or *Six* genes is able to dramatically affect the identity of a tissue in orchestrating complex developmental decisions and morphogenetic processes. This was made famous by the misexpression of the *Drosophila eyeless* gene (an orthologue of vertebrate *Pax6*), which results in ectopic eye development (Halder et al. 1995) involving ectopic activation of *Six* and *Eya* homologues (Halder et al. 1998a). Due to the employment of the network in multiple domains, however, particular fate changes can be obtained only in a certain subset of tissues. For instance, *eyeless* will lead to ectopic eye formation when misexpressed in the imaginal discs but not when misexpressed in other tissues of *Drosophila* (e.g., Gehring and Ikeo 1999; Wawersik and Maas 2000). Similar results have since been obtained in vertebrates after misexpression of *Pax6* or *Six3* (Chow et al. 1999; Loosli et al. 1999; reviewed in Wawersik and Maas 2000). It should be pointed out, however, that misexpression of other *Pax* genes typically has less dramatic effects (Underhill 2000).

Cell Type Modules

Whereas large multicellular organisms may contain millions or billions of cells, these belong to only a small number (on the order of 100–300) of specialized cell types. Most cell types are present in multiple copies in the organism. The determination and differentiation of a particular cell type is often controlled by a few genes that act as cell-type-specific selector genes or participate in small networks of genes that jointly regulate the expression of a large number of cell-type-specific proteins. This is perhaps best established for the myogenic genes, encoding a group of basic helix-loop-helix (bHLH) transcription factors including MyoD, Myf-5, and myogenin (reviewed, e.g., in Yun and Wold 1996;

Molkentin and Olson 1996; Perry and Rudnicki 2000). Mice lacking both MyoD and Myf-5 lack all skeletal muscles, suggesting that they are essential for the differentiation of skeletal muscle cells.

Similar networks, also involving many bHLH genes, are important for neuronal determination and differentiation (fig. 23.6, *A;* reviewed in Lee 1997; Guillemot 1999; Brunet and Ghysen 1999; Strähle and Blader 2004). Several neuronal determination genes (including *Neurogenin1* and *Neurogenin2*) converge on a few neuronal differentiation genes such as *NeuroD* which encode transcription factors that regulate many different target genes, thereby integrating various aspects of general neuronal differentiation. As in the regulation of myogenic differentiation, the genes involved in neurogenesis are not ordered in a simple linear hierarchy, but cross-regulate each other's expression in feedback loops (Koyano-Nakagawa et al. 2000). In mutants of neuronal determination or differentiation genes, neurogenesis is perturbed to various degrees (e.g., Ma et al. 1998; Fode et al. 1998; Morrow et al. 1999; Horton et al. 1999; Schwab et al. 2000). However, in the mutants analyzed so far, neurons are never completely absent, probably because of the existence of several genes with partially redundant or complementary functions.

Again, two kinds of observations support the idea that the network of factors regulating neurogenesis operates in a context-insensitive manner and hence acts as a module. First, one out of several neuronal determination genes is coexpressed with and presumably activates neuronal differentiation genes (such as *NeuroD*) in many different locations where neurogenesis takes place (fig. 23.6, *B*), including many areas in the central nervous system, all neurogenic placodes, and neural crest cells (e.g., Lee et al. 1995; Sommer et al. 1996; Korzh et al. 1998; Schlosser and Northcutt 2000). The same network is also important for the development of endocrine cells in the gut and pancreas (Naya et al. 1997; Gradwohl et al. 2000). Importantly, many of these multiple expression sites belong to different domains, judged by the fact that they independently activate the network for neurogenesis in unique spatiotemporal profiles and generate neurons that exhibit region-specific specializations in addition to their general neuronal features. For example, the same network responsible for primary neurogenesis in the spinal cord of amphibians also seems to govern the production of secondary neurons (Schlosser et al. 2002), although the latter are generated much later, are cytoarchitectonically distinct, and innervate different targets. Second, ectopic activation of the network results in neuronal differentiation even in cells not normally fated to become neurons (e.g., Lee et al. 1995; Ma et al. 1996; Blader et al. 1997; Perron et al. 1999), suggesting that it is able to operate normally in different cellular environments.

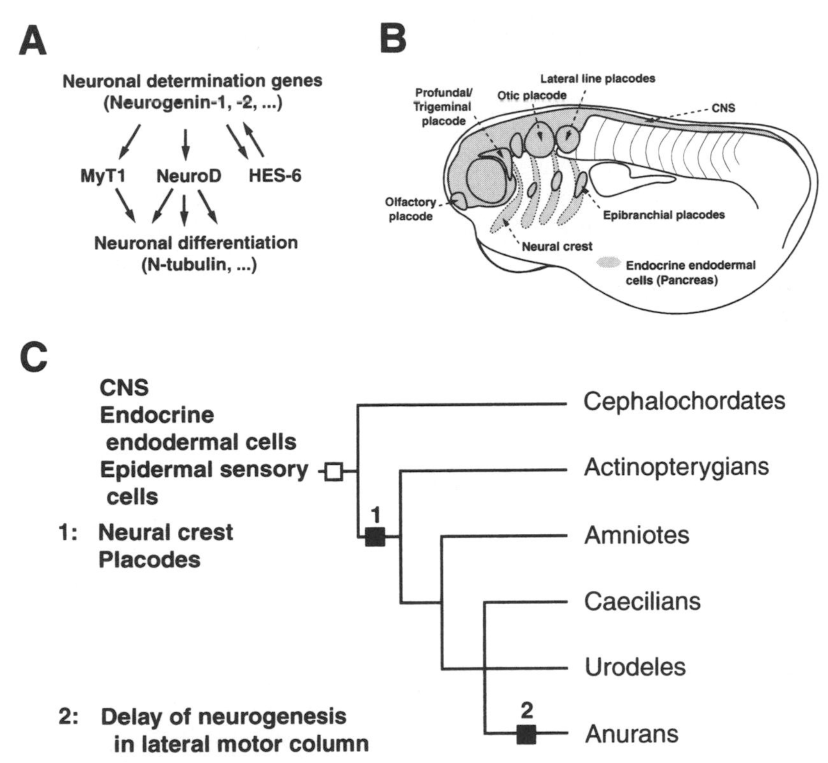

Fig. 23.6.—The network regulating vertebrate neurogenesis as a cell type module. *A*, Schematic diagram of interactions among various neuronal determination and differentiation genes (based on various sources, e.g., Koyano-Nakagawa et al. 1999, 2000). *B*, Schematic diagram of a vertebrate tail bud stage embryo depicting a subset of all the domains where *Neurogenins* and *NeuroD* (and in most cases other components of the network depicted in *A*) are known to be expressed. The employment of the network in multiple domains suggests that it acts as a relatively context-insensitive module in development. (Data are from Lee et al. 1995; Sommer et al. 1996; Korzh et al. 1998; Ma et al. 1996, 1998; Fode et al. 1998; Naya et al. 1997; Gradwohl et al. 2000; Schlosser and Northcutt 2000.) *C*, Outgroup comparison of vertebrates with cephalochordates (Holland et al. 2000) suggests that in the common ancestor of cephalochordates and vertebrates *Neurogenin* and probably other components of the network (whose expression in amphioxus is not yet known) was expressed in the CNS, endocrine endodermal cells, and epidermal sensory cells. During chordate evolution the network was repeatedly redeployed in additional domains or temporally shifted during development (only two out of probably many such events are indicated here), suggesting that it acted as a module in chordate evolution.

Organ Modules

The examples of modules presented so far concerned relatively small-scale units, such as networks of a few genes. However, larger-scale units, involving many different communicating cell types, such as organ primordia or even entire body regions like insect segments, can also act as modules. Organ primordia often develop as *morphogenetic fields* (Gilbert et al. 1996) with remarkable regulatory capacities. The underlying networks of developmental interactions are in most cases still incompletely understood, but in a few cases the robust behavior of such primordia can be simulated by relatively simple models (e.g., Thieffry and Sánchez 2004). For instance, it has recently been shown convincingly that a simulated network of *Drosophila* segmentation genes is relatively insensitive to perturbations or alterations of interaction parameters (von Dassow et al. 2000; von Dassow and Meir 2004).

The vertebrate limb bud is probably the most prominent example of a complexly integrated organ primordium that nonetheless develops relatively autonomously and, thus, constitutes a module (for review see, e.g., Raff 1996; Shubin et al. 1997; Tabin et al. 1999; Ng et al. 1999; Gilbert 2000). Many more examples exist, though, and I will focus here on the lateral line placodes of anamniotic vertebrates (fig. 23.7, *A*). These are local thickenings of the ectoderm that give rise to the mechano- and electroreceptive sense organs of the lateral line, as well as to the sensory neurons supplying them (e.g., Stone 1922; Northcutt 1992; Northcutt et al. 1994; Northcutt and Brändle 1995). Thus, the cell types that function together in lateral line sensory perception also share a common origin in development. Moreover, the activation of the Notch network in lateral line placodes (Smithers et al. 2000; Schlosser and Northcutt 2000; Itoh and Chitnis 2001) suggests that interactions between placodal cells are probably essential to generate the different cell types of the lateral line system (at least in the ear placode, cell fate decisions between similar cell types, namely, hair cells and supporting cells, depend on Notch signaling; e.g., Haddon et al. 1998; Riley et al. 1999).

Besides being tightly integrated, lateral line placodes develop in a relatively context-insensitive fashion. Even after transplantation to ectopic sites (fig. 23.7, *B*), they give rise to both neurons and lateral line primordia. The timing and polarity of migration of the latter, the type, size, spacing, and number of receptor organs formed, and the patterned differentiation of different cell types in these receptor organs, all proceed normally in ectopic sites (Stone 1928, 1929; Smith et al. 1990; Northcutt et al. 1994, 1995; Schlosser and Northcutt 2001). Only the pathways of migration for lateral line primordia seem to depend on external cues (Harrison 1904; Stone 1929; Smith et al. 1990).

Systemic Modules

Finally and perhaps counterintuitively, there may be modules that cannot be localized as a structural unit, but rather are distributed throughout the organism. This is typical for many hormonally mediated processes. Normally only a subset of cells is responsive to a particular hormone, but often these cells are found throughout the body

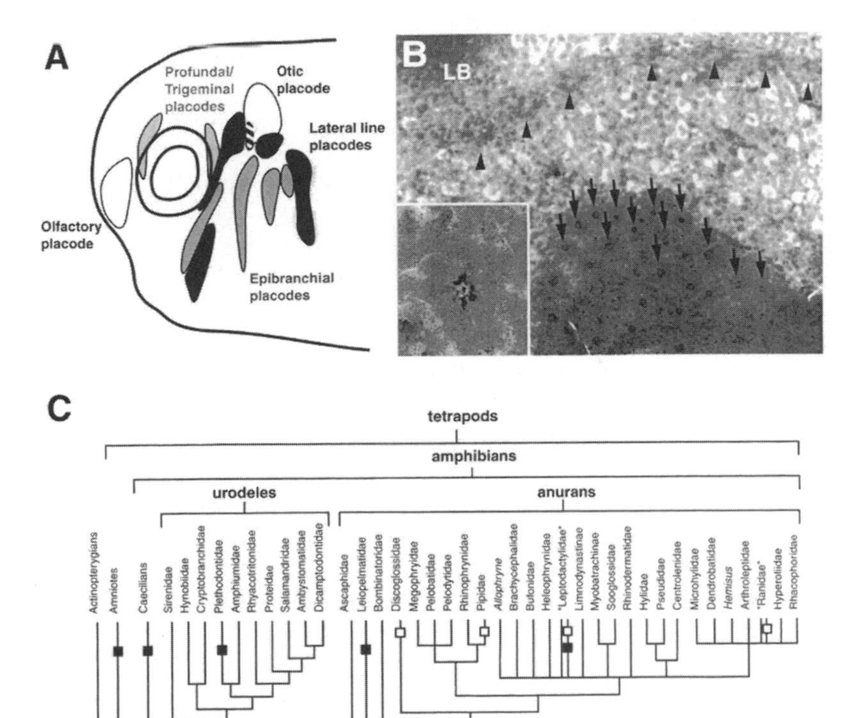

Fig. 23.7.—The vertebrate lateral line placodes as an organ module. *A,* Schematic diagram of different ectodermal placodes in a tail bud stage *Xenopus* embryo (after Schlosser and Northcutt 2000). Lateral line placodes are depicted in black. *B,* Lateral line placodes are modules that develop in a relatively context-insensitive manner. After transplantation of axolotl lateral line placodes from a pigmented donor embryo to the belly of an unpigmented host embryo, pigmented lateral line receptor organs called neuromasts (*arrows;* insert shows one neuromast at higher magnification) develop normally and are normally patterned in an ectopic location on the belly, ventral to the unpigmented neuromasts (*arrowheads*) of the host lateral lines (anterior is to the left; LB, limb bud). *C,* Phylogeny of tetrapods, including a detailed amphibian phylogeny based on Ford and Cannatella (1993) for anurans and Larson and Dimmick (1993) for urodeles. *Xenopus* belongs to the Pipidae, whereas the axolotl belongs to the Ambystomatidae. "Leptodactylidae" and "Ranidae" are considered paraphyletic and hence are listed in quotation marks. Out-group comparison of tetrapods with actinopterygians and elasmobranchs suggests that ancestral gnathostomes had a lateral line system, which was retained after metamorphosis. All components of the lateral line system were then completely lost several times in different lineages, and their fate during metamorphosis also changed repeatedly independently of other structures. There are possibly many more cases of complete loss in frogs, whose development has not yet been described. The fact that different components of the lateral line system show repeated coordinated changes during tetrapod evolution suggest that it acted as a module in tetrapod evolution. (Data are based on Lynn 1942; Stephenson 1951; Fritzsch and Wake 1986; Fritzsch et al. 1987; Wake et al. 1987; Fritzsch 1988, 1990; Roth et al. 1993; Schlosser and Roth 1997; Schlosser et al. 1999.)

intermingled with nonresponsive cells. Thyroid-hormone-dependent metamorphosis of amphibians provides a beautiful example (for another example see Davidson et al. 2004). Metamorphosis involves a plethora of well-orchestrated changes in many different tissues, including alterations in metabolic enzyme expression, extensive cell death, differentiation of new cell types in gut and epidermis, and remodeling of muscles, cartilage, and parts of the nervous system (reviewed in Dodd and Dodd 1976; Fox 1981; Duellman and Trueb 1986). Metamorphic changes appear to be confined to tissues that express thyroid hormone receptors and are initiated in a coordinated fashion by rising thyroid hormone concentrations at the end of larval life (reviewed, e.g., in Shi et al. 1996; Tata 1996, 1998; Su et al. 1999). Blocking hormone production in the thyroid gland completely arrests metamorphic changes (reviewed in Dodd and Dodd 1976; White and Nicoll 1981; see also Rose 1995a, 1995b, 1996; Schmidt and Roth 1996; Wakahara and Yamaguchi 1996). Because all metamorphic events depend on a common factor, they form a single domain, although they are not spatially contiguous. The spatial and temporal integration of such metamorphic events appears to be mainly due to the triggering of several stereotyped responses by a common factor without involving further interactions of the events thus triggered. Metamorphic events can be considered a relatively context-insensitive module because they can be initiated precociously by thyroid hormone injections or can be delayed by experimentally retarding thyroid hormone availability (reviewed, e.g., in Dodd and Dodd 1976; White and Nicoll 1981; Rose 1996), indicating that they do not depend on an exact temporal coordination with other developmental events (see also Schlosser in press).

Developmental Modularity Promotes the Capacity to Develop (Developability)

Development denotes a phase of the life cycle during which a complex multicellular organism is built from a small and often unicellular propagule like the zygote. Thus, development as it is normally understood is found only in multicellular organisms but has evolved several times independently (e.g., in plants and animals; Buss 1987; Maynard Smith and Szathmáry 1995). The capacity to develop in a reproducible fashion is promoted by a modular organization for two reasons. First, a system composed of many modules is at the same time robust (regarding the context-insensitive behavior of each module) and flexible (regarding the intermodular interactions), facilitating a stable and in each generation reproducible increase in complexity, relatively insensitive to perturbations (e.g., Gerhart and Kirschner 1997). Second, due to their relative autonomous behavior, modules are available for repeated employment in different contexts, allowing the rapid generation of dramatically complex systems by the combinatorial use of a relatively few basic building blocks (e.g., Bonner 1988).

That developmental modules are indeed employed in such a combinatorial fashion can be easily seen by considering the examples discussed above. They indicate that modules can be part of larger scale modules: the Notch module, the Pax-Six-Eya-Dach module, and the module of genes involved in neurogenesis are each involved in the development of placodes and hence are components of the placode module. However, the same smaller-scale modules in different combinations are also components of other large-scale modules. For instance, the Notch and Pax networks are also involved in pronephros development, but the neurogenesis module is not.

This combinatorial use of various modules appears to be greatly facilitated by the relatively free combinability of binding sites and/or enhancers in the cis-regulatory regions of genes discussed above (Gray et al. 1994; Kirchhamer et al. 1996a, 1996b; Arnone and Davidson 1997; Davidson 2001; Carroll et al. 2001). Several recent studies in *Drosophila* support this scenario (Halder et al. 1998b; Flores et al. 2000; Xu et al. 2000; Halfon et al. 2000; Guss et al. 2001; reviewed in de Celis 1999; Ghazi and Vijay Raghavan 2000; Nelson 2004) by demonstrating that particular cell-type-specific target genes are activated only after cooperative binding of various position-specific transcription factors and components of signaling pathways to particular binding sites in their enhancers. Moreover, different target genes may be activated by different combinations of the same transcription factors due to different combinations of binding sites in their enhancers. For example, the expression of *Pax2*, which is an important component of the network specifying cone cell identity in the *Drosophila* eye imaginal disc, is specifically activated in a subset of retinal cells by combination of the eye-specific transcription factor lozenge and inputs from the Notch and epidermal growth factor (EGF) signaling pathways, whereas *prospero,* which is important for R7 cell identity, also requires lozenge and EGF but not Notch input and hence is activated in a different subset of cells (Flores et al. 2000; Xu et al. 2000). Other combinatorial arrangements probably permit the same signaling pathways to induce the generation of different cell types in other imaginal discs.

Modules of Evolution

Unraveling the modularity of life cycles not only aids our understanding of development, but also has important implications for evolution. Many developmental modules that were first discovered in insects were later shown to be important for the development of other taxa as well (vertebrates, nematodes, etc.), suggesting that some modules of development tend to be evolutionarily conserved over surprisingly long periods. The general features of modules suggest an attractive hypothesis to explain this fact: the integrated and context-insensitive nature of modules may favor their preservation or coherent transformation during evolution but may permit their recombination and embedding in new contexts by modifications of intermodular interactions. According to this hypothesis, therefore, developmental modules will also form the *building blocks* of evolutionary transformation (Wagner 1995; Wagner and Altenberg 1996) or *units of evolution* (Schlosser 2002, in press). In other words, modules of development will also act as modules of evolution. In order to evaluate this claim, it will be necessary to spell out more precisely under which conditions several parts (or subprocesses) of a process form a *module of evolution* and whether modules of development tend to meet these conditions.

Modules of Evolution Are Units of Integrated and Context-Insensitive Evolutionary Change

Whereas developmental modules are units that behave in an integrated and context-insensitive fashion during development, modules of evolution are integrated and context-insensitive units of evolutionary change ("units of evolution" of Schlosser 2002). Obviously, this notion of a module of evolution can be applied only to processes or systems which evolve, that is, which reproduce themselves in a chain of generations, but which are open for the introduction of heritable variation, such as life cycles. And a subprocess or unit of interacting elements of such reproducing systems qualifies as a module of evolution only if it tends to make an integrated and context-insensitive contribution to its own reproduction in subsequent generations in the face of a particular class of perturbations, namely, heritable variations (e.g., mutations in genes of germ line cells).

Modules of evolution are, therefore, units of elements whose variants tend to have interdependent but relatively context-insensitive effects on fitness. Elements (such as genes) constituting a module of evolution tend to *coevolve*—that is, tend to change in a coordinated and interdependent fashion during evolution—because they reciprocally constrain the accumulation of each other's heritable modifications (integration), while their evolutionary modifications tend to be *dissociated* from changes in other elements of the life cycle, because their heritable variations are relatively unconstrained by heritable change in those other elements (context insensitivity). Two kinds of constraints have to be considered here: developmental and functional constraints. There are *developmental constraints* (see, e.g., Gould and Lewontin 1979;

Alberch 1980, 1982; Maynard Smith et al. 1985; Wake and Larson 1987; Schwenk 1994; Shubin and Wake 1996; Wagner and Schwenk 2000) when several elements (and hence their fitness) are not independently variable due to their common dependence on the same causal factors (e.g., common inducers). *Functional constraints,* in contrast, are due to the joint requirement of several elements for fulfilling some function, that is, for the generation of some vital state or behavior. As a consequence, the fitness of variants in any particular element will be highly sensitive to the presence of particular variants of the other elements (i.e., there will tend to be epistasis for fitness; for reviews of epistasis see Hedrick et al. 1978; Wade 1992; Whitlock et al. 1995; Fenster et al. 1997; Cheverud and Routman 1995; Wagner et al. 1998). This will result in internal selection, where the other elements of the unit rather than the external environment exert important selection pressures (reviewed in Schwenk 1994; Arthur 1997; Wagner and Schwenk 2000).

Again, it is important to bear in mind that integration and context insensitivity come in degrees and are not absolute properties. Modules of evolution *tend* to respond to heritable variations in a coherent way; that is, they may behave as a coherent unit with respect to many heritable variations of the life cycle but not with respect to all of them. In order for a unit of elements to behave as a coherent module in at least some evolutionary transformations, it is only required that the *probabilities* of linked variabilities and epistatic fitness interactions be higher among variants of components of the module than between variants of those components and other elements of the life cycle (for a detailed account see Schlosser 2002).[6]

Although modules of evolution have some affinity to the concept of a *unit of selection* in population genetics, there are important conceptual differences between the two notions that cannot be further explored here. I merely want to point out that a module of evolution is a broader notion that does not focus on single selection processes among simultaneously existing actual variants but rather considers probabilities of fitness interactions among possible variants that should be reflected in the long-term evolutionary fate of a unit (see Schlosser 2002).[7]

Do Modules of Development Act as Modules of Evolution?

Developmental modules are excellent candidates for modules of evolution: (1) they are reliably reproduced in each generation because they operate as integrated units that are relatively insensitive to perturbations of the context in which they are embedded, and (2) heritable variation is only a special kind of perturbation. However, there are two crucial differences between modules of development and modules of evolution which may lead to a mismatch between these two kinds of units (fig. 23.8). First, modules of evolution tend to make an integrated and context-insensitive contribution to their reproduction in subsequent generations and hence to their fitness, whereas modules of development tend to make an integrated and context-insensitive contribution to some developmental process. Second, modules of evolution are defined by their coherent response to heritable variations (i.e., by their "variational properties"; see Altenberg 1995; Wagner and Altenberg 1996; Wagner and Mezey 2004) and thus make up interdependent but relatively context-insensitive units of the "genotype-phenotype map" (Riedl 1975; Cheverud 1984, 1996; Wagner 1996; Wagner and Altenberg 1996; Mezey et al. 2000; Cheverud 2004; Wagner and Mezey 2004). In contrast, modules of development are defined by their coherent response to a different set of perturbations during development. Because heritable variations may have pleiotropic effects on the fitness of different developmental modules, they may link their evolutionary fate. Consequently, developmental modules will tend to act as modules of evolution only, when two additional conditions are met.

First, the integrated and context-insensitive behavior of a module during development should be reflected in a tendency for integrated and context-insensitive effects on fitness. However, this will be the case only when the behavior of a developmental module (i.e., its IOR) is not only relatively insensitive to perturbations of the context, but when the reverse is also true: the behavior of the module's surroundings must also be relatively insensitive to perturbations of the module (e.g., because the surroundings are also organized in a modular fashion); that is, the module must be relatively *dispensable* for the context in which it operates. The reason is that the fitness of any component of a life cycle is determined by the fitness of the entire life cycle in which it resides and thus depends on the fitness consequences of *all* its effects, including those side effects of a developmental module that are inconsequential for its coherent behavior (IOR) during development (see fig. 23.8). Thus, a developmental module is less likely to act as module of evolution if there is no reciprocal autonomy of the developmental module and its developmental context.[8]

For example, whereas lateral line placodes act as developmental modules that develop completely independently of pigment patterns in salamanders, the formation of particular pigment patterns depends strongly on how lateral line placodes migrate (Parichy 1996a, 1996b). As a consequence, the effects of heritable variations on the fitness of lateral line development and on pigment patterns are not independent of each other, and lateral line migration and pigment pattern formation may be tied together into a single module of evolution with a certain probability of coordinated evolutionary change due to developmental

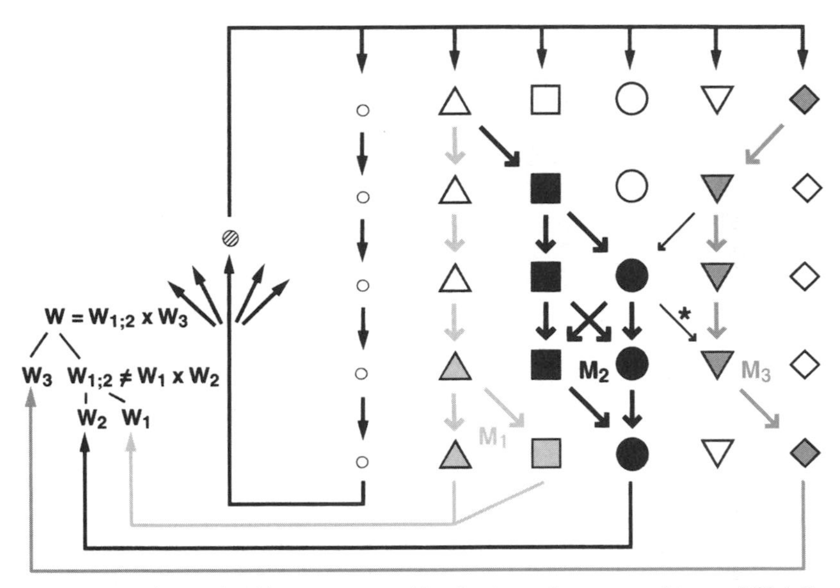

Fig. 23.8.—Developmental modules may act as modules of evolution when certain conditions are fulfilled. The schematic diagram shows several developmental modules (*light grey, black, dark grey*) embedded in a life cycle. The various symbols may, for instance, be interpreted as different cell types, the small circle representing the germ line and the hatched circle the propagule or zygote (in the case of sexual reproduction). The fitness W of a life cycle is its expected rate of reproduction relative to other variants of the life cycle in the population. It is assumed that different elements of the same developmental module strongly constrain each other's evolution, because either they are not independently variable or there is fitness epistasis between variants of these elements (i.e., the fitness of a particular combination of variants cannot simply be expressed as the product of fixed fitness values for each variant). However, similar constraints may also tie different developmental modules together. For instance, M_1 and M_2 represent different developmental modules that do not interact with each other and are to a certain degree independently perturbable during development (e.g., perturbation of the grey square late in development would affect the behavior only of M_1). However, they do not necessarily contribute independently to fitness, because the square element plays a role in both modules. Therefore, some heritable variations of the square may have pleiotropic effects on both circles and triangles and hence affect the fitness of M_1 as well as M_2. Because the fitness of squares may then depend epistatically on the fitness of circles as well as on triangles, their fitness contribution to M_1 may differ for different variants of circles. Consequently, fitness components W_1 (of M_1) and W_2 (of M_2) may not independently contribute to the fitness $W_{1;2}$ of the unit $M_{1;2}$ consisting of triangle, square and circle (i.e., $W_{1;2}$ cannot be calculated by simply multiplying the fitness components W_1 and W_2). If certain conditions are met, however, developmental modules tend to make independent fitness contributions. Assume, for example, that modules M_1 and M_2 are largely dispensable for M_3 (for example, it is assumed that variants of elements of M_2—circles or squares—are unlikely to modify the interaction marked with an asterisk in a way that alters the fitness of variants of elements of M_3 epistatically) and vice versa. Moreover, the elements of modules $M_{1;2}$ do not play additional roles in module M_3 and are assumed to show independent heritable variability (no pleiotropic effects on M_3) in congruence with their independent perturbability during development. As a consequence, M_3 and $M_{1;2}$ tend to make independent fitness contributions; that is, for most combinations of variants of M_3 and $M_{1;2}$ there will be no fitness epistasis (i.e., the fitness values of the life cycle W can be determined by multiplication of $W_{1;2}$ and W_3).

constraints (both processes cannot vary independently) and possibly functional constraints (certain migration patterns of placodes may be prohibited because they disrupt vital functions of pigmentation such as mate recognition or protection from ultraviolet light).

Second, developmental modules can be expected to act as modules of evolution only if the effects of heritable variations relatively faithfully mimic the effects of perturbations during development such as environmental fluctuations or somatic mutations. But such *congruence* of the effects of developmental perturbation and heritable variation (comparable to the "plastogenetic congruence" of Ancel and Fontana 2000) will not necessarily exist, because some heritable variations may have pleiotropic effects on multiple developmental interactions even when these are independently perturbable during development (fig. 23.8). Pleiotropic effects are particularly likely to occur after heritable variations of elements that play roles in multiple developmental modules, for example, in different domains of a particular type of module.

A germ line mutation in the coding region of a single gene, for example, will affect interactions of the encoded protein in all the domains where the gene is expressed, even when these domains operate as mutually independent, individualized developmental modules that are unlikely to be simultaneously affected by local developmental perturbations. Lateral inhibition by the Notch module, for instance, plays important roles during neurogenesis and hematopoiesis, to name only two of many domains. While the hematopoietic domain can be perturbed independently of the neurogenic domain by localized somatic mutations or experimental manipulations during development, many germ line mutations affecting intramodular interactions (e.g., mutations affecting the motif of the Notch protein binding to Suppressor of hairless) will affect all domains in which the Notch module is employed. Such heritable variations will therefore tie different individualized developmental domains (e.g., of a signaling pathway) together into a single module of evolution to the extent that these have the same genetic basis (i.e., involve orthologous and not paralogous genes).[9] Other heritable variations may pleiotropically affect interactions between different domains and their contexts (e.g., mutations affecting the binding of Suppressor of hairless to various domain-specific transcriptional coactivators), possibly resulting in additional functional or developmental constraints (see, e.g., Duboule and Wilkins 1998; Niehrs and Pollet 1999) that may link the evolutionary fate of a multiply instantiated developmental module with each of the contexts of its different domains.

In conclusion, modules of development will tend to coincide with modules of evolution only when (1) they are not only insensitive to, but also dispensable for, the developmental context in which they are embedded (reciprocal autonomy) and when (2) heritable variations have relatively few pleiotropic effects on different developmental modules that are individually and independently perturbable during development. Although we do not know how frequently these conditions are met during life cycle evolution, there is growing evidence for organizational features of life cycles that favor them. First, the existence of developmental modules is apparently not an exceptional feature of only some phases of development or some regions of the developing organism. Rather, modularity appears to be a pervasive architectural feature of living organisms (e.g., Raff 1996; Gerhart and Kirschner 1997). Therefore, the developmental context of a module may often itself be organized in a modular fashion favoring reciprocal autonomy. Second, the modularity of molecular interactions, for instance, modularity of gene regulation (see above) as well as modular interactions of multimotif proteins (e.g., Strähle and Blader 2004), greatly promotes the independent heritable variability of different developmental interactions of a multiply employed element (e.g., gene or protein) with few pleiotropic effects. For example, mutations in different cis-regulatory regions may have domain-specific effects on gene interactions, and mutations of different parts of the coding region of a gene may affect different protein domains, such as extra- or intracellular binding sites of a transmembrane receptor.

However, in order to assess the importance of developmental modules in evolution we must go beyond these general considerations and specify how claims that certain developmental modules act as modules of evolution can be tested empirically. The following section briefly sketches and illustrates a methodology for such tests.

Developmental Modules as Modules of Evolution: Examples from Different Levels

Characterizing a module of evolution as a unit of elements with relatively high probability of interdependent effects on fitness allows predictions to be made regarding the expected frequencies of coordinated evolutionary changes: evolutionary changes of different elements of the module should be repeatedly correlated with each other (coevolution) but should be uncorrelated with changes in other elements (dissociation). Consequently, elements belonging to a single module of evolution should exhibit frequent *dissociated coevolution* in the actual phylogeny of a lineage reflected in certain trends or parallelisms of evolutionary modifications (fig. 23.9).[10] Phylogenetic trends of recurrent dissociated coevolution of several elements thus indicate that these elements are likely to coevolve in a particular lineage (see also Alberch 1980; Alberch and Gale 1985; Shubin and Alberch 1986; Wake and Lar-

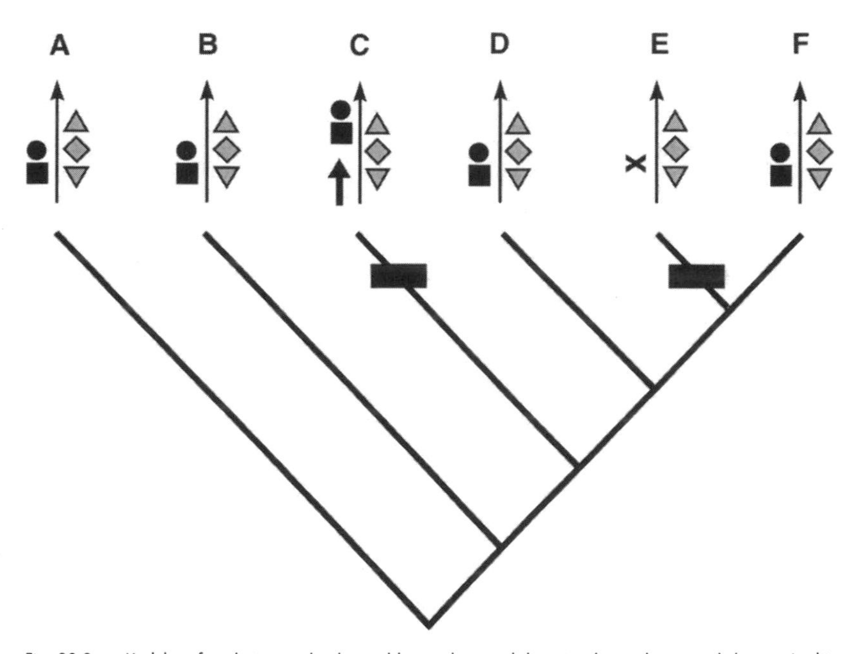

Fig. 23.9.—Modules of evolution can be detected by trends toward dissociated coevolution in phylogeny. In this example, repeated dissociated coevolution of characters is revealed by mapping sequences of developing characters (*symbols next to long arrows*) on a phylogenetic tree of species A–F. Out-group comparison among the species depicted allows us to infer two independent events (*black bars*) of dissociated coevolution of the black characters (*circle* and *square*) from the grey characters: a coordinated heterochronic shift (*short arrow*) in C, and a coordinated loss (*cross*) in E.

son 1987; Wake 1991; Wray and Bely 1994; Shubin and Wake 1996; Schlosser 2001, in press), assuming that recurrent environmental changes cannot account for the trends (e.g., Alberch 1980, 1983; Maynard Smith et al. 1985; Wake 1991; Shubin and Wake 1996; Wagner and Schwenk 2000).[11]

Dissociated coevolution of a unit of elements may manifest itself in many different ways. In particular, the unit may exhibit

1. Frequent losses of all its elements, without affecting the context
2. Frequent coordinated shifts in timing (heterochrony) or location (heterotopy) of developmental activity of all its elements relative to the context
3. Frequent redeployment of all its elements in new contexts (i.e., establishment of new domains) possibly accompanied by the evolution of novel functions

4. Other kinds of frequent coordinated evolution (e.g., in size, shape, activity) of all its elements (reflected, for instance, in a correlation of rates of evolutionary changes)

In order to test a claim that certain developmental modules act as modules of evolution in particular lineages, the frequency of dissociated coevolution of the components of the developmental module during phylogeny of these lineages can be elucidated using comparative phylogenetic methods (e.g., Felsenstein 1985; Maddison 1990; Harvey and Pagel 1991; Pagel 1999) such as out-group comparison (Fink 1982; Kluge and Strauss 1985; Northcutt 1990), assuming that the phylogenetic relations of the species in a certain lineage are already known from prior cladistic analysis (see Eldredge and Cracraft 1980; Wiley 1981). Because developmental modules do in fact often act as modules of evolution, comparative approaches such as phylogenetic profiling (e.g., Eisenberg et al. 2000) have recently gained importance in functional genomics as heuristic methods of *predicting* developmental modules from phylogenetic patterns of correlated change.

The following paragraphs will summarize some evidence for frequent dissociated coevolution (with particular emphasis on redeployment) of the developmental modules discussed above (under "Modules of Development") during metazoan or vertebrate phylogeny, suggesting that they in fact acted as modules of evolution in these lineages (see also Schlosser 2001, in press).

Modules of Gene Regulation as Modules of Evolution

The regulation of gene transcription and the sequence specific assembly of proteins not only constitute different developmental modules (see above); there is also evidence that they often evolve independently of each other. For example, in many cases where after gene duplication different paralogues have evolved divergent functions, nonetheless the proteins encoded by these paralogues are often functionally equivalent and can substitute for each other when expressed at appropriate times and locations (e.g., Li and Noll 1994; Zakany et al. 1996; Maconochie et al. 1996; Bouchard et al. 2000; Greer et al. 2000). This suggests that functional divergence in these cases is due exclusively to changes of gene regulation, either to sequence alterations of upstream transcription factors or to changes in cis-regulatory elements (for review see Purugganan 1998, 2000; Tautz 2000; Dover 2000; Stern 2000).

Moreover, different modules of gene regulation (involving different parts of the cis-regulatory regions of a gene, e.g., different enhancers or different binding sites in a single enhancer) often evolve in a mosaic fashion. This is supported by several observations. First, the evolutionary loss or modification of single binding sites (or enhancers) results in the selective loss of a subset of expression domains, suggesting loss of responsiveness to a particular transcription factor (or set of transcription factors), without affecting the responsiveness to others (Ross et al. 1994; Belting et al. 1998; Shashikant et al. 1998; Sucena and Stern 2000). For several cases of duplicate genes, the patterns of complementary losses of subsets of expression domains among paralogues in different lineages suggest that such events have repeatedly occurred during evolution, because each of the two paralogues has lost different cis-regulatory elements in different lineages (e.g., Sefton et al. 1998; Force et al. 1999, de Martino et al. 2000; Quint et al. 2000).

Second, as already noted, the same binding sites are found in the promoters or enhancers of multiple genes and confer similar types of responsiveness to transcription factors, wherever they occur (Arnone and Davidson 1997; Davidson 2001). As binding sites are small, part of this abundance may be explained by the independent generation of similar binding sites due to chance mutations (see Carroll et al. 2001; Force et al. 2004). However, there is also increasing evidence that binding sites or even larger cis-regulatory regions have been repeatedly duplicated and inserted at new sites during evolution (e.g., Britten 1996, 1997; Brosius 1999; Selinger and Chandler 1999; see below).

Third, enhancers are often surprisingly divergent, even when they govern similar expression domains of orthologous genes in different species and are able to functionally replace each other (Bonneton et al. 1997; Hancock et al. 1999; Takahashi et al. 1999; Ludwig et al. 2000). This appears to be due to the dissociated molecular coevolution of various binding sites within an enhancer: alterations of gene regulation due to changes in one part of the enhancer are compensated by the accumulation of changes in another part of the enhancer (coevolution), whereas there are no such compensatory changes across enhancer boundaries (dissociation) (e.g., Ludwig et al. 2000; see also Dover 2000; Stern 2000 for reviews).

Signaling Modules as Modules of Evolution

Dissociated coevolution of signaling pathways is evident in cases where correlated evolutionary changes occur specifically among the receptors, ligands, and other components of the pathway (see Goh et al. 2000 for an example). Dissociated coevolution is also supported by observations that signaling pathways are not only used multiply during development of a single species, but also have repeatedly acquired new domains with new functions during evolution. Typically, all components of a signaling pathway are jointly redeployed in these new domains (reviewed in Gerhart 1999; see Keys et al. 1999 for a nice example). The Notch module, for instance, has frequently acquired new domains during chordate evolution (fig. 23.4, *C*). While in amphioxus (Holland et al. 2001) and ascidians (Hori et al. 1997), where only the expression of Notch itself has been described so far, it is mainly employed in the central nervous system, somites, notochord, and endoderm, various new domains have evolved in vertebrates, for example, in the pronephros (McLaughlin et al. 2000), in lymphocytes (reviewed in Deftos and Bevan 2000; Rothenberg 2000), and in placodal (e.g., Haddon et al. 1998; Riley et al. 1999; Morrison et al. 1999; Schlosser and Northcutt 2000) and neural crest derivatives (Mitsiadis et al. 1998; Wakamatsu et al. 2000). During vertebrate evolution, novel domains evolved again several times, for example, in the development of feathers and hair in birds and mammals, respectively (Crowe et al. 1998; Powell et al. 1998; Lin et al. 2000), and in ciliated epidermal cells in amphibians (Deblandre et al. 1999).

Positional Modules as Modules of Evolution

Repeated redeployment of entire modules during evolution can also be documented for positional modules, such as the Pax-Six-Eya-Dach network discussed above (fig. 23.5, C). This network is evolutionarily ancient and is present in insects as well as in vertebrates (reviewed in Relaix and Buckingham 1999; Kawakami et al. 2000; Wawersik and Maas 2000; Kardon et al. 2004). In vertebrates, the network is clearly used in multiple domains (see above), and this is probably also the case for insects (see, e.g., Bonini et al. 1998). Focusing on chordate evolution, the repeated redeployment of the entire network is suggested by the evolution of novel domains of *Pax* gene expression, several of which are known to coexpress *Six, Eya,* and *Dach* genes in vertebrates. However, direct evidence for this scenario awaits the identification of *Six, Eya,* and *Dach* homologues in urochordates and cephalochordates. For example, *Pax3/7* in ascidians (Wada et al. 1996, 1997) is expressed in the neural tube, while in amphioxus additional expression domains are found in axial and paraxial mesoderm and nephridia (Holland et al. 1999). In vertebrates new expression domains have evolved in profundal placodes and ganglia and in hypaxial muscles, and these are also known to express *Six, Eya,* and *Dach* genes (e.g., Oliver et al. 1995; Bang et al. 1997; Stark et al. 1997; Ohto et al. 1998; Esteve and Bovolenta 1999; Davis et al. 1999, 2001; Heanue et al. 1999; Sahly et al. 1999; Kobayashi et al. 2000; Pandur and Moody 2000; David et al. 2001).

Similar changes are also observed for *Pax2/5/8* gene expression domains (fig. 23.5, C). In this case, however, the interpretation is slightly complicated by unclear homologies of ascidian expression domains (Wada et al. 1998): atrial expression in ascidians may correspond either to otic placodes (Wada et al. 1998) or to ectodermal gill slits (Kozmik et al. 1999a), whereas neural tube expression may be homologous to expression in the midbrain-hindbrain boundary (Wada et al. 1998) or rather to a general rhombospinal expression domain (Williams and Holland 1998; Holland and Holland 1999). Although the evolutionary scenarios differ for each of these possibilities, repeated evolution of new expression domains, such as in nephridia/kidney and optic support cells/optic stalk in amphioxus and vertebrates (e.g., Heller and Brändli 1997, 1999; Pfeffer et al. 1998; Kozmik et al. 1999a) and in B-lymphocytes in vertebrates (Adams et al. 1992; Morrison et al. 1998), can be inferred in each case. Again, *Six* and *Eya* genes are known to be coexpressed in overlapping domains in the kidney and retina (e.g., Xu et al. 1997; Kawakami et al. 2000; Wawersik and Maas 2000).

Cell Type Modules as Modules of Evolution

The origin of the neural crest and placodes in vertebrates (Northcutt and Gans 1983; Baker and Bronner-Fraser 1997) was associated with the evolution of several new types of neural crest and placodally derived neurons. Although these neurons differ from one another as well as from neurons in the central nervous system in the timing and location of neurogenesis and in their fate, neuronal differentiation in each case depends on similar genes (e.g., Lee et al. 1995; Sommer et al. 1996; Korzh et al. 1998; Ma et al. 1996, 1998; Fode et al. 1998; Schlosser and Northcutt 2000). This suggests that several new domains of the neuronal differentiation module evolved in vertebrates (fig. 23.6, *C*). The expression of amphioxus *neurogenin* (Holland et al. 2000) in the central nervous system (and some endodermal endocrine cells that may correspond to vertebrate pancreas) is compatible with this scenario, but its expression in epidermal sensory cells indicates that some ectodermal domains outside the central nervous system were already present in the vertebrate ancestor. Additional evidence for repeated dissociated coevolution of the neurogenesis module comes from repeated heterochronic shifts in the timing of development of particular neuronal cell populations that rely on this module for neuronal differentiation. For example, out-group comparison suggests that the timing of neurogenesis for neurons of the lateral motor column, which supply limb muscles, has repeatedly been altered relative to other developmental events (such as neurogenesis in cranial parts of the nervous system) in tetrapods, in parallel with temporal shifts in limb development (Richardson 1995; Richardson et al. 1997; Schlosser 2001). During the evolution of anurans, the development of forelimbs and of the lateral motor column neurons has been delayed from embryonic into larval stages, whereas some direct-developing frogs have secondarily reevolved early embryonic neurogenesis of the lateral motor column (Schlosser 2003) in parallel with early embryonic limb development (Elinson 1994; Richardson et al. 1998; Hanken et al. 2001).

Organ Modules as Modules of Evolution

Recurrent temporal shifts as well as complete losses of entire modules are also observed during the evolution of complex organ modules such as limb buds (see, e.g., discussion in Schlosser 2001, in press) or lateral line placodes. The entire lateral line system has, for example, been lost repeatedly in many direct-developing and viviparous amphibians (fig. 23.7, *C;* reviewed in Fritzsch 1989; Roth et al. 1993; Schlosser in press) without being obviously associated with modifications of other cranial structures, suggesting dissociated coevolution of all structures derived from lateral line placodes (receptor organs, nerves). This is further supported by frequent coordinated changes of the metamorphic fate of all these structures (fig. 23.7, *C;* Fritzsch and Wake 1986; Fritzsch 1988, 1990; Fritzsch et al. 1987; Roth et al. 1993). According to the most parsimonious scenario (Fritzsch 1990), ancestral amphibians retained the lateral line system after metamorphosis, but several groups of salamanders as well as the ancestors of frogs and caecilians secondarily evolved degeneration of the entire lateral line system at metamorphosis (Escher 1925; Fritzsch and Wake 1986; Wahnschaffe et al. 1987; Fritzsch 1988; Fritzsch et al. 1988; Roth et al. 1993). Metamorphic persistence of the lateral line system must then, however, have been reacquired by a few groups of frogs (Fritzsch et al. 1987) and caecilians (Fritzsch and Wake 1986; Roth et al. 1993).

Systemic Modules as Modules of Evolution

Finally, even systemic modules can act as coherent modules of evolution, as indicated by the repeated evolution of larval reproduction in urodeles (for reviews see Lynn 1961; Dent 1968; Wakahara 1996; Shaffer and Voss 1996; Rosenkilde and Ussing 1996; Rose 1999; Schlosser in press).[12] Urodeles with larval reproduction have essentially lost many thyroid-hormone-dependent metamorphic changes of their somatic tissues (with varying degrees of reduction in different taxa).[13] However, gonads develop normally, presumably because gonad maturation is independent of thyroid hormones in urodeles (Dodd and Dodd 1976; Wakahara 1996; Hayes 1997). Whereas in some salamanders larval reproduction is only facultative (i.e., occurs only under certain environmental conditions), in others it is obligatory. Among the latter, metamorphosis is still inducible by thyroid hormone treatments in some taxa but not in others. Although different alterations of developmental mechanisms probably underlie larval reproduction in different taxa, they all seem to involve some kind of disruption of the thyroid axis (Yaoita and Brown 1990; Tata et al. 1993; Shaffer and Voss 1996; Safi et al. 1997) and result in the coordinated failure of a wide variety of metamorphic events.

Developmental Modularity Promotes Evolutionary Modularity and Hence Evolvability

These examples—and many others could be cited—suggest that developmental modules indeed frequently act as modules of evolution: Developmental modularity, in conjunction with reciprocal context insensitivity and with sparseness of pleiotropic effects, restricts the fitness effects of heritable variation to small units, thereby promoting evolutionary modularity. This enhances evolvability (e.g., Wagner and Altenberg 1996; Gerhart and Kirschner 1997; Kirschner and Gerhart 1998) for two reasons.

First, developmental modularity may promote robustness (regarding the context-insensitive behavior of each module) and flexibility (regarding the intermodular interactions) not only relative to developmental perturbations, but also with respect to heritable variations. Therefore, it allows *quasi-independence* and *continuity* of selection on different units, two properties which Lewontin (1978) recognized as critical conditions for their separate adaptability and hence for the mosaic evolution of different traits (see also Brandon 1999; Schank and Wimsatt 2001). Quasi-independence requires units exhibiting a high probability of dissociated coevolution (i.e., units, whose evolutionary change is little constrained developmentally or functionally by elements that do not belong to the unit). Factually, developmental modules often behave like this (see above). Continuity demands that small variations have small effects on fitness. This is more likely in systems developing in a modular fashion because the number of fitness relevant interactions of any element is more restricted (mainly to other elements of the same module).

Second, due to their integrated and context-insensitive behavior, developmental modules can be embedded in new contexts during evolution without compromising their logical role. Thus, developmental modularity greatly promotes the evolution of complexity, because it facilitates the generation of novelty by the evolutionary redeployment and recombination of existing developmental modules (Schlosser 2002). This may create new opportunities for evolutionary change resulting in a kind of self-organizing process, a *modular ratcheting* similar to the "phylogenetic ratcheting" proposed by Katz (1987; see also Zuckerkandl 1997 for a similar suggestion): developmental modularity will tend to increase in evolution, because new domains of developmental modules can be added relatively easily, and with the accumulation of new domains and the generation of new combinations of domains the probability increases that one of them may become selectively advantageous later.[14]

As most heritable change occurs at the genetic level, the modularity of gene regulation itself is a crucial ingredient in this process of redeployment and recombination of domains, because it permits the independent heritable variation of inputs from outputs as well as of different inputs and outputs from each other in regulatory networks. Variations in cis-regulatory regions, in particular, will tend to specifically affect the regulatory input of an upstream transcription factor to a particular gene, whereas variations of the coding region of the upstream transcription factor would pleiotropically affect the regulation of all genes responsive to it (see also Stern 2000; Carroll 2000).

Moreover, modularity of gene regulation permits not only specific variation by local point mutations or deletions, but in conjunction with the locality and contiguity of many molecular interactions (such as transcription factor binding to a relatively small number of nucleotides that are located adjacent to each other in a strand of DNA), it facilitates the redeployment of gene activation in new domains by the recombination of existing inputs and outputs. This may happen simply by the shuffling around of cis-regulatory regions (fig. 23.10, C) like promoters, enhancers, or binding sites (as has already been suggested by Britten and Davidson 1969, 1971; Zuckerkandl 1994; García-Bellido 1996), similar to the exon shuffling that has been proposed as underlying the evolution of certain proteins that consist of modular functional motifs (e.g., Gilbert 1978; Wagner 1998; Patthy 1999; Gogarten and Olendzenski 1999; Hutter et al. 2000). Several mechanisms are known by which such enhancer shuffling may occur, including unequal crossing-over, gene conversion, slippage, and (retro)transposition (e.g., Henikoff et al. 1997; Gogarten and Olendzenski 1999; Eickbush 1999; Boeke and Pickeral 1999; Kazazian 2000; Dover 2000; Kidwell and Lisch 2001). The inpouring data from several genome projects indicate that such rearrangements are much more frequent than previously thought (Britten 1996, 1997; Henikoff et al. 1997; Brosius 1999; Fedoroff 1999; Deragon and Capy 2000; Prak and Kazazian 2000; Kidwell and Lisch 2001).

Importantly, the shuffling of cis-regulatory regions may require only the duplication of very small stretches of nucleotides, but it may allow the generation of new domains not only for the expression of a single gene, but potentially even for entire networks or developmental modules, without requiring the duplication of all the genes, which are now coordinately redeployed in the new domains (see also Niehrs and Pollet 1999; Niehrs 2004; Force et al. 2004). Creating a new domain of expression for a transcription factor due to the introduction of a new binding site in its cis-regulatory region (which may occur by shuffling or de novo generation due to series of neutral point mutations or deletions)

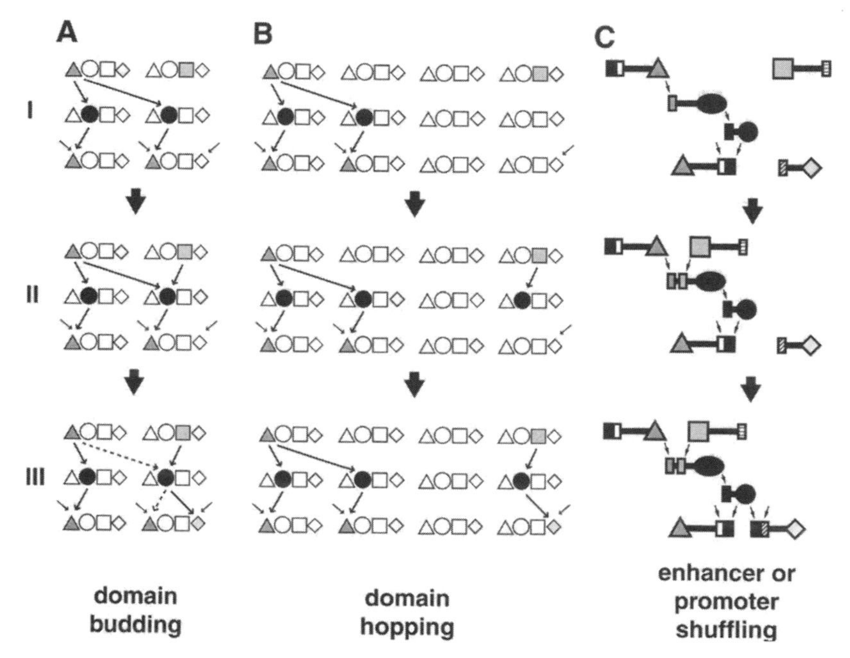

Fig. 23.10.—Two scenarios for the evolutionary redeployment of developmental modules in new domains. *A* and *B*, Three stages (I–III) of an evolutionary process leading to the establishment of a novel domain of a developmental module with a novel function. At each stage, the figure shows how a certain developmental module (*black circle*) is embedded in a sequence of developmental interactions of various kinds of elements in time (symbols may be interpreted as genes or gene networks, and quartets of symbols as cells; see fig. 23.2 for general explanation). It is assumed that the module is initially multiply instantiated, but all instantiations belong to a single domain because they are activated by the same upstream factor (*triangle*). In domain budding (*A*), a new domain of a module may bud off the original domain in two steps. In a first step (stages I–II), the module may acquire responsiveness to a new activator (*square*) while retaining its original function of activating the triangle (*long arrow*) in conjunction with a coactivator (*short arrow*). In the second step (stages II–III), it may acquire a new function (activation of the rhombus), for example, in conjunction with another context-specific coactivator (*small arrow*), and it may subsequently become spatiotemporally decoupled from the old domain, losing its original functions (*dashed arrows*). Domain hopping (*B*) also proceeds in two steps. In the first step (stages I–II), the module acquires responsiveness to a new activator that is active at a distant site. However, the new domain may not be able to fulfill its original function of activating the triangle there due to lack of coactivators. Therefore, it represents a spurious, functionally neutral domain. It may, however, acquire a new function (activation of the rhombus), for example, in conjunction with another context-specific coactivator (*small arrow*) in a second step (stages II–III). *C*, Both scenarios may proceed via the shuffling of cis-regulatory elements (*small rectangles*) in genes, whose coding regions (*large symbols*) remain unaltered. As in figure 23.3, those binding sites that are singly required and jointly sufficient forming an enhancer for gene activation are drawn adjacent to one another without a gap, whereas different enhancers are separated by a gap between them. For simplicity, the module of *A* and *B* is here represented by a simple chain of two genes (*black oval* and *circle*), but it may in fact be much more complex. Step 1 (stages I–II) may proceed via acquisition of a new regulatory element (*light grey rectangle*) of the black oval gene, while step 2 (stages II–III) may proceed via acquisition of a new regulatory element (*black rectangle*) in the downstream gene with the rhombic coding region.

may automatically create a new domain of expression for a whole battery of genes downstream from it. This allows evolution to "tinker" (Jacob 1977) with relatively large-scale modules of development, by locally cutting and pasting (or recreating) small fragments of DNA.

Two Scenarios for the Evolutionary Redeployment of Developmental Modules

In order to create functional evolutionary novelties by the mere redeployment and recombination of existing developmental modules (fig. 23.10), at least two conditions have to be satisfied. First, the function of the original domain of a developmental module must not be compromised. This will, for instance, be the case when the domain is duplicated, before it acquires different functions. As argued above, this may be achieved if one of the genes involved in the module acquires additional cis-regulatory regions; it does not require pervasive duplication of all genetic components of a module. Second, a novel function must be acquired by a new domain of the developmental module. It must be emphasized that acquisition of a new function will often necessitate two independent evolutionary changes (see also Sidow 1996), because in order to fulfill a novel function, the new domain must not only be activated by a novel combination of inputs, but must also typically feed into different kinds of downstream processes. For example, activation of the Notch module in the neural plate not only requires different activators, but also promotes different cell fate decisions—presumably by regulating different downstream genes—from those in the teeth. Because the context insensitivity of a developmental module will result in conserved intramodular interactions and therefore allow similar spatiotemporal and quantitative input-output relations—that is, a similar logical role of the module in each of its domains—the functional differences between domains may largely be due to *qualitative* differences in its input (e.g., their responsiveness to different upstream transcription factors) and output relations (e.g., their activation of different downstream genes).

The requirement for two independent evolutionary changes in the evolution of new functional domains has profound implications for the evolutionary process, which can only briefly be sketched here (fig. 23.10). It suggests that new domains of a developmental module with new functions might evolve either in a process that I will term *domain budding,* where intermediate stages are likely to be functional (but may involve redundancy), or in a process that I will term *domain hopping,* where intermediate stages are more likely to be neutral (Force et al. [2004] have independently developed a similar distinction between "subfunction fission" and "subfunction cooption"). In domain budding (fig. 23.10, *A, C*), a developmental module may become responsive to a new activator that is active in the vicinity of one of its domains. The domain is now responsive to two activators (which may differ in their relative importance for different parts of the domain) but initially continues to fulfill its old function by feeding into the same downstream processes. If the domain now acquires new outputs, new functional input-output relations may be established in the part of the domain responsive to the new activator. The new functional domain may then gradually become spatiotemporally decoupled and bud off from the old domain. In the alternative domain-hopping scenario (fig. 23.10, *B, C*), a developmental module becomes responsive to a new activator that is active somewhere spatiotemporally remote from one of its domains. This will result in the initiation of a new domain of the module. However, this new domain may be functionally neutral. It will often not be able to fulfill its old function, for example, because important coactivators of downstream genes in the old domain may not be available, and it will not yet have acquired a new function. It does, however, present a new opportunity for the generation of a novel function, for example, after acquiring new outputs (for similar ideas about the constructive role of neutral evolution see Huynen et al. 1996; Zuckerkandl 1997; Fontana and Schuster 1998; Force et al. 2004).

This is, of course, only a very simplistic account, and there are countless possibilities of modifications of the two scenarios. In particular, any domain may have multiple functions (or potential functions; e.g., there may be sets of genes that are responsive to the output of the module but cannot be activated, because other essential coactivators are not available) to begin with, which would increase the probability of establishing new domains with new functions either by domain hopping or by domain budding.

Both of these scenarios pose some problems for the operation of natural selection. In module budding, there is likely to be functional redundancy of the two domains in intermediate stages, whereas module hopping may even involve complete functional neutrality of one domain in intermediate stages. But redundancy and neutrality are predicted to be unstable under selection in most population genetic models, because disruptive mutations in redundant or neutral genes are not selected against (e.g., Ohta 1987; Walsh 1995; Nowak et al. 1997; Wagner 1998; Lynch and Force 2000). There are, however, two arguments supporting the view that this problem may be overestimated, in particular for the scenario of module budding. First, functional redundancy and neutrality in these scenarios do not require large-scale gene duplications but may be due to the duplication or convergent origin of very small sequences of nucleotides. This greatly reduces the probability of the accumulation of disruptive mutations (Krakauer and Nowak

1999). Second, functional redundancy (but not complete functional neutrality) may in addition be actively maintained under selection for developmental robustness (Nowak et al. 1997; Wilkins 1997; Krakauer and Nowak 1999) or by second-order selection for resistance against disruptive mutations (Wagner 2000).

I have suggested here that developmental modularity promotes the evolution of complex systems in a self-organizing process, because it facilitates the redeployment and recombination of modules, thereby providing new opportunities for evolutionary change. A potential argument against this proposal is that the adoption of new roles for any element (e.g., gene) or developmental module may slow down subsequent evolution and reduce evolvability, because multiple roles impose additional constraints (Duboule and Wilkins 1998; Force et al. 2004). According to this view, modules of development may be initially easy to redeploy and recombine, but this process will quickly grind to a halt, due to the increase in constraints it generates, unless the duplication of gene expression domains is paralleled by the duplication of all the genes thus redeployed. However, the fact that many signaling, cell type, or positional modules have acquired multiple new domains during evolution without concomitant gene duplication suggests that this argument does not hold generally. It tends to overlook that, like a transistor or a microchip, a developmental module plays the same logical role in each of its domains, so that alterations of intramodular interactions should affect different domains similarly. Therefore, as long as most heritable variations of the developmental module affect only intramodular interactions, and as long as only few variations have pleiotropic effects on its inputs or outputs in multiple contexts, employment in multiple domains may not necessarily bring with it a big increase in constraints or reduction of evolvability. Nonetheless, it remains true that constraints will be even smaller when different domains acquire completely independent genetic representations, for example, due to duplication of the genes involved followed by complementary loss of regulatory elements leading to the restriction of expression of each gene or set of genes to a single domain (Force et al. 1999, 2004).

Summary and Conclusions

The purpose of this chapter has been to make the notion of modularity more precise and to analyze whether and under what conditions modules of development can act as modules of evolution. Modules were introduced as units of interacting elements that make an integrated and relatively context-insensitive contribution to some containing process. Modules of *development* are units of interacting elements that make a relatively invariant contribution to the development of an organism due to their integrated and context-insensitive behavior in the face of local perturbations. Examples from gene regulation, signaling cascades, gene regulatory networks, cell types, and organ primordia demonstrated that developmental modules can be recognized at many different levels and that they may be components of larger-scale modules in a hierarchical fashion. Moreover, developmental modules are typically multiply instantiated in development, often in different domains (activated by different factors) and sometimes with different functions. Developmental modularity contributes to the "developability" of complex organisms, because it confers robustness and flexibility and allows the combinatorial increase of complexity during development.

Their integration and context insensitivity also make developmental modules good candidates for modules of *evolution*, that is, for units of elements that make an integrated and context-insensitive contribution to their own reproduction in subsequent generations in the face of heritable variations, thereby reciprocally constraining their evolutionary modifications. However, developmental modules will act as modules of evolution only when they fulfill additional conditions, namely, when they are not only relatively insensitive to but also dispensable for the context in which they are embedded, and when interactions that are independently perturbable during development also tend to be subject to independent heritable variations. Apparently, these conditions are often fulfilled, because there is evidence for frequent dissociated coevolution (e.g., recurrent loss, shifts in space and time, changes in the number of domains) of many developmental modules during phylogeny, suggesting that they indeed acted as modules of evolution in certain lineages. Thus, developmental modularity may promote evolvability, because it helps to delimit units of quasi-independent fitness contributions that can evolve in a mosaic fashion. In addition, developmental modularity may contribute to evolvability by facilitating the recombination of developmental modules, allowing an evolutionary increase in complexity. The modularity of gene regulation itself may be of outstanding importance during this process, because it allows heritable variation of the inputs and outputs of a developmental module to be generated without pleiotropic effects on intramodular interactions, thereby preserving the logical role of the development module (i.e., its spatiotemporal and quantitative input-output relations).

It seems that modularity does indeed play a central role for understanding development and evolution. However, this insight can be only the beginning of a new synthesis of developmental and evolutionary theory: it tells us where to look. We clearly need more empirical data (in particular, comparative data from a wide range of organisms) as much as further theoretical efforts in order to truly elucidate how frequently and under which conditions modules of development indeed act as modules of evolution.

Acknowledgments

I am grateful to Gerhard Roth, R. Glenn Northcutt, and Chris Kintner, in whose labs I conducted most of my own experimental work that drew my attention to modules. I also wish to thank Günter Wagner for many stimulating discussions about modularity in general and this chapter in particular, and Allan Force, George von Dassow, and Mario Wullimann for detailed comments on this chapter. This work was made possible by a scholarship of the Hanse Institute for Advanced Study, Delmenhorst, which also hosted and financially supported the symposium on modularity, which greatly contributed to the elaboration of many ideas presented in this chapter.

Notes

1. The IOR of a chain of interacting elements is not an additive superposition of the IORs of its components, because disruption of the output of a single element does not result in a proportional diminution of the output of the chain, but rather in its complete disruption.

2. One consequence of this is that the identity of a higher-order module may be retained in evolution, even though some of its component modules may be replaced by other modules with similar IORs (e.g., Roth 1991, 2001; Wagner 1995, 2001; Wagner and Altenberg 1996; von Dassow and Munro 1999; Wagner et al. 2000; Schlosser 2002). Evolutionary changes of the genetic or cellular basis of homologous structures provide examples of this (e.g., Striedter and Northcutt 1991; Dickinson 1995; Bolker and Raff 1996; Abouheif 1997).

3. However, where the limits are drawn depends on the kinds of perturbations considered: local retroviral infection, for example, may disrupt the hedgehog pathway in a single cell but not in its neighbor, whereas a germ line mutation in the hedgehog gene would disrupt hedgehog signaling throughout all cells of the body. Because local perturbations of development are possible (e.g., by perturbations of the immediate cellular environment or by somatic mutations), we often are able to recognize multiple individualized modules of a particular type, for instance, repeated employment of hedgehog signaling in different developmental contexts, even though these have the same genetic basis.

4. The question of whether an individualized developmental module employed in one domain is identical to a module employed in another domain (i.e., can be considered to belong to the *same type* of module), in fact raises very difficult questions that can only be sketched here briefly. Different developmental modules are identical in the most liberal sense when they can be characterized by the same or similar patterns of interacting elements and play identical or similar logical roles in each of the different contexts in which they are operating. When different developmental modules are similar with respect to some interaction parameters or some aspects of their behavior (IOR) but differ with respect to others, they may be considered to belong to different subtypes of a more broadly defined general type of module. In a more narrow sense, different developmen-

tal modules are identical when in addition these similarities are due to genealogical continuity, that is, due to a common evolutionary *origin* (modules could then be considered serial homologues). The main difficulty here stems from the fact that the common evolutionary origin of two developmental modules cannot simply be identified by the fact that they express homologous genes: while modules of common origin can be expected to initially express either orthologous or paralogous genes, later in evolutionary history there may be gradual replacements of some of their submodules by other submodules with a similar logical role but employing unrelated genes (see n. 2). Finally, different developmental modules are considered identical in the most specific and evolutionarily most significant sense only when they satisfy the additional condition that they cannot be reproduced (inherited) independently of each other, because they share the same genetic basis (i.e., they are not individualized relative to heritable variations such as germ line mutations). The latter condition is important to single out those types of developmental modules that have a common evolutionary *fate* and excludes modules with a similar connectivity, logical role, and possibly common ancestry but involving different (for instance, paralogous) genes. However, this last condition is often only partially satisfied, because modules may share some but not all of their components (the shared components forming a so-called synexpression group; see Niehrs and Pollet 1999; Niehrs 2004). For example, a signaling cascade in one cell may be activated when the ligand sonic hedgehog binds to the receptor patched1, whereas in another cell, patched1 is activated by binding of the paralogous banded hedgehog (see, e.g., Murone et al. 1999). This issue is important for understanding the relation between modules of development and modules of evolution discussed below ("Modules of Evolution").

5. As George von Dassow pointed out to me, every engineer knows that construction of a machine which is supposed to operate without human supervision specifically needs to include such parts that are designed to establish cross-talk between its different otherwise autonomous components.

6. Fitness epistasis should be rampant in tightly integrated developmental modules. However, some developmental modules achieve their context-insensitive behavior through mechanisms that allow for intramodular compensations of perturbations (canalization; see Waddington 1957; Wilkins 1997; Rutherford 2000; Gibson and Wagner 2000; von Dassow et al. 2000), and this kind of robustness should also make the module morc tolerant to heritable changes of its components and reduce the probability for fitness epistasis. Nevertheless, it should remain true that such a developmental module forms a coherent unit of evolutionary change only when the probability for fitness epistasis among variants of two components of a single module is higher than among a variant of a module component and a variant of the context. For an illustration of the notion of probability used here consider for example two variants of a receptor molecule that differ at their extracellular ligand-binding sites, so that variant R1 binds better to variant L1 of a ligand, whereas variant R2 binds better to variant L2 of the ligand. In this case, the relative fitness of R1 compared to R2 will depend on which variant of the ligand it encounters; that is, there will be fitness epistasis for receptor ligand interactions among a combination of variants R1, R2, L1, and L2. However, there may be another variant of the receptor R3 that differs from R1 with respect to its ability to interact with intracellular second messengers rather than with extracellular ligands. Accordingly, the relative fitness of R1 compared to R3 will not depend on which variant of ligand it encounters and there will be no fitness epistasis for receptor-ligand interactions among a combination of variants R1, R3, L1, and L2. Therefore, several elements (such as receptor and ligand) will belong to a single module of evolution only with a certain probability—termed *coevolution probability* here—and this probability depends on the frequency with which various possible variants of the elements (e.g., all one-mutant neighbors of genes) occur. Note that the same elements can belong to several modules of

evolution, each with a different coevolution probability. The spatiotemporal hierarchy of developmental modules, for instance, may be reflected in a hierarchy of modules of evolution with different coevolution probabilities (see Schlosser 2002 for details).

7. What counts as a *unit of selection* has long been a hotly debated question in evolutionary biology. While there is still some controversy, the importance of three elements is commonly recognized (e.g., Lewontin 1970, 1974; Wimsatt 1981; Sober 1981, 1984, 1987; Sober and Lewontin 1982; Dawkins 1982; Gould 1982; Brandon and Burian 1984; Eldredge 1985; Maynard Smith 1987; Lloyd 1988; Sterelny and Kitcher 1988; Brandon 1990, 1999; Waters 1991; Sober and Wilson 1998; Gould and Lloyd 1999; Schlosser 2002). In order to act as unit of selection, (1) an entity has to exist in several variants, and (2) these variants as well as (3) the fitness differences between them have to be reliably reproduced across generation boundaries (complications such as frequency-dependent selection must be ignored here for the sake of brevity; for a more detailed discussion see Schlosser 2002). Therefore, several elements belong to a single unit of selection when they each exist in several variants and when particular combinations of these variants as well as the fitness differences between them are reliably reproduced. Such a complex unit of selection is nondecomposable when all elements either are strictly linked or jointly and epistatically codetermine the fitness value of its variants (i.e., the fitness differences between the variants of one element differ, depending on which variants of the other elements they are associated with). Consider, for example, the case of two different nucleotides of a single gene, which are both polymorphic in a certain population. Often, these nucleotides will belong to a single nondecomposable unit of selection because they both code for amino acids that cooperate in determining the folding and functionality of a single protein. However, assuming that the gene codes for a protein with multiple domains that fold and function independently of each other, these two nucleotides may indeed belong to two separate units of selection, if the recombination rate between them is not too low and if each of the nucleotides codes for an amino acid in a different functional domain.

While the concept of "unit of selection" is very important in allowing us to understand the populational dynamics of single selection processes, it is of limited value for understanding why certain elements of the life cycle coevolve, because a unit of selection includes only elements which simultaneously exist in several variants during a particular selection process. Any attempt to define units of evolutionary change in terms of units of selection would drastically underestimate the number of elements that reciprocally constrain each other's evolution for two reasons. First, in case of low linkage between elements (e.g., different genes) in sexually reproducing species, reliable reproduction of particular combinations of variants is less likely (i.e., heritability is lower), the higher the number of independently variable elements is (e.g., Lewontin 1970; Maynard Smith 1987; Schlosser 2002), in particular, in large populations. Therefore, due to the low heritability of variants, a complex of a large number of polymorphic elements cannot act as a nondecomposable unit of selection under these conditions, even when these elements epistatically codetermine their fitness values (and therefore may constrain each other's future evolution; see below). Second, elements that are not simultaneously polymorphic in a particular selection process are by definition not part of a unit of selection. Nonetheless, they may reciprocally constrain each other's evolution, because the fitness differences between different variants of one element may depend on the outcome of *previous* selection processes among variants of the other element. For these reasons, a unit of reciprocal evolutionary constraints—what I refer to as module of evolution or unit of evolution (Schlosser 2002)—is a much more general notion than a unit of selection.

8. An important asymmetry between the epistasis of developmental effects and fitness epistasis should be noted here. When a component A of a developmental module affects

an element B of the context, but not vice versa, the developmental contributions of A to the input-output relation of the module will not differ after various perturbations of B; hence, the intramodular effects of A are not epistatically dependent on B; that is, they are context-insensitive. Nonetheless, the *fitness* of these intramodular effects of A may epistatically depend on B, because the fitness of any component of a life cycle is determined by the fitness of the entire life cycle in which it resides (see fig. 23.8) and this depends on B as well as on A. Under the simplifying assumptions that (1) each heritable variation—or mutation—will typically affect only one of several developmental interactions of an element and (2) fitness epistasis will be typically restricted to those variants of two elements that affect their functionally relevant interactions with each other, variations in an element of a module will have a low probability of showing fitness epistasis to variations in elements of the context when their functionally relevant interactions with each other represent only a minority of the total number of interactions of each element.

9. In a case in which several individualized domains of a particular type of developmental module (e.g., a certain signaling pathway) only partially share their genetic basis, because some of their interacting elements are paralogues and not orthologues (see n. 4), the probability of coherent evolutionary change across different domains will, of course, be reduced for the components that are not shared between different domains.

10. More precisely, the frequency of dissociated coevolution of a module of evolution should be higher than its disruption frequency (which is the frequency of dissociated coevolution of some of its elements with elements that are not part of the unit), thus indicating a coevolution probability that is higher than expected by the null hypothesis of uncorrelated change.

11. Developmental modules can act as modules of evolution only as long as they persist as developmental modules after heritable variation in the chain of generations, that is, as long as they exhibit *perseverance* in the face of iterated variation (see also Schlosser 2002). In other words, developmental modules will tend to act as modules of evolution in certain lineages for a significant period of evolutionary time, only when changes of the modular architecture (except those changes involving merely the recombination of existing modules; see below) are relatively unlikely. We still know very little about the conditions that favor this kind of robustness (but see, e.g., Wagner and Mezey 2000).

12. Larval reproduction is often referred to as "neoteny" (e.g., Wakahara 1996), but this is potentially misleading, because the latter term is also used in a more general sense for various kinds of retardation in differentiation processes (e.g., Gould 1977).

13. It should be noted, however, that at least a few metamorphic events may still occur in taxa with larval reproduction, presumably because they require only low doses of thyroid hormones (e.g., Ducibella 1974). Moreover, the degree of loss of metamorphic events differs in different taxa (e.g., in Amphiumidae many metamorphic events still take place), possibly due to different tissue sensitivities to thyroid hormones and varying degrees of decline in hormone levels (see Rose 1999 for review).

14. It remains unclear, however, why such a self-organizing process should be more pronounced in some lineages than in others.

References

Abouheif, E. 1997. Developmental genetics and homology: a hierarchical approach. *Trends. Ecol. Evol.* 12:405–408.

Adams, B., P. Dorfler, A. Aguzzi, Z. Kozmik, P. Urbanek, I. Maurer-Fogy, and M. Busslinger. 1992. Pax-5 encodes the transcription factor BSAP and is expressed in B lymphocytes, the developing CNS, and adult testis. *Genes Dev.* 6:1589–1607.

Akam, M. 1998. Hox genes: from master genes to micromanagers. *Curr. Biol.* 8: R676–R678.

Alberch, P. 1980. Ontogenesis and morphological diversification. *Am. Zool.* 20: 653–667.
Alberch, P. 1982. The generative and regulatory roles of development in evolution. In *Environmental adaptation and evolution,* ed. D. Mossakowski and G. Roth, 19–36. Stuttgart: Fischer.
Alberch, P. 1983. Morphological variation in the neotropical salamander genus *Bolitoglossa. Evolution* 37:906–919.
Alberch, P., and E. A. Gale. 1985. A developmental analysis of an evolutionary trend: digital reduction in amphibians. *Evolution* 39:8–23.
Altenberg, L. 1995. Genome growth and the evolution of the genotype-phenotype map. In *Evolution and biocomputation,* ed. W. Banzhaf and F. H. Eeckman, 205–259. Berlin: Springer.
Ancel, L. W., and W. Fontana. 2000. Plasticity, evolvability and modularity in RNA. *J. Exp. Zool. (Mol. Dev. Evol.)* 242–283.
Arnone, M. I., and E. H. Davidson. 1997. The hardwiring of development: organization and function of genomic regulatory systems. *Development* 124:1851–1864.
Arnosti, D. N., S. Barolo, M. Levine, and S. Small. 1996. The eve stripe 2 enhancer employs multiple modes of transcriptional synergy. *Development* 122:205–214.
Artavanis-Tsakonas, S., M. D. Rand, and R. J. Lake. 1999. Notch signaling: cell fate control and signal integration in development. *Science* 284:770–776.
Arthur, W. 1997. *The origin of animal body plans.* Cambridge: Cambridge University Press.
Axelrod, J. D., K. Matsuno, S. Artavanis-Tsakonas, and N. Perrimon. 1996. Interaction between wingless and notch signaling pathways mediated by dishevelled. *Science* 271:1826–1832.
Baker, C. V. H., and M. Bronner-Fraser. 1997. The origins of the neural crest. 2. An evolutionary perspective. *Mech. Dev.* 69:13–29.
Baker, N. E. 2000. Notch signaling in the nervous system: pieces still missing from the puzzle. *BioEssays* 22:264–273.
Balczarek, K. A., Z.-C. Lai, and S. Kumar. 1997. Evolution and functional diversification of the paired box (Pax) DNA-binding domains. *Mol. Biol. Evol.* 14:819–842.
Bang, A. G., N. Papalopulu, C. Kintner, and M. D. Goulding. 1997. Expression of Pax-3 is initiated in the early neural plate by posteriorizing signals produced by the organizer and by posterior non-axial mesoderm. *Development* 124:2075–2085.
Beatus, P., and U. Lendahl. 1998. Notch and neurogenesis. *J. Neurosci. Res.* 54:125–136.
Belting, H. G., C. S. Shashikant, and F. H. Ruddle. 1998. Modification of expression and cis-regulation of Hoxc8 in the evolution of diverged axial morphology. *Proc. Natl. Acad. Sci. U.S.A.* 95:2355–2360.
Bhalla, U. S., and R. Iyengar. 1999. Emergent properties of networks of biological signaling pathways. *Science* 283:381–387.
Blader, P., N. Fischer, G. Gradwohl, F. Guillemot, and U. Strähle. 1997. The activity of neurogenin1 is controlled by local cues in the zebrafish embryo. *Development* 124: 4557–4569.
Blaschuk, K. L., and C. Ffrench-Constant. 1998. Developmental neurobiology: notch is tops in the developing brain. *Curr. Biol.* 8:R334–337.
Boeke, J. D., and O. K. Pickeral. 1999. Retroshuffling the genomic deck. *Nature* 398: 108–111.
Bolker, J. A. 2000. Modularity in development and why it matters to evo-devo. *Am. Zool.* 40:770–776.
Bolker, J. A., and R. A. Raff. 1996. Developmental genetics and traditional homology. *BioEssays* 18:489–494.
Bonini, N. M., W. M. Leiserson, and S. Benzer. 1998. Multiple roles of the eyes absent gene in *Drosophila. Dev. Biol.* 196:42–57.

Bonner, J. T. 1988. *The evolution of complexity.* Princeton, N.J.: Princeton University Press.

Bonneton, F., P. J. Shaw, C. Fazarkerley, M. Shi, and G. A. Dover. 1997. Comparison of bicoid-dependent regulation of hunchback between *Musca domestica* and *Drosophila melanogaster. Mech. Dev.* 66:143–156.

Borsani, G., A. DeGrandi, A. Ballabio, A. Bulfone, L. Bernard, S. Banfi, C. Gattuso, M. Mariani, M. Dixon, D. Donnai, K. Metcalfe, R. Winter, M. Robertson, R. Axton, A. Brown, V. van Heyningen, and I. Hanson. 1999. EYA4, a novel vertebrate gene related to *Drosophila* eyes absent. *Hum. Mol. Genet.* 8:11–23.

Bouchard, M., P. Pfeffer, and M. Busslinger. 2000. Functional equivalence of the transcription factors Pax2 and Pax5 in mouse development. *Development* 127: 3703–3713.

Borycki, A.-G. 2004. Sonic hedgehog and Wnt signaling pathways during development and evolution . In *Modularity in development and evolution,* ed. G. Schlosser and G. P. Wagner. Chicago: University of Chicago Press.

Brandon, R. N. 1990. *Adaptation and environment.* Princeton, N.J.: Princeton University Press.

Brandon, R. N. 1999. The units of selection revisited: the modules of selection. *Biol. Philos.* 14:167–180.

Brandon, R. N., and R. M. Burian, eds.1984. Genes, organisms, populations: controversies over the units of selection. Cambridge, Mass.: MIT Press.

Brennan, K., T. Klein, E. Wilder, and A. M. Arias. 1999. Wingless modulates the effects of dominant negative notch molecules in the developing wing of *Drosophila. Dev. Biol.* 216:210–29.

Britten, R. J. 1996. DNA sequence insertion and evolutionary variation in gene regulation. *Proc. Natl. Acad. Sci. U.S.A.* 93:9374–9377.

Britten, R. J. 1997. Mobile elements inserted in the distant past have taken on important functions. *Gene* 205:177–182.

Britten, R. J., and E. H. Davidson. 1969. Gene regulation for higher cells: a theory. *Science* 165:349–357.

Britten, R. J., and E. H. Davidson. 1971. Repetitive and non-repetitive DNA sequences and a speculation on the origins of evolutionary novelty. *Q. Rev. Biol.* 46:111–138.

Brosius, J. 1999. Genomcs were forged by massive bombardments with retroelements and retrosequences. *Genetica* 107:209–238.

Brunet, J. F., and A. Ghysen. 1999. Deconstructing cell determination: proneural genes and neuronal identity. *BioEssays* 21:313–318.

Buss, L. W. 1987. The evolution of individuality. Princeton, N.J.: Princeton University Press.

Campos-Ortega, J. A. 1993. Mechanisms of early neurogenesis in *Drosophila melanogaster. J. Neurobiol.* 24:1305–1327.

Carroll, S. B. 2000. Endless forms: the evolution of gene regulation and morphological diversity. *Cell* 101:577–580.

Carroll, S. B., J. K. Grenier, and S. D. Weatherbee. 2001. From DNA to diversity. Malden: Blackwell Science.

Cheverud, J. M. 1984. Quantitative genetics and developmental constraints on evolution by selection. *J. Theor. Biol.* 110:155–171.

Cheverud, J. M. 1996. Developmental integration and the evolution of pleiotropy. *Am. Zool.* 36:44–50.

Cheverud, J. M. 2004. Modular pleiotropic effects of quantitative trait loci on morphological traits. In *Modularity in development and evolution,* ed. G. Schlosser and G. P. Wagner. Chicago: University of Chicago Press.

Cheverud, J. M., and E. J. Routman. 1995. Epistasis and its contribution to genetic variance components. *Genetics* 139:1455–1461.

Chitnis, A., D. Henrique, J. Lewis, D. Ish-Horowicz, and C. Kintner. 1995. Primary neurogenesis in *Xenopus* embryos regulated by a homologue of the *Drosophila* neurogenic gene Delta. *Nature* 375:761–766.

Chow, R. L., C. R. Altmann, R. A. Lang, and A. Hemmati-Brivanlou. 1999. Pax6 induces ectopic eyes in a vertebrate. *Development* 126:4213–4222.

Chuong, C. M., N. Patel, J. Lin, H. S. Jung, and R. B. Widelitz. 2000. Sonic hedgehog signaling pathway in vertebrate epithelial appendage morphogenesis: perspectives in development and evolution. *Cell. Mol. Life Sci.* 57:1672–1681.

Crowe, R., D. Henrique, D. Ish-Horowicz, and L. Niswander. 1998. A new role for Notch and Delta in cell fate decisions: patterning the feather array. *Development* 125: 767–775.

Dahl, E., H. Koseki, and R. Balling. 1997. Pax genes and organogenesis. *BioEssays* 19: 755–65.

David, R., K. Ahrens, D. Wedlich, and G. Schlosser. 2001. *Xenopus Eya1* demarcates all neurogenic placodes as well as migrating hypaxial muscle precursors. *Mech. Dev.* 103:189–192.

Davidson, B., M. W. Jacobs, and B. J. Swalla. 2004. The individual as a module: solitary-to-colonial transitions in metazoan evolution and development. In *Modularity in development and evolution,* ed. G. Schlosser and G. P. Wagner. Chicago: University of Chicago Press.

Davidson, E. H. 2001. *Genomic regulatory systems.* San Diego: Academic Press.

Davis, R. J., W. Shen, T. A. Heanue, and G. Mardon. 1999. Mouse Dach, a homologue of *Drosophila* dachshund, is expressed in the developing retina, brain and limbs. *Dev. Genes Evol.* 209:526–536.

Davis, R. J., W. Shen, Y. I. Sandler, T. A. Heanue, and G. Mardon. 2001. Characterization of mouse Dach2, a homologue of *Drosophila* dachshund. *Mech. Dev.* 102: 169–179.

Dawkins, R. 1982. *The extended phenotype.* Oxford: Oxford University Press.

Deblandre, G. A., D. A. Wettstein, N. Koyano-Nakagawa, and C. Kintner. 1999. A two-step mechanism generates the spacing pattern of the ciliated cells in the skin of *Xenopus* embryos. *Development* 126:4715–4728.

de Celis, J. 1999. The function of vestigial in *Drosophila* wing development: how are tissue-specific responses to signalling pathways specified? *BioEssays* 21: 542–545.

de Celis, J. 2004. The Notch signaling module. In *Modularity in development and evolution,* ed. G. Schlosser and G. P. Wagner. Chicago: University of Chicago Press.

Deftos, M. L., and M. J. Bevan. 2000. Notch signaling in T cell development. *Curr. Opin. Immunol.* 12:166–172.

De Martino, S., Y. L. Yan, T. Jowett, J. H. Postlethwait, Z. M. Varga, A. Ashworth, and C. A. Austin. 2000. Expression of sox11 gene duplicates in zebrafish suggests the reciprocal loss of ancestral gene expression patterns in development. *Dev. Dyn.* 217: 279–292.

Dent, J. N. 1968. Survey of amphibian metamorphosis. In *Metamorphosis: a problem in developmental biology,* ed. W. Etkin and L. I. Gilbert, 271–311. New York: Appleton-Century-Crofts.

Deragon, J. M., and P. Capy. 2000. Impact of transposable elements on the human genome. *Ann. Med.* 32:264–273.

Dickinson, W. J. 1995. Molecules and morphology: where's the homology? *Trends Genet.* 11:119–121.

Dodd, M. H. I., and J. M. Dodd. 1976. The biology of metamorphosis. In *Physiology of the amphibia,* vol. 3, ed. B. Lofts, 467–599. New York: Academic Press.

Dorsky, R. I., W. S. Chang, D. H. Rapaport, and W. A. Harris. 1997. Regulation of neuronal diversity in the *Xenopus* retina by Delta signaling. *Nature* 385:67–70.

Dorsky, R. I., D. H. Rapaport, and W. A. Harris. 1995. Xotch inhibits cell differentiation in the *Xenopus* retina. *Neuron* 14:487–496.
Dover, G. 2000. How genomic and developmental dynamics affect evolutionary processes. *BioEssays* 22:1153–1159.
Duboule, D., and A. S. Wilkins. 1998. The evolution of bricolage. *Trends Genet.* 14:54–59.
Ducibella, T. 1974. The occurrence of biochemical metamorphic events without anatomical metamorphosis in the axolotl. *Dev. Biol.* 38:175–186.
Duellman, W. E., and L. Trueb. 1986. *Biology of amphibians.* New York: McGraw-Hill.
Eickbush, T. 1999. Exon shuffling in retrospect. *Science* 283:1465–1466.
Eisen, M. B., P. T. Spellman, P. O. Brown, and D. Botstein. 1998. Cluster analysis and display of genome-wide expression patterns. *Proc. Natl. Acad. Sci. U.S.A.* 95:14863–14868.
Eisenberg, D., E. M. Marcotte, I. Xenarios, and T. O. Yeates. 2000. Protein function in the postgenomic area. *Nature* 405:823–826.
Eldredge, N. 1985. *Unfinished synthesis.* New York: Oxford University Press.
Eldredge, N., and J. Cracraft. 1980. *Phylogenetic patterns and the evolutionary process.* New York: Columbia University Press.
Elinson, R. P. 1994. Leg development in a frog without a tadpole (*Eleutherodactylus coqui*). *J. Exp. Zool.* 270:202–210.
Escher, K. 1925. Das Verhalten der Seitenorgane der Wirbeltiere und ihrer Nerven beim Übergang zum Landleben. *Acta Zool.* 6:307–414.
Esteve, P., and P. Bovolenta. 1999. cSix4, a member of the six gene family of transcription factors, is expressed during placode and somite development. *Mech. Dev.* 85:161–165.
Fedoroff, N. V. 1999. Transposable elements as a molecular evolutionary force. *Ann. N.Y. Acad. Sci.* 870:251–64.
Felsenstein, J. 1985. Phylogenies and the comparative method. *Am. Nat.* 125:1–15.
Fenster, C. B., L. F. Galloway, and L. Chao. 1997. Epistasis and its consequences for the evolution of natural populations. *Trends Ecol. Evol.* 12:282–286.
Fink, W. L. 1982. The conceptual relationship between ontogeny and phylogeny. *Paleobiology* 8:254–264.
Flores, G. V., H. Duan, H. J. Yan, R. Nagaraj, W. M. Fu, Y. Zou, M. Noll, and U. Banerjee. 2000. Combinatorial signaling in the specification of unique cell fates. *Cell* 103:75–85.
Fode, C., G. Gradwohl, X. Morin, A. Dierich, M. Lemeur, C. Goridis, and F. Guillemot. 1998. The bHLH protein neurogenin 2 is a determination factor for epibranchial placode-derived sensory neurons. *Neuron* 20:483–494.
Fontana, W., and P. Schuster. 1998. Continuity in evolution: on the nature of transitions. *Science* 280:1451–1455.
Force, A., M. Lynch, F. B. Pickett, A. Amores, Y.-L. Yan, and J. Postlethwait. 1999. Preservation of duplicate genes by complementary degenerative mutations. *Genetics* 151:1531–1545.
Force, A. G., W. A. Cresko, and F. B. Pickett. 2004. Informational accretion, gene duplication, and the mechanisms of genetic module parcellation. In *Modularity in development and evolution,* ed. G. Schlosser and G. P. Wagner. Chicago: University of Chicago Press.
Ford, L. S., and D. C. Cannatella. 1993. The major clades of frogs. *Herpetol. Monogr.* 7:94–117.
Fox, H. 1981. Cytological and morphological changes during amphibian metamorphosis. In *Metamorphosis: a problem in developmental biology,* ed. L. I. Gilbert and E. Frieden, 327–362. New York: Plenum Press.

Freeman, M. 2000. Feedback control of intercellular signalling in development. *Nature* 408:313–319.

Frisén, J., and U. Lendahl. 2001. Oh no, Notch again! *BioEssays* 23:3–7.

Fritzsch, B. 1988. The lateral-line and inner-ear afferents in larval and adult urodeles. *Brain Behav. Evol.* 31:325–348.

Fritzsch, B. 1989. Diversity and regression in the amphibian lateral line and electrosensory system. In *The mechanosensory lateral line,* ed. S. Coombs, P. Görner, and H. Münz, 99–114. New York: Springer.

Fritzsch, B. 1990. The evolution of metamorphosis in amphibians. *J. Neurobiol.* 21: 1011–1021.

Fritzsch, B., R. C. Drewes, and R. Ruibal. 1987. The retention of the lateral-line nucleus in adult anurans. *Copeia* 1987:127–135.

Fritzsch, B., U. Wahnschaffe, and U. Bartsch. 1988. Metamorphic changes in the octavolateralis system of amphibians. In *The evolution of the amphibian auditory system,* ed. B. Fritzsch, M. Ryan, W. Wilczynski, T. E. Hetherington, and W. Walkowiak, 359–376. New York: Wiley.

Fritzsch, B., and M. Wake. 1986. The distribution of ampullary organs in Gymnophiona. *J. Herpetol.* 20:90–93.

García-Bellido, A. 1975. Genetic control of wing disc development in *Drosophila. CIBA Found. Symp.* 29:161–182.

García-Bellido, A. 1996. Symmetries throughout organic evolution. *Proc. Natl. Acad. Sci. U.S.A.* 93:14229–14232.

Gehring, W. J., and K. Ikeo. 1999. Pax 6:- mastering eye morphogenesis and eye evolution. *Trends Genet.* 15:371–377.

Gellon, G., and W. McGinnis. 1998. Shaping animal body plans in development and evolution by modulation of Hox expression patterns. *BioEssays* 20:116–125.

Gerhart, J. 1999. Signaling pathways in development. *Teratology* 60:226–239.

Gerhart, J., and J. Kirschner. 1997. *Cells, embryos, and evolution.* Malden: Blackwell Science.

Ghazi, A., and K. Vijay Raghavan. 2000. Control by combinatorial codes. *Nature* 408: 419–420.

Gibson, G., and G. Wagner. 2000. Canalization in evolutionary genetics: a stabilizing theory? *BioEssays* 22:372–380.

Gilbert, S. F. 1998. Conceptual breakthroughs in developmental biology. *J. Biosci.* 23: 169–176.

Gilbert, S. F. 2000. *Developmental biology.* Sunderland: Sinauer.

Gilbert, S. F., and J. A. Bolker. 2001. Homologies of process and modular elements of embryonic construction. In *The character concept in evolutionary biology,* ed. G. P. Wagner, 559–579. San Diego: Academic Press.

Gilbert, S. F., J. M. Opitz, and R. A. Raff. 1996. Resynthesizing evolutionary and developmental biology. *Dev. Biol.* 173:357–372.

Gilbert, W. 1978. Why genes in pieces? *Nature* 271:501

Gogarten, J. P., and L. Olendzenski. 1999. Orthologs, paralogs and genome comparisons. *Curr. Opin. Genet. Dev.* 9:630–636.

Goh, C. S., A. A. Bogan, M. Joachimiak, D. Walther, and F. E. Cohen. 2000. Coevolution of proteins with their interaction partners. *J. Mol. Biol.* 299:283–293.

Gould, S. J. 1977. *Ontogeny and phylogeny.* Cambridge, Mass.: Harvard University Press.

Gould, S. J. 1982. Darwinism and the expansion of the evolutionary theory. *Science* 216:380–387.

Gould, S. J., and R. C. Lewontin. 1979. The spandrels of San Marco and the Panglossian paradigm: a critique of the adaptationist programme. *Proc. R. Soc. Lond. B Biol. Sci.* 205:581–598.

Gould, S. J., and E. A. Lloyd. 1999. Individuality and adaptation across levels of selection: how shall we name and generalize the unit of Darwinism? *Proc. Natl. Acad. Sci. U.S.A.* 96:11904–9.

Gradwohl, G., A. Dierich, M. Lemeur, and F. Guillemot. 2000. Neurogenin3 is required for the development of the four endocrine cell lineages of the pancreas. *Proc. Natl. Acad. Sci. U.S.A.* 97:1607–1611.

Gray, S., and M. Levine. 1996. Transcriptional repression in development. *Curr. Opin. Cell. Biol.* 8:358–364.

Gray, S., P. Szymanski, and M. Levine. 1994. Short-range repression permits multiple enhancers to function autonomously within a complex promoter. *Genes Dev.* 8: 1829–1838.

Greer, J. M., J. Puetz, K. R. Thomas, and M. R. Capecchi. 2000. Maintenance of functional equivalence during paralogous Hox gene evolution. *Nature* 403:661–665.

Guillemot, F. 1999. Vertebrate bHLH genes and the determination of neuronal fates. *Exp. Cell Res.* 253:357–364.

Guss, K. A., C. E. Nelson, A. Hudson, M. E. Kraus, and S. B. Carroll. 2001. Control of a genetic regulatory network by a selector gene. *Science* 292:1164–1167.

Haddon, C., Y. J. Jiang, L. Smithers, and J. Lewis. 1998. Delta-Notch signalling and the patterning of sensory cell differentiation in the zebrafish ear: evidence from the mind bomb mutant. *Development* 125:4637–4644.

Halder, G., P. Callaerts, S. Flister, U. Walldorf, U. Kloter, and W. J. Gehring. 1998a. *Eyeless* initiates the expression of both *sine oculis* and *eyes absent* during *Drosophila* compound eye development. *Development* 125:2181–2191.

Halder, G., P. Callaerts, and W. J. Gehring. 1995. Induction of ectopic eyes by targeted expression of the eyeless gene in *Drosophila. Science* 267:1788–1792.

Halder, G., P. Polaczyk, M. E. Kraus, A. Hudson, J. Kim, A. Laughon, and S. Carroll. 1998b. The vestigial and scalloped proteins act together to directly regulate wing-specific gene expression in *Drosophila. Genes Dev.* 12:3900–3909.

Halfon, M. S., A. Carmena, S. Gisselbrecht, C. M. Sackerson, F. Jimenez, M. K. Baylies, and A. M. Michelson. 2000. Ras pathway specificity is determined by the integration of multiple signal-activated and tissue-restricted transcription factors. *Cell* 103: 63–74.

Hall, B. K. 2000. Evo-devo or devo-evo: does it matter? *Evol. Dev.* 2:177–178.

Hammerschmidt, M., A. Brook, and A. P. Mc Mahon. 1997. The world according to hedgehog. *Trends Genet.* 13:14–21.

Hammond, K. L., I. M. Hanson, A. G. Brown, L. A. Lettice, and R. E. Hill. 1998. Mammalian and *Drosophila* dachshund genes are related to the Ski proto-oncogene and are expressed in eye and limb. *Mech. Dev.* 74:121–131.

Hancock, J. M., P. J. Shaw, F. Bonneton, and G. A. Dover. 1999. High sequence turnover in the regulatory regions of the developmental gene hunchback in insects. *Mol. Biol. Evol.* 16:253–265.

Hanken, J., T. F. Carl, M. K. Richardson, L. Olsson, G. Schlosser, C. K. Osabutey, and M. W. Klymkowsky. 2001. Limb development in a "nonmodel" vertebrate, the direct-developing frog *Eleutherodactylus coqui. J. Exp. Zool. (Mol. Dev. Evol.)* 291: 375–88.

Harrison, R. G. 1904. Experimentelle Untersuchungen über die Entwicklung der Sinnesorgane der Seitenlinie bei den Amphibien. *Arch. Mikrosk. Anat. Entwicklungsgesch.* 63:35–149.

Hartwell, L. H., J. J. Hopfield, S. Leibler, and A. W. Murray. 1999. From molecular to modular cell biology. *Nature* 402 (suppl.): C47–C52.

Harvey, P. H. and M. D. Pagel. 1991. *The comparative method in evolutionary biology.* Oxford: Oxford University Press.

Hayes, T. B. 1997. Hormonal mechanisms as potential constraints on evolution: examples from the Anura. *Am. Zool.* 37:482–490.

Heanue, T. A., R. Reshef, R. J. Davis, G. Mardon, G. Oliver, S. Tomarev, A. B. Lassar, and C. J. Tabin. 1999. Synergistic regulation of vertebrate muscle development by Dach2, Eya2, and Six1, homologs of genes required for *Drosophila* eye formation. *Genes Dev.* 13:3231–3243.

Hedrick, P., S. Jan, and L. Holden. 1978. Multilocus systems in evolution. *Evol. Biol. (N.Y.)* 11:101–184.

Heller, N., and A. Brändli. 1997. *Xenopus Pax-2* displays multiple splice forms during embryogenesis and pronephric kidney development. *Mech. Dev.* 69:83–104.

Heller, N., and A. W. Brändli. 1999. *Xenopus Pax-2/5/8* orthologues: novel insights into Pax gene evolution and identification of Pax-8 as the earliest marker for otic and pronephric cell lineages. *Dev. Genet.* 24:208–219.

Henikoff, S., E. A. Greene, S. Pietrokovski, P. Bork, T. K. Attwood, and L. Hood. 1997. Gene families: the taxonomy of protein paralogs and chimeras. *Science* 278: 609–614.

Henrique, D., J. Adam, A. Myat, A. Chitnis, J. Lewis, and D. Ish-Horowitz. 1995. Expression of a delta homologue in prospective neurons in the chick. *Nature* 375: 787–790.

Holland, L. Z., and N. D. Holland. 1999. Chordate origins of the vertebrate central nervous system. *Curr. Opin. Neurobiol.* 9:596–602.

Holland, L. Z., L. A. Rached, R. Tamme, N. D. Holland, H. Inoko, T. Shiina, C. Burgtorf, and M. Lardelli. 2001. Characterization and developmental expression of the amphioxus homolog of notch (AmphiNotch): evolutionary conservation of multiple expression domains in amphioxus and vertebrates. *Dev. Biol.* 232:493–507.

Holland, L. Z., M. Schubert, N. D. Holland, and T. Neuman. 2000. Evolutionary conservation of the presumptive neural plate markers AmphiSox1/2/3 and AmphiNeurogenin in the invertebrate chordate amphioxus. *Dev. Biol.* 226:18–33.

Holland, L. Z., M. Schubert, Z. Kozmik, and N. D. Holland. 1999. AmphiPax3/7, an amphioxus paired box gene: insights into chordate myogenesis, neurogenesis, and the possible evolutionary precursor of definitive vertebrate neural crest. *Evol. Dev.* 1:153–165.

Hori, S., T. Saitoh, M. Matsumoto, K. W. Makabe, and H. Nishida. 1997. Notch homologue from *Halocynthia roretzi* is preferentially expressed in the central nervous system during ascidian embryogenesis. *Dev. Genes Evol.* 207:371–380.

Horton, S., A. Meredith, J. A. Richardson, and J. E. Johnson. 1999. Correct coordination of neuronal differentiation events in ventral forebrain requires the bHLH factor MASH1. *Mol. Cell. Neurosci.* 14:355–369.

Hutter, H., B. E. Vogel, J. D. Plenefisch, C. R. Norris, R. B. Proenca, J. Spieth, C. Guo, S. Mastwal, X. Zhu, J. Scheel, and E. M. Hedgecock. 2000. Conservation and novelty in the evolution of cell adhesion and extracellular matrix genes. *Science* 287: 989–994.

Huynen, M. A., P. F. Stadler, and W. Fontana. 1996. Smoothness within ruggedness: the role of neutrality in adaptation. *Proc. Natl. Acad. Sci. U.S.A.* 93:397–401.

Itoh, M., and A. B. Chitnis. 2001. Expression of proneural and neurogenic genes in the zebrafish lateral line primordium correlates with selection of hair cell fate in neuromasts. *Mech. Dev.* 102:263–266.

Jacob, F. 1977. Evolution and tinkering. *Science* 196:1161–1166.

Jensen, J., R. S. Heller, T. Funder-Nielsen, E. E. Pedersen, C. Lindsell, G. Weinmaster, O. D. Madsen, and P. Serup. 2000a. Independent development of pancreatic alpha- and beta-cells from neurogenin3-expressing precursors: a role for the notch pathway in repression of premature differentiation. *Diabetes* 49:163–176.

Jensen, J., E. E. Pedersen, P. Galante, J. Hald, R. S. Heller, M. Ishibashi, R. Kageyama, F. Guillemot, P. Serup, and O. D. Madsen. 2000b. Control of endodermal endocrine development by Hes-1. *Nat. Genet.* 24:36–44.

Jiang, Y. J., B. L. Aerne, L. Smithers, C. Haddon, D. Ishhorowicz, and J. Lewis. 2000. Notch signalling and the synchronization of the somite segmentation clock. *Nature* 408:475–479.

Kardon, G., T. A. Heanue, and C. J. Tabin. 2004. The *Pax/Six/Eya/Dach* network in development and evolution. In *Modularity in development and evolution,* ed. G. Schlosser and G. P. Wagner. Chicago: University of Chicago Press.

Katz, M. J. 1987. Is evolution random? In *Development as an evolutionary process,* ed. R. A. Raff and E. C. Raff, 285–315. New York: Liss.

Kawakami, K., S. Sato, H. Ozaki, and K. Ikeda. 2000. Six family genes: structure and function as transcription factors and their roles in development. *BioEssays* 22: 616–626.

Kazazian, H. H. 2000. L1 retrotransposons shape the mammalian genome. *Science* 289:1152–1153.

Kermekchiev, M., M. Petterson, P. Matthias, and W. Schaffner. 1991. Every enhancer works with every promoter for all the combinations tested: could new regulatory pathways evolve by enhancer shuffling? *Gene Exp.* 1:71–81.

Kessler, D. S. 1997. Siamois is required for formation of Spemann's organizer. *Proc. Natl. Acad. Sci. U.S.A.* 94:13017–13022.

Keys, D. N., D. L. Lewis, J. E. Selegue, B. J. Pearson, L. V. Goodrich, R. L. Johnson, J. Gates, M. P. Scott, and S. B. Carroll. 1999. Recruitment of a hedgehog regulatory circuit in butterfly eyespot evolution. *Science* 283:532–534.

Kidwell, M. G., and D. R. Lisch. 2001. Transposable elements, parasitic DNA, and genome evolution. *Evolution* 55:1–24.

Kimble, J., and P. Simpson. 1997. The lin-12/notch signaling pathway and its regulation. *Annu. Rev. Cell Dev. Biol.* 13:333–361.

Kirchhamer, C. V., L. D. Bogarad, and E. H. Davidson. 1996a. Developmental expression of synthetic cis-regulatory systems composed of spatial control elements from two different genes. *Proc. Natl. Acad. Sci. U.S.A.* 93:13849–13854.

Kirchhamer, C. V., C.-H. Yuh, and E. H. Davidson. 1996b. Modular cis-regulatory organization of developmentally expressed genes: two genes transcribed territorially in the sea urchin embryo, and additional examples. *Proc. Natl. Acad. Sci. U.S.A.* 93: 9322–9328.

Kirschner, M., and J. Gerhart. 1998. Evolvability. *Proc. Natl. Acad. Sci. U.S.A.* 95: 8420–8427.

Kluge, A. G., and R. E. Strauss. 1985. Ontogeny and systematics. *Annu. Rev. Ecol. Syst.* 16:247–268.

Kobayashi, M., H. Osanai, K. Kawakami, and M. Yamamoto. 2000. Expression of three zebrafish Six4 genes in the cranial sensory placodes and the developing somites. *Mech. Dev.* 98:151–155.

Korzh, V., I. Sleptsova, J. Liao, J. He, and Z. Gong. 1998. Expression of zebrafish bHLH genes ngn1 and nrd defines distinct stages of neural differentiation. *Dev. Dyn.* 213: 92–104.

Koyano-Nakagawa, N., J. Kim, D. Anderson, and C. Kintner. 2000. Hes6 acts in a positive feedback loop with the neurogenins to promote neuronal differentiation. *Development* 127:4203–4216.

Koyano-Nakagawa, N., D. Wettstein, and C. Kintner. 1999. Activation of *Xenopus* genes required for lateral inhibition and neuronal differentiation during primary neurogenesis. *Mol. Cell. Neurosci.* 14:327–339.

Kozmik, Z., N. D. Holland, A. Kalousova, J. Paces, M. Schubert, and L. Z. Holland.

1999a. Characterization of an amphioxus paired box gene, AmphiPax2/5/8: developmental expression patterns in optic support cells, nephridium, thyroid-like structures and pharyngeal gill slits, but not in the midbrain-hindbrain boundary region. *Development* 126:1295–1304.

Kozmik, Z., P. Pfeffer, J. Kralova, J. Paces, V. Paces, A. Kalousova, and A. Cvekl. 1999b. Molecular cloning and expression of the human and mouse homologues of the *Drosophila* dachshund gene. *Dev. Genes Evol.* 209:537–45.

Krakauer, D. C., and M. A. Nowak. 1999. Evolutionary preservation of redundant duplicated genes. *Semin. Cell Dev. Biol.* 10:555–559.

Krumlauf, R. 1994. Hox genes in vertebrate development. *Cell* 78:191–201.

Lammert, E., J. Brown, and D. A. Melton. 2000. Notch gene expression during pancreatic organogenesis. *Mech. Dev.* 94:199–203.

Larson, A., and W. W. Dimmick. 1993. Phylogenetic relationships of the salamander families: an analysis of congruence among morphological and molecular characters. *Herpetol. Monogr.* 7:77–93.

Lee, J. E. 1997. Basic helix-loop-helix genes in neural development. *Curr. Opin. Neurobiol.* 7:13–20.

Lee, J. E., S. M. Hollenberg, L. Snider, D. L. Turner, N. Lipnick, and H. Weintraub. 1995. Conversion of *Xenopus* ectoderm into neurons by NeuroD, a basic helix-loop-helix protein. *Science* 268:836–844.

Lewin, B. 2000. *Genes VII.* New York: Oxford University Press.

Lewis, J. 1998. Notch signalling and the control of cell fate choices in vertebrates. *Semin. Cell Dev. Biol.* 9:583–589.

Lewontin, R. C. 1970. The units of selection. *Annu. Rev. Ecol. Syst.* 1:1–18.

Lewontin, R. C. 1974. *The genetic basis of evolutionary change.* New York: Columbia University Press.

Lewontin, R. C. 1978. Adaptation. *Sci. Am.* 239 (3): 156–169.

Li, X., and M. Noll. 1994. Evolution of distinct developmental functions of three *Drosophila* genes by acquisition of different cis-regulatory regions. *Nature* 367:83–87.

Lin, M. H., C. Leimeister, M. Gessler, and R. Kopan. 2000. Activation of the Notch pathway in the hair cortex leads to aberrant differentiation of the adjacent hair-shaft layers. *Development* 127:2421–2432.

Lloyd, E. 1988. *The structure and confirmation of evolutionary theory.* Princeton, N.J.: Princeton University Press.

Lockhart, D. J., and E. A. Winzeler. 2000. Genomics, gene expression and DNA arrays. *Nature* 405:827–836.

Loosli, F., S. Winkler, and J. Wittbrodt. 1999. Six3 overexpression initiates the formation of ectopic retina. *Genes Dev.* 13:649–654.

Ludwig, M. Z., C. Bergman, N. H. Patel, and M. Kreitman. 2000. Evidence for stabilizing selection in a eukaryotic enhancer element. *Nature* 403:564–567.

Lynch, M., and A. Force. 2000. The probability of duplicate gene preservation by subfunctionalization. *Genetics* 154:459–73.

Lynn, W. G. 1942. The embryology of *Eleutherodactylus nubicola,* an anuran which has no tadpole stage. *Contrib. Embryol.* 190:27–62.

Lynn, W. G. 1961. Types of amphibian metamorphosis. *Am. Zool.* 1:151–161.

Ma, Q. F., Z. F. Chen, I. D. Barrantes, J. L. de la Pompa, and D. J. Anderson. 1998. Neurogenin1 is essential for the determination of neuronal precursors for proximal cranial sensory ganglia. *Neuron* 20:469–482.

Ma, Q. F., C. Kintner, and D. J. Anderson. 1996. Identification of neurogenin, a vertebrate neuronal determination gene. *Cell* 87:43–52.

Maconochie, M., S. Nonchev, A. Morrison, and R. Krumlauf. 1996. Paralogous hox genes: function and regulation. *Annu. Rev. Genet.* 30:529–556.

Maddison, W. P. 1990. A method for testing the correlated evolution of two binary characters: are gains or losses concentrated on certain branches of a phylogenetic tree? *Evolution* 44:539–557.

Mannervik, M., Y. Nibu, H. Zhang, and M. Levine. 1999. Transcriptional coregulators in development. *Science* 284:606–609.

Mansouri, A., M. Hallonet, and P. Gruss. 1996. Pax genes and their roles in cell differentiation and development. *Curr. Opin. Cell Biol.* 8:851–857.

Maynard Smith, J. 1987. How to model evolution. In *The latest on the best,* ed. J. Dupré, 119–131. Cambridge, Mass.: MIT Press.

Maynard Smith, J., R. Burian, S. Kauffman, P. Alberch, J. Campbell, B. Goodwin, R. Lande, D. Raup, and L. Wolpert. 1985. Developmental constraints and evolution. *Q. Rev. Biol.* 60:265–287.

Maynard Smith, J., and E. Szathmáry. 1995. *The major transitions in evolution.* Oxford: Freeman.

McLaughlin, K. A., M. S. Rones, and M. Mercola. 2000. Notch regulates cell fate in the developing pronephros. *Dev. Biol.* 227:567–580.

McShea, D. W., and E. P. Venit. 2001. What is a part? In *The character concept in evolutionary biology,* ed. G. P. Wagner, 259–284. San Diego: Academic Press.

Mezey, J. G., J. M. Cheverud, and G. P. Wagner. 2000. Is the genotype-phenotype map modular? a statistical approach using mouse quantitative trait loci data. *Genetics* 156:305–311.

Mitsiadis, T. A., E. Hirsinger, U. Lendahl, and C. Goridis. 1998. Delta-Notch signaling in odontogenesis: correlation with cytodifferentiation and evidence for feedback regulation. *Dev. Biol.* 204:420–431.

Molkentin, J. D., and E. N. Olson. 1996. Defining the regulatory networks for muscle development. *Curr. Opin. Genet. Dev.* 6:445–453.

Morrison, A., C. Hodgetts, A. Gossler, M. H. Deangelis, and J. Lewis. 1999. Expression of Delta1 and Serrate1 (Jagged1) in the mouse inner ear. *Mech. Dev.* 84:169–172.

Morrison, A. M., S. L. Nutt, C. Thévenin, A. Rolink, and M. Busslinger. 1998. Loss- and gain-of-function mutations reveal as important role of BSAP (Pax-5) at the start and end of B cell differentiation. *Semin. Immunol.* 10:133–142.

Morrow, E. M., T. Furukawa, J. E. Lee, and C. L. Cepko. 1999. NeuroD regulates multiple functions in the developing neural retina in rodent. *Development* 126:23–36.

Mumm, J. S., and R. Kopan. 2000. Notch signaling: from the outside in. *Dev. Biol.* 228: 151–165.

Murone, M., A. Rosenthal, and F. J. deSauvage. 1999. Hedgehog signal transduction: from flies to vertebrates. *Exp. Cell Res.* 253:25–33.

Naya, F. J., H. P. Huang, Y. H. Qiu, H. Mutoh, F. J. DeMayo, A. B. Leiter, and M. J. Tsai. 1997. Diabetes, defective pancreatic morphogenesis, and abnormal enteroendocrine differentiation in beta2/neurod-deficient mice. *Genes Dev.* 11:2323–2334.

Nelson, C. 2004. Selector genes and the genetic control of developmental modules. In *Modularity in development and evolution,* ed. G. Schlosser and G. P. Wagner. Chicago: University of Chicago Press.

Ng, J. K., K. Tamura, D. Buscher, and J. C. Izpisuabelmonte. 1999. Molecular and cellular basis of pattern formation during vertebrate limb development. *Curr. Top. Dev. Biol.* 41:37–66.

Niehrs, C. 2004. Synexpression groups: genetic modules and embryonic development. In *Modularity in development and evolution,* ed. G. Schlosser and G. P. Wagner. Chicago: University of Chicago Press.

Niehrs, C., and N. Pollet. 1999. Synexpression groups in eukaryotes. *Nature* 402: 483–487.

Nishita, M., M. K. Hashimoto, S. Ogata, M. N. Laurent, N. Ueno, H. Shibuya, and

K. W. Y. Cho. 2000. Interaction between Wnt and TGF-beta signalling pathways during formation of Spemann's organizer. *Nature* 403:781–785.

Northcutt, R. G. 1990. Ontogeny and phylogeny: a re-evaluation of conceptual relationships and some applications. *Brain Behav. Evol.* 36:116–140.

Northcutt, R. G. 1992. The phylogeny of octavolateralis ontogenies: a reaffirmation of Garstang's hypothesis. In *The evolutionary biology of hearing,* ed. D. B. Webster, R. R. Fay, and A. N. Popper, 21–47. New York: Springer.

Northcutt, R. G., and K. Brändle. 1995. Development of branchiomeric and lateral line nerves in the axolotl. *J. Comp. Neurol.* 355:427–454.

Northcutt, R. G., K. Brändle, and B. Fritzsch. 1995. Electroreceptors and mechanosensory lateral line organs arise from single placodes in axolotls. *Dev. Biol.* 168: 358–373.

Northcutt, R. G., K. C. Catania, and B. B. Criley. 1994. Development of lateral line organs in the axolotl. *J. Comp. Neurol.* 340:480–514.

Northcutt, R. G., and C. Gans. 1983. The genesis of neural crest and epidermal placodes: a reinterpretation of vertebrate origins. *Q. Rev. Biol.* 58:1–28.

Noselli, S., and N. Perrimon. 2000. Are there close encounters between signaling pathways? *Science* 290:68–69.

Nowak, M. A., M. C. Boerlijst, J. Cooke, and J. M. Smith. 1997. Evolution of genetic redundancy. *Nature* 388:167–171.

Ohta, T. 1987. Simulating evolution by gene duplication. *Genetics* 115:207–213.

Ohto, H., S. Kamada, K. Tago, S. Tominaga, H. Ozaki, S. Sato, and K. Kawakami. 1999. Cooperation of Six and Eya in activation of their target genes through nuclear translocation of Eya. *Mol. Cell. Biol.* 19:6815–6824.

Ohto, H., T. Takizawa, T. Saito, M. Kobayashi, K. Ikeda, and K. Kawakami. 1998. Tissue and developmental distribution of Six family gene products. *Int. J. Dev. Biol.* 42: 141–148.

O'Kane, C. J., and W. J. Gehring. 1987. Detection in situ of genomic regulatory elements in *Drosophila. Proc. Natl. Acad. Sci. U.S.A.* 84:9123–7.

Oliver, G., R. Wehr, A. Jenkins, N. G. Copeland, B. N. R. Cheyette, V. Hartenstein, S. L. Zipursky, and P. Gruss. 1995. Homeobox genes and connective tissue patterning. *Development* 121:693–705.

Pagel, M. 1999. Inferring the historical patterns of biological evolution. *Nature* 401: 877–884.

Pandur, P. D., and S. A. Moody. 2000. *Xenopus* Six1 gene is expressed in neurogenic cranial placodes and maintained in differentiating lateral lines. *Mech. Dev.* 96: 253–257.

Parichy, D. M. 1996a. Pigment patterns of larval salamanders (Ambystomatidae, Salamandridae): the role of the lateral line sensory system and the evolution of pattern-forming mechanisms. *Dev. Biol.* 175:265–282.

Parichy, D. M. 1996b. Salamander pigment patterns: how can they be used to study developmental mechanisms and their evolutionary transformation? *Int. J. Dev. Biol.* 40:871–884.

Patthy, L. 1999. Genome evolution and the evolution of exon-shuffling: a review. *Gene* 238:103–114.

Perron, M., and W. A. Harris. 2000. Determination of vertebrate retinal progenitor cell fate by the Notch pathway and basic helix-loop-helix transcription factors. *Cell. Mol. Life Sci.* 57:215–23.

Perron, M., K. Opdecamp, K. Butler, W. A. Harris, and E. J. Bellefroid. 1999. X-ngnr-1 and Xath3 promote ectopic expression of sensory neuron markers in the neurula ectoderm and have distinct inducing properties in the retina. *Proc. Natl. Acad. Sci. U.S.A.* 96:14996–15001.

Perry, R. L., and M. A. Rudnicki. 2000. Molecular mechanisms regulating myogenic determination and differentiation. *Front. Biosci.* 5:D750–767.

Pfeffer, P. L., T. Gerster, K. Lun, M. Brand, and M. Busslinger. 1998. Characterization of three novel members of the zebrafish Pax2/5/8 family: dependency of Pax5 and Pax8 expression on the Pax2.1 (noi) function. *Development* 125:3063–3074.

Post, L. C., M. Ternet, and B. L. Hogan. 2000. Notch/Delta expression in the developing mouse lung. *Mech. Dev.* 98:95–98.

Powell, B. C., E. A. Passmore, A. Nesci, and S. M. Dunn. 1998. The Notch signalling pathway in hair growth. *Mech. Dev.* 78:189–192.

Prak, E. T., and H. H. Kazazian. 2000. Mobile elements and the human genome. *Nat. Rev. Genet.* 1:134–144.

Ptashne, M., and A. Gann. 1997. Transcriptional activation by recruitment. *Nature* 386:569–577.

Ptashne, M., and A. Gann. 1998. Imposing specificity by localization: mechanism and evolvability. *Curr. Biol.* 8:R812-R822

Purugganan, M. D. 1998. The molecular evolution of development. *BioEssays* 20: 700–711.

Purugganan, M. D. 2000. The molecular population genetics of regulatory genes. *Mol. Ecol.* 9:1451–1461.

Quint, E., T. Zerucha, and M. Ekker. 2000. Differential expression of orthologous Dlx genes in zebrafish and mice: implications for the evolution of the Dlx homeobox gene family. *J. Exp. Zool. (Mol. Dev. Evol.)* 288:235–241.

Raff, R. A. 1996. *The shape of life.* Chicago: University of Chicago Press.

Relaix, F., and M. Buckingham. 1999. From insect eye to vertebrate muscle: redeployment of a regulatory network. *Genes Dev.* 13:3171–3178.

Richardson, M. K. 1995. Heterochrony and the phylotypic period. *Dev. Biol.* 172: 412–421.

Richardson, M. K., T. F. Carl, J. Hanken, R. P. Elinson, C. Cope, and P. Bagley. 1998. Limb development and evolution: a frog embryo with no apical ectodermal ridge (AER). *J. Anat.* 192:379–390.

Richardson, M. K., J. Hanken, M. L. Gooneratne, C. Pieau, A. Raynaud, L. Selwood, and G. M. Wright. 1997. There is no highly conserved embryonic stage in the vertebrates: implications for current theories of evolution and development. *Anat. Embryol.* 196:91–106.

Riedl, R. 1975. *Die Ordnung des Lebendigen.* Hamburg: Parey.

Riley, B. B., M. Y. Chiang, L. Farmer, and R. Heck. 1999. The deltaA gene of zebrafish mediates lateral inhibition of hair cells in the inner ear and is regulated by pax2.1. *Development* 126:5669–5678.

Rose, C. S. 1995a. Skeletal morphogenesis in the urodele skull. 2. Effect of developmental stage in thyroid hormone-induced remodeling. *J. Morphol.* 223:149–166.

Rose, C. S. 1995b. Skeletal morphogenesis in the urodele skull. 3. Effect of hormone dosage in TH-induced remodeling. *J. Morphol.* 223:243–261.

Rose, C. S. 1996. An endocrine-based model for developmental and morphogenetic diversification in metamorphic and paedomorphic urodeles. *J. Zool.* 239:253–284.

Rose, C. S. 1999. Hormonal control in larval development and evolution: amphibians. In *The origin and evolution of larval forms,* ed. B. K. Hall, 167–216. San Diego: Academic Press.

Rosenkilde, P., and A. P. Ussing. 1996. What mechanisms control neoteny and regulate induced metamorphosis in urodeles? *Int. J. Dev. Biol.* 40:665–673.

Ross, J. L., P .P. Fong, and D. R. Cavener. 1994. Correlated evolution of the cis-acting regulatory elements and developmental expression of the *Drosophila* Gld gene in seven species from the subgroup *melanogaster. Dev. Genet.* 15:38–50.

Roth, G., K. C. Nishikawa, C. Naujoks-Manteuffel, A. Schmidt, and D. B. Wake. 1993. Paedomorphosis and simplification in the nervous system of salamanders. *Brain Behav. Evol.* 42:137–170.

Roth, V. L. 1991. Homology and hierarchies: problems solved and unresolved. *J. Evol. Biol.* 4:167–194.

Roth, V. L. 2001. Character replication. In *The character concept in evolutionary biology,* ed. G. P. Wagner, 81–107. San Diego: Academic Press.

Rothenberg, E. V. 2000. Stepwise specification of lymphocyte developmental lineages. *Curr. Opin. Genet. Dev.* 10:370–379.

Rutherford, S. L. 2000. From genotype to phenotype: buffering mechanisms and the storage of genetic information. *BioEssays* 22:1095–1105.

Safi, R., A. Begue, C. Hänni, D. Stehelin, J. R. Tata, and V. Laudet. 1997. Thyroid hormone receptor genes of neotenic amphibians. *J. Mol. Evol.* 44:595–604.

Sahly, I., P. Andermann, and C. Petit. 1999. The zebrafish *eya1* gene and its expression pattern during embryogenesis. *Dev. Genes Evol.* 209:399–410.

Schank, J. C., and W. C. Wimsatt. 2001. Evolvability: adaptation and modularity. In *Thinking about evolution,* ed. R. S. Singh, C. B. Krimbas, D. Paul, and J. Beatty, 322–335. Cambridge: Cambridge University Press.

Schlosser, G. 2001. Using heterochrony plots to detect the dissociated coevolution of characters. *J. Exp. Zool. (Mol. Dev. Evol.)* 291:282–304.

Schlosser, G. 2002. Modularity and the units of evolution. *Theory Biosci.* 121:1–80.

Schlosser, G. 2003. Mosaic evolution of neural development in anurans: acceleration of spinal cord development in the direct developing frog *Eleutherodactylus coqui. Anat. Embryol.* 206:215–227.

Schlosser, G. In press. Amphibian variations: the role of modules in mosaic evolution. In *Modularity: understanding the development and evolution of natural complex systems,* ed. W. Callebaut and D. Rasskin-Gutman. Cambridge, Mass.: MIT Press.

Schlosser, G., C. Kintner, and R. G. Northcutt. 1999. Loss of ectodermal competence for lateral line placode formation in the direct developing frog *Eleutherodactylus coqui. Dev. Biol.* 213:354–369.

Schlosser, G., N. Koyano-Nakagawa, and C. Kintner. 2002. Thyroid hormone promotes neurogenesis in the *Xenopus* spinal cord. *Dev. Dyn.* 225:485–498.

Schlosser, G., and R. G. Northcutt. 2000. Development of neurogenic placodes in *Xenopus laevis. J. Comp. Neurol.* 418:121–146.

Schlosser, G., and R. G. Northcutt. 2001. Lateral line placodes are induced during neurulation in the axolotl. *Dev. Biol.* 234:55–71.

Schlosser, G., and G. Roth. 1997. Evolution of nerve development in frogs. 2. Modified development of the peripheral nervous system in the direct-developing frog *Eleutherodactylus coqui* (Leptodactylidae). *Brain Behav. Evol.* 50:94–128.

Schlosser, G., and D. Thieffry. 2000. Modularity in development and evolution. *BioEssays* 22:1043–1045.

Schmidt, A., and G. Roth. 1996. Differentiation processes in the amphibian brain with special emphasis on heterochronies. *Int. Rev. Cytol.* 169:83–150.

Schwab, M. H., A. Bartholomae, B. Heimrich, D. Feldmeyer, S. Druffel-Augustin, S. Goebbels, F. J. Naya, S. Zhao, M. Frotscher, M. J. Tsai, and K. A. Nave. 2000. Neuronal basic helix-loop-helix proteins (NEX and BETA2/Neuro D) regulate terminal granule cell differentiation in the hippocampus. *J. Neurosci.* 20:3714–3724.

Schwenk, K. 1994. A utilitarian approach to evolutionary constraint. *Zoology* 98:251–262.

Sefton, M., S. Sanchez, and M. A. Nieto. 1998. Conserved and divergent roles for members of the Snail family of transcription factors in the chick and mouse embryo. *Development* 125:3111–3121.

Selinger, D. A., and V. L. Chandler. 1999. Major recent and independent changes in levels and patterns of expression have occurred at the b gene, a regulatory locus in maize. *Proc. Natl. Acad. Sci. U.S.A.* 96:15007–15012.

Shaffer, H. B., and S. R. Voss. 1996. Phylogenetic and mechanistic analysis of a developmentally integrated character complex: alternate life history modes in ambystomatid salamanders. *Am. Zool.* 36:24–35.

Shashikant, C. S., C. B. Kim, M. A. Borbely, W. C. Wang, and F. H. Ruddle. 1998. Comparative studies on mammalian Hoxc8 early enhancer sequence reveal a baleen whale–specific deletion of a cis-acting element. *Proc. Natl. Acad. Sci. U.S.A.* 95: 15446–15451.

Shi, Y. B., J. Wong, M. Puzianowska-Kuznicka, and M. A. Stolow. 1996. Tadpole competence and tissue-specific temporal regulation of amphibian metamorphosis: roles of thyroid hormone and its receptors. *BioEssays* 18:391–399.

Shubin, N., C. Tabin, and S. Carroll. 1997. Fossils, genes and the evolution of animal limbs. *Nature* 388:639–648.

Shubin, N., and D. Wake. 1996. Phylogeny, variation, and morphological integration. *Am. Zool.* 36:51–60.

Shubin, N. H., and P. Alberch. 1986. A morphogenetic approach to the origin and basic organization of the tetrapod limb. *Evol. Biol. (N.Y.)* 20:319–387.

Sidow, A. 1996. Gen(om)e duplications in the evolution of early vertebrates. *Curr. Opin. Genet. Dev.* 6:715–722.

Smith, S. C., M. J. Lannoo, and J. B. Armstrong. 1990. Development of the mechanorececeptive lateral-line system in the axolotl: placode specification, guidance of migration, and the origin of neuromast polarity. *Anat. Embryol.* 182:171–180.

Smithers, L., C. Haddon, Y.-J. Jiang, and J. Lewis. 2000. Sequence and embryonic expression of deltaC in the zebrafish. *Mech. Dev.* 90:119–123.

Sober, E. 1981. Holism, individualism and the units of selection. In *Proceedings of the 1980 biennial meeting of the Philosophy of Science Association,* vol. 2, ed. P. D. Asquith and R. N. Giere, 93–121.. East Lansing, Mich.

Sober, E. 1984. *The nature of selection.* Chicago: University of Chicago Press.

Sober, E. 1987. Comments on Maynard Smith's "How to model evolution." In *The latest on the best,* ed. J. Dupré, 133–149. Cambridge, Mass.: MIT Press.

Sober, E., and R. C. Lewontin. 1982. Artifact,cause and genic selection. *Philos. Sci.* 49: 157–180.

Sober, E., and D. S. Wilson. 1998. *Unto others.* Cambridge, Mass.: Harvard University Press.

Sommer, L., Q. Ma, and D. J. Anderson. 1996. *Neurogenins,* a novel family of atonal-related bHLH transcription factors, are putative mammalian neuronal determination genes that reveal progenitor heterogeneity in the developing CNS and PNS. *Mol. Cell. Neurobiol.* 8:221–241.

Somogyi, R., S. Fuhrman, G. Anderson, C. Madill, L. D. Greller, and B. Chang. 2004. Systematic exploration and mining of gene expression data provides evidence for higher-order, modular regulation. In *Modularity in development and evolution,* ed. G. Schlosser and G. P. Wagner. Chicago: University of Chicago Press.

Stark, M. R., J. Sechrist, M. Bronner-Fraser, and C. Marcelle. 1997. Neural tube-ectoderm interactions are required for trigeminal placode formation. *Development* 124:4287–4295.

Stephenson, W. G. 1951. Observations on the development of the amphicoelus frogs, *Leiopelma* and *Ascaphus. Zool. J. Linn. Soc.* 42:18–28.

Sterelny, K., and P. Kitcher. 1988. The return of the gene. *J. Philos.* 85:339–361.

Stern, D. L. 2000. Evolutionary developmental biology and the problem of variation. *Evolution* 54:1079–1091.

Stone, L. S. 1922. Experiments on the development of the cranial ganglia and the lateral line sense organs in *Amblystoma punctatum. J. Exp. Zool.* 35:421–496.
Stone, L. S. 1928. Primitive lines in *Amblystoma* and their relation to the migratory lateral-line primordia. *J. Comp. Neurol.* 45:169–190.
Stone, L. S. 1929. Experiments on the transplantation of placodes of the cranial ganglia in the amphibian embryo. 4. Heterotopic transplantation of the postauditory placodal material upon the head and body of *Amblystoma punctatum. J. Comp. Neurol.* 48:311–330.
Strähle, U., and P. Blader. 2004. The basic helix-loop-helix proteins in vertebrate and invertebrate neurogenesis. In *Modularity in development and evolution,* ed. G. Schlosser and G. P. Wagner. Chicago: University of Chicago Press.
Striedter, G. F., and R. G. Northcutt. 1991. Biological hierarchies and the concept of homology. *Brain Behav. Evol.* 38:177–189.
Su, Y., S. Damjanovski, Y. Shi, and Y. B. Shi. 1999. Molecular and cellular basis of tissue remodeling during amphibian metamorphosis. *Histol. Histopathol.* 14:175–183.
Sucena, E., and D. L. Stern. 2000. Divergence of larval morphology between *Drosophila sechellia* and its sibling species caused by cis-regulatory evolution of ovo/shaven-baby. *Proc. Natl. Acad. Sci. U.S.A.* 97:4530–4534.
Tabin, C. J., S. B. Carroll, and G. Panganiban. 1999. Out on a limb: parallels in vertebrate and invertebrate limb patterning and the origin of appendages. *Am. Zool.* 39: 650–663.
Takahashi, H., Y. Mitani, G. Satoh, and N. Satoh. 1999. Evolutionary alterations of the minimal promoter for notochord-specific Brachyury expression in ascidian embryos. *Development* 126:3725–3734.
Tata, J. R. 1996. Hormonal interplay and thyroid hormone receptor expression during amphibian metamorphosis. In *Metamorphosis. postembryonic reprogramming of gene expression in amphibian and insect cells,* ed. L. I. Gilbert, J. R. Tata, and B. G. Atkinson, 465–503. San Diego: Academic Press.
Tata, J. R. 1998. Amphibian metamorphosis as a model for studying the developmental actions of thyroid hormone. *Ann. Endocrinol.* 59:433–442.
Tata, J. R., B. S. Baker, I. Machuca, E. M. L. Rabelo, and K. Yamauchi. 1993. Autoinduction of nuclear receptor genes and its significance. *J. Steroid Biochem. Mol. Biol.* 46:105–119.
Tautz, D. 2000. Evolution of transcriptional regulation. *Curr. Opin. Genet. Dev.* 10: 575–579.
Thieffry, D., and D. Romero. 1999. The modularity of biological regulatory networks. *Biosystems* 50:49–59.
Thieffry, D., and L. Sánchez. 2004. Qualitative analysis of gene networks: toward the delineation of cross-regulatory modules. In *Modularity in development and evolution,* ed. G. Schlosser and G. P. Wagner. Chicago: University of Chicago Press.
Thomas, R. 1978. Logical analysis of systems comprising feedback loops. *J. Theor. Biol.* 73:631–656.
Thomas, R. 1991. Regulatory networks seen as asynchronous automata: a logical description. *J. Theor. Biol.* 153:1–23.
Thomas, R., D. Thieffry, and M. Kaufman. 1995. Dynamical behaviour of biological regulatory networks. 1. Biological role of feedback loops and practical use of the concept of the loop-characteristic state. *Bull. Math. Biol.* 57:247–276.
Underhill, D. A. 2000. Genetic and biochemical diversity in the Pax gene family. *Biochem. Cell Biol.* 78:629–38.
von Dassow, G. and E. Meir. 2004. Exploring modularity with dynamical models of gene networks. In *Modularity in development and evolution,* ed. G. Schlosser and G. P. Wagner. Chicago: University of Chicago Press.

von Dassow, G., E. Meir, E. M. Munro, and G. M. Odell. 2000. The segment polarity network is a robust development module. *Nature* 406:188–192.

von Dassow, G., and E. Munro. 1999. Modularity in animal development and evolution: elements of a conceptual framework for EvoDevo. *J. Exp. Zool. (Mol. Dev. Evol.)* 285:307–325.

Wada, H., P. W. H. Holland, and N. Satoh. 1996. Origin of patterning in neural tubes. *Nature* 384:123

Wada, H., P. W. H. Holland, S. Sato, H. Yamamoto, and N. Satoh. 1997. Neural tube is partially dorsalized by overexpression of HrPax-37: the ascidian homologue of Pax-3 and Pax-7. *Dev. Biol.* 187:240–252.

Wada, H., H. Saiga, N. Satoh, and P. W. H. Holland. 1998. Tripartite organization of the ancestral chordate brain and the antiquity of placodes: insights from ascidian Pax-2/5/8, Hox and Otx genes. *Development* 125:1113–1122.

Waddington, C. H. 1957. *The strategy of the genes.* London: George Allen and Unwin.

Wade, M. J. 1992. Epistasis. In *Keywords in evolutionary biology,* ed. E. Fox Keller and E. A. Lloyd, 87–91. Cambridge, Mass.: Harvard University Press.

Wagner, A. 1998. The fate of duplicated genes: loss or new function? *BioEssays* 20: 785–788.

Wagner, A. 2000. The role of population size, pleiotropy and fitness effects of mutations in the evolution of overlapping gene functions. *Genetics* 154:1389–1401.

Wagner, G. P. 1995. The biological role of homologues: a building block hypothesis. *Neues Jahrb. Geol. Paläontol. Abh.* 19:36–43.

Wagner, G. P. 1996. Homologues, natural kinds and the evolution of modularity. *Am. Zool.* 36:36–43.

Wagner,G. P. 2001. Characters, units and natural kinds: an introduction. In *The character concept in evolutionary biology,* ed. G. P. Wagner, 1–10. San Diego: Academic Press.

Wagner, G. P., and L. Altenberg. 1996. Complex adaptations and the evolution of evolvability. *Evolution* 50:967–976.

Wagner, G. P., C. H. Chiu, and M. Laubichler. 2000. Developmental evolution as a mechanistic science: the inference from developmental mechanisms to evolutionary processes. *Am. Zool.* 40:819–831.

Wagner, G. P., M. D. Laubichler, and H. Bagheri-Chaichian. 1998. Genetic measurement theory of epistatic effects. *Genetica* 102/103:569–580.

Wagner, G. P., and A. Mezey. 2000. Modeling the evolution of genetic architecture: a continuum of alleles model with pairwise A x A epistasis. *J. Theor. Biol.* 203:163–175.

Wagner, G. P., and J. G. Mezey. 2004. The role of genetic architecture constraints in the origin of variational modularity. In *Modularity in development and evolution,* ed. G. Schlosser and G. P. Wagner. Chicago: University of Chicago Press.

Wagner, G. P., and K. Schwenk. 2000. Evolutionarily stable configurations: functional integration and the evolution of phenotype stability. *Evol. Biol. (N.Y.)* 31:155–217.

Wahnschaffe, U., U. Bartsch, and B. Fritzsch. 1987. Metamorphic changes within the lateral-line system of Anura. *Anat. Embryol.* 175:431–442.

Wakahara, M. 1996. Heterochrony and neotenic salamanders: possible clues for understanding the animal development and evolution. *Zool. Sci.* 13:765–776.

Wakahara, M., and M. Yamaguchi. 1996. Heterochronic expression of several adult phenotypes in normally metamorphosing and metamorphosis-arrested larvae of a salamander *Hynobius retardatus. Zool. Sci.* 13:483–488.

Wakamatsu, Y., T. M. Maynard, and J. A. Weston. 2000. Fate determination of neural crest cells by NOTCH-mediated lateral inhibition and asymmetrical cell division during gangliogenesis. *Development* 127:2811–2821.

Wake, D. B. 1991. Homoplasy: the result of natural selection, or evidence of design limitations? *Am. Nat.* 138:543–567.

Wake, D. B., and A. Larson. 1987. Multidimensional analysis of an evolving lineage. *Science* 238:42–48.

Wake, D. B., G. Roth, and K. C. Nishikawa. 1987. The fate of the lateral line system in plethodontid salamanders. *Am. Zool.* 27:166A

Walsh, J. B. 1995. How often do duplicated genes evolve new functions? *Genetics* 139: 421–428.

Waters, K. 1991. Tempered realism about the force of selection. *Philos. Sci.* 58:553–573.

Wawersik, S., and R. L. Maas. 2000. Vertebrate eye development as modeled in *Drosophila. Hum. Mol. Genet.* 9:917–925.

Wen, X., S. Fuhrman, G. S. Michaels, D. B. Carr, S. Smith, J. L. Barker, and R. Somogyi. 1998. Large-scale temporal expression mapping of central nervous system development. *Proc. Natl. Acad. Sci. U.S.A.* 95:334–339.

White, B. A. and C. S. Nicoll. 1981. Hormonal control of amphibian metamorphosis. In *Metamorphosis: a problem in developmental biology,* ed. L. I. Gilbert and E. Frieden, 363–396. New York: Plenum Press.

Whitlock, M. C., P. C. Phillips, F. B.-G. Moore, and S. J. Tonsor. 1995. Multiple fitness peaks and epistasis. *Annu. Rev. Ecol. Syst.* 26:601–629.

Wiley, E. O. 1981. *Phylogenetics: the theory and practice of phylogenetic systematics.* New York: Wiley.

Wilkins, A. S. 1997. Canalization: a molecular genetic perspective. *BioEssays* 19: 257–262.

Wilkins, A. S. 1998. Evolutionary developmental biology: where is it going? *BioEssays* 20:783–784.

Williams, N. A., and P. W. H. Holland. 1998. Molecular evolution of the brain of chordates. *Brain Behav. Evol.* 52:177–185.

Wimsatt, W. C. 1981. Units of selection and the structure of the multilevel genome. In *Proceedings of the 1980 biennial meeting of the Philosophy of Science Association,* vol. 2, ed. P. D. Asquith and R. N. Giere, 122–183. East Lansing, Mich.

Winther, R. G. 2001. Varieties of modules: kinds, levels, origins, and behaviors. *J. Exp. Zool. (Mol. Dev. Evol.)* 291:116–129.

Wray, G. A. 1994. Developmental evolution: new paradigms and paradoxes. *Dev. Genet.* 15:1–6.

Wray, G. A., and A. E. Bely. 1994. The evolution of echinoderm development is driven by several distinct factors. *Development* 1994 (suppl.): 97–106.

Xu, C., R. C. Kauffmann, J. Zhang, S. Kladny, and R. W. Carthew. 2000. Overlapping activators and repressors delimit transcriptional response to receptor tyrosine kinase signals in the *Drosophila* eye. *Cell* 103:87–97.

Xu, P. X., J. Adams, H. Peters, M. C. Brown, S. Heaney, and R. Maas. 1999. Eya1-deficient mice lack ears and kidneys and show abnormal apoptosis of organ primordia. *Nat. Genet.* 23:113–117.

Xu, P.-X., I. Woo, H. Her, D. R. Beier, and R. L. Maas. 1997. Mouse Eya homologues of the *Drosophila* eyes absent gene require Pax6 for expression in lens and nasal placode. *Development* 124:219–231.

Yaoita, Y., and D. D. Brown. 1990. A correlation of thyroid hormone receptor gene expression with amphibian metamorphosis. *Genes Dev.* 4:1917–1924.

Yuh, C.-H., H. Bolouri, and E. H. Davidson. 1998. Genomic cis-regulatory logic: experimental and computational analysis of a sea urchin gene. *Science* 279:1896–1902.

Yun, K., and B. Wold. 1996. Skeletal muscle determination and differentiation: story of a core regulatory network and its context. *Curr. Opin. Cell Biol.* 8:877–889.

Zakany, J., M. Gerard, B. Favier, S. S. Potter, and D. Duboule. 1996. Functional equivalence and rescue among group 11 Hox gene products in vertebral patterning. *Dev. Biol.* 176:325–328.

Zuber, M. E., M. Perron, A. Philpott, A. Bang, and W. A. Harris. 1999. Giant eyes in *Xenopus laevis* by overexpression of XOptx2. *Cell* 98:341–352.
Zuckerkandl, E. 1994. Molecular pathways to parallel evolution. 1. Gene nexuses and their morphological correlates. *J. Mol. Evol.* 39:661–678.
Zuckerkandl, E. 1997. Neutral and nonneutral mutations: the creative mix: evolution of complexity in gene interaction systems. *J. Mol. Evol.* 44 (1 suppl.): S2–S8.

CONTRIBUTORS

Gary Anderson
Kingston, Ontario
E-mail: ganderson@computer.org

Patrick Blader
Centre de Biologie du Developpement
Université Paul Sabatier
Toulouse, France
E-mail: blader@pop.cict.fr

Anne-Gaëlle Borycki
University of Sheffield
Centre for Developmental Genetics
Sheffield, U.K.
E-mail: A.G.Borycki@sheffield.ac.uk

Bernard Chang
Scarborough, Ontario
E-mail: bchang@kos.net

James M. Cheverud
Department of Anatomy and Neurobiology
Washington University School of Medicine
St. Louis, Missouri
E-mail: cheverud@pcg.wustl.edu

José F. de Celis
Centro de Biología Molecular "Severo Ochoa"
Universidad Autónoma de Madrid
Madrid, Spain
E-mail: jfdecelis@cbm.uam.es

William A. Cresko
Institute of Neuroscience
University of Oregon
Eugene, Oregon
E-mail: wcresko@uoneuro.uoregon.edu

Brad Davidson
Department of Molecular and Cell Biology
University of California at Berkeley
Berkeley, California
E-mail: bandl@uclink.berkeley.edu

Marcus C. Davis
Department of Organismal Biology and Anatomy
University of Chicago
Chicago, Illinois
E-mail: marcusd@uchicago.edu

Allan G. Force
Virginia Mason Research Center
Benaroya Research Institute
Seattle, Washington
E-mail: force@vmresearch.org

Stefanie Fuhrman
Rockville, Maryland
E-mail: stefaniefuhrman@mac.com

Larry D. Greller
Biosystemix Ltd.
Sydenham, Ontario
E-mail: ldgphud@earthlink.net

Tiffany Heanue
Department of Molecular Neurobiology
National Institute for Medical Research
London, U.K.
E-mail: theanue@nimr.mrc.ac.uk

Molly Jacobs
Friday Harbor Laboratories
University of Washington
Friday Harbor, Washington
E-mail: mwjacobs@u.washington.edu

Gabrielle Kardon
Department of Genetics
Harvard Medical School
Boston, Massachusetts
E-mail: gkardon@genetics.med.harvard.edu

Chris Madill
Georgetown, Ontario
E-mail: chris_madill@yahoo.com

Eli Meir
SimBiotic Softward
Ithaca, New York
E-mail: meir@ecobeaker.com

Jason G. Mezey
Department of Biology
Florida State University
Tallahassee, Florida
E-mail: mezey@bio.fsu.edu

Richard E. Michod
University of Arizona
Department of Ecology and Evolutionary Biology
Tucson, Arizona
E-mail: michod@u.arizona.edu

Aurora M. Nedelcu
Department of Ecology and Evolutionary Ecology
University of Arizona
Tucson, Arizona
E-mail: nedelcua@u.arizona.edu

Craig Nelson
R. M. Bock Laboratories
University of Wisconsin—Madison
Madison, Wisconsin
E-mail: craignelson@facstaff.wisc.edu.

Christof Niehrs
Division of Molecular Embryology
Deutsches Krebsforschungszentrum
Heidelberg, Germany
E-mail: Niehrs@DKFZ-Heidelberg.de

F. Bryan Pickett
Department of Biology
Loyola University of Chicago
Chicago, Illinois
E-mail: fpicket@luc.edu

Luis Puelles
Department of Morphological Sciences
Faculty of Medicine
University of Murcia
Murcia, Spain
E-mail: puelles@um.es

Christoph Redies
Institute of Anatomy
University of Duisburg—Essen School of Medcine
Essen, Germany
E-mail: christoph.redies@uni-essen.de

Lucas Sánchez
Centro de Investigaciones Biológicas
Madrid, Spain
E-mail: lsanchez@cib.csic.es

Jeffery C. Schank
Department of Psychology and Animal Behavior Graduate Group
University of California, Davis
Davis, California
E-mail: jcschank@ucdavis.edu

Gerhard Schlosser
Brain Research Institute
University of Bremen
Bremen, Germany
E-mail: gschloss@uni-bremen.de

Urs Schmidt-Ott
Department of Organismal Biology and Anatomy
University of Chicago
Chicago, Illinois
E-mail: uschmid@uchicago.edu

Neil H. Shubin
Department of Organismal Biology and Anatomy
University of Chicago
Chicago, Illinois
E-mail: nshubin@uchicago.edu

Ralf J. Sommer
Max-Planck Institute for Developmental Biology
Department of Evolutionary Biology
Tübingen, Germany
E-mail: ralf.sommer@tuebingen.mpg.de

Roland Somogyi
Biosystemix, Ltd.
Sydenham, Ontario
E-mail: somogyi@kingston.net

Kim Sterelny
Department of Philosophy
Victoria University of Wellington
Wellington, New Zealand
E-mail: Kim.Sterelny@vuw.ac.nz
and
Philosophy Program

Research School of the Social Sciences
Australian National University
Canberra, Australia
E-mail: kimbo@coombs.anu.edu.au

Uwe Strähle
Institut de Génétique et de Biologie
Moléculaire et Cellulaire
Université Louis Pasteur
Strasbourg, France
E-mail: uwe@titus.u-strasbg.fr

Billie J. Swalla
Department of Biology
University of Washington
Seattle, Washington
and
Friday Harbor Laboratories
University of Washington
Friday Harbor, Washington
E-mail: bjswalla@u.washington.edu

Clifford J. Tabin
Department of Genetics
Harvard Medical School
Boston, Massachusetts
E-mail: tabin@rascal.med.harvard.edu

Denis Thieffry
ESIL-GBMA
Université de la Mediterranée
Marseille, France
E-mail: thieffry@esil.univ-mrs.fr

George von Dassow
Center for Cell Dynamics
Friday Harbor Laboratories
University of Washington
Friday Harbor, Washington
E-mail: dassow@u.washington.edu

Günter P. Wagner
Department of Ecology and Evolutionary
Biology
Yale University
New Haven, Connecticut
E-mail: gunter.wagner@yale.edu

Ernst A. Wimmer
Lehrstuhl für Genetik
Universität Bayreuth
Bayreuth, Germany
E-mail: Ernst.Wimmer@uni-bayreuth.de

William C. Wimsatt
Department of Philosophy
Committees on Evolutionary Biology
and Conceptual Foundations of
Science
University of Chicago
Chicago, Illinois
E-mail: w-wimsatt@uchicago.edu

Andrew Wuensche
Discrete Dynamics Inc.
Santa Fe, New Mexico
E-mail: wuensch@santafe.edu

INDEX